ANNUAL REVIEW OF MICROBIOLOGY

ANNUAL REVIEW OF MICROBIOLOGY

VOLUME 42, 1988

L. NICHOLAS ORNSTON, *Editor*

Yale University

ALBERT BALOWS, *Associate Editor*

Centers for Disease Control, Atlanta

PAUL BAUMANN, *Associate Editor*

University of California, Davis

ANNUAL REVIEWS INC.   4139 EL CAMINO WAY   P.O. BOX 10139   PALO ALTO, CALIFORNIA 94303-0897

ANNUAL REVIEWS INC.
Palo Alto, California, USA

International Standard Serial Number: 0066–4227
International Standard Book Number: 0–8243–1142-6
Library of Congress Catalog Card Number: 49-432

Annual Review and publication titles are registered trademarks of Annual Reviews Inc.

Annual Reviews Inc. and the Editors of its publications assume no responsibility for the statements expressed by the contributors to this *Review*.

Typesetting by Kachina Typesetting Inc., Tempe, Arizona; John Olson, President
Typesetting coordinator, Janis Hoffman

PRINTED AND BOUND IN THE UNITED STATES OF AMERICA

PREFACE

In recent prefaces to volumes in this series we have considered microbiology's affinity with the art forms of sculpture, ballet, and child-rearing. Poetry should not be neglected, because the concise expression of enduring thoughts is an art highly appropriate to a scientific review volume. One such thought is an appreciation that life is a dynamic state driven by free energy provided by the sun. The thermodynamics underlying this process was stated quite clearly by the seventeenth century poet Andrew Marvell, who declared that ". . . we cannot make our sun/Stand still, yet we will make him run." The thought is beautiful, but not quite general. Interesting exceptions to life supported by light are the complex living communities that derive their energy from geothermal vents under the sea. As described by Holger Jannasch, an explorer of life near the vents, the whole process is driven by microorganisms. In this analysis lies a powerful tool for generalization.

At the *Annual Review of Microbiology*, we are driven by deadlines, and we thank our authors for the care and promptness with which they prepared their manuscripts. Selection of topics and authors is the painful yet always delightful task of the Editorial Commitee, which for this volume was happily augmented by guests Carol Blair and Jack London. We will greatly miss the contributions of Frederick Murphy, who retired from the committee after the planning of this volume. Newly appointed and valued members of the Editorial Committee are Arnold Demain and Jonathan King. Central responsibility for preparation of the volume rested with Production Editor Andrea Perlis, who performed flawlessly. Responsibility for organization of the Editor remains with his extraordinary administrative associate, Susan Voigt, whose skill is rivaled only by her charm.

L. NICHOLAS ORNSTON
EDITOR

For the convenience of readers, a detachable order form/envelope is bound into the back of this volume.

Annual Review of Microbiology
Volume 42, 1988

CONTENTS

Indexes

OTHER REVIEWS OF INTEREST TO MICROBIOLOGISTS

From the *Annual Review of Biochemistry,* Volume 57 (1988)

Bacterial Electron Transport Chains, Yasuhiro Anraku
Posttranscriptional Regulatory Mechanisms in Escherichia coli, Larry Gold
DNA Polymerase III Holoenzyme of Escherichia coli, Charles S. McHenry
Viral Proteinases, Hans-Georg Kräusslich and Eckard Wimmer
Structure and Function of Bacterial Sigma Factors, John D. Helmann and Michael
   J. Chamberlin

From the *Annual Review of Biophysics and Biophysical Chemistry,* Volume
17 (1988)

Sensory Rhodopsins of Halobacteria, John L. Spudich and Roberto A. Bogomolni
Structure-Function Correlations in the Small Ribosomal Subunit From Escherichia
   coli, Peter B. Moore and Malcolm S. Capel
DNA Packing in Filamentous Bacteriophages, Loren A. Day, Christopher J. Mar-
   zec, Stephen A. Reisberg, and Arturo Casadevall

From the *Annual Review of Cell Biology,* Volume 4 (1988)

Conjugation in Saccharomyces cerevisiae, Fred Cross, Leland H. Hartwell, Cather-
   ine Jackson, and James B. Konopka
Environmentally Regulated Gene Expression for Membrane Proteins in Escherichia
   coli, Steven Forst and Masayori Inouye

From the *Annual Review of Genetics,* Volume 22 (1988)

Recombination Between Repeated Genes in Microorganisms, Thomas D. Petes and
   Charles W. Hill
Control of Antigen Gene Expression in African Trypanosomes, E. Pays and M.
   Steinert
Basic Processes Underlying Agrobacterium-*Mediated DNA Transfer to Plant Cells,*
   Patricia Zambryski
DNA Double-Chain Breaks in Recombination of Phage Lambda and of Yeast,
   Franklin W. Stahl and David S. Thaler
The Genetics of Bovine Papillomavirus Type 1, Paul Lambert, Carl Baker, and Peter
   M. Howley

From the *Annual Review of Medicine,* Volume 39 (1988)

Immunology of Respiratory Viral Infections, Robert C. Welliver and Pearay L. Ogra
Infectious Causes of Esophagitis, John S. Goff
Oral Antibiotic Therapy for Serious Infections, Arnold L. Smith
Transdermal Delivery of Drugs, Larry Brown and Robert Langer

Selective Manipulation of the Immune Response In Vivo by Monoclonal Antibodies,
    W. E. Seaman and D. Wofsy
Chlamydial Infections, Joseph Fraiz and Robert B. Jones
Genital Warts, Papillomaviruses, and Genital Malignancies, Keerti V. Shah and
    Joseph Buscema

From the *Annual Review of Phytopathology,* Volume 26 (1988)

Evolution of Concepts Associated With Soilborne Plant Pathogens, John L. Lock-
    wood
Perspectives on Progress in Plant Virology, Myron K. Brakke
Molecular Genetics of Pathogenicity in Phytopathogenic Bacteria, M. J. Daniels, J.
    M. Dow, and A. E. Osbourn
Expression and Function of Potyviral Gene Products, William G. Dougherty and
    James C. Carrington
Biological Control of Soilborne Plant Pathogens in the Rhizosphere with Bacteria,
    David M. Weller

From the *Annual Review of Plant Physiology and Plant Molecular Biology,*
Volume 39 (1988)

Genetic Analysis of Legume Nodule Initiation, Barry G. Rolfe and Peter M.
    Gresshoff

Ann. Rev. Microbiol. 1988. 42:1–34

A STRUCTURED LIFE

Robert G. E. Murray

Department of Microbiology and Immunology, University of Western Ontario, London, Ontario, Canada N6A 5C1

CONTENTS

> One of the profound lessons to be learned from science is not what it has done, but the way it goes about doing it.
>
> Hanbury Brown: *The Wisdom of Science*

I may not have been introduced to thinking about microbes at birth, but this was a likely event soon afterwards. I count myself lucky that bacteriology, its history, and science in general lived for me because my father, E. G. D. Murray (1890–1964) (88), was a bacteriologist whose teachers were young contemporaries of the giants of the early days. A few of that earlier generation were still teaching at Cambridge when I was there as a student, which emphasized the bridge between scientific generations. They were long past

1

0066-4227/88/1001-0001$02.00

retirement but had valuable things to say about science and the early days. This gives me a good feeling about being on postretirement appointment, as I am now, and still having contact with dedicated students. Between us, my father and I have enjoyed devotion to bacteriology for almost three quarters of this century.

It is hard to decide from biographies whether or not a career in science is assisted by having a parent a scientist. At the least, it gives a more defined set of prejudices as a basis for deciding what to do in life, or what to avoid. At best, it provides a strong set of interests, standards, stimuli and concerns which shape the inevitable explorations of the world around us and direct the neophyte's progress insensibly and in subtle ways from tentative to defined courses of action. A wealth of understanding gained informally provides the "gilt on the gingerbread" that amplifies capabilities.

BEGINNINGS

The Murrays who were my forebears lived in South Africa from 1835 and for the most part were dedicated to the land; my mother's family was English and of mercantile bent. My grandfather, G. A. E. Murray, departed from those norms by qualifying in medicine, and he set up for surgical practice and family life in the infant city of Johannesburg in 1888. My father followed in his footsteps and studied medicine at St. Bartholomew's Hospital, London. He did a degree at Cambridge (Christ's College, 1912), with a special interest in zoology, which I was to profit from in years ahead. While still a student at St. Bartholomew's he was much influenced by Dr. Mervyn Gordon, and the two worked together on studies of meningococcal meningitis. This interest continued during World War I after my father's qualification, marriage, and army service. He was in the Royal Army Medical College Vaccine Laboratories and was concerned with the making of vaccines when I was born in 1919 in Ruislip, a London suburb. The war was over and the bacteriological and epidemiological emergencies of wartime were displaced by the influenza pandemic with all its alarming complications. Soon after I arrived on the scene the family moved to Cambridge, where my father was appointed a research bacteriologist for the Medical Research Council (1920–1926). He revived his wartime interests in the meningococcus while supervising the production of therapeutic antisera. My very early memories include attending the bleeding of large and elegant horses at the Field Laboratories outside Cambridge.

My father was soon back to Cambridge academic life in Christ's College and in the Department of Pathology, where he became a lecturer. His chief, and later my teacher, was the redoubtable Professor H. R. Dean. We shared energetic summer vacations with his family for several years. Professor Dean was a pathologist with interests in bacteriology and immunology. (He introduced the concept of optimal proportions into serology, and he also studied complement.) It was my father's task to help him develop an honors course in pathology, which included medical microbiology in those days and still does. The first class graduated in 1925, and the nine students, now very distinguished British medical microbiologists, included A. A. Miles, C. H. Andrewes, and two who taught me later on, E. T. C. Spooner, who later returned to Cambridge, and Frederick Smith, who was to join my father at McGill. They were forerunners of a considerable company of medical scientists, and I was to join them myself.

My childhood was that of a homebody, but life changed considerably in 1927 when I was sent to Summer Fields, a boarding school at Oxford devoted to helping as many pupils as possible gain scholarships to enter one or another of the great "public" schools. It was a tough time because I had remarkably little experience of children my own age. It was a good school and there were miseries; but we worked hard and played hard as required by the school's motto, *"Mens sana in corpore sano."* I have every reason to bless the school for the basics (in everything but music, alas) that have served me well.

In 1930 my father became professor and head of the Department of Bacteriology and Immunology in the Faculty of Medicine of McGill University, Montreal. It was an exciting transition for all of us and one we embraced enthusiastically, even if I never managed more than minor modifications to an English accent. I was old enough to be more aware, and being at home for much of the year, I could now hear what was going on in academic life. So Montreal became my second home. As a great metropolis encompassing two cultures, the city provided an ambience of great importance to my development. The change in school systems was a traumatic experience, but happily I went to Lower Canada College, an English-type boy's school that offered a suitable program and games familiar to me, such as soccer and cricket. What helped me most in becoming Canadian without too many tears was the enthusiasm with which my parents took to the country and its pace, space, and natural beauties.

By 1936 I was ready for McGill University and completed the matriculation examinations, which then still required Latin. There was no science requirement, and indeed I did only a minimal combined chemistry and physics course at school. So McGill provided the grand and eagerly awaited start in the nuts and bolts of science and the opportunity to learn under senior and very capable teachers whom, in many instances, I already knew and admired. Largely because of my father's early interest in and retentive memory for zoological information, I set out with an intention to do zoology. However, all courses were interesting, and I performed well in the examinations for the two years 1936–1938.

There were omens of war in Europe, and evidently it would be now or never if I were to go to Cambridge University, which meant so much to my father and which I wanted with all my heart. Arrangements were made for me to enter Christ's College in October 1938. However, there was a summer to fill in, and I had the opportunity to go to a marine biological station for a course in invertebrate zoology at Salisbury Cove, near Bar Harbor in Maine, under Professor Ulric Dahlgren of Princeton University. The course was fun and influential; I was later to marry (in 1944) one of my classmates (Doris Marchand, also from a scientific family). Also, the scientific community provided much informal discourse with first class people such as Homer Smith and Gairdner Moment. After this very educational interlude I set off for England by ship during the Munich Crisis in September 1938.

Cambridge was a revelation to me, not just because of the quality of teaching but because of what was expected of the students. It was a matter of attitude and preparation. The attitude was expressed in a way by my tutor, C. P. Snow, who said (and he didn't say much more), " . . . you realize you have to present yourself for examination in May, two years hence." We had to pace ourselves both in and out of term, keep up, and do a wide range of independent reading, quite apart from lectures and laboratories. In this we were assisted by a supervisor of studies (I was assigned to John Yudkin, later a distinguished nutritionist, who then had interests in *Escherichia coli* physiology). We were expected to have the basics or quickly acquire them; I thought myself inordinately well prepared, but to my chagrin, my scholarly classmates just out of school had a remarkably well-honed understanding of basic physics and chemistry. The teaching also reflected attitude. For instance, the lectures and laboratories in introductory biochemistry were given by Ernest Baldwin. The first-term course was on comparative biochemistry, given in a fascinating fashion. Baldwin later went on to systematic biochemistry and the outlines of what was to become a distinguished textbook. It was good stuff and we learned a lot.

I was enrolled in medical studies, which meant the inclusion of anatomy, biochemistry, physiology, pathology (including bacteriology), and pharmacology in the first two years. Despite a rather miserable performance in anatomy, I was able to go on to a third year (Part II of the Natural Sciences Tripos) spent entirely in one disciplinary area, which was in my case pathology and bacteriology, still under Professor Dean. The Part II course in pathology was a combined effort coordinated for the most part by G. Williamson (pathology), E. T. C. Spooner, and A. W. Downie (medical microbiology). Downie was then running the Emergency Medical Service Laboratory in Cambridge. He was dealing with problems ranging from streptococcal diseases to leptospirosis, which he and his colleagues shared with the five of us students in exemplary fashion. Likewise, A. M. Barrett saw that we got all the tissue and bacteriological specimens that we needed from the autopsies he did for Addenbrooks Hospital. It was an extraordinarily thorough experience in the details and practice of medical bacteriology and pathology, full of interests to suit each of us. There were special additions: parasitology at the Molteno Institute and special series of lectures by senior (some very senior) colleagues such as G. S. Graham-Smith (toxins and venomous animals and insects), Louis Cobbett (tuberculosis), R. I. N. Greaves (immunology and the preservation of blood and sera), R. R. Race (blood grouping), and L. Faulds (cancer research). To me it was sad but not distressing that they excluded much general microbiology, which I heard a little about from a few fellow students in botany. The biochemistry department included some distinguished contributors to microbial biochemistry, and the group of us arranged an informal session or two with Marjory Stephenson and W. E. van Heynigen. As it turned out, I did not cover myself with glory and received a Class II overall, which was a bit humbling since my classmates John Stowers, Leo Wolman, and Alec Comfort all were Class I, as they have shown by their distinctions in later life. But after this year I understood what kind of advanced bacteriology course my father had developed at Cambridge (when Part II was initiated) and later introduced in refined form at McGill University.

Graduation in wartime brought the necessity of a clear track in medicine, in research, or in military service. My decision, encouraged by Alan Downie, came with the news of my acceptance at the McGill Medical School. The issuance of an exit permit and my assignment to a ship for the Atlantic crossing was slow, and I did not sail until late October 1941. I effectively spent the four-month interval on the wards of Addenbrooks Hospital learning physical diagnosis, which helped me considerably. I arrived back in Canada just before the invasion of Pearl Harbor changed the aspect of the war, after an exciting three-week voyage in convoy. We had few losses thanks to Canadian corvettes and the newly acquired lend-lease destroyer escorts supplied by the United States.

McGill had introduced an accelerated medical course with no vacations, and I settled down to a rather exhausting routine. Graduation came late in 1943, followed by internship at Royal Victoria Hospital. My internship was unconventional because it was based in the clinical laboratory of the Department of Bacteriology, which provided services for the hospital. This internship gave me clinical experience in most departments and laboratory experience in the fullest possible range of microbial diseases. The experience was all the better for the teaching of Fred Smith (who was holding the fort for my father while he was on war service) and for the practical clinical laboratory instruction of Gertrude Kalz, who really put the polish on my bacteriological training, for which I am ever grateful. Internship also brought me marriage and the start of family life.

The army called in 1944 with suggestions of using my laboratory competence, but predictably, after the usual training courses I ended up as a medical officer in tank training regiments. The war was all but over when an offer came from the dean of medicine at the University of Western Ontario.

INFLUENCES

Small things, thoughtful people, and lucky happenstances enrich life, teach us in subtle ways, establish values and interests, prepare the mind, and without intention guide the footsteps. Some of these influences I identify in what follows, with the certainty that others of equal importance are left out or were never recognized.

Nature was revealed to me in my family's series of well-tended gardens and country walks. My mother's keen eye for birds, beasts, and natural wonders, and my father's habit of collecting interesting specimens and sharing his knowledge about them were a part of many days' outings. Fishing was important to my father and it became so to me; his interest in the sport and his competence were infectious, and we fished together for many years. I acquired a great deal from the opportunity for gentle but serious conversation and for airing what we knew or thought we knew about life around us.

Microscopes had a pervasive, persuasive, and recurring role in my life because of my father's interest in microscopes and microscopy. Certainly I must have seen them early on just as I admired them on later occasions when visiting laboratories. However, I have a clear memory (identifiable by circumstance to the time when I was three or four years old) of seeing Leeuwenhoekian animalcules down an old brass microscope; their source was an inverted bell-jar aquarium full of weeds, water beetles, copepods, and other small forms. That same microscope was in more regular summertime use when I was seven to ten years old, and I still have it.

I did no formal biology until I went to McGill. Two kindnesses were of particular importance: I was attending N. J. Berrill's invertebrate zoology course and he asked me, "Would you like to have six feet of bench?" For sure I would. He gave me space in a corner, the use of a microscope, and facilities for fixing, paraffin-embedding, sectioning (an old "Cambridge rocker"), and staining preparations for microscopy. I enjoyed myself, learned by doing, and produced good sections of a frog kidney adenoma among other things. My father's kindness was no less formative; it was the present of a Leitz binocular microscope with appropriate optics and slide boxes of the preparations of microscopic invertebrates that he had made for himself as a student in 1909–1912. Thus I became familiar with preparative techniques and with an instrument that I have and use to this day, and which has provided all of my light micrographs to date.

I cannot remember life without books and discourse about books. My father was an inveterate book-buyer and reader of all sorts of subjects from Middle English to evolution. Books were treasured. Even the bad books were kept because the argument and the horrible example might be needed; they were annotated with marginalia, incisive comments on the title page if they were bad enough, and a personal index to interesting titbits noted on a back flyleaf. The home and the laboratory contained eclectic collections of books, journals, and reprints; as a result I became a card-carrying, second-generation pack rat and omnivorous reader. Fortunately few near disasters resulted, but a few there were, from such dilettantism. As long as I can remember the dinner table was a place to talk, to try out outrageous arguments, to remember interesting facts, and to embroider a good story. My mother had to contend with an impossibly elastic timetable for the course of a dinner.

I was lucky indeed that at home, at secondary school, and at the universities I was blessed with mentors who cared enormously about how we expressed ourselves in speech and in writing. At Summer Fields I took classes from L. A. G. Strong (a substantial novelist) and Cecil Day-Lewis (a major poet). At school in Montreal I was helped enormously by Hugh MacLennan (a major Canadian novelist and academic), D. S. Penton (a fine teacher), and V. C. Wansbrough (a scholarly headmaster), who were as much friends as teachers. Behind-the-scenes work on the school magazine, which was a rather elaborate annual, introduced me to the life of an editor. The interest must have been latent because I was to spend much time on the *McGill Medical Journal,* then a substantial quarterly, and one year as its editor. My involvement with the journal provided experience every step of the way from soliciting of contributions, to copy editing, proofreading, and production, to review writing. Only seven years later I was to be on the Editorial Board of the *Journal of Bacteriology,* enjoying the first of a number of appointments in scientific editing.

I was most fortunate to have done half of my schooling and half of my university training at first-class institutions on each side of the Atlantic. These opportunities were given at no little expense to my kind, generous parents, who gave me the best that they could of both worlds and all their support. I can only imagine a certain anguished concern at sending an only son as a student, not a serviceman, in the direction of crisis and war in 1938.

The final good fortune was to get started in teaching and research at a small medical school in a small university that had only one way to go and that was up, fast. There was infectious energy and ability at the top in the person of G. E. Hall, a fine scientist turned administrator, who knew what he wanted, usually got it, and made good decisions about people more of the time than do most administrators. He brought in able young professors to guide into the bright future the small (two–faculty member) departments, such as R. J. Rossiter (biochemistry) and A. C. Burton (biophysics), who became my very good friends and colleagues, as well as even younger faculty (of whom I was one) to back them up. His successor as dean of Medicine, J. B. Collip, a most distinguished endocrinologist, was even more influential and also pushed hard for productive involvement in basic research. Collip strengthened the school, with the help of Hall, then the university president, to a level to be proud of. There was no temptation to look for another job; promotion came fast, and the school was, and still is, an exciting place to be. As a result I lost no time along the way due to relocation and I received every possible encouragement. We prospered intellectually; we had to cooperate in all things because there were so few of us (in 1950 there were 18 full-time faculty in the Medical School), and we enjoyed each other's company.

A FLYING START IN ACADEME

The invitation to the University of Western Ontario Medical School was urgent because the professor of bacteriology and immunology, I. N. Asheshov, was in the hospital and was likely to be there for a few weeks. It was September 1945, and the course for 44 medical students had just started. Dean Hall, himself recently appointed after a senior posting in the Royal Canadian Air Force, knew how to pull strings. A brief weekend visit was enough for me to say yes to the post of lecturer in the department with duties to include teaching medical students, assisting in clinical bacteriology for the hospital, and involvement in research. Two weeks later I was seconded to the Wolsey Barracks, London, Ontario, for "special duties." My wife and infant daughter, with the help of friends, found us a temporary living place. Within a week of our being together again I was embarked on a 12-week stint of four lectures and four labs a week, in the midst of which I was demobilized. Many of this first class were ex-servicemen, and not a few were older than I was. We all survived the experience and I learned more from and with them in a shorter time than ever before or since then. Three of that class are senior colleagues still in this faculty.

Asheshov was out of the hospital before the end of the term, in time for a few lectures in his remarkable style, and he was marvelously helpful and supportive to me. He was a lively minded scientist, then engaged in seeking antibiotics, but at heart he was interested in bacteriophages, an area in which he had worked productively since 1922. He and I taught hard for five months of the year. It was a good old-fashioned bacteriology course topped off with minor dollops of virology, parasitology, and mycology. In many ways it was the oral examination rather than written papers that taught me the most about my subject, our teaching, and students. Whatever may be said about subjectivity in examinations, the hard work involved in these oral examinations was repaid in knowledge of the student and understanding of the strengths and weaknesses in teaching. Even more importantly, after the war we had mature veteran students who were remarkable people: dedicated, serious, accomplished, and worth every effort. They taught me a lot.

I had to get started on research in 1946 because there was time and I wanted to do it. Dr. Asheshov gave me every encouragement, without demanding any service or direct assistance in his own work, which was mighty kind of him. I was equally encouraged by Dean Hall and by Dean Collip (who succeeded Hall in 1948), who provided me with an initial grant for a phase microscope. Funds were very hard to get, so I had to make use of the sorts of things we worked with every day in the clinical laboratory. Each of my first four independent forays into research were important to my life and work in the laboratory. I write about them with that in mind.

The first project was prompted by the isolation of a mucoid Group A *Streptococcus pyogenes* from a chronic lung infection. This reminded me that when I was an intern I had seen similar mucoid colonies collapse flat in the neighborhood of some colonies of staphylococci and pneumococci. Knowing then that the capsular polysaccharide was hyaluronic acid provided a method for ready detection of hyaluronidase ("spreading factor") and also a diffusion assay for the enzyme, which was of great interest at that time as a component of virulence targeted on the intercellular matrix of connective tissue. Fortunately, R. H. Pearce, an equally young biochemist in the Department of Pathological Chemistry, was interested in acid mucopolysaccharides. The resulting joint paper was good experience and was accepted without demur by the *Canadian Journal of Research* (64). So I started out in collaborative research, and I am grateful to Pearce because I learned from him some of the many things I should have known had I had doctoral training in science. Entering PhD studies was an option, but I decided to get on with research because there was time and opportunity.

The next project arose from observing the remarkable motile colonies shown by a fortuitous isolation of *Bacillus circulans,* and it initiated a lifelong interest in swarming and motility. Both the rotating colonies and the curving paths of the bullet-shaped colonies showed an exact 2:1 ratio of counterclockwise to clockwise motion, which is still unexplained. For the first time I had the assistance, in the summers of 1947 and 1948, of a veteran medical student with laboratory experience, R. H. Elder, now a senior and respected clinical microbiologist in Ottawa. We wrote a neat paper which was accepted by the *Journal of Bacteriology* (60), and we made a short 16-mm movie, using a haywire rig, to illustrate a paper given at the 1948 ASM Meeting in Minneapolis. Most importantly, Carl Robinow introduced himself after that presentation, and we discussed swarming but also his observations on bacterial nuclei and endospores. So this project introduced me to a fascinating phenomenon which is still in mind, and to a dear colleague whose association with me and our students has been nothing but rewarding and a pleasure for 40 years.

Igor Asheshov taught me about phages and how to handle them just at the time the papers on the T-phages were coming out. The excitement of following the fate of the phage and the host components in these papers induced thoughts about research projects. Asheshov was advising Fred Heagy (an MD interested in science), a PhD candidate studying T2 phage infection, and suggested a parallel study of changes in the host cell in advance of lysis. Indeed, there were indications that these changes should be observable by microscopy (46). We tried Robinow's cytological techniques because I had been reading his papers and his "Addendum," which was a remarkable seminal chapter on bacterial cytology appended to *The Bacterial Cell* by René Dubos (75). We had hardly started on this in 1948 when Asheshov and his technical forces left to work on a March of Dimes project seeking antiviral agents at the Bronx Botanical Garden. So I was left willy-nilly to elaborate my own expertise and to supervise a doctoral candidate my own age. It was another learning experience, to use the language of educators, to become acting head of the department.

That summer and the next, Heagy and I were helped by D. H. Gillen, a medical student, and we made lots of preparations at timed intervals following infection of *Escherichia coli* B with phage T2. Our preparations were terrible but encouraging. The situation improved dramatically after Robinow paid us a visit and showed us in his inimitable fashion delightfully simple technical stratagems, which paid off in the quality of the preparations and the photomicrographs. There was a progression of cytological events involving a shift in the distribution of chromatin with eventual dissolution (host-cell DNA), gradual depletion of the cytoplasmic basophilia (host-cell RNA), and developing granularity in the last two thirds of the cycle due to synthesis of phage DNA. We published a good descriptive paper (62), and we were only slightly sorry to have been scooped by Luria & Human (45), who compared cytological events due to several of the T-phages. It was evident that the effect of phages on host cells was visible and was determined by the virus rather than the host. Our paper was appreciated and brought me into touch with S. E. Luria and his exciting group, paved the way for a collaboration with G. Bertani, and gained me an introduction to their colleagues at the University of Illinois, where laboratory visits were exhaustingly educational, night and day. The project was also the beginning of real cytological studies in our department because Dean Collip's unstinting support allowed me to invite Carl Robinow to join me in the department, which he did in 1949.

The fourth seminal research project followed the invitation by Charles Evans for me to spend the summer of 1949 in the Department of Microbiology, University of Washington, Seattle, as a sessional lecturer. There were fascinating people to learn from such as Erling Ordal and Evans. More to the point, I was able to put my newfound cytological competence to good use in helping Howard Douglas study the growth and division of an interesting budding and phototrophic bacterium, *Rhodomicrobium vanielii* (59). It was fun to find out that nuclear division took place in the mother cell and that one of the nuclei so formed was exported to the new bud through the long, 0.3-μm–diameter hyphal tube, which had the newly budded cell at its end. We spent happy hours observing fixed and stained preparations and taking time-lapse photomicrographs of living cells under phase contrast. I learned much that was useful from Douglas and his colleagues as well as from my teaching assistants, Wesley Volk and Quentin Myrvik. While I was there a telegram came from Dean Collip telling me I had been appointed professor and head of the department. This meant, I suppose, that the invitation from Seattle was a stimulus to decision. There was considerable rejoicing; I was just 30 years old and an expression of trust had been given by the university. As I see it, I was still drying behind the ears, as the saying goes.

RESPONSIBILITIES

I was lucky to be given from the outset as much responsibility as I could manage, such as the provision of a clinical microbiology service to the Children's Hospital and Victoria Hospital, our main teaching hospital, in which the department was then situated. I was responsible for this service from 1948 until 1965, when the department moved to the university campus and broadened its horizons. The clinical work kept us busy, and the major clinical problems came to me, day and night, in those days before we had physicians who specialized in infectious diseases. Antibiotics were with us and so were the developing problems concerning policy in treatment, spread of resistant strains within the hospitals, and infection control. This meant that research was a source of enjoyment and recreation in the midst of a busy life. Rightly or wrongly, the research projects grew more and more general and drifted further than might have seemed proper from the medical applications which were our clinical and teaching preoccupations.

Relief came in the late 1950s and early 1960s, when increased professional and technical forces were appointed to cope with the burgeoning use of clinical microbiology due to the introduction of a hospital insurance program in Ontario.

In those days there was much greater interdepartmental cooperation and we often had to teach what we could with any help we could get. I note with some astonishment that I gave courses in epidemiology (1948–1950), human genetics (1955–1958), and history and methods of medical science (for graduate students, 1958–1964). Luckily we were mostly young and energetic as well as communicative with each other owing to a lively lunchroom. This spirit and the efficiency and some of the enjoyment that went with it disappeared in a very few years with the increasing size and complexity of departments.

The move to the university campus in 1965 brought with it new faculty members and the integration of an honors course in microbiology and immunology into the biology programs. We were interlopers, and the introduction of six new programs by the basic medical sciences group was seen as a threat to the biologists and was not happily received. Time has resolved most of the problems, but increasing devotion of energy to administrative matters was required, which brought with it devotion to university affairs. I sat on the senate and many committees and spent a year as acting dean of science. So it was a degree of relief to give up being head of the department in 1974, after 25 years at it, and to administer thereafter nothing bigger than my own research group.

Another change came because of vastly increased faculty forces in medical microbiology, which drifted my teaching from medical microbiology to general bacteriology and systematics. It is not a bad thing to have such a shift in the midstream of a career and to be able to step away from administration before it becomes tedious or all-consuming. Too many good people regret losing their foothold in their academic discipline, and I have seen them get tired before their time. Life has to be lived with some care for what is best for the individual. Fortunately some good administrators and some good researchers and teachers find their appropriate niches and avoid the pitfalls of academic life; others are not as lucky as I was.

EXPLORING BACTERIAL CYTOLOGY

My feet were well set on cytological trails, but the diversity of my studies through the 1950s was needed to fix the major direction of my work. Observations on the cytology of phage infections continued with J. F. Whitfield and were extended by collaboration with G. Bertani to follow the cytological events during the establishment of a lysogenic state for P1 and P2 phages (94). Interesting as these phenomena were, perhaps the most fruitful derivative of questions about why changes in nuclear form accompanied phage infections turned out to be a study of the effect of the influx or efflux of cations on the conformation of the nucleoids of bacteria (95). As we and Kellenberger's group were to remember years later (35), an understanding of ionic effects on charged polymers is crucial to assembly and conformation of working structures and to lifelike preservation during cytological fixation of specimens. In fact, the crucial events involved in cellular structures and processes are nature's physicochemical experiments. So it was natural that my interest turned to trying to learn more about the structure of the host cell and that my studies broadened to the general cytology of bacteria. In this effort the daily discussions, suggestions, and joint efforts with Carl Robinow were determinative.

We started out in two directions, based much more on Robinow's experience than on mine. One was to do the best possible light microscopy of the components of what would now be called the cell envelope, i.e. the cytoplasmic membrane/cell wall complex (76). This was a necessary step because electron-microscope preparations of the time did not yet provide the contrast and differential staining in sections needed to demonstrate such membranes, although they soon would do so. The other approach, which was undertaken by P. C. Fitz-James (29), was to initiate studies of the biochemistry of a distinct cytological entity, the endospore of a *Bacillus* species. These studies of spores were influential then because they broke new ground and because they contributed to bacterial cytology (29). They indicated to us how powerful combined structural/biochemical research was in providing seminal data even with the restriction of resolution due to the wavelength of light. At this point it became obvious, despite our considerable pride at being able to attain excellent photomicrographs, that the potential of electron microscopy was developing rapidly and that high resolution was essential.

It is hard to realize now what we did not know about the structure of bacterial cells in 1950–1953. The microscopy of the day was giving a new dimension to understanding of cellular components and their functions by correlation with biochemical studies of many kinds of higher cells, and this was a stimulus to finding out what was hidden behind the somewhat undramatic facade of bacteria. The technical and intellectual essentials for effective electron microscopy were coming into focus. Metal shadowing (96) had been available for a few years and showed the topography of bacterial surfaces (36) and external structures such as flagella. Fractionation and chemical analysis of walls following disintegration of the cell was giving exciting data on the nature of bacterial walls in the capable hands of Salton & Horne (79) and could be monitored by electron microscopy even if the techniques were poorly developed. Sections thin enough for effective electron microscopy were obtainable from embeddings in methacrylate (70), and cytological fixatives such as osmium tetroxide (73) were being recognized as suitable preliminaries to embedding and sectioning tissues and bacteria for electron microscopy. I remember still my pleasure and excitement on viewing the first good micrographs of sections of a bacterial cell in the paper by Chapman & Hillier (16) submitted to me as an editorial board member for the *Journal of Bacteriology*. Importantly, the paper demonstrated that the images were recognizable in terms of the light-microscope cytological preparations that we and others were studying and relating to the biochemistry of fractions. Establishing the technical basis and accumulating the experience necessary for adequate preparation of bacterial cells and cell fractions for electron microscopy took more than a decade, and the techniques are still being refined.

Although we recognized the electron microscope as the coming instrument of cytology, Carl Robinow and I firmly believed then, as we do now, that the light microscope is here to stay. I remember brashly trying to persuade G. E. Palade that light microscopy could still do good things for understanding the insides of bacteria. Certainly, light microscopy gave us useful first approximations to the arrangement of the elements of the cell surface in studies of *Bacillus cereus* and *Bacillus megaterium*. A chance but very useful observation (65) showed us that when fixed cells were crushed on the cover slip and then stained, the walls were a rigid sleeve containing septa and the poles were hemispherical caps. The next step was based on Robinow's observation that these cells could be fixed in the plasmolyzed state to retain separation of the surface of the protoplast and the enveloping wall structures. Interpretation of the nature of the protoplasmic interface was greatly assisted by the partition of polar dyes (notably the Victoria blues) into an infinitely thin layer at the protoplasmic interface. This convinced us that this interface was an osmotic barrier, as had been known since 1895, and also was a differentiated structure equivalent to the cytoplasmic membrane of higher cells; the membrane profile was detected rather than resolved by light microscopy (76). We thus had an anatomical concept but were disappointed that the Chapman & Hillier (16) electron microscopic image and others of the time did not show a distinct cytoplasmic membrane at the surface of the protoplast. But most of the elements were there even if their topological features might be dim or distorted owing to inadequacies of fixation and processing; we could see a wall, a remarkably dense cytoplasm with some possible inclusions or spaces for them, cell wall septa accomplishing division, and something that could represent the Feulgen-positive nucleoids.

Salvation was around the corner. Robinow took his spore structure questions to K. R. Porter's laboratory at the Rockefeller Institute and found that sectioning was rewarding because it gave access to the hitherto inaccessible structure of the spore and revealed the cortex beneath the coats (75a). I took my interest in making preparations of infected host cells to R. W. G. Wyckoff's laboratory at the National Institutes of Health in Bethesda, Maryland, and found that whole-cell preparations were not going to take us far in phage or host-cell studies. These first experiences in electron microscopy and visits to James Hillier at the RCA Laboratories at Princeton, New Jersey, convinced us to seek funds for an electron microscope as an essential tool. A Philips EM-100 arrived early in 1954, which we shared for some years with our histologist friend and colleague, R. C. Buck. Learning was fun, but getting the very best out of that instrument (our only electron microscope until 1965) would have been slow indeed without the skill and experience of Aksel

Birch-Andersen (State Serum Institute, Copenhagen), who spent six months with us in 1955 and taught us more than the basics. Many tricks were needed then, including knowledge of when and where to kick an EM!

Carl Robinow's research interests turned to fungal cytology after 1953, and P. C. Fitz-James, now a department member, continued to utilize light and electron microscopy as a monitor and guide to recognizing the steps in spore morphogenesis and fractionation for biochemical studies. From then on each of us developed our own lines of inquiry, but they were close enough for mutual stimulus.

For my part, my graduate students and I explored the anatomy of a variety of bacteria including "blue-green algae." There was thought of taxonomic returns, then, as well as general insights into bacterial structure because J. P. Truant, J. W. Costerton, and I looked at a far wider range of physiological groups than was strictly necessary for the purposes of their theses. We worked toward better cellular descriptions for some bacterial genera such as *Moraxella* (67) and *Vitreoscilla* (19), with comments that applied to taxonomy. Obviously these experiences induced some broad intentions, because I wrote in 1960 (51): " . . . we can now glimpse the bare outline of what can be done to bring our knowledge of bacterial structure to a point where it can be incorporated into the whole picture we should like to have of the kingdoms of living things. . . . All the approaches to the unravelling of structure are bringing to light unique properties of bacteria that will stimulate more effective research." These forays were to be continued thereafter by increasingly focused inspections of *Lampropedia* (17), *Thioploca* (48), nitrifying bacteria (68), *Listeria* (31), and *E. coli* (66), each of which turned our thoughts to details of the cell wall, membranes, and protoplasmic structure.

At the end of all this work I had an unparalleled chance to assess the state of a large part of bacterial cytology by writing a chapter (51) for *The Bacteria,* a benchmark series edited by I. C. Gunsalus and R. Y. Stanier and published in 1960. I believe that volume, devoted entirely to structure and part of a still-significant series, put structural studies into the thinking of many microbiologists.

The major factors impeding the revelation and interpretation of cellular ultrastructure involved preparative techniques. Bacterial cells, particularly the nuclei, were sensitive to the ionic environment during fixation, and most fixation schedules gave unpredictable results. A suitable routine was not available until Ryter & Kellenberger (78) published in 1958 their still-useful method for using osmium tetroxide followed by uranyl acetate in a defined veronal buffer system. This method was not improved upon until dialdehydes, e.g. glutaraldehyde, were introduced in the mid 1960s and used in a prefixation step before osmium tetroxide postfixation. Embedding in a synthetic resin evaded the disturbing artefacts generated by differential polymerization rates in the methacrylates used up to that time.

The sections that we cut were good enough, but the structural contrast due to scattering of electrons by the embedded cell substance was inadequate and could be even less than that of the embedding plastic. Many microscopists realized gradually that scattering could be increased by treating tne sections with heavy metal salts to be taken up by ligands on structural polymers; for some time we favored lanthanum or uranyl salts rather than lead, which tended to precipitate. What was so surprising at the time was the discrimination of layers and substructure that became possible with these staining methods. An early outcome due to applying metal salts was our recognition in 1957 that a double-track unit membrane did indeed enclose the protoplast of the bacterial cell (50); not less important as a lesson was our then finding that earlier papers had shown such a membrane but it was not recognized. Generally we saw what we were prepared to see at each stage of our understanding of ultrastructure. We applied the technique usefully, for example in my study with Francombe and Mayall (61) of the remarkably direct effect of penicillin poisoning on the structure and integrity of staphylococcal cell walls, which helped to focus attention on the cell wall and peptidoglycan as the target of that remarkable antibiotic. Eventually in 1963 Reynolds (74) put into general use a method of combining lead citrate and uranyl acetate for staining sections, which became a standard treatment to give contrast.

We needed an effective method for following the fractionation of cells and cell walls and assessing the nature of the pieces of cells displayed after ballistic or sonic disruption. We used light microscopy to monitor cells and fractions negatively stained with nigrosin, a most useful technique then and now, but limited as to resolution; and we sectioned pellets to get profiles. Despite the amount of nigrosin that we used it never occurred to us that negative staining for electron microscopy was possible; the demonstration of this technique and practical applications by Brenner & Horne (7) was a scientific bombshell with repercussions continuing today. Brian Mayall and I tried it and got immediate results looking at a moiré pattern of an array on a wall fragment of a strange coccus, later identified as *Deinococcus (Micrococcus) radiodurans*. We described this in a review (52) and demonstrated it in a paper at the 1958 Microbiology Congress in Stockholm. It was exciting enough even if no formal publication resulted, but it did lead us into other applications and particularly studies of the nature of wall arrays in *Aquaspirillum serpens* (53), which have engaged us ever since.

These general and technical developments, which I hasten to say we participated in but did not originate, allowed us to enter into more detailed and complex studies. A worthwhile preliminary to our relation of structural studies and biochemical fractionation was determination of the location of the peptidoglycan component in the sectioned wall profile of *E. coli* and some other gram-negative bacteria (66). The location correlated exactly with the order established in the elegant biochemical dissection of the *E. coli* wall initiated by Weidel and colleagues (93). Then to carry this further I studied what happened to these layers during cell wall septum formation and cell division with Pamela Steed-Glaister, then doing her PhD studies, and we established an effective baseline for the wall studies to come (86). This work allowed us to understand the various forms of septum that we see. Septa incorporate a peptidoglycan layer as the primary component, which intrudes either as a double loop of peptidoglycan or as a single layer that thickens and differentiates into a doublet (laughingly attributed to "zipperase") to allow intrusion of outer layers and the ultimate separation of the cells. However, the constrictive divisions of *E. coli* and all but a few mutant enteric bacteria were a puzzle until we were able to show (first serendipitously, by my having let a water bath get too hot) some 10 years later that an appropriate fixation regime revealed septum formation of the basic type involving the peptidoglycan layer. We had to be confident and persuasive, thanks to the clear experiments of Ian Burdett (14), that this was an image that could not possibly be an artefact; yet the way some of our colleagues refer to septation in *E. coli* makes one wonder whether or not they are convinced. However, the structural details of these events are now being resolved in great detail for both gram-negative and gram-positive models (34, 38, 47).

Electron microscopy, although limited for cytochemical analysis, was adept at revealing structures in search of a function. Among these were the membranous intrusions often found near the site of septum formation and at the poles of the dividing cell. These lamellated structures were named mesosomes by Fitz-James (29) and were recognized as having continuity with the cytoplasmic membrane. These may have been among the sites identified earlier as bacterial mitochondria, because the granules exhibited strong oxidation-reduction reactions (49). However, cell fractionation indicates that respiratory activity is a general property of the cytoplasmic membrane. The polemical arguments died and left behind the interesting problems raised by the variety of membranous structures becoming visible in the cytoplasm. Some of these structures were obviously functional, as was supported by biochemical data, providing for photosynthesis or for complex metabolic processes requiring coordinated energy transfers such as nitrification.

Stanley Watson and I (68) were excited by the elaborate membranes of nitrifying species of *Nitrosocystis, Nitrosomonas,* and *Nitrobacter.* It was clear from these and other examples that there were membranous organelles, but in most cases they arose from and were still continuous with the cytoplasmic membrane. The exceptions, where the internal membranes separated from the peripheral membrane, still originated from the cytoplasmic membrane, however specialized the final function. Thylakoid membranes in cyanobacteria may yet prove to be the only bacterial membranes independent of the cytoplasmic membrane. In general, a major demand for a membrane-based function results in more membrane, and space for it has to be gained by internal projection, which may be structurally differentiated for whatever energy-linked process the cell may require. We now realize that not all membrane functions can be everywhere in the periphery and that a degree of area specialization is a necessity.

What is really involved in the formation of mesosomes is still far from certain; they may be artefacts of fixation because they are absent in freeze-cleaved preparations not exposed to a fixative (99). Yet they occur in persuasive sites, and the bits of membrane that give rise to them, even as artefacts, may be associated with some specific function. One concept that may not be too farfetched is that their function derives from their association with nucleoids (77) and the probability that DNA replication and nucleoid segregation requires an association with membrane sites.

VENTURING INTO MACROMOLECULAR ASSEMBLIES

An important outcome of the introduction of electron microscopy has been our ability to recognize specific cellular components and, by their structural characteristics, to recognize them also in the fractions generated after cell disintegration. This discriminatory function has been as essential to many cell biology structure/function studies as it has been to virology. Discrimination was made immensely more effective when negative staining with appropriate heavy metals (notably phosphotungstates, molybdates, and uranyl salts) was added to the techniques of metal shadowing of specimens and freeze-cleaved and etched preparations (7). What then became possible was the outlining of the shape and form of some kinds of macromolecules, mainly proteinaceous, to a level of detail limited by the resolution attainable on that type of specimen and the potential of that electron microscope. So an effective resolution of about 2.5 nm became possible on biological specimens, which cannot be infinitely thin. Negative staining was applied in many fields, but in microbiology there was a rapid evolution of structural studies on viruses, membranes, ribosomes, and in our case bacterial surface arrays.

I was very impressed by the images we obtained early in the 1960s from negatively stained specimens of cell wall fragments of *Deinococcus (Micrococcus) radiodurans* (52), *Lampropedia hyalina* (17), and *Aquaspirillum serpens* (53). The fragments showed as two-dimensional, hexagonal arrays of linked subunits with paracrystalline regularity that enveloped the external surface of the cells. The arrays had every possibility of being amenable to fractionation and analysis after the fashion of the techniques applied to *E. coli* by Weidel's group (93). Other possible arrays were assessed; M. V. Nermut, who was visiting in 1966, and I found that *B. polymyxa* possessed a tetragonal array (69). We explored the stability of this array on exposure to various chaotropes (e.g. types of detergents), which we thought might be of use in attempting isolation of the components. We found, as did Baddiley's group (32), that the fraction containing the array was mostly protein. We also took a first step toward image processing by enlisting the help of Klug's group at Cambridge. They found that each unit forming the tetragonal array was formed by four centers of mass arranged in p4 symmetry with the whole array (28), in which the lattice frequency was 10 nm. The units are now better resolved (15).

It was obvious that these arrays, or S-layers as they are now termed, were a common feature of bacteria in nature and, as a major protein component of the cell, would be retained on these cells for reasons of selective advantage. They needed serious study, and we chose to study the S-layer of *A. serpens* because it was a single layer and seemed to have properties useful for the manipula-tions required for isolation, as Pamela Steed-Glaister's studies of stability during growth in fluid media had shown (87). Francis Buckmire and I (11–13) isolated the *A. serpens* VHA S-layer protein and explored its properties, and these studies formed the basis of a study that I still continue in association with Susan Koval (42). The protein proved to be a large, 140-kd, acidic protein that can self-assemble into a lifelike array with the help of Ca^{2+} and the template provided by the outer membrane of the organism. This protein has been subjected recently to further biochemical and assembly studies (39–41), and its properties help to provide a basic description of many of the S-layer proteins (81, 82). I explored the variety of structures possible on the walls of a number of species of *Aquaspirillum* with T. J. Beveridge. Among them *A. putridiconchylium* showed p2 symmetry (4, 89), and several had two or more layers. *A. serpens* MW5 had a double layer and, in similar fashion, was subjected to image analysis in a collaboration with Murray Stewart (90). The MW5 array showed in freeze-etched preparations as a linear structure and in negative stains as a linear moiré. The image analysis resolved this moiré as two hexagonal arrays with similarly sized units superimposed but slipped a half interval along one three-fold axis. I studied the proteins with Marion Kist (37), and we found that the inner layer protein (150 kd) would self-assemble in vitro but would also form a directly linked array on the outer membrane. But the outer layer protein (125 kd) would assemble only on the formed inner layer, and these assemblies (like that of strain VHA) required either Ca^{2+} or Sr^{2+} even if the inner layer hexamer unit did not. The proteins have few or no sulfur-containing amino acids, and it is apparent in our model systems that divalent cations are required for both assembly and the conformation of the monomers in most cases (41). The structural studies of the

VHA S-layer were only two-dimensional or planar until recently; in 1986 some fortunate preparations allowed Robert Glaeser's group (24) to undertake filtered Fourier transform reconstructions from a tilt series of micrographs to give a three-dimensional model of the assembled units, each consisting of six monomers. However, other aspects of understanding are primitive or nonexistent, and these include molecular structure, transport, and regulation.

The other early examples of arrays that started us off have not been neglected: *D. radiodurans* turned out to have some considerable biochemical and taxonomic peculiarities, as I shall recount, and less attention was paid to the S-layers (43, 91, 100). However, Baumeister's group (2) has undertaken detailed structural analyses. *B. polymyxa* engaged me again in a comparative study with Stephen Burley (15), which served as an exercise in establishing our image-processing facility. Eventually, I have returned to the double S-layer that surrounds the cells of *Lampropedia hyalina* (1); this was much too complex a structure to consider before, and even now it is a challenge to work out the order of assembly and the identity of the multiple components of the outermost layer.

The considerable amount of information now available about S-layers is not susceptible to summation in this essay. They have been described as a component of the walls of about 200 species of bacteria from most major phylogenetic groups, and the phenomenon must be considered as a general attribute with varied functions. Fortunately there are several recent reviews (42, 81, 82) to provide the details. To a considerable extent my early work sparked an interest in these models of structural assembly of macromolecules, and the 25 years since then have involved me with 25 coworkers; sadly, it is not possible to give appropriate credit to all of them in this essay.

I had hoped at one time to be able to contribute to a structural and macromolecular description of the basal complex, the motor, of flagella in studies undertaken with James Coulton (20). It is more than a challenge to resolve components of this puzzling rotor-stator mechanism, which is an organelle not much more than 25 nm in total depth and diameter, effectively described in principle by DePamphilis & Adler (23). The closest we got to any macromolecular description of the operative part in the plasma membrane was to see that the central rod was made of triads forming a hole down the center, and to find a circlet of studs around the M-ring in the freeze-cleaved membrane. I remain sceptical, as are others (27), about the exact relationships of the structure to the plasma membrane, and I believe that we are missing the boundaries of a compartment and some part of the complex penetrating to the cytoplasm.

The bacterial wall is a remarkably complex and dynamic structure interposed between the cell and an often hostile environment. The export-import services and machinery such as flagella that traverse the wall put special demands on the maintenance of integrity and the assembly-disassembly capabilities of the entire complex. Enormous effort has been given to understanding the governance of the murein covalently cross-linked network because it is a target for antibiotics. The entropy-driven assemblies of membranes and porins and the integral transport mechanisms have had well-deserved attention in recent years. The S-layers are equal participants in the dynamics of walls in growth and division, and go by a simpler but no less sophisticated set of rules. Furthermore, they draw attention to functions other than strength (42), and the more complex of them may act as generalized models for the assembly of cell structures.

TAXONOMY IN TRANSITION

There were compelling reasons for my getting involved in taxonomy: My father was a trustee of *Bergey's Manual of Determinative Bacteriology* (1936–1964) and we talked about the problems; my work involved diagnostic bacteriology and wrestling with identification, and the study of the structure of bacteria inevitably drew attention to the inadequacies of descriptions. An early direct contact with the *Bergey's Manual* trust was an invitation from R. S. Breed in 1955 to join in discussing the description of the "Schizomycetes" and how it might be revised; my advice was not taken.

Work on structural aspects of bacteria had brought both Carl Robinow and me into contact with Roger Y. Stanier, whose thoughts about the nature of bacteria and approaches to taxonomy, sharpened by his association with C. B. van Niel, were particularly penetrating. His papers shaped my views and my intentions toward the definition of unique features of bacteria. The opportunity to do something about them came hot on the heels of my completing the review of structure for *The Bacteria* (51) with the invitation to contribute to the 1962 *Symposium of the Society for General Microbiology*. This essay, "Fine structure and taxonomy of bacteria" (52), had some good ideas, but the conclusions were not as strong and definitive as those of the almost coincident essay, "The concept of a bacterium," by Stanier & van Niel (85). Neither essay developed a formal taxonomic proposal, but both were disposed to accept any nomenclatural arrangement that recognized the relationship of blue-green algae and bacteria, that incorporated them both into a grouping distinct from all other microbes and macrobes, and that recognized their unique features of organization. Stanier & van Niel referred to them consistently as "prokaryotic organisms," which reflected Stanier's decision the year before (84) to describe bacteria as cells and to use Chatton's vernacular terminology, introduced in 1937 (18), for the major divisions of cellular organization, eukaryotes and prokaryotes. I feel that I rode on the shoulders of worthier colleagues because, almost inadvertently, I was the author of the formal name, the kingdom Procaryotae (54). I had something to do with establishing the generalizations that supported a position that became agreeable to most microbiologists.

It was clear to me that taxonomy was a worthwhile endeavor, even if it was still unpopular, because it required the use of all that one knew (or thought one knew) about organisms and because it drew attention to great gaps in basic knowledge. At that time objective means of checking the relationships in an arrangement of taxa or within any taxon were only just developing. Both the selection and the reliability of the phenotypic characters were largely a matter of faith. Mechanisms evolved over the past 30 years have given a scientific aspect to the assessment of characters and, far beyond the powers of serology, the definition of taxa. These mechanisms are now well known and came to include, successively, numerical (computer-assisted) taxonomy, murein (peptidoglycan) types, the G+C ratio in DNA, DNA/DNA homology, DNA/RNA hybridization, and the exploitation of the highly conserved RNA cistrons and the ribosomal RNA sequences. The power of these approaches to taxonomy has only been fully realized in the past decade, and the data supporting a phylogenetic assessment are still only partially complete.

I joined the *Bergey's Manual* Board of Trustees in 1964. The chairman, R. E. Buchanan, was strongly oriented to nomenclature and classical approaches. Most of the trustees were practical bacteriologists, but there was one grand heretic among them in the person of S. T. Cowan (22). A consensus in views was hard to accomplish, and arguments prolonged the gestation of the 8th edition, which did not appear until 1974. The turning point came in 1969, by which time R. Y. Stanier was also a member, with the decision to use vernacular names for the chapter headings of the 8th edition because the higher taxa were of dubious validity. However, our expert authors were by no means convinced, or if they were in principle, many traditional arrangements were nonetheless maintained.

We were by no means satisfied, scientifically, with what we had done, despite all the good intentions of N. E. Gibbons (the editor, following Buchanan's death) and the board of editor-trustees. I wrote at the time (55): "The future will have to bring a regrouping of higher taxa to express a more coordinate view. . . . Haste is unwise; all previous classifications seem to have suffered infinite rearrangement due to insufficient information. . . . The new insights are likely to come from a clear understanding, on a comparative level, of the components of the genome of procaryotic cells." That goal is closer now.

The ten years after 1974 saw some remarkable changes in the rules that govern the nomenclature that expresses the taxonomic conclusions of bacteriologists. The 1975 *International Code of Nomenclature of Bacteria* (44) allowed for clearing the records of the enormous list of synonyms and useless names by declaring January 1, 1980 to be the new starting date for bacterial nomenclature and arranging for the *Approved Lists of Bacterial Names* (80) to be available from that date. We owe this most unusual form of taxonomic housekeeping (which generated envy in taxonomists devoted to plants and animals) to the international efforts of V. B. D. Skerman and P. H. A. Sneath, in particular, through the Judicial Commission of the International Committee for Systematic Bacteriology. It was time for an assessment of taxonomic problems in the groupings of bacteria, and the *Bergey's Manual* trustees resolved to produce a systematic manual.

A new view of bacterial taxonomy has arisen since Carl Woese and his group proposed (30) that the computer-assisted comparison of T_1 ribonuclease–resistant oligonucleotide sequence catalogs of the 16S ribosomal RNA of representative bacteria gave clear evidence for the phylogenetic lineages emerging from the main stem of the clonal evolution of bacteria and, in fact, all living things (97). They all share the ribosomal mechanism for constructing proteins. The ribosomal RNA molecules are up to now the most powerful, universal, and practical biological semantides or molecular sequences documenting evolutionary history in the sense of Zuckerkandl & Pauling (101). So the past decade has seen the accumulation of data on some 500 representative strains from many but not all taxa, and many old taxonomic assumptions can be tested. There are several distinct high-level phylogenetic taxa in both of the major phylogenetic divisions, the Archaebacteria (at least three) and the Eubacteria (at least 10). Some of the rRNA groupings of Eubacteria (98) appear to pose few problems of phylogenetic and phenotypic interpretation (e.g. the spirochetes and the cyanobacteria); some provide associations that are hard for traditional bacteriologists to assimilate (e.g. among gram-positive bacteria, *Micrococcus* closely associated with *Arthrobacter*); and others show such a remarkable diversity of morphology and physiology that there will be difficulty in developing phenotypic consistency [e.g., notably, the purple photosynthetic bacteria and their relatives (92)]. One can be optimistic, as Woese is (97), that the bacteria will eventually fall into naturally (i.e. phylogenetically) defined taxa and that an appropriate search will reveal unifying phenotypic characters. I hope that we will soon discover other cistrons determining nearly universal, complex, and essential cellular functions that are highly conserved. We must try to cross-check the phylogenetic conclusions, now based almost entirely on the RNA cistrons, especially in the diverse and rapidly evolved groups to detect anomalies or lateral transfer. This is the more important because transfer of heterologous ribosomal RNA genes (*Proteus vulgaris* to *E. coli*) has been attained with a plasmid vector (71) and as a result 25% of ribosomes contained the heterologous rRNA. Genomic integration has also been claimed, and if this is true, a phylogenetic chimera is possible. Who knows what strange experiments have succeeded in nature's laboratory?

With all this ferment, the development of *Bergey's Manual of Systematic Bacteriology* in four volumes appearing serially 1984–1988 presented unusually difficult problems, which came during my time as chairman of the trustees, 1976–1988. The data were compelling but patchy. Of course, it is sad that one cannot instantly and dramatically reform the whole of bacterial taxonomy, as was the fervent hope of the main laborers in the field of molecular phylogeny (98). It was possible to institute no more than cosmetic changes in many groups, major revisions in only a few, and phylogenetic consistency only in the *Archaebacteria*. Now it can be recognized that it is important to express the principle of *festina lente* (make haste slowly), as exemplified by the approach to the problems in the *Bacillaceae* (83), and keep the complexities in reserve until a more complete survey and decisions are accomplished. This attitude is helpful for the retention of a useful framework of classification for groups, including several genera threatened with major surgery.

I have been interested in applying modern forms of taxonomy to the *Deinococcus* group (9), seemingly gram-positive cocci with gram-negative characteristics, which proved (10) to be a distinctive and ancient lineage but phenotypically deceptive. If it is an ancient set of clones there should be relatives which might be hard to recognize. The superficially distinctive feature of radiation resistance is a mutable character, so the usual selective mechanism of using radiation is likely to reveal only clones similar to the extant species. The distinctive polar lipid profile (21) may aid recognition, in addition to the ribosomal RNA sequences and select signatures, which have already identified a gram-negative relative in *Deinobacter grandis* (72). I am prepared to follow with great interest the next steps in developing a new natural taxonomy of bacteria.

Fortunately, what appears to be accepted in classificatory schemes is not immutable. Taxonomy has to represent the best of science and for that reason is bound to change with new knowledge. I have speculated on the need for an academic taxonomy as opposed to a practical classification to serve the bench worker who identifies bacteria (57). However, there is need for only one taxonomy (92), soundly based in phylogenetic terms and phenotypically recognizable at all ranks, expressing a best view of a natural order. The bench worker does not need a complete hierarchy but operates within a framework of experience and needs only a number of identifiable vernacular groupings. No doubt, we must eventually develop new molecular and phenotypic markers for the recognition of all taxa. The tests used should be available and practical.

There will be changes in approaches to identification, but it is to be hoped that whatever systems are used, they will not inhibit the recognition of new taxa. Almost every ecological niche includes species that have not been cultivated and recognized, which should dispel any feeling of diminishing returns in the study of natural populations or in taxonomy.

ODDS WITHOUT END

Research may owe its continuing support to high-profile projects, but the next generation of research and so-called innovation will owe a lot to the odds and ends undertaken out of sheer curiosity, to satisfy a hunch, to provide experience for a summer student, and for many other ostensibly trivial but important reasons. These items enliven what otherwise might be periods of diminishing returns and spawn unexpected new research; so they become "odds without end." All research programs should devote a proportion of their funds to free-wheeling exploration (What the X Foundation doesn't know isn't good for it), much as buildings need a proportion of cost devoted to art. Some experiments should be done with "controlled sloppiness," as S. E. Luria told me years ago in order to encourage the unexpected. (If you do the same old thing you will get the same old answer.)

When T. J. Beveridge was working with me on his doctorate and studying the wall structure of spirilla we had many an occasion to discuss the problem of revealing substructure by staining with metal salts. So we experimented with isolated *Bacillus subtilis* walls to see what sort of capabilities the durable cell-wall polymers and heteropolymers might have in capturing and sequestering metal ions from solutions of their salts. The results (5) showed that substantial amounts of many metals were taken out of solution (including those important to metal enzymes such as Fe^{3+}, Cu^{2+}, Mn^{2+}, and Zn^{2+}) but some were not absorbed (such as Li^+, Ba^{2+}, Co^{2+}, and Al^{3+}). Furthermore, if the walls were linked to a column and a series of metal-salt solutions was run through it, some metals were strongly bound (including Mg^{2+}, Ca^{2+}, Fe^{3+}, and Ni^{2+}) and others were displaced or replaced. We were interested in the effects of modifying ligands (6) or, for instance, of saturation with Mg^{2+} on the subsequent staining of structures with ruthenium (severe) or lead and uranyl acetate (minimal). But our geologist colleague, Professor W. S. Fyfe, was more interested in why many ore bodies have a high percentage of organic residues and particular selections of metals. The mediation of biopolymers provided a stimulating hypothetical mechanism. The observations were extended by collaboration on geological diagenesis (3) and have been explored further in recent years (26), as have the consequences for the physicochemical well-being of wall components (25). Of course, parallel work has been and is going on in many laboratories concerned with ore leaching and metal transport in waters, to which our observations have contributed.

A discussion with Carl Robinow, C. L. Hannay, and Philip Fitz-James about why *Bacillus laterosporus* spores were eccentric in the cell led to the recognition of parasporal bodies and to the rediscovery of the spore-associated crystals in *Bacillus thuringiensis* (33). This is certainly a fertile field of research and practical application today.

On a lesser scale, an early study with Aksel Birch-Andersen (58) on the nature of flagella basal structures was diverted somewhat by the finding of a decoration on the inside of the plasma membrane of *Aquaspirillum serpens,* which we called polar membrane because of its position surrounding the polar tuft of flagella. An exactly similar structure is associated with many lophotrichous bacteria. We have now tried to understand this "structure in search of a function" and have found that the polar membrane of *Campylobacter jejuni* is a close-packed array of ATPase (8). What is consequential is that this particular ATPase turned out to be unusual by being specifically directed to ATP, activated by Mg^{2+} but inhibited by Ca^{2+}, made of a single subunit, and serologically distinct from *E. coli* ATPase; it is also likely to be a phylogenetically distinct lineage of that highly conserved enzyme.

As at the beginning of my time in science, the odds and ends still prove to be a stimulus. The current example is the lucky finding that high growth temperature stimulates cyst formation by *Azospirillum brasilense* (63). Now we know that a heat shock is equally effective for cyst formation at a growth temperature that is usually ineffective.

SOME SERVICES TO SCIENCE

I was brought up to believe that science knew no boundaries and that it was a duty of scientists to communicate freely. So from the outset I joined with the major societies in my orbit and took part in the meetings of the Laboratory Section of the Canadian Public Health Association and the Society of American Bacteriologists, as they were known then. The friendly winter meetings of the former, with extensive discussions of medical microbiology, and the comprehensive coverage of microbiology each springtime at the meetings of the latter were just what I needed as a stimulus to get started. But I was soon (1950–1951) in the midst of organizing a Canadian Society of Microbiologists, a much-needed catalyst and unifier for the diverse applied and basic microbiologists of the country. It took off owing in large part to the efforts of N. E. Gibbons, the founding secretary. I was put by my senior colleagues in the position of chairman of the organizing committee and the inaugural meeting in Ottawa in 1951. This was unforgettable for me, not just because it gave me a fine experience and the confidence of my colleagues, but because my headache on the last day of that meeting was due to encephalitis, which bedded me in that city for three weeks and took seven months out of my working life. The charter members elected me the founding president, perhaps as an act of kindness. Societies and their journals have been important to me and my work ever since, particularly the Canadian and the American societies.

I had an interest in editing, but it is not clear to me why I was asked to join the editorial board of the *Journal of Bacteriology* by J. Roger Porter in 1951; the invitation was unexpected because I had published less than a handful of papers and only three in that journal. Membership on the editorial board was good experience for my appointment as founding editor of the *Canadian Journal of Microbiology* (1954–1960); this appointment was less mysterious because I had a role in persuading the National Research Council of Canada that the journal was justifiable. In these years I learned the basics of editing from experienced colleagues and by doing it. My most challenging assignment was as editor of *Bacteriological* (later *Microbiological*) *Reviews,* 1969–1979, when I learned properly that a scientific editor's crucial job was not just to adjudicate the reports of referees but was to help authors to do their best (56). It was no mean exercise in diplomacy to deal with the sensitivities of undoubted authorities in their fields, both authors and referees, who could produce not only an appalling text but also a distressing narrowness of view. This was one aspect of my education; the other was association with people of marvelous ability and judgment both on the editorial boards and on the Publications Board of the American Society for Microbiology, notably L. Leon Campbell and Robert A. Day, who really taught me about the management and production of scientific publications. My association with the International Committee for Systematic Bacteriology and the *International Journal of Systematic Bacteriology* has brought experience in the additional stringencies of monitoring the description of bacteria and the application of rules of bacterial nomenclature. There is a great need to help authors in a field unfamiliar to most of them. I feel strongly that an editor is not the savior of science; if the editor is any good, the role is more of a "friend in court" for the author, who is the only person really responsible for what is written. The editor is the author's most concerned critic and has the advantage of taking or refusing the comments of referees to formulate rational advice.

My involvement in *Microbiological Reviews* increased my interest in the affairs of the American Society for Microbiology and my interest in what societies do for science and society.

Being an editor of a first-rate journal has an enhancing effect on one's scientific profile. I was most honored to be a candidate and to be elected president of the American Society for Microbiology, 1972–1973, which was in all respects a remarkable experience. I am sure I owe this signal recognition by the membership as much to my activities as an editor as to my scientific contributions.

I regret no part of these varied experiences and recommend that all aspiring young scientists owe a duty to scientific societies and their journals. We have to be prepared to communicate and to promote the highest standards in both the meetings of scientists, in the broadest arenas possible, and the journals of science. It is not good practice to speak only to a group of connoisseurs. We must play the roles of critic and of supporting actor in the interplay of science and society.

ENVOI

Science has shifted in the past 50 years from being the province of scientists to being a crucial part of our culture, permeating all conditions of life and living. Microbiology may have been ahead of most disciplines in this matter because of the sensitivity to matters affecting human health. The shift should be encouraging because one might hope that wisdom should accompany or be encouraged by understanding and technical competence. But there has been no more than a minimal attainment of that goal, and man's place in nature's world (and vice versa) seems ever more insecure, much as we would prefer to interpret it otherwise. My own contributions to the fostering of wisdom have been minimal, especially on the public and political fronts, which are so critical. At one time I scorned taking on public enlightenment, but I am convinced now that it is the only route to the heart of the political animal, which is effectively indifferent to anything beyond votes and short-term gains. I now believe that scientific societies and institutions should balance their concerns for communication within the house of science with an equal concern for public understanding as a stimulus to political action. I am not proud of my inaction in the public arena, nor am I impressed with what we have managed to do in our universities and colleges to inculcate a high level of understanding in the 15–25% of the new generations that transit these institutions seeking enlightenment.

A major factor in fostering my scientific activity and productivity has been the continuing support by research grants thoughtfully administered, and I owe thanks to the old Medical Committee of the National Research Council of Canada and its successor, the Medical Research Council of Canada. The former did me a great service for ten of my early years by awarding a block grant, which had no strings attached and only required a letter at year's end asking that it be continued, with a minimal accounting. It was a fixed sum and I had to abandon the award due to inflation of costs, but it did a great deal of good. Life is not so easy for even the most fortunate of young scientists today, who must spend 10% or more of their year in supplicating for funds. From the way such applications are treated one wonders whether or not anything should be done for the first time.

An unwitting scientific reward resulted from my study of bacterial structure because understanding of the structure of cells was an essential accompaniment to the development of cell biology and modern microbiology. As I have pointed out, a part of the transformation of bacterial taxonomy involved the realization that the unique features of bacteria as cells are crucial to description, and this was equally true of the understanding of function. The microscopist was, therefore, in a fortunate position on the sidelines of the great game between the cell biologists and the molecular biologists, and had entry to the game in some of the interesting plays. I may not have contributed more than peripheral structure to the game, but some friends and colleagues in cytology were major players, such as Keith Porter and Edward Kellenberger, to each of whom we owe a lot of our understanding. The great importance of correlating structure and function will be no less tomorrow than it is today.

Many of my contemporaries joined in the great exploitation of *E. coli* and a few other "handmaidens of science," doing a great service to science and mankind in the doing. However, their lives had a hectic and unenviable pace compared to the life I led without the hot breath of competitors on my neck. My colleagues and I protected ourselves from that fate because, it may be noted, we usually worked on organisms that were not commonly exploited and we could work at our own rate. This attitude also allowed a diversity of research topics, from biochemical cytology to taxonomy, which could be pursued as and when competence allowed. There was opportunity for me to be helpful to colleagues and for them to be helpful to me, which provided scientific pleasures in good company.

Where is the bacteriology I once knew? It is changing in a technological fashion in the medical diagnostic laboratories where I worked in the past. I fancy I would be more comfortable now with the microbial ecologists and those interested in organisms and their associations; their studies give rise to vistas broader than the most direct route to diagnosis. Times change and attitudes must change with them, but our old bacteriologists have done their duty and assisted the birth and development of microbiology and a remarkable range of disciplines and subdisciplines of biology. Whatever happens to current developments, and despite all diminishing interdisciplinary distinctions, we must see that there is a strong disciplinary base of teaching about the life and interrelationships of microbes. Microbiology departments should not be indistinguishable from departments of biochemistry, however well we get on with each other. The biosphere has been explored only in part, and there is much that we need to learn about the life and nature of the microbes that are no small part of it.

It is obvious that I owe much to the people in my life, and I can mention only a few of them in the context of even fewer activities. The roles of some, such as my parents, have been made clear and must not be underestimated. However, my life would not have been half as productive without the understanding of my first wife, Doris, who was my helpmate for just short of 40 years, or, in the past few years, the supportiveness of my second wife, Marion. My life has been enlivened and the work in the laboratory has prospered because of the talent, energy, and adventurousness of the graduate students and postdoctoral fellows who have spent their time with us. I am ever grateful to a passing parade of friends in and around the laboratory; but a few, who are all colleagues whatever their calling, have contributed more to me over the years than I can possibly acknowledge: Igor N. Asheshov[†], Terrance J. Beveridge, Aksel Birch-Andersen, Myrtle Hall, Phyllis Hobson, Gertrude G. Kalz, Susan F. Koval, Marion I. Luney, John F. Marak, Dianne Moyles, Carl F. Robinow, Roger J. Rossiter[†], and Gertrude Vaughan-Dragon.

Microbiology, in its own ways, tells us about life as it is reflected in the microbial cells, which have been in existence for longer than the imagination can appreciate. It tells us about how and where life can be lived, about the extremes of survival, and something about how life has evolved; but how it came about and where it came from is among the mysteries. Mankind must listen to the messages.

> My dream is faded now, and I am through
> With dreaming . . . yet I know
> The iris still will keep its gorgeous hue.

Shushiki (Transl. C. H. Page)

Literature Cited

1. Austin, J. W., Murray, R. G. E. 1987. The perforate component of the regularly structured (RS) layer of *Lampropedia hyalina. Can. J. Microbiol.* 33(12):1039–45

2. Baumeister, W., Karrenberg, F., Engel, R. R., Tentteggeler, B., Saxton, W. D. 1982. The major cell envelope protein of *Micrococcus radiodurans* (R_1): Structural and chemical characterization. *Eur. J. Biochem.* 125:535–44

3. Beveridge, T. J., Meloche, J. D., Fyfe, W. S., Murray, R. G. E. 1983. Diagenesis of metals chemically complexed to bacteria. *Appl. Environ. Microbiol.* 45:1094–108

4. Beveridge, T. J., Murray, R. G. E. 1974. Superficial macromolecular arrays on the cell wall of *Spirillum putridiconchylium. J. Bacteriol.* 119:1019–38

5. Beveridge, T. J., Murray, R. G. E. 1976. Uptake and retention of metals by cell walls of *Bacillus subtilis. J. Bacteriol.* 127:1502–18

6. Beveridge, T. J., Murray, R. G. E. 1980. Sites of metal deposition in the cell wall of *Bacillus subtilis. J. Bacteriol.* 141:876–87

7. Brenner, S., Horne, R. W. 1959. A negative staining method for high resolution electron microscopy of viruses. *Biochim. Biophys. Acta* 34:103–10

8. Brock, F. M., Murray, R. G. E. 1988. The ultrastructure and ATPase nature of polar membrane in *Campylobacter jejuni. Can. J. Microbiol.* 34(5):In press

9. Brooks, B. W., Murray, R. G. E. 1981. Nomenclature for "*Micrococcus radiodurans*" and other radiation-resistant cocci: Deinococcaceae fam. nov. and *Deinococcus* gen. nov. including five species. *Int. J. Syst. Bacteriol.* 31:353–60

[†]Deceased.

10. Brooks, B. W., Murray, R. G. E., Johnson, J. L., Stackebrandt, E., Woese, C. R., et al. 1980. Red-pigmented micrococci: A basis for taxonomy. *Int. J. Syst. Bacteriol.* 30:627–46

11. Buckmire, F. L. A., Murray, R. G. E. 1970. Studies on the cell wall of *Spirillum serpens*. I. Isolation and partial purification of the outermost cell wall layer. *Can. J. Microbiol.* 16:883–87

12. Buckmire, F. L. A., Murray, R. G. E. 1973. Studies of the cell wall of *Spirillum serpens*. II. Chemical characterization of the outer structured layer. *Can. J. Microbiol.* 10:59–66

13. Buckmire, F. L. A., Murray, R. G. E. 1976. Substructure and *in vitro* assembly of the outer structural layer of *Spirillum serpens*. *J. Bacteriol.* 125:290–99

14. Burdett, I. D. J., Murray, R. G. E. 1974. Electron microscope study of septum formation in *Escherichia coli* strains B and B/r during synchronous growth. *J. Bacteriol.* 119:1039–56

15. Burley, S. K., Murray, R. G. E. 1983. Structure of the regular surface layer of *Bacillus polymyxa*. *Can. J. Microbiol.* 29:775–80

16. Chapman, G. B., Hillier, J. 1953. Electron microscopy of ultrathin sections of bacteria. I. Cellular division in *Bacillus cereus*. *J. Bacteriol.* 66:362–73

17. Chapman, J. A., Murray, R. G. E., Salton, M. R. J. 1963. The surface anatomy of *Lampropedia hyalina*. *Proc. R. Soc. London Ser. B.* 158:498–513

18. Chatton, E. 1937. *Titres et Travaux Scientifiques*. Sète, France: Sottano

19. Costerton, J. W. F., Murray, R. G. E., Robinow, C. F. 1961. Observations on the motility and the structure of *Vitreoscilla*. *Can. J. Microbiol.* 7:329–39

20. Coulton, J. W., Murray, R. G. E. 1978. Cell envelope associations of *Aquaspirillum serpens* flagella. *J. Bacteriol.* 136:1037–49

21. Counsell, T. J., Murray, R. G. E. 1986. Polar lipid profiles of the genus *Deinococcus*. *Int. J. Syst. Bacteriol.* 36: 202–6

22. Cowan, S. T. 1971. Sense and nonsense in bacterial taxonomy. *J. Gen. Microbiol.* 67:1–8

23. DePamphilis, M. L., Adler, J. 1971. Attachment of flagellar bodies to the cell envelope: specific attachment to the outer, lipopolysaccharide membrane and the cytoplasmic membrane. *J. Bacteriol.* 105:396–407

24. Dickson, M. R., Downing, K. H., Wu, W. H., Glaeser, R. M. 1986. Three-dimensional structure of the surface layer proteins of *Aquaspirillum serpens* VHA determined by electron crystallography. *J. Bacteriol.* 167:1025–34

25. Ferris, F. G., Beveridge, T. J. 1986. Physichochemical roles of soluble metal cations in the outer membrane of *Escherichia coli* K-12. *Can. J. Microbiol.* 32:594–601

26. Ferris, F. G., Beveridge, T. J., Fyfe, W. S. 1986. Iron-silica crystallite nucleation by bacteria in a geothermal sediment. *Nature* 320:609–11

27. Ferris, F. G., Beveridge, T. J., Marceau-Day, M. L., Larson, A. D. 1984. Structure and cell envelope associations of flagellar basal complexes of *Vibrio cholerae* and *Campylobacter fetus*. *Can. J. Microbiol.* 30:322–33

28. Finch, J. T., Klug, A., Nermut, M. V. 1967. Analysis of the fine surface structure of the macromolecular units on the cell wall of *Bacillus polymyxa*. *J. Cell Sci.* 2:587–90

29. Fitz-James, P. C. 1960. Participation of the cytoplasmic membrane in the growth and spore formation of bacilli. *J. Biophys. Biochem. Cytol.* 8:507–28

30. Fox, G. E., Pechman, K. R., Woese, C. R. 1977. Comparative cataloging of 16S ribosomal ribonucleic acid: molecular approach to procaryotic systematics. *Int. J. Syst. Bacteriol.* 27:44–57

31. Ghosh, B. K., Murray, R. G. E. 1967. Fine structure of *Listeria monocytogenes* in relation to protoplast formation. *J. Bacteriol.* 93:411–26

32. Goundry, J. A. L., Davison, A. R. A., Baddiley, J. 1967. The structure of the cell wall of *Bacillus polymyxa* (NCIB 4747). *Biochem. J.* 104:16

33. Hannay, C. L., Fitz-James, P. C. 1955. The protein crystals of *Bacillus thuringiensis* Berliner. *Can. J. Microbiol.* 1: 694–710

34. Higgins, M. L., Shockman, G. D. 1976. Study of a cycle of cell wall assembly in *Streptococcus faecalis* by three-dimensional reconstructions of thin sections of cells. *J. Bacteriol.* 127:1346–58

35. Hobot, J. A., Villiger, W., Escaig, J., Maeder, M., Ryter, A., Kellenberger, E. 1985. Shape and fine structure of nucleoids observed on sections of ultrarapidly frozen and cryosubstituted bacteria. *J. Bacteriol.* 162:960–71

36. Houwink, A. L. 1953. A macromolecular monolayer in the cell wall of *Spirillum* spec. *Biochim. Biophys. Acta* 10: 360–66

37. Kist, M. L., Murray R. G. E. 1984. Components of the regular surface array of *Aquaspirillum serpens* MW5 and their

assembly *in vitro. J. Bacteriol.* 157: 599–606

38. Koch, A. L., Doyle, R. J. 1986. The growth strategy of the gram-positive rod. *FEMS Microbiol. Rev.* 32:247–54

39. Koval, S. F., Murray, R. G. E. 1981. Cell wall proteins of *Aquaspirillum serpens. J. Bacteriol.* 146:1083–90

40. Koval, S. F., Murray, R. G. E. 1983. Solubilization of the surface protein of *Aquaspirillum serpens* by chaotropic agents. *Can. J. Microbiol.* 29:146–50

41. Koval, S. F., Murray, R. G. E. 1985. Effect of calcium on the *in vivo* assembly of the surface protein of *Aquaspirillum serpens* VHA. *Can. J. Microbiol.* 31:261–67

42. Koval, S. F., Murray, R. G. E. 1986. The superficial protein arrays on bacteria. *Microbiol. Sci.* 3:357–61

43. Lancy, P. Jr., Murray, R. G. E. 1978. The envelope of *Micrococcus radiodurans:* Isolation, purification, and preliminary analysis of the wall layers. *Can. J. Microbiol.* 24:162–76

44. Lapage, S. P., Sneath, P. H. A., Lessel, E. F., Skerman, V. B. D., Seeliger, H. P. R., Clarke, W. A., eds. 1975. *International Code of Nomenclature of Bacteria.* Washington, DC: Am. Soc. Microbiol.

45. Luria, S. E., Human, M. L. 1950. Chromatin staining of bacteria during bacteriophage infection. *J. Bacteriol.* 59:551–60

46. Luria, S. E., Palmer, J. L. 1946. Cytological studies of bacteria and bacteriophage growth. *Carnegie Inst. Washington Yearb.* 45:153–56

47. MacAlister, T. J., Cook, W. R., Wiegand, R., Rothfield, L. I. 1987. Membrane-murein attachment at the leading edge of the division septum: A second membrane-murein structure associated with morphogenesis of the gram-negative bacterial division septum. *J. Bacteriol.* 169:3945–51

48. Maier, S., Murray, R. G. E. 1965. The fine structure of *Thioploca ingrica* and a comparison with *Beggiatoa. Can. J. Microbiol.* 11:645–55

49. Mudd, S. 1953. The mitochondria of bacteria. *Bacterial Cytology, Symp. 6th Int. Congress Microbiol.* pp. 67–81, Suppl. Rend. Ist. Super. Sanità, Rome.

50. Murray, R. G. E. 1957. Direct evidence for a cytoplasmic membrane in sectioned bacteria. *Can. J. Microbiol.* 3:531–32

51. Murray, R. G. E. 1960. The internal structure of the cell. In *The Bacteria,* ed. I. C. Gunsalus, R. Y. Stanier, 1:35–96. New York: Academic

52. Murray, R. G. E. 1962. Fine structure and taxonomy of bacteria. *Symp. Soc. Gen. Microbiol.* 12:119–44

53. Murray, R. G. E. 1963. On the cell wall structure of *Spirillum serpens. Can. J. Microbiol.* 9:393–401

54. Murray, R. G. E. 1968. Microbial structure as an aid to microbial classification and taxonomy. *Spisy Prirodoved. Fak. Univ. J. E. Purkyne Brne* 43:249–52

55. Murray, R. G. E. 1974. A place for bacteria in the living world. In *Bergey Manual of Determinative Bacteriology,* ed. R. E. Buchanan, N. E. Gibbons, pp. 4–9. Baltimore, Md: Williams & Wilkins. 8th ed.

56. Murray, R. G. E. 1983. What *is* an Editor for? *CBE Views* 6:14–19

57. Murray, R. G. E. 1984. The higher taxa, or, a place for everything . . .? In *Bergey's Manual of Systematic Bacteriology,* ed. N. R. Krieg, J. G. Holt, pp. 31–34. Baltimore, Md: Williams & Wilkins.

58. Murray, R. G. E., Birch-Anderson, A. 1963. Specialized structure in the region of the flagella tuft in *Spirillum serpens. Can. J. Microbiol.* 9:393–401

59. Murray, R. G. E., Douglas, H. C. 1950. The reproductive mechanism of *Rhodomicrobium vanniellii* and the accompanying nuclear changes. *J. Bacteriol.* 59:157–67

60. Murray, R. G. E., Elder, R. H. 1949. The predominance of counterclockwise rotation during swarming of *Bacillus* species. *J. Bacteriol.* 58:351–59

61. Murray, R. G. E., Francombe, W. H., Mayall, B. H. 1959. The effect of penicillin on the cell structure of staphylococcal cell walls. *Can. J. Microbiol.* 5:641–48

62. Murray, R. G. E., Gillen, D. H., Heagy, F. C. 1950. Cytological changes in *Escherichia coli* produced by infection with phage T2. *J. Bacteriol.* 59:603–15

63. Murray, R. G. E., Moyles, D. 1987. Differentiation of the cell wall of *Azospirillum brasilense. Can. J. Microbiol.* 33:132–37

64. Murray, R. G. E., Pearce, R. H. 1949. The detection and assay of hyaluronidase by means of mucoid streptococci. *Can. J. Res. Sect. E* 27:254–64

65. Murray, R. G. E., Robinow, C. F. 1952. A demonstration of the disposition of the cell wall of *Bacillus cereus. J. Bacteriol.* 63:298–300

66. Murray, R. G. E., Steed, P., Elson, H. E. 1965. The location of the mucopeptide in sections of the cell wall of *Escherichia coli* and other gram-negative bacteria. *Can. J. Microbiol.* 11:547–60

67. Murray, R. G. E., Truant, J. P. 1954. The morphology, cell structure and taxonomic affinities of the *Moraxella*. *J. Bacteriol.* 67:13–22

68. Murray, R. G. E., Watson, S. W. 1965. Structure of *Nitrosocystis oceanus* and comparison with *Nitrosomonas* and *Nitrobacter*. *J. Bacteriol.* 80:1594–609

69. Nermut, M. V., Murray, R. G. E. 1967. The ultrastructure of the cell wall of *Bacillus polymyxa*. *J. Bacteriol.* 93:1949–65

70. Newman, S. B., Borysko, E., Swerdlow, M. 1949. New sectioning techniques for light and electron microscopy. *Science* 110:66–68

71. Niebel, H., Dorsch, M., Stackebrandt, E. 1987. Cloning and expression in *Escherichia coli* of *Proteus vulgaris* genes for 16S rRNA. *J. Gen. Microbiol.* 133:2401–9

72. Oyaizu, H., Stackebrandt, E., Schleifer, K. H., Ludwig, W., Pohla, H., et al. 1987. A radiation-resistant rod-shaped bacterium, *Deinobacter grandis* gen. nov., sp. nov., with peptidoglycan containing ornithine. *Int. J. Syst. Bacteriol.* 37:62–67

73. Palade, G. E. 1952. A study of fixation for electron microscopy. *J. Exp. Med.* 95:285–99

74. Reynolds, E. S. 1963. The use of lead citrate at high pH as an electron-opaque stain in electron microscopy. *J. Cell Biol.* 17:202–12

75. Robinow, C. F. 1945. Addendum: Nuclear apparatus and cell structure of rod-shaped bacteria. In *The Bacterial Cell*, by R. J. Dubos, pp. 355–77. Cambridge, Mass: Harvard Univ. Press

75a. Robinow, C. F. 1953. Spore structure as revealed by thin sections. *J. Bacteriol.* 66:300–11

76. Robinow, C. F., Murray, R. G. E. 1953. The differentiation of cell wall, cytoplasmic membrane and cytoplasm of gram positive bacteria by selective staining. *Exp. Cell. Res.* 4:390–407

77. Ryter, A. 1968. Association of the nucleus and the membrane of bacteria: A morphological study. *Bacteriol. Rev.* 32:39–54

78. Ryter, A., Kellenberger, E. 1958. Etude au microscope électronique de plasma contenant de l'acide désoxyribonucléique. I. Les nucléoides des bactéries

en croissance active. *Z. Naturforsch.* 133:597–605

79. Salton, M. R. J., Horne, R. W. 1951. Studies of the bacterial cell wall. II. Methods of preparation and some properties of cell walls. *Biochim. Biophys. Acta* 7:177–97

80. Skerman, V. B. D., McGowan, V., Sneath, P. H. A., eds. 1980. *Approved Lists of Bacterial Names*. Washington, DC: Am. Soc. Microbiol.

81. Sleytr, U. B., Messner, P. 1983. Crystalline surface layers on bacteria. *Ann. Rev. Microbiol.* 37:311–39

82. Smit, J. 1986. Protein surface layers of bacteria. In *Bacterial Outer Membranes as Model Systems*, ed. M. Inouye, pp. 343–76. Chichester, UK: Wiley

83. Stackebrandt, E., Ludwig, W., Weizenegger, M., Dorn, S., McGill, T. J., et al. 1987. Comparative 16S rRNA oligonucleotide analyses and murein types of round-spore-forming bacilli and non-spore-forming relatives. *J. Gen. Microbiol.* 133:2523–29

84. Stanier, R. Y. 1961. La place des bactéries dans le monde vivant. *Ann. Inst. Pasteur Paris* 101:297–312

85. Stanier, R. Y., van Niel, C. B. 1962. The concept of a bacterium. *Arch. Microbiol.* 42:17–35

86. Steed, P., Murray, R. G. E. 1966. The cell wall and cell division of gram-negative bacteria. *Can. J. Microbiol.* 12:263–70

87. Steed-Glaister, P. D. 1967. *A study of cell wall and division of gram-negative bacteria*. PhD thesis. Univ. Western Ontario, London, Ontario, Canada. 209 pp.

88. Stevenson, J. W., Cowan, S. T. 1967. Obituary notice: E. G. D. Murray. *J. Gen. Microbiol.* 46:1–21

89. Stewart, M., Beveridge, T. J., Murray, R. G. E. 1980. Structure of the regular surface layer of *Spirillum putridiconchylium*. *J. Mol. Biol.* 137:1–8

90. Stewart, M., Murray, R. G. E. 1982. Structure of the regular surface layer of *Aquaspirillum serpens* MW5. *J. Bacteriol.* 150:348–57

91. Thompson, B. G., Murray, R. G. E., Boyce, J. F. 1982. The association of the surface array and the outer membrane of *Deinococcus radiodurans*. *Can. J. Microbiol.* 28:1081–88

92. Wayne, L. G., Brenner, D. J., Colwell, R. R., Grimont, P. A. D., Kandler, O., et al. 1987. Report of the *ad hoc* committee on reconciliation of approaches to bacterial systematics. *Int. J. Syst. Bacteriol.* 37:463–64

93. Weidel, W., Frank, H., Martin, H. H. 1960. The rigid layer of the cell wall of *Escherichia coli* strain B. *J. Gen. Microbiol.* 22:158–66
94. Whitfield, J. F., Murray, R. G. E. 1954. A cytological study of the lysogenization of *Shigella dysenteriae* with P₁ and P₂ bacteriophages. *Can. J. Microbiol.* 1:216–26
95. Whitfield, J. F., Murray, R. G. E. 1956. The effects of the ionic environment on the chromatin structures of bacteria. *Can. J. Microbiol.* 2:245–60
96. Williams, R. C., Wyckoff, R. W. G. 1946. Applications of metallic shadow-casting to microscopy. *J. Appl. Phys.* 17:23
97. Woese, C. R. 1987. Bacterial evolution. *Microbiol. Rev.* 51:221–71
98. Woese, C. R., Blanz, P., Hahn, C. M. 1984. What isn't a pseudomonad: The importance of nomenclature in bacterial classification. *Syst. Appl. Microbiol.* 5:179–95
99. Woldringh, C. L., Nanninga, N. 1976. Organization of the nucleoplasm in *Escherichia coli* visualized by phase-contrast light microscopy, freeze fracturing, and thin sectioning. *J. Bacteriol.* 127:1455–64
100. Work, E., Griffiths, H. 1968. Morphology and chemistry of cell walls of *Micrococcus radiodurans*. *J. Bacteriol.* 95:641–57
101. Zuckerkandl, E., Pauling, L. 1965. Molecules as documents of evolutionary history. *J. Theor. Biol.* 8:357–66

Ann. Rev. Microbiol. 1988. 42:35–47

HOT NEWS ON THE COMMON COLD

D. A. J. Tyrrell

MRC Common Cold Unit, Harvard Hospital, Coombe Road, Salisbury, United Kingdom

CONTENTS

HOT NEWS ON THE COMMON COLD

As the title indicates, this review deals with topics that are exciting because they are undergoing rapid progress and may lead to important consequences. For that reason, and for lack of space, this is not a comprehensive update on all research on common colds, let alone on the many other closely related acute respiratory tract infections (e.g. pharyngitis, sinusitis, bronchitis, bronchiolitis, croup, and pneumonia). It focuses on those viral infections that cause a short, self-limiting rhinitis but that nevertheless are the most common viral infection of humans, affecting most of us several times a year.

Early microbiological research focused on establishing whether bacteria were the cause of the condition. They were not. Research then became a search for "the common cold virus." It was found that many different viruses can be detected in patients with colds and can cause colds when introduced into the nose of human volunteers. The most frequently recovered are rhinoviruses, of which over 100 serotypes are now recognized (16), and coronaviruses, of which there are few serotypes. Current estimates suggest that up

35

0066-4227/88/1001-0035$02.00

to one half of all colds are due to rhinoviruses and about one fifth to coronaviruses. Other viruses capable of causing colds apparently exist and remain to be cultured in the laboratory and identified (20).

In the past there have been excellent field studies of the etiology and epidemiology of colds. It has been established that there is a statistically significant association between infection with the viruses and the occurrence of cold symptoms (17), both in experimental infections in volunteers and in field studies of naturally acquired infections. Recent interesting work has been concerned with the mechanisms by which viruses cause colds. It is easy enough to say that we know that a virus causes colds, but we have not yet unraveled the events that occur after the arrival of a virus particle on the nasal mucosa and before the occurrence of the symptoms that characterize the illness, such as nasal discharge and obstruction, or the general symptoms of fever, malaise, and tiredness.

We always hope that studies will lead to effective prevention or treatment of colds, preferably by preventing the replication of the virus. There has been progress in this area. For example, intranasal sprays of interferons have effectively prevented virus replication and clinical symptoms not only in volunteers given experimental infections, but also in families with naturally acquired colds. However, these developments have been reviewed recently (33); I summarize here the trail that leads to possible antiviral treatment of a different kind.

INHIBITORS OF UNCOATING AS ANTIRHINOVIRUS DRUGS

In the 1950s, when I first became involved with antiviral research, much of our work was based on assumptions or on guesses, many of which were firmly denied by our colleagues. One such assumption was that we would be able to find a compound that would inhibit a biochemical process specific to the virus (by definition a compound nontoxic to the host) and that would be sufficiently stable metabolically and sufficiently well distributed within the body and in the target cells to be effective in an animal system and also in humans. We also discounted the idea of finding or using a substance that would neutralize the infectivity of a virus particle, as an antibody might do. We were probably influenced by the fact that known virus disinfectants were general cell poisons and that disinfectants had proved of little value in treating bacterial infections.

Early work showed that general metabolic inhibitors such as fluoroacetate could have an antiviral effect, but only at concentrations that were almost lethal for cells or experimental animals. Some of the early nucleoside analogs fell into that category too. In fact, one could regard blocking of viral growth as a sensitive indicator of the toxic effect of the molecule on cell metabolism. However, amantadine was found to be a specific and relatively nontoxic antiviral drug. Even though the introduction of this drug into clinical practice has been problematic, its discovery encouraged the belief that specific antiviral drugs could be developed (15).

It is now clear that amantadine has no direct effect on the virus particle, but interferes in a specific way with the early stages of infection and uncoating of influenza A viruses; it also has a less specific effect in raising the pH of cytoplasmic vesicles. The mechanism is not fully understood, but it apparently depends on the M2 peptide. Amantadine is also excreted into the respiratory tract after oral dosage; this property is both unusual and important. Unfortunately, the drug can also have some adverse effects on the CNS such as tremors, restlessness, or hallucinations. Nevertheless, clinical trials have established that amantadine is effective in prophylaxis and treatment of influenza A. Rimantadine, a secondary amine with the same cagelike ring, has the same mode of action (8) and seems to be marginally better in clinical use (13). ICI 130685 seems to be more active and less toxic, but is otherwise very similar (5).

Interferon has enormously high activity in vitro against many viruses and is nontoxic to cultured cells. Since it is a protein it is ineffective if given by mouth, but if administered locally into the respiratory tract it prevents rhinovirus colds in volunteers (33). However, on continued use it produces troublesome temporary adverse effects such as nasal inflammation, stuffiness, and bleeding, and it has no effect on a respiratory virus infection that is already clinically evident.

The highly effective antiherpes drugs and the partially successful drugs for respiratory infections all inhibit essential steps in the intracellular life cycle of the virus, such as the entry process (amantadine), nucleic acid synthesis (antiherpes drugs such as acyclovir), or the translation of viral messenger RNA to peptide (interferons). Thus there has been a common assumption that all useful antiviral drugs would act by blocking some virus-specific metabolic process within the cell.

However, another strand of antiviral research has been developing, begun as a result of studies of compounds active against enteroviruses. After early work at the Rockefeller Institute had shown that guanidine and hydroxybenzyl benzimidazole (HBB) blocked virus replication, Eggers (14) and Rosenwirth & Eggers (31) showed that rhodanine acts against echovirus 12 by preventing uncoating. For instance, it prevents the conversion of neutral red-treated virus from a light-sensitive to a light-insensitive state. It also stabilizes the free virus particle; for instance, it reduced the sensitivity of the particle to alkali. Eggers concluded that the drug probably interacts with the virus capsid and prevents it from opening up in the cell or in adverse environments.

Since the mid 1970s there have been systematic efforts to develop more satisfactory drugs (reviewed in 12). An early success was the development of arildone, a phenoxyalkyl diketone (9), which has a minimal inhibitory concentration (MIC) of 0.3 μg ml^{-1} against poliovirus type 2 and is active against other viruses. It stabilizes the virus to inactivation and prevents uncoating. The drug has an antiviral effect in animal models. Further development led to WIN 49321 (12), which has an isoxazole at one end of the alkyl chain in the middle of the molecule. Its in vitro activities resemble those of arildone; it was active against poliovirus type 2 in the mouse at a dose of 32 mg kg^{-1}. However, there is much interest in WIN 51711 (Figure 1). This drug is active at low concentration against rhinoviruses as well as enteroviruses. In vitro the MIC against poliovirus type 2 was 6×10^{-3} μg ml^{-1}; in the mouse the drug was active against the virus at a dose of 3.9 mg kg^{-1} daily. Its in vitro properties resemble those of arildone, but in addition its site of interaction with rhinovirus type 14 has been identified. Smith et al (35) soaked virus crystals in the drug and then used X-ray crystallography and other methods to deduce the structure of the capsid. They found that the drug molecule was inserted into the capsid as indicated in Figure 1. There is therefore little doubt now that a molecule that interacts directly with a capsid can produce the antiviral effects observed in vitro and in experimental animals.

While the above researchers were studying antienterovirus compounds, several other pharmaceutical laboratories were screening and developing drugs for preventing or treating rhinovirus colds. From these programs several compounds emerged that were of sufficient promise to warrant trials in volunteers at the MRC Common Cold Unit in Salisbury (Figure 2). These compounds were selected because they were more active in vitro than the

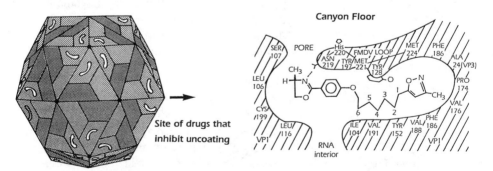

Figure 1   The location of a drug molecule in the capsid of human rhinovirus type 14. On the *left* the curved outline shows the position in the VP1. On the *right* it can be seen how the molecule is located at the bottom of the canyon that surrounds the fivefold axis. This canyon is believed to be the site of the interaction with the cell receptor that leads to uncoating of the particle. Adapted from Reference 32a with permission of the authors and Blackwell Scientific Publications.

benchmark compound enviroxime, which had weak activity in preventing and treating rhinovirus colds in volunteers but was effective in those volunteers who cleared the drug slowly (30). The efficacy of enviroxime was apparently dependent on local application. The drug did not produce statistically significant effects in volunteer trials elsewhere (18, 21), and it had no useful effect against natural colds. Nevertheless, results with this drug were sufficient to indicate that the MIC should be less than 1 μg ml^{-1} if a new compound were to be worth testing.

Dichloroflavan was developed by Wellcome Research Laboratories. It had high activity against some rhinoviruses, although certain serotypes were relatively unaffected. It was nontoxic, and if appropriately formulated it was well absorbed so that substantial and sustained levels were achieved in the

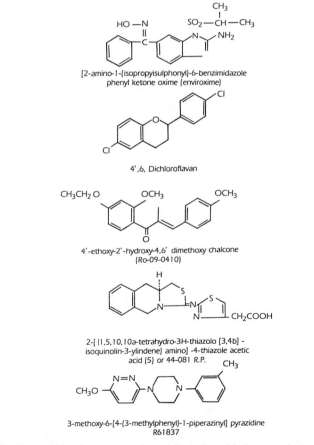

[2-amino-1-(isopropyisulphonyl)-6-benzimidazole
phenyl ketone oxime (enviroxime)

4',6, Dichloroflavan

4'-ethoxy-2'-hydroxy-4,6' dimethoxy chalcone
(Ro-09-0410)

2-[(1,5,10,10a-tetrahydro-3H-thiazolo [3,4b] -
isoquinolin-3-ylindene) amino] -4-thiazole acetic
acid (5) or 44-081 R.P.

3-methoxy-6-[4-(3-methylphenyl)-1-piperazinyl] pyrazidine
R61837

Figure 2   Five antirhinovirus compounds recently evaluated in human volunteers.

circulation. Nevertheless, when given by mouth in doses that were adequately absorbed it did not prevent rhinovirus infection. Phillpotts et al (29) concluded that the drug did not reach the susceptible cells, the ciliated nasal epithelium.

Researchers at Nippon Roche developed a chalcone with high antirhinovirus activity from an oriental herbal remedy (19). A phosphorylated prodrug was developed, which was absorbed from the alimentary tract and hydrolyzed to give high concentrations of the active molecule in the circulation. Nevertheless, like dichloroflavan, it did not prevent rhinovirus common colds when given orally (28).

It seemed particularly desirable to administer drugs intranasally. A drug prepared by Rhone Poulenc, 44 081 R.P., which was active at less than 1 μg ml^{-1} against at least a few rhinovirus serotypes and which was unsuitable for oral administration, was tested for efficacy when administered intranasally. The drug was relatively soluble, and was therefore thought likely to reach mucosal cells effectively. However, it produced no protective effect when given as an intranasal spray (42).

Dichloroflavan and the chalcone were also formulated for intranasal administration and given in maximum tolerated doses by repeated frequent intranasal sprays, but again they did not prevent infection or disease (3a, 6). This was particularly puzzling in the case of the chalcone, since fewer units of antiviral activity had been effective when interferon was administered in a previous series of experiments.

These molecules were studied in the laboratory for their mode of action. Although the details of their chemistry varied greatly, their basic effects were similar. 44 081 R.P. did not inactivate virus particles, but it prevented uncoating of virus adsorbed to the cell (3). DCF stablized particles against heat and low pH and bound to them to such an extent that they were noninfectious unless the drug was removed by shaking with chloroform. In the cell the drug prevented uncoating (36, 37). The chalcone had virtually the same properties as DCF; it was suggested that the two drugs even reacted at the same site on the virus capsid, since a resistant mutant of type 2 virus selected against one drug was resistant to the other (19, 26). However, other chalcone-resistant mutants of type 9 did not show cross-resistance against DCF (1). Furthermore, there was synergy between chalcone and DCF in antirhinovirus activity in vitro, which would not be expected if their sites of action were identical (2). It is most probable that these molecules interact at different sites within the structure of the virus capsid, so that although their effects are the same at the biological level they differ at the molecular level.

In recent years we have started work on a series of compounds produced by Janssen Laboratories (7). One of these, R61837 (Figure 2), has been studied in detail and has a similar pattern of in vitro activity but an even lower MIC than the compounds discussed above ($2 \times 10^{-3} \mu$g ml^{-1} against types 2 and 9). The drug binds so well to the virus that its effects are not reversed by chloroform extraction. In a first trial, it apparently reduced the clinical signs and symptoms for the first half of the cold, but treatments were stopped when the volunteers given placebo had suffered only half the duration of their colds. In the next trial drug sprays were continued until the controls recovered from their colds, and in this trial the drug completely suppressed the colds (4). Thus we have finally shown that a virus that inhibits uncoating can prevent colds. This was also the first occasion on which a low–molecular weight synthetic molecule has been shown to be effective against a rhinovirus-induced disease in humans.

There is clearly much more work to be done. We need trials to determine whether such a molecule is effective in treating as well as preventing colds, and we need to find out what characteristic of the spray made it effective when other applications were not. Was it the behavior of the molecule, the formulation, or possibly even some minor feature of the spray regime?

IMMUNITY AND PATHOGENESIS

Early work on rhinoviruses suggested that specific immunity was possible, although there were differing opinions on the relative importance of secretory and circulating antibody measured by neutralization or in a few cases by hemagglutination inhibition tests.

Recently it became possible to measure antibodies against coronaviruses by an enzyme-linked immunosorbent assay (ELISA) using whole virus as the antigen. This test measures mainly antibody against the surface peplomers of the virus, and the results correlate well with the results of neutralization tests. Volunteers were bled and nasal washings were collected, and then each volunteer was given intranasal coronavirus (of the 229E serotype). The results were monitored by clinical and virological means (10). The results were complex, as expected, but the role of each component of the antibody system could be separated by statistical analysis. This showed that both specific secretory immunoglobulin A (IgA) and circulating antibodies were important in protection. It was also found that total nasal secretion protein was correlated with resistance, as was originally observed many years ago with rhinoviruses (32). This correlation may be a manifestation of the antiviral activities of glycoproteins in secretions (23; K. G. Nicholson, unpublished). Furthermore, studies with volunteers, like population studies, showed that women and those who had not had a cold for over 6 mo were more susceptible to infection. It was found that individuals in these groups have less secretory and circulating antibody and also less nasal secretion protein.

We still have a limited understanding of the connection between the immediate effects of the virus and the symptoms and signs of disease when infection does occur. One group studied the site of virus replication after unilateral inoculation of a rhinovirus via either the eye or nares. The virus gradually appeared at several sites, apparently translocated across the mucosa. In addition, when symptoms appeared there was a marked reduction in the efficiency of the mucociliary transport from the nose to the throat (39). The cilia of biopsied cells still beat with normal frequency, but many cells had lost substantial numbers of cilia. The extent of loss was related to the degree of slowing of transport. Shed cells were found in the nasal secretions, and these cells contained viral RNA (detected by hybridization with cDNA) (C. Boyle, unpublished). It is not known whether the excess of nasal discharge is due to the loss of mucociliary transport or to an increase of secretion. I suspect that the latter occurs, but more experiments are needed to document this.

Recent work showed that swelling of the mucosa was associated with changes in blood flow in experimental coronavirus colds in volunteers (M. Bende, unpublished). However, Callow et al (11) have searched for clearcut changes in inflammatory mediators such as leukotrienes and histamine without success, even though volunteers with high total concentrations of IgE in the nasal secretion or circulation had worse colds than those with lower concentrations. However, we recently studied the effect of nedocromil, which was developed as a blocker of release of mediators from mast cells (and probably other cells as well). An intranasal spray developed for the treatment of nasal allergy had no effect on virus infection, but it reduced the total symptom score and the nasal secretion weight (D. A. J. Tyrrell, P. G. Higgins & W. Al Nakib, unpublished). We therefore have evidence that such mediators are involved in production of symptoms, but the details remain to be worked out.

In addition to inflammatory and other changes in the nose, colds are often accompanied by symptoms and signs of general illness such as malaise, tiredness, shivering, or fever. The occurrence of these symptoms suggests that some substance is released from the infected mucosa to produce effects in the CNS and elsewhere. Researchers have therefore looked for evidence of such a substance. It is believed that interleukin 1 induces the synthesis of serum amyloid A (SAA) protein in the liver; sera taken from volunteers a few days after their inoculation with common cold viruses showed an increased concentration of SAA (38). (The increase was even larger for volunteers inoculated with influenza viruses.) There is also evidence that alpha interferon (IFN-α) is produced in respiratory virus infections; if absorbed, this chemical can produce symptoms of general illness.

Feelings of illness, fever, and a sense that one cannot function properly all suggest that there may be mild alterations in CNS function during mild respiratory infections. On the other hand, inability to function properly is a subjective sensation that cannot be verified objectively. So are the changes in CNS function real? Smith et al (34) have used modern performance testing to answer this question. They found that volunteers with common colds showed impairment in their ability to perform tasks demanding manual dexterity and hand-and-eye coordination. For instance, when asked to use a joy stick to direct a symbol on a television screen to hit a moving target, volunteers with colds scored roughly half as many hits as when they were well, although in a task demanding attention they performed as well as when they were healthy. The effect was specific in that volunteers with influenza scored badly on the attention task but normally on the hand-and-eye coordination task. On the other hand, functions such as logical reasoning were unimpaired. The effect in influenza may be due to the mediation of interferon, since similar effects were produced by an intramuscular injection of purified recombinant human IFN-α (rIFN-α) (A. P. Smith, unpublished). The whole field is new and needs confirmation and elaboration. It has been found, for instance, that the CNS effects characteristic of common colds are produced by different viruses, e.g. rhinoviruses, coronaviruses, and respiratory syncytial virus. On the other hand, there are no data on whether these functional changes are important in everyday life, even though the changes in influenza, for example, are larger than those produced by blood alcohol concentrations above the legal limit for drivers of motor vehicles. It is now worth examining the possibility that respiratory virus infections are associated with accidents, performance at work or in athletics, and so on.

HOT AIR AND COMMON COLDS

Studies of the propagation and basic biology of rhinoviruses revealed that they grow optimally at about 33°C and that growth is impaired at normal body temperature, namely 37°C (17). It therefore seemed plausible that the temperature of the nasal mucosa might be an important factor in the pathogenesis of colds. It seems possible that increasing the temperature of the nasal mucosa, which normally oscillates around 33°C during inspiration and expiration, toward 37°C would limit virus replication. In the 1960s this possibility was explored in a largely unpublished series of studies at the MRC Common Cold Unit (D. A. J. Tyrrell, unpublished). The mucosal temperature was measured with a thermistor bead; in the early stages of a cold the mucosal temperature did not rise significantly, but when there was nasal obstruction the temperature rose to around 37°C. This effect could be mimicked in the normal nose by blocking the external nares. We considered the possibility of doing experiments to show whether blocking the nares would reduce the amount of virus replication, but even as a form of experimental therapy neither the investigators nor the staff were attracted to a treatment that might be at least as uncomfortable as the disease itself.

We therefore considered other ways of raising the nasal temperature, but found that inhalation of hot moist air, e.g. by breathing steam from a jug with a towel over the head, did not succeed, presumably because the nose is so efficient in adding heat and moisture to inspired air that it is almost impossible to stop the heat flux and thus the temperature gradient across the membrane. What was needed was a way of providing a constant flow of fully saturated warm air.

Lwoff (22) was noted for his work on the importance of the *ts* characteristic in attenuating polioviruses, which are members of the Picornaviridae closely related to rhinoviruses. On the basis of in vitro experiments he suggested that repeatedly heating the nasal mucosa to 43°C would not merely inhibit virus replication during the hyperthermia, but would block it permanently and thus cure the cold. He followed up this work by collaborating with a group in Israel that produced a device named the "Rhinotherm." This apparatus was irreverently called "the hair dryer" in our laboratory because of the fan and heater it contained, but it also contained a supply of water and a spraying device. It projected the humidified heated air into the two nostrils. It was evaluated in Israel for its ability to cure colds occurring in the community (41). Yerushalmi & Lwoff (41) claimed that 80% of those treated were better within a day. However, the trial used an easily detectable placebo (a machine with a greatly reduced airflow) and could not have been rigorously randomized. We therefore tried to repeat the study, but had difficulty first with the electrical safety of the apparatus and then with the acceptance of the machine by the volunteers (D. A. J. Tyrrell, unpublished). A small initial pilot trial was begun, but the obvious benefit claimed by Yerushalmi & Lwoff was not observed.

In the next phase (40), the Israeli group followed up an unexpected observation made in their common cold trial, namely that patients with allergic rhinitis derived benefit from the treatment. They therefore set up a placebo-controlled study, which showed that the original regime of three treatments of 30 min at 2-hr intervals (22) did alleviate the symptoms of the majority of a group of well-documented cases of nasal allergy. The researchers showed that the machine raised the temperature of the mucosa to 43°C, and they found that the benefits continued for up to a month (40). These results implied that the treatment had effects other than blocking of virus replication, but also indicated that the treatment could have a long-lasting beneficial effect on nasal pathology, presumably mediated by some modification of the response of the host.

Not knowing of this work, Beacham in South Africa was developing an apparatus to deliver hot humidified air to the nose for the relief of cold symptoms (A. Beacham, unpublished). A model machine was tried out on patients with colds, and although the trial was not controlled there were many reports of rapid improvement. Beacham moved to the United Kingdom and there further developed the apparatus, which is manufactured as the "Virotherm." It stands on a bench or table and delivers 40 liter min $^{-1}$ of fully humidified air at 43°C through an anesthetic mask. Initial studies in South Africa suggested that 43°C for 20 min was well tolerated and effective, so a clinical trial was organized in a National Health Service general practice operating from a health center in Andover, Hants, United Kingdom (D. A. J. Tyrrell, J. Arthur & A. Beacham, unpublished). The object was to determine the effect of such treatment on the course of clinically diagnosed colds. Volunteers were diagnosed on the basis of their history and a clinical examination, and were randomized to inhale under double-blind conditions from an active machine or from one delivering air at 30°C. Curiously, under such conditions it is not easy to distinguish between the two temperatures. The volunteers kept a daily symptom record and were followed up for five days. The groups were reasonably well balanced for sex and duration of illness, although by chance those given placebo had more symptoms than those given the treatment. Patients in both groups showed immediate improvement in their symptoms, but thereafter those given the treatment improved more rapidly than those given placebo, in that their total symptom scores declined significantly more rapidly. By four days after treatment 55% of the treated volunteers and 10% of the volunteers who had received placebo were asymptomatic. It may be estimated that the total number and severity of symptoms experienced over the five days after treatment are halved by the one treatment. The differences are large enough to be clinically interesting and are highly significant statistically, even allowing for the differences in the initial number of symptoms, so the method clearly deserves further investigation. The result must be confirmed by an independent study, which may be furnished by a recent Israeli trial using an improved version of the Rhinotherm (27). Although in my opinion the placebo was not satisfactory, the improvement was documented by an objective measurement of nasal obstruction. Ophir & Elad (27) used two treatments of 30 min and showed improvement in the course of the illness.

It is also important to obtain some plausible explanation for the beneficial effect. It could be that virus replication is directly inhibited, but there could also be some change in the host response, such as the local production of antibody or interferon or the resistance of cells to virus infection. Such changes may be due to activation of the recently described heat-shock genes (24, 25), which are likely to be involved in the regulation of various cell functions. Immune cells such as lymphocytes respond maximally to a stimulus of 43°C for 20 min in vitro. Studies to measure the clinical and virological response in volunteers with colds induced by experimental rhinovirus infection are now planned at the MRC Common Cold Unit.

This is clearly an unfinished story, but it should certainly be followed through, for it may yet lead to a useful way of limiting the symptoms and signs of infections of the upper respiratory tract. Analysis of the mechanism may also improve our understanding of the pathogenesis of these diseases.

Literature Cited

1. Ahmad, A. L. M., Dowsett, A. B., Tyrrell, D. A. J. 1987. Studies of rhinovirus resistance to an antiviral chalcone. *Antiviral Res.* 8:27–39
2. Ahmad, A. L. M., Tyrrell, D. A. J. 1986. Synergism between anti-rhinovirus antivirals: various human interferons and a number of synthetic compounds. *Antiviral Res.* 6:241–52
3. Alarcon, B., Zerial, A., Dupiol, C., Carrasco, L. 1988. Antirhinovirus compound 44 081 R.P. inhibits virus uncoating. *Antimicrob. Agents Chemother.* 30:31–34
3a. Al-Nakib, W., Higgins, P. G., Barrow, I., Tyrrell, D. A. J., Lenox-Smith, I., et al. 1988. Intranasal chalcone, Ro 09-0410 as prophylaxis against rhinovirus infection in human volunteers. *J. Antimicrob. Chemother.* 21:In press
4. Al-Nakib, W., Higgins, P. G., Tyrrell, D. A. J., Barrow, I. G., Taylor, N., et al. 1987. Tolerance and prophylactic efficacy of a new antirhinovirus compound, R61837. *Int. Virol. Congr., 7th, Edmonton,* Abstr. R32.3. Ottawa: Natl. Res. Counc. Can.
5. Al-Nakib, W., Higgins, P. G., Willman, J., Tyrrell, D. A. J., Swallow, D. L., et al. 1986. Prevention and treatment of experimental influenza A virus infections in volunteers with a new antiviral ICI 130,685. *J. Antimicrob. Chemother.* 18:119–29
6. Al-Nakib, W., Willman, J., Higgins, P. G., Tyrrell, D. A. J., Shepherd, W. M. 1987. Failure of intranasally administered 4',6-dichloroflavan to protect against rhinovirus infection in man. *Arch. Virol.* 92:255–60
7. Andries, K., Dewint, B., de Brabander, M., Stockbroeckx, R. 1987. In vitro activity of R61837, a new antirhinovirus compound. See Ref. 4, Abstr. R32.7
8. Belshe, R., Hay, A., Hall, C., Skehel, J. J. 1987. Genetic basis for resistance to rimantadine exhibited by influenza A viruses isolated from patients during treatment of influenza. See Ref. 4, Abstr. R32.5

9. Caliguiri, L. A., McSharry, J. J., Lawrence, G. W. 1980. Effect of arildone on modification of poliovirus in vitro. *J. Virol.* 105:86–93
10. Callow, K. A. 1985. Effect of specific humoral immunity and some nonspecific factors on resistance of volunteers to respiratory coronavirus infection. *J. Hyg.* 95:173–89
11. Callow, K. A., Tyrrell, D. A. J., Shaw, R. J., Fitzharris, P., Wardlaw, A. J., et al. 1988. The influence of atopy on the clinical manifestations of coronavirus infection in adult volunteers. *Clin. Allergy* In press
12. Diana, G. D., Otto, M. J., McKinlay, M. A. 1985. Inhibitors of picornavirus uncoating as antiviral agents. *Pharmacol. Ther.* 29:287–97
13. Dolin, R., Reichman, R. C., Madore, H. P., Maynard, R., Linton, P. M., et al. 1982. A controlled trial of amantadine and rimantadine in the prophylaxis of influenza infection. *N. Engl. J. Med.* 307:543–47
14. Eggers, H. J. 1977. Selective inhibition of uncoating of echovirus 12 by rhodanine. *Virology* 78:241–52
15. Galbraith, A. 1985. Influenza—recent developments in prophylaxis and treatment. *Br. Med. Bull.* 41:381–85
16. Hamparian, V. V., Colonno, R. J., Cooney, M. K., Dick, E. C., Gwaltney, J. M. Jr., et al. 1987. A collaborative report: rhinovirus—extension of the numbering system from 89 to 100. *Virology* 159:191–92
17. Hamre, D. 1968. *Rhinoviruses. Monogr. Virol.* Vol. 1. 88 pp.
18. Hayden, F. G., Gwaltney, J. M. Jr. 1982. Prophylactic activity of intranasal enviroxime against experimentally induced rhinovirus type 39 infection. *Antimicrob. Agents Chemother.* 21:892–917
19. Ishitsuka, H., Ninomiya, Y., Ohsawa, C., Fujiu, M., Suhara, Y. 1982. Direct and specific inactivation of rhinovirus by chalcone Ro 09-0410. *Antimicrob. Agents Chemother.* 22:617–21
20. Larson, H. E., Reed, S. E., Tyrrell, D.

A. J. 1980. Isolation of rhinoviruses and coronaviruses from 38 colds in adults. *J. Med. Virol.* 5:221–29

21. Levandowski, R. A., Pachucki, C. T., Rubenis, M., Jackson, G. G. 1982. Topical enviroxime against rhinovirus infection. *Antimicrob. Agents Chemother.* 22:1004–7

22. Lwoff, A. 1969. Death and transfiguration of a problem. *Bacteriol. Rev.* 33:390–403

23. Matthews, T. H. J., Nair, C. D. G., Lawrence, M. K., Tyrrell, D. A. J. 1976. Antiviral activity in milk of possible clinical importance. *Lancet* 2:1387–89

24. Morimoto, R., Fodor, E. 1984. Cell-specific expression of heat shock proteins in chicken reticulocytes and lymphocytes. *J. Cell Biol.* 99:1316–23

25. Neidhardt, F. C., Van Bogelen, R. A., Vaughn, V. 1984. The genetics and regulation of heat shock proteins. *Ann. Rev. Genet.* 18:295–329

26. Ninomiya, Y., Aoyama, M., Umeda, I., Suhara, Y., Ishitsuka, H. 1985. Comparative studies on the mode of action of the anti-rhinovirus agents Ro 09-0410, Ro 09-0179, RMI-15,731, 4',6-dichloroflavan, and enviroxime. *Antimicrob. Agents Chemother.* 27:595–99

27. Ophir, D., Elad, Y. 1987. Effect of steam inhalation on nasal patency and nasal symptoms in patients with the common cold. *Am. J. Otolaryngol.* 3:149–53

28. Phillpotts, R. J., Higgins, P. G., Willman, J. S., Tyrrell, D. A. J., Lenox-Smith, I. 1984. Evaluation of the anti-rhinovirus chalcone Ro 09-0415 given orally to volunteers. *J. Antimicrob. Chemother.* 14:403–9

29. Phillpotts, R. J., Wallace, J., Tyrrell, D. A. J., Freestone, D., Shepherd, W. 1983. Failure of oral 4',6-dichloroflavan to protect against rhinovirus infection in man. *Arch. Virol.* 75:115–21

30. Phillpotts, R. J., Wallace, J., Tyrrell, D. A. J., Tagart, V. B. 1983. Therapeutic activity of enviroxime against rhinovirus infection in volunteers. *Antimicrob. Agents Chemother.* 23:671–75

31. Rosenwirth, B., Eggers, H. J. 1979. Early processes of echovirus 12 infection: elution, penetration and uncoating under the influence of rhodanine. *Virology* 97:241–45

32. Rossen, R. P., Butler, W. T., Waldman, R. H., Alford, R. H., Hornick, R. B., et al. 1970. The proteins in nasal secretion. II. A longitudinal study of IgA and neutralizing antibody levels in nasal washings from men infected with influenza virus. *J. Am. Med. Assoc.* 211:1157–61

32a. Rossmann, M. G., Rueckert, R. R. 1987. What does the molecular structure of viruses tell us about viral function? *Microbiol. Sci.* 4(7):206

33. Scott, G. M., Tyrrell, D. A. J. 1985. Antiviral effects of interferon in man. In *Interferon,* Vol. 4, *In Vivo and Clinical Studies,* ed. N. B. Finter, R. K. Oldham, pp. 181–215. New York: Elsevier

34. Smith, A. P., Tyrrell, D. A. J., Coyle, K., Willman, J. S. 1987. Selective effects of minor illnesses on human performance. *Br. J. Psychol.* 78:183–88

35. Smith, T. J., Kremer, M. J., Luo, M., Vriend, G., Arnold, E., et al. 1986. The site of attachment in human rhinovirus 14 for antiviral agents that inhibit uncoating. *Science* 233:1286–93

36. Tisdale, M., Selway, J. W. T. 1983. Inhibition of an early stage of rhinovirus replication by dichloroflavan (BW 683C). *J. Gen. Virol.* 64:795–803

37. Tisdale, M., Selway, J. W. T. 1984. Effect of dichloroflavan (BW683C) on the stability and uncoating of rhinovirus type 1B. *J. Antimicrob. Chemother.* 14(Suppl. A):97–105

38. Whicher, J. T., Chambers, R. E., Higginson, J., Nashef, L., Higgins, P. G. 1985. Acute phase response of serum amyloid A protein and C reactive protein to the common cold and influenza. *J. Clin. Pathol.* 38:312–16

39. Wilson, R., Alton, E., Rutman, A., Higgins, P., Al Nakib, W., et al. 1987. Upper respiratory tract infection and mucociliary clearance. *Eur. J. Respir. Dis.* 70:272–79

40. Yerushalmi, A., Karman, S., Lwoff, A. 1982. Treatment of perennial allergic rhinitis by local hyperthermia. *Proc. Natl. Acad. Sci. USA* 79:4766–69

41. Yerushalmi, A., Lwoff, A. 1980. Traitement du coryza infectieux et des rhinites persistantes allergiques par la thermotherapie. *C. R. Acad. Sci.* 291:957–59

42. Zerial, A., Werner, G. H., Phillpotts, R. J., Willman, J. S., Higgins, P. G., et al. 1985. Studies on 44 081 R.P., a new antirhinovirus compound in cell cultures and in volunteers. *Antimicrob. Agents Chemother.* 27:846–50

Ann. Rev. Microbiol. 1988. 42:49–64

ECOLOGY OF COLORADO TICK FEVER

Richard W. Emmons

Viral and Rickettsial Disease Laboratory, Division of Laboratories, California State Department of Health Services, Berkeley, California 94704

CONTENTS

INTRODUCTION

Colorado tick fever (CTF) is a particularly suitable subject for a review on the ecology of a disease. Studies of this disease and its causative virus illustrate well the concept of *oikos* (Greek, "house" or "place to live"), the study of an organism "at home," i.e. its relationship to the environment and its functional status (niche) in the community of organisms and the ecosystem (1, 13, 50, 67). It is also a uniquely American disease, with a special place in the history and development of the western United States. Although the clinical,

49

epidemiological, and ecological aspects of the disease were extensively studied and described during the past century, some new information has accumulated which should stimulate further research. A review of current knowledge and some suggestions for research is therefore appropriate. Space does not allow a comprehensive summary of published work, but selected, pertinent sources are given to illustrate the principle points.

CTF was probably first recognized and described by physicians in the Rocky Mountain region in the 1850s as one of various mountain fevers, but for many decades it was poorly differentiated from typhus, typhoid fever, malaria, and other febrile illnesses. Published reports gradually accumulated and defined a disease that followed tick bite but was not typical of Rocky Mountain spotted fever (RMSF); it came to be called Colorado tick fever. Credit is usually given to Topping and coworkers (82) for the first detailed clinical description, and to Florio and coworkers (35, 36, 39, 40) for isolation in hamsters of the causative virus from human cases and *Dermacentor andersoni* ticks (See 18, 58 for reviews).

Koprowski & Cox (54–56) adapted the virus to grow in laboratory mice and chick embryos. The discovery that suckling mice were more suitable than hamsters for isolation of the virus (68) allowed field, laboratory, and clinical research on the disease to progress rapidly. Studies by Eklund, Kohls, Philip, Burgdorfer, Jellison, and others during the next two decades, particularly at the Rocky Mountain Laboratory in Hamilton, Montana (see section on natural history), further clarified the identity and role of tick vectors and natural vertebrate hosts and the clinical and epidemiologic aspects of the disease. The keenly observant, energetic, and dedicated pioneers revealed the basic features of CTF and its natural cycle within a short time despite relatively crude laboratory methods and limited resources.

THE VIRUS

Taxonomy

The virus that causes CTF is one of over 500 viruses in the heterogeneous group of arthropod-borne viruses (arboviruses). It is presently in the genus *Orbivirus* of the family Reoviridae, which includes at least 61 named viruses in 11 serogroups and one ungrouped set of viruses (51). However, the CTF virus genome contains 12 species or segments of double-stranded RNA, whereas other orbiviruses have 10, so that a revision of its taxonomic status may be needed (53). It was long thought that CTF virus had no serologic relatives, although minor antigenic differences among viral strains were noted. Ten strains isolated from humans and various mammalian hosts by RNA-RNA hybridization were quite homogeneous, in contrast to other members of orbivirus serogroups (3). However, a closely related virus

(Eyach) has been isolated from *Ixodes ricinus* ticks in the Federal Republic of Germany and in France (14, 74, 75) and from *Ixodes ventalloi* in France (14). In addition, there are virus strains in coastal California that are very similar, but not identical, to CTF (52, 57). The descriptions of these viruses indicate that CTF is a member of a group of viruses that may increase in number as more studies are done. Eyach virus cross-reacts with CTF virus by the complement-fixation test, but is distinct by neutralization tests. It may have a role in causing human disease in Europe, but confirmation requires further studies (14). The CTF-like virus isolated from the blood of a *Lepus californicus* jackrabbit in Mendocino County, California, and an apparently identical virus from a *Sciurus griseus* squirrel in San Luis Obispo County, California, also differ significantly from both Eyach and CTF viruses by serological tests (52, 57).

Morphology and Physicochemical Properties

The virion has an outer diameter of 80 nm, with a central core 50 nm in diameter containing the double-stranded RNA genome (43, 53, 65, 70). The apparent total molecular weight is 18×10^6, larger than other orbiviruses. The virion is nonenveloped, has cubic symmetry, and has a radial division of the capsomeres, which are thought to number 32. Virion development in infected cells is associated with cytoplasmic granular matrices and accumulation of distinctive cytoplasmic and nuclear fibrillar or filamentous structures (34, 64, 65, 70). Similar structures have been reported in Eyach virus–infected cells (14).

CTF virus is partially resistant to lipid solvents, but is inactivated at pH 3.0. The optimum pH for maintenance of viability is 7.5–7.8. The virus is relatively stable at room temperature and at 4°C, and it is preserved indefinitely by freezing or lyophilization if protected by a protein-containing medium (83). Virus protected within erythrocytes (RBC) in blood clots held at 4°C has remained viable for as long as 17–18 mo (30, 71).

Laboratory Hosts

Various cell culture types are susceptible (usually without overt cytopathic effects), including primary cultures or cell lines of avian, mammalian, and arthropod origin (52). When adapted to chick embryos or by numerous intracerebral (i.c.) passages to mice or hamsters, the virus becomes lethal to these hosts (54–56), but unadapted viral strains are lethal only for suckling mice. Transient depression of the peripheral leukocyte count occurs in infected animals. Experimentally infected rhesus monkeys had a depressed leukocyte count, prolonged viremia, and sometimes fever, but no other clinical signs (41).

In mammalian hosts the virus infects hemopoietic cells, resulting in a cell-associated viremia (H. N. Johnson, unpublished, cited in 28, 30). The viremia lasts several weeks to months, owing largely to the location of virus within erythrocytes, where it is protected from antibodies or other host immune defenses (29, 30, 34, 47, 69). The virus has been found in a wide variety of tissues, perhaps simply because of the blood contained in them; brain, spleen, other lymphoid tissues, bone marrow, and heart muscle of laboratory mice appear to support virus growth. Pathologic changes have been observed in these tissues and in liver, lungs, striated muscle, and other tissues in mice, hamsters, and guinea pigs (2, 22, 44, 63). Infection of pregnant mice caused resorption of fetuses, stillbirths, and neonatal deaths, and virus multiplied to high titer in placental and fetal tissues (17, 45). Such experimental studies, using artificial modes of infection and adapted strains of virus, may not accurately reflect the pathogenesis and tissue tropisms in natural hosts, however.

HUMAN INFECTION

Transmission and Pathogenesis

Infection is acquired by tick feeding and injection of tick saliva, although only about half of the patients actually find a tick attached, since the act of feeding is generally painless and even a brief feeding may still effectively transmit virus. The tick salivary glands are the main source of virus, but various tissues and fluids also contain virus and could be a source of infection by contamination of skin breaks or mucous membranes. Infection by contact with infected animal blood or tissues may also occur, such as in laboratory accidents, and person-to-person transmission via blood transfusion has been documented (73). The incubation period after virus inoculation is usually 3–6 days, but can be as long as 2 wk. The introduced virus presumably first replicates in erythropoietic cells and is then amplified in these cells and perhaps also in various lymphoid tissues, as suggested by the experimental animal studies described above (28, 34, 69).

Clinical Disease and Pathology

A brief review of the human disease follows; more detailed summaries are available elsewhere (20, 23, 24, 26, 31, 42, 58, 72, 79). Subclinical infections probably occur, but have not been adequately investigated. Most infections result in mild to moderately severe disease manifestations. Onset is typically sudden, with chilly sensations, high fever, severe headache, retrobulbar pain, photophobia, lethargy, myalgia, and arthralgia. The spleen and liver are sometimes palpable. There may also be anorexia, nausea, vomiting, abdominal pains, neurologic or encephalitic signs (disorientation, hallucinations, and stiff neck), and a variety of other rare or unusual complications.

Severe encephalitis and hemorrhagic manifestations have been limited to children. In earlier series of cases such complications were reported in more than 10% of the patients under 10 years of age, but more recent reports indicate that they are less common. Evidence of pericarditis and myocarditis has also been reported.

Symptoms and signs may persist for several days or for a few weeks. The fever is often biphasic (in about half the cases), with a 2- or 3-day initial phase, then 1–2 days of respite, followed by a second 2–3-day period of equal or higher fever and exacerbated symptoms. There may even be a third febrile period. Transient petechial or macular rash occurs in a small percentage of patients, but it is not like the persistent, severe, hemorrhagic, peripherally distributed rash that occurs in the majority of Rocky Mountain spotted fever (RMSF) cases. Nevertheless RMSF and CTF are often poorly differentiated because both are transmitted by ticks in the western United States, have many early clinical features in common, and require laboratory tests (sometimes not readily available) for definitive diagnosis.

The peripheral leukocyte count often drops dramatically during the acute illness to as low as 1000 μl^{-1}, with a relative lymphocytosis and a left shift of the granulocyte series. Metamyelocytes and myelocytes may appear in the peripheral blood. There may be a mild anemia and a transient but significant thrombocytopenia, presumably the cause of the rare cases of hemorrhage (49, 60). All these changes apparently reflect the transient cytopathic or depressant effects of the virus in hemopoietic tissues. As in other mammalian hosts, the characteristic persistent viremia is due largely to persistence of the virus within maturing and mature RBC. These cells are released into the peripheral circulation and survive their normal life span of about 120 days, carrying virus intracellularly, until they eventually become senescent and are eliminated from the blood. The virus-carrying RBC remain in the circulation long after the patient has recovered from symptoms and has established a normal antibody response (33, 34, 47, 69, 71). Various other examples of the coexistence in the body of persistent virus and circulating antibody are well known (e.g. with herpes simplex virus, varicella-zoster virus, Epstein-Barr virus, cytomegalovirus, hepatitis B virus, acquired immune deficiency syndrome virus), but these are usually thought to represent continual replication of virus in sites sheltered from the access of antibody or other immune mechanisms. In CTF there is no evidence that viral multiplication persists in bone marrow or elsewhere, although further study is needed (34, 41). Instead, virus produced during the initial amplification cycle appears to be stabilized and to persist in circulating RBC, despite the prolonged period at 37°C and the lack of cell metabolic machinery and substrates usually thought necessary to support viral replication. Further studies of this curious phenomenon might reveal interesting clues to mechanisms of viral persistence. There are numerous examples of viral carriage in the peripheral blood by leukocytes, RBC, or platelets, via attachment to blood cell surfaces, or of viral growth within supportive leukocytes. Examples of intracellular transport in RBC are few, but it is of interest that there is some evidence for this in other members of the family Reoviridae (46, 59).

Recovery from CTF is usually prompt, within 2 wk, but in a minority of patients convalescence is prolonged for months, with weakness, malaise, depression, and slow return to normal physical and mental functioning. Recovery is nearly always followed by a lasting immunity, although rare second attacks have been claimed. A second attack was documented in at least one case, a year following the first attack (42). Only a few fatalities have been recorded, but others may have occurred and been attributed to RMSF or other causes. In two fatal cases that were studied, hemorrhagic phenomena, disseminated intravascular coagulation syndrome, and pathologic changes in lung, lymph nodes, kidneys, spleen, liver, myocardium, intestines, and brain were found (16, 26). Despite the findings in pregnant laboratory mice (17, 45), infection of women during pregnancy is usually benign and does not cause significant fetal damage. However, one congenitally infected infant and one spontaneous abortion were reported (26, 72).

There is no specific treatment for CTF, although specific antiviral drugs are a future possibility (77). General supportive care is helpful, particularly for the rare serious complications. Most cases probably do not require hospitalization, but initial concern that the disease might be RMSF may prompt hospitalization for a period of observation or initial treatment with tetracycline or chloramphenicol until it can be determined that the patient has CTF instead. There is currently no commercial vaccine. An experimental vaccine for laboratory workers has been developed (80, 81), but its efficacy and safety have not been established; this vaccine is not available. Prevention consists of avoiding tick-infested areas and tick bites in the endemic zone. A patient should not donate blood for at least 6 mo after recovery, even if the patient is apparently completely well, because of the long period of cell-associated viremia (34, 71).

LABORATORY DIAGNOSIS

Since few specific clinical clues or routine laboratory tests aid in recognition of the disease, specific virologic diagnosis is necessary. This is readily accomplished in endemic areas where the disease is well known and laboratories have the experience and capability to do the necessary tests. Elsewhere, specimens must be mailed to a reference laboratory. The tests must be specifically requested, since few laboratories perform them as part of a routine battery of tests.

Isolation of the virus from blood by i.c. or intraperitoneal (i.p.) inoculation of 1–3-day old mice is the best method. Various cell culture systems could also be used, but there has been less experience with them. A suspension of acute-phase blood clot or a suspension of RBC from unclotted convalescent-phase blood, washed free of antibody-containing plasma, is used. The serum or plasma yields virus only during the first few days of illness. Viremia or antigenemia is regularly present for up to 90–120 days (47) and may last as long as 135 days or more (33, 34, 42). Inoculated mice sicken or die in 4–8 days. The virus isolation is confirmed by standard neutralization, by immunization-cross challenge tests, or more simply and quickly by immunofluorescence (IF) staining of the mouse brain or mouse blood smears (11, 33). Immunoperoxidase staining to detect antigen in cell cultures or mouse tissues has also been reported (17). The virus has also been isolated a few times from cerebrospinal fluid (38). Virus isolation from wildlife hosts and ticks is similar, but requires extra antibiotics and protective protein in the suspension medium to counteract bacterial contamination or toxicity of the specimens. To estimate the viral content of various tissues more accurately (as in pathogenesis studies), the blood must be rinsed or perfused from the tissue prior to testing it.

The IF staining procedure can also be used to detect viral antigen directly in RBC in peripheral blood smears as early as the first few days of symptoms or up to 4 mo later (20, 31, 33, 47, 71). Rapid, early confirmation of CTF in this way eliminates concern about RMSF, with which it may be clinically confused, and reduces unnecessary use of antibiotics and expensive hospitalization.

If virus isolation is not feasible, serologic tests showing a significant rise in antibody titer over the course of illness and recovery may be used. The indirect IF test is more sensitive and simple for this purpose than neutralization or complement fixation (32). An enzyme immunoassay method has also recently been described (12). Neutralization tests have been used for nearly all studies of immunity or pathogenesis of infection in wildlife hosts.

EPIDEMIOLOGY

The epidemiologic characteristics of CTF reflect its status as a zoonotic, vector-transmitted disease acquired in specific, focal ecologic niches. Cases occur primarily among residents of the endemic areas, but also among visitors who then return home, perhaps to distant states or even overseas. Cases also may occur anywhere outside of the endemic area if an infected tick is transported there, finds a victim, and takes a blood meal. Approximately 200–400 laboratory-verified cases per year are voluntarily reported in the United States (4). The true number of cases is undoubtedly much larger. The disease is so familiar in highly endemic areas that many cases may not be reported or submitted to laboratory tests. Also, since reporting is not required, it may be less complete than for other diseases, even for laboratory-confirmed cases. Cases also may be misdiagnosed as some other condition. In fact, it is probably the most common arthropod-transmitted, clinically overt, viral human disease in the United States. Mosquito-transmitted infections, such as California encephalitis, western equine encephalomyelitis, St. Louis encephalitis, and eastern equine encephalomyelitis, are more common, but are usually subclinical. The tick-transmitted disease Powassan virus encephalitis is rare. The more common tick-transmitted diseases, RMSF and Lyme disease, are caused by a rickettsia and a spirochete, respectively; and flea-transmitted murine typhus is caused by a rickettsia.

Human susceptibility to CTF is apparently universal. Variations by age group, gender, ethnic origin, or other catagories merely reflect relative risks of exposure to the tick vectors because of residence or occupational or recreational activities, not because of innate resistance or susceptibility. Thus male cases usually outnumber female cases two to three fold, and cases in males aged 20–40 predominate, reflecting exposure via fishing, hunting, lumbering, or similar activities. Seasonal occurrence is also correlated with the presence and host-seeking activities of adult *D. andersoni*. Cases have been documented from late February or early March until early November, but the majority are from May through July. The length of the season varies by region and from year to year, depending on the climate, cyclical abundance of vector ticks and mammalian hosts, and activities that bring people into the endemic areas and ecologic niches of the virus (4, 20, 23, 26, 42, 58, 79).

NATURAL HISTORY

Geographic Distribution

Evidence of the presence of CTF virus from human case records, virus isolations, and antibody surveys has indicated a sharply defined endemic zone encompassing mountainous and highland areas, from an altitude of about 4000 to over 10,000 ft (1220 to over 3050 m), in the Canadian provinces of British Columbia and Alberta and in at least 11 western states (California, Colorado, Idaho, Montana, Nevada, New Mexico, Oregon, South Dakota, Utah, Washington, and Wyoming) (4, 31). Sagebrush-juniper-pine vegetation, with moderate shrubs, herbs, and grass cover, provides the typical habitat in endemic areas for the mammalian hosts and ticks that maintain the virus cycle (13, 15, 50, 61).

The distribution of the virus coincides closely with the range of the principle vector, *D. andersoni* (23, 24, 26), commonly thought to be the major factor responsible for limiting the endemic region. The possible key role of a mammalian host, such as the Great Basin pocket mouse, *Perognathus parvus,* which has a similar limited distribution, has also been proposed (50). Exceptions to this limited distribution include a few virus isolations outside the known range of *D. andersoni* (24, 66; C. M. Clifford, cited in 71) in addition to occasional records of ticks transported to distant sites. A report by Florio and coworkers (37) of virus in *D. variabilis* ticks on Long Island, New York, could not be confirmed (62).

The distribution of virus and natural cycles of infection may be somewhat broader than the occurrence of human cases indicates, since there is some involvement of a few tick species that do not readily feed on humans and because the disease is not as likely to be suspected and verified at the edges of, or outside of, the endemic area described above. It remains to be seen if the recently discovered CTF-like virus strains (52, 57) will be considered as strain variants, extending the currently accepted range of CTF, or as distinct, new viruses which might cause separately designated human diseases that have not yet been documented.

These new findings also suggest that further studies should be done to follow up a report from eastern Canada (66) and perhaps even the earlier report from Long Island (37).

Mammalian Hosts

The virus is maintained by cycles of infection in various susceptible small mammals and in the tick species that feed on them, principally *D. andersoni.* The basic transmission cycle was described by Burgdorfer, Eklund, and coworkers (7–10, 21, 25, 26). Human cases do not contribute to the cycle, since they are rarely fed on by nymphal ticks. Direct animal-to-animal infection via ingestion of infected tissues and blood has been suggested as an alternative or supplementary mechanism (H. N. Johnson, personal communication), since it can be demonstrated in laboratory animals, but it is not known if it has an actual role in nature. As determined by field observations and laboratory experiments, infected animals undergo relatively prolonged periods of viremia with no recognized clinical or pathologic effects, presumably reflecting a long and successful evolutionary adaptation to the virus (5–8, 28, 29). The primary infection is followed by development of immunity and resistance to repeat infection. The annual production of susceptible young animals of the various short-lived species involved assures the availability of hosts that can become infected by feeding ticks and subsequently be sources of infection for ticks taking a blood meal from them. Large populations of both mammalian hosts and ticks are presumably necessary to maintain such an infectious cycle indefinitely, but the minimum sizes of these populations are not known.

The major naturally infected host species thus far identified include the golden-mantled ground squirrel, *Spermophilus (Citellus) lateralis tescorum;* Columbian ground squirrel, *Citellus columbianus columbianus;* yellow pine chipmunk, *Eutamias amoenus;* and least chipmunk, *Eutamias minimus* (6, 9, 10, 13, 15, 26, 61). Various other species are less frequently involved, including the pine squirrel, *Tamiasciurus hudsonicus richardsoni;* tassel-eared squirrel, *Sciurus aberti;* Richardson ground squirrel, *Spermophilus (Citellus) richardsoni;* deer mouse, *Peromyscus maniculatus;* piñon mouse, *Peromyscus truei;* California ground squirrel, *Spermophilus (Citellus) be-echeyi;* meadow vole, *Microtus pennsylvanicus;* California vole, *Microtus californicus;* boreal redback vole, *Clethrionomys gapperi;* kangaroo rat, *Dipodomys californicus;* Great Basin pocket mouse, *Perognathus parvus;* bushytail woodrat, *Neotoma cinerea cinerea;* porcupine, *Erethizon dorsatum epixanthum;* mountain cottontail, *Sylvilagus nuttalli;* snowshoe hare, *Lepus americanus;* black-tailed jackrabbit, *Lepus californicus;* marmot, *Marmota flaviventris;* elk, *Cervus canadensis;* coyote, *Canis latrans;* mule deer, *Odocoileus hemionus;* and domestic sheep and horses (6, 9, 10, 15, 25, 31, 50, 57, 61).

The key mammalian species involved in the cycle may vary from one locality to another. Only a few species in each locality have sufficient abundance and population stability, are regularly fed upon by both larval and nymphal ticks, are highly susceptible to infection, and maintain a sufficiently high and sustained level of viremia (5, 6, 9, 10, 13, 21, 25, 61, 78). Experimental infection studies of various species have shown that *E. minimus, E. umbrinus, S. lateralis, P. maniculatus, S. richardsoni, E. amoenus, S. columbianus,* and *E. dorsatum epixanthum,* all of which are readily fed upon by *D. andersoni* immature stages, are the best virus hosts, particularly as juveniles. In chipmunks, viremia levels of up to $10.^{6.2}$ mouse i.c. LD_{50} μl^{-1} have been recorded; such levels are far higher than the minimum necessary to infect feeding ticks (see below) (5–8, 28, 29). The relatively long viremia and the persistence of transmissible virus in the long-lived tick vector contribute to stable and persistent foci for the virus, as in certain other vector-transmitted diseases (e.g. Lyme disease, tick-borne relapsing fever). In contrast, diseases like bubonic plague tend to occur in epizootic waves, with shifting focal distributions within the overall major endemic zones, since susceptible hosts are temporarily used up in each focal area.

Various species of ground-feeding birds are hosts for larval and nymphal *D. andersoni* ticks and other tick species and so are theoretically possible hosts for the virus, which is known to be infectious for domestic chicks. However, the possible role of birds has not been adequately studied.

Nearly all the evidence of infected mammals was obtained before closely related strains of virus were known, so the currently accepted list of host ranges and vector tick species might have to be modified based upon more specific serologic surveys and virus identification.

Overwintering of the virus in hibernating ground squirrels and chipmunks is an interesting possibility (28, 29). Under artificial conditions, viremia was shown to persist for as long as 160 days in *S. lateralis* squirrels that had hibernated for 4 mo; the virus persisted for up to 47 days after their arousal and transfer to room temperature. This suggests that ticks seeking such hosts in early spring might pick up virus. Limited attempts to infect *D. andersoni* nymphs by having them feed on the viremic posthibernation animals were not successful, but further studies should be done. However, it is clear that the main mechanism for virus survival through the winter is persistence in dormant nymphal and adult ticks.

Tick Vectors and Environmental Factors

Although CTF virus has been isolated from at least eight species of ticks, *D. andersoni* has yielded by far the most strains and is the only proven vector for humans. Mosquitoes and other blood-feeding vectors are not involved. The distribution of human cases and of infected mammalian hosts matches closely the distribution of *D. andersoni,* as discussed earlier (23, 24, 26). Other infected tick species found within the range of *D. andersoni* or very near to it include *Dermacentor parumapertus, Dermacentor occidentalis, Dermacentor albipictus, Otobius lagophilus, Haemaphysalis leporispalustris, Ixodes spinipalpus,* and *Ixodes sculptus* (4, 23, 24, 26, 31). Earlier reports of naturally infected *D. variabilis* (37) could not be verified (62), but this species was readily infected experimentally, as was *Ornithodoros savignyi* (48).

The tick is a significant reservoir for the virus, since transstadial transmission of virus (from larva to nymph to adult) is a regular feature and the tick remains infected and infectious for life. Survival of infected ticks for up to three years has been shown (19). Evidence of transovarial transmission, which is a regular feature of *D. andersoni* transmission of *Rickettsia rickettsii* and various other rickettsiae, has been claimed (36), but could not be confirmed (27). Therefore, persistence of CTF virus requires, in addition to the tick, a constant availability of susceptible mammalian hosts. The prolonged viremia in these hosts contributes to the stability of the cycle, since the chance for virus to be transmitted to other feeding ticks is enhanced. Both male and female ticks can acquire the virus during feeding and are found infected in nature with about equal frequency. The virus is amplified and maintained by growth in various tissues. It is transmitted via the saliva after the tick develops to the next life stage, seeks a new host, and takes a blood meal. The *D. andersoni* life cycle usually requires two years, but sometimes only one, or three or more, depending on environmental conditions and the availability of blood meals. Usually the tick utilizes three separate hosts and host species to complete the life cycle. A blood meal is needed for the tick to develop and molt to each successive life stage and for the adult female to produce eggs. Adult ticks begin host-seeking activity in late February or early March. They prefer larger hosts, such as porcupines, deer, elk, and humans. Eggs laid by the adult female in June or July hatch to release the six-legged larvae, which then feed from June through September on mice, ground squirrels, chipmunks, or other small hosts, if the larvae can find them. Nymphs, which either develop from larvae during the summer and remain dormant through the winter or develop during the spring, are active from April through September. They usually feed on chipmunks, ground squirrels, and larger hosts such as hares, rabbits, and porcupines. Simultaneous or sequential feeding by larvae and nymphs on the same animals allows the transfer of virus to the tick via ingested blood or to the susceptible animals via the tick saliva (4, 9, 10, 13, 15, 19, 21, 24, 36, 78).

The dynamics of infection in *D. andersoni* and transmission to laboratory animals (76) and various natural hosts (9, 10, 76) were studied in detail. A minimum viremia titer of about 10^2 mouse i.c. LD_{50} μl^{-1} was found necessary to sufficiently infect feeding ticks so they would maintain the virus through subsequent molts and life stages. The ingested virus was amplified in the tissues of ticks fed as larvae or nymphs, reaching a titer of 10^2–10^3 mouse i.c. LD_{50} ml^{-1} following the molt to the next stage. Infected adult ticks maintained virus titer of 10^2–10^5 mouse i.c. LD_{50} ml^{-1} for up to 10–13 mo (76).

Ticks are usually more abundant on grass and brush near streams and rocky hillsides, particularly on south-facing slopes, where soil depth, temperature, moisture, shelter, ground cover, food sources, and other conditions are favorable for them and for the hosts on which they feed. Since they are restricted in ability to travel far, dispersal of larvae, nymphs, or adult ticks depends mostly on the movements of the hosts to which they attach. The distribution of ticks infected with CTF virus is even more restricted and irregular and can vary widely from place to place, even among adjacent sites. The interactions of hosts, environmental conditions, and tick populations determine whether or not the virus will survive and be perpetuated in a specific site. Nevertheless, because of factors discussed above, these conditions are often so stable and predictable that viremic hosts and infected ticks can often be found year after year, decade after decade, in the same microhabitat and focus, at highly predictable times of the year.

DIRECTIONS FOR FUTURE STUDIES

As suggested in this review, several research areas may be productive, not only for understanding CTF, but for lessons that might be applied to other diseases and disease organisms: a search for new virus strains or relatives and for new or changing natural cycles; development of laboratory models for disease pathogenesis studies; determination of the mechanisms of virus persistence in the vector and host animal and of prolonged RBC-associated viremia; development of more sensitive and rapid virus and viral-antibody detection methods; and discovery of new and better methods for prevention and treatment of CTF and similar viral diseases.

Much has been learned about this fascinating disease using simple techniques and keen observation. The present availability of sophisticated laboratory tools will certainly lead to important new information and concepts, but a plea should be made not to retreat entirely into the laboratory and study only the molecular biology and chemistry of the virus. Colorado tick fever is a disease in nature that should also continue to be studied in its natural setting, its ecological niche. The continuing collaboration of vertebrate zoologists, acarologists, botanists, ecologists, meteorologists, and statisticians as well as physicians, biochemists, molecular biologists, and other research scientists will certainly be needed.

Literature Cited

1. Audy, J. R. 1958. The localization of disease with special reference to the zoonoses. Trans. R. Soc. Trop. Med. Hyg. 52:308–34
2. Black, W. C., Florio, L., Stewart, M. O. 1947. A histologic study of the reaction in the hamster spleen produced by the virus of Colorado tick fever. Am. J. Pathol. 23:217–24
3. Bodkin, D. K., Knudson, D. L. 1987. Genetic relatedness of Colorado tick fever virus isolates by RNA-RNA blot hybridization. J. Gen. Virol. 68:1199–204
4. Bowen, G. S. 1988. Colorado tick fever. In The Epidemiology of Arthropod-Borne Viral Diseases, ed. T. P. Monath. Boca Raton, Fla: CRC. In press
5. Bowen, G. S., McLean, R. G., Shriner, R. B., Francy, D. B., Pokorny, K. S., et al. 1981. The ecology of Colorado tick fever in Rocky Mountain National Park in 1974. II. Infection in small mammals. Am. J. Trop. Med. Hyg. 30(2):490–96
6. Bowen, G. S., Shriner, R. B., Pokorny, K. S., Kirk, L. J., McLean, R. G. 1981. Experimental Colorado tick fever virus infection in Colorado mammals. Am. J. Trop. Med. Hyg. 30(1):224–29
7. Burgdorfer, W. 1959. The behavior of CTF virus in the porcupine. J. Infect. Dis. 104:101–4
8. Burgdorfer, W. 1960. Colorado tick fe-ver. II. The behavior of Colorado tick fever virus in rodents. J. Infect. Dis. 107:384–88
9. Burgdorfer, W., Eklund, C. M. 1959. Studies on the ecology of Colorado tick fever virus in western Montana. Am. J. Hyg. 69(2):127–37
10. Burgdorfer, W., Eklund, C. M. 1960. Colorado tick fever. I. Further ecological studies in western Montana. J. Infect. Dis. 107(3):379–83
11. Burgdorfer, W., Lackman, D. 1960. Identification of the virus of Colorado tick fever in mouse tissues by means of fluorescent antibodies. J. Bacteriol. 80:131–36
12. Calisher, C. H., Poland, J. D., Calisher, S. B., Warmoth, L. A. 1985. Diagnosis of Colorado tick fever virus infection by enzyme immunoassays for immunoglobulin M and G antibodies. J. Clin. Microbiol. 22:84–88
13. Carey, A. B., Mclean, R. G., Maupin, G. O. 1980. The structure of a Colorado tick fever ecosystem. Ecol. Monogr. 50(2):131–51
14. Chastel, C., Main, A. J., Couatarma-nac'h, A., Le Lay, G., Knudson, D. L., et al. 1984. Isolation of Eyach virus (Reoviridae, Colorado tick fever group) from Ixodes ricinus and I. ventalloi ticks in France. Arch. Virol. 82:161–71
15. Clark, G. M., Clifford, C. M., Fadness,

L. V., Jones, E. K. 1970. Contributions to the ecology of Colorado tick fever virus. *J. Med. Entomol.* 7(2):189–97

16. Dawson, D. L., Vernon, T. M. 1972. Colorado tick fever. Colorado. *Morb. Mortal. Wkly. Rep.* 21(44):374

17. Desmond, E. P., Schmidt, N. J., Lennette, E. H. 1979. Immunoperoxidase staining for detection of Colorado tick fever virus, and a study of congenital infection in the mouse. *Am. J. Trop. Med. Hyg.* 28(4):729–32

18. Drevets, C. C. 1957. Colorado tick fever. Observations on eighteen cases and review of the literature. *J. Kans. Med. Soc.* 58:448–55

19. Eads, R. B., Smith, G. C. 1983. Seasonal activity and Colorado tick fever virus infection rates in Rocky Mountain wood ticks, *Dermacentor andersoni* (Acari: Ixodidae), in north-central Colorado, USA. *J. Med. Entomol.* 20(1):49–55

20. Earnest, M. P., Breckinridge, J. C., Barr, R. J., Francy, D. B., Mollohan, C. S. 1971. Colorado tick fever. Clinical and epidemiologic features and evaluation of diagnostic methods. *Rocky Mount. Med. J.* 68(2):60–62

21. Eklund, C. M. 1963. Role of mammals in maintenance of arboviruses. *Proc. 7th Int. Congr. Trop. Med. Malaria. An. Microbiol.* 11A:99–105

22. Eklund, C. M., Kennedy, R. C. 1962. Preliminary studies of pathogenesis of Colorado tick fever virus infection of mice. In *Biology of Viruses of the Tick-Borne Encephalitis Complex. Proc. Symp. Smolenice, Czechoslovakia, 1960,* ed. H. Libiková, pp. 286–93. Prague: Czech. Acad. Sci.

23. Eklund, C. M., Kennedy, R. C., Casey, M. 1961. Colorado tick fever. *Rocky Mount. Med. J.* 58:21–25

24. Eklund, C. M., Kohls, G. M., Brennan, J. M. 1955. Distribution of Colorado tick fever and virus carrying ticks. *J. Am. Med. Assoc.* 157:335–37

25. Eklund, C. M., Kohls, G. M., Jellison, W. L. 1958. Isolation of Colorado tick fever virus from rodents in Colorado. *Science* 128:413

26. Eklund, C. M., Kohls, G. M., Jellison, W. L., Burgdorfer, W., Kennedy, R. C., et al. 1959. The clinical and ecological aspects of Colorado tick fever. *Proc. 6th Int. Congr. Trop. Med. Malaria, Lisbon,* 5:197–203. Lisbon: Inst. Med. Trop.

27. Eklund, C. M., Kohls, G. M., Kennedy, R. C. 1961. Lack of evidence of transovarial transmission of Colorado tick fever virus in *Dermacentor andersoni.* See Ref. 22, pp. 401–9

28. Emmons, R. W. 1965. *Experimental Colorado tick fever infection in wild rodents: viremia and antibody response in active and hibernating animals.* PhD thesis. Univ. Calif., Berkeley. 82 pp.

29. Emmons, R. W. 1966. Colorado tick fever: prolonged viremia in hibernating *Citellus lateralis. Am. J. Trop. Med. Hyg.* 15(3):428–33

30. Emmons, R. W. 1967. Colorado tick fever along the Pacific slope of North America. *Jpn. J. Med. Sci. Biol.* 20(Suppl):166–70

31. Emmons, R. W. 1988. Orbivirus (Colorado tick fever). In *Laboratory Diagnosis of Infectious Diseases: Principles and Practice.* ed. A. Balows, W. J. Hausler, E. H. Lennette. New York/Berlin/Heidelberg/Tokyo: Springer-Verlag. In press

32. Emmons, R. W., Dondero, D. V., Devlin, V., Lennette, E. H. 1969. Serologic diagnosis of Colorado tick fever. A comparison of complement-fixation, immunofluorescence, and plaque-reduction methods. *Am. J. Trop. Med. Hyg.* 18(5):796–802

33. Emmons, R. W., Lennette, E. H. 1966. Immunofluorescent staining in the laboratory diagnosis of Colorado tick fever. *J. Lab. Clin. Med.* 68(6):923–29

34. Emmons, R. W., Oshiro, L. S., Johnson, H. N., Lennette, E. H. 1972. Intraerythrocytic location of Colorado tick fever virus. *J. Gen. Virol.* 17(2):185–95

35. Florio, L., Miller, M. S. 1948. Epidemiology of Colorado tick fever. *Am. J. Public Health* 38:211–13

36. Florio, L., Miller, M. S., Mugrage, E. R. 1950. Colorado tick fever. Isolation of the virus from *Dermacentor andersoni* in nature and a laboratory study of the transmission of the virus in the tick. *J. Immunol.* 64(4):257–63

37. Florio, L., Miller, M. S., Mugrage, E. R. 1950. Colorado tick fever. Isolation of virus from *Dermacentor variabilis* obtained from Long Island, New York, with immunological comparisons between eastern and western strains. *J. Immunol.* 64(4):265–72

38. Florio, L., Miller, M. S., Mugrage, E. R. 1952. Colorado tick fever: recovery of virus from human cerebrospinal fluid. *J. Infect. Dis.* 90:285–89

39. Florio, L., Steward, M. O., Mugrage, E. R. 1944. The experimental transmission of Colorado tick fever. *J. Exp. Med.* 80:165–87

40. Florio, L., Steward, M. O., Mugrage,

E. R. 1946. The etiology of Colorado tick fever. *J. Exp. Med.* 83:1–10

41. Gerloff, R. K., Larson, C. L. 1959. Experimental infection of rhesus monkeys with Colorado tick fever virus. *Am. J. Pathol.* 35(5):1043–54

42. Goodpasture, H. C., Poland, J. D., Francy, D. B., Bowen, G. S., Horn, K. A. 1978. Colorado tick fever: clinical, epidemiologic, and laboratory aspects of 228 cases in Colorado in 1973–1974. *Ann. Intern. Med.* 88(3):303–10

43. Green, I. J. 1970. Evidence for the double-stranded nature of the RNA of Colorado tick fever virus, an ungrouped arbovirus. *Virology* 40:1056–59

44. Hadlow, W. J. 1957. Histopathologic changes in suckling mice infected with the virus of Colorado tick fever. *J. Infect. Dis.* 101:158–67

45. Harris, R. E., Morghan, P., Coleman, P. 1975. Teratogenic effects of Colorado tick fever. *J. Infect. Dis.* 131(4):397–402

46. Hoff, G. L., Trainer, D. O. 1973. Experimental infection in North American elk with epizootic hemorrhagic disease virus. *J. Wild Dis.* 9:129–32

47. Hughes, L., Casper, E. A., Clifford, C. 1974. Persistence of Colorado tick fever virus in red blood cells. *Am. J. Trop. Med. Hyg.* 23:530–32

48. Hurlbut, H. S., Thomas, J. I. 1960. The experimental host range of the arthropod-borne animal viruses in arthropods. *Virology* 12:391–407

49. Johnson, E. S., Napoli, V. M., White, W. C. 1960. Colorado tick fever as a hematologic problem. *Am. J. Clin. Pathol.* 34(2):118–24

50. Johnson, H. N. 1963. The ecological approach to the study of small mammals in relation to arboviruses. *Proc. 7th Int. Congr. Trop. Med. Malaria. An. Microbiol.* 11A:107–9

51. Karabatsos, N., ed. 1985. *International Catalogue of Arboviruses Including Certain Other Viruses of Vertebrates*. San Antonio, Tex: Am. Soc. Trop. Med. Hyg. 1147 pp.

52. Karabatsos, N., Poland, J. D., Emmons, R. W., Mathews, J. H., Calisher, C. H., et al. 1987. Antigenic variants of Colorado tick fever virus. *J. Gen. Virol.* 68(Pt. 5):1463–69

53. Knudson, D. L. 1981. Genome of Colorado tick fever. *Virology* 112:361–64

54. Koprowski, H., Cox, H. R. 1946. Adaptation of Colorado tick fever virus to mouse and developing chick embryo. *Proc. Soc. Exp. Biol. Med.* 62:320–22

55. Koprowski, H., Cox, H. R. 1947. Colorado tick fever. I. Studies on mouse brain adapted virus. *J. Immunol.* 57:239–53

56. Koprowski, H., Cox, H. R. 1947. Colorado tick fever. II. Studies on chick embryo adapted virus. *J. Immunol.* 57:255–62

57. Lane, R. S., Emmons, R. W., Devlin, V., Dondero, D. V., Nelson, B. C. 1982. Survey for evidence of Colorado tick fever virus outside of the known endemic area in California. *Am. J. Trop. Med. Hyg.* 31(4):837–43

58. Lloyd, L. W. 1951. Colorado tick fever. *Med. Clin. North Am.* 2:587–92

59. Luedke, A. J. 1970. Distribution of virus in blood components during viremia of bluetongue. *Proc. Ann. Meet. US Anim. Health A* 74:9–21

60. Markovitz, A. 1963. Thrombocytopenia in Colorado tick fever. *Arch. Intern. Med.* 111(3):307–8

61. McLean, R. G., Francy, D. B., Bowen, G. S., Bailey, R. E., Calisher, C. H., et al. 1981. The ecology of Colorado tick fever in Rocky Mountain National Park in 1974. *Am. J. Trop. Med. Hyg.* 30(2):483–89

62. Miller, J. K. 1960. Colorado tick fever: failure to isolate the virus from Dermacentor variabilis on Long Island, N.Y. *Public Health Lab.* 18(3):53–54

63. Miller, J. K., Tomkins, V. N., Sieracki, J. C. 1961. Pathology of Colorado tick fever in experimental animals. *Arch. Pathol.* 72(2):149–57

64. Murphy, F. A., Borden, E. C., Shope, R. E., Harrison, A. 1971. Physicochemical and morphological relationships of some arthropod-borne viruses to bluetongue virus—a new taxonomic group. Electron microscopic studies. *J. Gen. Virol.* 13:273–88

65. Murphy, F. A., Coleman, P. H., Harrison, A. K., Gray, G. W. Jr. 1968. Colorado tick fever virus: an electron microscopic study. *Virology* 35:28–40

66. Newhouse, V. F., McKiel, J. A., Burgdorfer, W. 1964. California encephalitis, Colorado tick fever and Rocky Mountain spotted fever in eastern Canada. *Can. J. Public Health* 55:257–61

67. Odum, E. P., Odum, H. T. 1959. *Fundamentals of Ecology*. Philadelphia/London: Saunders. 546 pp. 2nd ed.

68. Oliphant, J. W., Tibbs, R. O. 1950. Colorado tick fever. Isolation of virus strains by inoculation of suckling mice. *Public Health Rep.* 65(15):521–22

69. Oshiro, L. S., Dondero, D. V., Emmons, R. W., Lennette, E. H. 1978. The development of Colorado tick fever

virus within cells of the haemopoietic system. *J. Gen. Virol.* 39:73–79

70. Oshiro, L. S., Emmons, R. W. 1968. Electron microscopic observations of Colorado tick fever virus in BHK-21 and KB cells. *J. Gen. Virol.* 3:279–80

71. Philip, R. N., Casper, E. A., Cory, J., Whitlock, J. 1975. The potential for transmission of arboviruses by blood transfusion with particular reference to Colorado tick fever. In *Transmissible Disease & Blood Transfusion,* ed. T. J. Greenwalt, G. A. Jamieson, pp. 175–95. New York/San Francisco/London: Grune & Stratton. 298 pp.

72. Poland, J. D. 1985. Colorado tick fever. In *Current Diagnosis 7,* ed. R. B. Conn, pp. 195–97. Philadelphia: Saunders. 1364 pp.

73. Randall, W. H., Simmons, J., Casper, E. A., Philip, R. N. 1975. Transmission of Colorado tick fever virus by blood transfusion—Montana. *Morb. Mortal. Wkly. Rep.* 24(50):422–23, 27

74. Rehse-Küpper, B., Casals, J., Danielova, V., Ackermann, R. 1979. Eyach virus: the first relative of Colorado tick fever virus isolated in Germany. *Recent Adv. Acarol.* 2:233–38

75. Rehse-Küpper, B., Casals, J., Rehse, E., Ackerman, R. 1976. Eyach—an arthropod-borne virus related to Colorado tick fever virus in the Federal Republic of Germany. *Acta Virol.* 20: 339–42

76. Rozeboom, L. E., Burgdorfer, W. 1959. Development of Colorado tick fever virus in the Rocky Mountain wood tick, *Dermacentor andersoni. Am. J. Hyg.* 69(2):138–45

77. Smee, D. F., Sidwell, R. W., Clark, S. M., Barnett, B. B., Spendlove, R. S. 1981. Inhibition of bluetongue and Colorado tick fever orbiviruses by selected antiviral substances. *Antimicrob. Agents Chemother.* 20:533–38

78. Sonenshine, D. E., Yunker, C. E., Clifford, C. M., Clark, G. M., Rudbach, J. A. 1976. Contributions to the ecology of Colorado tick fever virus. 2. Population dynamics and host utilization of immature stages of the Rocky Mountain wood tick, *Dermacentor andersoni. J. Med. Entomol.* 12(6):651–56

79. Spruance, S. L., Bailey, A. 1973. Colorado tick fever. A review of 115 laboratory confirmed cases. *Arch. Intern. Med.* 131:288–93

80. Thomas, L. A., Eklund, C. M., Philip, R. N., Casey, M. 1963. Development of a vaccine against Colorado tick fever for use in man. *Am. J. Trop. Med. Hyg.* 12:678–85

81. Thomas, L. A., Philip, R. N., Patzer, E., Casper, E. 1967. Long duration of neutralizing-antibody response after immunization of man with a formalinized Colorado tick fever vaccine. *Am. J. Trop. Med. Hyg.* 16(1):60–62

82. Topping, N. H., Cullyford, J. S., Davis, G. E. 1940. Colorado tick fever. *Public Health Rep.* 55:2224–37

83. Trent, D. W., Scott, L. V. 1966. Colorado tick fever virus in cell culture. II. Physical and chemical properties. *J. Bacteriol.* 91:1282–88

Ann. Rev. Microbiol. 1988. 42:65–95

GENETICS AND REGULATION OF CARBOHYDRATE CATABOLISM IN *BACILLUS*

A. F. Klier and G. Rapoport

Department of Biotechnology, Institut Pasteur, 25, rue du Docteur Roux, 75724 Paris Cedex 15, France

CONTENTS

INTRODUCTION

Scientists have amassed a considerable body of knowledge concerning the genetics, biochemistry, physiology, and molecular biology of bacilli. Problems such as DNA-mediated transformation and sporulation have been successfully addressed in *Bacillus* species. Bacilli are also of major importance in the fermentation industry, since they produce a variety of useful proteins and antibiotics. Their ability to secrete degradative enzymes in large amounts

65

makes them very attractive for commercial exploitation as well as for basic research. Among the enzymes secreted are those able to hydrolyze high–molecular weight polymers of sugars such as starch, cellulose, xylan, or levan. The metabolism of the resulting low–molecular weight compounds and other mono- or oligosaccharides requires specific transport systems, which are found not only in bacilli but also in other bacteria, both gram-positive and gram-negative. Similarly, the subsequent steps in the metabolic pathways of sugars do not seem to be specific to bacilli. The study of the regulation of carbohydrate metabolism is only beginning, and only a few systems have been examined in detail.

This contribution is not a complete survey of the literature on sugar catabolism in bacilli, but is rather an effort to focus on points relevant to the current status of research in the field.

The review deals essentially with *Bacillus subtilis,* which is the most studied organism among the bacilli. The choice of the genetic systems that are analyzed is based on the data available in the recent compilations concerning the genetic linkage map of *B. subtilis* (117, 162). About 15 different systems involved in the carbohydrate pathways have been listed. They can be divided into two classes: the systems in which the relevant loci are defined only by fine-structure genetic mapping and the systems that include genes that have already been cloned and possibly some that have been sequenced.

The review is divided according to the sugars considered: monosaccharides, disaccharides, and polysaccharides. For each genetic system we describe the isolation and properties of the mutants, the mapping of the corresponding loci on the *B. subtilis* chromosome, the characterization of the cloned gene products (if data are available), and the demonstration of any linkage of structural genes in an operon structure (e.g. in gluconate and sucrose). As a consequence, emphasis is focused on those systems for which comprehensive knowledge has been acquired, such as the α-amylase family and the sucrose metabolic system.

The interaction of the different gene products is analyzed to give an overall view of the regulation of these systems. Information is also given for some known related systems in other bacilli (for instance, α-amylases, β-glucanases, and xylanases).

Finally, the different systems are discussed with reference to homologies and novel types of regulation.

CATABOLISM OF MONO- AND DISACCHARIDES

Glycerol

B. subtilis 168 has two known metabolic pathways for the utilization of glycerol. Each pathway leads to the formation of dihydroxyacetone phosphate (82, 92, 126).

Glycerol is first taken up by a specific permease induced by glycerol or glycerol-3-phosphate (G3P). The first pathway involves a NAD-dependent glycerol dehydrogenase and a dihydroxyacetone kinase. The second pathway includes a glycerol kinase and a NAD-independent glycerol phosphate dehydrogenase.

Dissimilation of glycerol in *B. subtilis* proceeds primarily via the second pathway (82, 92, 126). Only the enzymes of the second pathway are inducible by glycerol or G3P. Enzymes of both pathways are repressed by glucose (82, 126).

A mutant affected in glycerol permeation that has all the enzymes required for the degradation of endocellular glycerol has been obtained, and in this mutant repression of sporulation by glycerol is absent (125). Mutants in the main pathway have been described in detail (81–83). Mutants in the glycerophosphate kinase locus *(glpK)* fail to grow on glycerol but utilize G3P. Those deficient in G3P dehydrogenase activity *(glpD)* are inhibited by glycerol owing to the accumulation of toxic G3P.

Pleiotropic negative glycerol mutants $(glpP_I)$ were also found with low glycerophosphate kinase and G3P dehydrogenase activities (82, 83). These mutants are deficient in G3P transport and are resistant to the antibiotic fosfomycin, which is normally taken up by the G3P transport system. It is probable that this class of mutants is affected in the control of the expression of the glycerol-catabolizing enzymes. The $glpP_I$ mutations were mapped by density transfer experiments, phage PBS1–mediated transduction, and three-factor transformation crosses. The order of the loci on the *B. subtilis* chromosome is *glpP-glpK-glpD-gtaC* (82). The results suggest that $glpP_I$ codes for a protein with a positive regulatory function. They also indicate that the structural genes *glpK* and *glpD* do not belong to the same operon, since their synthesis is not coordinated.

Two other types of fosfomycin-resistant mutants affected in G3P transport have been isolated (81). Mutants of the first type (GlpT) are resistant to fosfomycin but can grow on glycerol as sole carbon source. Mutants of the second type are less resistant to the antibiotic and transport G3P as efficiently as wild-type cells. The *glpT* mutations map by transduction in the *cysA-aroI* region of the *B. subtilis* chromosome. Recently, the *glpK* and *glpD* genes have been cloned (L. Rutberg, personal communication).

Arabinose

The metabolic pathway of L-arabinose in *B. subtilis* 168 was first identified by Lepesant & Dedonder (74). It includes the following enzymatic reactions:

$$\text{L-arabinose} \overset{1}{\rightleftharpoons} \text{L-ribulose} \overset{2}{\rightleftharpoons} \text{L-ribulose-5-phosphate}$$
$$\overset{3}{\rightleftharpoons} \text{D-xylulose-5-phosphate} \rightarrow \text{intermediary metabolism.}$$

Reactions 1, 2, and 3 are catalyzed by L-arabinose isomerase, ATP-dependent L-ribulose kinase, and L-ribulose-5-phosphate-4-epimerase, respectively. The expression of these enzymes is induced by L-arabinose, but constitutive mutants for the three enzymes have been isolated after mutagenesis (75). Mutants of *B. subtilis* unable to utilize L-arabinose have been isolated and characterized (116). Genes for arabinose utilization are located in at least three different regions of the *B. subtilis* chromosome: between *hisA* and *cysB* (region I) at 294°, between *aroG* and *leuA* (region II) at 256°, and between *gltA* and *thyA* (region III) at 172°. The structural genes *araA, araB,* and *araD,* coding for L-arabinose isomerase, L-ribulose kinase, and L-ribulose-5-phosphate-4-epimerase, respectively, were mapped between *aroG* and *leuA* (H. de Lencastre, personal communication) and probably correspond to region *ara* II. Mutants in another locus, *araC,* located in the region *hisA–cysB* (at 300° and probably corresponding to region I), are defective in isomerase and ribulose kinase activities. Constitutive mutations for L-arabinose isomerase were isolated and mapped precisely in the *araC* region, which suggests the regulatory nature of this locus (H. de Lencastre, personal communication). Two recombinant plasmids containing the *aroA, aroB,* and *aroD* genes were isolated from a *B. subtilis* DNA bank and identified by complementation of the corresponding mutations of *B. subtilis* (H. de Lencastre, personal communication).

Phosphotransferase System

The involvement of phosphotransferase systems (PTS) in the transport of several sugars (e.g. glucose, fructose, sucrose) in *B. subtilis* 168 has been shown by biochemical and genetic studies (43, 45, 76, 87, 88, 99).

A soluble cellular fraction containing an Enzyme I–type activity (EI) has been demonstrated; the enzyme catalyzes the following reaction, first described by Kundig et al (64):

$$PEP + HPr \xrightarrow{\text{EI, } Mg^{2+}} P\text{-}HPr + pyruvate$$

The heat-resistant histidine-containing phosphocarrier protein (HPr), which is an obligate intermediary in the transfer of the phosphoryl group from PEP, has been purified from *B. subtilis* (86) and is similar to that found in *Staphylococcus aureus*. The sequence around the unique histidine residue, the site of phosphorylation, was compared to the sequences found in HPr from *S. aureus* and *Streptococcus faecalis* (1). The amino acid sequence Gly-Ile-His-Ala-Arg-Pro-Ala-Thr is conserved in all three microorganisms.

Gonzy-Tréboul & Steinmetz (46) have recently cloned a DNA fragment from *B. subtilis* that complemented mutations affecting the gene coding for the Enzyme I of *Escherichia coli (ptsI)*. This DNA fragment has also been shown to complement mutations in the HPr locus (*ptsH*) in *E. coli* (46). In *E. coli, ptsI* and *ptsH* are organized in an operon. In *B. subtilis* the Enzyme I gene *(ptsI)* was mapped at 120° on the chromosome (43, 99) and is therefore adjacent to the *ptsH* loci; thus it is likely that the two genes in this microorganism, too, belong to the same operon.

Until recently, no Enzyme III had been described from *B. subtilis*. This cytoplasmic protein found in other gram-positive bacteria is specific to a given sugar (118). The protein transfers the phosphoryl group from the phosphorylated form of HPr to the specific Enzyme II, which will finally phosphorylate its corresponding sugar. However, Gonzy-Tréboul & Steinmetz (46) have recently obtained complementation of defective Enzyme III in *E. coli* (*crr* mutants) with a *B. subtilis* DNA fragment allowing growth on glucose. This fragment carries *ptsI* and *ptsH,* and the order of the markers appears to be *crr*(?)*-ptsH-ptsI,* which is different from that observed in *E. coli* (*ptsH-ptsI-crr*). It would be interesting to know if these genes are also organized in an operon in *B. subtilis* as they are in *E. coli.*

Regulation by the PTS has been shown by pleiotropic mutants of *ptsI* unable to grow not only on sugars normally transported by PTS but also on other sugars transported by other systems, such as glycerol, arabinose, and glucitol (43). The transport of glycerol is inhibited by α-methylglucoside, and the level of inhibition depends upon the balance between the Enzyme I and Enzyme II activities, as shown by the use of a *ptsI* thermosensitive mutant (99, 119). Other catabolite functions are also impaired in PTS mutants, such as the induction of fructose-1-phosphate kinase and mannitol-1-phosphate dehydrogenase by glucitol, which is completely repressed (43), and the regulation of sporulation by glucose (29). We stress that not all *ptsI* mutants are affected in these functions. For example, of the two phenotypic classes of thermosensitive *ptsI* mutants, one class is only defective in the transport of PTS sugars, while the other is impaired in different catabolite regulations (99).

Mannitol

D-Mannitol-1-phosphate dehydrogenase activity was first described in *B. subtilis* by Horwitz & Kaplan (57).

Gay (40a) subsequently obtained mutants in D-mannitol metabolism, which were affected in either the transport of hexitol *(mtlA),* which involves a phosphotransferase system, or mannitol-1-phosphate dehydrogenase *(mtlB),* which catalyzes the formation of fructose-6-phosphate. Both mutations map in the same region of the *B. subtilis* chromosome near the *dal* marker (at 34–35°) (80). The involvement of a PTS in the transport of D-mannitol was confirmed by the discovery of pleiotropic mutants of the Enzyme I system that are unable to take up several sugars, including D-mannitol (40a).

The *mtl* system is induced by D-mannitol and also, at a lower level, by D-sorbitol (43, 57).

Fructose

D-Fructose is transported into *B. subtilis* by way of a PTS that includes a specific component *(fruA)*. This transport results mainly in the intracellular synthesis of fructose-1-phosphate (44). Fructose-1-phosphate is then metabolized through the action of fructose-1-phosphate kinase *(fruB)*. Both *fruA* and *fruB* map at the same position on the *B. subtilis* chromosome (120°) (41). Mutations of the *fruA* locus affect only the transport of fructose when the cells are subjected to catabolite repression, for instance in the presence of sugars such as glucose, glycerol, or D-mannitol (44). When grown in the presence of slowly metabolized substrates such as succinate and glutamate, these mutants were able to transport fructose. This specific phenotypic suppression, which is still dependent upon an active Enzyme I of the PTS, may proceed through the derepression of other genes, giving rise to intracellular fructose-1-phosphate or fructose-6-phosphate.

Growth of mutants altered in the *fruB* gene was inhibited by the presence of fructose in the medium; these mutants were shown to accumulate fructose-1-phosphate (45). In wild-type cells fructose-1-phosphate can be dephosphorylated and the free intracellular fructose can be converted to fructose-6-phosphate by the *fruC* gene product fructokinase (22).

It has also been shown that fructose can be excreted from the cell into the medium and recaptured by the PTS (21). Furthermore, phosphorylation of intracellular fructose by the PTS has been reported, but seems to be devoid of a physiological role (21).

Glucitol

The catabolic pathway of D-glucitol (sorbitol) in *B. subtilis* 168 includes the following steps (18). First, the hexitol is transported by a specific permease, which is affected by *gutA* mutations. The hexitol permease delivers unmodified D-glucitol inside the cells, which is then oxidized by a D-glucitol dehydrogenase, the product of the *gutB* gene, leading to intracellular D-fructose. This compound is phosphorylated by the PTS either at the C-1 position or at the C-6 position as mentioned before (see section on metabolism of fructose, above). Fructose can also be phosphorylated by a fructokinase (fructose-ATP-6-phosphotransferase) which is affected by the *fruC* mutations (located at 51° on the *B. subtilis* chromosome).

In both wild-type and *gutA* mutants the transport of D-glucitol can also be mediated by the D-mannitol PTS, but the D-glucitol-6-phosphate formed cannot be further catabolized by *B. subtilis* (18). The permease is induced by D-glucitol and to a lesser extent by D-mannitol (18).

Horwitz & Kaplan (57) described a NAD-dependent D-glucitol dehydrogenase that is able to oxidize D-xylitol and L-iditol in addition to D-glucitol. The enzyme is induced only by D-glucitol (18). The system is not induced by D-xylitol, which is not metabolized in the wild-type strain. However, mutants affected in another locus, *gutR*, are able to utilize D-xylitol as sole carbon source and synthesize constitutively both the permease and the dehydrogenase (18). The *gutA, gutB,* and *gutR* loci were mapped by PBS1 transduction and DNA-mediated transformation between the *purB* and *dal* markers on the *B. subtilis* chromosome at 50°, in the order *gutR-gutB-gutA*. It is likely that these loci belong to the same operon (42). The *fruC* locus was also mapped near the *gut* cluster, but it does not belong to the same system. The catabolic pathway in *B. subtilis* is therefore different from that of Enterobacteriaceae, in which vectorial phosphorylation by the PTS seems to be the general rule for the utilization of D-glucitol (72, 136).

Gluconate

As *B. subtilis* lacks the Entner-Doudoroff pathway, the only way for *B. subtilis* cells to metabolize D-gluconate is through the pentose cycle. Two inducible enzymes are needed for the utilization of gluconate: gluconate permease, which transports gluconate into the cell, and gluconate kinase, which phosphorylates gluconate (34, 35, 85). The induction is repressed by rapidly metabolized sugars such as glucose (102). Several mutants that are unable to grow when gluconate is the sole carbon source have been isolated (34, 35). All these mutations were localized between *iol6* and *fdp74* on the *B. subtilis* chromosome (mapped at 344°). Among these mutations, most were localized at either the permease locus or the kinase locus. One mutation *(gnt9)* affected synthesis of both permease and kinase, which suggests that the two structural genes belong to a single operon (35).

The gluconate operon of *B. subtilis* has been initially cloned in the temperate phage rho11 (32) as a 7-kb *Eco*RI fragment. It was shown that this fragment can correct all the *gnt* mutations. Analysis of the sequence revealed four open reading frames, each of which was preceded by a ribosome binding site (33). The order of these open reading frames is *gntR-gntK-gntP-gntZ*. Deletions and insertional inactivation experiments demonstrated that the *gntK* and *gntP* gene products (consisting of 513 and 448 amino acids, respectively) correspond to gluconate kinase and permease. The *gntP* product was found to be very hydrophobic, which suggests that this protein might have a membrane location. The function of the *gntZ* gene product is not known. The first gene of the operon, *gntR*, encodes a polypeptide of 243 amino acids, which has a regulatory function on the expression of the operon. An overlap' of 5 bases was found between the coding sequences of *gntR* and *gntK*. Such cistron overlapping has recently been reported in the *B. subtilis trp* operon (49) and between the *sacP* and *sacA* genes (see below). The overlap of the *gnt* genes may have an important function in regulating their expression, probably at the translational level.

Analysis of the transcripts by S1 nuclease mapping experiments indicated that the *gnt* operon was mainly transcribed as a polycistronic mRNA (33). However, this giant molecule has never been detected. The major promoter, located 40 base pairs (bp) upstream from the coding sequence of the *gntR* gene, was identified by S1 analysis and by subcloning in a promoter probe vector (30). This region is responsible for induction by gluconate and for repression by glucose. Interestingly, two minor constitutive promoters are localized between the *gntP* and *gntZ* genes, which suggests that the *gntZ* product might be required for an unknown function besides gluconate catabolism. The *gnt* transcripts terminate about 45 bp downstream of the *gntZ* gene (33).

A 223-bp fragment corresponding to the promoter and the upstream part of the *gnt* operon, on a high–copy number plasmid, titrated a regulator of the operon, which led to constitutive expression of the structural genes located on the chromosome (31). This regulator is encoded by the *gntR* gene, and insertion of a 4-bp sequence within its coding sequence led to constitutive expression of the permease and kinase genes (31). As the level of mRNA initiated at the promoter was not changed by the insertional mutation in the presence or absence of the inducer, it was concluded that the *gntR* gene codes for a negative transcriptional regulator of the *gnt* operon (31). Recently, the repressor has been purified and footprinting experiments have allowed the localization of the *gnt* operator at the transcriptional start site of the operon (Y. Fujita, personal communication). Interestingly, cloning of the promoter and the *gntR* gene in *E. coli* demonstrated that this inducible system can operate in a heterologous host (93).

Sucrose Metabolic System

Sucrose metabolism of *Bacillus subtilis* has been studied extensively (77, 78). Two saccharolytic enzymes, sucrase and levansucrase, can be detected in crude extracts of *B. subtilis* 168 after induction by sucrose. Sucrase is an intracellular enzyme, whereas levansucrase is secreted. Both enzymes act as β-D-fructofuranosidases, and levansucrase also catalyzes the formation of the high–molecular weight fructose polymer levan. A third structural gene, *sacP*, codes for an inducible membrane component of the PTS of sucrose transport. Another exocellular saccharolytic enzyme, levanase, which is able to hydrolyze levans and inulin, was characterized after the isolation of a class of mutants referred to as *sacL* (66). Sucrase has been purified and has a molecular weight of 55×10^3 (68), whereas levansucrase, whose low-resolution crystallographic structure has been determined (70), has a molecular weight of 50×10^3 (20a, 140).

The structural gene of levansucrase, *sacB*, maps between *cysB* and *hisA* on the *B. subtilis* chromosome at 305° (79). Several classes of regulatory mutations that affect its expression have been identified (78). A first class of mutations, which lead to constitutive synthesis of levansucrase (*sacR^c*), are tightly linked to the *sacB* locus. The *sacR^c* mutations did not affect the inducibility of the intracellular sucrase. The sucrase structural gene, *sacA*, is linked to the *sacP* locus, and both map to the left of *purA* at 335° (73). A mutation termed *sacT* constitutive (*sacT^c*) was also shown to be linked to the *sacA* and *sacP* loci and leads to the constitutive ability of *B. subtilis* to hydrolyze sucrose. Another regulatory locus, *sacS*, affects the synthesis of both sucrase and levansucrase (78). This locus is linked to the *sacA-sacP-sacT* cluster by PBS1 transduction, but it is completely separated from this group. Three types of mutations have been defined at this locus. *SacS^c* mutations lead to a constitutive synthesis of both saccharolytic enzymes, whereas *sacS$^-$* and *sacS^h* mutations affect only levansucrase production, leading to insignificant and high-level synthesis, respectively.

The systematic isolation of mutants affected in levansucrase synthesis led to the identification of two other loci, *sacQ* and *sacU* (67). The *sacU* locus was originally defined by two series of mutants referred to as *sacU^h* and *sacU$^-$*, which exhibit respectively a hyperproduction (Lvs^h phenotype) and a hypoproduction (Lvs$^-$ phenotype) of levansucrase. This locus maps between *uvrA* and *gtaB*. So far, a single *sacQ* mutant has been isolated. It displays a Lvs^h phenotype, and the locus is located to the left of *thrA* on the *B. subtilis* chromosome. The strains carrying the *sacU* and *sacQ* mutations are affected in the production of other extracellular enzymes (α-amylase, proteases) and are also impaired in their motility and their transformation by exogenous DNA.

The *sacA* and *sacP* genes have been cloned and expressed in *E. coli* (26–28). The deduced amino acid sequence of the *sacA* gene product gives a protein with molecular weight of 54,827, which is consistent with that of the purified sucrase. The amino acid sequence of the protein shows homology with that of yeast invertase (*SUC2* gene product). The sequence of the *B. subtilis* levanase gene *(sacC)* product has recently been determined (89). The deduced protein has a molecular weight of 75,866, and its sequence displays large similarities to the sequences of sucrase and invertase. A cysteine residue has been implicated in the active site of these three enzymes, which are β-D-fructofuranosyl transferases. Interestingly, each of these sequences contains only one cysteine residue, and the region around this residue is highly conserved. It is therefore tempting to postulate that the sequence (Tyr/Trp)-Glu-Cys-Pro-(Gly/Asp)-Leu corresponds to the peptide containing the active SH group of the three enzymes. The deduced amino acid sequence of *sacP* gives a molecular weight of 48,945 for the corresponding polypeptide (26). Fouet et al (26) demonstrated that *E. coli* cells harboring the *B. subtilis sacP* gene transported and phosphorylated sucrose. Thus the *sacP* gene encodes the Enzyme II (EII^{Suc}) of the sucrose PTS of *B. subtilis*. *E. coli* cells harboring *sacA* and *sacP* genes grew well in medium containing sucrose as the sole carbon source, demonstrating that the EII^{Suc} interacts with the other nonspecific components of the *E. coli* PTS. The amino acid sequence of the *B. subtilis* EII^{Suc} was compared to that of other EIIs from enteric bacteria, EII^{Scr} and EII^{Mtl} from *Salmonella typhimurium* and EII^{Bgl} and EII^{Nag} from *E. coli*. At the amino acid level, the EII^{Suc} encoded by the *sacP* gene shows a very high homology (52%) with the EII^{Scr} from *S. typhimurium* (23). A strong similarity was also found between EII^{Suc} from *B. subtilis* and EII^{Bgl} from *E. coli*, which suggests that these genes are derived from a common ancestor (12, 26). Moreover, the comparison of the EII amino acid sequences shows several blocks of similarity along the sequences, which reflect the common properties of these membrane proteins (13, 26). The nucleotide sequences encoding EII^{Suc} and sucrase overlap at the sequence ATGA, where ATG is the initiation codon of the sucrase gene and TGA is the stop codon of the *sacP* gene. This observation supports the hypothesis that *sacP* and *sacA* are part of an operon in which *sacA* is distal to *sacP*. If this assumption is correct, the promoter is located upstream from the *sacP* gene and could correspond to part of the *sacT* locus. This locus would be the target of the inducer.

The sequence of the *sacB* gene and its upstream *cis*-acting control region, *sacR*, has been reported (25, 140). The deduced protein is a 473-residue polypeptide, which includes a 444–amino acid sequence almost identical to that already determined for the mature levansucrase. The 29 amino acid residues located upstream from the mature sequence are typical of a signal sequence with a hydrophilic region followed by a hydrophobic stretch with an Ala residue at position −1. The existence of the precursor was demonstrated in an *E. coli* minicell-producing strain (25). In this strain a significant proportion of the mature enzyme is recovered in the supernatant.

The *sacR* locus contains the *sacB* promoter and the targets of the products of *sacU*, *sacQ*, and *sacS* (3, 62, 131). S1 nuclease mapping of the *sacB* promoter defined the transcription start site 199 bp upstream of the *sacB* coding sequence (131). Between the promoter and the *sacB* coding sequence a region of dyad symmetry of approximately 2 × 30 nucleotides was shown to act as a transcriptional terminator structure (131). Deletions of this termination structure or single base changes that destabilize the structure lead to constitutive synthesis of levansucrase. The *sacR*^{c} mutations have been shown to be changes of the nucleotide sequence in the putative stem-and-loop structure (137). Analysis of transcripts demonstrated constitutive expression from the *sacB* promoter (132). However, these transcripts did not extend into the *sacB* structural gene in the absence of sucrose. Transcription continues beyond the region of dyad symmetry only in the presence of the inducer (sucrose). This termination structure preceding the *sacB* gene suggests that *sacB* expression might be positively regulated by an antitermination process similar to that regulating the *B. subtilis trp* operon (132).

Fusions in which different parts of the *sacR* region were linked to heterologous genes allowed identification of the targets of different regulatory gene products (3, 50, 62, 131, 163). A region located between the promoter and the ribosome binding site of the *sacB* gene was shown to be the target of the *sacS* gene product. This region harbors the putative stem-and-loop structure. Deletion of either the left or the right part of the palindromic structure abolished the induction by sucrose and led to constitutive expression of the *sacB* gene promoter. Consequently, the palindromic structure or a region located just upstream from it could be the actual target of the antiterminator protein, which is encoded by the *sacS* gene. Another region of *sacR* located near or just upstream from the promoter was shown to be the target of the *sacU* and *sacQ* gene products. Fusions between the *sacB* promoter and heterologous genes were constructed and gave constitutive expression of these genes. The levels of expression were increased by the presence of the *sacQ*^{h} and *sacU*^{h} alleles (62, 163). Moreover, strains carrying these alleles exhibited increased steady-state levels of transcripts from the *sacB* promoter (131). Recently, a second target of the *sacU* gene product has been hypothesized (50). The second target could correspond to the *sacS* gene target, located downstream of the promoter, which would suggest that the *sacS* gene expression is under the control of *sacU* (50). The different fusions used have not allowed discrimination between the targets of the *sacU* and *sacQ* gene products. One possible explanation is that these two regulatory products have the same target. A second possibility is that *sacU* acts on a target upstream of the promoter and *sacQ* acts indirectly by affecting the action of *sacS* and *sacQ*.

The *sacS* locus was cloned in *E. coli* (4, 20). As the *sacS*^{c} allele, placed on a multicopy plasmid, was dominant over the wild-type chromosomal *sacS*$^+$ allele, it was proposed that the *sacS* gene encoded a positive regulatory protein (20). Genetic and sequence analyses (4, 138) of the *sacS* locus suggested that it contained two genes. One (termed *sacY*) encoded a positive regulator of levansucrase synthesis. As deletion of *sacS* abolished levansucrase synthesis and as the mutant $sacR^c$ $\Delta sacS$ is constitutive for $sacB$ expression, it was proposed that $sacY$ encodes the antiterminator protein. This conclusion was further confirmed by the strong similarities at the amino acid level between the $sacY$ gene product and the *E. coli bglC* gene product, which is known to be a posivite regulator of the *bgl* operon and which acts as an antiterminator (4, 128).

A second gene was detected upstream from the $sacY$ gene in the $sacS$ locus after insertion of the transposon Tn*917 lac* (138). The strains harboring the transposon showed constitutive $sacB$ expression. The newly found gene was called $sacX$, and its corresponding product could be involved in negative control of $sacY$. Inactivation of the protein by insertion or mutation of $sacX$ led to constitutive $sacB$ synthesis. Interestingly, recent data (M. Zukowski, personal communication) indicate that the amino acid sequence of part of the protein encoded by $sacX$ is highly homologous to the C-terminal sequence of the EII^{Suc}, which is involved in the transport of sucrose (26). This homology could imply the involvement of the $sacX$ gene product in sucrose transport.

As induction of $sacA$ by sucrose was not affected by a deletion of $sacY$ and part of $sacX$, Steinmetz et al (138) suggested that the $sacS$ locus was not essential for $sacA$ inducibility. However, $sacS$ mutations or insertions make $sacA$ constitutive. This paradox could be explained by the involvement of two parallel pathways in the induction of saccharolytic enzymes by sucrose (138). Each of them would be preferential for one saccharolytic enzyme but could also allow the expression of the other at a low level. If this hypothesis is true, an as yet unidentified gene product similar to that encoded by $sacY$ is responsible for $sacA$ induction.

Although the promoter sequences of several genes whose expression is affected by the $sacU^h$ and $sacQ^h$ mutations have been determined, there are no obvious sequence similarities to indicate target sites. The exact nature of the $sacU^h$ mutations is not known at this time. The $sacQ$ gene encodes a 46–amino acid polypeptide. The $sacQ^h$ mutation is a single base change upstream from the coding sequence, which leads to a higher level of expression (155). Genes leading to similar phenotypic alterations have been cloned from *Bacillus licheniformis* (2) and *Bacillus amyloliquefaciens* (107). The $sacQ$ product from *B. licheniformis* has shown the greatest stimulatory activity. The same nucleotide change has been identified at the same position upstream from the coding sequence of the $sacQ$ genes isolated from both bacteria. Moreover, Amory et al (2) have compared the polypeptides by inserting the coding sequences of each $sacQ$ gene downstream of a heterologous promoter. A higher stimulatory effect was observed with the coding sequence from *B. licheniformis*, which demonstrated that the difference in efficiency was maintained. This means that the higher efficiency of the *B. licheniformis* gene is at once due to the structure of the $sacQ$ gene product and to the regulation of its transcription. Moreover, the insertion of the gene into high–copy number plasmids led to a stimulation of the synthesis rate of a large number of secreted enzymes, such as protease(s), β-glucanase(s), and xylanase (2). Furthermore, substitution of the 3' end of the *sacQ* gene by a foreign coding sequence led to a more active fusion protein, which demonstrated that the C-terminal end of the polypeptide is not essential for the stimulatory activity (2).

It was demonstrated that the *sacQ* polypeptide stimulates the transcription of the target genes. However, partial chromosomal deletion of the *sacQ* gene had no discernable phenotypic effect on *sacB* expression (155). Amory et al (2) have shown that the *sacU* gene product is epistatic to the *sacQ* gene product, and they postulated that the *sacQ*^{h} mutation caused an increase in the production of the *sacQ* polypeptide, leading to an increased effect of the *sacU* gene on *sacB* expression. One may speculate that the function of the *sacQ* gene is to adapt the *sacB* expression level to different growth conditions.

The *prtR* genes of *Bacillus natto* and *B. subtilis,* cloned on a high–copy number plasmid in *B. subtilis,* also stimulate production of levansucrase as well as alkaline and neutral proteases (145, 157). They encode identical 60–amino acid polypeptides. Genetic mapping localized the *prtR* gene near *metB* at 200°, a location distinct from that of other pleiotropic genes that cause similar effects. Deletions of the gene from the *B. subtilis* chromosome had no obvious phenotypic effect. The *prtR* polypeptide, like the *sacQ* polypeptide, affects the target genes at the transcriptional level. No amino acid homologies between the predicted sequences of the *sacQ* and *prtR* genes have been detected (157). However, the predicted secondary structures of both polypeptides have some similarities, suggesting a common mechanism of action (157).

A new gene, called *sacV,* was recently identified in *B. subtilis;* it encodes a 64–amino acid polypeptide (90). This gene led to levansucrase overproduction when it was present on a high–copy number plasmid. It probably acts by increasing the transcription of the *sacB* gene. Interestingly, Martin et al (90) have noted a homology at the amino acid level with the product of the sporulation-inhibition gene sequence *(sin),* which apparently encodes a DNA-binding protein (40). This new gene has been located on the *B. subtilis* chromosome near the *dal* locus (40°) and defines a new locus involved in sucrose metabolism.

How genes encoding enzymes involved in degradation processes are controlled by small regulatory molecules is not yet fully understood. The polypeptides affect the rate of transcription of these genes. Several hypotheses concerning the mechanism of action of these peptides, at the DNA level or at the RNA polymerase level, can be made (65). Interestingly, this kind of regulation seems to be unique in the Bacillaceae.

CATABOLISM OF POLYSACCHARIDES

Xylanases

Xylan is a polymer consisting of a β-1.4 linked xylose backbone with branches formed by pentoses including xylose, hexoses, and uronic acids. This compound can represent up to one third of the total sugar content of plant biomass (146). Microbial degradation of xylan is similar to that of starch, the initial step being degradation to oligosaccharides by an extracellular endo-acting xylanase. Free pentoses are subsequently generated by the action of a β-xylosidase (53). Among the many microorganisms known to produce these enzymes, Bacillaceae are excellent sources.

Genes controlling xylan utilization by *B. subtilis* were first identified by Roncero (122). She obtained a set of mutants after *N*-methyl-*N*'-nitro-*N*-nitrosoguanidine (NTG) mutagenesis that were unable to metabolize xylan. The eight mutations analyzed are linked on the chromosome to the left of *purB33* and distal to *tre12* (55–60°). She concluded that two genes, which have been named *xynA* and *xynB,* code for an extracellular β-xylanase and for a cell-associated β-xylosidase, respectively. The two genes have been cloned from *B. subtilis* PAP115 (6, 7). *B. coagulans* 26 (24), and the hyperproducing *B. pumilus* IPO (94, 115).

In *B. pumilus,* the *xynA* and *xynB* are closely linked on the chromosome within a 14.4-kbp fragment (114). The *xynA* gene was located 4.6 kbp downstream of the 3' end of *xynB*. Analysis of the transcripts by Northern blot experiments strongly suggested that the corresponding mRNAs were independently produced. Additional proof came from the identification of putative hairpin structures downstream from the termination codon of both genes. Probable promoter-like sequences, identified by S1 nuclease mapping experiments, were found upstream of both the *xynA* and *xynB* coding regions. The −10 regions for promoter sequences approximate to the consensus sequence for *Bacillus* sigma 43 RNA polymerase, whereas the −35 regions do not.

The sequences of both genes were determined and the amino acid sequences were deduced (39, 114). A 684-bp open reading frame for the xylanase gene was observed, and the amino acid sequence of the N-terminal region of mature xylanase was determined to be Arg-Thr-Ile-Thr, which suggests processing. A signal sequence consisting of 27 amino acids, including three basic residues at the N-terminal end followed by a stretch of 15 hydrophobic residues, terminates with an Ala residue (39).

The subunit of β-xylosidase, a homodimer enzyme, consists of 539 amino acid residues (94). The active site of xylanase was supposed to be a lysozyme type from amino acid alignments (95). Xylanase crystals have been obtained and analyzed to 2.5-Å resolution (95). The xylanase molecule contains two structural domains. Between them, a long crevice was observed, believed to be the active site in which Asp98 and Glu124 could be the two active residues (95).

The *xyn* genes were cloned from *B. subtilis* strain PAP115 (6–8). The *xynA* gene was first isolated by shotgun cloning in *E. coli,* and the intracellular xylanase was purified. Its electrophoretic mobility indicated a molecular weight of 22×10^3. The *xynB* gene was cloned on a 3.2-kbp DNA fragment and further subcloned in the plasmid harboring the *xynA* gene. The two genes were expressed in *E. coli* (6).

Alkalophilic *Bacillus* sp. strain C125 produced two types of xylanases (N and A) whose molecular weights were estimated as 43×10^3 and 16×10^3, respectively (55). The gene encoding the *xynA* enzyme was located on a 4.6-kbp DNA fragment that was cloned in *E. coli.* One surprising result was that more than 80% of the total activity expressed in *E. coli* was detected in the culture medium (54).

To use xylose, the microorganism must convert it into xylulose and then into xylulose-5-phosphate. In *B. subtilis,* this pathway involves two enzymes, the D-xylose isomerase and the xylulose kinase, coded by the genes *xylA* and *xylB,* respectively. Both genes from *B. subtilis* W23 and 168 were cloned in *E. coli* either by complementation of Xyl⁻ mutants (148) or by detection of the activity of the *xynB* gene (48), which is closely linked. Expression of the *xylA* gene was obtained only when an insertion element, IS5, was fortuitously inserted in front of the gene, which suggests that a repressor is involved (148). The gene *xylA* was sequenced and the deduced amino acid sequence corresponded to a molecular weight of 49,680. Comparison with the amino acid sequence of the corresponding *E. coli* enzyme indicated 50% identity over the entire polypeptide (149). Although the protein was expressed at a high level in *E. coli* and yeast, the enzyme was inactive in both cases (53).

It has been shown that on the *B. subtilis* 168 chromosome *xynB, xylA,* and three other genes belonging to the xylose regulon, *xynC, xylR,* and *xylB,* are located on a 7.5-kb fragment (48). All of these genes were cloned in *E. coli,* first by shotgun cloning with partially cut DNA by screening for β-xylosidase activity and secondly by inserting a recombinant replication-minus plasmid into the *B. subtilis* chromosome. Homologous recombination and isolation of clones with larger DNA fragments by the marker rescue technique recovered the genes. The five genes were sequenced, and it was shown that *xylA* and *xylB* encoded the xylose isomerase and the xylulose kinase, respectively. The *xynC* gene coded for a protein of 463 amino acids, with long stretches of hydrophobic residues suggesting a membrane localization. This protein could be the permease for xylose oligomers. It was also shown that expression of the operon *xynC-xynB* was dependent on a 430-bp upstream DNA sequence. The *xylR* gene, which is the repressor of the four other genes, is transcribed in the opposite sense from the other genes. Mutations inside the coding sequence of *xylR* led to a constitutive phenotype.

The promoters of the two operons were identified by S1 nuclease mapping experiments, and the start sites were found within the region of homology of both operons. This sequence was found 300 and 100 bp upstream of *xynC* and *xylA*, respectively, and corresponded to a 10-bp inverted repeat, which is typical for operator sequences. This operator could be the target of the *xylR* gene product. The xylose regulon appears therefore to be controlled by a negative process. It is tempting to speculate that the *xynA* gene, which is nearby, may contain the same operator sequence.

β-Glucanase Family

Cellulose and related β-glucans are the most abundant carbohydrates in plant biomass. Although bacilli are not truly cellulolytic organisms, several strains secrete endoglucanases, including β-1.4 glucanases, β-1.3-1.4 glucanases or lichenases, and β-1.3 glucanases. These glucanases have been increasingly studied because of their potential use in the brewing industry (9) and in the bioconversion of agricultural waste material to useful products (124). They also provide a model for the study of mechanisms involved in the export of enzymes involved in polysaccharide degradation.

Information on *Bacillus* β-glucanases has largely been provided by molecular cloning of the corresponding genes. At present, in total 15 β-glucanase genes from five different *Bacillus* species have been cloned, and most have been expressed in heterologous hosts such as *E. coli*, *B. subtilis*, and yeasts (17). Only one gene has been mapped on the *B. subtilis* chromosome. This gene, *bgl*, encodes a lichenase of 22 kd which is linked to other loci in the order *ctrA-sacA-bgl-purA* (11, 108). It appears that the other β-glucanase genes are widely distributed on the chromosome.

Several lichenases from *Bacillus* species have been purified and their molecular weights have been estimated as $24–27 \times 10^3$ (10, 142). The cloning of four *Bacillus* lichenase genes has been reported (11, 15, 51, 52). The sequences of two of these genes have been determined and indicate a molecular weight of approximately 27×10^3 for each gene product (52, 96). This is the size of the precursor form of each protein, which is processed during secretion to give the mature enzyme of about 24 kd. Analysis of the sequences revealed that the minor differences between them are typical of the divergence of homologous genes in related species. Moreover, the sequences can be aligned with few gaps, and the amino acid sequences of the encoded proteins are not dramatically changed by the nucleotide differences, which are largely silent mutations.

Carboxymethyl (CM-) cellulases are secreted by different *Bacillus* species, and include endo-acting enzymes that hydrolyze only the β-1.4-D-glucosidic linkages. Seven *Bacillus* CM-cellulase genes have been identified by molecular cloning and expression in *E. coli*, and six of them have been fully sequenced (17). Most of these enzymes have similar size of about 45–55 kd, as deduced from the nucleotide sequences of their corresponding genes. The β-1.4-glucanase of *B. subtilis* IF01034 is synthesized as a precursor protein of 55 kd, which is processed to give the active extracellular form of 51 kd (98). Interestingly, the processing of the precursor protein coded by the *B. subtilis* DLG β-glucanase gene gave an unusually small final product of 35 kd (120, 121). It has been reported that for the CM-cellulase precursor of *B. subtilis* PAP115 (55.2 kd) two processing steps occur sequentially (17, 84). The first step is the classical removal of the signal peptide, which gives rise to an inactive proenzyme (53 kd). This product is activated by removal of the C-terminal third of the molecule, which leads to an active enzyme of 36 kd.

The alkalophilic *Bacillus* sp. strain 1139 synthesizes an unusually large CM-cellulase of 92 kd (37). Fukumori et al (37) identified the corresponding gene by molecular cloning and showed that the removal of the first 29 or 30 amino acid residues, the signal sequence, is the only processing step known to occur in vivo. Interestingly, it has been reported that the C-terminal half of this protein can be removed without affecting the activity or the pH optimum of the enzyme (36). Several differently sized forms of the enzyme have been found within a single strain. For example, *B. subtilis* N4, an alkalophilic strain, has two CM-cellulases of different molecular weights (38).

In a recent review, Cantwell et al (17) reported that four genes encoding CM-cellulases from *B. subtilis* ATCC 6633 (130), PAP115 (84), DLG (121), and IFO3034 (98) are closely related. Most DNA sequence changes do not result in a change in the amino acid sequence of the mature protein. However, the signal sequences show less homology than the mature proteins. Comparison of the DNA sequences of three genes from alkalophilic *Bacillus* species indicated that the two genes from strain N4 are closely related (38). One of them shows a duplication of the last 80 codons. The third gene, from strain 1139, showed only partial similarity to either gene from strain N4 (37). Interestingly, the CM-cellulases encoded by *B. subtilis* are closely related in amino acid sequence to those from the alkalophilic strain N4. One explanation could be that the number of CM-cellulase genes is limited and that all the different genes isolated are derived from the same gene, with minor alterations. No obvious similarity has been observed between the CM-cellulases and lichenases of *Bacillus* species. Apparently, there is no great similarity between the β-glucanases from the genus *Bacillus* and other known cellulases. However, it has been reported that the β-1.4-glucanase from *B. subtilis* PAP115 has a small region of similarity to the protein encoded by the *Clostridium thermocellum celB* gene (84).

The cloned β-glucanase offers an efficient system for the study of protein export both in *Bacillus* species and in *E. coli*. Surprisingly, in most cases, when a β-glucanase gene has been overexpressed in *E. coli* a significant part of the enzyme activity has been recovered in the culture supernatant. However, the controls have not been performed to distinguish between exportation and lysis of the cells. In the case of lichenase from *B. subtilis* C120, the hypothesis is that the observed export could be due to an alteration of the *E. coli* outer membrane (17). A different secretion mechanism is reported for the *B. subtilis* PAP115 CM-cellulase cloned in *E. coli* (17). Seventeen percent of the total activity was recovered in the extracellular medium, and leakage of the *E. coli* outer membrane did not seem to be responsible. The CM-cellulase from *B. subtilis* DLG did not seem to be exported by *E. coli* (120). However, two active forms of the enzyme are synthesized. One form is located in the cytoplasm, whereas the second is associated with the cell in an unknown manner. Robson & Chambliss (120) have proposed that the latter may represent an incorrectly processed form of the protein.

Cantwell et al (14) have obtained expression of the lichenase gene from *B. subtilis* 168 in *Saccharomyces cerevisiae* by inserting the yeast alcohol dehydrogenase promoter (ADH1) in front of the cellulase gene. However, a high level of secretion could be demonstrated only by fusing the promoter and the signal sequence of the yeast *MFα1* gene to the sequence encoding the mature lichenase (16, 69). In this case more than 87% of the total activity was recovered in the yeast culture supernatant (69).

No obvious induction mechanism regulating these enzymes has been identified. However, a region of homology with *sacR* was found upstream from the coding sequence of the lichenase genes from *B. subtilis* and *B. amyloliquefaciens* (52). This region consists of an inverted repeat sequence that could form a stem-and-loop structure. As in the case of the *sacR–sacB* region, where transcriptional regulation via antitermination has been demonstrated, it has been proposed that this structure could be involved in the inducibility of expression of these genes.

It has been shown that production of lichenase in *B. subtilis* is stimulated by the presence of the $sacQ^h$ mutation (2). Similarly, it has been shown that the lichenase gene from *B. amyloliquefaciens* BE 20/78, cloned in *B. subtilis*, was expressed more strongly in the presence of the $sacU^h$ mutation. This mutation, which normally increases the synthesis of extracellular enzyme, also increased the glucanase activity 40-fold (10). This indicates that these genes are also under the control of the *sacU* and *sacQ* loci.

α-Amylase Family

Extensive work has been carried out on the α-amylase genes of various bacilli, from a basic point of view as well as for industrial purposes. The most studied species have been *B. subtilis* and *B. amyloliquefaciens*. The α-amylases are useful as model systems in research on the regulation of gene expression, the mechanism of protein secretion, and the relationship between structure and function.

α-AMYLASE FROM *BACILLUS SUBTILIS* MARBURG   α-Amylase–defective (Amy⁻) mutants of *B. subtilis* 168 were described two decades ago (160). By transformation of Amy⁻ strains by wild-type DNA, *amy* markers were identified that were closely linked to the *aro116* marker. Yuki (160) also postulated that the rate of α-amylase production is controlled by a specific locus located near the *amy* gene, which is probably the *amyR1* locus. Other mutants affected in α-amylase expression were characterized (152), such as mutants carrying thermosensitive, cross-reacting material and suppressible mutations, which mapped in the *amyR–amyE* region near the *aroI* marker (25° on the *B. subtilis* chromosome). The α-amylase level of *B. subtilis* Marburg was increased about fivefold by introducing the *amyR2* marker by transformation from the related overproducer *B. subtilis* natto 1212 (151).

Another type of mutation *(pap)* led to increased α-amylase production (by two to three fold) compared to that of the parental *B. subtilis* Marburg strain (158). This phenotype was accompanied by other phenotypic changes: increased protease production (by five to sixteen fold), lower competence for DNA transformation, formation of filamentous cells, and a decrease in autolytic activity. All these characters could be transferred by transformation and presumably result from a point mutation (5).

Other pleiotropic mutations have also been described, including the *amyB* mutation, which affects not only α-amylase production but also production of serine and metal proteases (129). This locus did not seem to be linked to the structural gene of α-amylase *amyE*. Subsequently it has been shown that it maps at the same locus as the *pap* and *sacU*^{h} mutations (139).

The α-amylase gene from *B. subtilis* has been isolated by cloning, and it conferred the Amy⁺ phenotype to a transformed Amy⁻ strain of *B. subtilis*. Its nucleotide sequence was determined, including the transcriptional and translational signals (156). The deduced molecular weight of the gene product is 72.8×10^3, and the putative signal sequence has 32 amino acids.

Nicholson & Chambliss (100, 101) have isolated *cis*-acting mutations in *B. subtilis* 168 that confer resistance to catabolite repression of α-amylase synthesis by glucose (100, 101). The *cis*-acting alleles *gra5* and *gra10* and also the previously described allele *amyR2* (151) were cloned by gene conversion transformation. The *gra10* mutation was mapped to the vicinity of the structural gene *amyE* by transduction and transformation. It did not affect the postexponential activation of *amyE* expression (100). The sequences corresponding to the two independently isolated mutations *gra5* and *gra10* were determined, and each was a transition from G·C to A·T at a position 5 bp downstream from the putative promoter (101a). These catabolite repression–resistant mutations affect transcription levels of α-amylase and are located in a sequence that shares homology with operators from *E. coli*.

α-AMYLASE FROM *BACILLUS SUBTILIS* NATTO   The expression of α-amylase in this strain has been studied in detail (5, 47, 91, 104–106, 127, 135, 143, 153, 154, 159). As in *B. subtilis* Marburg, the structural gene *amyE* is controlled by the proximal locus *amyR*. The allele *amyR1* of strain NA20-22 is equivalent to the *amyR1* allele of *B. subtilis* Marburg, and its presence gives rise to low-level α-amylase activity. On the other hand, the locus *amyR2*, which is present in strains NA20 and NA1212, allows higher production of the enzyme (159).

Enhanced production of α-amylase was also obtained by chemical mutagenesis and stepwise genetic transformation of *B. subtilis* natto. The increase in enzyme activity reached 1500–2000-fold. The resulting high-production strain T2N26 contains at least six kinds of regulatory genes (91, 153).

Hyperproduction of α-amylase has also been observed in a tunicamycin-resistant mutant of *B. subtilis* NA64. This mutant produces five times as much enzyme as the parental strain. The locus involved, *tmrA*, is linked to the *amyR–amyE* cluster (104, 127). Increased production of the enzyme is due to gene amplification of this region, which can be transferred by DNA transformation. The amplification consists of about ten copies of a 16-kb tandemly repeated unit (47).

Yamazaki et al (154) obtained the complete nucleotide sequence of *amyE⁺–amyR2* from *B. subtilis* NA64. This sequence is very similar to the sequence found in *B. subtilis* 168 (156) and includes a signal sequence of 41 amino acids (106). The authors found an A·T-rich palindromic structure upstream from the promoter and suggested that it corresponds to either the *amyR2* locus or the transcription termination of a neighboring gene located upstream of *amyE* (154). Takano et al (143) studied the role of this palindromic structure using a gene fusion between the β-lactamase *(bla)* gene of *E. coli* and the DNA coding for the signal sequence of *amyE* (143). The palindrome is absent from the corresponding *amyR1* region of *B. subtilis* 168, which has low α-amylase activity. The presence of such a structure was shown to activate the transcription of the α-amylase gene after the addition of starch. Expression was only partially repressed by the presence of glucose. Takano et al (143) suggested that the palindromic sequence of *amyR2* acts as the transcription terminator signal for the proximate gene of *amyE* and may be the target of regulatory genes such as *sacU (pap)* or *sacQ* gene products. In the case of *sacR* (see above) a similar structure is also present, but it is not directly involved in regulation by the *sacQ* and *sacU* loci (3, 50, 62, 131, 163).

α-AMYLASE FROM *BACILLUS AMYLOLIQUEFACIENS*   This α-amylase sytem has been analyzed extensively by Palva (110), who has cloned and expressed the *B. amyloliquefaciens* gene in *B. subtilis* using the vector pUB110. Expression of α-amylase activity in the heterologous system was about 2500 times higher than that in the wild-type *B. subtilis* and about five times higher than that in the parental strain. The activity was essentially found in the culture medium. The nucleotide sequence of the promoter and the α-amylase gene indicated that a signal sequence of 31 amino acids precedes the native enzyme, a protein of 483 amino acids with a molecular weight of 54,778. As is typical for prokaryotic terminators, palindromic sequences with stretches of T residues were found in the flanking 5' and 3' regions (112, 144).

The transcription initiation and termination sites of α-amylase were determined by S1 mapping. The sequence of the promoter appears to be that recognized by the sigma 43 of vegetative RNA polymerase of *B. subtilis*. The terminator is probably the rho-independent type, including a stretch of U residues; it is not, however, transcribed in vivo in *B. subtilis* (71).

The palindromic structure located upstream from the *B. amyloliquefaciens* α-amylase gene is similar to that described in *B. subtilis* natto (60). This inverted repeat region functions as a termination signal of an as yet unidentified upstream operon of 2.2 kb. The removal of this structure led to a dramatic decrease in α-amylase production and the concomitant formation of a readthrough transcript starting at the promoter of the 2.2-kb operon. Kallio (60) therefore suggested that such a structure does not enhance the production of the enzyme, but rather prevents the inhibition of α-amylase expression resulting from the readthrough transcription arising from the upstream promoter.

The production of α-amylase from the *B. amyloliquefaciens* gene cloned on a multicopy vector (pUB110) in *B. subtilis* Marburg was not increased by the presence of positive regulatory loci such as *amyR* or *pap* (134a). Integration of the α-amylase gene into the *B. subtilis* chromosome allowed the amplification of the gene, by antibiotic selection, to up to eight copies for one genome. The gene dosage was reflected in the amount of the corresponding mRNA and led to an eightfold increase in the α-amylase activity (61).

α-AMYLASE FROM OTHER BACILLI   The expression of α-amylase has also been examined in other *Bacillus* species such as *B. licheniformis, B. stearothermophilus, B. coagulans,* and *B. circulans,* which are of industrial importance for the production of α-amylase. The corresponding α-amylase genes have been cloned.

Bacillus licheniformis   The thermostable α-amylase gene from *B. licheniformis* FD02 was cloned directly in *B. subtilis* by the marker rescue technique.

The cloned gene product was expressed and secreted by *B. subtilis* (109). The amino acid sequence inferred from the 5' region presents a strong homology with the beginning of the mature protein of *B. amyloliquefaciens,* while the signal peptides are unrelated (141). The N-terminal sequence is almost identical to that of the α-amylase gene from *B. licheniformis* ATCC 14580, with a putative signal peptide 29 amino acids long (63, 134).

A similar thermostable α-amylase gene from another strain of *B. licheniformis* (ATCC 6598) was cloned and expressed in *E. coli* and *B. subtilis* (58). The recombinant plasmid was stably maintained only in *E. coli.* Joyet et al (58) showed that the gene product was located in the periplasmic space in *E. coli,* while in *B. subtilis* it was recovered in the culture supernatant. High thermostability of the heterologous protein was observed only in *E. coli* (58). This α-amylase gene was also integrated into the *B. subtilis* chromosome by a Campbell-type mechanism. When the gene was put under the control of the *sacB* promoter (see section above on the sucrose metabolic system), induction in the wild-type strain was obtained only in the presence of sucrose; moreover, the level of α-amylase activity was dependent upon the genetic background of the host, such as whether it was an SacS^c (constitutive) or SacU^h (hyperproducing) strain. Activity could be increased by gene amplification under increased antibiotic selective pressure (59).

Yuuki et al (161) have reported the complete nucleotide sequence of a gene coding for a thermostable and alkali-resistant α-amylase from strain ATCC 27811. The mature enzyme consists of 483 amino acid residues, giving a molecular weight of 55.2×10^3, with a signal sequence of 29 amino acids. Identification of the transcription initiation site by S1 nuclease mapping indicated the position of the promoter, which is comparable to the consensus sequence recognized by the sigma 43 RNA polymerase of *B. subtilis.*

The regulation of the expression of α-amylase from *B. licheniformis* 5A1 has recently been reexamined (123). Contradicting previous results, Rothstein et al (123) showed that in *B. licheniformis* the α-amylase is produced essentially during growth and not during the stationary phase. However, catabolite repression by glucose was confirmed, and the replacement of glucose by slowly metabolized carbon sources such as glutamate or citrate caused a strong derepression of α-amylase synthesis. It was also shown that this regulation takes place at the transcriptional level.

Bacillus stearothermophilus   A thermophilic α-amylase gene from *B. stearothermophilus* was cloned and stably maintained in *E. coli.* The 61-kd gene product was expressed and was efficiently exported to the periplasmic space of this host. Its enzymatic properties and thermostability were the same as those observed in the parental strain (147).

The α-amylase gene that codes for a thermostable enzyme was obtained from another strain of *B. stearothermophilus*, A631 *(amyT631)* (135). It was fused to the promoter and the signal sequence of the α-amylase gene of *B. subtilis*. Fusions were made in phase at different sites in the signal sequence of the *B. stearothermophilus* gene, and the constructs were introduced into *B. subtilis*. The cells synthesized preproenzymes, which were processed in at least two steps to mature extracellular enzymes; the N-terminal amino acid sequences of the mature proteins were the same as that of the parental α-amylase, in spite of the different lengths and amino acid composition of the preprosequences.

Bacillus coagulans   The α-amylase gene from a *B. coagulans* strain was isolated in phage λ and subcloned in pBR322. The enzyme was expressed in *E. coli* as a 60-kd polypeptide and accumulated in the periplasmic space (19). When this gene was cloned on a multicopy plasmid, the *E. coli* host grew very poorly in media containing maltose or glycerol as carbon sources (150).

Bacillus circulans   An α-amylase gene from *B. circulans* was cloned in *B. subtilis* in a plasmid derived from pUB110, and its nucleotide sequence was determined. The precursor form of the enzyme consists of 528 amino acids, corresponding to a molecular weight of 58,776 (103). Regions homologous to other α-amylases from bacilli were observed (97). The promoter, which is located 250 bp upstream from the starting codon, is typical of the sigma 43 vegetative RNA polymerase and is also functional in *E. coli*.

USE OF α-AMYLASE GENES AS SECRETION VECTOR SYSTEMS IN *BACILLUS SUBTILIS*   Several host vector systems have been developed in *B. subtilis* for the expression and secretion of homologous and heterologous gene products using α-amylase genes. *E. coli* β-lactamase (105, 113), mouse beta interferon (IFN-β), and human IFN-α_2 and IFN-β (56, 111, 133) have each been secreted from *B. subtilis* in an active form when vectors of this type have been used.

CONCLUDING REMARKS

In recent years, more information about the regulation of carbohydrate catabolism has become available. The emphasis has been on genetics and the regulation of genes encoding enzymes of these pathways. Cloning, sequencing, and expression in homologous and heterologous hosts have facilitated the detailed analysis of regulatory functions and have led to the identification of novel regulatory systems and proteins.

The hydrolysis of high–molecular weight polymers of sugars by bacilli involves several steps and is usually performed by several enzymes. The corresponding genes are often linked on the chromosome and their expression may be coordinated. Both negative and positive regulatory systems that control operons encoding enzymes of carbohydrate metabolism have been identified. An example of a negatively controlled system is the *gnt* operon of *B. subtilis*. The regulation is mediated by a repressor, which is displaced from its target site on the DNA by the inducer; this allows transcription of the operon. In contrast, the expression of the *sacB* gene is mediated by a positive regulator, the *sacS* gene product. This protein is an antiterminator, which allows transcription through a terminator upstream of the *sacB* structural gene. The only other known examples of this type of positive antitermination regulation are regulation of the *bgl* operon of *E. coli* and regulation of phage λ infection by the N protein. However, the molecular details of this antitermination process are not well understood in the case of *B. subtilis*.

Although many of the carbohydrate-degrading enzymes are organized in operons, regulons involving several genes at different positions on the chromosome have been identified, e.g. the genes of the arabinose, sucrose, and xylan catabolic pathways.

Several pleiotropic regulatory loci that control the level of expression of degradative enzymes have been identified. Some of the corresponding genes code for small polypeptides, and these small polypeptides are not all related at the amino acid sequence level. How genes encoding the degradative enzymes are controlled by such small molecules is not understood. These polypeptides affect the transcription rate of the gene targets. Several hypotheses concerning the mechanism of action at the DNA or at the RNA polymerase levels can be made. Interestingly, this kind of regulation by small polypeptides seems to be unique to *Bacillus* species.

The enzymes involved in the early stages of the catabolic pathways of sugar polymers are secreted enzymes, synthesized as precursors, with typical signal sequences of 20–40 amino acid residues. Although the mechanism of secretion is not clearly understood, several signal sequences have been used to export heterologous proteins from *Bacillus*. Use of *Bacillus* for this type of process is going to be of increasing industrial importance.

ACKNOWLEDGMENTS

We would like to thank the following colleagues for providing information and preprints: B. Cantwell, G. H. Chambliss, H. de Lencastre, Y. Fujita, S. Hastrup, D. J. Henner, C. Hollenberg, I. Palva, L. Rutberg, M. Steinmetz, K. Yamane, and M. Zukowski. We are grateful to A. Edelman for correcting the manuscript. The research from our laboratory (directed by R. Dedonder) reported in this review was supported by Institut Pasteur, Centre National de la Recherche Scientifique, Université de Paris VII, and Fondation pour la Recherche Médicale.

Literature Cited

1. Alpert, C. A., Frank, R., Stüber, K., Deutscher, J., Hengstenberg, W. 1985. Phosphoenolpyruvate-dependent protein kinase enzyme I of *Streptococcus faecalis:* purification and properties of the enzyme and characterization of its active center. *Biochemistry* 24:955–64
2. Amory, A., Kunst, F., Aubert, E., Klier, A., Rapoport, G. 1987. Characterization of the *sacQ* genes from *Bacillus licheniformis* and *Bacillis subtilis. J. Bacteriol.* 169:324–33
3. Aymerich, S., Gonzy-Tréboul, G., Steinmetz, M. 1985. 5′-Noncoding region *sacR* is the target of all identified regulation affecting the levansucrase gene in *Bacillus subtilis. J. Bacteriol.* 166:993–98
4. Aymerich, S., Steinmetz, M. 1987. Cloning and preliminary characterization of the *sacS* locus from *Bacillus subtilis,* which controls the regulation of the exoenzyme levansucrase. *Mol. Gen. Genet.* 208:114–20
5. Ayusawa, D., Yoneda, Y., Yamane, K., Maruo, B. 1975. Pleiotropic phenomena in autolytic enzyme(s) content, flagellation, and simultaneous hyperproduction of extracellular α-amylase and protease in a *Bacillus subtilis* mutant. *J. Bacteriol.* 124:459–69
6. Bernier, R. Jr., Desrochers, M. 1985. Molecular cloning of a β-xylosidase gene from *Bacillus subtilis. J. Gen. Appl. Microbiol.* 31:513–18
7. Bernier, R. Jr., Driguez, R., Desrochers, M. 1983. Molecular cloning of a *Bacillus subtilis* xylanase gene. *Gene* 26:59–65
8. Bernier, R. Jr., Rho, D., Arcano, Y., Desrochers, M. 1985. Partial characterization of an *Escherichia coli* strain harboring a xylanase encoding plasmid. *Biotechnol. Lett.* 7:797–802
9. Borris, R., Baumlein, H., Hofemeister, J. 1985. Expression in *Escherichia coli* of a cloned β-glucanase gene from *Bacillus amyloliquefaciens. Appl. Microbiol. Biotechnol.* 22:63–71
10. Borris, R., Hofemeister, J. 1985. Mapping and cloning of the gene coding for β-glucanase in various bacilli. *Mol. Genet. Microbiol. Virol.* 2:21–26 (In Russian)
11. Borris, R., Süss, M., Manteuffel, R., Hofemeister, J. 1986. Mapping and properties of *bgl* (β-glucanase) mutants of *Bacillus subtilis. J. Gen. Microbiol.* 132:431–42
12. Bramley, H. F., Kornberg, H. L. 1987. Nucleotide sequence of *bglC,* the gene

specifying enzyme II[Bgl] of the PEP:sugar phosphotransferase system in *Escherichia coli* K-12 and overexpression of the gene product. *J. Gen. Microbiol.* 133:563–73
13. Bramley, H. F., Kornberg, H. L. 1987. Sequence homologies between proteins of bacterial phospho*enol*pyruvate-dependent sugar phosphotransferase systems: identification of possible phosphate-carrying histidine residues. *Proc. Natl. Acad. Sci. USA* 84:4777–80
14. Cantwell, B. A., Brazil, G., Murphy, N., McConnell, D. J. 1985. Expression of the gene for the endo β-1.3-1.4 glucanase from *Bacillus subtilis* in *Saccharomyces cerevisiae. Eur. Brew. Conv. Proc. Congr.* 20:259–66
15. Cantwell, B. A., McConnell, D. J. 1983. Molecular cloning and expression of a *Bacillus subtilis* β-glucanase gene in *Escherichia coli. Gene* 23:211–19
16. Cantwell, B. A., Ryan, T., Hurley, J. C., Doherty, M., McConnell, D. J. 1987. Degradation of barley β-glucan by brewers' yeast. In *Brewers Yeast, Eur. Brew. Conv. Monogr.* 12:186–98
17. Cantwell, B. A., Sharp, P. M., Gormley, E., McConnell, D. J. 1988. Molecular cloning of *Bacillus* β-glucanases. In *Biochemistry and Genetics of Cellulose Degradation,* ed. J.-P. Aubert, P. Béguin, J. Millet, pp. 181–201. London: Academic
18. Chalumeau, H., Delobbe, A., Gay, P. 1978. Biochemical and genetic study of D-glucitol transport and catabolism in *Bacillus subtilis. J. Bacteriol.* 134:920–28
19. Cornelis, P., Digneffe, C., Willemot, K. 1982. Cloning and expression of a *Bacillus coagulans* amylase gene in *Escherichia coli. Mol. Gen. Genet.* 186:507–11
20. Débarbouillé, M., Kunst, F., Klier, A., Rapoport, G. 1987. Cloning of the *sacS* gene encoding a positive regulator of the sucrose regulon in *Bacillus subtilis. FEMS Microbiol. Lett.* 41:137–40
20a. Delfour, A. 1981. *Approche tactique pour une analyse par spectrophotométrie de masse de la structure primaire des proteines; application à la levane-saccharase de* Bacillus subtilis. PhD thesis. Univ. Paris
21. Delobbe, A., Chalumeau, H., Claverie, J.-M., Gay, P. 1976. Phosphorylation of intracellular fructose in *Bacillus subtilis* mediated by phosphoenolpyruvate-1-fructose phosphotransferase. *Eur. J. Biochem.* 66:485–91

22. Delobbe, A., Chalumeau, H., Gay, P. 1975. Existence of two alternative pathways for fructose and sorbitol metabolism in *Bacillus subtilis* Marburg. *Eur. J. Biochem.* 51:503–10

23. Ebner, R., Lengeler, J. W. 1987. DNA sequence of the gene *scrA* encoding the sucrose transport protein Enzyme II^{Scr} of the phosphotransferase system from enteric bacteria: homology of the Enzyme II^{Scr} and Enzyme II^{Bgl} proteins. *Mol. Microbiol* 2:9–17

24. Esteban, R., Chordi, A., Villa, T. G. 1983. Some aspects of a 1-4-β-D-xylanase and a β-D-xylosidase secreted from *Bacillus coagulans* strain 26. *FEMS Microbiol. Lett.* 17:163–66

25. Fouet, A., Arnaud, M., Klier, A., Rapoport, G. 1984. Characterization of the precursor form of the exocellular levansucrase from *Bacillus subtilis*. *Biochem. Biophys. Res. Commun.* 119:795–800

26. Fouet, A., Arnaud, M., Klier, A., Rapoport, G. 1987. *Bacillus subtilis* sucrose-specific Enzyme II of the phosphotransferase system: expression in *Escherichia coli* and homology to Enzymes II from enteric bacteria. *Proc. Natl. Acad. Sci. USA* 84:8773–77

27. Fouet, A., Klier, A., Rapoport, G. 1982. Cloning and expression in *Escherichia coli* of the sucrase gene from *Bacillus subtilis*. *Mol. Gen. Genet.* 186:399–404

28. Fouet, A., Klier, A., Rapoport, G. 1986. Nucleotide sequence of the sucrase gene of *Bacillus subtilis*. *Gene* 45:221–25

29. Freese, E., Klofat, W., Galliers, E. 1970. Commitment to sporulation and induction of glucose-phosphoenolpyruvate-transferase. *Biochim. Biophys. Acta* 222:265–89

30. Fujita, Y., Fujita, T. 1986. Identification and nucleotide sequence of the promoter region of the *Bacillus subtilis* gluconate operon. *Nucleic Acids Res.* 14:1237–52

31. Fujita, Y., Fujita, T. 1987. The gluconate operon *gnt* of *Bacillus subtilis* encodes its own transcriptional negative regulator. *Proc. Natl. Acad. Sci. USA* 84:4524–28

32. Fujita, Y., Fujita, T., Kawamura, F., Saito, H. 1983. Efficient cloning of genes for utilization of D-gluconate of *Bacillus subtilis* in phage ρ11. *Agric. Biol. Chem.* 47:1679–82

33. Fujita, Y., Fujita, T., Miwa, Y., Nihashi, J., Aratani, Y. 1986. Organization and transcription of the gluconate operon, *gnt*, of *Bacillus subtilis*. *J. Biol. Chem.* 261:13744–53

34. Fujita, Y., Ichi, J., Nihashi, J., Fujita, T. 1985. Organization and cloning of a gluconate *(gnt)* operon of *Bacillus subtilis*. In *Molecular Biology of Microbial Differentiation*, ed. J. A. Hoch, P. Setlow, pp. 203–8. Washington, DC: Am. Soc. Microbiol.

35. Fujita, Y., Ichi, J., Nihashi, J., Fujita, T. 1986. The characterization and cloning of a gluconate *(gnt)* operon of *Bacillus subtilis*. *J. Gen. Microbiol.* 132:161–69

36. Fukumori, F., Kudo, T., Horikoshi, K. 1987. Truncation analysis of an alkaline cellulase from an alkalophilic *Bacillus* species. *FEMS Microbiol. Lett.* 4:311–14

37. Fukumori, F., Kudo, T., Narahashi, Y., Horikoshi, K. 1986. Molecular cloning and nucleotide sequence of the alkaline cellulase gene from the alkalophilic *Bacillus* sp. strain 1139. *J. Gen. Microbiol.* 132:2329–35

38. Fukumori, F., Sashihara, N., Kudo, T., Horikoshi, K. 1986. Nucleotide sequence of two cellulase genes from alkalophilic *Bacillus* sp. N4 and their strong homology. *J. Bacteriol.* 168:479–85

39. Fukusaki, E., Panbangred, N., Shinmyo, A., Okada, H. 1984. The complete nucleotide sequence of the xylanase gene *(xynA)* of *Bacillus subtilis*. *FEBS Lett.* 171:197–201

40. Gaur, N. K., Dubnau, E., Smith, I. 1986. Characterization of a cloned *Bacillus subtilis* gene that inhibits sporulation in multiple copies. *J. Bacteriol.* 168:860–69

40a. Gay, P. 1979. *Le metabolisme vectoriel des glucides et le catabolisme du fructose chez* Bacillus subtilis *Marburg: étude génétique et biochimique*. PhD thesis. Univ. Paris

41. Gay, P., Carayon, A., Rapoport, G. 1970. Isolement et localisation génétique de mutants du système métabolique du fructose chez *Bacillus subtilis*. *C. R. Acad. Sci. Ser. D* 271:263–66

42. Gay, P., Chalumeau, H., Steinmetz, M. 1983. Chromosomal localization of *gut*, *fruC*, and *pfk* mutations affecting genes involved in *Bacillus subtilis* D-glucitol catabolism. *J. Bacteriol.* 153:1133–37

43. Gay, P., Cordier, P., Marquet, M., Delobbe, A. 1973. Carbohydrate metabolism and transport in *Bacillus subtilis*. A study of *ctr* mutations. *Mol. Gen. Genet.* 121:355–68

44. Gay, P., Delobbe, A. 1977. Fructose

transport in *Bacillus subtilis*. *Eur. J. Biochem.* 79:363–73

45. Gay, P., Rapoport, G. 1970. Etude des mutants dépourvus de fructose-1-phosphate kinase chez *Bacillus subtilis*. *C. R. Acad. Sci. Ser. D* 271:374–77

46. Gonzy-Tréboul, G., Steinmetz, M. 1987. Phosphoenolpyruvate:sugar phosphotransferase system of *Bacillus subtilis:* cloning of the region containing the *ptsH* and *ptsI* genes and evidence for a *crr*-like gene. *J. Bacteriol.* 169:2287–90

47. Hashiguchi, K., Tanimoto, A., Nomura, S., Yamane, K., Yoda, K., et al. 1986. Amplification of the *amyE–tmrB* region on the chromosome in tunicamycin-resistant cells of *Bacillus subtilis*. *Mol. Gen. Genet.* 204:36–43

48. Hastrup, S. 1987. Analysis of the *Bacillus subtilis* xylose regulon. In *Genetics and Biotechnology of Bacilli*, ed. A. T. Ganesan, J. Hoch. New York: Academic. In press

49. Henner, D. J., Band, L., Shimotsu, H. 1984. Nucleotide sequence of the *Bacillus subtilis trp* operon. *Gene* 34:169–77

50. Henner, D. J., Yang, M., Band, L., Shimotsu, H., Ruppen, M., Ferrari, E. 1988. Genes of *Bacillus subtilis* that regulate the expression of degradative enzymes. In *5th Int. Symp. Genet. Ind. Microorg., Split, Yugoslavia, 1986*, ed. M. Alacevic, D. Hranueli, Z. Toman. Zagreb, Yugoslavia: Pliva. In press

51. Hinchliffe, E. 1984. Cloning and expression of a *Bacillus subtilis* endo-1.3-1.4-β-D-glucanase gene in *Escherichia coli* K-12. *J. Gen. Microbiol.* 130:1285–91

52. Hofemeister, J., Kurtz, A., Borris, R., Knowles, J. 1986. The β-glucanase gene from *Bacillus amyloliquefaciens* shows extensive homology with that of *Bacillus subtilis*. *Gene* 49:177–87

53. Hollenberg, C. P., Wilhelm, M. 1987. New substrates for old organisms. *Biotec* 1:21–31

54. Honda, H., Kudo, T., Horikoshi, K. 1985. Molecular cloning and expression of the xylanase gene of alkalophilic *Bacillus* sp. strain C125 in *Escherichia coli*. *J. Bacteriol.* 161:784–85

55. Honda, H., Kudo, T., Ikura, Y., Horikoshi, K. 1985. Two types of xylanases of alkalophilic *Bacillus* sp. No C125. *Can. J. Microbiol.* 31:538–42

56. Honjo, M., Akaoka, A., Nakayama, A., Shimada, H., Mita, I., et al. 1986. Construction of secretion vector and secretion of hIFN-β. In *Bacillus Molecular Genetics and Biotechnology Applica-*

tions, ed. A. T. Ganesan, J. A. Hoch, pp. 89–100. New York: Academic

57. Horwitz, S. B., Kaplan, N. O. 1964. Hexitol dehydrogenases of *Bacillus subtilis*. *J. Biol. Chem.* 239:830–38

58. Joyet, P., Guérineau, M., Heslot, H. 1984. Cloning of a thermostable α-amylase gene from *Bacillus licheniformis* and its expression in *Escherichia coli* and *Bacillus subtilis*. *FEMS Microbiol. Lett.* 21:353–58

59. Joyet, P., Levin, D., de Louvencourt, L., Le Révérent, B., Heslot, H. 1986. Expression of a thermostable alpha-amylase gene under the control of levansucrase inducible promoter from *Bacillus subtilis*. See Ref. 56, pp. 479–91

60. Kallio, P. 1986. The effect of the inverted repeat structure on the production of the cloned *Bacillus amyloliquefaciens* α-amylase. *Eur. J. Biochem.* 158:491–95

61. Kallio, P., Palva, A., Palva, I. 1987. Enhancement of α-amylase production by integrating and amplifying the α-amylase gene of *Bacillus amyloliquefaciens* in the genome of *Bacillus subtilis*. *Appl. Microbiol. Biotechnol.* 27:64–71

62. Klier, A., Fouet, A., Débarbouillé, M., Kunst, F., Rapoport, G. 1987. Distinct control sites located upstream from the levansucrase gene of *Bacillus subtilis*. *Mol. Microbiol.* 1:233–41

63. Kuhn, H., Fietzek, P. P., Lampen, J. O. 1982. N-Terminal amino acid sequence of *Bacillus licheniformis* α-amylase: comparison with *Bacillus amyloliquefaciens* and *Bacillus subtilis* enzymes. *J. Bacteriol.* 149:372–73

64. Kundig, W., Ghosh, S., Roseman, S. 1964. Phosphate bound to histidine in a protein as an intermediate in a novel phosphotransferase system. *Proc. Natl. Acad. Sci. USA* 52:1067–74

65. Kunst, F., Amory, A., Débarbouillé, M., Martin, I., Klier, A., Rapoport, G. 1987. Polypeptides activating the syntheses of secreted enzymes. See Ref. 48, In press

66. Kunst, F., Lepesant, J.-A., Dedonder, R. 1977. Presence of a third sucrose hydrolyzing enzyme in *Bacillus subtilis:* constitutive levanase synthesis by mutants of *Bacillus subtilis* 168. *Biochimie* 59:287–92

67. Kunst, F., Pascal, M., Lepesant, J., Lepesant, J.-A., Billault, A., Dedonder, R. 1974. Pleiotropic mutations affecting sporulation conditions and the syntheses of extracellular enzymes in *Bacillus subtilis* 168. *Biochimie* 56:1481–89

68. Kunst, F., Pascal, M., Lepesant, J.-A.,

Walle, J., Dedonder, R. 1974. Purification and some properties of an endocellular sucrase from a constitutive mutant of *Bacillus subtilis* Marburg 168. *Eur. J. Biochem.* 42:611–20

69. Lancashire, W. E., Wilde, R. J. 1988. Expression of glucanase genes in yeasts. *Eur. Brew. Conv. Proc. Congr.* 21:In press

70. Lebrun, E., Van Rapenbusch, R. 1980. The structure of *Bacillus subtilis* levan-sucrase at 3-Å resolution. *J. Biol. Chem.* 255:12034–36

71. Lehtovaara, P., Ulmanen, I., Palva, I. 1984. *In vivo* transcription initiation and termination sites of an α-amylase gene from *Bacillus amyloliquefaciens* cloned in *Bacillus subtilis*. *Gene* 30:11–16

72. Lengeler, J., Lin, E. C. C. 1972. Reversal of the mannitol-sorbitol diauxie in *Escherichia coli*. *J. Bacteriol.* 112:840–48

73. Lepesant, J.-A., Billault, A., Lepesant, J., Pascal, M., Kunst, F., Dedonder, R. 1974. Identification of the structural gene for sucrase in *Bacillus subtilis* Marburg. *Biochimie* 56:1465–70

74. Lepesant, J.-A., Dedonder, R. 1967. Métabolisme du L-arabinose chez *Bacillus subtilis* Marburg Ind⁻ 168. *C. R. Acad. Sci. Ser. D* 264:2683–86

75. Lepesant, J.-A., Dedonder, R. 1967. Isolement de mutants du système métabolique du L-arabinose chez *Bacillus subtilis* Marburg Ind⁻ 168. *C. R. Acad. Sci. Ser. D* 264:2832–34

76. Lepesant, J.-A., Dedonder, R. 1968. Transport du saccharose chez *Bacillus subtilis*. *C. R. Acad. Sci. Ser. D* 267:1109–12

77. Lepesant, J.-A., Kunst, F., Lepesant-Kejzlarova, J., Dedonder, R., 1972. Chromosomal location of mutations affecting sucrose metabolism in *Bacillus subtilis* Marburg. *Mol. Gen. Genet.* 118:135–60

78. Lepesant, J.-A., Kunst, F., Pascal, M., Lepesant-Kejzlarova, J., Steinmetz, M., Dedonder, R. 1976. Specific and pleiotropic regulatory mechanisms in the sucrose system of *Bacillus subtilis* 168. In *Microbiology 1976*, ed. D. Schlessinger, pp. 58–69. Washington, DC: Am. Soc. Microbiol.

79. Lepesant, J.-A., Lepesant-Kejzlarova, J., Pascal, M., Kunst, F., Billault, A., Dedonder, R. 1974. Identification of the structural gene of levansucrase in *Bacillus subtilis*. *Mol. Gen. Genet.* 128:213–21

80. Lepesant-Kejzlarova, J., Lepesant, J.-A., Walle, J., Billault, A., Dedonder, R. 1975. Revision of the linkage map of *Bacillus subtilis* 168: indications for circularity of the chromosome. *J. Bacteriol.* 121:823–34

81. Lindgren, V. 1978. Mapping of a genetic locus that affects glycerol 3-phosphate transport in *Bacillus subtilis*. *J. Bacteriol.* 133:667–70

82. Lindgren, V., Rutberg, L. 1974. Glycerol metabolism in *Bacillus subtilis*: gene-enzyme relationships. *J. Bacteriol.* 119:431–42

83. Lindgren, V., Rutberg, L. 1976. Genetic control of the *glp* system in *Bacillus subtilis*. *J. Bacteriol.* 127:1047–57

84. MacKay, R., Lo, A., Withich, G., Züker, M., Band, S., et al. 1986. Structure of a *Bacillus subtilis* endo-β-1.4-glucanase gene. *Nucleic Acids Res.* 14:9159–70

85. MacKillen, M. N., Rountres, J. H. 1973. Gluconate transport in *Bacillus subtilis*. *Biochem. Soc. Trans.* 1:442–45

86. Marquet, M., Creignou, M.-C., Dedonder, R. 1976. The phosphoenolpyruvate:methyl-α-D-glucoside phosphotransferase system in *Bacillus subtilis* Marburg 168: purification and identification of the phosphocarrier protein (HPr). *Biochimie* 58:435–41

87. Marquet, M., Creignou, M.-C., Delobbe, A., Gay, P., Rapoport, G. 1970. Mise en évidence de systèmes de phosphotransférases dans le transport du glucose, du fructose et du saccharose chez *Bacillus subtilis*. *C. R. Acad. Sci. Ser. D* 271:449–52

88. Marquet, M., Wagner, M.-C., Dedonder, R. 1971. Separation of components of the phosphoenolpyruvate–glucose phosphotransferase system from *Bacillus subtilis* Marburg. *Biochimie* 53:1131–34

89. Martin, I., Débarbouillé, M., Ferrari, E., Klier, A., Rapoport, G. 1987. Characterization of the levanase gene of *Bacillus subtilis* sharing homologies with the yeast invertase. *Mol. Gen. Genet.* 208:177–84

90. Martin, I., Débarbouillé, M., Klier, A., Rapoport, G. 1987. Identification of a new locus, *sacV*, involved in the regulation of levansucrase synthesis in *Bacillus subtilis*. *FEMS Microbiol. Lett.* 44:39–43

91. Maruo, B., Yamane, K., Yoneda, Y., Hitotsuyanagi, K. 1978. Stepwise genetic transformation of *Bacillus subtilis* with enhancement of productivity of α-amylase. *Proc. Jpn. Acad. Ser. B* 54:435–39

92. Mindich, L. 1968. Pathway for oxidative dissimilation of glycerol in *Bacillus subtilis*. *J. Bacteriol.* 96:565–66

93. Miwa, Y., Fujita, Y. 1987. Efficient utilization and operation of the gluconate inducible operon of the promoter of the *Bacillus subtilis gnt* operon in *Escherichia coli. J. Bacteriol.* 169:5333–35

94. Moriyama, H., Fukusaki, F., Cabrera Crespo, J., Shinmyo, A., Okada, H. 1987. Structure and expression of genes coding for xylan depending enzymes of *Bacillus pumilus. Eur. J. Biochem.* 166:539–45

95. Moriyama, H., Hata, Y., Yamaguchi, H., Shinmyo, A., Sato, M., et al. 1987. Crystallization and preliminary X-ray studies of *Bacillus pumilus* IPO xylanase. *J. Mol. Biol.* 193:237–38

96. Murphy, N., McConnell, D. J., Cantwell, B. A. 1984. The DNA sequence of the gene and genetic control sites for the excreted *Bacillus subtilis* β-glucanase. *Nucleic Acids Res.* 12:5355–67

97. Nakajima, R., Imanaka, T., Aiba, S. 1986. Comparison of aminoacid sequence of eleven different α-amylases. *Appl. Microbiol. Biotechnol.* 23:355–60

98. Nakamura, A., Uozimi, T., Beppu, T. 1987. Nucleotide sequence of a cellulase gene of *Bacillus subtilis. Eur. J. Biochem.* 164:317–20

99. Niaudet, B., Gay, P., Dedonder, R. 1975. Identification of the structural gene of the PEP-phosphotransferase Enzyme I in *Bacillus subtilis* Marburg. *Mol. Gen. Genet.* 136:337–49

100. Nicholson, W. L., Chambliss, G. H. 1985. Isolation and characterization of a *cis*-acting mutation conferring catabolite repression resistance to α-amylase synthesis in *Bacillus subtilis. J. Bacteriol.* 161:875–81

101. Nicholson, W. L., Chambliss, G. H. 1986. Molecular cloning of *cis*-acting regulatory alleles of the *Bacillus subtilis amyR* region by using gene conversion transformation. *J. Bacteriol.* 165:663–70

101a. Nicholson, W. L., Park, Y.-K., Henkin, T. M., Won, M., Weickert, M. J., et al. 1987. Catabolite repression–resistant mutations of the *Bacillus subtilis* alpha-amylase promoter affect transcription levels and are in an operator-like sequence. *J. Mol. Biol.* 198:609–18

102. Nihashi, J., Fujita, Y. 1984. Catabolite repression of inositol dehydrogenase and gluconate kinase syntheses in *Bacillus subtilis. Biochim. Biophys. Acta* 798:88–95

103. Nishizawa, M., Ozawa, F., Hishinuma, F. 1987. Molecular cloning of an amylase gene of *Bacillus circulans. DNA* 6:255–65

104. Nomura, S., Yamane, K., Sasaki, T., Yamasaki, M., Tamura, G., Maruo, B. 1978. Tunicamycin-resistant mutants and chromosomal locations of mutational sites in *Bacillus subtilis. J. Bacteriol.* 136:818–21

105. Ohmura, K., Nakamura, K., Yamazaki, H., Shiroza, T., Yamane, K., et al. 1984. Length and structural effect of signal peptides derived from *Bacillus subtilis* α-amylase on secretion of *Escherichia coli* β-lactamase in *B. subtilis* cells. *Nucleic Acids Res.* 12:5307–19

106. Ohmura, K., Yamazaki, H., Takeichi, Y., Nakayama, A., Otozai, K., et al. 1983. Nucleotide sequence of the promotor and NH$_2$-terminal signal peptide region of *Bacillus subtilis* α-amylase gene cloned in pUB110. *Biochem. Biophys. Res. Commun* 112:678–83

107. Okada, J., Shinogaki, H., Murata, K., Kimura, A. 1984. Cloning of the gene responsible for the extracellular proteolytic activities of *Bacillus licheniformis. Appl. Microbiol. Biotechnol.* 20:406–12

108. O'Kane, C., Cantwell, B. A., McConnell, D. J. 1985. Mapping of the gene for endo-β-1.3-1.4-glucanase of *Bacillus subtilis. FEMS Microbiol. Lett.* 29:135–39

109. Ortlepp, S. A., Ollington, J. F., McConnell, D. J. 1983. Molecular cloning in *Bacillus subtilis* of *Bacillus licheniformis* gene encoding a thermostable alpha amylase. *Gene* 23:267–76

110. Palva, I. 1982. Molecular cloning of α-amylase gene from *Bacillus amyloliquefaciens* and its expression in *B. subtilis. Gene* 19:81–87

111. Palva, I., Lehtovaara, P., Kääriäinen, L., Sibakov, M., Cantell, K., et al. 1983. Secretion of interferon by *Bacillus subtilis. Gene* 22:229–35

112. Palva, I., Petterson, R. F., Kalkkinen, N., Lehtovaara, P., Sarvas, M., et al. 1981. Nucleotide sequence of the promoter and NH$_2$-terminal signal peptide region of the α-amylase gene from *Bacillus amyloliquefaciens. Gene* 15:43–51

113. Palva, I., Sarvas, M., Lehtovaara, P., Sibakov, M., Kääriäinen, L. 1982. Secretion of *Escherichia coli* β-lactamase from *Bacillus subtilis* by the aid of α-amylase signal sequence. *Proc. Natl. Acad. Sci. USA* 79:5582–86

114. Panbangred, N., Fukusaki, E., Epifanic, E., Shinmyo, A., Okada, H. 1985. Expression of a xylanase gene of *Bacillus pumilus* in *Escherichia coli* and in *Bacillus subtilis. Appl. Microbiol. Biotechnol.* 22:259–64

94   KLIER & RAPOPORT

115. Panbangred, N., Kondo, T., Negoro, S., Shinmyo, A., Okada, H. 1983. Molecular cloning of the genes for xylan degradation of *Bacillus pumilus* and their expression in *Escherichia coli*. *Mol. Gen. Genet.* 192:335–41

116. Paveia, H., Archer, L. 1980. Location of genes for arabinose utilization in the *Bacillus subtilis* chromosome. *Brotéria Genét.* 76:169–76

117. Piggot, P. J., Hoch, J. A. 1985. Revised genetic linkage map of *Bacillus subtilis*. *Microbiol. Rev.* 49:158–79

118. Postma, P. W., Lengeler, J. W. 1985. Phosphoenolpyruvate:carbohydrate phosphotransferase system of bacteria. *Microbiol. Rev.* 49:232–69

119. Reizer, J., Novotny, M. J., Stuiver, I., Saier, M. H. Jr. 1984. Regulation of glycerol uptake by the phosphoenolpyruvate–sugar phosphotransferase system in *Bacillus subtilis*. *J. Bacteriol.* 159:243–50

120. Robson, L. M., Chambliss, G. H. 1986. Cloning of the *Bacillus subtilis* DLG β-1.4-glucanase gene and its expression in *Escherichia coli* and *Bacillus subtilis*. *J. Bacteriol.* 165:612–19

121. Robson, L. M., Chambliss, G. H. 1987. Endo-β-1.4-glucanase gene of *Bacillus subtilis* DLG. *J. Bacteriol.* 169:2017–25

122. Roncero, M. I. G. 1983. Genes controlling xylan utilization by *Bacillus subtilis*. *J. Bacteriol.* 156:257–63

123. Rothstein, D. M., Devlin, P. E., Cate, R. L. 1986. Expression of α-amylase in *Bacillus licheniformis*. *J. Bacteriol.* 168:839–42

124. Ryu, D. D. Y., Mandels, M. 1980. Cellulases: biosynthesis and applications. *Enzyme Microb. Technol.* 2:91–102

125. Saheb, S. A. 1972. Perméation du glycérol et sporulation chez *Bacillus subtilis*. *Can. J. Microbiol.* 18:1307–13

126. Saheb, S. A. 1972. Etude de deux mutants du métabolisme du glycérol chez *Bacillus subtilis*. *Can. J. Microbiol.* 18:1315–25

127. Sasaki, T., Yamasaki, M., Maruo, B., Yoneda, Y., Yamane, K., et al. 1976. Hyperproductivity of extracellular α-amylase by a tunicamycin resistant mutant of *Bacillus subtilis*. *Biochem. Biophys. Res. Commun.* 70:125–31

128. Schnetz, K., Tologzky, C., Rak, B. 1987. β-Glucoside *(bgl)* operon of *Escherichia coli* K-12: Nucleotide sequence, genetic organization and possible evolutionary relationships to regulatory components of two *Bacillus subtilis* genes. *J. Bacteriol.* 169:2579–70

129. Sekiguchi, J., Takada, N., Okada, H. 1975. Genes affecting the productivity of α-amylase in *Bacillus subtilis* Marburg. *J. Bacteriol.* 121:688–94

130. Seo, Y. S., Young, O. L., Um, U. P. 1985. Molecular cloning of β-glucanase gene from *Bacillus subtilis* and its expression in *Escherichia coli*. *Korean Biochem. J.* 18:367–76

131. Shimotsu, H., Henner, D. J. 1986. Modulation of *Bacillus subtilis* levansucrase gene expression by sucrose and regulation of the steady state mRNA level by *sacU* and *sacQ* genes. *J. Bacteriol.* 168:380–88

132. Shimotsu, H., Kuroda, M., Yanofski, C., Henner, D. J. 1986. Novel form of transcription attenuation regulates expression of the *Bacillus subtilis trp* operon. *J. Bacteriol.* 166:467–71

133. Shiroza, T., Nakazawa, K., Tashiro, N., Yamane, K., Yanagi, K., et al. 1985. Synthesis and secretion of biologically active mouse interferon-β using a *Bacillus subtilis* α-amylase secretion vector. *Gene* 34:1–8

134. Sibakov, M., Palva, I. 1984. Isolation and the 5'-end nucleotide sequence of *Bacillus licheniformis* α-amylase gene. *Eur. J. Biochem.* 145:567–72

134a. Sibakov, M., Sarvas, M., Palva, I. 1983. Increased secretion of α-amylase from *Bacillus subtilis* caused by multiple copies of α-amylase gene from *B. amyloliquefaciens* is not further increased by genes enhancing the basic level of secretion. *FEMS Microbiol. Lett.* 17:81–85

135. Sohma, A., Fujita, T., Yamane, K. 1987. Protein processing to form extracellular thermostable α-amylases from a gene fused in a *Bacillus subtilis* secretion vector. *J. Gen. Microbiol.* 133:3271–77

136. Solomon, E., Lin, E. C. C. 1972. Mutations affecting the dissimilation of mannitol by *Escherichia coli* K-12. *J. Bacteriol.* 111:566–74

137. Steinmetz, M., Aymerich, S. 1986. Analyse génétique de *sacR*, régulateur en *cis* de la synthèse de la lévanesaccharase de *Bacillus subtilis*. *Ann. Microbiol. Inst. Pasteur* 137A:3–14

138. Steinmetz, M., Aymerich, S., Gonzy-Tréboul, G., Le Coq, D. 1987. Levansucrase induction in *Bacillus subtilis* involves an antiterminator. Homology with the *Escherichia coli bgl* operon. See Ref. 48, In press

139. Steinmetz, M., Kunst, F., Dedonder, R. 1976. Mapping of mutations affecting synthesis of exocellular enzymes in *Bacillus subtilis*. Identity of the *sacU^h*, *amyB* and *pap* mutations. *Mol. Gen. Genet.* 148:281–85

140. Steinmetz, M., Le Coq, D., Aymerich, S., Gonzy-Tréboul, G., Gay, P. 1985. The DNA sequence of the gene for the secreted *Bacillus subtilis* enzyme levansucrase and its genetic control sites. *Mol. Gen. Genet.* 200:220–28

141. Stephens, M. A., Ortlepp, S. A., Ollington, J. F., McConnell, D. J. 1984. Nucleotide sequence of the 5' region of the *Bacillus licheniformis* α-amylase gene: comparison with the *B. amyloliquefaciens* gene. *J. Bacteriol.* 158:369–72

142. Suzuki, H., Kaneko, H. 1976. Degradation of barley glucan and lichenan by a *Bacillus pumilus* enzyme. *Agric. Biol. Chem.* 40:577–86

143. Takano, J., Kinoshita, T., Yamane, K. 1987. Modulation of *Bacillus subtilis* α-amylase promoter activity by the presence of a palindromic sequence in front of the gene. *Biochem. Biophys. Res. Commun.* 146:73–79

144. Takkinen, K., Pettersson, R. F., Kalkkinen, N., Palva, I., Söderlund, H., Kääriäinen, L. 1983. Amino acid sequence of α-amylase from *Bacillus amyloliquefaciens* deduced from the nucleotide sequence of the cloned gene. *J. Bacteriol.* 258:1007–13

145. Tanaka, T., Kawata, M., Nagami, Y., Uchiyama, H. 1987. *prtR* enhances the mRNA level of the *Bacillus subtilis* extracellular proteases. *J. Bacteriol.* 169:3044–50

146. Timell, T. E. 1967. Recent progress in the chemistry of wood hemicellulose. *Wood Sci. Technol.* 1:45–70

147. Tsukagoshi, N., Ihara, H., Yamagata, H., Udaka, S. 1984. Cloning and expression of a thermophilic α-amylase gene from *Bacillus stearothermophilus* in *Escherichia coli*. *Mol. Gen. Genet.* 193:58–63

148. Wilhelm, M., Hollenberg, C. P. 1984. Selective cloning of *Bacillus subtilis* xylose isomerase and xylulokinase genes in *Escherichia coli* by IS5-mediated expression. *EMBO J.* 3:2555–60

149. Wilhelm, M., Hollenberg, C. P. 1985. Nucleotide sequence of the *Bacillus subtilis* xylose isomerase gene: extensive homology between the *Bacillus* and *E. coli* enzymes. *Nucleic Acids Res.* 13:5717–12

150. Willemot, K., Cornelis, P. 1983. Growth defects of *Escherichia coli* cells which contain the gene of an α-amylase from *Bacillus coagulans* on a multicopy plasmid. *J. Gen. Microbiol.* 129:311–19

151. Yamaguchi, K., Nagata, Y., Maruo, B. 1974. Genetic control of the rate of α-amylase synthesis in *Bacillus subtilis*. *J. Bacteriol.* 119:410–15

152. Yamaguchi, K., Nagata, Y., Maruo, B. 1974. Isolation of mutants defective in α-amylase from *Bacillus subtilis*: genetic analyses. *J. Bacteriol.* 119:416–24

153. Yamane, K., Shinomiya, S. 1982. *Bacillus subtilis* α-amylases: regulation of production and molecular cloning. In *Microbiology 1982*, ed. D. Schlessinger, pp. 8–11. Washington, DC: Am. Soc. Microbiol.

154. Yamazaki, H., Ohmura, K., Nakayama, A., Takeichi, Y., Otozai, K., et al. 1983. α-Amylase genes (*amyR2* and *amyE*+) from an α-amylase–hyperproducing *Bacillus subtilis* strain: molecular cloning and nucleotide sequences. *J. Bacteriol.* 156:327–37

155. Yang, M., Ferrari, E., Chen, E., Henner, D. J. 1986. Identification of the pleiotropic *sacQ* gene of *Bacillus subtilis*. *J. Bacteriol.* 166:113–19

156. Yang, M., Galizzi, A., Henner, D. 1983. Nucleotide sequence of the amylase gene from *Bacillus subtilis*. *Nucleic Acids Res.* 11:237–49

157. Yang, M., Shimotsu, H., Ferrari, E., Henner, D. J. 1987. Characterization and mapping of the *Bacillus subtilis prtR* gene. *J. Bacteriol.* 169:434–37

158. Yoneda, Y., Maruo, B. 1975. Mutation of *Bacillus subtilis* causing hyperproduction of α-amylase and protease, and its synergistic effect. *J. Bacteriol.* 124:48–54

159. Yoneda, Y., Yamane, K., Yamaguchi, K., Nagata, Y., Maruo, B. 1974. Transformation of *Bacillus subtilis* in α-amylase productivity by deoxyribonucleic acid from *B. subtilis* var. *amylosacchariticus*. *J. Bacteriol.* 120:1144–50

160. Yuki, S. 1968. On the gene controlling the rate of amylase production in *Bacillus subtilis*. *Biochem. Biophys. Res. Commun.* 31:182–87

161. Yuuki, T., Nomura, T., Tezuka, H., Tsuboi, A., Yamagata, H., et al. 1985. Complete nucleotide sequence of a gene coding for heat- and pH-stable α-amylase of *Bacillus licheniformis*: comparison of the amino acid sequences of three bacterial liquefying α-amylases deduced from the DNA sequences. *J. Biochem. Tokyo* 98:1147–56

162. Zeigler, D. R., Dean, D. H. 1985. Revised genetic map of *Bacillus subtilis* 168. *FEMS Microbiol. Rev.* 32:101–34

163. Zukowski, M. M., Miller, L. 1986. Hyperproduction of an intracellular heterologous protein in a *sacU*h mutant of *Bacillus subtilis*. *Gene* 46:247–55

Ann. Rev. Microbiol. 1988. 42:97–125

PROTEIN PHOSPHORYLATION IN PROKARYOTES

Alain J. Cozzone

Laboratoire de Biologie Moléculaire, Université de Lyon, 69622 Villeurbanne, France

CONTENTS

INTRODUCTION

Phosphorylation is one of the covalent modifications that proteins are liable to undergo in a posttranslational process. Although it was originally detected almost a hundred years ago, the idea that it might have an important role not only in the structure but also in the functioning of proteins in cellular

97

0066-4227/88/1001-0097$02.00

metabolism was slow to take root. Its direct involvement in a metabolic pathway was first demonstrated in 1956 by Krebs & Fischer in a eukaryotic system (89). These authors showed that the activity of rabbit skeletal muscle glycogen phosphorylase, the rate-limiting enzyme in glycogenolysis, is regulated by reversible phosphorylation: The dephosphorylated inactive *b*-form of the enzyme can be converted to the phosphorylated active *a*-form by a specific protein kinase, and conversely, it can be regenerated from the *a*-form by a dephosphorylation reaction catalyzed by a specific protein phosphatase. During the next ten years, two additional phospho-dephospho enzymes of glycogen metabolism, phosphorylase kinase and glycogen synthase, were discovered (57, 90). Then the field started to move rapidly with some important developments such as the discovery of cyclic AMP–dependent protein kinase in rabbit muscle (200) and the observation that histones and other nuclear proteins from rat liver and thymus undergo phosphorylation-dephosphorylation (137), which extended the applicability of this mechanism to nonenzymic proteins. In recent years, a seemingly exponentially increasing number of phosphoenzymes and phosphoproteins have been characterized in a wide variety of eukaryotic systems from fungi to mammals, and protein phosphorylation has been definitively recognized as a major dynamic process involved in a large array of cellular functions. Numerous reviews (44, 77, 83, 88, 172, 209) and monographs (1, 25, 170, 185) provide excellent coverage of the vast literature currently available in the field. The reader interested in detailed information pertaining to the multiple aspects of protein phosphorylation in eukaryotes is encouraged to consult these.

For prokaryotes, the interest in protein phosphorylation took even longer to gather momentum than for eukaryotes. In fact, even the occurrence of this chemical modification in microorganisms was a matter of controversy for many years. The first attempt to characterize a protein kinase activity in bacteria was made in 1969 by Kuo & Greengard (93). The authors reported the presence in *Escherichia coli* extracts of a cyclic AMP–dependent enzyme that could catalyze the phosphorylation by ATP of histones, which are exogenous basic proteins. Soon thereafter, two different protein kinases, regulated in a reciprocal fashion by cyclic AMP, were described in oral streptococci, also phosphorylating histones and protamines (82), and a few more reports were published on this topic (66, 94, 149). But still no definite conclusion could be drawn on the existence of a protein kinase activity in prokaryotes, namely because of the unreproducibility of certain results, as in the case of ribosomal protein phosphorylation (66, 95), or because of the incomplete chemical characterization of the phosphorylated moiety of the proteins. The latter point was critical, since bacteria are known to contain some kinds of kinases that are quite different from protein kinases and whose in vitro activity is stimulated especially by basic proteins. For example, *E.*

coli extracts have been shown to harbor a polyphosphate kinase that transfers phosphate from ATP to polyphosphate and requires histones or protamines for maximum activity (105). Also, the eubacterium *Caulobacter crescentus* contains a histone-stimulated enzyme that catalyzes the transfer of the γ-phosphoryl group of ATP to a protein acceptor, with the formation of an acyl phosphate rather than a phosphate ester bond (3). From such observations the originally described phosphorylating activities were soon found to be questionable and in many cases were ascribed to kinases other than protein kinases (105, 143).

The first clear demonstration of a protein kinase in a prokaryotic system came from an analysis of virus-infected bacteria. Rahmsdorf et al (152, 153) showed that a cyclic nucleotide–independent protein kinase appears in *E. coli* cells following infection with bacteriophage T7. The induced enzyme can phosphorylate both endogenous and exogenous proteins. The products of the reaction have the chemical characteristics of phosphoserine and phosphothreonine, distinguishing the enzyme from acyl phosphate and polyphosphate kinases. However, the authors simultaneously showed that this protein kinase is, in fact, coded for by a specific viral gene, since its appearance is prevented by ultraviolet irradiation of the phage genome but not by that of the host genome (153). This gene was mapped in the early region of the T7 genome, and the kinase was purified and characterized (141, 142). In view of these data it was generally assumed for several years that uninfected bacteria did not carry any protein kinase activity and, therefore, that protein phosphorylation was restricted to eukaryotic cells (143, 153). In the past decade, however, this concept has been reversed by conclusive evidence that bacteria do contain specific protein kinases. The aim of this review is to summarize the decisive findings and to assess the present status of knowledge of bacterial phosphoproteins and phosphorylating-dephosphorylating enzymes. It also illuminates the role of protein phosphorylation in the regulation of bacterial physiology.

Research in protein phosphorylation is rather unconventional. The problem is most often entered in the middle, by first targeting a phosphorylated protein in a gel slice or in a chromatography column eluate. Then one proceeds both upstream to characterize the enzyme(s) responsible for its phosphorylation and downstream to determine the functional consequences of its modification. The next sections of this article are presented in this same order.

PHOSPHORYLATED PROTEINS

Evidence for phosphorylation of proteins by protein kinases was first obtained simultaneously, and independently, in both *E. coli* (61, 114) and *Salmonella typhimurium* (202). Since then phosphorylation has been demonstrated in a number of other bacterial species. In most cases phosphoproteins have been detected through their ability to incorporate radioactivity in vivo from ^{32}P-labeled orthophosphate or in vitro from [γ-^{32}P]ATP, and the nature of the phosphorylated moiety has been characterized chemically and enzymatically. Some of the phosphoproteins have been identified.

Phosphoproteins of Escherichia coli and Salmonella typhimurium

The bacterial phosphoproteins most extensively studied so far are those from *E. coli* (30). Early determinations indicated that their number within the cell ranged from a dozen to about 40 (33, 50, 53, 114). A recent, tentatively comprehensive catalog indicates that *E. coli* K-12 contains 130 different phosphoprotein species (29, 165). This catalog has been established by analyzing the effects on protein phosphorylation of various culture conditions, including growth on different carbon sources in either the exponential or the stationary phase, treatment of cells with ethanol, heat shock, and amino acid starvation. With a reasonable estimate of 2500 proteins as the possible genomic coding capacity of *E. coli* (144), it therefore appears that only about 5% of the total proteins undergo phosphorylation, which means that this modification is rather specific. Using the O'Farrell technique (135, 136), the 130 phosphoproteins have been characterized individually by both molecular weight and isoelectric point, as well as by the extent of their labeling in vivo in the presence of radioactive orthophosphate (29, 165). Only some of them are phosphorylated when bacteria are grown anaerobically with nitrate as electron acceptor (151). Analysis of various subcellular preparations has shown that most of the phosphoproteins are located in the cytoplasmic fraction of cells; none of them are within ribosomes or nucleoids, and only three are associated with membranes (29).

Fewer endogenous proteins are phosphorylated in vitro than in vivo. In cellular extracts prepared from exponentially growing *E. coli* only six major proteins and a few minor components were significantly phosphorylated at the expense of ATP (31, 50, 116). No qualitative difference in the phosphorylation pattern was observed when cellular lysates were prepared from stationary-phase cultures, but some quantitative variations in the extent of protein labeling did appear (50). Most of the proteins that are phosphorylated in vitro exhibit electrophoretic properties identical to those of proteins phosphorylated in vivo, which suggests that they are not artifacts of cell-free systems (50, 116).

In *S. typhimurium* only 10 phosphoproteins have been detected in vivo (202, 203). In contrast, more than 24 proteins have been reported to undergo phosphorylation in vitro; some of these proteins presumably correspond to covalent intermediates of enzymes that catalyze phosphoryl-transfer reactions

(203). Considering the well-known homology between *E. coli* and *S. typhimurium*, it seems probable that the real number of total phosphoproteins in *S. typhimurium* has been underestimated.

Phosphoproteins of Other Bacterial Species

Protein phosphorylation appears to be a universal phenomenon in prokaryotes. Since its discovery in *E. coli* and *S. typhimurium*, it has been revealed in a variety of organisms that belong to the two bacterial kingdoms, the eubacteria and archaebacteria, as well as to the line of cyanobacteria (20, 55). All together, these organisms comprise 26 different species from 13 distinct families or groups (listed in Table 1) and present a great diversity of morphological, physiological, and biochemical features. Phosphoproteins are found in cells that are rod-shaped, coccal, endospore-forming, and microcyst-forming; gram-negative and gram-positive; aerobic, facultative, and anaerobic; chemoorganotrophic and chemolithotrophic; photosynthetic and nonphotosynthetic. The number of polypeptides that are significantly phosphorylated varies from five (107) to more than 25 (191). Since most of the information showing that phosphoproteins are ubiquitous in bacteria has been obtained only in the past three years, it can be anticipated that more examples supporting this concept will be forthcoming before long.

Identification of Specific Phosphoproteins

A possible physiological role for protein kinases in prokaryotic systems (see below) can only be determined when the phosphorylated proteins have been identified and when the effects of phosphorylation on the functional properties of these proteins have been studied. In fact only a limited number of individual phosphoproteins have been identified as yet, and much work remains to be done along this line.

The NADP-dependent isocitrate dehydrogenase (IDH) of *E. coli* was the first bacterial phosphoprotein clearly identified (61, 62, 78). This enzyme is a dimer of identical subunits of 42–53 kd each, with isoelectric point of 5.3. It catalyzes the metabolism of isocitrate at an important branch point between the Krebs cycle and the glyoxylate bypass (13, 71, 130). Depending on the nature of the energy/carbon source provided to the bacterium, IDH is either phosphorylated or not; the phosphorylation reaction leads to its complete inactivation (61, 126, 127). The enzyme has been purified to homogeneity (62, 195), and its two forms, active and inactive, have been separated on the basis of their difference in charge (16). Inactive phosphorylated IDH is much less susceptible to proteolysis than the active form and is unable to bind NADP (60). The inactivation of the enzyme is mediated by the negative charge of the phosphate (187).

Another phosphoprotein in *E. coli* is the *dnaK* protein, which is acidic (pI

Table 1 Prokaryotic organisms subject to protein phosphorylation[a]

Family/group	Species	References
Eubacteria		
Enterobacteriaceae	*Escherichia coli* K-12	50, 53, 61, 114, 126, 197, 206
	E. coli B	[b]
	E. coli ML308	13, 15, 16
	Salmonella typhimurium	27, 53, 202, 204, 206
Halobacteriaceae	*Halobacterium halobium*	182, 183
Myxococcaceae	*Myxococcus xanthus*	85
Neisseriaceae	*Acinetobacter calcoaceticus*	156, [b]
Pseudomonadaceae	*Pseudomonas fluorescens*	[b]
Rhodospirillaceae	*Rhodospirillum rubrum*	70, 72, 192
	Rhodomicrobium vannielii	191
	Rhodocyclus gelatinosus	10
Bacillaceae	*Bacillus megaterium*	[b]
	B. sphaericus	[b]
	B. subtilis	84, 157
	Clostridium sphenoides	8, 9
	C. thermohydrosulfuricum	107
Corynebacteriaceae	*Arthrobacter globiformis*	[b]
Lactobacillaceae	*Lactobacillus brevis*	157
Micrococcaceae	*Staphylococcus aureus*	157
Streptococcaceae	*Streptococcus faecalis*	36, 39, 41
	S. lactis	37
	S. mutans	122
	S. pyogenes	36, 40, 159, 160
	S. salivarius	205
Archaebacteria		
Thiobacillaceae	*Sulfolobus acidocaldarius*	180
Cyanobacteria		
	Synechococcus 6301	4
Bacteriophages		
T4,T7		64, 142, 152, 153, 176
T3		142
φCd1		69

[a]The nomenclature used is from References 20, 55, and 91.
[b]M. Dadssi & A. J. Cozzone, unpublished data.

5.1) and largely monomeric; it has an apparent mass of 72 and 78.4 kd under denaturing and native conditions, respectively (12, 212). This protein modulates the heat-shock response of bacteria (124, 188) and is required for the replication of cellular DNA (80, 140, 173). Moreover, it is essential for both bacteriophage lambda and M13 in vitro DNA replication systems that are dependent on the lambda O and P proteins (58, 103, 104). The *dnaK* protein is phosphorylated in vivo in exponentially growing bacteria (165, 197) and, to a larger extent, in phage-infected cells (165). It can also be phosphorylated in vitro, presumably as a result of an autophosphorylation reaction (213).

Evidence has been presented that both glutaminyl-tRNA synthetase and threonyl-tRNA synthetase are phosphorylated in wild-type strains of *E. coli* K-12, but not in temperature-sensitive *dnaK* and *dnaJ* mutants grown at restrictive temperature (197). No other component of the protein-synthesizing machinery appears to be phosphorylated. In particular, immunochemical analysis of total proteins has ruled out the possibility of phosphorylation of the initiation and elongation factors (73, 153). For many years, much controversy has surrounded the phosphorylation of ribosomal proteins (66, 94, 153). It is now clear, however, that none of these proteins are modified by protein kinases in normally growing cells, even though some phosphate-containing molecules are attached to ribosomes (27, 66). After infection with bacteriophages T7 and T4, a few *E. coli* ribosomal proteins become phosphorylated by a viral kinase activity (152, 153). The same type of virus-induced modification applies to the β and β' subunits of host DNA–dependent RNA polymerase (141, 211). Infection of the dimorphic bacterium *C. crescentus* by the small lytic phage ϕCd1 also leads to the phosphorylation of host ribosomal proteins and the β' subunit of RNA polymerase (69).

Protein HPr is phosphorylated in several gram-positive bacteria (reviewed in 157). This protein is one of the phosphate carrier enzymes (42, 169) of the phosphoenolpyruvate (PEP)-dependent phosphotransferase system (PTS). The other proteins involved are the soluble enzyme I and enzyme III and the membrane-bound enzyme II, which form a phosphorylation chain with HPr (68, 148). For the sugar uptake reaction and the concomitant phosphorylation of the sugar, HPr is phosphorylated by enzyme I at the N-1 position of a single histidine residue at the expense of PEP (14, 206, 207). In *Streptococcus salivarius* a protein kinase activity that catalyzes the formation of phosphohistidine at the expense of ATP has been described (205). In various streptococci, *Staphylococcus aureus, Bacillus subtilis,* and *Micrococcus brevis*, protein HPr is phosphorylated, in addition, at a serine residue in an ATP-dependent protein kinase–catalyzed reaction (38, 40, 157, 205). This reaction has not been observed in gram-negative bacteria.

Several proteins from *E. coli* are in vitro substrates for exogenous mammalian or viral protein kinases. For example, the sigma factor of RNA polymerase (119), the initiation factor IF2 (52), and a few proteins from the small and large subunits of ribosomes (189) are phosphorylated by rabbit skeletal muscle protein kinase, while certain basic histone-like proteins can be modified by either beef heart protein kinase (92) or Rauscher murine leukemia virus kinase (184). Although such reactions are very specific with regard to the nature of the phosphorylatable substrates, their biological significance is difficult to assess, especially because most, if not all, of the proteins involved are not phosphorylated in bacteria in vivo.

Recent work has indicated that the most abundant phosphopolypeptide in the purple nonsulfur bacterium *Rhodomicrobium vannielii* is the large subunit of ribulose 1,5-bisphosphate carboxylase/oxygenase (N. H. Mann & A. M. Turner, unpublished results).

SITES OF PHOSPHORYLATION OF PROTEIN SUBSTRATES

To demonstrate the occurrence of protein phosphorylation it is important to show that phosphoryl groups are covalently bound to certain amino acids of protein substrates and to characterize the nature of the linkage involved.

General Aspects

Phosphorylated amino acid residues in proteins are commonly classified (120) into three main groups. (*a*) *O*-Phosphates or *O*-phosphomonoesters are formed by phosphorylation of the hydroxyamino acids serine, threonine, and tyrosine; they all resist acid or hydroxylamine treatment, but all except phosphotyrosine are cleaved by alkali. (*b*) *N*-Phosphates or phosphoramidates are produced by phosphorylation of the basic amino acids arginine, histidine, and lysine; they are relatively base stable (except for phosphoarginine in hot alkali), but they are hydrolyzed by hydroxylamine and acid. (*c*) Acyl phosphates or phosphate anhydrides are generated by phosphorylation of the acidic amino acids aspartic acid and glutamic acid; they are extremely acid and base labile, and they are destroyed by hydroxylamine and reductive cleavage.

Two types of intracellular phosphoproteins are known: enzymes that are intermediately phosphorylated, usually at their active sites, and proteins that are phosphorylated by protein kinases (83, 88). It is generally assumed that phosphorylated intermediates in enzymatic mechanisms are either phosphoramidates (14, 75, 123, 198) or acyl phosphates (7, 23, 56, 199), but in some cases, such as *E. coli* phosphoglucomutase and alkaline phosphatase (51, 81, 174), the active site of the enzyme is modified at a serine residue. On the other hand, phosphorylation by protein kinases is considered to take place essentially on hydroxyamino acids (166, 172, 209), but the existence of enzymes specific for other amino acids cannot be excluded (59, 120).

In prokaryotes, as in eukaryotes, the more commonly employed assay procedures and isolation methods in the investigations of phosphoproteins involve the use of acid (53, 59, 120). As mentioned above, acid conditions are precluded in investigations of proteins containing *N*-phosphates or acyl phosphates. Since acid treatment of phosphoproteins during purification and determination of phosphoamino acids is routine, almost only O-phospho-monoesters have been studied, and the existence of acid-labile phosphates has been largely overlooked. The latter class of phosphates may well be more widespread than is generally believed.

Errors can be made in the characterization of phosphoamino acids when phosphoryl groups act as bridging moieties in phosphodiester linkages of nucleotide to protein (120). This type of modification has been found, for instance, in adenylated glutamine synthetase (21, 177) and in uridylated P_{II} regulatory protein (2, 164) from $E.\ coli$. It has also been detected during the formation of covalent complexes between DNA and DNA topoisomerases of $E.\ coli$ and $Micrococcus\ luteus$ (19, 106, 190). Partial acid digests of these phosphodiesters produce O-phosphates (5, 171). Consequently, treatment of phosphoprotein with phosphatases yielding free inorganic phosphate is required to demonstrate an O-phosphomonoester linkage.

Nature of Phosphorylated Amino Acids

The overall amount of phosphoamino acids in prokaryotes is much lower than that in eukaryotic organisms (54, 154). The small amount renders their detection and characterization relatively difficult and may explain, at least in part, the earlier lack of conclusive data on the presence of protein kinase activity in bacteria (66, 82, 93).

Experiments performed on total phosphoproteins of $E.\ coli$ radioactively labeled in vivo have shown that about 96% of the acid-stable phosphoamino acids are phosphoserine residues. The rest are phosphothreonine and, unexpectedly, phosphotyrosine residues (29). This analysis was carried out using a two-dimensional analytical system with high resolving power (175). Such equipment is imperative to separate phosphoamino acids accurately not only from one another but also from contaminating phosphate-containing molecules such as the nucleoside monophosphates arising from RNA hydrolysis or the pentose monophosphates produced by depurination of nucleotides (117). Alternative methods that are equally accurate and sensitive have been developed (22, 210). The large proportion of phosphoserine in $E.\ coli$ phosphoproteins has been confirmed by a report (154) indicating that the ratio of phosphothreonine to phosphoserine is $1:38$; the concentration of phosphoserine in bacteria is around 0.02 μmol g^{-1} wet weight (54). On the average, phosphoserine is the phosphoamino acid most represented in proteins not only in $E.\ coli,$ but also in the majority of the other bacterial species hitherto analyzed (see Table 1), as determined from both in vivo and in vitro experiments.

A detailed analysis of the phosphoamino acid content of some individual $E.\ coli$ phosphoproteins has been made (29). Among the 20 proteins examined, all except one are phosphorylated at serine. Three proteins, termed PP21,

PP71, and PP108, contain a significant amount of phosphothreonine in addition to phosphoserine, while one protein, PP40, is phosphorylated exclusively at tyrosine. The latter protein is associated with the membrane/ribosome fraction of the cell, but it is not an intrinsic constituent of either ribosomes or membranes (27); it seems rather to be located at their interface.

The presence of phosphotyrosine in bacteria is not confined to the *E. coli* cells in which it was originally discovered (27). Since then it has been detected in *S. typhimurium* (27), *Rhodospirillum rubrum* (72, 192), *Clostridium thermohydrosulfuricum* (107), *R. vannielii* (191), *Acinetobacter calcoaceticus, Arthrobacter globiformis,* and *Pseudomonas fluorescens* (M. Dadssi & A. J. Cozzone, unpublished data). In eukaryotic systems tyrosine phosphorylation is well documented, and its involvement in a number of crucial metabolic processes including viral transformation and growth control has been established (76, 77). The challenge now is to determine its physiological significance in bacteria.

Amino Acid Sequence of Phosphorylation Sites

The sites of phosphorylation of two bacterial proteins, isocitrate dehydrogenase and protein HPr, have been investigated in detail. Analysis of a tryptic digest obtained from *E. coli* IDH labeled in vivo established the phosphorylation site as Ser(P)-Leu-Asn-Val-Ala-Leu-Arg (110). Independent experiments on the enzyme labeled in vitro and digested with chymotrypsin indicated that the phosphorylated peptide is Thr-Thr-Pro-Val-Gly-Gly-Gly-Ile-Arg-Ser(P)-Leu-Asn-Val-Ala(Asx,Glx,Glx,Pro,Leu,His,Lys,Tyr)   (15). These data are in good agreement, since the first five residues of the former tryptic peptide, Ser(P)-Leu-Asn-Val-Ala, are identical to residues 10–14 of the chymotryptic peptide. Neither threonine nor tyrosine, when substituted for the serine at the phosphorylation site, is detectably phosphorylated (187). The sequence on the N-terminal side is unlike the sequences surrounding the sites phosphorylated by most eukaryotic cyclic AMP–dependent protein kinases, which usually contain two basic residues separated from the phosphorylated amino acid by one or two intervening residues (24, 145). This observation is consistent with the finding that IDH cannot be phosphorylated by cyclic AMP–dependent protein kinase (15). The location of an acidic amino acid two residues away from a serine on the C-terminal side appears to be characteristic for recognition of the phosphorylation site by several cyclic AMP–independent protein kinases in eukaryotes (146, 181). However, IDH, which does not contain such an acidic residue in its phosphorylation site, is nonetheless phosphorylated by a cyclic AMP–independent kinase (31, 109). An unusual feature of the sequence of the phosphopeptide of IDH is the presence of one proline and a run of three glycine residues in a five-residue stretch on the N-terminal side of the phosphorylation site. This region of the polypeptide is likely to form a flexible random coil and may have a role in the enzyme-substrate recognition process (15).

From comparative analysis of active and inactive forms of IDH it has been suggested that the phosphorylation of the enzyme may occur close to or at its coenzyme (NADP) binding site (60, 126). In this context it is noted that the sequence Arg-Ser-Leu-Asn found in the phosphopeptide of IDH is also present in another NADP-binding enzyme, dihydrofolate reductase from chicken liver, in a region very close to the NADP binding site (196). The suggestion above was based on the assumption that IDH is phosphorylated at a serine residue with a stoichiometry of one per subunit (99, 131). However, this estimate has recently been questioned in a report showing that IDH can exist in at least three different phosphorylated states involving the modification of threonine as well as serine residues (28).

The second bacterial protein for which the amino acid sequence surrounding the phosphorylation site was determined was protein HPr from *Streptococcus faecalis* (39). ATP-dependent phosphorylation occurs at Ser46 within this 89-residue protein, in a segment with the structure Val-Asn-Leu-Lys-Ser(P)-Ile-Met. This sequence presents some similarity with the sequence surrounding the phosphorylatable serine residue of *E. coli* IDH (15, 110). The phosphorylation site of HPr also resembles the reactive center of lima bean trypsin inhibitor (34) and one of the phosphorylatable sites of troponin I of rabbit skeletal muscle (74). A cyclic AMP–dependent protein kinase isolated from rabbit cardiac muscle phosphorylates two serine residues in troponin I, one of which is located at position 19 (208). The sequence Leu-Lys-Ser-Val-Met around Ser19 of troponin I is nearly identical to the site of ATP-dependent phosphorylation in HPr.

The amino acid sequences of HPr from *E. coli* and from *S. typhimurium* have been determined and are almost identical (32, 150). In particular, both possess the sequence Lys-Ser-Leu spanning residues 45 to 47. This is similar to the sequence Lys-Ser-Ile in Hpr from *S. faecalis* (39) and identical to the sequence in HPr from *S. aureus* (14). However, homology is restricted to these three amino acids, which may explain why HPr from *E. coli* is not a substrate of the ATP-dependent HPr kinase isolated from either *S. faecalis* or *Streptococcus pyogenes* (36, 45).

Multiplicity of Phosphorylation Sites on Individual Proteins

Evidence has been presented that some bacterial proteins can be phosphorylated at different sites along their primary structure and that differential phosphorylation of these sites may occur depending on the environmental conditions.

One example, already mentioned, concerns protein HPr from streptococcal cells. In most strains, HPr can be phosphorylated at the N-1 position of His15 in a PEP-dependent reaction catalyzed by enzyme I of the sugar phosphotrans-ferase system (14, 179, 207). A second phosphorylation can occur at Ser46 catalyzed by an ATP-dependent protein kinase (39, 40). The latter phosphorylation is stimulated by glycolytic intermediates and is severely inhibited by inorganic phosphate (36, 159). The reaction strongly reduces the ability of HPr to be phosphorylated at histidine in the PEP-dependent reaction (159). In *S. salivarius,* HPr is doubly phosphorylated through a particular, still unclear, mechanism that involves two distinct ATP-dependent protein kinases (205). In this case the histidine kinase is activated by glucose-6-phosphate, but it is inhibited by other glycolytic intermediates.

Another protein subject to differential phosphorylation is the *dnaK* protein, which is phosphorylated exclusively at serine residues during exponential growth of *E. coli* cells and essentially at threonine residues after infection of bacteria with bacteriophage M13 (165). This qualitative change in the nature of the phosphorylated amino acids is accompanied by a sevenfold increase in the extent of phosphorylation of the protein following infection. Phosphoryla-tion of the *dnaK* protein in vitro occurs only at threonine residues (213).

Isocitrate dehydrogenase of *E. coli* also contains multiple phosphorylation sites, as evidenced by isoelectric focusing analysis of its various forms, peptide mapping, and identification of the different phosphorylatable amino acids (28, 62). As in the case of the *dnaK* protein, there is again a difference in the nature of the amino acids that are phosphorylated in vivo and in vitro: In growing cells IDH is modified at both serine and threonine, whereas in an acellular system it is modified only at serine (28, 131).

PROTEIN KINASES AND PHOSPHOPROTEIN PHOSPHATASES

The extent of phosphorylation of most proteins is dependent on a dynamic equilibrium between the phosphorylation reaction catalyzed by a protein kinase from a nucleoside-triphosphate (usually ATP) and a dephosphorylation reaction promoted by a phosphoprotein phosphatase. The balance between the phosphorylated and dephosphorylated states of a protein can be shifted by stimulation or inhibition of the enzymatic activity of either the kinase or the phosphatase.

In comparison with the wealth of information amassed concerning the nature and properties of protein kinases and phosphatases in eukaryotic systems (44, 88, 170, 172), there has been a paucity of reports dealing with these enzymes in prokaryotes. In a few cases, however, detailed analyses of their structural and functional characteristics have been performed based on both biochemical and genetic approaches.

Nature and Properties of Protein Kinases

E. coli cells contain two main classes of protein kinases, which phosphorylate different polypeptides; one is in the cytoplasmic fraction and the other is attached to the ribosome/membrane fraction (114, 116, 149). From a crude RNA polymerase preparation a 120-kd protein kinase has been purified to apparent homogeneity (50). This enzyme, which phosphorylates an un-identified 90-kd protein, is composed of two subunits in the form of either a heterodimer of the 61- and 66-kd polypeptides or a homodimer of one of these polypeptides. From the same crude preparation another protein kinase of 100 kd with a self-phosphorylation activity has been isolated (50). The *dnaK* protein of *E. coli,* which also possesses an autophosphorylating activity, has been purified to homogeneity (213). This protein exhibits, in addition to its kinase activity, a weak but significant DNA-independent ATPase activity, which is encoded by the same *dnaK* gene. A recent report indicates that the protein participates, together with the other heat shock protein, encoded by *dnaJ,* in the in vivo phosphorylation of at least five endogenous proteins (197). Whether it acts directly as a protein kinase for these proteins or indirectly by stimulating some other protein kinases in the cell remains to be elucidated.

In *S. typhimurium* four different protein kinases have been detected (202, 203). They can be distinguished by their substrate specificity, chromatographic behavior, and sensitivity to various inhibitors such as AMP, GTP, and pyrophosphate. A tyrosine kinase activity from the photosynthetic bacterium *R. rubrum* has been partially purified (192). This kinase phosphorylates three endogenous proteins as well as a number of synthetic peptides, namely angiotensin derivatives, that only contain a single tyrosine residue as phosphate acceptor. The thermophile *C. thermohydrosulfuricum* harbors several protein kinases in both soluble and particulate fractions which phosphorylate distinct substrate proteins (107). Their activity in vitro is modulated by fructose-1,6-bisphosphate, glucose-1,6-bisphosphate, hexose monophosphates, and longer-chain polyamines, especially the odd-numbered polyamines characteristic of thermophiles (138, 139). Another effector seems to be an endogenous calmodulin-like protein similar to that found in other bacterial species (79, 102).

An enzyme that catalyzes the ATP-dependent phosphorylation of protein HPr has been purified from *S. faecalis* (36). It is capable of phosphorylating not only HPr from this bacterium but also HPr from other gram-positive species such as *S. pyogenes, Streptococcus lactis, S. aureus, Staphylococcus carnosus, Lactobacillus casei,* and *B. subtilis;* this suggests that phosphorylation occurs at a preserved region of the protein. By contrast, HPr of *E. coli* is not a substrate of this enzyme (see above), which raises the question of whether or not phosphorylation of HPr at a serine residue is restricted to gram-positive bacteria. The kinase of *S. faecalis*, which has a mass of about 65 kd, is stimulated by fructose-1,6-bisphosphate, but is largely inhibited by inorganic phosphate and EDTA. Mg^{2+} and Mn^{2+} overcome inhibition by EDTA (36). A similar ATP-dependent protein kinase has been purified from *S. pyogenes* (40, 159). The soluble form of the enzyme, released from the particulate fraction of the cells with concentrated salt in the presence of a protease inhibitor, has an approximate mass of 60 kd. Its activity is dependent on divalent cations as well as several metabolites. Of the latter compounds, fructose-1,6-bisphosphate is most active, while gluconate-6-phosphate, 2-phosphoglycerate, 2,3-diphosphoglycerate, PEP, and pyruvate have lower stimulatory effects. Enzyme activity is unaffected by *p*-chloromercuribenzo-ate, *N*-ethylmaleimide, or iodoacetate, but is strongly inhibited by di-ethylpyrocarbonate (159). In the oral pathogen *Streptococcus mutans*, ATP-dependent protein kinase activities have been detected in both membrane and cytoplasmic fractions (122). These kinases phosphorylate different polypeptides in the two fractions. Those phosphorylated in cytoplasm include protein HPr and a 61-kd protein that exhibits an autophosphorylating activity.

The viral protein kinase whose synthesis is induced in *E. coli* upon infec-tion with phage T7 has been purified approximately 5000-fold from the ribosomal wash fraction of infected cells (141). It consists of a single 37-kd peptide chain that requires low ionic strength for optimal activity and is inhibited by *p*-chloromercuribenzoate. It can phosphorylate histones and protamines in vitro, but its best substrate is, uniquely, egg-white lysozyme (142). The protein kinase induced by phage T3 resembles the T7-induced enzyme in all respects tested (142).

The capacity of prokaryotic protein kinases to phosphorylate exogenous proteins in vitro seems to be specific to viral enzymes. None of the known bacterial kinases is able to phosphorylate proteins such as casein, histones, protamines, or phosvitin, which are, on the other hand, readily targeted by eukaryotic kinases (115, 202). Considering the high selectivity of the pro-karyotic kinases, it is understandable that early searches employing these exogenous proteins as substrates failed to demonstrate the presence of kinases in bacteria (82, 93). *E. coli* extracts not only fail to phosphorylate histones or casein, but moreover impair the phosphorylation of these proteins by specific eukaryotic enzymes (beef heart histone kinase and rat liver casein kinase, respectively) (118). This effect is not due to a phosphoprotein phosphatase, ATPase, or protease activity in bacterial preparations. It is rather attributable to a protein kinase inhibitor (118). Bacterial protein kinase inhibitors have also been observed in *S. typhimurium* (203) and *C. thermohydrosulfuricum* (107).

The activity of prokaryotic protein kinases is cyclic nucleotide–independent. The pattern of protein phosphorylation in vitro was identical whether or not cyclic AMP or cyclic GMP was present during incubation of cellular extracts from *E. coli* (31, 116), *R. rubrum* (72), *Sulfolobus acidocaldarius* (180), and all other species tested (see Table 1) as well as from T7-infected *E. coli* (142). When protein phosphorylation in vivo in wild-type cells was compared with that in adenylate cyclase deletion mutants of *E. coli* (31, 53, 109) or *S. typhimurium* (203), similar results were obtained. One can thus imagine that in bacteria cyclic nucleotides, namely cAMP, are specialized for the regulation of protein synthesis (108, 143), while protein phosphorylation regulates specific metabolic controls and is itself controlled by metabolites (88, 203), even though some exceptions to this rule are presented below.

Nature and Properties of Phosphoprotein Phosphatases

Protein kinases in prokaryotes are complex enough, but, as in eukaryotes (178), the phosphoprotein phosphatases are even less well defined in terms of their structure, regulation, and substrate specificity. *S. typhimurium* contains at least two soluble protein phosphatase activities, which differentially dephosphorylate specific endogenous phosphoproteins in vitro (203). These enzymes differ from the nonspecific acid phosphatases harbored by the bacterium (which lacks alkaline phosphatase) in their insensitivity to sodium fluoride. In *C. thermohydrosulfuricum* two phosphatases that dephosphorylate distinct substrate proteins have been found, one in the cytoplasm and the other in the particulate fraction of the cell (107).

The phosphate group of the phosphoserine residue borne by protein HPr of *S. pyogenes* can be hydrolyzed by a soluble phosphatase (70 kd), which has been partially purified (40). The type I enzymes of the PTS and PEP, separately, do not exert any influence over the activity of this phosphatase in vitro. However, they inhibit it drastically when they are added together to the incubation medium, and they can thus phosphorylate the histidine residue of HPr. Doubly phosphorylated HPr is therefore a poor substrate for the phosphatase (40). A similar phosphatase activity (75 kd), which is capable of splitting the seryl phosphate bond of HPr from *S. faecalis,* has been purified and characterized (38). It is stimulated by inorganic phosphate, an inhibitor of the ATP-dependent HPr kinase, but is insensitive to fructose-1,6-bisphosphate and other glycolytic intermediates. ATP, but not ADP, severely inhibits this enzyme. EDTA is also a potent inhibitor, but divalent cations such as Mg^{2+}, Mn^{2+}, and Co^{2+} overcome its inhibitory action. The phosphatase is able to hydrolyze the seryl phosphate bond in HPr of *S. lactis, S. aureus, B. subtilis, S. pyogenes,* and *L. casei* in addition to that of *S. faecalis*. Besides this cytoplasmic enzyme, a membrane-bound phosphoprotein phosphatase with identical functional properties has been detected (38).

Isocitrate Dehydrogenase Kinase/Phosphatase

The kinase and phosphatase activities that modify IDH in *E. coli* exhibit an unusual feature: They are physically associated with the same protein (98). This bifunctional enzyme gives a single band in denaturing gel electrophoresis; its native form, however, is a dimer, but no separate monofunctional kinase or phosphatase exists (131). The association of the two activities with the same molecule was initially suggested by their coelution from ion exchange, gel filtration, and affinity columns (98, 111) and was further confirmed by genetic studies. It has been shown that IDH kinase and IDH phosphatase are coded for by a single gene *(aceK)* and that mutants of *E. coli* that are defective in kinase activity are simultaneously devoid of phosphatase activity (97, 100). Recently, the complete nucleotide sequence of the *aceK* gene has been determined (26). This is the only bacterial kinase (and phosphatase) gene whose primary structure has been determined. The sequence comprises 1731 nucleotides encoding a protein with a molecular weight of 66,528 (577 amino acid residues). Expression of the gene is essential for growth on minimal acetate selective medium (48, 100).

The IDH kinase and phosphatase located on the same protein, although unusual, is not the only known bifunctional regulatory enzyme. The activities that catalyze the adenylation and deadenylation of glutamine synthetase in *E. coli* are located on a single polypeptide chain (21, 163, 164). Similarly, in eukaryotes, the enzymes responsible for the synthesis and breakdown of fructose-2,6-bisphosphate in rat liver are physically associated (46, 47). The discovery of such systems raises intriguing questions as to the possible advantages of having opposing activities localized on the same polypeptide chain (63, 98).

Interestingly, many of the metabolites that inhibit IDH kinase also stimulate IDH phosphatase (98, 99, 132). This is true of AMP, ADP, isocitrate, 3-phosphoglycerate, oxaloacetate, 2-oxoglutarate, PEP, and pyruvate; exceptions include ATP, which is a substrate for the kinase and an activator of the phosphatase, and NADPH, which inhibits both activities (17, 49, 128). ATP ($K_m = 100~\mu M$) is a better substrate for IDH phosphatase than ADP ($K_m = 400~\mu M$). It therefore seems that the phosphatase activity is not simply the reverse of the kinase reaction, since ADP, not ATP, would then be the superior cosubstrate. This conclusion is supported by the finding that the ^{32}P released from [^{32}P]phospho-IDH by the phosphatase appears not in ATP but as inorganic phosphate (98).

BIOLOGICAL ROLE OF PHOSPHORYLATION

The discovery that protein kinases and phosphoprotein phosphatases coexist within prokaryotic cells was an indication that the reversible phosphorylation of proteins in prokaryotes could represent, by analogy with that in eukaryotes, a regulatory device evolved by lower organisms for controlling their cellular functions. Actually, there have been several reports that give evidence or strongly suggest that in a number of cases protein function is indeed controlled by phosphorylation-dephosphorylation.

Intracellular Carbon Flux

The role of phosphorylation in the control of *E. coli* isocitrate dehydrogenase activity has been investigated with the most thoroughness (reviewed in 126, 127). When acetate is the sole carbon source, the bacteria can oxidize it by way of the Krebs cycle to produce energy efficiently, but the two carbon atoms are totally converted to CO_2, and no material is then left for the synthesis of cellular components (86, 96, 130). The bacteria adjust by inducing two new enzymes, isocitrate lyase and malate synthase, which divert some of the carbon flux through the glyoxylate cycle, providing energy and four-carbon intermediates for the biosynthetic pathways (71, 194). Thus when acetate must serve as the energy/carbon source, the channeling of metabolites from the Krebs cycle into the glyoxylate bypass occurs at isocitrate. Under these conditions, the activity of IDH declines drastically, and the enzyme becomes phosphorylated concomitantly with this decrease in activity (13, 61, 201). Conversely, when *E. coli* cells are cultured on carbon sources, such as glucose, that do not employ the glyoxylate cycle as an anaplerotic sequence, IDH is fully active and dephosphorylated. Similar behavior has been described in *S. typhimurium* (113, 204). Therefore the phosphorylation of IDH is generally thought to control the flux of carbon between the competing Krebs and glyoxylate cycles (126), even though this interpretation has been questioned (155, 156). It is interesting, in this context, that *E. coli* isocitrate lyase, one of the two glyoxylate bypass enzymes, has lately been found to be phosphorylated at histidine (167, 168).

The activity of IDH is regulated by a number of metabolites, cited above, and the extent of its phosphorylation is determined by the kinetic behavior of the opposing kinase and phosphatase reactions (98, 99, 132). Analysis of the control of IDH phosphorylation under steady-state conditions has shown that its sensitivity to regulation is greatly enhanced when the substrate saturates the converter enzyme (99, 101). The possible physiological consequences of this phenomenon, termed "zero-order ultrasensitivity," have been discussed (65, 87).

The genes coding for malate synthase *(aceB)*, isocitrate lyase *(aceA)*, and IDH kinase/phosphatase *(aceK)* are present, in that order, in the same *ace* operon (23a, 100, 112, 193) located at 90 min on the *E. coli* K-12 linkage map (11). The expression of the *ace* operon is under the transcriptional control of two genes: *iclR*, which is adjacent to the operon, and *fadR*, which maps at 25 min and is also involved in the regulation of the fatty acid degradation *(fad)* regulon (18, 134). The inclusion of the *aceK* gene in the *ace* operon obviously provides a mechanism for coordinating its expression with that of glyoxylate bypass enzymes. However, the cellular concentration of IDH kinase/phosphatase is approximately 1000-fold less than that of isocitrate lyase (100). In this respect, recent determination of the nucleotide sequence of the 5'-flanking region of the *aceK* gene has revealed an original intercistronic structural pattern consisting of two consecutive long dyad symmetries that can yield very stable stem-loop units (26). These regulatory structures are probably responsible for the very low level of expression of the *aceK* gene compared to that of the *aceA* gene (26).

Sugar Accumulation

At least five distinct mechanims regulating the uptake of external sugars have been demonstrated in bacteria (reviewed in 42, 148). A novel regulatory mechanism, operating by exclusion of external sugar and expulsion of intracellular sugar phosphate, has been proposed for gram-positive bacteria (160–162). This mechanism responds to variations in energy levels and metabolite concentrations (ATP, sugar phosphates, and/or inorganic phosphate) by changing the direction and rate of sugar transport (129, 159, 186). This response appears to be achieved by the ATP-dependent phosphorylation of one of the PTS proteins, HPr (37, 40). The primary function of the PEP-dependent phosphorylation of the active histidine residue in HPr is to drive concomitant sugar uptake and phosphorylation (35, 42). In contrast to the histidine-bound phosphoryl group, the serine-bound phosphoryl group of HPr cannot be used for sugar phosphorylation (40, 148). Phosphorylation at the histidine of phospho-seryl HPr of *S. lactis* by PEP and enzyme I is about 5000 times slower than that of free HPr (37). Therefore an important aspect of the regulatory mechanism of sugar accumulation is the inhibition of the PTS-mediated uptake of carbohydrates by conditions that favor the formation of phospho-seryl HPr (122, 158). However, it is not yet clear how the phosphorylation of HPr promotes sugar expulsion.

Enzymes of type III are able to relieve the inhibition of PEP-dependent phosphorylation. For example, enzyme I, which does not appreciably phosphorylate phospho-seryl-HPr, can form a complex with the gluconate enzyme III from *S. faecalis* and thus stimulate the transfer of phosphate from phospho-histidyl enzyme I to the histidine residue of phospho-seryl HPr by several orders of magnitude (37). A tentative role for this reaction, whose efficiency depends on the type of enzyme III involved, has been presented (37, 129, 158).

Bacteriophage Infection

The involvement of protein phosphorylation in the regulatory transcriptional and translational mechanisms triggered by phage infection has been documented. After infection of *E. coli* with phage T7, the synthesis of host RNA and of various early T7 RNA species is rapidly arrested (147). This early transcriptional control is defective, however, when bacteria are infected with mutants lacking protein kinase activity (211). Phosphorylation of the β and β' subunits of *E. coli* RNA polymerase following infection coincides with the appearance of viral protein kinase and with the shutoff of RNA synthesis. It is therefore likely that the phosphorylation of RNA polymerase represents the molecular basis of early transcriptional control (153, 211). A similar phenomenon occurs upon infection of *E. coli* with phage T3 (142) or upon infection of *C. crescentus* with phage ϕCd1 (6, 69). However, when *E. coli* is infected with phage T4, the α subunit of RNA polymerase is then subject to ADP ribosylation (64, 176), which indicates that different bacteriophages may employ different strategies to induce the shutoff of macromolecular synthesis. The protein kinase of phage T7 seems to be implicated, moreover, in the posttranscriptional regulation of T7 gene expression. Indeed, RNase III, which in *E. coli* processes the T7 early polycistronic mRNA synthesized by the host RNA polymerase (43), is stimulated fourfold when bacteria are infected with wild-type T7, but is not stimulated when cells are infected with a T7 protein kinase mutant (121). Furthermore, the fact that phage infection promotes the phosphorylation of several ribosomal proteins and other host proteins suggests that alterations in the host also occur at the level of translation or post translation (69, 152, 153).

Protein phosphorylation may have a role in DNA synthesis. The phosphorylation of the *E. coli dnaK* protein, like that of several other proteins, is greatly increased upon infection with the filamentous single-stranded DNA phage M13, while its rate of synthesis is not significantly changed (165). Since the *dnaK* protein is required for the formation of the M13 duplex replicative form (58, 103, 104), it has been proposed that the increase in its phosphorylation may reflect the increase in its activity which is necessary to sustain viral DNA replication (165).

Other Systems

In several other cases a possible regulatory role of protein phosphorylation has been envisaged, but as yet no conclusive evidence has been presented in terms of identification of the phosphorylated substrates and/or phosphorylating-dephosphorylating enzymes and precise characterization of the molecular effectors involved. For example, phosphorylation may be involved in the control of bacterial photosynthesis, as indicated by experiments with the extreme halophile *Halobacterium halobium,* which employs retinal-containing protein pigments for both photoenergy and photosensory transduction (182, 183), the nonsulfur purple bacterium *R. rubrum,* in which bacteriochlorophyll serves as light-harvesting pigment (70, 72), and the cyanobacterium *Synechococcus* strain 6301, in which light is harvested largely by extrinsic phycobiliprotein complexes (4). Other examples include the modification of citrate lyase ligase in the anaerobic bacterium *Clostridium sphenoides* (8), of the arginine-ornithine periplasmic transport protein in *E. coli* (23), and of certain proteins of *Myxococcus xanthus* during spore or fruiting-body formation (85). Protein phosphorylation has also been implicated in the transcriptional control of the synthesis of glutamine synthetase in *E. coli* (133), the regulation of nitrogenase activity in *Chromatium vinosum* (67), and the regulation of enzyme III^{Glc} of the PTS in *S. typhimurium* (125).

CONCLUDING REMARKS

The observations summarized in this review conclusively demonstrate that, contrary to an early concept, protein phosphorylation is not confined to eukaryotes but occurs in prokaryotes as well. This modification therefore seems to be of ancient evolutionary origin, like a number of other posttranslational modifications such as methylation, acetylation, or ADP ribosylation, which are found in almost all forms of life. The reversible phosphorylation of proteins in growing or virus-infected bacteria resembles that in higher organisms in the sense that it regulates various cellular functions at different levels, including carbohydrate and energy metabolism, transport processes, gene expression, and possibly DNA replication. However, the tuning of physiological reactions to the requirements of the prokaryotic cell is operated by phosphorylating-dephosphorylating enzymes that differ in many respects from the corresponding eukaryotic enzymes and, for this reason, constitute an interesting new field to investigate. In this regard, the restricted substrate specificity of bacterial protein kinases and their general insensitivity to cyclic nucleotides are striking. Also, the coexistence of two opposing activities, IDH kinase and IDH phosphatase, on the same protein and the phosphorylation of protein HPr on two distinct amino acid residues by two distinct protein phosphotransferases utilizing two different phosphoryl donors are quite unusual aspects of modifying enzymes that raise puzzling questions and await further studies.

Bacteria provide much simpler biological systems for studying protein phosphorylation than most eukaryotes. For instance, by simply changing the composition of the culture medium one can analyze the cellular response to environmental variations in connection with the regulation of metabolism by phosphorylation. In addition, with bacteria it is easier to carry out a direct study of the metabolism of an intact whole organism in its natural environment. By contrast, investigations of eukaryotic systems often start with in vitro studies to simplify the experimental approach, but the results thus obtained are frequently difficult to correlate with the in vivo reality. A further advantage of utilizing bacteria is that biochemical analysis of the nature and function of the modifying enzymes and their protein substrates can be completed rather easily by genetic studies. Considering the impressive progress and exciting discoveries made in recent years, one can predict that prokaryotes will continue to be of great help for our understanding of the basic processes in living organisms.

ACKNOWLEDGMENTS

The interest of a large number of colleagues who provided me with both published and unpublished information is gratefully acknowledged. I also thank Dr. P. Donini for critical reading of the manuscript. Work from my laboratory was supported by the Centre National de la Recherche Scientifique (UA1176), Fondation pour la Recherche Médicale, and Institut pour la Santé et la Recherche Médicale (contract 841007).

Literature Cited

1. Abraham, A. K., Eikhom, T. S., Pryme, I. F., eds. 1983. *Protein Synthesis Translational and Post-Translational Events.* Clifton, NJ: Humana. 477 pp.
2. Adler, S. P., Durich, D., Stadtman, E. R. 1975. Cascade control of *Escherichia coli* glutamine synthetase. Properties of the P_{II} regulatory protein and the uridyltransferase-uridylyl-removing enzyme. *J. Biol. Chem.* 250:6264–72
3. Agabian, N., Rosen, O. M., Shapiro, L. 1972. Characterization of a protein acyl kinase from *Caulobacter crescentus*. *Biochem. Biophys. Res. Commun.* 49:1690–98
4. Allen, J. F., Sanders, C. E., Holmes, N. G. 1985. Correlation of membrane protein phosphorylation with excitation energy distribution in the cyanobacterium *Synechococcus* 6301. *FEBS Lett.* 193:271–75
5. Ambros, V., Baltimore, D. 1978. Protein is linked to the 5' end of poliovirus RNA by a phosphodiester linkage to tyrosine. *J. Biol. Chem.* 253:5263–66
6. Amemiya, K., Raboy, B., Shapiro, L. 1980. Involvement of the host RNA polymerase in the early transcription program of *Caulobacter crescentus* bacteriophage φCd1 DNA. *Virology* 104:109–16
7. Anthony, R. S., Spector, L. B. 1972.

Phosphorylated acetate kinase. Its isolation and reactivity. *J. Biol. Chem.* 247:2120–25
8. Antranikian, G., Herzberg, C., Gottschalk, G. 1985. Covalent modification of citrate lyase ligase from *Clostridum sphenoides* by phosphorylation/dephosphorylation. *Eur. J. Biochem.* 153:413–20
9. Antranikian, G., Herzberg, C., Gottschalk, G. 1985. In vivo phosphorylation of proteins in *Clostridium sphenoides*. *FEMS Microbiol. Lett.* 27:135–38
10. Averhoff, B., Antranikian, G., Gottschalk, G. 1986. Phosphorylation and nucleotidylation of proteins in *Rhodocyclus gelatinosus*. *FEMS Microbiol. Lett.* 33:299–304
11. Bachmann, B. J., Low, K. B. 1980. Linkage map of *Escherichia coli* K-12, edition 6. *Microbiol. Rev.* 44:1–56
12. Bardwell, J. C. A., Craig, E. A. 1984. Major heat shock gene of *Drosophila* and the *Escherichia coli* heat-inducible *dnaK* gene are homologous. *Proc. Natl. Acad. Sci. USA* 81:848–52
13. Bennett, P. M., Holms, W. H. 1975. Reversible inactivation of the isocitrate dehydrogenase of *Escherichia coli* ML308 during growth on acetate. *J. Gen. Microbiol.* 87:37–51

14. Beyreuther, K., Raufuss, K. H., Schrecker, O., Hengstenberg, W. 1977. The phosphoenolpyruvate-dependent phosphotransferase system of *Staphylococcus aureus:* amino acid sequence of the phosphocarrier protein HPr. *Eur. J. Biochem.* 75:275–86

15. Borthwick, A. C., Holms, W. H., Nimmo, H. G. 1984. Amino acid sequence round the site of phosphorylation in isocitrate dehydrogenase from *Escherichia coli* ML308. *FEBS Lett.* 174:112–15

16. Borthwick, A. C., Holms, W. H., Nimmo, H. G. 1984. Isolation of active and inactive forms of isocitrate dehydrogenase from *Escherichia coli* ML308. *Eur. J. Biochem.* 141:393–400

17. Borthwick, A. C., Holms, W. H., Nimmo, H. G. 1984. The phosphorylation of *Escherichia coli* isocitrate dehydrogenase in intact cells. *Biochem. J.* 222:797–804

18. Brice, C. G., Kornberg, H. L. 1968. Genetic control of isocitrate lyase activity in *Escherichia coli*. *J. Bacteriol.* 96:2185–86

19. Brown, P. O., Peebles, C. L., Cozzarelli, N. R. 1979. A topoisomerase from *Escherichia coli* related to DNA gyrase. *Proc. Natl. Acad. Sci. USA* 76:6110–14

20. Buchanan, R. E., Gibbons, N. E., eds. 1974. *Bergey's Manual of Determinative Bacteriology*. Baltimore, Md: Williams & Wilkins. 1246 pp.

21. Caban, C. E., Ginsburg, A. 1976. Glutamine synthetase adenylyltransferase from *Escherichia coli:* purification and physical and chemical properties. *Biochemistry* 15:1569–80

22. Capony, J. P., Demaille, J. G. 1983. A rapid microdetermination of phosphoserine, phosphothreonine and phosphotyrosine in proteins by automatic cation exchange on a conventional amino acid analyzer. *Anal. Biochem.* 128:206–12

23. Celis, R. T. F. 1984. Phosphorylation in vivo and in vitro of the arginine-ornithine periplasmic transport protein of *Escherichia coli*. *Eur. J. Biochem.* 145:403–11

23a. Chung, T., Klumpp, D. J., LaPorte, D. C. 1988. Glyoxylate bypass operon of *Escherichia coli:* cloning and determination of the functional map. *J. Bacteriol.* 170:386–92

24. Cohen, P. 1980. Protein phosphorylation and the coordinated control of intermediary metabolism. *Mol. Aspects Cell. Regul.* 1:255–68

25. Cohen, P., ed. 1980. *Recently Discovered Systems of Enzyme Regulation by Reversible Phosphorylation*. *Mol. Aspects Cell. Regul.* Vol 1. 273 pp.

26. Cortay, J. C., Bleicher, F., Rieul, C., Reeves, H. C., Cozzone, A. J. 1988. Nucleotide sequence and expression of the *aceK* gene coding for isocitrate dehydrogenase kinase/phosphatase in *Escherichia coli*. *J. Bacteriol.* 170:89–97

27. Cortay, J. C., Duclos, B., Cozzone, A. J. 1986. Phosphorylation of an *Escherichia coli* protein at tyrosine. *J. Mol. Biol.* 187:305–8

28. Cortay, J. C., Reeves, H. C., Cozzone, A. J. 1986. Multiplicity of phosphorylation sites on *Escherichia coli* isocitrate dehydrogenase. *Curr. Microbiol.* 13:251–54

29. Cortay, J. C., Rieul, C., Duclos, B., Cozzone, A. J. 1986. Characterization of the phosphoproteins of *Escherichia coli* cells by electrophoretic analysis. *Eur. J. Biochem.* 159:227–37

30. Cozzone, A. J. 1984. Protein phosphorylation in bacteria. *Trends Biochem. Sci.* 9:400–3

31. Dadssi, M., Cozzone, A. J. 1985. Cyclic AMP independence of *Escherichia coli* protein phosphorylation. *FEBS Lett.* 186:187–90

32. De Reuse, H., Roy, A., Danchin, A. 1985. Analysis of the *ptsH-ptsI-crr* region in *Escherichia coli* K-12: nucleotide sequence of the *pts* gene. *Gene* 35:199–207

33. Desmarquets, G., Cortay, J. C., Cozzone, A. J. 1984. Two-dimensional analysis of proteins phosphorylated in *E. coli* cells. *FEBS Lett.* 173:337–41

34. Deutscher, J. 1985. Unusual resemblance. *Trends Biochem. Sci.* 10:232

35. Deutscher, J. 1985. Phosphoenolpyruvate-dependent phosphorylation of a 55-kDa protein of *Streptococcus faecalis* catalyzed by the phosphotransferase system. *FEMS Microbiol. Lett.* 29:237–43

36. Deutscher, J., Engelmann, R. 1984. Purification and characterization of an ATP-dependent protein kinase from *Streptococcus faecalis*. *FEMS Microbiol. Lett.* 23:157–62

37. Deutscher, J., Kessler, U., Alpert, C. A., Hengstenberg, W. 1984. Bacterial phosphoenolpyruvate-dependent phosphotransferase system: P-Ser-HPr and its possible regulatory function. *Biochemistry* 23:4455–60

38. Deutscher, J., Kessler, U., Hengstenberg, W. 1985. Streptococcal phosphoenolpyruvate:sugar phosphotransferase system. Purification and characterization of a phosphoprotein phosphatase which hydrolyzes the phosphoryl

bond in seryl-phosphorylated histidine-containing protein. *J. Bacteriol.* 163:1203–9

39. Deutscher, J., Pevec, B., Beyreuther, K., Kiltz, H. H., Hengstenberg, W. 1986. The streptococcal phosphoenol-pyruvate:sugar phosphotransferase system: amino acid sequence and site of ATP-dependent phosphorylation of HPr. *Biochemistry* 25:6543–50

40. Deutscher, J., Saier, M. H. Jr. 1983. ATP-dependent protein kinase–catalyzed phosphorylation of a seryl residue in HPr, a phosphate carrier protein of the phosphotransferase system in *Streptococcus pyogenes*. *Proc. Natl. Acad. Sci. USA* 80:6790–94

41. Deutscher, J., Sauerwald, H. 1986. Stimulation of dihydroxyacetone and glycerol kinase activity in *Streptococcus faecalis* by phosphoenolpyruvate-dependent phosphorylation catalyzed by enzyme I and HPr of the phosphotransferase system. *J. Bacteriol.* 166:829–36

42. Dills, S. S., Apperson, A., Schmidt, M. R., Saier, M. H. Jr. 1980. Carbohydrate transport in bacteria. *Microbiol. Rev.* 44:385–418

43. Dunn, J. J., Studier, F. W. 1973. T7 early RNAs and *Escherichia coli* ribosomal RNAs are cut from large precursor RNAs in vivo by ribonuclease III. *Proc. Natl. Acad. Sci. USA* 70:3296–300

44. Edelman, A. M., Blumenthal, D. K., Krebs, E. G. 1987. Protein serine/threonine kinases. *Ann. Rev. Biochem.* 56:567–613

45. El-Kabbani, O. A. L., Waygood, E. B., Delbaere, L. T. J. 1987. Tertiary structure of histidine-containing protein of the phosphoenolpyruvate:sugar phosphotransferase system of *Escherichia coli*. *J. Biol. Chem.* 262:12926–29

46. El-Maghrabi, M. R., Claus, T. H., Pilkis, J., Fox, E., Pilkis, S. J. 1982. Regulation of rat liver fructose 2,6-bisphosphatase. *J. Biol. Chem.* 257:7603–7

47. El-Maghrabi, M. R., Fox, E., Pilkis, J., Pilkis, S. J. 1982. Cyclic AMP-dependent phosphorylation of rat liver 6-phosphofructo 2-kinase/fructose 2,6-bisphosphatase. *Biochem. Biophys. Res. Commun.* 106:794–802

48. El-Mansi, E. M. T., MacKintosh, C., Duncan, K., Holms, W. H., Nimmo, H. G. 1987. Molecular cloning and overexpression of the glyoxylate bypass operon from *Escherichia coli* ML308. *Biochem. J.* 242:661–65

49. El-Mansi, E. M. T., Nimmo, H. G., Holms, W. H. 1985. The role of isoci-trate in control of the phosphorylation of isocitrate dehydrogenase in *Escherichia coli* ML308. *FEBS Lett.* 183:251–55

50. Enami, M., Ishihama, A. 1984. Protein phosphorylation in *Escherichia coli* and purification of a protein kinase. *J. Biol. Chem.* 259:526–33

51. Engström, L., Ekman, P., Humble, E., Ragnarsson, U., Zetterqvist, O. 1984. Detection and identification of substrates for protein kinases: use of proteins and synthetic peptides. *Methods Enzymol.* 107:130–54

52. Fakunding, J. L., Traugh, J. A., Traut, R. R., Hershey, J. W. B. 1972. Phosphorylation of initiation factor IF-2 from *Escherichia coli* with skeletal muscle kinase. *J. Biol. Chem.* 247:6365–67

53. Ferro-Luzzi Ames, G., Nikaido, K. 1981. Phosphate-containing proteins of *Salmonella typhimurium* and *Escherichia coli*. Analysis by a new two-dimensional gel system. *Eur. J. Biochem.* 115:525–31

54. Forsberg, H., Zetterqvist, O., Engström, L. 1969. Protein-bound phosphoserine in different issues and organisms. *Biochim. Biophys. Acta* 181:171–75

55. Fow, G. E., Stackebrandt, E., Hespell, R. B., Gibson, J., Maniloff, J., et al. 1980. The phylogeny of prokaryotes. *Science* 209:457–63

56. Fox, D. K., Roseman, S. 1986. Isolation and characterization of homogeneous acetate kinase from *Salmonella typhimurium* and *Escherichia coli*. *J. Biol. Chem.* 261:13487–97

57. Friedman, D. I., Larner, J. 1963. Studies on UDPG-α glucan transglucosylase. III. Interconversion of two forms of muscle UDPG-α glucan transglucosylase by a phosphorylation-dephosphorylation reaction sequence. *Biochemistry* 2:669–75

58. Friedman, D. I., Olson, E. R., Georgopoulos, C., Tilly, K., Herskowitz, I., Banuett, F. 1984. Interactions of bacteriophage and host macromolecules in the growth of bacteriophage λ. *Microbiol. Rev.* 48:299–325

59. Fujitaki, J. M., Smith, R. A. 1984. Techniques in the detection and characterization of phosphoramidate-containing proteins. *Methods Enzymol.* 107:23–36

60. Garland, D., Nimmo, H. G. 1984. A comparison of the phosphorylated and unphosphorylated forms of isocitrate dehydrogenase from *Escherichia coli* ML 308. *FEBS Lett.* 165:259–64

61. Garnak, M., Reeves, H. C. 1978. Phosphorylation of isocitrate dehy-

drogenase of *Escherichia coli. Science* 203:1111–12

62. Garnak, M., Reeves, H. C. 1979. Purification and properties of phosphorylated isocitrate dehydrogenase of *Escherichia coli. J. Biol. Chem.* 254:7915–20

63. Gillies, R. J. 1983. Can enzymes contain two active sites? *Trends Biochem. Sci.* 8:301

64. Goff, C. G. 1974. Chemical structure of a modification of the *Escherichia coli* ribonucleic acid polymerase α polypeptides induced by bacteriophage T4 infection. *J. Biol. Chem.* 249:6181–90

65. Goldbeter, A., Koshland, D. E. Jr. 1984. Ultrasensitivity in biochemical systems controlled by covalent modification. Interplay between zero-order and multistep effects. *J. Biol. Chem.* 259:14441–47

66. Gordon, J. 1971. Determination of an upper limit to the phosphorus content of polypeptide chain elongation factor and ribosomal proteins in *Escherichia coli. Biochem. Biophys. Res. Commun.* 44:579–86

67. Gotto, J. W., Yoch, D. C. 1985. Regulation of nitrogenase activity by covalent modification in *Chromatium vinosum. Arch. Microbiol.* 141:40–43

68. Hengstenberg, W. 1977. Enzymology of carbohydrate transport in bacteria. *Curr. Top. Microbiol. Immunol.* 77:97–126

69. Hodgson, D., Shapiro, L., Amemiya, K. 1985. Phosphorylation of the β' subunit of RNA polymerase and other host proteins upon ϕCd1 infection of *Caulobacter crescentus. J. Virol.* 55:238–41

70. Holmes, N. G., Allen, J. F. 1986. Protein phosphorylation as a control for excitation energy transfer in *Rhodospirillum rubrum. FEBS Lett.* 200:144–48

71. Holms, W. H., Bennett, P. M. 1971. Regulation of isocitrate dehydrogenase activity in *Escherichia coli* on adaptation to acetate. *J. Gen. Microbiol.* 65:57–68

72. Holuigue, L., Lucero, H. A., Vallejos, R. H. 1985. Protein phosphorylation in the photosynthetic bacterium *Rhodospirillum rubrum. FEBS Lett.* 181:103–8

73. Howe, J. G., Hershey, J. W. B. 1982. Immunochemical analysis of molecular forms of protein synthesis initiation factors in crude cell lysates of *Escherichia coli. Arch. Biochem. Biophys.* 214:446–51

74. Huang, T. S., Bylund, D. B., Stull, J. T., Krebs, E. G. 1974. The amino acid sequences of the phosphorylation sites in

troponin-I from rabbit skeletal muscle. *FEBS Lett.* 42:249–52

75. Huebner, V. D., Matthews, H. R. 1985. Phosphorylation of histidine in proteins by a nuclear extract of *Physarum polycephalum* plasmodia. *J. Biol. Chem.* 260:16106–13

76. Hunter, T. 1982. Phosphotyrosine—a new protein modification. *Trends Biochem. Sci.* 7:246–49

77. Hunter, T., Cooper, J. A. 1985. Protein-tyrosine kinases. *Ann. Rev. Biochem.* 54:897–930

78. Hy, M., Burke, W. F., Reeves, H. C. 1978. NADP$^+$-dependent isocitrate dehydrogenase protein kinase in *E. coli. Bacteriol. Abstr.* 1978:146

79. Isawa, Y., Yonemitsu, K., Matsui, K. 1981. Calmodulin-like activity in the soluble fraction of *Escherichia coli. Biochem. Biophys. Res. Commun.* 98:656–60

80. Itikawa, H., Ryu, J. I. 1979. Isolation and characterization of a temperature-sensitive *dnaK* mutant in *Escherichia coli. J. Bacteriol.* 138:339–44

81. Kelley, P. M., Neumann, P. A., Shriefer, K., Cancedda, F., Schlesinger, M. J., Bradshaw, R. A. 1973. Amino acid sequence of *Escherichia coli* alkaline phosphatase. Amino- and carboxyl-terminal sequences and variations between two isozymes. *Biochemistry* 12:3499–503

82. Khandelwal, R. L., Spearman, T. N., Hamilton, I. R. 1973. Protein kinase activity in cariogenic and non-cariogenic oral streptococci: activation and inhibition by cyclic AMP. *FEBS Lett.* 31:246–50

83. Knowles, J. R. 1980. Enzyme-catalyzed phosphoryl transfer reactions. *Ann. Rev. Biochem.* 49:877–919

84. Köhler, E., Antranikian, G. 1988. Covalent modification of proteins in *Bacillus subtilis* during the process of sporulation and germination. *Arch. Microbiol.* In press

85. Komano, T., Brown, N., Inouye, S., Inouye, M. 1982. Phosphorylation and methylation of proteins during *Myxococcus xanthus* spore formation. *J. Bacteriol.* 151:114–18

86. Kornberg, H. 1966. The role and control of the glyoxylate shunt in *Escherichia coli. Biochem. J.* 99:1–11

87. Koshland, D. E. Jr. 1987. Switches, thresholds and ultrasensitivity. *Trends Biochem. Sci.* 12:225–29

88. Krebs, E. G., Beavo, J. A. 1979. Phosphorylation-dephosphorylation of enzymes. *Ann. Rev. Biochem.* 48:923–59

89. Krebs, E. G., Fischer, E. H. 1956. The phosphorylase *b* to *a* converting enzyme of rabbit skeletal muscle. *Biochim. Biophys. Acta* 20:150–57

90. Krebs, E. G., Graves, D. J., Fischer, E. H. 1959. Factors affecting the activity of muscle phosphorylase *b* kinase. *J. Biol. Chem.* 234:2867–73

91. Krieg, N. R., Holt, J. G., eds. 1984. *Bergey's Manual of Systematic Bacteriology.* Baltimore, Md: Williams & Wilkins. 964 pp.

92. Kuo, C. H., August, J. T. 1972. Histone or bacterial basic protein required for replication of bacteriophage RNA. *Nature New Biol.* 237:105–8

93. Kuo, J. F., Greengard, P. 1969. An adenosine 3',5'-monophosphate-dependent protein kinase from *Escherichia coli. J. Biol. Chem.* 244:3417–19

94. Kurek, E., Grankowski, N., Gasior, E. 1972. On the phosphorylation of *Escherichia coli* ribosomes. I. An *in vivo* labeling of ribosomes. *Acta Microbiol. Pol.* 4:171–76

95. Kurek, E., Grankowski, N., Gasior, E. 1972. On the phosphorylation of *Escherichia coli* ribosomes. II. Reaction in cell-free system. *Acta Microbiol. Pol.* 4:177–83

96. Lakshmi, T., Helling, R. 1978. Acetate metabolism in *Escherichia coli. Can. J. Microbiol.* 24:149–53

97. LaPorte, D. C., Chung, T. 1985. A single gene codes for the kinase and phosphatase which regulate isocitrate dehydrogenase. *J. Biol. Chem.* 260:15291–97

98. LaPorte, D. C., Koshland, D. E. Jr. 1982. A protein with kinase and phosphatase activities involved in regulation of tricarboxylic acid cycle. *Nature* 300:458–60

99. LaPorte, D. C., Koshland, D. E. Jr. 1983. Phosphorylation of isocitrate dehydrogenase as a demonstration of enhanced sensitivity in covalent regulation. *Nature* 305:286–90

100. LaPorte, D. C., Thorsness, P. E., Koshland, D. E. Jr. 1985. Compensatory phosphorylation of isocitrate dehydrogenase. A mechanism for adaptation to the intracellular environment. *J. Biol. Chem.* 260:10563–68

101. LaPorte, D. C., Walsh, K., Koshland, D. E. Jr. 1984. The branch point effect. Ultrasensitivity and subsensitivity to metabolic control. *J. Biol. Chem.* 259:14068–75

102. Leadlay, P. F., Roberts, G., Walker, J. E. 1984. Isolation of a novel calcium-binding protein from *Streptomyces erythreus. FEBS Lett.* 178:157–60

103. LeBowitz, J. H., McCaken, R. 1984. The bacteriophage lambda *O* and *P* protein initiators promote the replication of single-stranded DNA. *Nucleic Acids Res.* 12:3069–88

104. LeBowitz, J. H., Zylicz, M., Georgopoulos, C., McCaken, R. 1985. Initiation of DNA replication on single-stranded DNA templates catalyzed by purified replication proteins of bacteriophage lambda and *Escherichia coli. Proc. Natl. Acad. Sci. USA* 82:3988–92

105. Li, H. C., Brown, G. G. 1973. Orthophosphate and histone dependent polyphosphate kinase from *E. coli. Biochem. Biophys. Res. Commun.* 53:875–81

106. Little, J. W., Zimmerman, S. B., Oshinsky, C. K., Gellert, M. 1967. Enzymatic joining of DNA strands. An enzyme-adenylate intermediate in the DPN-dependent DNA ligase reaction. *Proc. Natl. Acad. Sci. USA* 58:2004–11

107. Londesborough, J. 1986. Phosphorylation of proteins in *Clostridium thermohydrosulfuricum. J. Bacteriol.* 165:595–601

108. Markman, R. S., Sutherland, E. Q. 1965. Adenosine 3',5'-phosphate in *Escherichia coli. J. Biol. Chem.* 240:1309–14

109. Malloy, P. J., Reeves, H. C. 1983. Cyclic AMP-independent phosphorylation of *Escherichia coli* isocitrate dehydrogenase. *FEBS Lett.* 151:59–62

110. Malloy, P. J., Reeves, H. C., Spiess, J. 1984. Amino acid sequence of the phosphorylation site of isocitrate dehydrogenase from *Escherichia coli. Curr. Microbiol.* 11:37–42

111. Malloy, P. J., Robertson, E. F., Reeves, H. C. 1985. Purification of the isocitrate dehydrogenase kinase/phosphatase. *Microbiol. Abstr.* 1985:181

112. Maloy, S. R., Nunn, W. D. 1982. Genetic regulation of the glyoxylate shunt in *Escherichia coli* K-12. *J. Bacteriol.* 149:173–80

113. Maloy, S. R., Wilson, R. B. 1987. Isolation and characterization of *Salmonella typhimurium* glyoxylate shunt mutants. *J. Bacteriol.* 169:3029–34

114. Manaï, M., Cozzone, A. J. 1979. Analysis of the protein kinase activity of *Escherichia coli* cells. *Biochem. Biophys. Res. Commun.* 91:819–26

115. Manaï, M., Cozzone, A. J. 1979. Mise en évidence d'une activité protéine-kinase chez *Escherichia coli. C. R. Acad. Sci. Ser. D* 289:367–70

116. Manaï, M., Cozzone, A. J. 1982.

Endogenous protein phosphorylation in *Escherichia coli* extracts. *Biochem. Biophys. Res. Commun.* 107:981–88

117. Manaï, M., Cozzone, A. J. 1982. Two-dimensional separation of phosphoamino acids from nucleoside monophosphates. *Anal. Biochem.* 124:12–18

118. Manaï, M., Duclos, B., Cozzone, A. J. 1984. Inhibitory effect of bacterial extracts on the activity in vitro of eukaryotic histone kinase. *FEMS Microbiol. Lett.* 23:319–23

119. Martelo, O. J., Savio, L. C., Reimann, E. R., Davie, E. W. 1970. Effect of protein kinase on ribonucleic acid polymerase. *Biochemistry* 9:4807–13

120. Martensen, T. M. 1984. Chemical properties, isolation and analysis of O-phosphates in proteins. *Methods Enzymol.* 107:3–23

121. Mayer, J. E., Schweiger, M. 1983. RNase III is positively regulated by T7 protein kinase. *J. Biol. Chem.* 258:5340–43

122. Mimura, C. S., Poy, F., Jacobson, G. R. 1987. ATP-dependent protein kinase activities in the oral pathogen *Streptococcus mutans*. *J. Cell. Biochem.* 33:161–171

123. Narindrasorasak, S., Bridger, W. A. 1977. Phosphoenolpyruvate synthetase of *Escherichia coli*. Molecular weight, subunit composition, and identification of phosphohistidine in phosphoenzyme intermediate. *J. Biol. Chem.* 252:3121–27

124. Neidhardt, F. C., VanBogelen, R. A., Lau, E. T. 1983. Molecular cloning and expression of a gene that controls the high-temperature regulon of *Escherichia coli*. *J. Bacteriol.* 153:597–60

125. Nelson, S. O., Schuitema, A. R. J., Postma, P. W. 1986. The phosphoenolpyruvate:glucose phosphotransferase system of *Salmonella typhimurium*. The phosphorylated form of III^{Glc}. *Eur. J. Biochem.* 154:337–41

126. Nimmo, H. G. 1984. Control of *Escherichia coli* isocitrate dehydrogenase: an example of protein phosphorylation in a prokaryote. *Trends Biochem. Sci.* 9:475–78

127. Nimmo, H. G. 1984. The control of bacterial isocitrate dehydrogenase by phosphorylation. *Mol. Aspects Cell. Regul.* 3:123–41

128. Nimmo, H. G. 1986. Kinetic mechanism of *Escherichia coli* isocitrate dehydrogenase and its inhibition by glyoxylate and oxaloacetate. *Biochem. J.* 234:317–23

129. Nimmo, H. G. 1987. Regulation of bacterial metabolism by protein phosphorylation. *Essays Biochem.* 23:1–27

130. Nimmo, H. G., Borthwick, A. C., El-Mansi, E. M. T., Holms, W. H., MacKintosh, C., Nimmo, G. A. 1988. Regulation of the enzymes at the branchpoint between the citric acid cycle and the glyoxylate bypass in *Escherichia coli*. *Biochem. Soc. Symp.* 54:93–101

131. Nimmo, G. A., Borthwick, A. C., Holms, W. H., Nimmo, H. G. 1984. Partial purification and properties of isocitrate dehydrogenase kinase/phosphatase from *Escherichia coli* ML 308. *Eur. J. Biochem.* 141:401–8

132. Nimmo, G. A., Nimmo, H. G. 1984. The regulatory properties of isocitrate dehydrogenase kinase and isocitrate dehydrogenase phosphatase from *Escherichia coli* ML308 and the roles of these activities in the control of isocitrate dehydrogenase. *Eur. J. Biochem.* 141:409–14

133. Ninfa, A. J., Magasanik, B. 1986. Covalent modification of the *glnG* product, NR$_I$, by the *glnL* product, NR$_{II}$, regulates the transcription of the *glnALG* operon in *Escherichia coli*. *Proc. Natl. Acad. Sci. USA* 83:5909–13

134. Nunn, W. D. 1986. A molecular view of fatty acid catabolism in *Escherichia coli*. *Microbiol. Rev.* 50:179–92

135. O'Farrell, P. H. 1975. High resolution two-dimensional electrophoresis of proteins. *J. Biol. Chem.* 250:4007–21

136. O'Farrell, P. Z., Goodman, H. M., O'Farrell, P. H. 1977. High resolution two-dimensional electrophoresis of basic as well as acidic proteins. *Cell* 12:1133–42

137. Ord, M. G., Stocken, L. A. 1966. Metabolic properties of histones from rat liver and thymus gland. *Biochem. J.* 98:888–97

138. Oshima, T. 1975. Thermine: a new polyamine from an extreme thermophile. *Biochem. Biophys. Res. Commun.* 63:1093–98

139. Oshima, T. 1982. A pentamine is present in an extreme thermophile. *J. Biol. Chem.* 257:9913–14

140. Paek, K. H., Walker, G. C. 1987. *Escherichia coli dnaK* null mutants are inviable at high temperature. *J. Bacteriol.* 169:283–90

141. Pai, S. H., Ponta, H., Rahmsdorf, H. J., Hirsch-Kauffmann, M., Herrlich, P., Schweiger, M. 1975. Protein kinase of bacteriophage T7. 1. Purification. *Eur. J. Biochem.* 55:299–304

142. Pai, S. H., Rahmsdorf, H. J., Ponta, H.,

Hirsch-Kauffmann, M., Herrlich, P., Schweiger, M. 1975. Protein kinase of bacteriophage T7. 2. Properties, enzyme synthesis in vitro and regulation of enzyme synthesis and activity in vivo. *Eur. J. Biochem.* 55:305–14

143. Pastan, I., Adhya, S. 1976. Cyclic adenosine 5'-monophosphate in *Escherichia coli. Bacteriol. Rev.* 40:527–51

144. Philipps, T. A., Bloch, P. L., Neidhardt, F. C. 1980. Protein identifications on O'Farrell two-dimensional gels: location of 55 additional *Escherichia coli* proteins. *J. Bacteriol.* 144:1024–33

145. Pinna, L. A., Agostinis, P., Ferrari, S. 1986. Selectivity of protein kinases and protein phosphatases: a comparative analysis. *Adv. Protein Phosphatases* 3:327–68

146. Pinna, L. A., Meggio, F., Donella-Deana, A. 1980. Structure of the sites of substrate proteins undergoing phosphorylation by protein kinases, with special reference to liver casein kinases. See Ref. 185, pp. 8–16

147. Ponta, H., Rahmsdorf, H. J., Pai, S. H., Hirsch-Kauffmann, M., Herrlich, P., Schweiger, M. 1974. Control of gene expression in bacteriophage T7: transcriptional controls. *Mol. Gen. Genet.* 134:281–87

148. Postma, P. W., Lengeler, J. W. 1985. Phosphoenolpyruvate:carbohydrate phosphotransferase system of bacteria. *Microbiol. Rev.* 49:232–69

149. Powers, D. M., Ginsburg, A. 1973. Protein kinase activity in *Escherichia coli.* In *Metabolic Interconversions of Enzymes,* ed. E. H. Fischer, E. G. Krebs, E. R. Stadtman, pp. 131–43. New York: Springer-Verlag

150. Powers, D. A., Roseman, S. 1984. The primary structure of *Salmonella typhimurium* HPr, a phosphocarrier protein of the phosphoenolpyruvate:glycose phosphotransferase system. A correction. *J. Biol. Chem.* 259:15212–14

151. Quentmeier, A., Antranikian, G. 1986. Covalent modification of proteins in *Escherichia coli* growing anaerobically with nitrate as electron acceptor. *FEMS Microbiol. Lett.* 34:231–35

152. Rahmsdorf, H. J., Herrlich, P., Pai, S. H., Schweiger, M., Wittmann, H. G. 1973. Ribosomes after infection with bacteriophage T4 and T7. *Mol. Gen. Genet.* 127:259–71

153. Rahmsdorf, H. J., Pai, S. H., Ponta, H., Herrlich, P., Roskoski, R. Jr. et al. 1974. Protein kinase induction in *Escherichia coli* by bacteriophage T7.

Proc. Natl. Acad. Sci. USA 71:586–89

154. Rask, L., Walinder, O., Zetterqvist, O., Engström, L. 1970. Protein-bound phosphorylthreonine in some tissues and organisms. *Biochim. Biophys. Acta* 221:107–13

155. Reeves, H. C., Malloy, P. J. 1983. Phosphorylation of isocitrate dehydrogenase in *Escherichia coli* mutants with a non-functional glyoxylate cycle. *FEBS Lett.* 158:239–42

156. Reeves, H. C., O'Neil, S., Weitzman, P. D. J. 1983. Modulation of isocitrate dehydrogenase activity in *Acinetobacter calcoaceticus* by acetate. *FEBS Lett.* 163:265–68

157. Reizer, J., Deutscher, J., Grenier, F., Thompson, J., Hengstenberg, W., Saier, M. H. Jr. 1988. The phosphoenolpyruvate:sugar phosphotransferase system in gram-positive bacteria: properties, mechanism and regulation. *CRC Crit. Rev. Microbiol.* In press

158. Reizer, J., Deutscher, J., Sutrina, S., Thompson, J., Saier, M. H. Jr. 1985. Sugar accumulation in gram-positive bacteria : exclusion and expulsion mechanisms. *Trends Biochem. Sci.* 10:32–35

159. Reizer, J., Novotny, M. J., Hengstenberg, W., Saier, M. H. Jr. 1984. Properties of ATP-dependent protein kinase from *Streptococcus pyogenes* that phosphorylates a seryl residue in HPr, a phosphocarrier protein of the phosphotransferase system. *J. Bacteriol.* 160:333–40

160. Reizer, J., Novotny, M. J., Panos, C., Saier, M. H. Jr. 1983. Mechanism of inducer expulsion in *Streptococcus pyogenes*: a two-step process activated by ATP. *J. Bacteriol.* 156:354–61

161. Reizer, J., Panos, C. 1980. Regulation of β-galactoside phosphate accumulation in *Streptococcus pyogenes* by an expulsion mechanism. *Proc. Natl. Acad. Sci. USA* 77:5497–501

162. Reizer, J., Saier, M. H. Jr. 1983. Involvement of lactose enzyme II of the phosphotransferase system in rapid expulsion of free galactosides from *Streptococcus pyogenes. J. Bacteriol.* 156:236–42

163. Rhee, S. G., Park, R., Chock, P. B., Stadtman, E. R. 1978. Allosteric regulation of monocyclic interconvertible enzyme cascade systems: use of *Escherichia coli* glutamine synthetase as an experimental model. *Proc. Natl. Acad. Sci. USA* 75:3138–42

164. Rhee, S. G., Park, S. C., Koo, J. H. 1985. The role of adenylyltransferase and

uridylyltransferase in the regulation of glutamine synthetase in *Escherichia coli. Curr. Top. Cell. Regul.* 27:221–32

165. Rieul, C., Cortay, J. C., Bleicher, F., Cozzone, A. J. 1987. Effect of bacteriophage M13 infection on phosphorylation of *dnaK* protein and other *Escherichia coli* proteins. *Eur. J. Biochem.* 168:621–27

166. Roach, P. J. 1984. Protein kinases. *Methods Enzymol.* 107:81–101

167. Robertson, E. F., Hoyt, J. C., Reeves, H. C. 1987. In vitro phosphorylation of *Escherichia coli* isocitrate lyase. *Curr. Microbiol.* 15:103–5

168. Robertson, E. F., Reeves, H. C. 1987. Purification and characterization of isocitrate lyase from *Escherichia coli. Curr. Microbiol.* 14:347–50

169. Robillard, G. T. 1982. The enzymology of the bacterial phosphoenolpyruvate-dependent sugar transport system. *Mol. Cell. Biochem.* 46:3–24

170. Rosen, O. M., Krebs, E. G., eds. 1981. *Protein Phosphorylation.* Cold Spring Harbor, NY: Cold Spring Harbor Lab.

171. Rothberg, P. G., Harris, T. J. R., Nomoto, A., Wimmer, E. 1978. O^4-(5'-uridylyl)tyrosine is the bond between the genome-linked protein and the RNA of poliovirus. *Proc. Natl. Acad. Sci. USA* 75:4868–72

172. Rubin, C. S., Rosen, O. M. 1975. Protein phosphorylation. *Ann. Rev. Biochem.* 44:831–87

173. Saito, H., Uchida, H. 1977. Initiation of the DNA replication of bacteriophage lambda in *Escherichia coli* K-12. *J. Mol. Biol.* 113:1–25

174. Schwartz, J. H., Lipmann, F. 1961. Phosphate incorporation into alkaline phosphatase of *E. coli. Proc. Natl. Acad. Sci. USA* 47:1996–2005

175. Sefton, B. M., Hunter, T., Beemon, K., Eckhart, W. 1980. Evidence that the phosphorylation of tyrosine is essential for cellular transformation by Rous sarcoma virus. *Cell* 20:807–16

176. Seifert, W., Qasba, P., Walter, G., Palm, P., Schachner, M., Zillig, W. 1969. Kinetics of the alteration and modification of DNA-dependent RNA polymerase in T4-infected *E. coli* cells. *Eur. J. Biochem.* 9:319–24

177. Shapiro, B. M., Stadtman, E. R. 1968. 5'-adenylyl-*O*-tyrosine. The novel phosphodiester residue of adenylylated glutamine synthetase from *Escherichia coli. J. Biol. Chem.* 243:3769–71

178. Shenolikar, S., Ingebritsen, T. S. 1984. Protein (serine and threonine) phosphate phosphatases. *Methods Enzymol.* 107:102–29

179. Simoni, R. D., Hays, J. B., Nakazawa, T., Roseman, S. 1973. Sugar transport. VI. Phosphoryl transfer in the lactose phosphotransferase system of *Staphylococcus aureus. J. Biol. Chem.* 248:957–65

180. Skorko, R. 1984. Protein phosphorylation in the archaebacterium *Sulfolobus acidocaldarius. Eur. J. Biochem.* 145:617–22

181. Sparks, J. W., Brautigan, D. L. 1986. Molecular basis for substrate specificity of protein kinases and phosphatases. *Int. J. Biochem.* 18:497–504

182. Spudich, E. N., Spudich, J. L. 1982. Measurement of light-regulated phosphoproteins of *Halobacterium halobium. Methods Enzymol.* 88:213–16

183. Spudich, J. L., Stoeckenius, W. 1980. Light-regulated retinal-dependent reversible phosphorylation of *Halobacterium* proteins. *J. Biol. Chem.* 255:5501–3

184. Strand, M., August, J. T. 1971. Protein kinase and phosphate acceptor proteins in Rauscher murine leukaemia virus. *Nature New Biol.* 233:137–40

185. Thomas, G., Podesta, E. J., Gordon, J., eds. 1980. *Protein Phosphorylation and Bio-Regulation.* Basel: Karger. 232 pp.

186. Thompson, J., Saier, M. H. Jr. 1981. Regulation of methyl-β-D-thiogalactopyranoside 6-phosphate accumulation in *Streptococcus lactis* by exclusion and expulsion mechanisms. *J. Bacteriol.* 146:885–94

187. Thorsness, P. E., Koshland, D. E. Jr. 1987. Inactivation of isocitrate dehydrogenase by phosphorylation is mediated by the negative charge of the phosphate. *J. Biol. Chem.* 262:10422–25

188. Tilly, K., McKittrick, N., Zylicz, M., Georgopoulos, C. 1983. The *dnaK* protein modulates the heat-shock response of *Escherichia coli. Cell* 34:641–46

189. Traugh, J. A., Traut, R. R. 1972. Phosphorylation of ribosomal proteins of *Escherichia coli* by protein kinase from rabbit skeletal muscle. *Biochemistry* 11:2503–9

190. Tse, Y. C., Kirkegaard, K., Wang, J. C. 1980. Covalent bonds between protein and DNA. Formation of phosphotyrosine linkage between certain DNA topoisomerases and DNA. *J. Biol. Chem.* 255:5560–65

191. Turner, A. M., Mann, N. H. 1986. Protein phosphorylation in *Rhodomicrobium vannielii. J. Gen. Microbiol.* 132:3433–40

192. Vallejos, R. H., Holuigue, L., Lucero,

H. A., Torruella, M. 1985. Evidence of tyrosine kinase activity in the photosynthetic bacterium *Rhodospirillum rubrum. Biochem. Biophys. Res. Commun.* 126:685–91

193. Vanderwinkel, E., DeVlieghere, M. 1968. Physiologie et génétique de l'isocitritase et des malate synthases chez *Escherichia coli. Eur. J. Biochem.* 5:81–90

194. Vanderwinkel, E., Liard, P., Ramos, F., Wiame, J. M. 1963. Genetic control of the regulation of isocitritase and malate synthase in *Escherichia coli* K-12. *Biochem. Biophys. Res. Commun.* 12:157–62

195. Vasquez, B., Reeves, H. C. 1979. NADP-specific isocitrate dehydrogenase of *Escherichia coli.* Purification by chromatography on Affi-gel blue. *Biochim. Biophys. Acta* 578:31–40

196. Volz, K. W., Mathews, D. A., Alden, R. A., Freer, S. T., Hansch, C., et al. 1982. Crystal structure of avian dihydrofolate reductase containing phenyltriazine and NADPH. *J. Biol. Chem.* 257:2528–36

197. Wada, M., Sekine, K., Itikawa, H. 1986. Participation of the *dnaK* and *dnaJ* gene products in phosphorylation of glutaminyl-tRNA synthetase and threonyl-tRNA synthetase of *Escherichia coli* K12. *J. Bacteriol.* 168:213–20

198. Walinder, O. 1968. Identification of a phosphate-incorporating protein from bovine liver as nucleoside diphosphate kinase and isolation of 1-^{32}P-phosphohistidine, 3-^{32}P-phosphohistidine, and *N*-ϵ-^{32}P-phospholysine from erythrocytic nucleoside diphosphate kinase, incubated with adenosine triphosphate-^{32}P. *J. Biol. Chem.* 243:3947–52

199. Walsh, C. T. Jr., Spector, L. B. 1971. A phosphoenzyme intermediary in phosphoglycerate kinase action. *J. Biol. Chem.* 246:1255–61

200. Walsh, D. A., Perkins, J. P., Krebs, E. G. 1968. An adenosine 3',5'-monophosphate-dependent protein kinase from rabbit skeletal muscle. *J. Biol. Chem.* 243:3763–74

201. Walsh, K., Koshland, D. E. Jr. 1985. Branch point control by the phosphorylation state of isocitrate dehydrogenase. *J. Biol. Chem.* 260:8430–37

202. Wang, J. Y. J., Koshland, D. E. Jr. 1978. Evidence for protein kinase activities in the prokaryote *Salmonella typhimurium. J. Biol. Chem.* 253:7605–8

203. Wang, J. Y. J., Koshland, D. E. Jr. 1981. The identification of distinct protein kinases and phosphatases in the prokaryote *Salmonella typhimurium. J. Biol. Chem.* 256:4640–48

204. Wang, J. Y. J., Koshland, D. E. Jr. 1982. The reversible phosphorylation of isocitrate dehydrogenase of *Salmonella typhimurium. Arch. Biochem. Biophys.* 218:59–67

205. Waygood, E. B., Mattoo, R. L., Erickson, E., Vadeboncoeur, C. 1986. Phosphoproteins and the phosphoenolpyruvate:sugar phosphotransferase system of *Streptococcus salivarius.* Detection of two different ATP-dependent phosphorylations of the phosphocarrier protein HPr. *Can. J. Microbiol.* 32:310–18

206. Waygood, E. B., Reiche, B., Hengstenberg, W., Lee, J. S. 1987. Characterization of mutant histidine-containing proteins of the phosphoenolpyruvate:sugar phosphotransferase system of *Escherichia coli* and *Salmonella typhimurium. J. Bacteriol.* 169:2810–18

207. Weigel, N., Powers, A. D., Roseman, S. 1982. Sugar transport by the bacterial phosphotransferase system. Primary structure and active site of a general phosphocarrier protein (HPr) from *Salmonella typhimurium. J. Biol. Chem.* 257:14499–509

208. Wilkinson, J. M., Grand, R. J. A. 1975. The amino acid sequence of troponin I from rabbit skeletal muscle. *Biochem. J.* 149:493–96

209. Wold, F. 1981. In vivo chemical modification of proteins (post-translational modification). *Ann. Rev. Biochem.* 50:783–814

210. Yang, J. C., Fujitaki, J. M., Smith, R. A. 1982. Separation of phosphohydroxyamino acids by high-performance liquid chromatography. *Anal. Biochem.* 122:360–63

211. Zillig, W., Fujiki, H., Blum, W., Janekovic, D., Schweiger, M., et al. 1975. In vivo and in vitro phosphorylation of DNA-dependent RNA polymerase of *Escherichia coli* by bacteriophage-T7–induced protein kinase. *Proc. Natl. Acad. Sci. USA* 72:2506–10

212. Zylicz, M., Georgopoulos, C. 1984. Purification and properties of the *Escherichia coli dnaK* replication protein. *J. Biol. Chem.* 259:8820–25

213. Zylicz, M., LeBowitz, J. H., McCaken, R., Georgopoulos, C. 1983. The *dnaK* protein of *Escherichia coli* possesses an ATPase and autophosphorylating activity and is essential in an in vitro DNA replication system. *Proc. Natl. Acad. Sci. USA* 80:6431–35

Ann. Rev. Microbiol. 1988. 42:127–50

GENETIC DETERMINANTS OF
SHIGELLA PATHOGENICITY[1]

Anthony T. Maurelli[2]

Department of Microbiology, Uniformed Services University of the Health Sciences, Bethesda, Maryland 20814-4799

Philippe J. Sansonetti

Service des Entérobactéries, Unité 199, Institut National de la Santé et de la Recherche Médicale, Institut Pasteur, 75724 Paris Cedex 15, France

CONTENTS

[1]The US Government has the right to retain a nonexclusive, royalty-free license in and to any copyright covering this paper.

[2]The opinions or assertions of ATM contained herein are the private ones of the author and are not to be construed as official or reflecting the views of the Department of Defense or the Uniformed Services University of the Health Sciences.

INTRODUCTION

Although endemic throughout the world, shigellosis or bacillary dysentery is of special concern in developing areas, where poor sanitation and low hygiene standards account for a large number of cases. Children are the major target of this enteric disease. Prevalent serotypes in these areas belong to the *Shigella flexneri* species, while *Shigella sonnei*, *Shigella boydii*, and enteroinvasive *Escherichia coli* (EIEC) are occasionally isolated. *Shigella dysenteriae* 1 (Shiga bacillus) is rarely encountered but may cause devastating epidemics. *S. sonnei* is prevalent in western countries. The disease ranges from a mild diarrhea to a severe dysenteric syndrome with blood, mucus, and pus in stools.

The essential step in the pathogenesis of shigellosis is invasion of the human colonic mucosa (49). The invasive process encompasses complex features that include penetration into epithelial cells, intracellular multiplication, and spreading to adjacent cells and to the connective tissue of intestinal villi. These events lead to a strong inflammatory reaction, which causes abscesses and ulcerations of the colon. Bacillary dysentery does not usually affect the small intestine (15). Although severe, the infection process is limited to the mucosal surface and does not spread significantly from the lamina propria to the submucosa (75, 76, 98). The intensity of the inflammatory reaction may prevent systemic dissemination of the pathogen, thus accounting for the low frequency of bacteremia. Therefore, the overall invasive process can be considered as the integration of two steps, invasion of the individual epithelial cells and invasion of connective tissue. In the tissue, the infection process may encompass both an extracellular stage and an intracellular stage within phagocytes.

Although dysentery is particularly characteristic of shigellosis, watery diarrhea is also common and often precedes dysentery. Studies of intestinal perfusions performed on monkeys infected with shigellae have shown that transport abnormalities occur within the colon and correlate with the degree of bacterial invasion (75). However, fluid secretion is also observed at the level of the jejunum without evidence of invasion, and monkeys challenged intracecally develop only dysentery (46). Such data indicate that dysentery is purely colonic, whereas diarrhea is the result of a jejunal secretion that cannot be reabsorbed owing to colonic abnormalities. Shiga toxin, which is produced at high levels by *S. dysenteriae* 1, is both a potent cytotoxin and an enterotoxin (16) and is thus a likely candidate for enterotoxicity. Isolates of *S. sonnei* and *S. flexneri* produce low levels of a Shiga-like toxin (SLT) (45, 65, 66), which may also account for diarrhea. Shiga toxin and SLT are the subject of a recent review (63).

Several laboratory models are used to assess different aspects of *Shigella* virulence (19). Virulent microorganisms invade cultured mammalian cell monolayers by a process that appears similar to invasion of intestinal epithelial cells in vivo (49). This is the simplest system for the study of host cell–bacteria relationships and the least stringent, requiring only that the bacteria be capable of invasion (33, 37). Epithelial and nonepithelial cell lines (e.g. HeLa cells) are commonly used. A modification of the standard tissue culture assay allows quantitation of the microorganisms' capacity to invade cells, multiply intracellularly, and spread to contiguous cells (62). Under the assay conditions shigellae cause a cytopathic effect on confluent monolayers and form clear plaques in the monolayer. The Serény test (92) assesses the virulence of *Shigella* by measuring the ability of the bacteria to cause a keratoconjunctivitis in guinea pigs or rabbits. The efficient invasion and multiplication of shigellae within the corneal epithelium (72) mimics their behavior within intestinal epithelium.

The ligated rabbit ileal loop is a more definitive assay for intestinal virulence in which invasive organisms are introduced into the lumen (23). In this model, qualitative and quantitative evaluation of the invasive process can be performed by examination of fluid accumulation, the nature of the fluid (presence of blood, pus, and mucus), and the appearance of mucosal surfaces. Microscopic examination can also be performed. Invasive microorganisms and their effects can be visualized by standard biological staining, immunofluorescence, and electron microscopy. In contrast with tissue culture assays, animal models also permit exploration of the capacity of invasive bacteria to survive host-cell defenses such as attack by phagocytic cells and the bactericidal activity of serum.

This review emphasizes the genetic and molecular basis of invasiveness. Interactions of shigellae with individual eukaryotic cells and tissue infection are particularly considered.

OVERVIEW OF EARLY WORK

Studies undertaken at the Walter Reed Army Institute of Research in the 1960s took advantage of the close genetic relatedness between *E. coli* K-12 and *Shigella* to provide our first insights into the genes involved in *Shigella* virulence. Genetic material from an *E. coli* K-12 Hfr donor was transferred into *S. flexneri* 2a via conjugation, and the effect on virulence properties was determined. Transfer of approximately 50% of the *E. coli* chromosome into *Shigella* had no effect on virulence. A transconjugant that had inherited the *xyl-rha* region of the *E. coli* chromosome expressed a reduced ability to multiply intracellularly (18). This transconjugant proved to be Serény test positive and was capable of penetration of intestinal epithelium. However, the organism failed to cause a fatal infection in starved, opiated guinea pigs (24) and did not cause disease when fed to rhesus monkeys (25). Thus the ability to multiply intracellularly after penetration was established as an essential step in *Shigella* pathogenicity, and the *xyl-rha* region was implicated in expression of this phenotype.

Hfr matings using *E. coli* as a donor revealed a second chromosomal locus that is essential for *Shigella* virulence. *S. flexneri* 2a recipients that had inherited the *lac-gal* region from the *E. coli* donor lost the ability to provoke keratoconjunctivitis in guinea pigs. Transductional mapping showed the genetic locus, termed *kcpA* for keratoconjunctivitis provocation, to be cotransducible with *purE* (21).

A third chromosomal region implicated in *Shigella* virulence was similarly identified by screening His⁺ hybrids following a mating between an *E. coli* K-12 Hfr donor and an *S. flexneri* 2a recipient. The recombinants lacked the group- and type-specific *S. flexneri* somatic antigens and could not invade the intestinal mucosa of experimental animals or produce a positive Serény test (22). Thus the genes controlling *Shigella* O-antigen biosynthesis map close to *his,* a chromosomal region analogous to the site of lipopolysaccharide (LPS) biosynthetic genes *(rfb)* in *E. coli* and *Salmonella.* A gene(s) mapping near *pro* is responsible for the type specificity of the O-antigen of *S. flexneri* (20). As we discuss later, some species of *Shigella* also employ plasmid-encoded genes for the synthesis of their particular O-antigens.

An *S. flexneri* Hfr donor was used to demonstrate that *Shigella* genes could be transferred and stably maintained and expressed in *E. coli* hybrids (89). In addition to confirming the similar gene order of the *E. coli* and *Shigella* chromosome maps, these experiments made possible the transfer of genes in both directions between the two organisms. Matings using an *S. flexneri* Hfr donor to transfer virulence determinants into *E. coli* K-12 were undertaken as part of a program to develop live oral vaccines against shigellosis. Unfortunately, none of the recombinants had any protective value as vaccines. Even more interesting was the fact that all attempts to confer *Shigella* virulence properties on a laboratory strain of *E. coli* K-12 by classical gene transfer techniques failed (22, 26). This result underscored the fact that the pathogenicity of *Shigella* involves a number of different genes dispersed around the chromosome. What was not known at the time was that a plasmid has a key role in *Shigella* virulence.

INTERACTION WITH INDIVIDUAL EPITHELIAL CELLS

Role of the Virulence Plasmid

The irreversible generation of noninvasive variants of *Shigella* isolates upon subculture led to the idea that plasmid-borne genes were responsible for the invasive phenotype. Initial observations were made in *S. sonnei.* A 180-

kilobase (kb) plasmid present in invasive Form I strains was absent from noninvasive Form II strains (47, 79). This plasmid encodes the Form I somatic antigen and when reintroduced into Form II bacteria restores both the invasive phenotype and the somatic specificity (47, 83). A 220-kb plasmid was then shown to be involved in the invasive ability of *S. flexneri* (84). This observation was extended to other species of *Shigella* (80) as well as to EIEC strains (40, 81). The large virulence plasmids are functionally interchangeable among these different species (81). Although their restriction endonuclease patterns vary considerably from one serotype to another, hybridization experiments indicate a high level of homology (39, 80). Therefore, these virulence plasmids are probably descended from a common ancestor. This view has recently been reinforced by replicon typing studies in which hybridization with specific probes demonstrated that all virulence plasmids tested belonged to incompatibility group FII (R. M. Silva, S. Saudi, W. K. Maas, submitted for publication). The major role of the virulence plasmid in controlling invasion of eukaryotic cells by shigellae is emphasized by the results of experiments in which the Tn5-labeled virulence plasmid pWR110 of *S. flexneri* serotype 5 was mobilized into *E. coli* K-12. Transconjugants carrying pWR110 expressed high invasive potential for HeLa cells (82).

Regulation of the Invasive Phenotype by Growth Temperature

Enteric pathogens such as *Shigella* undergo a dramatic shift in life-style in passing from the outside environment to their mammalian host. To survive passage through the gastrointestinal tract to the colon, to invade epithelial cells, and to kill these cells, *Shigella* most likely needs specialized functions that it does not require when it is outside the host. It would be prudent for a bacterium to economize its protein-synthesizing energies and regulate production of gene products essential for virulence until they are required by the bacterium. In addition, the bacterium should have a system to coordinate expression of the widely scattered genes essential for the virulent phenotype. *Shigella* seems to have solved these two regulatory problems by using temperature as a signal for controlling expression of its virulence genes. The virulent phenotype of *Shigella* spp. is temperature regulated such that strains grown at 37°C are fully invasive and penetrate mammalian cells, while the same strains when grown at 30°C fail to invade the target cells (55). This loss of invasive capacity is reversible. Full virulence is restored after the growth temperature is shifted up to 37°C and the bacteria are permitted to synthesize proteins and to continue to grow. Therefore one or more virulence genes of *Shigella* are subject to regulation by growth temperature. The temperature response is regulated at the level of transcription, as Maurelli & Curtiss (57) demonstrated by fusing a plasmid-encoded virulence gene with the gene for β-galactosidase, creating a *vir*::*lacZ* operon fusion. In this fusion, expression of the *lacZ* gene was driven by the promoter of the virulence gene. Thus β-galactosidase levels reflected the temperature regulation seen for the virulent phenotype, i.e. they were high after growth at 37°C and greatly reduced after growth at 30°C. This fusion was exploited to isolate mutants defective in temperature regulation of virulence-gene expression. One such regulatory mutant, derived by Tn*10* insertion mutagenesis, had no temperature control and constitutively expressed the virulent phenotype at both 30 and 37°C (58). The mutation was not on the 220-kb plasmid, but mapped to the chromosome close to *galU*. The inactivated gene, called *virR*, is postulated to encode a repressor of virulence-gene expression. The wild-type gene has been cloned and was shown to complement the mutant phenotype in *trans* (58).

The regulation of *Shigella* virulence by growth temperature has enabled us to define a virulence regulon—a network of diverse, unlinked genes that share a common regulatory signal. Among the phenotypes associated with virulence, the ability to invade HeLa cells, the production of a positive Serény test, the binding of Congo red (55), the production of contact-mediated hemolysin (86), and the expression of several invasive plasmid antigens [*ipaA, B, C,* and *D* (38, 54)] are all regulated by growth temperature and controlled by *virR*. Further discussion of these individual virulence-associated properties is found in subsequent sections of this review.

Strategies for Studying the 220-kb Virulence Plasmid

After the discovery of virulence-associated plasmids in *Shigella* spp. and EIEC, concerted efforts in several laboratories were directed at identifying the plasmid genes responsible for virulence. These efforts followed two genetic approaches: cloning and transposon mutagenesis. It was hoped that by generating transposon insertions that abolished the invasive capacity of the organism, plasmid genes required for penetration could be identified and the corresponding sequences in the wild-type strain could be cloned. Unfortunately, this tactic did not yield cloned sequences that could restore invasiveness to a *Shigella* strain that had been cured of the virulence plasmid, because multiple, unlinked genes were involved in the penetration process (54). To maximize the likelihood of cloning all of the required genes on a single large fragment of virulence-plasmid DNA, a cosmid cloning strategy was employed. This resulted in the cloning of a plasmid sequence of about 37 kb (Figure 1), which was sufficient to enable a plasmidless mutant of *S. flexneri* to invade HeLa cells (54). The cosmid clone, pHS4108, also expressed the invasive phenotype in a temperature-regulated manner. However, for reasons to be discussed later, this clone did not produce a positive Serény test or create plaques in confluent monolayers of HeLa cells. Maurelli et al (54), concluded that pHS4108 carried the genes essential for penetration of mammalian cells but lacked a function(s) required for efficient intracellular multiplication.

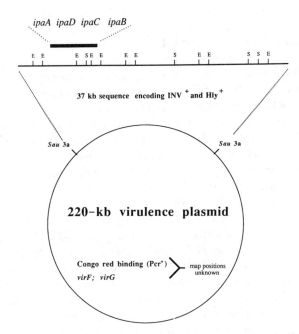

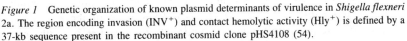

Figure 1   Genetic organization of known plasmid determinants of virulence in *Shigella flexneri* 2a. The region encoding invasion (INV$^+$) and contact hemolytic activity (Hly$^+$) is defined by a 37-kb sequence present in the recombinant cosmid clone pHS4108 (54).

Transposon mutagenesis of this cloned fragment defined at least five noncontiguous regions spanning 20 kb in which insertions abolished or altered the invasive phenotype (1, 54). This suggests that at least five nonlinked gene loci contribute to epithelial cell penetration.

The virulence plasmid of *S. sonnei* was also analyzed by transposon mutagenesis for regions essential for invasion, and four separate *Hind*III fragments were implicated (102). A restriction map of about 20 kb of plasmid sequences in this region showed a striking similarity to the right-hand side of the restriction map of the *S. flexneri* virulence plasmid cloned in pHS4108 (Figure 1) (1, 102). This similarity provided further confirmation of the genetic relatedness of the virulence plasmids among *Shigella* species and supports the argument that the plasmids arose from a common ancestor.

Sasakawa et al (87) constructed a deletion map of a virulence plasmid from *S. flexneri* 2a by analyzing *Sal*I digests of 39 deletion derivatives. The measured virulence properties of over 300 independent Tn5 insertion mutants defined three regions of virulence genes: *Sal*I fragment G; *Sal*I fragment F; and a group of four contiguous *Sal*I fragments (B, P, H, and D) (88). Insertions in *Sal*I fragment G were unusual in that they caused a negative mouse Serény test but did not alter invasiveness for monkey epithelial cells in culture or Congo red binding. Insertions in fragments B, P, H, D, and F resulted in a pleiotropic loss of Congo red binding ability and invasiveness for both the mouse Serény test and cells in tissue culture. Preliminary results indicated that the insertions in fragments B, P, D, and H lay within a region covering 33 kb of the four contiguous fragments. Yet this region is separated by more than 30 kb from fragment F, which is also required for invasion. Thus one would assume that a minimum contiguous region of 63 kb of plasmid DNA is necessary for invasion in tissue culture cells. However, as we have already indicated, a 37-kb fragment of DNA from the virulence plasmid of S. flexneri 5 was sufficient to encode the invasive phenotype (54). The reason for this discrepancy is unknown, but it may be related to sequence divergence among strains of S. flexneri.

Identification of specific virulence plasmid–encoded peptides was first undertaken by examining minicell-producing derivatives of S. flexneri 2a. These anucleate bacterial cells, containing only plasmid DNA, retained the capability of penetrating HeLa cells (39). Sodium dodecyl sulfate–polyacrylamide gel electrophoresis (SDS-PAGE) of the polypeptides expressed in outer membranes of minicells isolated from invasive strains of S. flexneri revealed a complement of 9–16 polypeptides ranging in size from 12 to 64 kd (39). Seven polypeptides, designated a–g, were identified by two-dimensional gel electrophoresis as unique to virulence plasmids of S. flexneri 2a and 5 and EIEC serogroup O143 (38). Expression of these virulence plasmid–specific polypeptides was also partially repressed at 30°C, the non-permissive temperature for the invasive phenotype. Polypeptides a and b were not detected at all after growth at 30°C, while c–g were weakly expressed. Polypeptides a–d were also expressed in a temperature-regulated manner by strains containing the cosmid clone pHS4108 (54). Their sizes were estimated to be 78, 62, 43, and 38 kd, respectively. Several of the virulence plasmid–specific polypeptides displayed extremely basic isoelectric points, which suggests that they may act as regulatory proteins capable of binding to DNA or as anchorage or transport proteins in the outer membrane. Some of the plasmid-encoded polypeptides necessary for the invasive phenotype may not be detectable in such minicell experiments. Conversely, not all of the polypeptides detected are necessarily involved in virulence, as later studies revealed.

Western blot analysis using rabbit antisera raised against individual plasmid-encoded polypeptides or serum from a monkey that had been infected with shigellae established the immunological similarity of polypeptides b–d among strains of Shigella and EIEC (38). Although a discussion of the serum immune response to Shigella infection is beyond the scope of this review (for a review of attempts at developing anti-Shigella vaccines, see Reference 26), we note that polypeptides a–d, as well as an additional plasmid-encoded polypeptide of 140 kd, are predominant *Shigella* antigens recognized in immunoblots by most convalescent human and monkey sera (61). That these plasmid-encoded polypeptides are the principal antigens that induce a serum immune response during *Shigella* infection does not necessarily implicate them as essential virulence antigens, but this property has been useful in studies of these polypeptides.

The genetic organization of the immunogenic peptides a–d has recently been determined by two groups. Buysse et al (5) constructed a λgt11 expression library from *S. flexneri* 5 strain M90T carrying the Tn5-tagged virulence plasmid pWR110. They used rabbit antisera specific for the immunogenic polypeptide antigens b, c, and d to screen for recombinant phages expressing these antigens. In this way they cloned invasion plasmid antigen *(ipa)* genes *ipaB, C,* and *D* and identified a fourth gene, *ipaH.* The protein product of this last gene had a molecular mass similar to that of IpaB (58 kd) but was immunologically distinct. Southern blot hybridization of various recombinants revealed that three *ipa* genes mapped to contiguous *Hind*III fragments in the order *ipaBCD.* The restriction map of this region of the virulence plasmid was identical to the left-hand end of the cosmid clone pHS4108 (Figure 1), with the exception of a single missing *Bam*HI site. The location of *ipaH* was not determined. Buysse et al (5) did not clone the gene encoding polypeptide a because they lacked specific antiserum for screening for such a clone. However, strain M90T carrying pWR110 was both invasive in HeLa cells and positive in the Serény test (82), yet it did not synthesize detectable levels of this 78-kd polypeptide. This suggested that polypeptide a may not be essential for invasion.

Baudry et al (1) generated Tn5 insertions in the cosmid clone pHS4108 and analyzed mutants that were altered or blocked in invasion of HeLa cells for expression of immunogenic peptides a–d. The results of immunoblots using convalescent monkey antiserum showed that insertions that reduced the expression of polypeptides a, b, and d greatly decreased the invasive ability of the organism. The role of polypeptide c in invasion could not be determined because no Tn5 insertions that abolished expression of c were obtained. Two insertions that altered expression of polypeptide a had no effect on invasive potential, which confirmed that polypeptide a was not essential for invasion of HeLa cells. Two-dimensional gel electrophoresis revealed that one of the Tn5 mutants with reduced invasive ability no longer expressed a 21-kd polypeptide, which was tentatively identified as the polypeptide g described earlier (38). Subcloning of fragments that expressed polypeptides a–d and mapping of the Tn5 insertions within pHS4108 permitted Baudry et al (1) to deduce a gene order of *ipaBCDA* from right to left on the map of pHS4108 (Figure 1). The gene order and location within pHS4108 are in agreement with the conclusions of Buysse et al (5).

Based on expression data from the λgt11 clones, Buysse et al (5) postulated discrete transcriptional units for *ipaB, C,* and *D*. However, Baudry et al (1) interpreted the results of the Tn*5* insertion mutations to indicate the existence of an operon encompassing, in order, the 21-kd peptide gene, *ipaB, C, D,* and *A*. Both groups are currently sequencing this region of plasmid DNA, and the contrasting models of transcription should be resolved in the near future.

The virulence plasmid of *S. sonnei* also encodes genes for expression of polypeptides that are similar in size and immunological cross-reactivity to the immunogenic peptides a–d described above in *S. flexneri* (38). Subclones of two contiguous *Hind*III fragments expressed at least four polypeptides of 80, 47, 41, and 38 kd. There is no evidence to date that these peptides are the ones that are similar to the peptides a–d expressed in *S. flexneri*. They are likely to be new, unrelated peptides, since the region to which these *Hind*III fragments map lies at the opposite end of pHS4108 from the sequences known to encode for polypeptides a–d of *S. flexneri* (1, 102).

ANALYSIS OF INTERACTION OF BACTERIA WITH CELLS

Entry

Experiments with M90T, an invasive isolate of *S. flexneri* serotype 5, have indicated that its virulence plasmid pWR100 participates at every essential step of the invasive process in the HeLa cell model, including entry into cells, rapid intracellular multiplication, and early killing of host cells. Preliminary experiments indicated that shigellae entered mammalian cells through endocytosis. No leakage of macromolecules from recipient cells could be observed during entry (34). Cytochalasin D, which inhibits microfilament functions, blocked the entry process (37). Recent evidence indicates that *S. flexneri* enters HeLa cells via directed phagocytosis (10). Use of an anti-myosin monoclonal antibody and of 7-nitrobenz-2-oxa-1,3-diazole phallacidin, a fluorescent dye that stains polymerized F-actin, demonstrated accumulation of myosin and F-actin, two major components of the cell cytoskeleton, underneath the cytoplasmic membrane at the site of bacterial entry. The nature of the plasmid-encoded bacterial product, as well as that of the transmembrane signaling system that triggers the phagocytic process, remains unknown. As mentioned earlier, Tn*5* mutagenesis of recombinant plasmid pHS4108 revealed at least five regions within a 20-kb portion of the insert that are implicated in this phenotype (1). The observation that *S. flexneri* requires expression from several different gene loci to penetrate cells emphasizes the complexity of the strategy that has evolved in this microorganism. It differs from that of *Yersinia pseudotuberculosis*, in which only one gene appears necessary for promoting entry into cells (41).

Intracellular Multiplication

Plasmid genes are also needed for efficient intracellular multiplication (86). Plasmid pWR100 endows *E. coli* K-12 with the same replication potential as wild-type *S. flexneri*. A sequential electron-microscopic study of infected HeLa cells demonstrated that invasive *S. flexneri* induced lysis of the phagocytic membrane shortly after penetration into cells. By 30 min after centrifugation-induced penetration, all bacteria were lying free within the cytoplasm of host cells (86). Similar invasiveness was observed with *E. coli* K-12 carrying pWR100. On the other hand, *Salmonella typhimurium*, whose lysis of the phagocytic membrane is late and inefficient, grows poorly intracellularly. A plasmid-mediated contact-hemolytic activity demonstrated in virulent shigellae provides a likely mechanism for lysis of the phagosome (8, 86). Molecular characterization of this contact hemolysin is not yet available, although preliminary evidence indicates that the gene product(s) that triggers entry also accounts for hemolysis. This proposal is based on the observation that all Tn5 mutations in the virulence plasmid that eliminate the entry phenotype also eliminate contact-hemolytic activity (1). By contrast, no correlation has been observed between rapid intracellular growth of shigellae and the level of Shiga toxin or SLT production (9, 86). During the infectious process both *S. dysenteriae* 1 and *S. flexneri* 2a block host-cell protein synthesis; the block leads to preferential incorporation of labeled amino acids into bacterial proteins (35). Although secretion of Shiga toxin or SLT should give these invasive pathogens an additional advantage by precipitating cell killing, it does not seem to account for rapid intracellular growth. Another factor, the iron chelator aerobactin, may also be critical for bacterial replication within cells in which iron is immobilized by ferritin. Independent studies have recently shown that mutants of *S. flexneri* that no longer produce aerobactin demonstrate no significant alteration in their capacity to multiply intracellularly (50, 60). Therefore, only early lysis of the phagocytic vacuole so far appears to be a major factor for the rapid intracellular growth that is characteristic of *Shigella*. Metabolic pathways that lead to products such as aromatic components that are not available within eukaryotic cells will also probably turn out to be major factors for intracellular growth.

Early Killing of Host Cells

Plasmid genes are also involved in early killing of host cells. In a study of the intracellular fate of both an invasive strain and a noninvasive, plasmidless derivative of *S. flexneri,* plasmid pWR100 appeared to mediate killing of host cells (the continuous macrophage cell line J774) within 4 hr (9). For expression of this activity bacteria had to be intracellular, since macrophages were protected by cytochalasin D. Although both strains produced equivalent levels of SLT and inhibited protein synthesis of macrophages within 2 hr, only invasive bacteria were able to kill host cells. Damage to macrophages correlated with the ability of invasive bacteria to rapidly and efficiently lyse the membrane of the phagocytic vacuole (9). Metabolic events that mediate early killing have recently been demonstrated to include a rapid drop in the intracellular concentration of ATP, an increase in pyruvate concentration, and arrest of lactate production (85). The molecular basis of such inhibition of host-cell respiration and fermentation is unknown. It is not yet clear whether plasmid genes are directly responsible for this effect or whether plasmid-mediated lysis of the phagocytic vacuole allows diffusion of toxic products encoded by the chromosome into the cytosol.

Continuous Reinfection of Adjacent Cells

A region (virG) of the S. flexneri virulence plasmid is considered to be necessary for continuous reinfection of adjacent cells (53). virG mutants can invade cells and multiply intracellularly but do not spread to adjacent cells. Within epithelia, bacteria tend to localize within the cytoplasm and convert to a spherical morphology before being eliminated. The precise alteration of the invasive phenotype of the virG mutants has not been characterized. We speculate that they are either altered in their capacity to destroy the membrane of the phagocytic vacuole or unable to resist lysosomal killing. The location of virG with respect to the essential invasion genes contained on pHS4108 (Figure 1) is not known.

To summarize, the identification of the four major phenotypes (entry into cells, intracellular multiplication, early killing of host cells, and continuous reinfection of adjacent cells) represents a solid basis for an analytical approach to the invasive process of individual epithelial cells.

INTERACTION OF BACTERIA WITH TISSUES

Shiga and Shiga-like Toxins

Although Shiga toxin is one of the most potent bacterial toxins, its actual role in the pathogenesis of shigellosis is still unclear (for a review on Shiga toxin and SLT, see Reference 63). S. dysenteriae 1, which produces the most severe shigellosis, also produces much more Shiga toxin than other species of Shigella (44, 45, 64–66). This toxin is composed of two subunits. Subunit A (32-kd) possesses the biological activities. It is combined with five molecules of the B subunit (7.7 kd), which are responsible for binding to cell-surface receptors (14, 90). Some strains of S. flexneri and S. sonnei produce low levels of Shiga-like toxin, which is neutralizable by anti–Shiga toxin sera (45, 66). This toxin has not yet been purified.

Shiga toxin has three main biological activities: enterotoxicity, neurotoxicity, and cytotoxicity (16). The toxin inhibits protein synthesis in eukaryotic cells (4, 99) by acting on the 60S ribosomal subunit (3, 67, 74). The molecular basis of Shiga toxin cytotoxicity has recently been clearly established (17). Like the plant lectin ricin, Shiga toxin depurinates adenine 4324 near the 3' end of 28S RNA. Therefore, the A catalytic chain of Shiga toxin is a highly specific N-glycosidase.

The contribution of Shiga toxin to virulence is unclear. It may contribute to or even be responsible for the diarrheal component of shigellosis (75). However, how Shiga toxin acts as an enterotoxin is unknown as yet. The role of Shiga toxin in the pathogenesis of dysentery is not clear either. Volunteers fed a low toxin–producing mutant of *S. dysenteriae* 1 (29) showed milder symptoms than those fed the parental strain (52). This would indicate that the toxin has a role at some stage of the infection process. However, Shiga toxin and SLT do not appear to be involved in penetration into cells, intracellular multiplication, or early cell killing based on in vitro models (9, 86). This view is confirmed by recent experiments in which a Tox⁻ mutant of *S. dysenteriae* 1 still invaded HeLa cells and caused a positive Serény test (91). Three possibilities may be viewed at present: (*a*) Free toxin within the intestinal lumen [which can be detected during *S. dysenteriae* 1 infection (13)] may bind to specific receptors on the surface of enterocytes and kill those cells by inhibition of protein synthesis; Shiga toxin has been shown to be cytotoxic for primary cultures of human colonic epithelial cells (59). (*b*) Toxin released within infected enterocytes or within connective tissue of the lamina propria may diffuse and alter adjacent cells, including phagocytic cells (if sensitive to Shiga toxin), thus increasing the severity of the lesions. (*c*) The third possibility, which we favor at present, is derived from recent evidence that the hemolytic uremic syndrome (HUS) observed after *S. dysenteriae* 1 or *E. coli* O157:H7 infection (30, 42, 48, 70, 73) may in part be due to systemic dissemination of the toxin. Shiga toxin appears to mediate vascular damage, since vascular injury observed in HUS (48) resembles that observed in cerebral vessels of animals inoculated with Shiga toxin (2, 6). We thus speculate that Shiga toxin produced during colonic infection may increase the severity of the disease by causing ischemia of the intestinal tissue by either local or systemic diffusion of the toxin. Alteration of vasa nervosum may also account for more severe symptoms, including toxic megacolon. The recent production of a transposon mutant of *S. dysenteriae* 1 that does not produce Shiga toxin should allow precise evaluation of the role of this toxin in the infection process (91).

Information on the genetics of Shiga toxin production has been scarce and contradictory. Initial reports localized a region of the *S. flexneri* chromosome necessary for fluid production in rabbit ileal loops to the *rha-mtl* region, near the lysine decarboxylase–negative locus (82). However, no evidence was provided that the ability to cause fluid accumulation was due to the Shiga-like toxin of *S. flexneri*. In another study, Timmis et al (100) constructed an R-prime plasmid that contained the *argE* region of the chromosome of *S. dysenteriae* 1 and encoded production of Shiga toxin. The discrepancy between the location of the toxin genes in the two species may be explained by a different chromosome map or the presence of several copies of the gene in *S. dysenteriae* 1. However, Sekizaki et al (91) recently reported that *E. coli* K-12 was able to produce high levels of Shiga toxin after conjugation with *S. dysenteriae* 1 Hfr derivatives. P1 transduction analysis allowed mapping of the gene encoding Shiga toxin *(stx)* near *pyrF* at 30 min on the linkage map of *E. coli* K-12 (91). This series of experiments seems to solve the problem of the location of the Shiga toxin gene on the chromosome of *S. dysenteriae*. A similar approach did not reveal the location of the Shiga-like toxin gene of *S. flexneri* (T. Sekizaki, personal communication). The structural gene for Shiga toxin has recently been cloned. The sequence differs from that of the SLT-I toxin gene of *E. coli* by three base pairs with a single amino acid change (94).

Role of Lipopolysaccharide in Shigella Virulence

The lipopolysaccharide (LPS) layer of gram-negative bacteria is an important component of the cell surface and contributes to the virulence of many pathogens by providing resistance to certain host defenses such as serum killing and phagocytosis. Its importance has been demonstrated by the observation that smooth-colony clinical isolates readily segregate from rough-colony variants upon passage in the laboratory. These rough mutants are altered in their LPS structure and have lost their virulence in animal models. Rough mutants of *S. flexneri* 2a are still capable of invading HeLa cells but do not produce keratoconjunctivitis in guinea pigs (69). Mutants representing a range of rough chemotypes from Ra to Re (no O-antigen but complete to incomplete cores) also retained the ability to invade HeLa cells, which ruled out a role for *Shigella* O-antigens in the invasion step in vitro (68). These mutants were uniformly negative in the Serény test, which suggests that a complete LPS structure is essential in either bacterial multiplication after penetration of the epithelial cell or bacterial survival in an animal host before invasion.

The importance of a specific O-antigen structure for *Shigella* virulence was shown in early experiments involving construction of intergeneric hybrids. *S. flexneri* hybrids expressing the *E. coli* O-25 antigen retained the ability to penetrate epithelial cells and produce keratoconjunctivitis in guinea pigs (28). On the other hand, hybrids expressing *E. coli* O-8 somatic antigen lost their invasive capacity. The avirulence of the O-8 hybrids was attributed to the chemical structure of the O-8 antigen, which was less similar than that of the O-25 antigen to the structure of *Shigella* somatic antigen (for a detailed review of the chemistry of *S. flexneri* O-antigens, see Reference 93). Thus possession of a smooth LPS alone would not appear to be sufficient for virulence, and a possible role for the chemical composition of the particular *Shigella* somatic antigen is implicated. Further evidence of the need for a specific type of O-antigen for virulence is the strong immunological relatedness of the majority of the O-antigen serotypes expressed by EIEC to the O-antigens of various *Shigella* spp. (7). The role of the O-antigen was further clarified in reconstruction experiments involving sequential transfer of *S. flexneri* chromosomal DNA into an *E. coli* K-12 recipient carrying the 220-kb virulence plasmid (82). Expression of the *Shigella* group antigen in conjunction with the virulence plasmid and the *kcp* locus was important in producing a positive Serény test and inflammation in rabbit ileal loops (82). However, construction of the intergeneric hybrids involved gene transfer by conjugation, and one cannot rule out the possibility that other virulence-associated genes may be closely linked to *his* and may have contributed to the observed behavior of these hybrids.

The genetic information required for O-antigen biosynthesis in *S. flexneri* is chromosomally located, and the virulence plasmid does not appear to encode any genes for production of the LPS (84). However, *S. sonnei* and *S. dysenteriae* are different. Virulent smooth (Form I) isolates of *S. sonnei* irreversibly dissociate to rough (Form II) variants at a high frequency, and these variants are unable to invade epithelial cells in the Serény test (47). The irreversible nature of the transition from Form I to Form II suggested a plasmid role in O-antigen biosynthesis. A 180-kb plasmid was subsequently identified in Form I isolates of *S. sonnei* from diverse geographical sources. Form II isolates had lost the plasmid. When the 180-kb plasmid was mobilized back into Form II *S. sonnei*, the resulting transconjugants again expressed the characteristic *S. sonnei* O-antigen. Mobilization of the 180-kb plasmid into *S. flexneri* and *Salmonella typhi* also resulted in expression of *S. sonnei* O-antigen by the transconjugants, which confirmed that the plasmid carried all of the genes necessary for the biosynthesis of the *S. sonnei* O-antigen. As noted earlier, this 180-kb plasmid of *S. sonnei* is also functionally and genetically analogous to the virulence plasmid of *S. flexneri* and is required for expression of the invasive phenotype (83, 101).

A small 9-kb plasmid in strains of *S. dysenteriae* 1 was found to be associated with production of O-antigen. Mutants that had lost the plasmid expressed a rough phenotype and reduced virulence in HeLa cells and the Serény test. Reconstruction experiments introducing the 9-kb plasmid back into the plasmid-cured rough mutant restored both *S. dysenteriae* O-antigen production and virulence (104). Cloning and transposon mutagenesis localized the plasmid-borne determinant involved in O-antigen synthesis, and minicell experiments demonstrated that the gene, designated *rfp*, produced a 41-kd protein (103). The cloned *rfp* gene, when introduced into a *S. dysenteriae* 1 strain cured of the native 9-kb plasmid, restored normal O-antigen production and virulence. When introduced into *E. coli* K-12, the cloned *rfp* gene modified the *E. coli* LPS core by the addition of a galactose residue, the first sugar of the *S. dysenteriae* 1 O-antigen repeat unit (95). However, synthesis of a complete *S. dysenteriae* 1 O-antigen in *E. coli* K-12 requires the chromosomal *rfb* gene cluster as well (36). This region, spanning 6.4–7.5 kb near *his*, specifies at least six determinants for O-antigen production, including the synthetases and transferases required for adding two rhamnose residues to the galactose to complete the O-antigen side chain (96).

kcpA *Locus*

The *kcpA* locus, located between the *lac* and the *gal* genes on the *Shigella* chromosome, is necessary for production of keratoconjunctivitis in guinea pigs (21). Transduction with phage P1 demonstrated that this locus cotransduced with *purE*. Although the function of the *kcpA* locus in disease is unclear, *E. coli* K-12–*S. flexneri* hybrid strains that had received the *kcpA* locus appeared more invasive in the rabbit ligated ileal loop assay, which suggests that this locus encodes a function necessary for tissue invasiveness (82).

Miscellaneous Phenotypes Associated With Virulence

AEROBACTIN   Microorganisms have evolved efficient high-affinity iron uptake systems in order to grow in iron-limited environments. The ability of a microorganism to compete for iron in an infected host is considered a virulence factor for many pathogens (105). Certain *E. coli* ColV strains produce the hydroxamate siderophore aerobactin, and this iron uptake system is associated with the strain's ability to cause generalized extraintestinal infections (106). Most strains of *S. flexneri* and *S. boydii* also utilize the aerobactin system for chelating iron, while many *S. sonnei* and *S. dysenteriae* strains employ only the enterobactin system (51). The chromosomal region *arg-mtl*, which was shown by conjugational gene transfer experiments to be associated with *S. flexneri* virulence (82), was subsequently found to contain the gene for aerobactin and a 76-kd iron-regulated outer-membrane protein, which is probably the aerobactin receptor protein (32). A formal test of the hypothesis that aerobactin is a virulence factor for *Shigella* was undertaken when defined transposon mutants of the aerobactin gene *(iuc)* was isolated. Lawlor et al (50) assayed an *iuc*::Tn5 mutant constructed in vitro for invasiveness in HeLa cells and the Serény test. They found the mutant unaffected in invasive capacity and ability to multiply intracellularly. The *iuc*::Tn5 mutant was also screened for virulence in the chicken embryo model (71). It was found to be lethal for chicken embryos, but at LD_{50} values 10–100 times higher than those of the wild-type parent. This higher LD_{50} was attributed to a slower growth rate for the *iuc*::Tn5 mutant in the allantoic fluid.

A Tn*10* insertion in the aerobactin gene of a *S. flexneri* 5 strain was similarly tested in a separate study (60). Like the *iuc*::Tn5 mutant, the *iuc*::Tn*10* mutant was unaltered in its ability to invade and multiply within HeLa cells and killed the host cells with the same kinetics as the parent strain. However, an inoculum-dependent difference was observed in the Serény test; the wild-type strain produced a positive reaction with 10^6 bacteria, while the mutant required a 10-fold higher inoculum to produce a similar reaction. Virulence was also tested by infecting ligated rabbit ileal loops with the aerobactin mutant and evaluating fluid production and histopathology. An inoculum-dependent effect was again observed with the *iuc*::Tn*10* mutant. When an inoculum of 10^7 bacteria was used, ileal loops infected with the mutant displayed markedly fewer mucosal lesions and lower fluid production than those infected with the parent strain. Taken together, these results suggest that aerobactin production may have a role in bacterial growth in the extracellular compartment within host tissues. It should also be noted that aerobactin does not appear to be preferred over enterobactin as a siderophore for *Shigella,* since *S. dysenteriae* 1, which causes the most severe form of dysentery among *Shigella* spp., has no aerobactin genes and apparently produces only enterochelin (51).

CONGO RED BINDING ABILITY   The ability to bind the dye Congo red (referred to as Pcr^+ or Crb^+ in the literature) was originally recognized as a phenotype that differentiated between virulent and avirulent derivatives of *Yersinia pestis.* The observation was subsequently extended to include *Shigella* and several other gram-negative pathogens (71). Spontaneous mutants of *S. flexneri* that are no longer capable of binding Congo red (Pcr^-) arise at a frequency of $\sim 10^{-4}$, are noninvasive, and frequently have either lost the virulence plasmid or suffered deletions in the plasmid (56). Conversely, screening of noninvasive mutants of *S. flexneri* revealed that they had also lost the ability to bind Congo red. This observation demonstrates a close association of the Pcr^+ phenotype with invasive ability and the presence of the virulence plasmid, which suggests Congo red binding as a possible virulence factor. In addition, as mentioned earlier, the Pcr^+ phenotype is expressed in a temperature-regulated fashion, as are the virulence genes of *Shigella* (55). Confirmation of the role of the plasmid in Congo red binding came with the cloning of a gene(s) for this phenotype from the virulence plasmid (11, 77). According to one report (11), a cloned 9.0-kb *Bam*HI fragment from the virulence plasmid of *S. flexneri* 1b conferred Congo red binding ability on *E. coli* and restored the Pcr^+ phenotype to a plasmidless *S. flexneri* 1b. Congo red binding by the recombinant was not fully temperature regulated in *S. flexneri* 1b, possibly because of the high copy number of the cloning vector. The 9.0-kb fragment alone was not sufficient, however, to restore virulence to a plasmidless *S. flexneri* 1b.

Another study found that a 1-kb sequence of DNA (the *vir*F region) from the virulence plasmid of *S. flexneri* 2a was sufficient for expression of the Pcr$^+$ phenotype in *E. coli* K-12 (77). However, unlike the previous clone, a plasmidless *S. flexneri* 2a containing the *vir*F clone did not become Pcr$^+$. It is not clear why this clone expressed Pcr$^+$ in *E. coli* but not in *S. flexneri*. The authors reported finding several other regions of the virulence plasmid that expressed the Pcr$^+$ phenotype in *E. coli,* yet these clones were not tested in *S. flexneri*. It is difficult to evaluate the relevance of these plasmid regions to Congo red binding because the proper phenotype could not be restored in the parent. The *vir*F region has been sequenced and was determined to specify three peptides of 30, 27, and 21 kd in minicells (78). A computer-generated secondary structure prediction of the product of an open reading frame in this sequence indicated a protein rich in β-sheets. This may represent a structure capable of direct interaction with Congo red, which is known to bind to such a structure (31).

The function of the Congo red binding gene(s) in *Shigella* is unknown. Congo red binding in other bacterial systems has been correlated with absorption of hemin [*Y. pestis* (97)] and protoporphyrin IX [*Aeromonas salmonicida* (43)]. Daskaleros & Payne (12) observed this correlation in *Shigella* as well and found that absorption of these iron-containing compounds was growth-temperature dependent (12). They also found that bacteria that had prebound hemin showed increased invasiveness in HeLa cells, which suggested a possible involvement of the Congo red binding gene(s) in bacterial attachment to mammalian cells.

CONCLUSIONS AND PROSPECTS

During the six years since the first report of the involvement of a plasmid in invasion of mammalian cells by *Shigella,* research in defining the plasmid virulence determinants has moved rapidly. Application of new techniques for creating protein fusions using *phoA* and *lacZ* will further help define plasmid virulence genes and their role in invasion. Study of the immunogenic peptides encoded by the plasmid should yield information that will be useful for the design of effective vaccines against shigellosis. The DNA sequence of several *ipa* genes will soon be available (B. Baudry, M. Kaczorek, P. Sansonetti, manuscript in preparation), and this will certainly give us clues as to the possible roles of their gene products in virulence.

Genetic studies will need to focus again on the chromosomal determinants of *Shigella* virulence. The gene product and the specific function of the *kcpA* locus, which was first described in 1971, still remain to be uncovered. The genetic basis for the colony-morphology transition from transparent to opaque, which leads to loss of virulence and a host of pleiotropic changes (27, 49), is unknown. The interaction between chromosomal and plasmid genes should also be considered, as the pathogenicity of *Shigella* is dependent on the coordinate expression of genes on both genetic elements. Research on the molecular mechanisms underlying the temperature regulation of *Shigella* virulence will be important in understanding these interactions.

Future research will probably examine the intracellular life-style of the bacteria, specifically intracellular growth and host-cell killing. For example, the availability of defined Tox$^-$ mutants of *S. dysenteriae* 1 will make it possible to examine definitively the role of toxin in pathogenesis at the cellular and tissue levels. However, many of the genetic studies currently under way to determine the genes and their products that trigger entry may provide only partial answers. These studies should be complemented by experiments that approach the question from the point of view of membrane biochemistry and cell biology. This approach is already beginning to reveal the biochemical events involved in the host-parasite interactions in *Shigella* infection. Future studies will surely benefit from this multidisciplinary approach.

ACKNOWLEDGMENTS

The authors wish to acknowledge the outstanding contributions of Sam Formal to our understanding of the genetics of *Shigella* pathogenicity and cite his considerable influence in our own research. We also wish to thank T. L. Hale, P. Clerc, B. Baudry, and A. D. O'Brien for helpful discussions. Research on the genetics of *Shigella* virulence in the laboratory of ATM is currently supported by USUHS protocol RO-7385.

Literature Cited

1. Baudry, B., Maurelli, A. T., Clerc, P., Sadoff, J. C., Sansonetti, P. J. 1987. Localization of plasmid loci necessary for *Shigella flexneri* entry into HeLa cells, and genetic organization of one locus encoding four immunogenic polypeptides. *J. Gen. Microbiol.* 133:3403–13
2. Bridgewater, F. A. J., Morgan, R. S., Rowson, K. E. K., Wright, G. P. 1955. The neutrotoxin of *Shigella shigae*. Morphological and functional lesions produced in the central nervous system of rabbits. *Br. J. Exp. Pathol.* 36:447–53
3. Brown, J. E., Obrig, T. G., Ussery, M. A., Moran, T. P. 1986. Shiga toxin from *Shigella dysenteriae* 1 inhibits protein synthesis in reticulocyte lysates by inactivation of aminoacyl-tRNA binding. *Microb. Pathog.* 1:325–34
4. Brown, J. E., Rothman, S. W., Doctor, B. P. 1980. Inhibition of protein synthesis in intact HeLa cells by *Shigella dysenteriae* 1 toxin. *Infect. Immun.* 29:98–107
5. Buysse, J. M., Stover, C. K., Oaks, E. V., Venkatesan, M., Kopecko, D. J. 1987. Molecular cloning of invasion plasmid antigen *(ipa)* genes from *Shigella flexneri:* analysis of *ipa* gene products and genetic mapping. *J. Bacteriol.* 169:2561–69
6. Cavanagh, J. B., Howard, J. G., Whitby, J. L. 1956. The neurotoxin of *Shigella shigae*. A comparative study of the effects produced in various laboratory animals. *Br. J. Exp. Med.* 37:272–78
7. Cheasty, T., Rowe, B. 1983. Antigenic relationships between the enteroinvasive *Escherichia coli* O antigens O28ac, O112ac, O124, O136, O143, O144, O152, and O164 and *Shigella* O antigens. *J. Clin. Microbiol.* 17:681–84
8. Clerc, P., Baudry, B., Sansonetti, P. J. 1986. Plasmid-mediated contact haemo-

lytic activity in *Shigella* species: correlation with penetration into HeLa cells. *Ann. Inst. Pasteur Microbiol.* 137A:267–78

9. Clerc, P., Ryter, A., Mounier, J., Sansonetti, P. J. 1987. Plasmid mediated early killing of eukaryotic cells by *Shigella flexneri* as studied by infection of J774 macrophages. *Infect. Immun.* 55:521–27

10. Clerc, P., Sansonetti, P. J. 1987. Entry of *Shigella flexneri* into HeLa cells: evidence for directed phagocytosis involving actin polymerization and myosin accumulation. *Infect. Immun.* 55:2681–88

11. Daskaleros, P. A., Payne, S. M. 1985. Cloning the gene for Congo red binding in *Shigella flexneri*. *Infect. Immun.* 48:165–68

12. Daskaleros, P. A., Payne, S. M. 1987. Congo red binding phenotype is associated with hemin binding and increased infectivity of *Shigella flexneri* in the HeLa cell model. *Infect. Immun.* 55:1393–98

13. Donahue-Rolfe, A., Kelley, N. A., Bennish, M., Keusch, G. T. 1986. Enzyme-linked immunosorbant assay for *Shigella* toxin. *J. Clin. Microbiol.* 24:65–68

14. Donohue-Rolfe, A., Keusch, G. T., Edson, C., Thorley-Lawson, D., Jacewicz, M. 1984. Pathogenesis of *Shigella* diarrhea. IX. Simplified high yield purification of *Shigella* toxin and characterization of subunit composition and function by the use of subunit-specific monoclonal and polyclonal antibodies. *J. Exp. Med.* 160:1767–81

15. Dupont, H. L., Formal, S. B., Hornick, R. B., Snyder, M., Libonati, J. P., et al. 1971. Pathogenesis of *Escherichia coli* diarrhea. *N. Engl. J. Med.* 285:1–9

16. Eiklid, K., Olsnes, S. 1983. Animal toxicity of *Shigella dysenteriae* cytotoxin: evidence that the neurotoxic, enterotoxic and cytotoxic activities are due to one toxin. *J. Immunol.* 130:380–84

17. Endo, Y., Tsurugi, K. 1987. RNA *N*-glycosidase activity of ricin A-chain. *J. Biol. Chem.* 262:8128–30

18. Falkow, S., Schneider, H., Baron, L. S., Formal, S. B. 1963. Virulence of *Escherichia-Shigella* genetic hybrids for the guinea pig. *J. Bacteriol.* 86:1251–58

19. Formal, S. B., Dupont, H. L., Hornick, R., Snyder, M. J., Libonati, J., LaBrec, E. H. 1971. Experimental models in the investigation of the virulence of dysentery bacilli and *Escherichia coli*. *Ann. NY Acad. Sci.* 176:190–96

20. Formal, S. B., Gemski, P. Jr., Baron,

L. S., LaBrec, E. H. 1970. Genetic transfer of *Shigella flexneri* antigens to *Escherichia coli* K-12. *Infect. Immun.* 1:279–87

21. Formal, S. B., Gemski, P. Jr., Baron, L. S., LaBrec, E. H. 1971. A chromosomal locus which controls the ability of *Shigella flexneri* to evoke keratoconjuctivitis. *Infect. Immun.* 3:73–79

22. Formal, S. B., Hornick, R. B. 1978. Invasive *Escherichia coli*. *J. Infect. Dis.* 137:641–44

23. Formal, S. B., Kundel, D., Schneider, H., Kuneu, N., Spunz, H. 1961. Studies with *Vibrio cholerae* in the ligated loop of the rabbit intestine. *Br. J. Exp. Pathol.* 42:504–10

24. Formal, S. B., LaBrec, E. H., Kent, T. H., Falkow, S. 1965. Abortive intestinal infection with an *Escherichia coli–Shigella flexneri* hybrid strain. *J. Bacteriol.* 89:1374–82

25. Formal, S. B., LaBrec, E. H., Palmer, A., Falkow, S. 1965. Protection of monkeys against experimental shigellosis with attenuated vaccines. *J. Bacteriol.* 90:63–68

26. Formal, S. B., Levine, M. M. 1983. Shigellosis. In *Bacterial Vaccines*, ed. R. Germanier, pp. 167–86. New York: Academic

27. Gemski, P. Jr., Formal, S. B. 1975. Shigellosis: an invasive infection of the gastrointestinal tract. In *Microbiology 1975*, ed. D. Schlessinger, pp. 165–69. Washington DC: Am. Soc. Microbiol.

28. Gemski, P. Jr., Sheahan, D. G., Washington, O., Formal, S. B. 1972. Virulence of *Shigella flexneri* hybrids expressing *Escherichia coli* somatic antigens. *Infect. Immun.* 6:104–11

29. Gemski, P. Jr., Takeuchi, A., Washington, O., Formal, S. B. 1972. Shigellosis due to *Shigella dysenteriae* 1: relative importance of mucosal invasion versus toxin production in pathogenesis. *J. Infect. Dis.* 126:523–30

30. Gianantonio, C., Vitacco, M., Mendilaharzu, F., Rutty, A., Mendilaharzu, J. 1964. The hemolytic-uremic syndrome. *J. Pediatr.* 64:478–91

31. Glenner, G. G. 1980. Amyloid deposits and amyloidosis. *N. Engl. J. Med.* 302:1283–92

32. Griffiths, E., Stevenson, P., Hale, T. L., Formal, S. B. 1985. Synthesis of aerobactin and a 76,000-dalton iron-regulated outer membrane protein by *Escherichia coli* K-12–*Shigella flexneri* hybrids and by enteroinvasive strains of *Escherichia coli*. *Infect. Immun.* 49:67–71

33. Hale, T. L., Bonventre, P. F. 1979.

Shigella infection of Henle intestinal epithelial cells: role of the bacteria. *Infect. Immun.* 24:879–86

34. Hale, T. L., Formal, S. B. 1980. Cytotoxicity of *Shigella dysenteriae* 1 for cultured mammalian cells. *Am. J. Clin. Nutr.* 33:2485–90

35. Hale, T. L., Formal, S. B. 1981. Protein synthesis in HeLa or Henle 407 cells infected with *Shigella dysenteriae* 1, *Shigella flexneri* 2a, or *Salmonella typhimurium* W118. *Infect. Immun.* 32:137–44

36. Hale, T. L., Guerry, P., Seid, R. C., Kapfer, C., Wingfield, M. E., et al. 1984. Expression of lipopolysaccharide O antigen in *Escherichia coli* K-12 hybrids containing plasmid and chromosomal genes from *Shigella dystenteriae* 1. *Infect. Immun.* 46:470–75

37. Hale, T. L., Morris, R. E., Bonventre, P. F. 1979. *Shigella* infection of Henle intestinal epithelial cells: role of the host cell. *Infect. Immun.* 24:887–94

38. Hale, T. L., Oaks, E. V., Formal, S. B. 1985. Identification and antigenic characterization of virulence-associated, plasmid-coded proteins of *Shigella* spp. and enteroinvasive *Escherichia coli*. *Infect. Immun.* 50:620–29

39. Hale, T. L., Sansonetti, P. J., Schad, P. A., Austin, S., Formal, S. B. 1983. Characterization of virulence plasmids and plasmid-associated outer membrane proteins in *Shigella flexneri*, *Shigella sonnei*, and *Escherichia coli*. *Infect. Immun.* 40:340–50

40. Harris, J. R., Wachsmuth, I. K., Davis, B. R., Cohen, M. L. 1982. High molecular weight plasmid correlates with *Escherichia coli* enteroinvasiveness. *Infect. Immun.* 37:1295–98

41. Isberg, R. R., Falkow, S. 1985. A single genetic locus encoded by *Yersinia pseudotuberculosis* permits invasion of cultured animal cells by *Escherichia coli* K-12. *Nature* 317:262–64

42. Karmali, M. A., Petric, M., Lim, C., Fleming, P. C., Arbus, G. S., Lior, H. 1985. The association between idiopathic hemolytic uremic syndrome and infection by Verotoxin-producing *Escherichia coli*. *J. Infect. Dis.* 151:775–82

43. Kay, W. W., Phipps, B. M., Ishiguro, E. E., Trust, T. J. 1985. Porphyrin binding by the surface array virulence protein of *Aeromonas salmonicida*. *J. Bacteriol.* 164:1332–36

44. Keusch, G. T., Donohue-Rolfe, A., Jacewicz, M. 1982. *Shigella* toxin(s): description and role in diarrhea and dysentery. *Pharmacol Ther.* 15:403–38

45. Keusch, G. T., Jacewicz, M. 1977. The pathogenesis of *Shigella* diarrhea. VI. Toxin and antitoxin in *S. flexneri* and *S. sonnei* infections in humans. *J. Infect. Dis.* 135:552–56

46. Kinsey, M. D., Formal, S. B., Dammin, G. J., Giannella, R. A. 1976. Fluid and electrolyte transport in rhesus monkeys challenged intracecally with *Shigella flexneri* 2a. *Infect. Immun.* 14:368–71

47. Kopecko, D. J., Washington, O., Formal, S. B. 1980. Genetic and physical evidence for plasmid control of *Shigella sonnei* Form I cell surface antigen. *Infect. Immun.* 29:207–14

48. Koster, F., Levin, J., Walker, L., Tung, K. S. K., Gilman, R. H., et al. 1977. Hemolytic-uremic syndrome after shigellosis. Relation to endotoxemia and circulating immune complexes. *N. Engl. J. Med.* 298:927–33

49. LaBrec, E. H., Schneider, H., Magnani, T. J., Formal, S. B. 1964. Epithelial cell penetration as an essential step in the pathogenesis of bacillary dysentery. *J. Bacteriol.* 88:1503–18

50. Lawlor, K. M., Daskaleros, P. A., Robinson, R. E., Payne, S. M. 1987. Virulence of iron transport mutants of *Shigella flexneri* and utilization of host iron compounds. *Infect. Immun.* 55:594–99

51. Lawlor, K. M., Payne, S. M. 1984. Aerobactin genes in *Shigella* spp. *J. Bacteriol.* 160:266–72

52. Levine, M. M., Dupont, H. L., Formal, S. B., Hornick, R. B., Takeuchi, A., et al. 1973. Pathogenesis of *Shigella dysenteriae* 1 (Shiga) dysentery. *J. Infect. Dis.* 127:261–70

53. Makino, S., Sasakawa, C., Kamata, K., Kurata, T., Yoshikawa, M. 1986. A genetic determinant required for continuous reinfection of adjacent cells on large plasmid in *Shigella flexneri* 2a. *Cell* 46:551–55

54. Maurelli, A. T., Baudry, B., d'Hauteville, H., Hale, T. L., Sansonetti, P. J. 1985. Cloning of virulence plasmid DNA sequences involved in invasion of HeLa cells by *Shigella flexneri*. *Infect. Immun.* 49:164–71

55. Maurelli, A. T., Blackmon, B., Curtiss, R. III. 1984. Temperature-dependent expression of virulence genes in *Shigella* species. *Infect. Immun.* 43:195–201

56. Maurelli, A. T., Blackmon, B., Curtiss, R. III. 1984. Loss of pigmentation in *Shigella flexneri* 2a is correlated with loss of virulence and virulence-associated plasmid. *Infect. Immun.* 43:397–401

57. Maurelli, A. T., Curtiss, R. III. 1984. Bacteriophage Mu d1 (Ap^r lac) generates vir-lac operon fusions in Shigella flexneri 2a. Infect. Immun. 45:642–48

58. Maurelli, A. T., Sansonetti, P. J. 1988. Identification of a chromosomal gene controlling temperature regulated expression of Shigella virulence. Proc. Natl. Acad. Sci. USA In press

59. Moyer, M. P., Dixon, P. S., Rothman, S. W., Brown, J. E. 1987. Cytotoxicity of Shiga toxin for primary cultures of human colonic and ileal epithelial cells. Infect. Immun. 55:1533–35

60. Nassif, X., Mazert, M. C., Mounier, J., Sansonetti, P. J. 1987. Evaluation with an iuc : : Tn10 mutant of the role of aerobactin production in the virulence of Shigella flexneri. Infect. Immun. 55: 1963–69

61. Oaks, E. V., Hale, T. L., Formal, S. B. 1986. Serum immune response to Shigella protein antigens in rhesus monkeys and humans infected with Shigella spp. Infect. Immun. 53:57–63

62. Oaks, E. V., Wingfield, M. E., Formal, S. B. 1985. Plaque formation by virulent Shigella flexneri. Infect. Immun. 48: 124–29

63. O'Brien, A. D., Holmes, R. K. 1987. Shiga and Shiga-like toxins. Microbiol. Rev. 51:206–20

64. O'Brien, A. D., LaVeck, G. D. 1982. Immunochemical and cytotoxic activities of Shigella dysenteriae 1 (Shiga) and Shiga-like toxins. Infect. Immun. 35: 1151–54

65. O'Brien, A. D., LaVeck, G. D., Thompson, M. R., Formal, S. B. 1982. Production of Shigella dysenteriae type 1–like cytotoxin by Escherichia coli. J. Infect. Dis. 146:763–69

66. O'Brien, A. D., Thompson, M. R., Gemski, P., Doctor, B. P., Formal, S. B. 1977. Biological properties of Shigella flexneri 2a toxin and its serological relationship to Shigella dysenteriae 1 toxin. Infect. Immun. 15:796–98

67. Obrig, T. G., Moran, T. P., Colinas, R. J. 1985. Ribonuclease activity associated with the 60S ribosome-inactivating proteins ricin A, phytolaccin and Shiga toxin. Biochem. Biophys. Res. Commun. 130:879–84

68. Okamura, N., Nagai, T., Nakaya, R., Kondo, S., Murakami, M., Hisatsune, K. 1983. HeLa cell invasiveness and O-antigen of Shigella flexneri as separate and prerequisite attributes of virulence to evoke keratoconjunctivitis in guinea pigs. Infect. Immun. 39:505–13

69. Okamura, N., Nakaya, R. 1977. Rough mutant of Shigella flexneri 2a that penetrates tissue culture cells but does not evoke keratoconjunctivitis in guinea pigs. Infect. Immun. 17:4–8

70. Pai, C. H., Gordon, R., Sims, H. V., Bryan, L. E. 1984. Sporadic cases of hemolytic colitis associated with Escherichia coli 0157 : H7. Ann. Intern. Med. 101:738–42

71. Payne, S. M., Finkelstein, R. A. 1977. Detection and differentiation of iron-responsive mutants on Congo red agar. Infect. Immun. 18:94–98

72. Piéchaud, M., Szturm-Rubinstein, S., Piéchaud, D. 1958. Evolution histologique de la kératoconjonctivite à bacilles dysentériques du cobaye. Ann. Inst. Pasteur 94:298–309

73. Raghupathy, P., Date, A., Shastry, J. C. M., Sudarsanam, A., Jadhav, M. 1978. Haemolytic-uraemic syndrome complicating shigella dysentery in South Indian children. Br. Med. J. 1:1518–21

74. Reisbig, R., Olsnes, S., Eiklid, K. 1981. The cytotoxic activity of Shigella toxin. Evidence for catalytic inactivation of the 60S ribosomal subunit. J. Biol. Chem. 256:8739–44

75. Rout, W. R., Formal, S. B., Gianella, R. A., Dammin, G. J. 1975. The pathophysiology of Shigella diarrhea in the rhesus monkey; intestinal transport, morphology and bacteriological studies. Gastroenterology 68:270–78

76. Rutgeerts, P., Geboes, K., Porelte, E., Caremaus, G., Vautroffer, G. 1982. Acute infective colitis caused by endemic pathogens in Western Europe. Endoscopic features. Endoscopy 141: 212–19

77. Sakai, T., Sasakawa, C., Makino, S., Kamata, K., Yoshikawa, M. 1986. Molecular cloning of a genetic determinant from Congo red binding ability which is essential for the virulence of Shigella flexneri. Infect. Immun. 51: 476–82

78. Sakai, T., Sasakawa, C., Makino, S., Yoshikawa, M. 1986. DNA sequence and product analysis of the virF locus responsible for Congo red binding and cell invasion in Shigella flexneri 2a. Infect. Immun. 54:395–402

79. Sansonetti, P. J., David, M., Toucas, M. 1980. Corrélation entre la perte d'ADN plasmidique et le passage de la phase I virulente a la phase II avirulente chez Shigella sonnei. C. R. Acad. Sci. Ser. D. 290:879–82

80. Sansonetti, P. J., d'Hauteville, H., Ecobichon, C., Pourcel, C. 1983. Molecular comparison of virulence plasmids in Shigella and enteroinvasive Es-

cherichia coli. Ann. Microbiol. (Inst. Pasteur) 134A:295–318

81. Sansonetti, P. J., d'Hauteville, H., Formal, S. B., Toucas, M. 1982. Plasmid-mediated invasiveness of "*Shigella*-like" *Escherichia coli. Ann. Inst. Pasteur Microbiol.* 132A:351–55

82. Sansonetti, P. J., Hale, T. L., Dammin, G. J., Kapfer, C., Collins, H. H. Jr., Formal, S. B. 1983. Alterations in the pathogenicity of *Escherichia coli* K-12 after transfer of plasmid and chromosomal genes from *Shigella flexneri. Infect. Immun.* 39:1392–1402

83. Sansonetti, P. J., Kopecko, D. J., Formal, S. B. 1981. *Shigella sonnei* plasmids: evidence that a large plasmid is necessary for virulence. *Infect. Immun.* 34:75–83

84. Sansonetti, P. J., Kopecko, D. J., Formal, S. B. 1982. Involvement of a plasmid in the invasive ability of *Shigella flexneri. Infect. Immun.* 35:852–60

85. Sansonetti, P. J., Mounier, J. 1987. Metabolic events mediating early killing of host cells infected by *Shigella flexneri. Microb. Pathog.* 3:53–61

86. Sansonetti, P. J., Ryter, A., Clerc, P., Maurelli, A. T., Mounier, J. 1986. Multiplication of *Shigella flexneri* within HeLa cells: lysis of the phagocytic vacuole and plasmid-mediated contact hemolysis. *Infect. Immun.* 51:461–69

87. Sasakawa, C., Kamata, K., Sakai, T., Murayama, S. Y., Makino, S., Yoshikawa, M. 1986. Molecular alteration of the 140-megadalton plasmid associated with loss of virulence and Congo red binding activity in *Shigella flexneri. Infect. Immun.* 51:470–75

88. Sasakawa, C., Makino, S., Kamata, K., Yoshikawa, M. 1986. Isolation, characterization, and mapping of Tn5 insertions into the 140-megadalton invasion plasmid defective in the mouse Serény test in *Shigella flexneri* 2a. *Infect. Immun.* 54:32–36

89. Schneider, H., Falkow, S. 1964. Characterization of an Hfr strain of *Shigella flexneri. J. Bacteriol.* 88:682–89

90. Seidah, N. G., Donohue-Rolfe, A., Lazure, C., Auclair, F., Keusch, G. T., Chrétien, M. 1986. Complete amino acid sequence of *Shigella* toxin B-chain. A novel polypeptide containing 69 amino acids and one disulfide bridge. *J. Biol. Chem.* 261:13928–31

91. Sekizaki, T., Harayama, S., Brazil, G. M., Timmis, K. N. 1987. Localization of *stx,* a determinant essential for high level production of Shiga toxin by *Shigella dysenteriae* serotype 1, near

pyrF and generation of *stx* transposon mutants. *Infect. Immun.* 55:2208–14

92. Serény, B. 1957. Experimental keratoconjuctivitis shigellosa. *Acta Microbiol. Acad. Sci. Hung.* 4:367–76

93. Simmons, D. A. R. 1971. Immunochemistry of *Shigella flexneri* O-antigens: a study of structural and genetic aspects of the biosynthesis of cell-surface antigens. *Bacteriol. Rev.* 35:117–48

94. Strockbine, N. A., Jackson, M. P., Sung, L. M., Holmes, R. K., O'Brien, A. D. 1988. Cloning and sequencing of the genes for Shiga toxin from *Shigella dysenteriae* type 1. *J. Bacteriol.* In press

95. Sturm, S., Jann, B., Jann, K., Fortnagel, P., Timmis, K. N. 1986. Genetic and biochemical analysis of *Shigella dysenteriae* 1 O antigen polysaccharide biosynthesis in *Escherichia coli* K-12: 9 kb plasmid of *S. dysenteriae* 1 determines addition of a galactose residue to the lipopolysaccharide core. *Microb. Pathog.* 1:299–306

96. Sturm, S., Jann, B., Jann, K., Fortnagel, P., Timmis, K. N. 1986. Genetic and biochemical analysis of *Shigella dysenteriae* 1 O antigen polysaccharide biosynthesis in *Escherichia coli* K-12: structure and functions of the *rfb* gene cluster. *Microb. Pathog.* 1:307–20

97. Surgalla, M. J., Beesley, E. D. 1969. Congo red agar plating medium for detecting pigmentation of *Pasteurella pestis. Appl. Microbiol.* 18:834–37

98. Takeuchi, A., Sprinz, H., LaBrec, E. H., Formal, S. B. 1965. Experimental acute colitis in the rhesus monkey following peroral infection with *Shigella flexneri. Am. J. Pathol.* 52:503–29

99. Thompson, M. R., Steinberg, M. S., Gemski, P., Formal, S. B., Doctor, B. P. 1976. Inhibition of *in vitro* protein synthesis by *Shigella dysenteriae* 1 toxin. *Biochem. Biophys. Res. Commun.* 71:783–88

100. Timmis, K. N., Clayton, C. L., Sekizaki, T. 1985. Localization of Shiga toxin gene in the region of *Shigella dysenteriae* 1 chromosome specifying virulence functions. *FEMS Microbiol. Lett.* 30:301–5

101. Watanabe, H., Nakamura, A. 1985. Large plasmids associated with virulence in *Shigella* species have a common function necessary for epithelial cell penetration. *Infect. Immun.* 48:260–62

102. Watanabe, H., Nakamura, A. 1986. Identification of *Shigella sonnei* Form I plasmid genes necessary for cell invasion and their conservation among *Shigella* species and enteroinvasive *Es-*

cherichia coli. *Infect. Immun.* 53:352–58

103. Watanabe, H., Nakamura, A., Timmis, K. N. 1984. Small virulence plasmid of *Shigella dysenteriae* 1 strain W30864 encodes a 41,000-dalton protein involved in formation of specific lipopolysaccharide side chains of serotype 1 isolates. *Infect. Immun.* 46:55–63

104. Watanabe, H., Timmis, K. N. 1984. A small plasmid in *Shigella dysenteriae* 1 specifies one or more functions essential for O antigen production and bacterial virulence. *Infect. Immun.* 43:391–96

105. Weinberg, E. D. 1978. Iron and infection. *Microbiol. Rev.* 42:45–66

106. Williams, P. H. 1979. Novel iron uptake system specified by ColV plasmids: an important component in the virulence of invasive strains of *Escherichia coli. Infect. Immun.* 26:925–32

Ann. Rev. Microbiol. 1988. 42:151–76

ENZYMES AND GENES FROM THE *lux* OPERONS OF BIOLUMINESCENT BACTERIA

Edward A. Meighen

Department of Biochemistry, McGill University, Montreal, Quebec, Canada, H3G 1Y6

CONTENTS

INTRODUCTION

Bioluminescence has intrigued mankind over many centuries owing to its intrinsic beauty, the wonder of its biological and chemical origin, and the

151

0066-4227/88/1001-0151$02.00

realization of the potential application of this naturally occuring phenomenon. Direct observations of bioluminescence were recorded over 2000 years ago in China and in Greece. In the fourth century BC Aristotle recognized that bioluminescence was a process of light emission without heat.

Systematic scientific studies on bioluminescence were published over 300 years ago by Robert Boyle, who reported the requirement of air for light emission by rotten wood and dead fish (9). This discovery essentially recognized the need for O_2 for light emission by fungi and bacteria but predated the discovery of O_2 and microbes by well over 100 years. During the intervening years a multitude of luminescent organisms were recognized, including bacteria, fireflies, fish, squid, clams, worms, and algae, with light emission extending from blue to red to yellow (40). In recent years significant advances have been made in our knowledge of the biochemistry and molecular biology of light emission in these organisms, including identification of the DNA coding for a number of luminescence-related proteins (22, 29, 81). These advances have laid the foundation for the application of bioluminescence to an expanding range of problems (62, 80).

In the last few years the genes from the bioluminescence operons of a number of marine bacteria have been cloned (6, 7, 18, 21, 29, 70), and the specific functions of the polypeptide products of the structural genes (*luxA–E)* have been identified (Table 1). The *luxA* and *B* genes code for the α and β subunits of luciferase (17, 47). The *luxC, D,* and *E* genes code for the

Table 1 Enzymes and genes from the bioluminescence systems of marine bacteria[a]

Gene	Length (bp)	Polypeptide	Mass[b] (kd)	Function
luxA	1062[d]–1065[c]	Luciferase α subunit	40–44	Catalysis of the bioluminescence reaction (FMNH$_2$ +
luxB	972[c], 978[d]	Luciferase β subunit	35–40	O_2 + aldehyde $\rightarrow$ light)
luxC	1431[c]	Reductase	53–58	NADPH-dependent reduction of activated fatty acyl groups to aldehyde
luxD	915[c], 918[e]	Acyl-transferase	32–34	Generation of fatty acids (tetradecanoic acid) for the luminescence system
luxE	1122[e]	Synthetase	42–50	ATP-dependent activation of fatty acids

[a]Identified in *V. harveyi, V. fischeri,* and *P. phosphoreum.* Two regulatory genes, *luxI* and *luxR,* have also been identified in *V. fischeri,* and their nucleotide sequences have been determined (33).
[b]Range from SDS-PAGE for *lux* polypeptides of *V. harveyi, V. fischeri,* and *P. phosphoreum.*
[c]For *V. harveyi lux* genes (17, 47, 71a; C. Miyamoto, unpublished data).
[d]For *V. fischeri lux* genes (34a).
[e]For *P. phosphoreum lux* genes (S. Ferri & R. Soly, unpublished data).

polypeptides of the fatty acid reductase system (reductase, acyl-transferase, and synthetase) (8, 30) responsible for synthesis of the fatty aldehyde substrate (15, 84, 87, 89). This review focuses on the recent advances in the biochemistry and molecular biology of bacterial bioluminescence. Particular attention is given to the identity, function, and regulation of the enzymes and genes coded on the *lux* operon(s) and their role in diversion of fatty acids from lipid metabolism into the bacterial bioluminescence reaction.

Bioluminescent Bacteria

Most luminescent bacteria are marine in origin and can be isolated from seawater or certain luminous species in the ocean. Currently, luminous bacteria are classified into four major genera: *Vibrio, Photobacterium, Alteromonas,* and *Xenorhabdus* (4, 44). Only bacteria of the latter genus can infect terrestrial organisms.

Detailed biochemical studies have only been conducted on the luminescence systems of marine *Vibrio* and *Photobacterium,* which can be distinguished by the presence of sheathed and unsheathed flagella, respectively. The properties and classification of marine bacteria have been well documented by Baumann and coworkers (4). Their primary life-style appears to be some form of symbiotic relationship with certain marine species; most strains are associated with specific organs of luminescent fish (74). Indeed, luminescence in most but not all marine fish can be attributed to the coexistence of luminescent bacteria. The vast majority (up to 96%) and most species of deep sea fish ($\geqslant 50$ %) are reported to be luminescent (40). However, some luminescent bacteria appear to exist in only a saprophytic or parasitic relationship with their hosts (43).

Various advantages have been proposed for marine organisms coexisting with light-emitting bacteria, including attraction of prey, diversion of predators, and communication (43, 74). However, the persistence of the luminescence system is difficult to explain, and the advantage to the bacteria is still unknown. There are literally hundreds of different strains of luminescent bacteria, which have maintained this energy-expending system of light emission over many generations, even though dark mutants can readily be obtained.

BIOLUMINESCENCE REACTION

The bioluminescence reaction in light-emitting organisms is catalyzed by enzymes designated as luciferases. In bioluminescent bacteria the luciferase-catalyzed reaction involves the oxidation of a long-chain aldehyde and $FMNH_2$ and results in the emission of a blue-green light (42, 106). The primary source of energy for the light (60 kcal Einstein^{-1}) is supplied by the conversion of the aldehyde to the corresponding fatty acid (24). Oxidation of the $FMNH_2$ accounts for the remaining energy in the reaction

$$FMNH_2 + O_2 + R\text{--}CO\text{--}H \rightarrow FMN + R\text{--}COOH + H_2O + light. \quad 1.$$

Aldehyde Specificity

Long-chain aldehydes were discovered to be essential for the luminescence reaction when they were identified as the stimulatory component in extracts of kidney cortex (19). Lipid extracts of luminescent bacteria, however, only contain sufficient amounts of aldehyde, primarily tetradecanal, to maintain luminescence for a few seconds in vivo (94). Both *Vibrio harveyi* and *Photobacterium phosphoreum* luciferases show a marked preference for tetradecanal at low, nonsaturating concentrations (65), although nonanal and decanal (45, 61) and some unsaturated aldehydes (62) can in some cases give high levels of light. Based on in vivo and in vitro data, tetradecanal appears to be the natural substrate for the luminescence reaction.

Reduced Flavin Specificity

The bioluminescence reaction catalyzed by bacterial luciferase is highly specific for reduced riboflavin phosphate ($FMNH_2$) (41, 44). Alteration of the flavin ring or removal of the phosphate decreases the activity significantly (64). In the latter case, substitution of a carboxyl group or exogenous addition of high concentrations of phosphate or sulfate, which act as autosteric effectors, can sometimes partially restore activity.

Some controversy exists over whether more than one reduced flavin molecule is required for a single turnover of the enzyme (5, 55, 56, 63, 106). Regardless of the relative validity of the different arguments, the participation of more than one molecule of a given substrate, except in a few condensation reactions, would be very unusual.

Reaction Mechanism and the Excited State

The first chemical step in the mechanism involves reaction of $FMNH_2$ with O_2 to form a 4a-peroxyflavin intermediate (41, 44, 96). Although some questions have been raised about its identification (56), hydroperoxy flavins have been shown to emit light with aldehydes in a nonenzymatic reaction (52). Moreover, FMN and H_2O_2 can be substituted for $FMNH_2$ and O_2 in the luciferase-catalyzed reaction to give low levels of luminescence (46). A mechanism in which a peroxyflavin aldehyde adduct is cleaved across the peroxy group with oxidation of the aldehyde and emission of light has been proposed (27, 41).

A number of different flavin intermediates have been considered as possible emitters in the excited state (106). The possibility that the flavin or another molecule acts as a secondary emitter to which the energy of the initial excited state has been transferred has also been proposed (41, 56, 106). Emission of yellow light by a *Vibrio fischeri* strain is a dramatic example of secondary emission by a yellow fluorescence protein (92). In *Photobacterium,* proteins containing bound 6,7-dimethyl-8-ribityllumazine affected the wavelength and efficiency of light emission in the reaction in vitro (79). Since there was a blue shift in the emitted light as well as an increase in intensity, it has been proposed that the lumazine protein competes with flavin for energy transfer from a primary emitter (56). Alternatively, the blue shift could arise from a change in the microenvironment of the primary chromophore (e.g. flavin) on interaction of luciferase with the lumazine protein.

Although the lumazine protein affects light emission, a number of questions must still be resolved concerning its role in the luminescence system. Very high concentrations of lumazine and luciferase are needed for observable effects on spectral emission. Consequently, the lumazine protein would be expected to be under the same regulation as the other luminescence-related proteins and to be present in high amounts in all strains, including those of *Vibrio.* Determination of the precise role of lumazine and related proteins will be important in elucidating the mechanism of emission from the excited state in bacterial bioluminescence.

Structure and Properties of Luciferases

All bacterial luciferases contain two nonidentical subunits, α and β, of 40–44 kd and 35–40 kd, respectively (37, 42, 61, 106). Luciferases ($\sim$80 kd) have been purified from a number of different strains including *V. fischeri, V. harveyi, P. phosphoreum,* and *Photobacterium leiognathi.* Comparisons of the amino acid and nucleotide sequences have demonstrated close homologies among the different luciferases as well as between the α and β subunits (3, 17, 47), which indicate that the α and β subunits have arisen by gene duplication (106).

Hybridization experiments and mutant analyses have indicated that the active site is located primarily on the α subunit (16, 66). However, the role of specific amino acid chains on the α subunit implicated in or near the active site is still unknown (20, 77). The β subunit, however, is still essential for the light-emitting reaction, and some evidence suggests that this subunit can affect the interaction of the enzyme with the reduced flavin (61). The properties and mechanism of bacterial luciferase and the bioluminescence reaction have been summarized comprehensively in Reference 106.

BIOSYNTHESIS OF FATTY ALDEHYDES

The mechanism of synthesis of the aldehyde substrate of the luminescence reaction was not deduced until over 30 years after its discovery as an essential component for bacterial luminescence. Analyses of dark mutants of *V. harveyi* showed that light emission could be stimulated by fatty acids in some mutants as well as by aldehyde, indicating that fatty acid was a precursor to the aldehyde (97, 98). Maximum stimulation of light was obtained with tetradecanoic acid; other fatty acids gave much lower light levels (100).

Analysis of extracts of *P. phosphoreum* revealed an enzyme that catalyzed the synthesis of aldehyde on incubation with ATP, NADPH, and fatty acid, providing direct in vitro evidence that fatty acids are converted to aldehyde (60, 82). Following this discovery a fatty acid reductase complex involved in producing aldehyde was purified (84), and its properties were characterized.

Fatty Acid Reductase Reaction

The reduction of fatty acids and long-chain acyl-CoAs has been demonstrated in a wide variety of biological organisms including plants, insects, mammals, and bacteria (83). Although acyl-CoA reductases have been partially purified from a number of tissues, the only system capable of reducing free fatty acids that has been resolved from crude extracts and purified is the fatty acid reductase complex from the luminescent bacterium *P. phosphoreum* NCMB 844. The net reaction involves reduction of fatty acid to aldehyde with oxidation of NADPH and cleavage of ATP to AMP and PP_i:

$$R–COOH + ATP + NADPH \rightarrow R–CHO + AMP + PP_i + NADP. \quad 2.$$

Maximum activity, as measured by the rate of release of AMP (86), is obtained with tetradecanoic acid, confirming earlier studies using a luciferase-coupled assay (82). The reaction is highly specific for ATP and NADPH, which give apparent K_ms of 20 nM and 1 μM, respectively, at saturating concentrations of the other substrates. A 1:1 stoichiometry between the amount of AMP, aldehyde, and NADP produced and the fatty acid consumed demonstrates a tight coupling of the multistep enzyme reaction (86).

Two different protein components are necessary for in vitro reconstitution of this enzyme activity: acyl-protein synthetase, [s], which is responsible for the ATP-dependent activation of the fatty acid (89), and the reductase component, [r], which catalyzes the NADPH-dependent reduction of the activated fatty acid (87). A third component, acyl-transferase, [t], present in the partially purified fatty acid reductase complex, is involved in the generation of fatty acids (15), and its properties are described later. The specific steps outlined below, which are catalyzed by the *P. phosphoreum* fatty acid reductase, may provide the basis for a general model for the reduction of fatty acids in biological organisms.

FATTY ACID ACTIVATION   Metabolism of fatty acids, including biosynthesis, degradation, reduction, and transport, in almost all instances requires activation of the carboxyl groups. In the bacterial luminescence system this step is catalyzed by the acyl-protein synthetase component, either alone or as part of the fatty acid reductase complex, and results in the synthesis of acyl-AMP (85):

$$R\text{–}COOH + ATP \rightleftharpoons R\text{–}CO\text{–}AMP + PP_i. \qquad\qquad 3.$$

The reverse reaction can be easily measured by the incorporation of $^{32}PP_i$ into ATP.

The mechanism appears to be analogous to the activation of amino acids by aminoacyl-tRNA synthetases. The acyl-AMP product is so tightly bound to the enzyme that it can be resolved from the substrates by gel filtration (85). Acyl-AMP intermediates have been detected during the synthesis of acetyl-CoA and butyryl-CoA, and it seems likely that acyl-AMP is a common intermediate in most fatty acid activating systems.

ACYLATION OF THE SYNTHETASE   Reaction of bound acyl-AMP with the synthetase subunit,

$$R\text{–}CO\text{–}AMP + [s] \xrightarrow{[r]} R\text{–}CO\text{–}[s] + AMP, \qquad\qquad 4.$$

readily occurs in the fatty acid reductase complex, but occurs to only a small extent in the absence of the reductase subunit (89). Interestingly, three other proteins involved in aldehyde metabolism in luminescent bacteria, the acyl-transferases from *P. phosphoreum* (15) and *V. harveyi* (13) and the *V. harveyi* aldehyde dehydrogenase (12), can also stimulate acylation of the synthetase. Since the reductase subunit can also transfer acyl groups from the synthetase to thiols, the level of acylation can be significantly increased by reducing the concentration of free thiols in the assay (105) or by blocking the active site of the reductase with *N*-ethylmaleimide (88) and preventing acyl turnover. Apparently, interaction between the reductase and the synthetase mediates a conformational change in the synthetase, and even a nonfunctional reductase can stimulate acylation.

A cysteinyl residue has been proposed as the site of acylation of the synthetase based on the susceptibility of thioesters and the acyl group to cleavage by neutral hydroxylamine (88). One mole of *N*-ethylmaleimide inactivated the synthetase and on partial proteolysis generated the same modified peptides as the acylated enzyme. Since the reductase can protect against proteolytic cleavage of the synthetase as well as stimulate acylation, it is possible that the acylation site on the synthetase is constrained specifically toward the bound acyl-AMP on interaction with the reductase. This model would also be consistent with the direct intersubunit transfer of the acyl group from the synthetase to the reductase and would provide a highly efficient mechanism for channeling the activated fatty acid to aldehyde.

FATTY ACYL TRANSFER   Turnover of the acyl group on the synthetase arises from transfer of the acyl group to thiol acceptors (XSH) via the reductase subunit:

$$R\text{--}CO\text{--}[s] \rightleftharpoons R\text{--}CO\text{--}[r] \rightleftharpoons R\text{--}CO\text{--}SX. \qquad\qquad 5.$$

By limiting the concentration of β-mercapotoethanol or dithiothreitol in the reaction, high levels of acylation of the reductase as well as the synthetase can be obtained (0.4–0.8 mol fatty acid mol^{-1} polypeptide) on incubation with fatty acid and ATP (105). Since acyl transfer is reversible, the synthetase can also be labeled with acyl-CoA if the latter is mixed with the reductase. Other activated acyl esters, including acyl–acyl carrier protein (acyl-ACP) do not appear to be as effective as acyl-CoA in direct acylation of the reductase subunit. An increase in mobility on SDS-PAGE on acylation of the reductase provides a very useful probe for determining the degree of labeling of the active site (105).

REDUCTION TO ALDEHYDE   The reduction of the acylated reductase is specific for NADPH, with a much lower activity with NADH (87):

$$R\text{--}CO\text{--}[r] + NADPH \rightleftharpoons NADP + R\text{--}CHO + [r]. \qquad\qquad 6.$$

NADPH or NADP will block transfer of the acyl groups to thiols, which indicates that these molecules bind at the same site on the enzyme.

Although the reductase was originally identified by its ability to reduce acyl-CoA and was designated as an acyl-CoA reductase, its function appears to be more directly related to reduction of fatty acids activated by the synthetase. These results raise the possibility that acyl-CoA reductases in other organisms may actually have a very different role in cellular metabolism than is defined by their present kinetic properties (83). Purification and characterization of the fatty acid reductases from other biological systems would be necessary to determine if the acyl-CoA reductases are components of similar complexes in these organisms.

Channeling of Fatty Acids Into the Luminescence System

The generation of fatty acids for the fatty acid reductase system is catalyzed by an acyl-transferase specifically associated with the luminescence system. In *P. phosphoreum*, the acyl-transferase can either be copurified with the fatty acid reductase complex and then be resolved from the other components in the complex or be purified from crude extracts as a separate entity (15). A luminescence-related acyl-transferase has also been purified from *V. harveyi* (13).

The acyl-transferase can react with different esters including acyl-CoA, acyl-*S*-mercaptoethanol, acyl-*p*-nitrophenol, acyl-glycerol, and acyl-ACP; the reaction results in transfer of the acyl group to the enzyme (13–15). The acyl acceptor can be either water, generating free fatty acid, or different alcohol or thiol acceptors (A) including β-mercaptoethanol, glycerol, and ethyleneglycol, which stimulate cleavage of acyl-CoA, acyl-*p*-nitrophenol, and acyl-ACP, forming the corresponding ester:

$$R\text{--}CO\text{--}X \rightleftharpoons R\text{--}CO\text{--}[t] \underset{A}{\overset{H_2O}{\rightleftharpoons}} \begin{matrix} RCOOH. \\ \\ R\text{--}CO\text{--}A. \end{matrix} \qquad 7.$$

The final acceptors appear to be determined to some degree by the nature of the acyl donor. Acyl-*S*-ACP is cleaved primarily to fatty acid, and acyl-CoA is cleaved to acyl-*S*-mercaptoethanol. Since the next step in aldehyde biosynthesis is activation of the fatty acid by the synthetase, this result might suggest that acyl-ACP is the natural substrate in vivo (14). However, free fatty acid can also be produced from acyl-CoA, and this activity can be stimulated by the synthetase at high protein concentrations (15), which indicates that protein interactions may affect the specificity of the transferase.

Maximum activity has been observed for acyl chain lengths of 14 carbons for esters of ACP, *p*-nitrophenol, and coenzyme A (S. Ferri, unpublished data). This result is in agreement with the observations that tetradecanoic acid was the preferred substrate for conversion into aldehyde by the fatty acid reductase (86) and that less tetradecanoic acid was produced in a transferase-deficient mutant (11). Similarly, tetradecanal has been demonstrated to be the preferred substrate for luciferase (65).

The specific utilization of acyl-ACP by the acyl-transferase has been demonstrated in extracts of *V. harveyi* (14). Acyl-ACP could not be turned over in extracts of a mutant lacking a functional acyl-transferase or in extracts of *E. coli* (14); this result indicates that the luminescence-related acyl-transferases are the only enzymes that can cleave acyl-ACP at a high rate in marine bacteria or *Escherichia coli*.

The current data suggest that the acyl-transferases are involved in diverting fatty acids into the luminescence system. The transferases would preferentially release fatty acids from acyl-ACP with a chain length of 14 carbons and would thus prevent further elongation of tetradecanoyl-ACP to palmitic acid by the fatty acid synthetase. The release of shorter-chain fatty acids from the fatty acid biosynthetic pathway by specific esterases has been demonstrated in a number of biological systems (90, 95). Although the luminescence-related transferases do not appear to cause a significant disruption in lipid biosynthesis, an elongation system that can convert the released tetradecanoic acid into palmitic acid has recently been identified in luminous bacteria (D. Byers, personal communication).

Structure of the Fatty Acid Reductase Complex in Photobacterium phosphoreum

The fatty acid reductase complex from *P. phosphoreum* is a weakly interacting multienzyme system that dissociates at low protein concentrations. Although the complex has been purified to only 80% homogeneity, each of the polypeptide components has been resolved by Blue A Sepharose (84) and purified to homogeneity (15, 87, 89), and the enzyme complex has been reconstituted.

The basic complex with fatty acid reductase activity ($\sim 4 \times 10^5$ kd) contains four subunits each of the reductase (58 kd) and synthetase (50 kd) components (103). The central core consists of reductase polypeptides that can be resolved from the complex as a tetramer (2×10^5 kd) with acyl-CoA reductase activity, whereas the synthetase is isolated as a monomer of 50 kd. The synthetase, however, still retains its ability to catalyze the fatty acid–dependent PP_i-ATP exchange and formation of acyl-AMP (85).

The partially purified fatty acid reductase complex also contains low amounts of the transferase polypeptides. Complementation experiments as well as cross-linking studies have demonstrated a specific interaction of the transferase with the synthetase in a 1:1 stoichiometry (103). The amount of transferase found in the fatty acid complex may be lower because of partial dissociation during purification.

Fatty Acid Reductase Polypeptides in Other Luminescent Species

Much less is known about the properties of fatty acid reductases in other luminescent marine bacteria. Fatty acid reductase activity has been detected in extracts of a number of *Photobacterium* species but not in *Vibrio* (58, 60, 89, 101). In contrast, the acyl-ACP cleavage activity of the acyl-transferase component ([t]) has been detected in all luminescent species so far tested. The absence of fatty acid reductase activity in *Vibrio* extracts may reflect the decoupling of the fatty acid–activating and reducing enzymes ([s] and [r]) on lysis of the cells due to dissociation upon dilution. The polypeptides of the fatty acid reductase complex in the different luminescent strains can, however, be specifically identified by acylation with fatty acids even in crude extracts (89). Three acylated polypeptides, corresponding to the transferase, synthetase, and reductase polypeptides, are induced during the development of luminescence in *V. harveyi* (102) and *V. fischeri* (7) as well as in *P. phosphoreum* (104).

Figure 1 gives the characteristic patterns on SDS-PAGE for the fatty acid reductase polypeptides ([r], [s], and [t]) labeled in vivo with [³H]tetradecanoic acid and in vitro with [³H]tetradecanoyl-CoA for five luminescent strains. Changes in the acylation patterns due to alteration of the experimental conditions (e.g. time, thiol concentration) or acylation procedures (e.g. in vitro with fatty acid) can be used to identify the specific polypeptides. The major bands labeled in vivo correspond to the [r] and [s] polypeptides. The labeled *V. harveyi* reductase polypeptide has a higher molecular weight (~57 kd) on SDS-PAGE than the reductase polypeptides from the other strains (53–54 kd), whereas both the labeled *Vibrio* synthetases (~42 kd) migrate faster than the *Photobacterium* synthetase polypeptides (48–50 kd). The observation that the molecular weight of the labeled *P. phosphoreum* reductase is lower than that of the unmodified polypeptide (58 kd) is consistent with an increase in mobility on SDS-PAGE on acylation of the reductase polypeptide (105).

On in vitro acylation with [³H]tetradecanoyl-CoA the transferase poly-

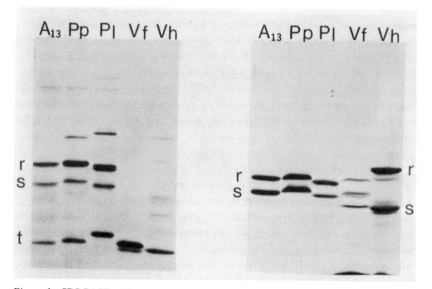

Figure 1 SDS-PAGE and autoradiography patterns of extracts of *P. phosphoreum* A13, *P. phosphoreum* NCMB 844, *P. leiognathi* ATCC 25521, *V. fischeri* ATCC 7744, and *V. harveyi* B392 after acylation in vitro with [³H]tetradecanoyl-CoA *(left)* and in vivo with [³H]tetradecanoic acid *(right)*. [Reductase (r), synthetase (s), transferase (t).] (Courtesy of R. Szittner, unpublished data.)

peptides (32–34 kd), as well as the *Photobacterium* reductase and synthetase polypeptides, are easily detected. In vitro, longer exposure times and reaction with [^{3}H]tetradecanoic acid plus ATP are generally needed to detect the *Vibrio* reductase and synthetase polypeptides by acylation. The fatty acid reductase system consisting of [t], [s], and [r] subunits thus appears to be present in all luminescent marine bacteria.

CLONING AND EXPRESSION OF THE *lux* GENES

The generation of light-emitting *E. coli* on transformation with cloned genes coding for the luminescence functions (29) has advanced our understanding of the gene organization and regulation of luminescence in marine bacteria. Moreover, the creation of new light-emitting organisms (54, 80) has stimulated a wide range of interest in the application of luminescence to applied and basic research.

Luciferase Genes, luxA *and* B

E. coli that produce high levels of luminescence in the presence of exogenous long-chain aldehyde have been obtained by transformation with DNA from a number of luminescent marine bacteria. For aldehyde-dependent luminescence, only a <2.5-kbp DNA fragment with the *luxA* and *B* genes coding for the luciferase α and β subunits need be transformed into *E. coli* under a suitable promoter (1, 2, 6, 32, 38, 51, 68, 69). The nucleotide sequences for the *luxA* and *B* genes are known for *V. harveyi* luciferase (17, 47) as well as for *V. fischeri* (34a), *P. leiognathi,* and *P. phosphoreum* luciferases (T. Baldwin, personal communication). Both genes are approximately 1 kbp and they are located immediately adjacent to one another and are transcribed in the same direction.

The *luxA* and *B* genes have been expressed in bacteria other than *E. coli* (34, 51, 57) and even in plant cells (54). The high sensitivity of the assay for light emission, coupled with the absence of endogenous luciferase in the cells to be transformed, permits the detection of very low levels of the luciferase product. Since the FMNH$_2$ substrate appears to be sufficiently abundant in most bacteria, and since aldehyde can readily cross the cellular membrane, the luminescence assay can be conducted and the light can be quantitated without cell disruption. These properties provide considerable potential for the use of the luciferase genes and their protein products in the development of rapid screening systems, e.g. for the detection of bacterial promoters (2, 32, 34, 57, 68). The luciferase genes have also been expressed in eukaryotic cells using separate promoters for the *luxA* and *B* genes, although it was necessary to lyse the cells and light emission was relatively low in the in vitro assay (54).

Fatty Acid Reductase Genes, luxC, D *and* E

To obtain light emission without addition of aldehyde, the genes coding for aldehyde synthesis *(luxC–E)* must also be transformed and expressed in *E. coli.* Completely luminescent aldehyde-independent *E. coli* have been produced on transformation with DNA from *V. fischeri* MJ1 and 7744 (7, 29), *V. harveyi* (70), *P. leoignathi* (21), and *P. phosphoreum* (J. Mancini, M. Boylan, R. Soly, A. F. Graham, & E. A. Meighen, manuscript in preparation). Figure 2 gives the restriction maps for the DNA from *P. phosphoreum, V. harveyi,* and *V. fischeri.* The structural genes *lux C–E*, encoding the three polypeptides of the fatty acid reductase complex ([r], [t], and [s]), required for aldehyde synthesis, flank the luciferase genes *(luxA* and *B)* in all luminescent systems. The *luxC* and *D* genes are located immediately upstream of *luxA* and *B,* and *luxE* is located within 1 kbp downstream; transcription is from left to right. In *P. phosphoreum* a small gene (<1 kbp) separates *luxB* and *E* (J. Mancini & R. Soly, unpublished data). The lux structural genes in *V. fischeri* have been shown by transposon mutagenesis to be in the same operon (29).

Fatty acid reductase activity has been detected in extracts of transformed *E. coli* containing the complete *V. fischeri* luminescence system, but not in extracts of the other cloned systems (7). However, individual polypeptides of the *luxC, E,* and *D* genes have been identified in *E. coli* transformed with *V. fischeri,, V. harveyi,* or *P. phosphoreum* DNA by acylation in vivo with fatty acid or in vitro with fatty acid (+ATP) or acyl-CoA (8). The *luxD* gene

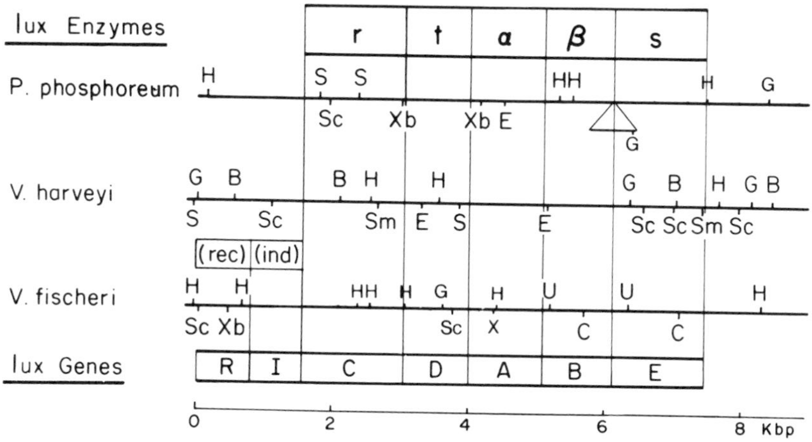

Figure 2   Restriction maps and location of the *lux* enzymes and genes in *V. harveyi* B392, *V. fischeri* ATCC 7744, and *P. phosphoreum* NCMB 844. All restriction sites for *Bam*HI (B), *Bgl*II (G), *Eco*RI (E), *Hind*III (H), *Sac*I (Sc), *Sal*I (S), *Sma*I (Sm), *Xba*I (Xb), and *Xho*I (X) are given except for *Xba*I sites in *V. harveyi* DNA. *Cla*I (C) and *Pvu*II (U) sites are shown only for *V. fischeri* DNA. As depicted, ~ 0.7 kbp of DNA are between *luxB* and *E* in *P. phosphoreum* (R. Soly, unpublished data).

product can also be quantitated in transformed *E. coli* by its specific capability to cleave acyl-ACP.

Regulatory Genes

In addition to the structural genes, two regulatory genes, *luxI* and *R,* separated by a 218-bp noncoding region (33), have been identified in *V. fischeri* MJ1 (29). The *luxI* gene in *V. fischeri,* consisting of 579 bp, is located on the same operon (the right) as the structural genes *luxA–E,* whereas the *luxR* gene (750 bp) is transcribed in the opposite direction on the left operon (29, 33).

The *V. fischeri luxI* and *R* genes were identified by complementation experiments in *E. coli* transformed with mutated *lux* DNA (30, 31). Light emission could be restored to *luxI* but not *luxR* mutants by the addition of the *V. fischeri* autoinducer. Consequently, it was proposed that *luxR* encodes a receptor protein that responds to an autoinducer synthesized by the *luxI* gene. Polypeptides of 25 and 27 kd on SDS-PAGE (30, 31) have been identified in *E. coli* minicells on expression of the *luxI* and *luxR* genes, respectively. The *luxR* gene product has been isolated from transformed *E. coli,* in which it was required for expression of luminescence (49). However, the purified protein did not bind specifically to *lux* DNA or the autoinducer, which suggests that receptor function was lost during purification or that binding conditions were inappropriate. Alternatively, the *luxR* protein may have a different function.

Although comparable regulatory genes would be expected in other luminescent bacteria in view of the similarities in organization of the *lux* structural genes and the regulation of luminescence expression, cross-hybridization of *lux* DNA of *V. harveyi, V. fischeri,* and *P. phosphoreum* has shown that the close homologies in the DNA do not extend upstream past the *luxC* gene (71a; E. P. Greenberg, personal communication) where the *luxI* and *R* genes of *V. fischeri* are located (Figure 2). In *V. fischeri* MJ1, the 579-bp *luxI* gene ends 56 bp before the start of the *luxC* gene (33). In contrast, an open reading frame greater than 40 codons with an ATG initiation codon is not present within 600 bp upstream of the start of the *luxC* gene in *V. harveyi* (71a) or *P. phosphoreum* (R. Szittner & E. A. Meighen, unpublished data). As the nearest gene is transcribed in the opposite direction to *luxC,* a regulatory gene cannot be the first gene in an operon containing the structural genes in *V. harveyi* or *P. phosphoreum.* Consequently, it appears that the organization of *lux* regulatory genes and/or the basic mechanism controlling the induction of luminescence must have diverged in marine bacteria.

Only 6.3 kbp of *V. harveyi* DNA, coding for *luxA–E,* are needed to obtain aldehyde-independent luminescence in mutant *E. coli* cells (70), whereas approximately 8.0 kbp of *V. fischeri* DNA are needed, encompassing the regulatory (*luxR* and *I*) and structural genes *(luxA–E).* This difference may reflect the ability of the *E. coli* mutant cells to bypass whatever regulatory mechanisms control the expression of the *V. harveyi* luminescence system. The level of expression depends on the particular set of cloned genes, the *E. coli* host, and the growth or incubation conditions. The *V. fischeri* system appears to have the widest *E. coli* host range for expression (7, 21, 29, 70). In contrast, the *V. harveyi* luminescence system is expressed only at high levels in the mutant *E. coli* cells (70). Only extremely low levels of expression have been observed for the cloned *P. phosphoreum* system compared to those for the other luminescence systems (68). However, the low levels may reflect a high dependence on growth conditions and the *E. coli* host, as has been demonstrated for the *P. leiognathi* system (21).

The difference in expression may also largely reflect the capability of *E. coli* RNA polymerase to recognize the initiation and termination sequences in the *lux* DNA from marine bacteria. In this regard, results obtained from *E. coli* transformed with the *lux* system must be interpreted with reasonable caution. The transfer of cloned *lux* genes back into dark mutants of luminescent bacteria (39) may allow a more definitive study of regulation of luminescence in the parent strain.

mRNA and Transcription

A set of polycistronic mRNAs of 8, 7, 4, and 2.6 kb, detected by hybridization with *lux* DNA (71), extends to the left of the transferase gene *(luxD)* and at least 2–3 kb downstream past the end of the synthetase gene *(luxE)* both in *V. harveyi* and in the *V. harveyi* system cloned in *E. coli* (Figure 3). Cloned DNA that was truncated immediately after the synthetase gene and missing the large mRNAs (7 and 8 kb) also gave high levels of light in *E. coli*. It is possible that this downstream region codes for proteins involved in some other aspect of luminescence that is not essential for expression of light, at least in *E. coli*. A new, less abundant set of *V. harveyi* mRNAs, starting just above the *luxC* gene and extending downstream 1.5–10 kb (Figure 3), has also recently been detected and suggests that a promoter for the *lux* structural genes may be located in front of *luxC*. The polycistronic mRNAs could arise from different promoter and termination sites and/or specific in vivo processing or degradation of the mRNA.

REGULATION

The intensity of light emission from luminescent marine bacteria depends upon a number of parameters, including the nutritional state of the cell, the oxygen and salt concentrations, the temperature, and the level of catabolites and cAMP (43). In addition, a wide range of compounds including environmental pollutants, drugs, anesthetics, and mutagens can affect in vivo luminescence (44), providing the basis for sensitive light-emitting bioassays.

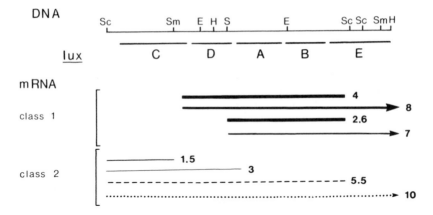

Figure 3   Location of the polycistronic mRNAs of the *V. harveyi* luminescence system (courtesy of C. Miyamoto, unpublished data). The relative levels of mRNA are indicated by the intensities of the lines. *Arrowheads* indicate that mRNA extends downstream ($\sim$ 3 kb) after *luxE*.

Even under optimal nutritional conditions for expression of luminescence, the light emission per cell in most luminescent bacteria is very low at the early stages of cellular growth (74, 75).

Changes in the intensity of luminescence could arise by a number of different mechanisms that affect the availability of luciferase or the substrates for the luminescence reaction. Modulation of the activities of the *lux* enzymes or the processes involved in transcription or translation of the genes coding for the α and β subunits of luciferase or the [r], [s], and [t] polypeptides of the fatty acid reductase system can change the intensity of in vivo luminescence. In addition, the levels of substrates (e.g. $FMNH_2$), precursors, or regulatory molecules could be changed directly or by affecting the expression of other genes or proteins. Since luminescence can easily and continuously be quantitated at very low levels without cell disruption, the luminous systems have a significant advantage over other systems for the study of the regulation of gene and enzyme expression inside the cell. However, interpretation of the effects of different conditions on luminescence may be complicated, and there is some danger of misinterpreting changes in the intensity of in vivo luminescence in view of the relatively extensive range of possible mechanisms.

Induction of Luminescence

In many luminescent bacteria, light emission lags remarkably behind cellular growth and then increases several thousand fold at higher cell densities (75). Both subunits of luciferase (67) and the [r], [s], and [t] polypeptides of the fatty acid reductase system (104) are specifically synthesized during the induction of luminescence. Since the genes for these polypeptides are encoded on the same operon in *V. fischeri* and their relative location is similar in other luminescent strains (Figure 2), a mechanism involving the coordinate regulation of their expression would appear to be soundly based. The synthesis of other soluble (67) and membrane (76) polypeptides, including a longchain aldehyde dehydrogenase (67), has been associated with the development of bioluminescence in *V. harveyi*. However, the induced aldehyde dehydrogenase was recently shown to be differentially regulated from the *lux* enzymes in minimal media (10).

The growth-dependent lag in luminescence occurs because the cells must produce an inducer and remove inhibitors from the media for expression of luminescence (25, 53). This lag is greatly reduced in media conditioned by prior growth of the same luminescent strain. The early induction of luminescence upon addition of low amounts of conditioned media to cells inoculated in fresh media provided strong evidence that a positive factor (autoinducer) was being produced. Under conditions where the autoinducer cannot accumulate in the media, luminescence is repressed (91, 99). Consequently, luminous bacteria living free in the ocean would emit very low levels of light compared to bacteria living in a confined environment.

The autoinducer for the *V. fischeri* luminescence system is *N*-β-ketocaproylhomoserine lactone (26). It can induce luminescence in *V. fischeri* and *V. logei* but not in other species (29). A heat-sensitive autoinducer from *V. harveyi* (25) appears to be produced by a number of other marine bacteria (36), which brings into question the species specificity for induction of luminescence by any one autoinducer. Although different autoinducers might be expected to be structurally or metabolically related if a common mechanism regulates luminescence in different species, analogs of the *V. fischeri* autoinducer could not stimulate luminescence in other species (28). Further experiments to establish the structure and properties of autoinducers from other species are clearly needed.

The enzyme function responsible for synthesis of the autoinducer of *V. fischeri*, *N*-β-ketocaproylhomoserine lactone, has not yet been identified. Engebrecht et al (29) have proposed that the *luxI* product is responsible for synthesis of the autoinducer in *V. fischeri* and in *E. coli* transformed with *lux* DNA from this species, suggesting that the precursor(s) must be present in both bacteria. The autoinducer is composed of an amino acid and a fatty acid metabolite and thus possibly provides a signaling mechanism for the nutritional state in the cell. Possible precursors could be homoserine and β-ketocaproyl-ACP or β-ketocaproyl-CoA. The last two compounds are intermediates in fatty acid biosynthesis and degradation, respectively, and thus could be indicators of the relative levels of lipid metabolism in the cell. An increase in the level of these potential precursors could be important for induction of a system that involves diversion of the fatty acids away from normal lipid metabolism and into the luminescent system.

A proposed mechanism for the rapid induction of bioluminescence in *V. fischeri* involves both positive and negative feedback regulation (29, 31). The autoinducer, which freely diffuses across the cell membrane (48), would be produced initially at a low constitutive rate. When enough autoinducer had accumulated during cellular growth, it would interact with the *luxR* gene product and turn on transcription of the right operon. This interaction would lead to further synthesis of the *luxI* gene product and the autoinducer and would result in an exponential increase in synthesis of the *lux* enzymes required for bioluminescence. High levels of the receptor-inducer complex would then turn off expression of the *luxR* gene. This mechanism would explain why luminescence becomes constant or declines at higher cell density.

This model would predict that synthesis of the autoinducer (via *luxI* expression) as well as expression of the structural genes *(luxA–E)* would increase during growth. Although synthesis of the autoinducer appears to be constitutive (72) and luciferase and fatty acid reductase activities increase with growth (7), recent experiments have shown that the mRNA for the operon containing *luxI* is induced (33). A different regulatory mechanism may be needed to account for induction of luminescence in *V. harveyi* and *P. phosphoreum*, which lack a gene in the location of the *luxI* gene of *V. fischeri*.

Catabolite Repression

The expression of the bioluminescence system of *V. harveyi* is partially controlled by catabolite repression (59, 73). Expression of the luciferase and fatty acid reductase polypeptides is decreased by the addition of glucose, and these functions are partially restored on addition of cAMP (10, 73). In *V. fischeri,* catabolite repression by glucose has been observed in phosphate-limited chemostat cultures prior to induction of luminescence (35). Addition of cAMP or autoinducer relieves this repression. Using *E. coli* mutants lacking cAMP receptor protein (CRP) or cAMP and transformed with the *V. fischeri lux* system, cAMP and CRP were found to be necessary for high expression of the left operon, containing the *luxR* gene, whereas expression of the right operon was inhibited in the absence of the *luxR* gene product (23). These results may explain why both cAMP and autoinducer relieve repression of luminescence by glucose (35), since expression of the right operon, containing the *lux* structural genes, would be increased by higher levels of the cAMP-stimulated *luxR* gene product as well as by the autoinducer.

Differential Expression of the lux Structural Genes

In *P. phosphoreum* up to 20% of the soluble protein is luciferase, whereas the level of fatty acid reductase, constituted by the *luxC* and *E* gene products ([r]

and [s]) in 1:1 molar ratio, is at least fivefold lower (104). As the genes for aldehyde biosynthesis *(luxC–E)* flank the luciferase genes *(luxA* and *B)*, expression of the *lux* genes must be closely regulated. By placing the luminescence system under the control of the T7 phage promoter and analyzing expression of the lux enzymes in *E. coli,* differences in transcription can be eliminated. Figure 4 shows that the synthesis of the luciferase polypeptides coded by *luxA* and *B* of *V. fischeri* is still much higher than that of the fatty acid reductase polypeptides coded by *luxC* and *E* even under the T7 promoter. Differential control of expression of the *lux* genes must therefore be exerted after gene transcription owing to differences in rates of translation, degradation, or processing of the mRNA.

Turnover of Fatty Acid in the Luminescence System

A general model relating the enzymes in the luminescent system to fatty acid metabolism is given in Figure 5. The transferase enzyme diverts fatty acids from the pathway leading to lipid biosynthesis with release of tetradecanoic acid. Since the synthetase interacts with the transferase (103), the released fatty acid may be preferentially activated to fatty acyl-AMP by the synthetase. Interaction with the reductase is then required for efficient acylation of the synthetase, followed by acyl transfer to the reductase and reduction to aldehyde. Consequently, formation of the fatty acid reductase complex would control the flow of fatty acid into aldehyde. Channeling of aldehyde to luciferase through a direct interaction between the fatty acid reductase and luciferase subunits may also occur; this would prevent aldehyde loss via other

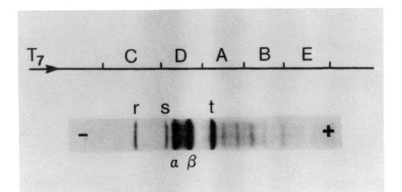

Figure 4 Specific expression in *E. coli* of the *lux* structural genes under the T7 phage promoter. *V. fischeri* DNA containing the *lux* structural genes was inserted after the T7 promoter and transformed into *E. coli* containing the T7 RNA polymerase gene on a compatible plasmid. Addition of rifampycin, which inhibits *E. coli* but not T7 RNA polymerase, allowed exclusive expression of the *lux* genes. The polypeptides were labeled with a mixture of tritiated amino acids and subjected to SDS-PAGE followed by autoradiography (courtesy of M. Boylan, unpublished data).

metabolic pathways and maintain the aldehyde concentration at a low level. Fatty acid regenerated in the luminescence reaction could then be recycled back to aldehyde by the fatty acid reductase reaction (Figure 5).

The regulation of the flux of fatty acid and aldehyde through the fatty acid reductase system is critical in determining the final level of light emission in luminescent bacteria. One milliliter of bright *P. phosphoreum* cells, which emit light at a rate of $\sim 1 \times 10^{13}$ quantum sec^{-1} [based on the luminol standard (78)], would require a minimum rate of aldehyde synthesis of 0.9 nmol min^{-1} assuming 100% quantum efficiency in vivo. Although 10–20% quantum efficiency has been measured for aldehyde turnover in the in vitro bioluminescence reaction (24, 93), analyses of the oxygen consumption and ATP levels in the luminescent cells have indicated that the quantum efficiency in vivo may be much higher (50). A maximum rate of 1 nmol min^{-1} for reduction of tetradecanoic acid was measured for an extract of 1 ml of bright *P. phosphoreum* cells (E. Meighen, unpublished data). Thus the fatty reductase system can catalyze the synthesis of sufficient aldehyde to maintain luminescence providing that the chemical reaction and light emission catalyzed by luciferase are tightly coupled in vivo.

In contrast, the maximum rate of cleavage of tetradecanoyl-ACP to form tetradecanoic acid is only 0.2 nmol min^{-1} for an extract of 1 ml of bright *P. phosphoreum* cells. This rate is too low for the acyl-transferase enzyme to be the exclusive source of fatty acid for reduction to aldehyde unless the fatty acid produced in the luminescence reaction can be recycled back through the fatty acid reductase system (Figure 5). Recycling of the tetradecanoic acid would minimize the amount of fatty acid needed to be diverted from the normal metabolic pathways involved in lipid metabolism. Regulation of

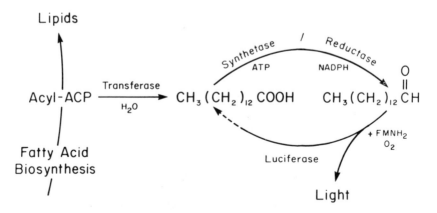

Figure 5 Scheme depicting the relationship between lipid metabolism and the turnover of tetradecanoic acid and tetradecanal in the luminescence system.

expression of the *lux* structural genes and the respective functions of the luciferase and products must be closely coordinated so that lipid turnover and luminescence can be maintained without generating high intracellular levels of fatty acid and aldehyde.

SUMMARY

Over the past five years, major advances have been made in our knowledge of the function and organization of the enzymes and genes in the luminescent bacteria. The *lux* structural genes have now been identified and have been shown to be transcribed and located in the order *luxCDABE,* although an additional gene is located between *luxB* and *E* in *P. phosphoreum.* The structural genes code for the two luciferase subunits, α and β (*luxA* and *B*), and the three fatty acid reductase polypeptides, r, s, and t (*luxC, E,* and *D*), involved in the diversion of fatty acids from lipid biosynthesis into aldehyde required for the luminescence reaction. Tetradecanoic acid is specifically produced by the transferase, with the most probable source being tetradeca-noyl-ACP; activation by the synthetase, followed by reduction after transfer of the acyl group to the reductase, converts the fatty acid to tetradecanal, the preferred substrate for luciferase. Aldehyde and fatty acid turnover by the luciferase and fatty acid reductase enzymes must be closely coupled to control the level and flux of lipid in the luminescence system. Characterization of the specific chemical mechanisms of the luciferase and fatty acid reductase enzymes, the interactions between these proteins, the mechanism of channel-ling the lipid substrate, and the interfacing of the *lux* system with the other metabolic pathways will be needed to understand more closely how the expression of the luminescence system is controlled at the enzyme level.

What mechanism(s) regulates the growth-dependent induction of lumines-cence? Excreted autoinducers stimulate light emission in a number of species. Do the *lux* regulatory genes in *V. fischeri, luxI* and *R,* believed to be responsible for autoinducer synthesis and reception, respectively, exist in other species and code for these or similar functions? If so, the organization of the *lux* regulatory genes must have diverged in the different species. Moreov-er, the bacterial luminescence systems differ in glucose repression. Evolution of different regulatory mechanisms in marine bacteria to limit luminescence at low cell density may reflect a critical need to prevent utilization of energy or specific metabolites under these conditions. These control mechanisms must also allow for the coordinate expression of the luciferase genes at a higher level than the flanking fatty acid reductase genes. The mechanisms would require differences in translation rates, mRNA processing or degradation, or protein turnover as well as transcriptional regulation. Expression of the *lux* system in marine bacteria appears to be as complex, varied, and finely coordinated as that of any prokaryotic or eukaryotic system. The ability to manipulate and anaylze the *lux* system will allow a more in-depth understanding of the regulatory mechanisms controlling expression of biological systems.

ACKNOWLEDGMENTS

The author wishes to thank his coworkers over the past five years, particularly Michael Boylan, David Byers, Jack Cao, Jilly Evans, Stefano Ferri, Angus Graham, Joe Mancini, Carol Miyamoto, David Morse, Denis Riendeau, Miguel (Angel) Rodriguez, Robert Soly, Elana Swartzman, Rose Szittner, and Lee Wall for their ideas and research contributions. I am also indebted to Tom Baldwin, Peter Greenberg, John Lee, Dennis O'Kane, and Michael Silverman, who generously provided current information prior to publication. The patience and skill of Josie D'Amico in the preparation of the manuscript are also greatly appreciated. The author gratefully acknowledges the support of grants from the Medical Research Council of Canada.

Literature Cited

1. Baldwin, T. O., Berends, T., Bunch, T. A., Holzman, T. F., Rausch, S. K., et al. 1984. Cloning of the luciferase structural genes from *Vibrio harveyi* and expression of bioluminescence in *Escherichia coli*. *Biochemistry* 23:3663–67
2. Baldwin, T. O., Devine, J. H., Lin, J.-W., Legocki, R., Szalay, A., et al. 1987. Applications of the cloned bacterial luciferase genes *luxA* and *luxB* to the study of transcriptional promoters and terminators. In *Bioluminescence and Chemiluminescence. New Perspectives*, ed. J. Schölmerich, R. Andreesen, A. Kapp, M. Ernst, W. G. Woods, pp. 373–76. New York: Wiley. 600 pp.
3. Baldwin, T. O., Ziegler, M. M., Powers, D. A. 1979. Covalent structure of subunits of bacterial luciferase: NH_2-terminal sequence demonstrates subunit homology. *Proc. Natl. Acad. Sci. USA* 76:4887–89
4. Baumann, P., Baumann, L., Woolkalis, M., Bang, S. 1983. Evolutionary relationships in *Vibrio* and *Photobacterium*. A basis for a natural classification. *Ann. Rev. Microbiol.* 37:369–98
5. Becvar, J., Hastings, J. W. 1975. Bacterial luciferase requires one reduced flavin for light emission. *Proc. Natl. Acad. Sci. USA* 72:3374–76
6. Belas, R., Mileham, R., Cohn, D., Hilmen, M., Simon, M., Silverman, M. 1982. Bacterial bioluminescence: isolation and expression of the luciferase genes from *Vibrio harveyi*. *Science* 218:791–93
7. Boylan, M., Graham, A. F., Meighen, E. A. 1985. Functional identification of the fatty acid reductase components encoded in the luminescence operon of *Vibrio fischeri*. *J. Bacteriol.* 163:1186–90
8. Boylan, M., Miyamoto, C., Wall, L., Graham, A. F., Meighen, E. A. 1988. *luxC, D* and *E* genes of the *V. fischeri* luminescence operon code for the reductase, transferase and synthetase enzymes involved in aldehyde biosynthesis. *Photochem. Photobiol.* In press
9. Boyle, R. 1667. Experiments concerning the relation between light and air. *Philos. Trans. R. Soc. London Ser. B* 2:202–15
10. Byers, D. M., Bognar, A., Meighen, E. A. 1988. Differential regulation of enzyme activities involved in aldehyde metabolism in the luminescent bacterium *Vibrio harveyi*. *J. Bacteriol.* 170:967–71
11. Byers, D. M., Cook, H., Meighen, E. A. 1987. The source of fatty acid for bacterial luminescence. See Ref. 2, pp. 405–8
12. Byers, D. M., Meighen, E. A. 1984. *Vibrio harveyi* aldehyde dehydrogenase. Partial reversal of aldehyde oxidation and its possible role in the reduction of fatty acids for the bioluminescence reaction. *J. Biol. Chem.* 259:7109–14

13. Byers, D. M., Meighen, E. A. 1985. Purification and characterization of a bioluminescence-related fatty acyl esterase from *Vibrio harveyi*. *J. Biol. Chem.* 260:6938–44

14. Byers, D. M., Meighen, E. A. 1985. Acyl-acyl carrier protein as a source of fatty acids for bacterial bioluminescence. *Proc. Natl. Acad. Sci. USA* 82:6085–89

15. Carey, L. M., Rodriguez, A., Meighen, E. A. 1984. Generation of fatty acids by an acyl esterase in the bioluminescent system of *Photobacterium phosphoreum*. *J. Biol. Chem.* 259:10216–21

16. Cline, T. W., Hastings, J. W. 1972. Mutationally altered bacterial luciferase. Implications for subunit functions. *Biochemistry* 11:3359–70

17. Cohn, D. H., Mileham, A. J., Simon, M. I., Nealson, K. H., Rausch, S. K., et al. 1985. Nucleotide sequence of the *luxA* gene of *Vibrio harveyi* and the complete amino acid sequence of the α subunit of bacterial luciferase. *J. Biol. Chem.* 260:6139–46

18. Cohn, D. H., Ogden, R. C., Abelson, J. N., Baldwin, T. O., Nealson, K. H., et al. 1983. Cloning of the *Vibrio harveyi* luciferase genes: use of a synthetic oligonucleotide probe. *Proc. Natl. Acad. Sci. USA* 80:120–23

19. Cormier, M. J., Strehler, B. L. 1953. The identification of KCF: requirement of long chain aldehydes for bacterial extract luminescence. *J. Am. Chem. Soc.* 75:4864–65

20. Cousineau, J., Meighen, E. A. 1976. Chemical modification of bacterial luciferase with ethoxyformic anhydride. Evidence for an essential histidyl residue. *Biochemistry* 15:4992–5000

21. Delong, E. F., Steinhauer, D., Israel, A., Nealson, K. H. 1987. Isolation of the *lux* genes from *Photobacterium leiognathi* and expression in *Escherichia coli*. *Gene* 54:203–10

22. DeWet, J., Wood, K. V., Helinski, D. R., Deluca, M. 1985. Cloning of firefly luciferase cDNA and the expression of active luciferase in *Escherichia coli*. *Proc. Natl. Acad. Sci. USA* 82:7870–73

23. Dunlap, P. V., Greenberg, E. P. 1985. Control of *Vibrio fischeri* luminescence gene expression in *Escherichia coli* by cyclic AMP and cyclic AMP receptor protein. *J. Bacteriol.* 164:45–50

24. Dunn, D. K., Michaliszyn, G. A., Bogacki, I. B., Meighen, E. A. 1973. Conversion of aldehyde to acid in the bacterial bioluminescent reaction. *Biochemistry* 12:4911–18

25. Eberhard, A. 1972. Inhibition and activation of bacterial luciferase synthesis. *J. Bacteriol.* 109:1101–5

26. Eberhard, A., Burlingame, A. L., Eberhard, C., Kenyon, G. L., Nealson, K. H., Oppenheimer, N. J. 1981. Structural identification of autoinducer of *Photobacterium fischeri* luciferase. *Biochemistry* 20:2444–49

27. Eberhard, A., Hastings, J. W. 1972. A postulated mechanism for the bioluminescent oxidation of reduced flavin mononucleotide. *Biochem. Biophys. Res. Commun.* 47:348–53

28. Eberhard, A., Widrig, C. A., Mcbath, P., Schineller, J. B. 1986. Analogs of the autoinducer of bioluminescence in *Vibrio fischeri*. *Arch. Microbiol.* 146:35–40

29. Engebrecht, J., Nealson, K. H., Silverman, M. 1983. Bacterial bioluminescence: isolation and genetic analysis of functions from *Vibrio fischeri*. *Cell* 32:773–81

30. Engebrecht, J., Silverman, M. 1984. Identification of genes and gene products necessary for bacterial bioluminescence. *Proc. Natl. Acad. Sci. USA* 81:4154–58

31. Engebrecht, J., Silverman, M. 1986. Regulation of expression of bacterial genes for bioluminescence. *Genet. Eng.* 8:31–44

32. Engebrecht, J., Silverman, M. 1986. Techniques for cloning and analyzing bioluminescence from marine bacteria. *Methods Enzymol.* 133:83–98

33. Engebrecht, J., Silverman, M. 1987. Nucleotide sequence of the regulatory locus controlling expression of bacterial genes for bioluminescence. *Nucleic Acids Res.* 15:10455–67

34. Engebrecht, J., Simon, M., Silverman, M. 1985. Measuring gene expression with light. *Science* 227:1345–47

34a. Foran, D. R., Brown, W. M. 1988. Nucleotide sequence of the *luxA* and *luxB* genes of the bioluminescent marine bacterium *Vibrio fischeri*. *Nucleic Acids Res.* 16:777

35. Friedrich, W. F., Greenberg, E. P. 1983. Glucose repression of luminescence and luciferase in *Vibrio fischeri*. *Arch. Microbiol.* 134:87–91

36. Greenberg, E. P., Hastings, J. W., Ultizur, S. 1979. Induction of luciferase synthesis in *Beneckea harveyi* by other marine bacteria. *Arch. Microbiol.* 120:87–91

37. Gunsalus-Miguel, A., Meighen, E. A., Nicoli, M. Z., Nealson, K. H., Hastings, J. W. 1972. Purification and properties of bacterial luciferases. *J. Biol. Chem.* 247:398–404

38. Gupta, S. C., O'Brien, D., Hastings, J. W. 1985. Expression of the cloned subunits of bacterial luciferase from separate replicons. *Biochem. Biophys. Res. Commun.* 127:1007–11

39. Gupta, S. C., Reese, C. P., Hastings, J. W. 1986. Mobilization of cloned luciferase genes into *Vibrio harveyi* luminescence mutants. *Arch. Microbiol.* 143:325–29

40. Harvey, E. N. 1952. *Bioluminescence.* New York: Academic. 649 pp.

41. Hastings, J. W. 1986. Bioluminescence in bacteria and dinoflagellates. In *Light Emission by Plants and Bacteria*, ed. J. Govindjee, J. Amesz, D. C. Fork, pp. 363–98. New York: Academic

42. Hastings, J. W., Baldwin, T. O., Nicoli, M. Z. 1978. Bacterial luciferase: assay purification and properties. *Methods Enzymol.* 57:135–52

43. Hastings, J. W., Nealson, K. H. 1977. Bacterial bioluminescence. *Ann. Rev. Microbiol.* 31:549–95

44. Hastings, J. W., Potrikas, C. J., Gupta, S. C., Kurfurst, M., Makemson, J. C. 1985. Biochemistry and physiology of bioluminescent bacteria. *Adv. Microbiol. Physiol.* 26:235–91

45. Hastings, J. W., Spudich, J., Malnic, G. 1963. The influence of aldehyde chain length upon the relative quantum yield of the bioluminescent reaction of *Achromobacter fischeri.* *J. Biol. Chem.* 238:3100–5

46. Hastings, J. W., Tu, S.-C., Becvar, J. E., Presswood, R. P. 1979. Bioluminescence from the reaction of FMN, H_2O_2 and long chain aldehyde with bacterial luciferase. *Photochem. Photobiol.* 29:383–87

47. Johnston, T. C., Thompson, R. B., Baldwin, T. O. 1986. Nucleotide sequence of the *luxB* gene of *Vibrio harveyi* and the complete amino acid sequence of the β subunit of bacterial luciferase. *J. Biol. Chem.* 261:4805–11

48. Kaplan, H. B., Greenberg, E. P. 1985. Diffusion of autoinducer is involved in regulation of the *Vibrio fischeri* luminescence system. *J. Bacteriol.* 163:1210–14

49. Kaplan, H. B., Greenberg, E. P. 1987. Overproduction and purification of the *luxR* gene product: the transcriptional activation of the *Vibrio fischeri* luminescence system. *Proc. Natl. Acad. Sci. USA* 84:6639–43

50. Karl, D. M., Nealson, K. H. 1980. Regulation of cellular metabolism during synthesis and expression of the luminous system in *Beneckea* and *Photobacterium.* *J. Gen. Microbiol.* 117:357–68

51. Karp, M. T., Meyer, Lähde, M., Lahti, R., Mantsala, P. I. 1987. Construction of a shuttle plasmid containing bacterial luciferase genes for gene expression in *Escherichia coli* and *Bacillus subtilis.* See Ref. 2, pp. 361–64

52. Kemal, C., Chan, T. W., Bruice, T. C. 1977. Chemiluminescent reactions and electrophilic oxygen donating ability of 4a-hydroperoxy flavins: general synthetic method for the preparation of N^5-alkyl-1,5-dihydroflavins. *Proc. Natl. Acad. Sci. USA* 74:405–9

53. Kempner, E. S., Hanson, F. E. 1968. Aspects of light production by *Photobacterium fischeri.* *J. Bacteriol.* 95:975–79

54. Koncz, C., Olsson, O., Langbridge, W. H., Schell, J., Szalay, A. A. 1987. Expression and assembly of functional bacterial luciferase in plants. *Proc. Natl. Acad. Sci. USA* 84:131–35

55. Lee, J. 1972. Bacterial bioluminescence. Quantum yields and stoichiometry of the reactants, reduced flavin mononucleotide, dodecanal, and oxygen and of a product, hydrogen peroxide. *Biochemistry* 11:3350–59

56. Lee, J. 1985. The mechanism of bacterial luminescence. In *Chemi- and Bioluminescence*, ed. J. Burr, pp. 401–37. New York: Dekker

57. Legocki, R. P., Legocki, M., Baldwin, T. O., Szalay, A. A. 1986. Bioluminescence in soybean root nodules: demonstration of a general approach to assay gene expression in vivo by using bacterial luciferase. *Proc. Natl. Acad. Sci. USA* 83:9080–84

58. Lummen, P., Winkler, U. 1986. Bioluminescence of outer membrane defective mutants of *Photobacterium* phosphoreum. *FEMS Microbiol. Lett.* 37:293–98

59. Makemson, J. C., Hastings, J. W. 1982. Iron represses bioluminescence and affects catabolite repression of luminescence in *Vibrio harveyi.* *Curr. Microbiol.* 7:181–86

60. Meighen, E. A. 1979. Biosynthesis of aliphatic aldehydes for the bacterial bioluminescent reaction. Stimulation by ATP and NADDH. *Biochem. Biophys. Res. Commun.* 87:1080–86

61. Meighen, E. A., Bartlet, I. 1980. Complementation of subunits from different bacterial luciferases. Evidence for the role of the β subunit in the bioluminescent mechanism. *J. Biol. Chem.* 255:11181–87

62. Meighen, E. A., Grant, G. G. 1985. Bioluminescence analysis of long chain aldehydes. Detection of insect pheromones. In *Bioluminescence and Chemiluminescence: Instruments and Applications*, ed. K. Van Dyke, 2:253–68. Boca Raton, Fla: CRC. 276 pp.

63. Meighen, E. A., Hastings, J. W. 1971. Binding site determination from kinetic data; reduced flavin mononucleotide binding to bacterial luciferase. *J. Biol. Chem.* 246:7666–74

64. Meighen, E. A., MacKenzie, R. E. 1973. Flavine specificity of enzyme-substrate intermediates in the bacterial bioluminescent reaction. *Biochemistry* 12:1482–91

65. Meighen, E. A., Slessor, K. N., Grant, G. G. 1982. Development of a bioluminescent assay for aldehyde pheromones of insects. Sensitivity and specificity. *J. Chem. Ecol.* 8:911–21

66. Meighen, E., Ziegler, M., Hastings, J. W. 1971. Hybridization of bacterial luciferase with a variant produced by chemical modification. *Biochemistry* 22:4062–68

67. Michaliszyn, G. A., Meighen, E. A. 1976. Induced polypeptide synthesis during the development of bacterial bioluminescence. *J. Biol. Chem.* 251:2541–49

68. Miyamoto, C. M., Boylan, M., Graham, A. F., Meighen, E. A. 1986. Cloning and expression of the genes from the bioluminescent system of marine bacteria. *Methods Enzymol.* 133:70–83

69. Miyamoto, C. M., Byers, D. M., Boylan, M., Graham, A. F., Meighen, E. A. 1987. Vector and cell dependent expression in *Escherichia coli* of the genes responsible for bacterial bioluminescence. See Ref. 2, pp. 389–92

70. Miyamoto, C. M., Byers, D., Graham, A. F., Meighen, E. A. 1987. Expression of bioluminescence in *Escherichia coli* by recombinant *Vibrio harveyi* DNA. *J. Bacteriol.* 169:247–53

71. Miyamoto, C. M., Graham, A. D., Boylan, M., Evans, J. F., Hasel, K. W., et al. 1985. Polycistronic mRNAs code for polypeptides of the *Vibrio harveyi* luminescence systems. *J. Bacteriol.* 161:995–1001

71a. Miyamoto, C. M., Graham, A. F., Meighen, E. A. 1988. Nucleotide sequence of the *luxC* gene and the upstream DNA from the bioluminescent system of *Vibrio harveyi*. *Nucleic Acids Res.* 16:1551–62

72. Nealson, K. H. 1977. Autoinduction of bacterial luciferase; occurrence, mechanism and significance. *Arch. Microbiol.* 112:73–79

73. Nealson, K. H., Eberhard, A., Hastings, J. W. 1972. Catabolite repression of bacterial luminescence: functional implications. *Proc. Natl. Acad. Sci. USA* 69:1073–76

74. Nealson, K. H., Hastings, J. W. 1979. Bacterial bioluminescence: its control and ecological significance. *Microbiol. Rev.* 43:496–18

75. Nealson, K. H., Platt, T., Hastings, J. W. 1970. Cellular control of the synthesis and activity of the bacterial luminescent system. *J. Bacteriol.* 104:313–22

76. Ne'eman, Z., Ulitzur, S., Branton, D., Hastings, J. W. 1977. Membrane polypeptides co-induced with the bacterial bioluminescent system. *J. Biol. Chem.* 252:5150–54

77. Nicoli, M. Z., Meighen, E. A., Hastings, J. W. 1974. Bacterial luciferase: chemistry of the reactive sulfhydryl. *J. Biol. Chem.* 249:2385–92

78. O'Kane, D. J., Ahmad, M., Mathesan, I., Lee, J. 1986. Purification of bacterial luciferase by high performance liquid chromatography. *Methods Enzymol.* 133:109–28

79. O'Kane, D. J., Lee, J. 1986. Purification and properties of lumazine proteins from *Photobacterium* strains. *Methods Enzymol.* 133:149–72

80. Ow, D. W., Wood, K. V., Deluca, M., DeWet, J. R., Helinski, D. R., Howell, S. H. 1986. Transient and stable expression of the firefly luciferase gene in plant cells and transgenic plants. *Science* 234:856–59

81. Prasher, D., McCann, R., Longiara, M., Cormier, M. J. 1987. Sequence comparisons of complementary DNAs encoding aequorin isotypes. *Biochemistry* 26:1326–32

82. Riendeau, D., Meighen, E. A. 1979. Evidence for a fatty acid reductase catalyzing the synthesis of aldehydes for the bacterial bioluminescent reaction. *J. Biol. Chem.* 154:7488–90

83. Riendeau, D., Meighen, E. A. 1985. Enzymatic reduction of fatty acids and acyl-CoAs to long chain aldehydes and alcohols. *Experientia* 41:707–13

84. Riendeau, D., Rodriguez, A., Meighen, E. A. 1982. Resolution of the fatty acid reductase from *Photobacterium phosphoreum* into acyl–protein synthetase and acyl–CoA reductase activities. Evidence for an enzyme complex. *J. Biol. Chem.* 257:6908–15

85. Rodriguez, A., Meighen, E. A. 1985.

Fatty acyl-AMP as an intermediate in fatty acid reduction to aldehyde in luminescent bacteria. *J. Biol. Chem.* 260:771–74

86. Rodriguez, A., Nabi, I. R., Meighen, E. A. 1985. ATP turnover by the fatty acid reductase complex of *Photobacterium phosphoreum. Can. J. Biochem.* 63:1106–11

87. Rodriguez, A., Riendeau, D., Meighen, E. A. 1983. Purification of the acyl–CoA reductase component from a complex responsible for the reduction of fatty acids in bioluminescent bacteria. Properties and acyl-transferase activity. *J. Biol. Chem.* 258:5233–38

88. Rodriguez, A., Wall, L., Raptis, S., Zarkadas, C. G., Meighen, E. A. 1988. Different sites for fatty acid activation and acyl transfer by the synthetase subunit of fatty acid reductase. Acylation of a cysteinyl residue. *Biochim. Biophys. Acta* 964:266–75

89. Rodriguez, A., Wall, L., Riendeau, D., Meighen, E. A. 1983. Fatty acid acylation of proteins in bioluminescent bacteria. *Biochemistry* 22:5604–11

90. Rogers, L., Kollatakudy, P. E., de-Renobales, M. 1982. Purification and characterization of *S*-acyl fatty acid synthase and thioester hydrolase which modifies the product specificity of fatty acid synthase in the uropygial gland of mallard. *J. Biol. Chem.* 257:880–85

91. Rosson, R. A., Nealson, K. H. 1981. Autoinduction of bacterial bioluminescence in a carbon limited chemostat. *Arch. Microbiol.* 129:299–304

92. Ruby, E.-G., Nealson, K. H. 1977. A luminous protein that emits yellow light. *Science* 196:432–34

93. Shimomura, O., Johnson, F. H., Kohama, Y. 1972. Reactions involved in bioluminescence systems of limpet *(Latia neritoides)* and luminous bacteria. *Proc. Natl. Acad. Sci. USA* 69:2086–89

94. Shimomura, O., Johnson, F. H., Morise, H. 1974. The aldehyde content of luminous bacteria and of an "aldehyde-less" dark mutant. *Proc. Natl. Acad. Sci. USA* 71:4666–69

95. Smith, S., Libertini, L. J. 1978.

Purification and properties of a thioesterase from lactating rat mammary gland which modifies the product specificity of fatty acid synthetase. *J. Biol. Chem.* 253:1393–401

96. Tu, S.-C. 1986. Bacterial luciferase 4a-hydroperoxyflavin intermediates: stabilization, isolation and properties. *Methods Enzymol.* 133:128–39

97. Ulitzur, S., Hastings, J. W. 1978. Myristic acid stimulation of bacterial luminescence in "aldehyde" mutants. *Proc. Natl. Acad. Sci. USA* 75:266–69

98. Ulitzur, S., Hastings, J. W. 1979. Evidence for tetradecanal as the natural aldehyde in bacterial bioluminescence. *Proc. Natl. Acad. Sci. USA* 76:265–67

99. Ulitzur, S., Hastings, J. W. 1979. Autoinduction of luminous bacterium: a confirmation of the hypothesis. *Curr. Microbiol.* 2:345–48

100. Ulitzur, S., Hastings, J. W. 1979. Control of aldehyde synthesis in the luminous bacterium *Beneckea harveyi*. *J. Bacteriol.* 137:854–59

101. Ulitzur, S., Hastings, J. W. 1980. Reversible inhibition of bacterial bioluminescence by long chain fatty acids. *Curr. Microbiol.* 3:295–300

102. Wall, L. A., Byers, D. M., Meighen, E. A. 1984. *In vivo* and *in vitro* acylation of polypeptides in *Vibrio harveyi*: identification of proteins involved in aldehyde production for bioluminescence. *J. Bacteriol.* 159:720–24

103. Wall, L. A., Meighen, E. A. 1986. Subunit structure of the fatty acid reductase complex from *Photobacterium phosphoreum. Biochemistry* 25:4315–21

104. Wall, L., Rodriguez, A., Meighen, E. A. 1984. Differential acylation *in vitro* with tetradecanoyl-CoA and tetradecanoic Acid (+ATP) of three polypeptides shown to have induced synthesis in *Photobacterium phosphoreum. J. Biol. Chem.* 259:1409–14

105. Wall, L., Rodriguez, A., Meighen, E. 1986. Intersubunit transfer of fatty acyl groups during fatty acid reduction. *J. Biol. Chem.* 261:16018–25

106. Ziegler, M. M., Baldwin, T. O. 1981. Biochemistry of bacterial bioluminescence. *Curr. Top. Bioenerg.* 12:65–113

NOTE ADDED IN PROOF   The gene located between *luxB* and *luxE* in *P. phosphoreum,* now designated as *luxF*, codes for a protein of 231 residues with approximately 30% homology to the β-subunit of luciferase (R. Soly, J. Mancini, M. Boylan, S. Ferri & E. Meighen, manuscript in preparation). The amino-terminal sequence is identical to that of the nonfluorescent flavoprotein found in *Photobacterium* sp. (D. J. O'Kane & J. Lee, personal communication).

Ann. Rev. Microbiol. 1988. 42:177–99

BACULOVIRUSES AS GENE EXPRESSION VECTORS

Lois K. Miller

Departments of Genetics and Entomology, University of Georgia, Athens, Georgia 30602

CONTENTS

INTRODUCTION

Although baculoviruses have been of interest for many years by virtue of their impact on pest insect populations, the recent burgeoning interest in these viruses stems from their usefulness as helper-independent viral vectors for the high-level expression of foreign genes in eukaryotes. The popularity of baculoviruses as gene expression vectors is directly related to the success that numerous academic and commercial laboratories have had in obtaining substantial quantities of biologically active products from a variety of eukaryotic genes, especially those that posed difficulties in lower eukaryotic or prokaryotic expression systems. Baculovirus-based expression is achieved rapidly using conventional virological techniques. As with other expression systems, however, the ability of the baculovirus system to provide abundant quantities of any particular foreign gene must be determined empirically, since many aspects of foreign gene expression and protein stability are not known at this time.

All currently available baculovirus expression vectors provide the advantage of a natural viral gene regulation phenomenon, namely the very late but highly abundant expression of the polyhedrin gene. This gene is expressed during occlusion, the second of two stages of the infection cycle. Each stage produces a biochemically and functionally distinct form of the virus: The extracellular budded form of the virus is produced during the first stage, and the occluded form of the virus, composed primarily of polyhedrin protein, is produced during the second stage. Knowledge of the biology and the molecular basis of the baculovirus infection process allows for a fuller understanding and appreciation of the advantages and power of the expression vector system.

This review first surveys the basic biology of baculoviruses at the organismal and molecular levels, with specific emphasis on those features relevant to the use of the viruses as expression vectors. An outline of the available baculovirus expression vectors is followed by a brief discussion of some factors that affect the quantity and biological quality of the gene products; of particular interest in this regard is the nature of posttranslational modifications in insect cells. Methods for cell propagation, recombinant virus construction, and protein production are considered elsewhere (44, 68). Prior reviews relevant to baculovirus expression vectors (37a, 44, 45, 48) and baculovirology in general (9, 12, 16, 17, 33, 67, 70) are available.

BIOLOGY OF BACULOVIRUSES

Structure and Classification

Viruses within the family Baculoviridae possess a single molecule of circular supercoiled double-stranded DNA of 80–220 kb. The DNA is packaged as rod-shaped nucleocapsids, which acquire membrane envelopes either by budding through the plasma membrane of the cell or by a nuclear envelopment process. In the subgroup A baculoviruses known as nuclear polyhedrosis viruses (NPVs), those virions that obtain an envelope by an intranuclear envelopment process can be occluded within a paracrystalline protein matrix, forming large (1–5 μm) occlusion bodies of polyhedral morphology containing multiple virions. NPVs are further distinguished on the basis of whether they contain a single nucleocapsid (SNPV) or multiple nucleocapsids (MNPV) per envelope.

Baculoviruses have been isolated from invertebrates only. Over 400 baculoviruses have been reported in the literature, but less than 20 have been studied at the molecular level. The most intensively studied at this level is an MNPV originally isolated from *Autographa californica,* a lepidopteran noctuid (which in its adult stage is a nocturnal moth) commonly known as the alfalfa looper. The virus is known as AcMNPV or AcNPV.

Although AcMNPV has been used most extensively for gene expression vector purposes, other NPVs may also be developed as vectors based on the experience from the AcMNPV system. For example, the NPV of *Bombyx mori* (silkworm) (BmNPV) has been developed as a gene expression system (40) even though only a single gene (the highly conserved polyhedrin gene) has been identified in this virus.

The Two Infectious Forms of Nuclear Polyhedrosis Viruses

The two forms of an NPV, the occluded virus (OV) and the extracellular budded virus (EV) have distinct roles in infection (reviewed in 18). OVs are involved in the horizontal transmission of virus infection from insect to insect. Insects usually acquire the disease by consuming food contaminated with OV. The paracrystalline matrix of OVs is comprised primarily of a 29-kd protein, polyhedrin. This matrix apparently has two roles in horizontal transmission: (*a*) to protect the embedded virions from inactivation during the interval of transmission between host organisms and (*b*) to effect the release of the virions at the primary site of infection, the midgut epithelial cells, by dissolving in the high-pH (10.5) environment of the midgut lumen. Systemic infection of the insect (as well as infection in cell cultures) is mediated by EVs, which bud from infected cells into the hemolymph of the insect.

The baculovirus vectors that are routinely used are based on the substitution of the polyhedrin gene with the foreign gene of interest (40, 44, 48, 49, 68). Substitution of the polyhedrin gene interferes with the production of OV but has minimal effect on the production of EV (62). Thus the polyhedrin-based substitution vectors are helper-independent vectors for gene expression in cell cultures but are defective in per os transmission from insect to insect. The lack of OV production is a visually selectable plaque phenotype and provides a rapid means of identifying recombinant viruses (44, 62, 68). Recombinant EVs are defective only in horizontal transmission among insects via the oral infection route.

In a synchronous infection of a permissive cell line such as *Spodoptera frugiperda* (fall armyworm) IPLB-SF-21, the production of the bulk of progeny EV precedes the production of OV. The delay in OV production can be accounted for by a delay in the production of stable transcripts from occlusion-specific genes such as polyhedrin (reviewed in 14). Whereas the bulk of EV is produced within the first 18 hr postinfection (p.i.), polyadenylated transcripts from the polyhedrin gene begin to accumulate in the cytoplasm of the cells at 18–24 hr p.i. Translation products appear shortly thereafter and accumulate through 70 hr p.i. Thus in a polyhedrin-based baculovirus expression system, the production of foreign gene products is expected to have minimal impact on the ability of the recombinant virus to replicate. There should be minimal selection pressure to eliminate the foreign gene insert, providing insert stability, and it should be possible to express at least some foreign gene products that adversely affect cell viability. Although insert stability is usually observed, the latter feature has not been rigorously tested yet. A trick that might be used to facilitate the isolation of recombinant viruses expressing cytotoxic products is to utilize a host cell that supports EV production but fails to express polyhedrin [e.g. AcMNPV-mediated expression in *B. mori* cells (69)].

MOLECULAR BIOLOGY OF AcMNPV

EVs apparently enter cells by adsorptive endocytosis (73). The rod-shaped nucleocapsids enter the nucleus, where the nucleoprotein core is released from the capsid (71). The core of AcMNPV is comprised of the 128-kb supercoiled DNA associated with a protamine-like, arginine-rich 6.9-kd protein (74). Although very early events are difficult to study, some data suggest that the viral DNA may be released from this core and assume a nucleosome-like structure using host-encoded histones. Late in infection, the viral DNA, but not the host DNA, assumes a unique nucleoprotein structure as observed by micrococcal nuclease protection experiments (75). At least three phases of AcMNPV gene expression can be distinguished in AcMNPV-infected *S. frugiperda* cells: (*a*) an early phase, (*b*) a late phase, and (*c*) a very late or occlusion-specific phase.

Early Phase of Gene Expression

During the early phase of gene expression (from 0 to approximately 6 hr p.i.), which precedes viral DNA replication, a number of early genes that are distributed throughout the AcMNPV genome are expressed (reviewed in 14). There may be two classes of early gene transcription: immediate early and delayed early. Transient expression assays suggest that the product of the IE-1 gene can act as a *trans*-acting transcriptional activator of other early genes (22). It remains to be demonstrated genetically whether IE-1 has such a role in early gene expression or whether it is a component of the virion. The effects of IE-1 activation in transient assays are dramatic; thousandfold increases in gene expression are observed when a "homologous region" sequence is present in *cis* with the reporter gene (23). There are six short (400–900 bp) homologous regions (hr1, hr2, hr3, hr4l, hr4r, and hr5) interspersed through the genome of AcNPV (13, 21); the nucleotide sequences of these regions reveal the reiteration of an approximately 60-bp sequence encompassing a highly conserved 26-bp imperfect palindrome, the center of which is an *Eco*RI site (21, 37). These sequences, in conjunction with IE-1, resemble enhancers (21, 23) and, by analogy with other DNA viruses, may serve as origins of viral DNA replication.

Late Phase of Gene Expression

The late phase of AcMNPV replication, extending from approximately 6 through 18 hr p.i., is characterized by extensive replication of viral DNA and the formation of progeny EV. This phase encompasses the production of structural proteins of nucleocapsids, including the major capsid and the core proteins, which have been sequenced recently (74; S. M. Thiem & L. K. Miller, unpublished data). Another abundant late protein is a 64-kd glycoprotein that is found in the envelope of EVs, but not OVs, and that has a role in receptor-mediated entry of EV into cells (73).

Very Late Phase of Gene Expression

The final phase of gene expression, from approximately 20 through 72 hr p.i., is referred to as the occlusion phase or very late phase of gene expression. This phase is characterized by abundant polyhedrin synthesis and the formation of OV within the nucleus. By 70 hr p.i., wild-type AcMNPV produces an average of 70 OVs per nucleus; each OV is approximately 2 μm in diameter and contains numerous enveloped nucleocapsids. Polyhedrin becomes the most predominant protein of the cell, composing 25–50% of the total stainable protein of the cell by 70 hr p.i. The expression of at least one other protein of AcMNPV, the abundant p10 protein, is regulated in a similar fashion to that of polyhedrin (reviewed in 13, 14). Although p10 is not essential for the formation of refractive occlusions, it is thought to be involved in the maturation of OVs and possibly in the cytoskeletal structure (P. Faulkner & G. Rohrmann, personal communication).

Late Promoters

The two late phases of gene transcription are apparently mediated by a virus-induced α-amanitin–resistant RNA polymerase (15, 20). It is not yet known whether the virus encodes any portion of this induced polymerase. The polymerase probably recognizes, directly or indirectly, a consensus sequence that serves as the initiation site for RNA synthesis for all known late and occlusion phase–specific genes characterized to date (61). Based on the sequences of the 5' start sites of eight different major late RNAs of AcMNPV [polyhedrin, p10, the 6.9-kd core protein, the 39-kd late start site, two transcripts of the major capsid protein (S. M. Thiem & L. K. Miller, unpublished data), and the long terminal repeats (LTRs) of AcMNPV-borne transposable element D], the following consensus sequence has emerged: a-a/t-a/t-y-a/t-a/t-A-T-A*-A-G-a-a-a/t-t-a/t-a/y-a/t-a/t-t. The capital letters ATAAG represent invariant nucleotides. The bases represented by small letters are usually found in seven of eight sequences (y represents a pyrimidine). The asterisk notes the position of the 5' start site that my laboratory has determined for six of the eight abundant late transcripts (polyhedrin, the 6.9-kd core protein, the capsid transcripts, and the LTRs of transposable element D).

The effects of deletions in the region upstream of the polyhedrin coding region have suggested that this consensus sequence and additional downstream sequences are intimately, and possibly exclusively, involved in the abundant expression of this gene (25, 43), although distant enhancers such as the homologous regions may have a role. In the most extensive study reported to date, a series of AcMNPV expression vectors carrying the infuenza virus hemagglutinin (HA) gene in lieu of the polyhedrin gene was constructed such that the lengths of the RNA leaders differed (43). Recombinants lacking 7, 11, 14, 16, 27, 31, 46, 51, and 58 bases immediately upstream of the original polyhedrin ATG were assayed for HA activity. A significant reduction in HA activity was observed between -16 and -27. The HA assays may not have been sensitive enough to discern subtle differences between -1 and -27. SDS–polyacrylamide gel analysis (43) provided evidence that the region from -1 to -8 is essential in obtaining maximum levels of at least one foreign gene product.

Sequences upstream of -70 contribute minimally if at all to polyhedrin expression. Large insertions of nonviral DNA at the *Eco*RV site at -92 suggested that the region upstream, with the possible exception of distant enhancers, is unessential for high-level expression (56a). Successive deletions from the *Eco*RV site toward the ATG have suggested that the region from approximately -69 to -92 has little or no effect on polyhedrin promoter-driven gene expression (56a). These results are consistent with experiments in which the 92 bp upstream of the polyhedrin ATG, a chloramphenicol acetyl-transferase (CAT) gene fusion, and the polyadenylation site of the transferase gene were excised from a pEV55-based plasmid vector (Figure 1) and placed in opposite orientation within the plasmid vector (C. Rankin, B.-G. Ooi & L.

K. Miller, manuscript in preparation). Transient expression of plasmids carrying the CAT gene in reverse orientation resulted in higher levels of CAT activity than is observed for the normal orientation following cotransfection of recombinant plasmids with wild-type viral DNA.

We have recently constructed a series of pEV55-based, CAT-containing plasmids with linker-scanning mutations through the -1 to -92 upstream polyhedrin promoter/leader region (C. Rankin, B.-G. Ooi & L. K. Miller, manuscript in preparation). Transient expression assays of CAT activity following cotransfection of the plasmids with wild-type viral DNA indicated that linker replacement in the late promoter conserved consensus sequence (see above) inactivates the promoter. Virtually no effect on CAT expression was observed for linker replacement upstream of -60. Severalfold decreases were observed for some linker replacements in the -1 to -40 region. Initial experiments with recombinant viruses carrying these linker-scanning mutations confirm the results observed by transient expression assays.

The region from -1 to -60, encompassing the entire 50-base leader region, appears to be sufficient, in the context of the viral genome, for abundant polyhedrin expression. As noted below, vectors that lack any portion of the leader sequence may not provide optimal levels of foreign gene expression. It is not yet known how addition of nucleotides to the leader sequence may influence transcription and/or translation of foreign genes, although expression of genes with long leaders has been reported. It is recommended that such additional nucleotides be kept to a minimum during vector construction. Since the polyhedrin leader is $A+T$ rich, the inclusion of $G+C$-rich sequences in the leader should be avoided, if possible.

To date, there is no convincing evidence to suggest that fusions to polyhedrin coding sequences are more highly transcribed than genes inserted at $+1$ of the polyhedrin coding region. Such fusions are unessential for the abundant expression of at least some genes; expression of the lymphocytic choriomeningitis virus (LCMV) N gene, when the gene is inserted at $+1$ and employs its own ATG, reaches levels representing 50% of the total stainable protein of the infected cell (43). Much remains to be learned about RNA stability and translational efficiency in baculovirus-infected cells. We do know that not all genes will be expressed efficiently even if inserted at the $+1$ position.

Signal sequences or polyhedrin coding region fusions may be useful in providing stability to otherwise unstable proteins by altering cellular location or increasing resistance to proteolysis. In a recent study, expression of a gene encoding a 36–amino acid scorpion toxin was not detectable even though the synthetic gene was inserted so that its ATG was six nucleotides downstream from the -1 position (L. F. Carbonell, M. R. Hodge, M. D. Tomalski & L. K. Miller, manuscript submitted). However, attachment of a signal sequence

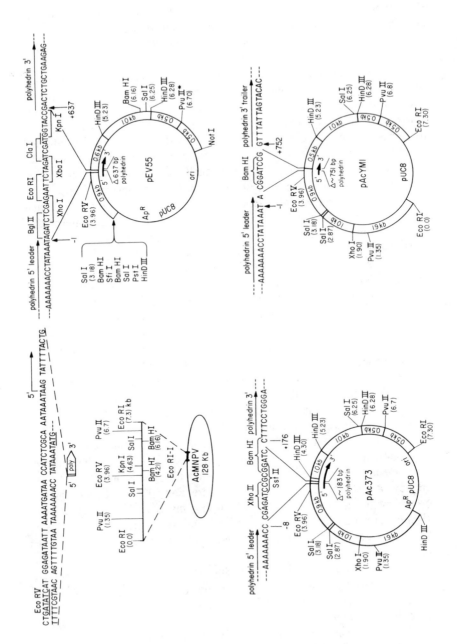

in-frame with the gene resulted in expression that was detectable by [^{35}S]methionine labeling. Fusion of the 36-codon gene to the N-terminal 58 codons of polyhedrin allowed synthesis that was detectable by Coomassie blue staining. Although levels of foreign gene transcripts for all three recombinants were equivalent to those of polyhedrin RNA in wild-type infections, foreign protein levels were far lower than those of polyhedrin. Thus expression of this peptide is apparently limited at the level of translational efficiency and/or protein stability.

A codon bias in the highly expressed late baculovirus proteins has been noted (61). It is unlikely that this codon bias seriously affects the translational efficiency of most genes because very high levels of β-galactosidase gene expression have been observed (55), and this large gene contains many codons that are underrepresented in late, highly expressed baculovirus genes. Attention should be given to the context of the initiating ATG of the foreign gene. The initiating ATG of late, highly expressed baculovirus genes is usually AANATG.

Gene and Transcript Organization

The mapping of the transcripts of AcMNPV and the sequencing of approximately one third of the AcMNPV DNA genome (reviewed in 13, 14; recently extended in 24, 37, 52, 74; L. K. Miller, M. D. Tomalski, J.-G. Wu, S. M. Thiem & D. R. O'Reilly, unpublished data) have revealed the following. (a) Coding regions are arranged in linear nonoverlapping units, which are usually separated by A+T-rich regions. (b) Tandem or bidirectional arrangements of open reading frames are observed. (c) The transcriptional initiation signals and polyadenylation sites often overlap in the A+T-rich regions. (d) Transcriptional motifs include multiple overlapping RNAs with common 5' or 3' termini; it is not uncommon for one open reading frame to be transcribed in both directions into early- and late-abundance polyadenylated RNAs. (e) Early and late transcripts often overlap, perhaps for regulatory purposes. (f)

Figure 1   The design of AcMNPV transplacement plasmids. *Upper left:* Location of the polyhedrin gene of AcMNPV within the 7.3-kb *Eco*RI-I fragment of the 128-kb genome (72). The position and direction of the polyhedrin gene within this *Eco*RI fragment are shown above. The sequence of the promoter/leader region beginning at the *Eco*RV site at −92 is provided. The start site for polyhedrin gene transcription is indicated *(5' arrow). Upper right, lower left, lower right:* AcMNPV transplacement plasmids pEV55 (44), pAc373 (64, 68), and pAcYM1 (43). All are very similar in their basic design. Regions of the *Eco*RI-I fragment of AcMNPV are shown as double-lined plasmid segments. All three plasmids have portions of the polyhedrin gene deleted; the size of the deletion is indicated within the circle. In place of these deletions, multicloning sites have been inserted. The sequence of the sites and their positions relative to the polyhedrin leader are shown. The numbers (e.g. −1, +637) are relative to the original polyhedrin ATG (+1, +2, +3). Key restriction sites in the plasmids are indicated. All plasmids are pUC-based and confer resistance to ampicillin.

Early, late, and very late genes are interspersed in the genome; no pattern for gene arrangement has been discerned yet. (*g*) All characterized coding regions, including early, late, and very late genes, lack introns; splicing of natural viral transcripts has not been observed to date.

It is recommended that all foreign genes to be expressed abundantly in the baculovirus expression system be inserted without introns (e.g. as cDNAs). Introns of SV40 tumor antigens were reported to be removed from RNA transcripts in recombinant-infected cells (31). However, large T-antigen expression, which is dependent on correct splicing, was not observed (31). Small t antigen was expressed, but its synthesis is not dependent on correct splicing. We have observed excellent synthesis of large T antigen from a recombinant carrying an intronless version of this SV40 gene (D. R. O'Reilly & L. K. Miller, manuscript submitted). Although it is possible that foreign genes with introns may be expressed in baculovirus-infected cells, it is unlikely that introns will augment expression and probable that they will inhibit expression.

TRANSPLACEMENT PLASMIDS FOR RECOMBINANT VIRUS CONSTRUCTION

Owing to the large size of baculovirus genomes, most recombinant virus construction relies on in vivo recombination to replace a viral allele with the gene of interest (40, 46, 49, 55, 62, 64). Transplacement plasmids contain the site for foreign gene insertion as well as flanking viral sequences, which provide homologous sequences for recombination. Cotransfection of cells with viral and recombinant plasmid DNAs allows cell-mediated allelic replacement of the target viral gene with the plasmid-borne foreign gene.

Vectors for High-Level Expression of a Single Gene

The polyhedrin promoter and its flanking sequences are usually employed for allelic replacement into the AcMNPV genome because the polyhedrin promoter/leader region provides abundant transcription and because replacement of the polyhedrin gene with a foreign gene allows visual selection of recombinants on the basis of their occlusion-negative (occ⁻) phenotype. Three similar AcMNPV transplacement plasmids are schematically presented in Figure 1; analogous vectors are available for BmNPV (25, 50). All of these vectors have linkers with one or more unique restriction sites in place of all or part of the polyhedrin coding region. They also have sufficient flanking viral sequences for efficient recombination into the viral genome. The pAc373 vector, in contrast to pEV55 or pAcYM1, has a linker positioned at −8 rather than −1 relative to the original polyhedrin ATG (+1, +2, +3). Since these eight nucleotides are implicated as important components for optimal expression (43; see above), optimum levels of expression of some genes may not be achieved with this vector.

The three vectors differ in the nature of the linker. In addition to unique sites that can be used to position genes for expression using their own ATG, pEV55 has a terminal *Kpn*I site that can be filled in with a polymerase to provide a blunt end containing a translational initiation codon (ATG) to allow expression of cDNAs lacking their own initiation signal (37a, 44). Other vector systems are available to provide in-frame fusions to varying lengths of the polyhedrin N-terminus (37a, 68). Although it is not clear that such fusions provide additional advantage to a +1 vector system, they are useful for expression of genes lacking their own ATG and may provide protein stability (see above). In expressing a synthetic gene using BmNPV, a polyhedrin fusion protein was produced abundantly, and the polypeptide was subsequently liberated by cyanogen bromide cleavage; biological activity of this product was not reported (41).

The variation in the length of the deletion of polyhedrin coding sequences (Figure 1) presumably has no effect on foreign gene expression; the largest deletion (13 nucleotides of the trailer region in pAcYM1) can provide levels of expression equivalent to those of polyhedrin in the case of at least one gene (43). All the vectors include the polyadenylation signal of polyhedrin. It is not known whether this is an important feature; foreign genes containing their own polyadenylation signals can be efficiently expressed.

Expression of Two or More Foreign Genes

For many applications, it is of interest to express two or more foreign genes simultaneously in the same cells (e.g. multisubunit proteins). One simple way to achieve this goal is to construct separate recombinant viruses (e.g. using the vectors described above) so that each virus expresses one of the protein subunits. The viruses can be mixed in appropriate proportion and used for cell inoculation. As long as the multiplicity of infection for each virus is at least 5 plaque-forming units (PFU) per cell, over 95% of the cells will be infected with at least one of each of the viruses, which ensures coexpression of the genes in the same cell. This coinfection technique was applied in a study of complex formation between two subunits of the influenza virus polymerase (65) and in a study demonstrating complex formation between SV40 T antigen and mouse p53 protein (D. R. O'Reilly & L. K. Miller, manuscript submitted).

Transplacement vectors specifically designed to allow high-level coexpression of two or more genes are in the development stage. One approach has been to insert a foreign gene (the LCMV N gene), under polyhedrin promoter control, upstream of and in opposite orientation to the polyhedrin gene (10).

The insertion site utilized was the *Eco*RV site, -92 bp upstream of the polyhedrin ATG (43). The occ$^+$ recombinant thus has two polyhedrin promoters in opposite orientations, one driving polyhedrin expression and one driving LCMV N expression. The next step would be to replace the polyhedrin gene with a second foreign gene of interest, presumably by selecting for a polyhedrin occ$^-$ phenotype. One important question regarding this system, which remains to be tested by Southern blot analysis, is whether the duplication of the polyhedrin promoter leads to recombinant virus instability. Since only approximately 60 nucleotides of the polyhedrin promoter appear to be necessary for high-level gene expression (see above), it is probable that very convenient systems for expression can be developed that will minimize promoter homology. Other dual expression systems may be developed to exploit the very late p10 promoter. Additional methods for identifying recombinant viruses are available besides selection for occlusion body phenotype; both plaque hybridization to DNA probes and β-galactosidase (*lacZ*) gene marker systems can be used (44, 68).

Another system for dual gene expression utilized a polyhedrin/*lacZ* fusion as a marker for recombinant virus selection (6), while the CAT gene, under the control of the Rous sarcoma virus long terminal repeat (RSV-LTR), was inserted downstream of the *lacZ* gene. The occ$^-$, blue-plaque recombinant virus expressed β-galactosidase and CAT in a temporally distinct fashion. The transplacement plasmid may be used for inserting genes under control of their own promoter. A gene under polyhedrin or p10 promoter control could be inserted into this vector, although the stability of a construct containing a duplicated promoter must be determined as noted above. Replacement of the polyhedrin/*lacZ* fusion gene with a foreign gene can then be achieved with an appropriately constructed transplacement plasmid, and recombinant viruses can be selected as non-blue-plaque viruses in the presence of a chromogenic indicator.

Baculovirus Pesticide Applications

Foreign gene expression in baculoviruses is also of interest in the development of more effective baculovirus pesticides (6, 7). Most applications of baculovirus expression vectors have been aimed at producing foreign proteins as abundantly as possible and have therefore utilized polyhedrin or related promoters to drive gene expression. For pesticide applications it may be preferable to utilize an early promoter to drive the expression of the foreign gene (i.e. an insect-specific neurotoxin) and to retain polyhedrin expression so that OVs are produced (6, 7). A recombinant AcMNPV containing a marker oligonucleotide was recently field-tested in the United Kingdom. Virus persistence and spread in the environment were monitored with a view toward future development and release of genetically improved pesticides (3).

SELECTION AND MAINTENANCE OF VIRUS VECTORS AND HOSTS

The nature of the virus and the host used for expression deserve consideration. It is beyond the scope of this review to deal with these aspects in depth, but virus stability and hosts are briefly considered.

Virus Stability

Baculoviruses, like most other animal viruses, undergo genomic alteration upon serial passage in cell culture (reviewed in 47; see also 35). The most obvious manifestation of such changes is a general decline in the ability of NPVs to produce OVs after serial passage of EVs by routes that eliminate the need for OVs (i.e. cell culture or hemolymph injection of larvae). Although for historical reasons this phenomenon has been documented most extensively for passage of AcMNPV and its closely related variant *Trichoplusia ni* MNPV (TnMNPV) in *T. ni,* the same phenomenon probably also occurs for other viruses and other hosts. Serially passaged EV stocks usually exhibit the few-polyhedra (FP) phenotype, characterized by the production of fewer than 10 polyhedra per infected cell. FP mutants eventually predominate in serially passaged stocks owing to a selective growth advantage in cell culture; at least some FP viruses produce a higher titer of EV than the wild-type viruses (reviewed in 47, 57, 58). The FP phenotype of at least some FP mutants is apparently due to a defect in the nuclear envelopment of nucleocapsids, which is a prerequisite to occlusion. The lack of nuclear envelopment provides a rationale for the overproduction of EVs and the selective growth advantage of the FP phenotype in serial passages that avoid the oral route of infection. Whether there is less polyhedrin synthesis in FP than in wild-type infected cells is somewhat controversial; FP mutants do make high, but possibly not optimum, levels of polyhedrin.

Generation of FP virus might have the following impacts on the use of the virus for expression purposes. (*a*) If viral DNA from serially passaged virus stocks is used in the initial cotransfection for recombinant DNA selection, the visual selection for the occ$^-$ phenotype will be obscured by the FP phenotype. (*b*) Amplification of a recombinant virus by long-term serial passage can result in a decline in expression. Both problems can be minimized, if not solved, by careful attention to the quality of the virus stock. For wild-type stocks, a wild-type plaque can be visually selected by its many-polyhedra (MP) phenotype. The virus can be amplified into large (500 ml) stocks within two to three passages; most of the second- or third-passage stock can be reserved for generation of larger third-, fourth-, or fifth-passage stocks. Stocks can be monitored by plaque assay for the integrity of the wild-type phenotype. For recombinant virus stocks whose MP phenotype cannot be checked easily, the best approach is again to limit the number of serial passages of the virus to six or less and to initiate new stocks from early-passage stocks of the characterized recombinant.

Production in Cell Culture Versus Insect Larvae

Although most expression work has thus far been conducted in cell culture, it is likely that many laboratories will desire large quantities of product and will begin to explore avenues of expression that will minimize costs of production. One possibility that has been suggested is production in insect larvae (40, 50). The decision of whether to produce a foreign protein in larvae rather than in cell culture may depend on the amount of product required, the ease of purification from larval carcasses, the purity of the product necessary, the relative convenience of the two modes of production, and the comparative costs. The questions of cost, convenience, product stability, and purity may need to be balanced for each product and may even depend on the application of each product.

The use of insect larvae for production purposes has been promoted because levels of protein products in hemolymph exceed those in cell culture (40, 50). In two of three cases (40, 41, 50), hemolymph contained a tenfold higher concentration of product (units or mg ml^{-1}) than cell culture media; approximately 0.5 ml of hemolymph are obtained per silkworm, whereas 5 ml of media are used per 1.5×10^6 cells. In one of those cases, 3.6 mg of a polyhedrin/insulin-like growth factor II (IGF-II) fusion protein were recovered per larva, and 0.3 mg protein were produced per milliliter of culture medium (41). At these levels it would be necessary either to inject 300 larvae or to infect 3 liters of cells to obtain 1 g of purified product. In the third case, that of mouse interleukin-3 (IL-3) expression (50), there was a significantly higher (i.e. approximately 500 fold) level of biologically active protein in hemolymph than in cell culture media. It is not clear why IL-3 expression differs from that of the other two proteins studied; it is possible that some proteins are more susceptible to proteolysis in cell culture media (which usually contain fetal bovine serum) than in insect hemolymph. In the case of IL-3, it would appear that production in larvae is preferable. IL-3 is secreted into the hemolymph, and purification from hemolymph is easily achieved. Whether collection of hemolymph can be performed on a large-scale basis remains to be demonstrated; if not, purification from entire larvae would be necessary, as for proteins that are not secreted into the hemolymph. This may prove to be more challenging.

Marumoto et al (41) advocate expression in silkmoths because of their large size, the long history of sericulture, and computerized mass-rearing systems. The difference in larval size may be a consideration to the bench scientist who is considering injecting EV into hemocoels and collecting hemolymph. *T. ni* larvae, which support AcMNPV vectors, are approximately one third the size of *B. mori* larvae. Although *T. ni* have not been reared by humans for 3000 years, laboratory strains are conveniently reared in large numbers and are routinely available for research purposes. Rearing of silkmoths is considered expensive by those in the United States who have experience in mass rearing of insects for viral pesticide production (28; C. M. Ignoffo, personal communication). Such economic considerations, however, are primarily within the realm of industrial-scale production.

The development of vectors that allow dual expression of both polyhedrin and foreign genes so that OV are produced (see above) may make the use of insect larvae for large-scale protein production more convenient; an occ$^+$ virus can be introduced as a contaminant of the food supply, while occ$^-$ viruses must be individually injected into the hemocoel. Occlusion of occ$^-$ recombinant viruses can currently be achieved by coinfection of cultured cells with wild-type and recombinant virus with a multiplicity of infection (MOI) of 5 or more PFU per cell; thus both genes are expressed in the same cell, and the polyhedrin, supplied by the wild type, can occlude the recombinant virions. It remains to be determined whether the coexpression of polyhedrin interferes with the expression and/or purification of the foreign protein. Possible difficulties of purifying protein from larvae remain.

Another avenue currently being explored by some laboratories is reduction in the cost of insect cell culture media. Most media for the culture of lepidopteran cells in the laboratory are supplemented with fetal bovine serum. For a number of years government and industrial laboratories have undertaken to reduce the cost of media for mass production of baculoviruses for pesticide purposes; methods are often retained as proprietary. Much work remains to be done, but rapid progress is likely based on experience from mammalian cell culture work.

Host Range of AcMNPV

AcMNPV has a relatively wide host range for a baculovirus; it is known to infect productively at least 33 different species within 10 different families of Lepidoptera (19). It is not known to infect productively any species outside this order. The IPLB-SF-21 cell line of S. frugiperda is usually used for expression work because it performs excellently in both monolayer and suspension culture.

Using a dual marker gene recombinant of AcMNPV, it was possible to demonstrate that AcMNPV enters dipteran cells (e.g. Drosophila and Aedes) and efficiently expresses the CAT gene under the control of the RSV-LTR, which behaves at least in part as an early promoter in permissive S. frugiperda cells (6). Late genes, however, are not expressed in these recombinant virus–infected dipteran cells, and the infection is not productive (6, 60). Expression of the RSV-LTR CAT gene is not observed in recombinant AcMNPV-infected mammalian cells, which indicates that there is a block in the ability of AcMNPV to deliver its DNA genome to the mammalian nucleus

(7). Virus entry into the cytoplasm of mammalian cells has been confirmed by recombinant virus studies (7), but the efficiency of entrance appears to be low, and the progress of the virus is apparently blocked, perhaps in the lysosomes, prior to nuclear entry or uncoating. These observations support the view that baculovirus vector systems are relatively safe with respect to mammals.

POSTTRANSLATIONAL MODIFICATION AND BIOLOGICAL ACTIVITY OF BACULOVIRUS-PRODUCED PROTEINS

A wide variety of genes have been expressed in baculovirus expression vectors. These include genes for (*a*) immune regulatory proteins such as human alpha interferon (25, 39, 40), beta interferon (64), IL-2 (63), and IL-3 (50); (*b*) DNA-binding proteins such as *Drosophila* Kruppel gene product (53), *Neurospora* qa-1f activator (2), human T-cell leukemia virus type 1 (HTLV-1) p40^x transactivator protein (30), c-*myc* (51), and the large T antigens of SV40 and polyomavirus (59; D. R. O'Reilly & L. K. Miller, manuscript in preparation); (*c*) virus structural proteins such as LCMV N and G proteins (42, 43), influenza hemagglutinin proteins (36, 56), rotavirus VP6 (11), bluetongue virus VP2 (29), human immunodeficiency virus (HIV) *gag* (38) and *env* proteins (27), phlebovirus N protein (54), parainfluenza virus HA neuraminidase (HN) (8), and hepatitis B virus surface antigen (32); and (*d*) other proteins such as SV40 small t antigen (31), influenza virus polymerase genes (65), IGF-II (41), phlebovirus NS$_s$ protein (54), β-galactosidase (55), and CAT (6). Expression of many other genes is under way in numerous academic and industrial laboratories.

Many researchers are using eukaryotic expression vectors, rather than prokaryotic or lower eukaryotic vectors, to optimize the likelihood that the product will be biologically active. Many factors may affect the biological activity of a eukaryotic protein, including posttranslational modifications (e.g. glycosylation, proteolysis, phosphorylation, ADP-ribosylation, acylation, sulfation) and the tertiary structure (e.g. disulfide bond formation) or quaternary structure (e.g. oligomerization or complex formation). Since numerous genes that are being expressed with baculovirus vectors are of mammalian origin, the question of how posttranslational modification and cellular localization in baculovirus-infected insect cells will compare with those observed in mammalian cells becomes particularly important. Several of these factors are considered below.

Proteolysis, Including Signal Sequence Cleavage

Insect cells apparently recognize and cleave mammalian signal sequences that direct proteins to the endoplasmic reticulum (ER). The signals of human alpha and beta interferons (40, 64) and interleukins 2 and 3 (50, 63) are all recognized and cleaved in insect cells as in mammalian cells. It thus seems likely that most, if not all, proteins with appropriate mammalian signals will be transported to the ER and cleaved in a normal fashion. Whether similar signal recognition will occur in lysosomal or mitochondrial protein transport remains to be determined.

Certain other specific proteolytic cleavages have also been observed for other baculovirus-produced proteins. The HA protein of at least one influenza virus was cleaved, albeit slowly, in baculovirus-infected cells (36); in another report cleavage was not observed (56). The HA cleavage is a biologically important one associated with fusion activity and viral pathogenicity; the differences in HA susceptibility to cleavage by a host cell appear to be correlated with pathogenicity of the virus. The HIV envelope protein is also cleaved slowly, but apparently accurately (27). Both HA and env cleavage involve an Arg-Lys rich recognition site. The HIV *gag* gene product is also proteolytically cleaved in baculovirus-infected cells; one cleavage is gag-protease related, but host-mediated cleavage is apparently responsible for a second proteolytic event (38).

Glycosylation

Protein glycosylation is thought to have a variety of different roles, including protein stabilization, cellular interaction, and intracellular protein localization; posttranslational modifications of this sort may be important in biological activity of some but not necessarily all proteins. The most common and most thoroughly characterized type of mammalian glycosylation is N-linked glycosylation in the ER, where a common oligosaccharide is attached to asparagine in a reaction that is mediated by a phospholipid carrier. N-Linked glycosylation is sensitive to inhibition by tunicamycin in both mammalian cells and baculovirus-infected insect cells (34, 66).

From one careful comparative study, it appears that the sites that are targeted for glycosylation in insect cells are the same as those of mammalian cells (26). However, this study also demonstrated that the nature of the oligosaccharide at these sites differs in insect- and mammal-derived protein. Mammalian cells extensively modify the core oligosaccharide in terminal glycosylation events involving the transfer of glucosamine-galactose and sialic acid residues to form complex oligosaccharides. Insect cells appear to lack galactose and sialic acid transferases (4, 5) and trim the oligosaccharide to a central core of $GlcNAc_2Man_3$ (26). Whether this is generally the case remains to be demonstrated.

A variety of baculovirus-produced mammalian gene products are known to be glycosylated, e.g. influenza HA (36), parainfluenza HN (8), beta interferon (64), HIV *env* protein (27), and IL-3 (50). Glycosylation is sensitive to tunicamycin inhibition and can be detected with a [^{3}H]mannose label (61).

Differences in the nature of the glycosylation were suggested by differences in the electrophoretic mobility of HA and IL-3 (36, 50), but no difference in biological activity was observed. No glycosylation of IL-2 in insect cells was observed (63). However, glycosylation patterns in mammalian cells also differ; 40% of the IL-2 from one human T-cell line was not glycosylated.

It is likely that differences between insect-mediated glycosylation and mammalian glycosylation will be observed. Whether such differences influence biological activity of the proteins in a positive, neutral, or negative fashion remains to be determined for individual expression products. Indeed, the baculovirus expression system may be ideal for addressing some of the questions regarding the role of glycosylation in protein structure and function.

Phosphorylation

Three nuclear proteins are reported to be phosphorylated in baculovirus-infected insect cells: the Kruppel gene product of *Drosophila* (53), the HTLV-1 p40^x protein (30), and c-*myc* (51). Thus far, the only data reported indicate that some phosphorylation occurs, but it is not known whether phosphorylation occurs in a position identical to that in proteins isolated from the natural host cell. One report (31) has alluded to unpublished data indicating that the phosphorylation patterns are "virtually identical." More information is necessary on the efficiency and accuracy of insect phosphorylation.

Oligomerization and Complex Formation

It is likely that baculovirus-infected cells will provide an adequate in vivo setting for complex formation for most proteins as long as all required factors for complex formation are present and their cellular location is appropriate for complex formation to occur. The rotavirus VP6 protein exhibits disulfide bond–dependent oligomerization; this virus structural protein also shows evidence of morphological unit assembly (11). In vivo complex formation between two of the three subunits of the influenza virus polymerase has been reported (65). However, the failure of the third polymerase subunit to form a complex may reflect either the lack of a necessary factor for successful complex formation in this system or simply the inability of insect cells to provide an adequate environment or modification necessary for complex formation. The baculovirus expression system might prove useful for exploring factors or modifications necessary for such complex formation.

Cellular Location

From the cellular localization studies undertaken thus far, it appears that mammalian proteins segregate in insect cells as they would in mammalian cells. The c-*myc* gene product, for example, is found in the nucleus of baculovirus-infected cells (51); it remains to be formally demonstrated that insect cells recognize the same nuclear transfer signals as mammalian cells. The influenza virus HA and parainfluenza HN proteins are located at the cell surface (8, 36, 56) as expected for these membrane proteins. The LCMV N gene product forms cytoplasmic inclusions, which suggests that this viral structural protein forms natural aggregates, whereas a glycoprotein of LCMV, GPC, is directed to the cell surface (42, 43); hyperexpression of GPC results in a highly vacuolated cytoplasm and dense ER (43). It remains to be determined whether this is a general cellular response to hyperexpression of glycoproteins. Efficient secretion of hepatitis B virus S-antigen, human alpha interferon, beta interferon, IL-2, and IL-3 are observed. The ability to secrete IL-3 in soluble form is considered an advantage over prokaryotic expression and a key to obtaining a product with high specific activity (50). Some proteins that are not normally secreted may also be observed in the media of infected cells; for example, rotavirus VP6 is observed in media as well as in cells (11). Whether this extracellular location is due to premature cell lysis or facile entry into the ER remains to be determined.

Biological Activity

The interleukins and interferons expressed with baculovirus vectors are biologically active (40, 50, 63, 64). The most extensive study of this activity indicated that IL-3 has specific activity equivalent to that produced in mammalian cells (50); the specific activity of IL-2 is also similar to that of mammal-derived material (63). Baculovirus-produced HTLV-1 p40^x exhibited transactivation of an HTLV-1 LTR-promoted globin gene in insect cells (30). The *Neurospora crassa* qa-1f activator and polyomavirus T antigen exhibited specific DNA-binding properties in vitro (2, 59). Baculovirus-derived influenza HA protein has full biological activity despite a difference in the nature of the glycosylation (36). This HA protein elicits neutralizing and protective antibodies. For viral proteins, it is important that major epitopes be displayed if they are to be used for diagnostic or vaccine purposes. Thus far baculovirus-derived antibodies raised against simian rotavirus VP6, parainfluenza virus HN, bluetongue virus VP2, HIV *gag,* and HIV *env* proteins have been useful as diagnostic reagents. In appropriate cases they have provided experimental animals with protection from virus challenge. One study noted that antisera raised to the baculovirus-derived antigen did not react with mammalian cell proteins (11). In another study a low cutoff value and a high signal-to-background ratio were obtained (27). Baculovirus-derived proteins may thus require less purification prior to being used as antigens. Baculovirus-derived HIV *env* protein is being tested in humans as a vaccine to protect against AIDS. The vaccine was developed by MicroGeneSys, Inc (West Haven, Connecticut).

SUMMARY AND FUTURE PROSPECTS

Baculoviruses are useful vectors for the high-level expression of genes in eukaryotes. The success in constructing vectors that produce biologically active protein products using relatively easy and rapid technology makes this vector system a popular one from a research perspective. Initial studies have shown that posttranslational modifications such as signal cleavage and glycosylation are similar in insect and mammalian cells, but glycosylation apparently varies in detail. Although it is currently possible to coexpress two or more genes in the same cell, new expression vectors with additional capabilities including high-level multigene expression are being developed. Scaleup of production will benefit from additional attention to the costs of media and the improved design of vectors for expression in insect larvae. It will be fascinating to determine the mechanism(s) that regulate the expression of baculovirus genes.

ACKNOWLEDGMENTS

I would like to acknowledge those individuals from my laboratory who contributed substantially to the early development of baculoviruses as expression vectors: Dr. Gregory D. Pennock (55), Dr. Michael J. Adang (1, 49), and Dr. David W. Miller (49). Continuous support for my baculovirus research was provided by Public Health Service grants from the National Institutes of Allergy and Infectious Diseases over the past 11 years. Current research is funded by NIH 1R37 AI 23719. I thank Dr. David O'Reilly, Dr. Luis F. Carbonell, and Ms. Naomi A. Miller for reading this review in manuscript form and providing helpful comments.

Literature Cited

1. Adang, M. J., Miller, L. K. 1982. Molecular cloning of DNA complementary to mRNA of the baculovirus *Autographa californica* nuclear polyhedrosis virus: location and gene products of RNA transcripts found late in infection. *J. Virol.* 44:782–93
2. Baum, J. A., Geever, R., Giles, N. H. 1987. Expression of qa-1f activator protein: identification of upstream binding sites in the QA gene cluster and localization of DNA-binding domains. *Mol. Cell. Biol.* 7:1256–66
3. Bishop, D. H. L. 1986. UK release of genetically marked virus. *Nature* 323:496
4. Butters, T. D., Hughes, R. C. 1981. Isolation and characterization of mosquito cell membrane glycoproteins. *Biochim. Biophys. Acta* 640:655–71
5. Butters, T. D., Hughes, R. C., Visher,

P. 1981. Steps in the biosynthesis of mosquito cell membrane glycoproteins and the effects of tunicamycin. *Biochim. Biophys. Acta* 640:672–86
6. Carbonell, L. F., Klowden, M. J., Miller, L. K. 1985. Baculovirus-mediated expression of bacterial genes in dipteran and mammalian cells. *J. Virol.* 56:153–60
7. Carbonell, L. F., Miller, L. K. 1987. Baculovirus interaction with nontarget organisms: a virus-borne reporter gene is not expressed in two mammalian cell lines. *Appl. Environ. Microbiol.* 53:1412–17
8. Coelingh, K. L. V. W., Murphy, B. R., Collins, P. L., Lebacq-Verheyden, A.M., Battey, J. F. 1987. Expression of biologically active and antigenically authentic parainfluenza type 3 virus hemagglutinin-neuraminidase glycopro-

tein by a recombinant baculovirus. *Virology* 160:465–72

9. Doerfler, W., Bohm, P. 1986. *The Molecular Biology of Baculoviruses.* Berlin/Heidelberg/New York: Springer-Verlag. 168 pp.

10. Emery, V. C., Bishop, D. H. L. 1987. The development of multiple expression vectors for high level synthesis of eukaryotic proteins: expression of LCMV N and AcNPV polyhedrin protein by a recombinant baculovirus. *Protein Eng.* 1:359–66

11. Estes, M. K., Crawford, S. E., Penaranda, M. E., Petrie, B. L., Burns, J. W., et al. 1987. Synthesis and immunogenicity of the rotavirus major capsid antigen using a baculovirus expression system. *J. Virol.* 61:1488–94

12. Faulkner, P. 1981. Baculovirus. In *Pathogenesis of Invertebrate Microbial Diseases,* ed. E. W. Davidson pp. 3–37. Totowa, NJ: Allanheld, Osmun. 265 pp.

13. Faulkner, P., Carstens, E. B. 1986. An overview of the structure and replication of baculoviruses. See Ref. 9, pp. 1–17

14. Friesen, P. D., Miller, L. K. 1986. The regulation of baculovirus gene expression. See Ref. 9, pp. 31–49

15. Fuchs, L. Y., Woods, M. S., Weaver, R. F. 1983. Viral transcription during *Autographa californica* nuclear polyhedrosis virus infection: a novel RNA polymerase induced in infected *Spodoptera frugiperda* cells. *J. Virol.* 48:641–46

16. Granados, R. R., Federici, B. A. 1986. *The Biology of Baculoviruses,* Vol. 1, *Biological Properties and Molecular Biology.* Boca Raton, Fla: CRC. 275 pp.

17. Granados, R. R., Federici, B. A. 1986. *The Biology of Baculoviruses,* Vol. 2, *Practical Application for Insect Control.* Boca Raton, Fla: CRC. 276 pp.

18. Granados, R. R., Williams, K. A. 1986. In vivo infection and replication of baculoviruses. See Ref. 16, pp. 90–127

19. Gröner, A. 1980. Specificity and safety of baculoviruses. See Ref. 16, pp. 178–202

20. Grula, M. A., Buller, P. L., Weaver, R. F. 1981. Amanitin-resistant viral RNA synthesis in nuclei isolated from nuclear polyhedrosis virus–infected *Heliothis zea* larvae and *Spodoptera frugiperda* cells. *J. Virol.* 38:916–21

21. Guarino, L. A., Gonzalez, M. A., Summers, M. D. 1986. Complete sequence and enhancer function of the homologous DNA regions of *Autographa californica* nuclear polyhedrosis virus. *J. Virol.* 60:224–29

22. Guarino, L. A., Summers, M. D. 1986. Functional mapping of a *trans*-activating

gene required for expression of a baculovirus delayed-early gene. *J. Virol.* 57:563–71

23. Guarino, L. A., Summers, M. D. 1986. Interspersed homologous DNA of *Autographa californica* nuclear polyhedrosis virus enhances delayed-early gene expression. *J. Virol.* 60:215–23

24. Guarino, L. A., Summers, M. D. 1987. Nucleotide sequence and temporal expression of a baculovirus regulatory gene. *J. Virol.* 61:2091–99

25. Horiuchi, T., Marumoto, Y., Saeki, Y., Sato, Y., Furusawa, M., et al. 1987. High-level expression of the human α-interferon gene through the use of an improved baculovirus vector in the silkworm, *Bombyx mori. Agric. Biol. Chem.* 51:1573–80

26. Hsieh, P., Robbins, P. W. 1984. Regulations of asparagine-linked oligosaccharide processing. *J. Biol. Chem.* 259:2375–82

27. Hu, S., Kowoski, S. G., Schaaf, K. F. 1987. Expression of envelope glycoproteins of human immunodeficiency virus by an insect virus vector. *J. Virol.* 61:3617–20

28. Ignoffo, C. M. 1966. In *Insect Colonization and Mass Production,* ed. C. N. Smith pp. 501–30. New York: Academic. 125 pp.

29. Inumaru, S., Roy, P. 1987. Production and characterization of the neutralization antigen VP2 of bluetongue virus serotype 10 using a baculovirus expression vector. *Virology* 157:472–79

30. Jeang, K., Giam, C., Nerenberg, M., Khoury, G. 1987. Abundant synthesis of functional human T-cell leukemia virus type I p40^x protein in eucaryotic cells by using a baculovirus expression vector. *J. Virol.* 61:708–13

31. Jeang, K., Holmgren-Konig, M., Khoury, G. 1987. A baculovirus vector can express intron-containing genes. *J. Virol.* 61:1761–64

32. Kang, C. Y., Bishop, D. H. L., Seo, J. S., Matsuura, Y., Cheo, M. H. 1988. Secretion of human hepatitis B virus surface antigen from *Spodoptera frugiperda* cells. *J. Gen. Virol.* 68:2607–13

33. Kelly, D. C. 1982. Baculovirus replication. *J. Virol.* 63:1–13

34. Kelly, D. C., Lescott, T. 1983. Baculovirus replication: glycosylation of polypeptides synthesized in *Trichoplusia ni* nuclear polyhedrosis virus–infected cells and the effect of tunicamycin. *J. Gen. Virol.* 64:1915–26

35. Kumar, S., Miller, L. K. 1987. Effects of serial passage of *Autographa californica* nuclear polyhedrosis virus in cell culture. *Virus Res.* 7:335–49

36. Kuroda, K., Hauser, C., Rott, R., Klenk, H., Doefler, W. 1986. Expression of the influenza virus haemagglutinin in insect cells by a baculovirus vector. *EMBO J.* 5:1359–65

37. Liu, A., Qin, J., Rankin, C., Hardin, S. E., Weaver, R. F. 1986. Nucleotide sequence of a portion of the *Autographa californica* nuclear polyhedrosis virus genome containing the *Eco*RI site-rich region (hr$_5$) and an open reading frame just 5' of the p10 gene. *J. Gen. Virol.* 67:2565–70

37a. Luckow, V. A., Summers, M. D. 1988. Trends in the development of baculovirus expression vectors. *Biotechnology* 6:47–55

38. Madisen, L., Travis, B., Hu, S.-L., Purchio, A. F. 1987. Expression of the human immunodeficiency virus *gag* gene in insect cells. *Virology* 158:248–50

39. Maeda, S., Kawai, T., Obinata, M., Chika, T., Horiuchi, T., et al. 1984. Characteristics of human interferon-α produced by a gene transferred by a baculovirus vector in the silkworm, *Bombyx mori*. *Proc. Jpn. Acad.* 60:423–26

40. Maeda, S., Kawai, T., Obinata, M., Fujiwara, H., Horiuchi, T., et al. 1985. Production of human α-interferon in silkworm using a baculovirus vector. *Nature* 315:592–94

41. Marumoto, Y., Sato, Y., Fujiwara, H., Sakano, K., Saeki, Y., et al. 1988. Hyperproduction of polyhedrin-IGF-II fusion protein in silkworm larvae infected with recombinant *Bombyx mori* nuclear polyhedrosis virus. *J. Gen. Virol.* In press

42. Matsuura, Y., Possee, R. D., Bishop, D. H. L. 1986. Expression of the S-coded genes of lymphocytic choriomeningitis arenavirus using a baculovirus vector. *J. Gen. Virol.* 67:1515–29

43. Matsuura, Y., Possee, R. D., Overton, H. A., Bishop, D. H. L. 1987. Baculovirus expression vectors: the requirements for high level expression of proteins, including glycoproteins. *J. Gen. Virol.* 68:1233–50

44. Miller, D. W., Safer, P., Miller, L. K. 1986. An insect baculovirus host-vector for high-level expression of foreign genes. In *Genetic Engineering*, ed. J. K. Setlow, A. Hollaender, 8:277–98. New York: Plenum

45. Miller, L. K. 1981. A vector for genetic engineering in invertebrates. In *Genetic Engineering in the Plant Sciences*, ed. N. J. Panopoulos, pp. 203–35. New York: Praeger. 271 pp.

46. Miller, L. K. 1981. Construction of a genetic map of the baculovirus *Autographa californica* nuclear polyhedrosis virus by marker rescue of temperature-sensitive mutants. *J. Virol.* 39:973–76

47. Miller, L. K. 1986. The genetics of baculoviruses. See Ref. 16, pp. 217–38

48. Miller, L. K. 1987. Baculoviruses for foreign gene expression in insect cells. In *Vectors: A Survey of Molecular Cloning Vectors and Their Uses*, ed. D. Denhardt, F. Rodriguez, pp. 457–65. Stoneham, Mass: Butterworth. 578 pp.

49. Miller, L. K., Miller, D. W., Adang, M. J. 1983. An insect virus for engineering: developing baculovirus polyhedrin substitution vectors. In *Genetic Engineering in Eukaryotes*, ed. P. F. Lurquin, A. Kleinhofs, pp. 89–98. New York: Plenum. 282 pp.

50. Miyajima, A., Schreurs, J., Otsu, K., Kondo, A., Arai, K., Maeda, S. 1987. Use of the silkworm, *Bombyx mori*, and an insect baculovirus vector for high-level expression and secretion of biologically active mouse interleukin-3. *Gene* 58:273–81

51. Miyamoto, C., Smith, G. E., Farrell-Towt, J., Chizzonite, R., Summers, M. D., Ju, G. 1985. Production of human c-*myc* protein in insect cells infected with a baculovirus expression vector. *Mol. Cell. Biol.* 5:2860–65

52. Oellig, C., Happ, B., Muller, T., Doerfler, W. 1987. Overlapping sets of viral RNAs reflect the array of polypeptides in the *Eco*RI J and N fragments (map positions 81.2 to 85.0) of the *Autographa californica* nuclear polyhedrosis virus genome. *J. Virol.* 61:3048–57

53. Ollo, R., Maniatis, T. 1987. *Drosophila* Kruppel gene product produced in a baculovirus expression system is a nuclear phosphoprotein that binds to DNA. *Proc. Natl. Acad. Sci. USA* 84:5700–4

54. Overton, H. A., Ihara, T., Bishop, D. H. L. 1987. Identification of the N and NS$_s$ proteins coded by the ambisense S RNA of Punta Toro phlebovirus using monospecific antisera raised to baculovirus expressed N and NS$_s$ proteins. *Virology* 157:338–50

55. Pennock, G. D., Shoemaker, C., Miller, L. K. 1984. Strong and regulated expression of *Escherichia coli* β-galactosidase in insect cells with a baculovirus vector. *Mol. Cell. Biol.* 4:399–406

56. Possee, R. D. 1986. Cell-surface expression of influenza virus haemagglutinin in insect cells using a baculovirus vector. *Virus Res.* 5:43–59

56a. Possee, R. D., Howard, S. C. 1987. Analysis of the polyhedrin gene promoter of the *Autographa californica* nuclear polyhedrosis virus. *Nucleic Acids Res.* 15:10233–48

57. Potter, K. N., Faulker, P., MacKinnon, E. A. 1976. Strain selection during serial passage of *Trichoplusia ni* nuclear polyhedrosis virus. *J. Virol.* 18:1040–50

58. Potter, K. N., Jaques, R. P., Faulkner, P. 1978. Modification of *Trichoplusia ni* nuclear polyhedrosis virus passaged *in vivo*. *Intervirology* 9:76–85

59. Rice, W. C., Lorimer, H. E., Prives, C., Miller, L. K. 1987. Expression of polyomavirus large T antigen by using a baculovirus vector. *J. Virol.* 61:1712–16

60. Rice, W. C., Miller, L. K. 1986. Baculovirus transcription in the presence of inhibitors and in nonpermissive *Drosophila* cells. *Virus Res.* 6:155–72

61. Rohrmann, G. F. 1986. Polyhedrin structure. *J. Gen. Virol.* 67:1499–513

62. Smith, G. E., Fraser, M. J., Summers, M. D. 1983. Molecular engineering of the *Autographa californica* nuclear polyhedrosis virus genome: deletion mutations within the polyhedrin gene. *J. Virol.* 46:584–93

63. Smith, G. E., Ju, G., Ericson, B. L., Moschera, J., Lahm, H.-W., et al. 1985. Modification and secretion of human interleukin 2 produced in insect cells by a baculovirus expression vector. *Proc. Natl. Acad. Sci. USA* 82:8404–8

64. Smith, G. E., Summers, M. D., Fraser, M. J. 1983. Production of human beta interferon in insect cells infected with a baculovirus expression vector. *Mol. Cell. Biol.* 3:183–92

65. St. Angelo, C. S., Smith, G. E., Summers, M. D., Krug, R. M. 1987. Two of the three influenza viral polymerase (P) proteins expressed using baculovirus vectors form a complex in insect cells. *J. Virol.* 61:361–65

66. Stiles, B., Wood, H. A. 1983. A study of the glycoproteins of *Autographa californica* nuclear polyhedrosis virus (AcNPV). *Virology* 131:230–41

67. Summers, M. D. 1978. Baculoviruses (Baculoviridae). In *The Atlas of Insect and Plant Viruses*, ed. K. Maramorosch, pp. 3–33. New York: Academic. 320 pp.

68. Summers, M. D., Smith, G. E. 1987. *A Manual of Methods for Baculovirus Vectors and Insect Cell Culture Procedures*. *Tex. Agric. Exp. Stn. Bull.* Vol. 1555. 56 pp.

69. Summers, M. D., Volkman, L. E., Hsieh, C.-H. 1978. Immunoperoxidase detection of baculovirus antigens in insect cells. *J. Gen. Virol.* 40:545–57

70. Tanada, Y., Hess, R. T. 1984. The cytopathology of baculovirus infections in insects. In *Insect Ultrastructure*, Vol. 2, ed. R. C. King, H. Akai, pp. 517–56. New York: Plenum. 620 pp.

71. Tweeten, K. A., Bulla, L. A. Jr., Consigli, R. A. 1980. Characterization of an extremely basic protein derived from granulosis virus nucleocapsids. *J. Virol.* 33:866–76

72. Vlak, J. M., Smith, G. E. 1982. Orientation of the genome of *Autographa californica* nuclear polyhedrosis virus: a proposal. *J. Virol.* 41:1118–21

73. Volkman, L. E. 1986. The 64K envelope protein of budded *Autographa californica* nuclear polyhedrosis virus. See Ref. 9, pp. 103–18

74. Wilson, M. E., Mainprize, T. H., Friesen, P. D., Miller, L. K. 1987. Location, transcription, and sequence of a baculovirus gene encoding a small arginine-rich polypeptide. *J. Virol.* 61:661–66

75. Wilson, M. E., Miller, L. K. 1986. Changes in the nucleoprotein complexes of baculovirus DNA during infection. *Virology* 151:315–28

Ann. Rev. Microbiol. 1988. 42:201–30

COMPLEMENT EVASION BY BACTERIA AND PARASITES[1]

Keith A. Joiner

Laboratory of Parasitic Diseases, National Institute of Allergy and Infectious Diseases, National Institutes of Health, Bethesda, Maryland 20892

CONTENTS

INTRODUCTION

Microorganisms that are pathogenic for humans possess virulence determinants that allow evasion of host defense mechanisms. One of the principal components of the humoral host defense system is the complement cascade. This paper reviews the complement cascade and then elucidates, for those

---

bacterial and parasitic systems for which data are available, the mechanisms and microbial strategies involved in complement evasion. This review explicitly concentrates on published studies investigating the mechanisms of complement evasion from the complement standpoint. In general, the studies cited contain data on either the amount, the location, or the form of complement for the pathogen of interest. Many excellent studies that have analyzed the general issue of complement activation by microorganisms and microbial constituents are therefore not discussed. Several recent papers (16, 25, 41, 67, 68, 145, 164, 167), most notably the comprehensive review of Taylor (164), provide an outline of the microbial determinants responsible for complement evasion.

THE COMPLEMENT SYSTEM: AN OVERVIEW

In its simplest form the complement system can be thought of as two distinct but related pathways that lead to the generation of opsonic, chemotactic, and lytic functions against invading pathogens. On a more specific level, the two pathways, the classical and the alternative, contain 25 separate activation and regulatory molecules and at least seven complement receptors. These components interact in a complex cascade of events including enzymatic cleavage of molecules and dramatic conformational changes generating biologically active molecules. A simplified version of this latter construct is shown in Figure 1 in the context of interaction of complement with bacteria and parasites.

The alternative pathway is generally considered to be the older of the two pathways phylogenetically, is usually activated in the absence of antibody, and is therefore thought of as an early warning system that allows the naive host to respond to an invading pathogen before specific antibody has developed. In contrast, the classical pathway typically requires specific antibody for activation to be triggered. These concepts are clearly overly simplistic; many circumstances are now defined in microbial systems in which classical pathway activation occurs in the absence of specific antibody or in which specific antibody is necessary for alternative pathway activation.

Not all features of the complement cascade are of equal importance in the discussion of microbe-complement interactions. The initial discussion of the complement system is focused on those aspects of the complement cascade that are most relevant to the published studies on complement evasion by bacteria and parasites. These areas include C1 binding and activation, C3 chemistry, control of alternative pathway activation, C3 receptors, C5b-9 and poly C9 formation, and regulatory molecules. Excellent detailed reviews (38, 45, 105, 138) are available describing the complement system and complement receptors, and the reader is referred to these papers for additional information.

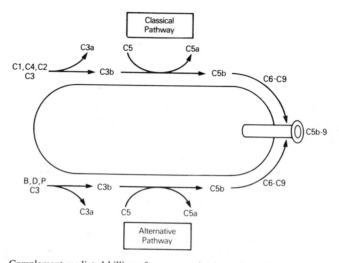

Figure 1 Complement-mediated killing of gram-negative bacteria and parasites. A simplified scheme is shown, illustrating the major components of the cascade.

C1 Activation and Control

Whole C1 consists of a pentameric complex between one C1q subunit and two subunits each of C1r and C1s (reviewed in 27). Activation of C1, which can occur spontaneously in the fluid phase or following binding of whole C1 to specific activating ligands, involves the sequential proteolytic cleavage (activation) of C1r and C1s. The activated C$\overline{1}$s moiety then has the capacity to cleave sequentially additional molecules in the cascade, specifically C4 followed by C2, whose products combine to form a complex protease, C14b2a, the classical pathway C3 convertase.

The process of C1 activation is not synonymous with C1 binding. Surfaces that can bind either the purified C1q subcomponent of C1 or even whole, nonactivated C1 may be incapable of supporting C1 activation in the presence of the normal serum inhibitor of this step, C1 esterase inhibitor. This heavily glycosylated molecule dissociates the activated C$\overline{1}$r and C$\overline{1}$s subunits from C1q, thereby blocking further enzymatic activity of C1 and simultaneously exposing the collagenous portion of C1q to specific receptors on a variety of cells. Specific antibody, when bound to antigen, is the prototypic activator of C1 and provides a protected site for C1 activation that cannot be overridden by C1 inhibitor. As discussed below, a growing number of microbial constituents are known to activate intact C1 directly in the absence of antibody.

C3 Chemistry

The classical and alternative pathways converge at the step of C3 cleavage and deposition. A knowledge of C3 chemistry is critical to a fundamental understanding of complement activation and control (reviewed in 45, 105). Enzymatic cleavage of C3 by the C3 convertase of either the classical pathway (C14b2a) or the alternative pathway (PC3bBb) liberates the small anaphylatoxic C3a molecule and results in exposure and activation of an internal thioester group within the remaining C3b fragment. The labile thioester group either undergoes hydrolysis by H_2O, forms a covalent ester bond with hydroxyl groups on nearby acceptor molecules, or forms a covalent amide bond with amino groups on acceptor molecules (Figure 2). Ester linkages are preferred at physiologic pH; this preference for ester bonds in large part explains the propensity of C3 to react with carbohydrate residues. Ester linkages, unlike amide bonds, are susceptible to spontaneous cleavage in a time-, temperature-, and pH-dependent fashion. Such cleavage leads to the release of intact C3b. Proteolytic cleavage of bound C3b may also occur, mediated by the active complement enzyme factor I and either of the necessary cofactors, factor H or CR1 (discussed below; Figure 3). Formation of iC3b and further cleavage of this fragment to C3d or C3dg block further C3 or

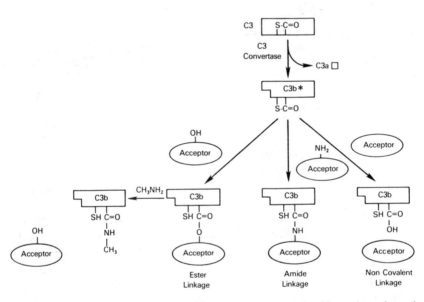

Figure 2 Mechanisms of C3 interaction with acceptor surfaces. Native C3 contains an internal thioester group, which is exposed when the molecule is cleaved to C3b by the C3 convertase of either the classical pathway or the alternative pathway. The internal thioester may be hydrolyzed by water *(right)*, may form a covalent amide bond between C3b and exposed amino groups on acceptor molecules *(center)*, or may form a covalent ester bond between the C3b and exposed hydroxyl groups on acceptor molecules *(left)*. Ester linkages are susceptible to cleavage by nucleophiles such as methylamine, which liberates the acceptor from the covalently bound C3b fragment *(far left)*.

C5 convertase formation, abruptly delimiting the direct microbicidal activity of the cascade. Lytically inactive C3 fragments are not inert, however, since distinct receptors exist for the C3b, iC3b, and C3d or C3dg fragments of complement (see below).

Control of Alternative Pathway Activation

Undetectable activation of the alternative pathway occurs continuously in serum, but it is limited by normal control processes. Low-level deposition of C3b or of water-hydrolyzed C3, which is C3b-like, occurs continually in the fluid phase and on both activating and nonactivating surfaces (reviewed in 45, 105) (Figure 3). The affinity and extent of the subsequent interaction of C3b with either factor H or factor B determines whether activation will occur. On a nonactivating surface, the C3b fragment is rapidly cleaved to iC3b by factors H and I, and no further activation ensues (Figure 3). Detectable activation occurs when these normal control processes are overridden. Thus on an activating surface the affinity of B for C3b exceeds the affinity of H, and activation ensues with amplification of C3b deposition and formation of a C5 convertase. In the best studied system, erythrocytes, the affinity of H for C3b is low on activating surfaces and high on nonactivating, sialic acid–bearing

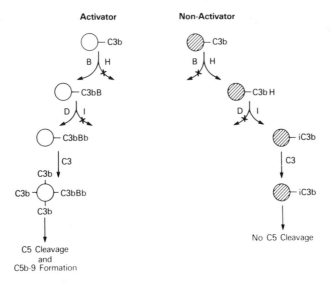

Figure 3 Control of alternative pathway activation. C3b or C3b-like molecules bind randomly to the surface of both activating and nonactivating particles. On activating surfaces, factor B binds to C3b in preference to factor H, permitting the amplification pathway to proceed *(left)*. Additional molecules of C3b are deposited and a C5 convertase forms. In contrast, on nonactivating surfaces *(right)* factor H binds to C3b in preference to factor B, leading to factor I-mediated cleavage of C3b to the hemolytically inactive iC3b fragment. The lytic cascade stops at this point.

surfaces, while the affinity of B for C3b does not vary on the two surfaces. In microbial systems, this scenario does not always apply.

C3 Receptors

Four receptors for fragments of C3 have been defined (43, 173, 174; reviewed in 38, 138). The cellular distribution, chain structure, and ligand specificity of the receptors are shown in Table 1. The best studied receptors are CR1 (or the C3b receptor) and CR3 (or the iC3b receptor). CR1 has a dual function, since it serves not only as a receptor for attachment and adherence of complement-coated ligands but also as a cofactor for proteolytic cleavage of C3b to iC3b. CR3 is a member of a family of receptors, the LFA-1, p150,95, Mo-1 family, that share a common β-chain (CD18) of 95 kd and have variable α-chain size. The identification of patients lacking the entire family of molecules has delineated multiple functions for this receptor group (reviewed in 2). Neutrophils from these patients display markedly deficient adherence and aggregation and an abnormal oxidative burst and phagocytic capacity for iC3b-coated targets, but not for soluble stimuli. Binding of the natural ligand iC3b to CR3 occurs through a sequence containing Arg-Gly-Asp in iC3b (184). The presence of multiple ligand binding sites in CR3, each mediating different cellular responses, has been suggested (137). Two receptors with different ligand specificities have been termed CR4 (43, 108, 173, 174). The molecule that predominates on macrophages (Table 1) and binds iC3b may be p150,95 (108).

C5b-9 and Poly C9 Formation

Formation of the C5b-9 complex is identical whether the alternative or classical pathway is activated. Assembly of the lytic C5b-9 complex involves the sequential high-affinity association of C5b, C6, C7, C8, and C9, thus generating an amphiphilic complex capable of inserting into membranes. This

Table 1 C3 receptors

Receptor	Cells[a]	Chain structure and mass	Ligand specificity
CR1	RBC, PMN, Mono, M0, all B lymphocytes, mast cells, some T cells	160–250 kd single chain	C3b>iC3b
CR2	B lymphocytes	140-kd single chain	C3d>C3dg>iC3b
CR3	PMN, Mono, M0, LGL	165-kd α chain 95-kd β chain	iC3b
CR4	PMN, M0	150-kd α chain 95-kd β chain	iC3b

[a]RBC, erythrocytes; PMN, polymorphonuclear leukocytes; Mono, monocytes; M0, macrophages; LGL, large granular lymphocytes.

capacity can be blocked by the incorporation of the soluble serum control factor, S protein. Although the stoichiometry of the complex was debated earlier, it is now recognized that 1–16 molecules of C9 can bind to the C5b-8 complex on a membrane, leading to progressive enlargement of the channel created by the complex (reviewed in 186). Binding of multiple C9 molecules also leads to formation of the classic "doughnut" lesion visualized by electron microscopy, which represents a ring-polymerized molecule with detergent and protease resistance (poly C9). These attributes of poly C9 may be important in cytolysis of protease-rich targets and for direct complement killing of microorganisms with complex cell walls such as gram-negative bacteria (18). Of recent interest is the demonstration of sequence homology and structural similarity among C7, C8, and C9, as well as antigenic relatedness between C9 and the cytotoxic protein molecule of similar molecular weight isolated from granules of cytotoxic large granular lymphocytes (186).

Control Proteins

A variety of control mechanisms influence the process of complement activation. Soluble control proteins of the complement cascade include factor H, C4-binding protein, and C1 esterase inhibitor, all of which have been discussed briefly above and can limit complement in the fluid phase or on surfaces. Two membrane proteins have gained attention recently: decay-accelerating factor (DAF) and homologous restriction factor (HRF). These molecules are unique because they limit complement activation and lysis on cells by homologous, but not heterologous, complement components.

DAF is a 70-kd glycoprotein with a wide cellular distribution and multiple functions, including limiting formation and accelerating the normal intrinsic decay of classical and alternative pathway C3 convertases (96, 110). DAF thus limits complement activation by limiting the extent of C3 cleavage and deposition. Although not proven, it is presumed that the molecule binds to C3b and C4b and inhibits their interaction with factor B, C2, or other C3b- and C4b-binding molecules. This hypothesis is supported by the observation that the DAF sequence contains four repeats (21) of the 60–63–amino acid partial-homology sequence shared by all of the evolutionarily related C3b- and C4b-binding molecules (H, B, C2, C4-binding protein, CR1, DAF) (132).

HRF, also termed C8-binding protein, was more recently described (155, 187). This integral membrane glycoprotein limits C9 polymerization, apparently by binding to both C8 and C9 from only homologous species. HRF thus limits the extent of C9 binding and membrane insertion. Where polymerized C9 is necessary for effective cytolysis or where the extent of C9 polymerization affects the efficiency of C5b-9 endocytosis or shedding, HRF may have an important regulatory role.

STRATEGIES OF COMPLEMENT EVASION

To be successful pathogens, microorganisms must avoid destruction by the opsonic, chemotactic, and lytic functions of the complement cascade. Therefore, complement evasion may be mediated at a variety of steps in the cascade (Figure 4). Evasion may occur on the surface of the organism because (*a*) the complement cascade is not activated, (*b*) the cascade is blocked before C5b-9 formation, or (*c*) a C5b-9 complex forms that does not lyse the organism. Alternatively, microorganisms may shed molecules that cause fluid-phase depletion of complement or actively destroy complement proteins in the vicinity. Finally, organisms may subvert host complement proteins and receptors to gain access to safe intracellular confines. The remainder of the paper reviews specific examples of each process.

Failure to Activate Complement

CLASSICAL PATHWAY   Direct classical pathway activation by microorganisms is unusual. Although a variety of microbial constituents, prototypically gram-negative bacterial lipopolysaccharide (LPS), can bind C1q and whole C1, C1 binding does not generally result in C1 activation. In a well studied system, Tenner et al (168) demonstrated that whole, reconstituted C1 bound to isolates of *Escherichia coli* bearing either truncated (rough) or complete

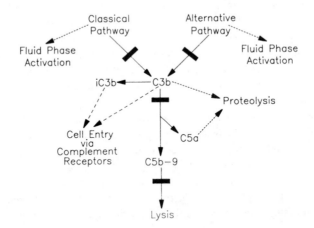

Figure 4 Strategies for complement evasion by bacteria and parasites. Organisms may evade the complement system because complement is not activated, because the cascade is blocked before completion of C5b-9 formation, or because the assembled C5b-9 complex is not lytic (*solid arrows*). Alternatively, organisms may nonspecifically inactivate complement components in the local vicinity by fluid-phase complement activation or proteolysis, or they may produce more specific inactivators of complement molecules (*dotted arrows*). Finally, some organisms subvert complement receptors to gain safe access to obligatory intracellular confines (*dashed arrows*).

(smooth) LPS and that activation ensued in both cases in the absence of C1 esterase inhibitor. Addition of C1 inhibitor, which reproduces the situation in serum, blocked C1 activation on the pathogenic, serum-resistant organism bearing smooth LPS, but not on the avirulent, serum-sensitive strain bearing rough LPS. Tenner et al (168) postulated that either C1s or C1r may have a binding site on the surface of the latter organisms, which leads to a high-affinity two-site interaction (C1q and C1r or C1s). Evidence for C1 binding to the major outer-membrane porin proteins has recently been presented (159). Aubert et al (3), after examining a deep, rough, heptoseless mutant of *E. coli* containing only 2-keto-3-deoxyoctulosonic acid (KDO) in the core polysaccharide of LPS, concluded that the C1 inhibitor–resistant association of C1 was mediated exclusively through the collagenous portion of C1q and did not involve a second-site interaction. In *Neisseria gonorrhoeae,* which bears lipooligosaccharide (LOS) containing KDO and only a limited number of additional sugars, C1 activation leading to C3 cleavage and deposition occurs even in the presence of C1 inhibitor (151). A rough mutant of *N. gonorrhoeae* containing only KDO in the LOS consumed C1, C4, and C3 from antibody-free serum, which implied that C1 inhibitor-resistant activation occurred (153). Vukajlovich et al (175) have shown that the presence of D-mannoheptose in addition to KDO in the core polysaccharide of rough LPS converts the molecule from a classical pathway activator to an alternative pathway activator in serum. Taken as a whole, these results suggest that the polysaccharide components of the LPS core inhibit direct C1 activation by deeper LPS core structures, possibly by blocking a second-site C1 interaction or by enhancing C1 inhibitor binding.

Other investigators have examined the binding and activation of C1 by a variety of gram-negative and gram-positive bacteria and by several protozoan and helminthic parasites. Although direct classical pathway activation has been demonstrated with several rough strains of *E. coli* and with purified LPS (6, 23, 87, 99), these studies were not done with C1 inhibitor. Conclusions can therefore not be drawn with regard to whether the reactions will occur in serum. Both *Schistosoma mansoni* (147) and *Trypanosoma brucei* (107) can bind C1q, although the biological consequences of this process are not clear.

Binding of C1 or C1q may facilitate cell attachment and entry for some organisms, potentially in conjunction with fibronectin, thereby circumventing a host-protective role for C1 (4, 156). *Trypanosoma cruzi* vector-stage epimastigotes and vertebrate-stage tissue-culture trypomastigotes bind but degrade C1q as well as native C1, and this process is not controlled by C1 inhibitor (136). The presence of either intact C1q or collagenous tail fragments of C1q facilitates the entry of trypomastigotes, but not epimastigotes, into both human monocytes and human monocyte-derived macrophages.

ALTERNATIVE PATHWAY   Microorganisms may also evade complement activation through the alternative pathway, by presenting structures to the environment that are not complement-activating surfaces. Bacterial polysaccharide capsules provide one of the clearest examples. Generally, neither gram-positive bacterial capsules nor acidic gram-negative bacterial capsules activate the alternative pathway. Although the polysaccharide capsule is permeable to complement molecules and although subcapsular structures may efficiently activate the alternative pathway (17, 125, 160, 163, 180, 181), the complement fragments that are deposited do not have access to complement receptors on phagocytic cells (17, 180). Thus the deposited complement molecules are ineffective in mediating ingestion by phagocytic cells.

The molecular basis for the inability of capsular polysaccharides to activate the alternative pathway has been studied with sialic acid–containing capsules, with the type 7 *Streptococcus pneumoniae* capsule, and with the polyribose phosphate capsule of *Haemophilus influenzae*. Sialic acid–containing capsules of the *E. coli* K-1 group, of *Neisseria meningitidis* type B, and of type III group B streptococci are poor activators of the alternative complement pathway in the absence of specific antibody. The exact molecular mechanism by which capsules containing sialic acid inhibit alternative pathway activation is not known. By way of analogy, however, sialic acid on erythrocytes limits alternative pathway activation by facilitating factor H binding to C3b on the sialic acid–bearing surface (37). Desialation of erythrocytes decreases the extent of factor H binding and allows activation to proceed. In an analogous system, deposition of C3 and factor B increased on an enzymatically desialated isolate of group B *N. meningitidis* (62). With type III group B streptococci, which do not cause C3 cleavage in nonimmune serum, removal or chemical modification of the sialic acid allows complement activation and C3 cleavage to occur. Opsonic anticapsular antibody apparently has a similar role, to effectively mask sialic acid residues (33, 34). With *E. coli* K-1 capsules, the presence of sialic acid limits the amount of complement consumed in comparison with that consumed by organisms bearing capsules of a different composition (119, 160). Finally, pathogenic *Treponema pallidum*, which contains nearly four times as much surface sialic acid as nonpathogenic *Treponema vicentii*, activates the alternative pathway less efficiently than the latter, as reflected by C3a generation. Although the role of a capsular structure is unclear, removal of sialic acid from *T. pallidum* markedly enhanced alternative pathway activation (42). A comparison of factor H and factor B binding to C3b on the surface of the above organisms has not been made, but it is likely that the presence of sialic acid enhances H binding.

Capsular polysaccharides of *S. pneumoniae* are often poor activators of the alternative and classical pathways in the absence of specific antibody (47, 121, 182). Deposited fragments may be rapidly cleaved to iC3b and C3dg

(56), although two studies revealed predominantly C3b on the capsule (18, 109). Brown et al (18) have examined the control of alternative pathway activation on the capsule of type 7 *S. pneumoniae*. The alternative pathway is not activated because of inefficient binding of factor B to C3b on the capsule.

Finally, with *H. influenzae* type b, the polyribose phosphate capsule is incapable of serving as an acceptor for covalent C3 deposition (88). Nonetheless, subcapsular structures activate complement efficiently in the absence of anticapsular antibody (125, 158, 163). Therefore, with three different polysaccharide capsules, three different mechanisms apparently limit alternative pathway activation: enhanced H binding, poor B binding, and poor C3 binding. The mechanisms by which other capsular structures inhibit activation (55, 172) remain to be defined.

Noncapsular polysaccharides may also limit alternative pathway activation. The greater virulence for mice of *Salmonella typhimurium* than of *Salmonella enteritidis* and *Salmonella montevideo* has been ascribed to poor complement activation by *S. typhimurium*. The rate and extent of microbial uptake into peritoneal macrophages is dictated solely by the extent of C3 deposition on these species, which in turn is directly related to the carbohydrate composition of the O-antigen within the LPS (50, 51, 91; reviewed in 92). Jimenez-Lucho et al (63) have shown that the extent of C3 deposition is controlled at the level of factor B binding to C3b. The only difference between the isogenic *S. typhimurium* and *S. enteritidis* strains is a change from abequose to tyvulose within the LPS O-antigen polysaccharide. These sugars are epimers of one another; thus the different behavior of the two strains illustrates the exquisite sensitivity of the alternative pathway C3 convertase to subtle changes in the nature of the activating surface.

Outer membrane proteins also limit C3 deposition on the bacterial surface (1, 14, 60, 169). The mechanism for this effect is now understood with *Campylobacter fetus* (14) and with certain group A streptococci bearing M protein (54), as discussed further below.

Microbial Shedding of Molecules That Activate or Destroy Complement

Microorganisms may shed constituents that activate the classical or alternative pathway (Figure 4). One consequence may be that the microbial surface becomes inert with respect to the complement cascade. For example, mechanical schistosomula of *S. mansoni* incubated in serum-free media shed molecules capable of consuming C3 from serum (93) and consequently become nonactivators of the alternative pathway (93, 128, 144). Shed molecules may include residual cercarial glycocalyx, which is known to be a C3 acceptor (143). Residence of the schistosomula within the host also results in loss of the capacity to bind homologous C3 (139, 140), although the capacity to activate complement from heterologous species is not totally abrogated (30, 146).

The second consequence of microbial shedding of complement-activating or complement-inactivating molecules is that organisms may deplete complement components in the vicinity of or even at a distance from the microbial cell membrane. Thus complement deposition on the microbial surface is blocked. Conceptually, any microbial cell wall constituent that activates complement and that is released normally or with cell death could block complement deposition in this way. Although a review of all such constituents is beyond the scope of this paper, several examples are more completely understood than others at the molecular level. Levy has suggested that the released type Ia group B streptococcal capsule may reduce C1 in this fashion (90). The copious mucoexopolysaccharide slime from *Pseudomonas aeruginosa* activates the alternative pathway with C3 and factor B cleavage (86). *Staphylococcus aureus* produces and sheds a decomplementation antigen (7), recently identified as soluble techoic acid (S. Bhakdi, personal communication), which consumes early but not terminal classical pathway components from serum in conjunction with antibody.

Complement molecules may also be inactivated or destroyed by microbial products through a process other than conventional complement activation. Hostetter et al (57) have suggested that local high concentrations of ammonia produced by *P. aeruginosa* may inactivate C3 by nucleophilic attack. Microbial proteases may also inactivate complement components. One of the best known examples is elastase from *P. aeruginosa,* which cleaves a number of complement molecules (150). Most of the proteases described are nonspecific, readily attacking a variety of complement and noncomplement substrates (22, 94, 111, 150, 162, 176). Proteases from *Serratia, Pseudomonas,* and group A streptococci may generate chemotactic fragments from native C5, a process which would appear to be of dubious protective value to the pathogen (150, 176). Recently, however, a specific C5a inactivator produced by group A streptococci has been identified, purified, and characterized. This predominantly cell bound molecule, which cleaves a six-residue peptide from the carboxyl-terminal end of C5a, destroys the chemotactic activity of C5a (113, 178, 179).

Synthesis or Acquisition of Regulatory Molecules

A sophisticated method for evasion of alternative pathway activation is the microbial synthesis of protein molecules that have functional similarity to regulatory molecules of the complement cascade. Recent studies have shown that trypomastigotes of *Trypanosoma cruzi* synthesize a complement-regulatory molecule, thus evading lysis by serum. The interaction with complement has been compared for epimastigote and trypomastigote stages of the parasite. Epimastigotes are the noninfective form of the parasite, found within the foregut and midgut of the insect vector. Metacyclic trypomastigotes (CMT) are the infective stages, found within the hindgut and feces of the reduviid bug vector. Bloodstream trypomastigotes, analogous to tissue culture trypomastigotes (TCT), circulate in the blood and invade tissues within the vertebrate host. Epimastigotes, which never come in contact with the vertebrate complement system, lyse when incubated in normal human serum (106). Trypomastigotes, which during the normal life cycle are exposed to host complement, are serum resistant (112). This complement resistance is a preadaptation of infective and vertebrate stages commonly observed in protozoan parasites. It reflects developmental regulation that optimizes parasite evasion of vertebrate host defenses (39, 44, 112).

Initial studies showed that six- to eightfold more C3 and C9 bound to epimastigotes than to either CMT or TCT during alternative pathway activation in serum (65) (Figure 5). C3b predominated on epimastigotes, whereas iC3b predominated on CMT and TCT (66). Formation of the alternative pathway C3 convertase on CMT and TCT was inefficient owing to poor binding of factor B to C3b on the parasite surface. Treatment with pronase, and to a lesser extent endoglycosidase, rendered CMT susceptible to complement killing, increased B binding to C3b, and resulted in a threefold increase in C3 and C9 binding (154). It was subsequently demonstrated that CMT and TCT, but not epimastigotes, produce and shed molecules that interfere with formation and accelerate decay of classical and alternative pathway C3 convertases, whether these convertases are assembled on inert particles or on the biologically relevant epimastigote surface (135). The responsible activity has been characterized and is contained in a 87–93-kd band intrinsically labeled with ^{35}S, which is destroyed by papain and removed on concanavalin A (75). The *T. cruzi* complement inhibitor, which does not enhance factor I–mediated cleavage of C3b to iC3b, is thus more analogous to decay-accelerating factor than to other soluble regulatory molecules of the complement cascade such as factor H or C4-binding protein. Nonetheless, there is no antigenic cross-reactivity between the molecule produced by trypomastigotes and any of the C3b- and C4b-binding molecules of the human complement cascade. Kipnis et al (82) have also shown that *T. cruzi* trypomastigotes produce molecules that accelerate decay of the classical pathway C3 convertase. In contrast, Schenkman et al (148) suggested that control of alternative pathway activation on *T. cruzi* strain G CMT resulted from more extensive binding of factor H to C3b on CMT than on epimastigotes.

Instead of synthesizing regulatory molecules, organisms may bind or acquire regulatory molecules from the complement system of the host. The first examples of this process have recently been reported. Strains of group A streptococci containing M protein are more virulent than those lacking M

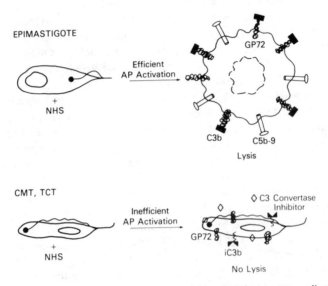

Figure 5 Model of the interaction of complement with epimastigotes, metacyclic trypomasti-gotes (CMT), and tissue culture trypomastigotes (TCT) of *Trypanosoma cruzi*. Epimastigotes are lysed by the alternative complement pathway in nonimmune serum, whereas both CMT and TCT resist lysis. The major C3 fragment on the epimastigote surface is C3b, which attaches ex-clusively to the developmentally regulated glycoprotein gp72. Deposition of C3b leads to C5 convertase formation, C5b-9 generation, and lysis of the parasite. On CMT and TCT, six- to eightfold less C3 is deposited on the parasite in nonimmune serum. Most bound C3b is cleaved to iC3b, which cannot participate in C5 convertase formation or lead to C5b-9. Inefficient C3 deposition on CMT and TCT results from the production of a molecule that has functional similarity to human decay-accelerating factor, as discussed in the text.

protein. C3 deposition on M protein–bearing strains is limited, and the accessibility of bound C3 to exogenous C3 receptors is blocked as well (10, 60, 61, 118). Direct binding of factor H to purified M protein has recently been demonstrated; this binding leads to efficient cleavage of C3b to iC3b on the M protein–bearing surface (54). Lastly, it has been speculated for some time that schistosomula, which are known to incorporate host molecules into their membranes, could acquire host regulatory molecules. Recently, pre-liminary evidence for the presence of host decay-accelerating factor on the schistosomal membrane has been provided (126; E. Pearce & B. F. Hall, personal communication).

Blockade of Activation Before C5b-9 Formation

Some microorganisms initiate complement activation but block the cascade at a specific step before opsonic or lytic complement fragments are deposited on the microbial surface (Figure 4). A general mechanism for blocking activation at a single step or at multiple steps is the production and shedding of microbial molecules that either bind or degrade complement proteins. This process has been described above.

More specific blockade of the activation cascade on the microbial surface has also been described in several instances. A deep, rough *Salmonella* mutant activated C1 and C4 but failed to bind C4b in the absence of an additional serum factor (24). A dissociation between consumption and deposition of C3 may also occur. Schiller & Joiner (149) compared C3 deposition on a rough, serum-sensitive isolate of *Pseudomonas aeruginosa* and a smooth, serum-resistant isolate derived from the rough strain by serial passage in normal human serum. In the face of greater consumption of C3 by the resistant strain, approximately four- to fivefold less C3 binding occurred on the surface of the resistant isolate than on the serum-sensitive isolate. Blaser et al (13) have shown that serum-resistant isolates of *Campylobacter fetus*, but not serum-sensitive isolates, contain high–molecular weight proteins of 100 and 125 kd, which appear to interfere with efficient C3 deposition on the bacterial surface (14). Thus, despite equivalent C3 consumption by strains bearing and lacking the high–molecular weight proteins, the extent of C3 deposition, C5 convertase formation, and C9 deposition on the serum-resistant isolates is four- to fivefold lower than on the serum-sensitive strains. The mechanism limiting C3 deposition is not yet clear.

Related processes may be operative in the serum resistance of the bloodstream trypomastigotes of *Trypanosoma brucei gambiense* and *Trypanosoma lewisi*. Although C3 is deposited on the parasite surface during incubation in normal human serum, few C5 or C9 molecules are bound. Bound C3 on *T. lewisi* is rapidly cleaved to iC3b and possibly to C3dg, which cannot participate in C5 convertase formation (32, 161). Although this cleavage may be sufficient to explain the small amounts of C5 found on the parasite surface, it is also possible that C5b-9 complexes form but are incapable of penetrating the confluent variable surface glycoprotein surface layer in the absence of high-titer antibody. This type of resistance would be analogous to the serum resistance in smooth strains of *Salmonella* spp. and *E. coli*, as described below.

Although lipopolysaccharide of gram-negative organisms is typically thought of as a protected site for alternative pathway C3 convertase formation, rapid cleavage of C3b to iC3b has been noted on some enteric organisms. This cleavage provides a mechanism for serum and phagocytosis resistance at an intermediate step in the cascade (48, 85). A similar process may be operative with tachyzoites of *Toxoplasma gondii* (46).

With *Leishmania donovani* promastigotes, rapid cleavage of C3b to iC3b occurs on the parasite surface, and the bound iC3b is then released through proteolytic cleavage by a parasite-derived protease (122). The contribution of the major surface protease of *Leishmania,* gp63 (35), to this event is under investigation. A correlation between C3 release and levels of gp63 protease activity has been noted by one laboratory (D. M. Mosser, personal communication).

Isolates of *Bacteroides fragilis* contain proteases capable of cleaving C3b to iC3b, providing a potential mechanism for serum resistance (11). Finally, Densen & McRill (31) have recently shown that the terminal complement complex can accelerate decay of the alternative pathway C3 convertase on *Neisseria gonorrhoeae* by facilitating the release of bound factor B from the C3 convertase, C3bBb.

Formation of a Nonlytic C5b-9 Complex

SHEDDING OF C5b-9   Formation of a C5b-9 complex that deposits in non-microbicidal locations is a common mechanism of complement evasion (Figure 4). The clearest examples of this phenomenon are with smooth nonencapsulated strains of *Salmonella* spp. and *E. coli.* Early studies indicated that consumption and binding of early components were equivalent on serum-resistant and serum-sensitive *Salmonella* strains, suggesting a block later in the cascade (40, 133). Several years ago, Joiner et al (73) showed that terminal components were completely consumed from serum incubated with resistant isolates of *Salmonella* spp. and *E. coli.* Nonetheless, binding of C9 to the resistant isolates was minimal, and the C5b-9 that did bind was released with continued incubation (Figure 6). Subsequent experiments suggested that the bulk of C5b-9 on serum-sensitive organisms was intimately associated with hydrophobic portions of the outer membrane, since the complex was not released by ionic manipulations or proteolytic attack (74, 77). In contrast, the C5b-9 on resistant isolates was bound by weak, salt-susceptible bonds and was not inserted into hydrophobic domains of the outer membrane unless bactericidal antibody was added (69–71). Complement activation, C3 deposition, and by inference C5b-9 deposition occurred only on those LPS molecules bearing more than 40–50 O-antigen subunits (72). This population of molecules represents only 2–3% of the LPS in the cell and contains eightfold more O-antigen subunits than the average O-antigen side chain. Experiments were also performed to study the precise distribution and length necessary for O-polysaccharide side chains to mediate serum resistance. There was a striking relationship between the percentage of LPS molecules bearing more than 13 O-antigen subunits (long-chain LPS) and serum sensitivity. When more than 20% of the LPS molecules were long-chain LPS, the strain was serum resistant; when less than 20% of the LPS molecules were long-chain LPS, the strain was serum sensitive (52). Molecular modeling of these data supports the idea that long-chain LPS imparts a steric barrier to the access of C5b-9 to hydrophobic domains of the outer membrane.

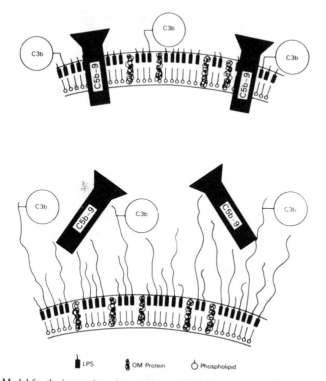

Figure 6   Model for the interaction of complement with isolates of *Salmonella* spp. and *E. coli.* *Top:* Complement is activated close to the hydrophobic surface on rough, serum-sensitive strains, leading to C5b-9 insertion into the outer membrane. *Bottom:* Complement is activated exclusively by long-chain LPS on smooth, serum-resistant strains because of steric blockade of shorter-chain molecules to host complement proteins. C5b-9 forms but cannot insert into hydrophobic domains, is shed, and thus is not bactericidal.

Related observations have been made about complement deposition on other gram-negative organisms bearing smooth LPS, including *P. aeruginosa* (149), *C. fetus* (14), and *E. coli* (77, 166). The contention of steric hindrance is also supported by studies on antibody and bacteriophage accessibility to outer membrane proteins and LPS core structures (170, 171), as well as by a recently published model using LPS incorporated into liposomes (120).

A similar process may be operative with some serum-resistant *Leishmania* species and stages. A number of investigators have demonstrated C3 deposition by immunofluorescence or direct binding on both serum-sensitive and serum-resistant promastigotes and amastigotes of *Leishmania donovani* (44, 122, 183), *Leishmania major* (100, 101, 104, 123), and *Leishmania mexicana* (141). In general, the amount of C3 bound does not correlate with the lytic susceptibility. In fact, Wozencraft et al (183) suggested that serum-resistant promastigotes of *L. donovani* bound more C3 during incubation in serum than did serum-susceptible promastigotes. Russell (141) reached the opposite conclusion in experiments with *L. mexicana*. Puentes et al (124) have recently demonstrated that serum-resistant, infective, metacyclic promastigotes of *L. major* activate complement efficiently, bear C3b on their surface, and efficiently consume C5 and C9 from serum. Nonetheless, minimal C9 deposition occurs, and the small amount of bound C9 is readily released as a C5b-9 complex (124). Serum-sensitive, noninfective promastigotes also activate complement efficiently, with formation of C3b and a C5 convertase; however, in contrast, C9 binds stably in large amounts and the parasites are killed. Changes in the prominent cell-surface glycolipid, excreted factor, are critical in the transition from the noninfective to the infective form (142), appear to alter the form of C3 deposition (123), and may also alter the interaction of C5b-9 with the promastigote surface. Alternatively, C5b-9 may be shed from the parasite surface by active exocytosis in metacyclic promastigotes.

The uniform resistance of gram-positive bacteria to the lytic effects of the serum complement cascade presumably reflects an inability of the formed C5b-9 complex to penetrate the thick, rigid peptidoglycan layer. Bound C5b-9 may be released or may stay associated with the cell wall via hydrophobic interaction (64).

STABLE INSERTION OF C5b-9   Stable deposition of nonlytic C9 on the microbial surface is the most enigmatic of all reported mechanisms for serum resistance. Stable and quantitatively equivalent binding of C5b-9 to the surface of resistant and susceptible strains of *Neisseria gonorrhoeae* occurs during incubation in serum (53, 79). *N. gonorrhoeae* possesses only LOS and lacks conventional O-antigen side chains within LPS. Thus, long-chain O-antigen polysaccharide molecules cannot sterically hinder C5b-9 insertion into the outer membrane as observed for *Salmonella* spp. and *E. coli*. The sedimentation profile and membrane associations of C5b-9 extracted from resistant strains differ from those of C5b-9 extracted from sensitive isolates (79, 80). The relative levels of bactericidal and blocking antibodies determine the ultimate location and configuration of C5b-9 on the bacterial surface (49, 76, 79, 81, 134). Although these results emphasize the critical role of antibody in determining the bactericidal efficiency of C5b-9, the precise molecular correlations are not known.

Strains of *E. coli* containing the *traT* outer membrane protein are more resistant to serum killing than are isogenic *traT* mutants (98). Strains bearing *traT* and their more serum-sensitive counterparts lacking the *traT* protein consume and stably bind equivalent amounts of C9 (9, 19, 114). The mechanism for the enhanced resistance conferred by *traT* is therefore uncertain.

Virulent strains of *Entamoeba histolytica* are serum resistant (58, 97, 131).

Serum-resistant as well as serum-susceptible strains of *E. histolytica* activate complement efficiently, with stable deposition of C9 on the trophozoite surface (129, 130). Although the molecular mechanism of serum resistance is unknown in this instance, the resistance could arise from a block in C9 polymerization or insertion due to the presence of a molecule analogous to homologous restriction factor.

Hypothetical Mechanisms for Complement Resistance at the C5b-9 Step

At least four additional mechanisms for complement resistance at the C5b-9 step can be considered based on published data: (*a*) The C9:C5b-8 ratio may be limited on a particular organism. Multimeric C9 within the C5b-9 complex (C9:C5b-8 $\geqslant$ 3:1) is needed for killing of rough strains of *E. coli* and *Salmonella* spp. (8, 15, 78). At least one potential mechanism that may limit the C9:C5b-8 ratio and hence block direct bacterial killing is the production by microorganisms of a molecule analogous to homologous restriction factor, which limits C9 polymerization or insertion. (*b*) If proteolysis of C9 rather than formation of poly C9 is required for bacterial killing, as has been suggested by Dankert & Esser (28, 29), the lack of production or proper location of the responsible protease might mediate serum resistance. (*c*) Since bacterial metabolic processes are required for a lethal effect of the complement cascade on gram-negative organisms (84, 165), the absence of the necessary metabolic pathway may lead to serum resistance. (*d*) Active endocytosis or exocytosis of bound C5b-9, as shown for complement resistance in nucleated eukaryotic cells (127), may occur in some microbial systems, but has not yet been demonstrated.

Microbial Use of Complement and Complement Receptors to Gain Access to Obligatory Intracellular Locations

Complement receptors on phagocytic cells have conventionally been considered important for the uptake and destruction of microbial pathogens. However, some microorganisms use complement receptors and complement fragments to gain safe access to obligatory intracellular confines. This represents a form of complement evasion, since the organisms parasitize complement receptor–ligand systems for uptake into cells. A presumed advantage for the microbe is that uptake of complement-coated particles may not trigger a respiratory burst upon phagocytosis (20, 185). Hence the organism is not exposed to the lethal effects of toxic oxygen products such as hydroxyl radical (OH·), hydrogen peroxide (H_2O_2), singlet oxygen (O_2·), and products of the myeloperoxidase-halide system. Other receptors, in particular the Fc receptor, in the ligated state and during the process of phagocytosis, are potent triggers for generation of the toxic oxygen radicals produced by phagocytes.

Complement receptors and complement components were first implicated in intracellular uptake of the rat parasite *Babesia rhodaini*. Jack & Ward (59) presented data suggesting that merozoites of *B. rhodaini* possessed receptors for complement fragments that facilitated uptake of the parasites into rat red cells, and that complement-coated *B. rhodaini* entered rat erythrocytes through C3b receptors on the red cells. A puzzling aspect of this study is the general observation from other work that only primate erythrocytes bear receptors for C3b (i.e. CR1 receptors). Studies with *Babesia* spp. in murine systems (89, 152) and experimental studies with malarial parasites have not demonstrated a role for C3 receptors in parasite uptake into erythrocytes (177). Although one study suggested that human erythrocytes infected with *Plasmodium falciparum* become activators of the alternative pathway at the trophozoite stage (157), the consequences of such activation were not defined.

The best studied situation in which complement receptors and complement fragments participate in establishing intracellular infection is with promastigotes of *Leishmania* species. Blackwell and coworkers (12) showed that uptake of promastigotes of *L. donovani* into murine peritoneal macrophages was blocked by monoclonal antibodies to the iC3b (CR3) receptor. Inhibition of uptake was described in a serum-free system. The authors have provided evidence that complement components produced locally by the macrophages are bound to the parasite (183) and are necessary for parasite attachment and infectivity through the CR3 receptor. A similar process of "local opsonization" had been described earlier by Ezekowitz et al (36), who used zymosan as the target particle. Infective and noninfective promastigotes may use different receptor binding sites on the CR3 molecule (26). Mosser & Edelson (102) confirmed the above results with murine macrophages using a visceralizing strain of *L. major*. They recently suggested, using a mixed population of infective and noninfective *L. major* promastigotes, that C3 deposition is necessary not only for cellular attachment and uptake, but also for infectivity within cells (103). da Silva et al (R. da Silva, B. F. Hall, K. A. Joiner & D. L. Sacks, manuscript in preparation) recently showed that uptake of serum-incubated *L. donovani* promastigotes into human macrophages and monocytes was inhibited with monoclonal antibodies to CR3 but not to CR1. In contrast, monoclonal antibodies to CR1 but not to CR3 effectively inhibited cell attachment and entry of serum-incubated infective metacyclic promastigotes of *L. major* but not noninfective log-phase *L. major* promastigotes, despite the presence of C3b on the surface of both forms. Taken in toto, these results suggest that entry into cells via complement receptors is not sufficient, although it may be necessary, for effective establishment of intracellular infection.

An increasing number of organisms that utilize complement receptors and complement fragments to gain access to an intracellular locale is being recognized. For example, *Legionella pneumophila*, the causative agent of Legionnaires' disease, apparently enters monocytes and macrophages through both CR1 and CR3 receptors (117). Complement deposition occurs on the organism during incubation in serum, at least in part on the 60-kd major outer membrane protein, and leads to deposition of iC3b and perhaps C3d despite serum resistance (5). Thus, *L. pneumophila* resembles *Leishmania* spp. in activating complement efficiently, which leads to deposition of complement fragments mediating cellular infectivity, while it remains serum resistant.

Additional organisms that use complement receptors to gain or enhance entry into cells are metacyclic trypomastigotes of *Trypanosoma cruzi* (R. da Silva & A. Sher, manuscript in preparation) and *Mycobacterium tuberculosis* (116). It is certain that this list will grow as the ligand-receptor systems used in cell attachment and entry are examined for more organisms. Similarly, the presence of complement receptors on the microbial surface (83, 95, 115) and their contribution to virulence are fruitful areas for future study.

CONCLUSIONS

Pathogenic bacteria and parasites successfully evade destruction mediated by the opsonic, chemotactic, and lytic components of the complement system. This review has characterized the mechanisms involved in this process of complement evasion. A variety of microbial strategies are used to circumvent the destructive effects of the complement cascade, and it is likely that new mechanisms will be defined. The study of parasites is particularly instructive, since comparisons can be made of different developmental stages that vary in the capacity and the necessity for complement evasion. One of the most intriguing issues for continued study involves the definition of specific complement-regulatory molecules that are either synthesized by the invading organisms or acquired from the host. Nonetheless, the most critical area for continued study, and one that has not been reviewed here, is the further delineation of those host defense mechanisms capable of rendering bacteria and parasites susceptible to the destructive effects of complement.

ACKNOWLEDGMENTS

The author acknowledges the helpful comments of B. F. Hall, M. Marques, and A. Sher in the preparation of the manuscript and the excellent editorial assistance of Jean Carolan and Rochelle Howard.

Literature Cited

1. Aguero, M. E., Aron, L., DeLuca, A. G., Timmis, K. N., Cabello, F. C. 1984. A plasmid-encoded outer membrane protein, TraT, enhances resistance of *Escherichia coli* to phagocytosis. *Infect. Immun.* 46:740–46

2. Anderson, D. C., Springer, T. A. 1987. Leukocyte adhesion deficiency: an inherited defect in the Mac-1, LFA-1, and p150,95 glycoproteins. *Ann. Rev. Med.* 38:175–94

3. Aubert, G., Chesne, S., Arlaud, G. J., Colomb, M. G. 1985. Antibody-independent interaction between the first component of human complement, C1, and the outer membrane of *Escherichia coli* D31 m4. *Biochemistry* 232:513–19

4. Baughn, R. E. 1986. Antibody-independent interactions of fibronectin, C1q, and human neutrophils with *Treponema pallidum*. 54:456–64

5. Bellinger-Kawahara, C. G., Horwitz, M. A. 1987. The major outer membrane protein is a prominent acceptor molecule for complement component C3 on *Legionella pneumophila*. *Clin. Res.* 35:468A

6. Betz, S. J., Isliker, H. 1981. Antibody-independent interactions between *E. coli* J5 and human complement components. *J. Immunol.* 127:1748–54

7. Bhakdi, S., Muhly, M. 1985. Decomplementation antigen, a possible determinant of staphylococcal pathogenicity. *Infect. Immun.* 47:41–46

8. Bhakdi, S., Tranum-Jensen, J. 1986. C5b-9 assembly: average binding of one C9 molecule to C5b-8 without poly-C9 formation generates a stable transmembrane pore. *J. Immunol.* 136:2999–3006

9. Binns, M. W., Mayden, J., Levine, R. P. 1982. Further characterization of complement resistance, conferred on *Escherichia coli* by the plasmid genes traT of R100 and iss of ColV, I-K94. *Infect. Immun.* 35:654–59

10. Bisno, A. L. 1979. Alternative complement pathway activation by group A streptococci: role of M-protein. *Infect. Immun.* 26:1172–76

11. Bjornson, A. G., Magnafichi, P. I., Schreiber, R. D., Bjornson, H. S. 1987. Opsonization of *Bacteroides* by the alternative complement pathway reconstructed from isolated plasma proteins. *J. Exp. Med.* 164:777–98

12. Blackwell, J. M., Ezekowitz, R. A. B., Roberts, M. B., Channon, J. Y., Sim, R. B., et al. 1985. Macrophage complement and lectin-like receptors bind *Leishmania* in the absence of serum. *J. Exp. Med.* 162:324–31

13. Blaser, M. J., Smith, P. F., Hopkins, J. A., Heinzer, I., Bryner, J. H., Wang, W.-L. L. 1987. Pathogenesis of *Campylobacter fetus* infections: serum resistance associated with high-molecular-weight surface proteins. *J. Infect. Dis.* 155:696–706

14. Blaser, M. J., Smith, P. F., Repin, J. E., Joiner, K. A. 1988. Pathogenesis of *Campylobacter fetus* infections. II. Failure of encapsulated *Campylobacter fetus* to bind C3b explains serum and phagocytosis resistance. *J. Clin. Invest.* 81:1434–44

15. Bloch, E. F., Schmetz, M. A., Foulds, J., Hammer, C. H., Frank, M. M., Joiner, K. A. 1987. Multimeric C9 within C5b-9 is required for inner membrane damage to *Escherichia coli* J5 during complement killing. *J. Immunol.* 138:842–48

16. Brown, E. J. 1985. Interaction of gram-positive microorganisms with complement. *Curr. Top. Microbiol. Immunol.* 121:159–87

17. Brown, E. J., Hosea, S. W., Hammer, C. H., Burch, C. G., Frank, M. M. 1982. A quantitative analysis of the interactions of antipneumococcal antibody and complement in experimental pneumococcal bacteremia. *J. Clin. Invest.* 69:85–98

18. Brown, E. J., Joiner, K. A., Gaither, T. A., Hammer, C. H., Frank, M. M. 1983. The interaction of C3b bound to pneumococci with factor H (β1H globulin), factor I (C3b/C4b inactivator), and properdin factor B of the human complement system. *J. Immunol.* 131:409–15

19. Cabello, F. 1988. Plasmid mediated complement and phagocytosis resistance in *E. coli*. In *Bacteria, Complement and the Phagocytic Cell*, ed. F. Cabello, C. Pruzzi. Heidelberg: Springer-Verlag. In press

20. Cain, J. A., Newman, S. L., Ross, G. D. 1987. Role of complement receptor type three and serum opsonins in the neutrophil response to yeast. *Complement* 4:75–86

21. Caras, I. W., Davitz, M. A., Rhee, L., Weddell, G., Martin, D. W. Jr., et al. 1987. Cloning of decay-accelerating factor suggests novel use of splicing to generate two proteins. *Nature* 325:545–49

22. Catanese, J., Kress, L. F. 1984. Enzymatic inactivation of human plasma C1-inhibitor and α_1-antichymotrypsin by

Pseudomonas aeruginosa proteinase and elastase. *Biochim. Biophys. Acta* 789:37–43

23. Clas, F., Loos, M. 1981. Antibody-independent binding of the first component of complement (C1) and its subcomponent C1q to the S and R forms of *Salmonella minnesota*. *Infect. Immun.* 31:1138–44

24. Clas, F., Loos, M. 1982. Requirement for an additional serum factor essential for antibody-independent activation of the classical complement sequence by gram-negative bacteria. *Infect. Immun.* 37:935–39

25. Clas, F., Schmidt, G., Loos, M. 1985. The role of the classical pathway for the bactericidal effect of normal sera against gram-negative bacteria. *Curr. Top. Microbiol. Immunol.* 121:19–72

26. Cooper, A., Wozencraft, A. O., Roach, T. I. A., Blackwell, J. M. 1987. Different epitopes of the macrophage type three complement receptor (CR3) are used to bind *Leishmania* promastigotes harvested at different phases of their growth cycle. *Proc. NATO-ASI Symp. Leishmaniasis, Zakynthos, Greece*

27. Cooper, N. R. 1985. The classical complement pathway: activation and regulation of the first complement component. *Adv. Immunol.* 37:151–216

28. Dankert, J. R., Esser, A. F. 1986. Complement-mediated killing of *Escherichia coli*: dissipation of membrane potential by a C9-derived peptide. *Biochemistry* 25:1094–100

29. Dankert, J. R., Esser, A. F. 1987. Bacterial killing by complement: C9-mediated killing in the absence of C5b-8. *Biochem. J.* 244:393–99

30. da Silva, W. D., Kazatchkine, M. D. 1980. *Schistosoma mansoni*: activation of the alternative pathway of human complement by schistosomula. *Exp. Parasitol.* 50:278–86

31. Densen, P., McRill, C. M. 1987. C5b-8 promotes alternative pathway C3 convertase decay on *Neisseria*. *Fed. Proc.* 46:1196

32. Devine, D. V., Falk, R. J., Balber, A. E. 1986. Restriction of the alternative pathway of human complement by intact *Trypanosoma brucei* subsp. *gambiense*. *Infect. Immun.* 52:223–29

33. Edwards, M. S., Kasper, D. L., Jennings, H. J., Baker, C. J., Nicholson-Weller, A. 1982. Capsular sialic acid prevents activation of the alternative complement pathway by type III, group B streptococci. *J. Immunol.* 128:1278–83

34. Edwards, M. S., Nicholson-Weller, A., Baker, C. J., Kasper, D. L. 1980. The role of specific antibody in alternative complement pathway–mediated opsonophagocytosis of type III, group B *Streptococcus*. *J. Exp. Med.* 151:1275–87

35. Etges, R., Bouvier, J., Bordier, C. 1986. The major surface protein of *Leishmania* promastigotes is a protease. *J. Biol. Chem.* 20:9098–101

36. Ezekowitz, R. A. B., Sim, R. B., Hill, M., Gordon, S. 1984. Local opsonization by secreted macrophage complement components. Role of receptors for complement in uptake of zymosan. *J. Exp. Med.* 159:244–60

37. Fearon, D. T. 1978. Regulation by membrane sialic acid of β1H-dependent decay-dissolution of amplification C3 convertase of the alternative complement pathway. *Proc. Natl. Acad. Sci. USA* 75:1971–75

38. Fearon, D. T., Wong, W. W. 1983. Complement ligand-receptor interactions that mediate biological responses. *Ann. Rev. Immunol.* 1:243–71

39. Ferrante, A., Allison, A. C. 1983. Alternative pathway activation of complement by African trypanosomes lacking a glycoprotein coat. *Parasite Immunol.* 5:491–98

40. Fierer, J., Finley, F. 1979. Lethal effect of complement and lysozyme on polymyxin-treated, serum-resistant, gram-negative bacilli. *J. Infect. Dis.* 140:581–88

41. Fine, D. P. 1981. *Complement and Infectious Diseases*. Boca Raton, Fla: CRC. 157 pp.

42. Fitzgerald, T. J. 1987. Activation of the classical and alternative pathways of complement by *Treponema pallidum* subsp. *pallidum* and *treponema vincentii*. *Infect. Immun.* 55:2066–73

43. Frade, R., Myones, B. L., Barel, M., Krikorina, L., Charriaut, C., Ross, G. D. 1985. gp140, a C3b-binding membrane component of lymphocytes, is the B cell C3dg/C3d receptor (CR2) and is distinct from the neutrophil C3dg receptor (CR4). *Eur. J. Immunol.* 15:1192–97

44. Franke, E. D., McGreevy, P. B., Katz, P., Sacks, D. L. 1985. Growth cycle–dependent generation of complement-resistant *Leishmania* promastigotes. *J. Immunol.* 134:2713–17

45. Fries, L. F., Frank, M. M. 1987. Molecular mechanisms of complement action. In *The Molecular Basis of Blood Diseases*, ed. G. Stamatoyannopoulos, A. Nienhuis, P. Leder, P. Majerus, pp. 450–98. Philadelphia: Saunders

224   JOINER

46. Fuhrman, S. A., Joiner, K. A. 1987. Antibody-independent binding of the third complement component (C3) by *Toxoplasma gondii*. *Complement* 4:156

47. Giebink, G. S., Grebner, J. V., Kim, Y., Quie, P. G. 1978. Serum opsonic deficiency produced by *Streptococcus pneumoniae* and by capsular polysaccharide antigens. *Yale J. Biol. Med.* 51:527–38

48. Gordon, D. L., Rice, J. 1987. Deposition and degradation of opsonic components of C3 on bacteria. *Complement* 4:160

49. Griffiss, J. M., Broud, D. D., Bertram, M. A. 1975. Bactericidal activity of meningococcal antisera: blocking by IgA of lytic antibody in human convalescent sera. *J. Immunol.* 114:1779–84

50. Grossman, N., Joiner, K. A., Frank, M. M., Leive, L. 1986. C3b binding but not its breakdown is affected by the structure of the O-antigen polysaccharide in lipopolysaccharide from salmonellae. *J. Immunol.* 136:2208–12

51. Grossman, N., Leive, L. 1984. Complement activation via the alternative pathway by purified *Salmonella* lipopolysaccharide is affected by its structure but not its O-antigen length. *J. Immunol.* 132:376–85

52. Grossman, N., Schmetz, M. A., Foulds, J., Klima, E. N., Jiminez, V., et al. 1987. Lipopolysaccharide size and distribution determine serum resistance in *Salmonella montevideo*. *J. Bacteriol.* 169(2):856–63

53. Harriman, G. R., Podack, E. R., Braude, A. L., Corbeil, L. C., Esser, A. F., Curd, J. G. 1982. Activation of complement by serum-resistant *Neisseria gonorrhoeae*. Assembly of the membrane attack complex without subsequent cell death. *J. Exp. Med.* 156:1235–44

54. Horstmann, R. D., Fishetti, V. A. 1987. M protein, the major virulence factor of group A streptococci, inhibits alternative pathway activation by selectively binding factor H. *Complement* 4:170

55. Horwitz, M. A., Silverstein, S. C. 1980. Influence of the *Escherichia coli* capsule on complement fixation and on phagocytosis and killing by human phagocytes. *J. Clin. Invest.* 65:82–94

56. Hostetter, M. K. 1986. Serotypic variations among virulent pneumococci in deposition and degration of covalently bound C3b: Implications for phagocytosis and antibody production. *J. Infect. Dis.* 153:682–93

57. Hostetter, M. K., Johnson, G. M., Etsi-nas, E. M. 1986. Amidation of C3 by mucoid *Pseudomonas aeruginosa*: a mechanism for opsonic failure and phagocytic activation in the cystic fibrosis lung. *Clin. Res.* 34:520A

58. Huldt, G., Davies, P., Allison, A. G., Schorlemmer, H. U. 1979. Interactions between *Entamoeba histolytica* and complement. *Nature* 277:214–16

59. Jack, R. M., Ward, P. A. 1980. The role in vivo of C3 and the C3b receptor in babesial infection in the rat. *J. Immunol.* 124:1574–78

60. Jacks-Weis, J., Kim, Y., Cleary, P. P. 1982. Restricted deposition of C3 on M^+ group A streptococci: correlation with resistance to phagocytosis. *J. Immunol.* 128:1897–902

61. Jacks-Weis, J., Law, S. K., Levine, R. P., Cleary, P. P. 1985. Mechanism of resistance to phagocytosis by group A streptococci: failure of deposited complement opsonins to interact with cellular receptors. *J. Immunol.* 134:500–5

62. Jarvis, G. A., Vedros, N. A. 1987. Sialic acid of group B *Neisseria meningitidis* regulates alternative complement pathway activation. *Infect. Immun.* 55:174–80

63. Jimenez-Lucho, V. E., Joiner, K. A., Foulds, J., Frank, M. M., Leive, L. 1987. C3b generation is affected by the structure of the O-antigen polysaccharide in lipopolysaccharide from salmonellae. *J. Immunol.* 139:1253–59

64. Joiner, K., Brown, E., Hammer, C., Warren, K., Frank, M. 1983. Studies on the mechanism of bacterial resistance to complement-mediated killing. III. C5b-9 deposits stably on rough and type 7 *S. pneumoniae* without causing bacterial killing. *J. Immunol.* 130:845–49

65. Joiner, K., Heiny, S., Kirchhoff, L. V., Sher, A. 1985. gp72, the 72 kilodalton glycoprotein, is the membrane acceptor site for C3 on *Trypanosoma cruzi*. *J. Exp. Med.* 161:1196–212

66. Joiner, K., Sher, A., Gaither, T., Hammer, C. 1986. Evasion of alternative complement pathway by *Trypanosoma cruzi* results from inefficient binding of factor B. *Proc. Natl. Acad. Sci. USA* 83:6593–97

67. Joiner, K. A. 1985. Studies on the mechanism of bacterial resistance to complement-mediated killing and on the mechanism of action of bactericidal antibody. *Curr. Top. Microbiol. Immunol.* 121:99–133

68. Joiner, K. A., Brown, E. J., Frank, M. M. 1984. Complement and bacteria: chemistry and biology in host defense. *Ann. Rev. Immunol.* 2:461–91

69. Joiner, K. A., Fries, L. F., Schmetz, M. A., Frank, M. M. 1985. IgG bearing covalently bound C3b has enhanced bactericidal activity for *Escherichia coli* 0111. *J. Exp. Med.* 162:877–89

70. Joiner, K. A., Goldman, R. C., Hammer, C. H., Leive, L., Frank, M. M. 1983. Studies on the mechanism of bacterial resistance to complement-mediated killing. V. IgG and F(ab')₂ mediate killing of *E. coli* 0111B4 by the alternative complement pathway without increasing C5b-9 deposition. *J. Immunol.* 131:2563–69

71. Joiner, K. A., Goldman, R. C., Hammer, C. H., Leive, L., Frank, M. M. 1983. Studies on the mechanism of bacterial resistance to complement-mediated killing. VI. IgG increases the bactericidal effiency of C5b-9 for *E. coli* 0111B4 by acting a step before C5 cleavage. *J. Immunol.* 131:2570–75

72. Joiner, K. A., Grossman, N., Schmetz, M., Leive, L. 1986. C3 binds preferentially to long-chain lipopolysaccharide during alternative pathway activation by *Salmonella montevideo*. *J. Immunol.* 136(2):710–15

73. Joiner, K. A., Hammer, C. H., Brown, E. J., Cole, R. J., Frank, M. M. 1982. Studies on the mechanism of bacterial resistance to complement-mediated killing. I. Terminal complement components are deposited and released from *Salmonella minnesota* S218 without causing bacterial death. *J. Exp. Med.* 155:797–804

74. Joiner, K. A., Hammer, C. H., Brown, E. J., Frank, M. M. 1982. Studies on the mechanism of bacterial resistance to complement-mediated killing. II. C8 and C9 release C5b67 from the surface of *Salmonella minnesota* S218 because the terminal complex does not insert into the bacterial outer membrane. *J. Exp. Med.* 155:809–15

75. Joiner, K. A., Rimoldi, M. T., Kipinis, T., Dias da Silva, W., Hammer, C. H., Sher, A. 1987. Trypomastigotes of *Trypanosoma cruzi* produce molecules which accelerate the decay of complement C3 convertases. *Complement* 5:175

76. Joiner, K. A., Scales, R., Warren, K. A., Frank, M. M., Rice, P. 1985. Mechanism of action of blocking IgG for *Neisseria gonorrhoeae*. *J. Clin. Invest.* 76:1765–72

77. Joiner, K. A., Schmetz, M. A., Goldman, R. C., Leive, L., Frank, M. M. 1984. Mechanism of bacterial resistance to complement-mediated killing: inserted C5b-9 correlates with killing for *Escherichia coli* 0111B4 varying in O-antigen capsule and O-polysaccharide coverage of lipid A core oligosaccharide. *Infect. Immun.* 45:113–17

78. Joiner, K. A., Schmetz, M. A., Sanders, M. E., Murray, T. G., Hammer, C. H., et al. 1985. Multimeric complement component C9 is necessary for killing of *Escherichia coli* J5 by terminal attack complex C5b-9. *Proc. Natl. Acad. Sci. USA* 82:4808–12

79. Joiner, K. A., Warren, K. A., Brown, F. J., Swanson, J., Frank, M. M. 1983. Studies on the mechanism of bacterial resistance to complement-mediated killing. IV. C5b-9 forms high molecular weight complexes with bacterial outer membrane constituents on serum resistant but not on serum sensitive *Neisseria gonorrhoeae*. *J. Immunol.* 131:1443–51

80. Joiner, K. A., Warren, K. A., Hammer, C., Frank, M. M. 1985. Bactericidal but not nonbactericidal C5b-9 is associated with distinctive outer membrane proteins in *Neisseria gonorrhoeae*. *J. Immunol.* 134:1920–25

81. Joiner, K. A., Warren, K. A., Tam, M., Frank, M. M. 1985. Monoclonal antibodies directed against gonococcal protein I vary in bactericidal activity. *J. Immunol.* 134:3411–19

82. Kipnis, T. L., Tambourgi, D. V., Sucupira, M., Dias da Silva, W. 1986. Effect of *Trypanosoma cruzi* membrane components on the formation of the classical pathway C3 convertase. *Braz. J. Med. Biol. Res.* 19:271–78

83. Krettli, A. U., Pontes de Carvalho, L. C. 1985. Binding of C3 fragments to the *Trypanosoma cruzi* surface in the absence of specific antibodies and without activation of the complement cascade. *Clin. Exp. Immunol.* 62:270–77

84. Kroll, H. P., Bhakdi, S., Taylor, P. W. 1983. Membrane changes induced by exposure of *Escherichia coli* to human serum. *Infect. Immun.* 42:1055–66

85. Kubens, B. S., Wettstein, M., Opferkuch, W. 1985. A new mechanism of serum resistance in *Escherichia coli*. *Complement* 2:47

86. Lambris, J., Papmichail, M., Ioannidis, C., Dimitracoppoulos, G. 1982. Activation of the alternative pathway of human complement by the extracellular slime glycolipoprotein of *Pseudomonas aeruginosa*. *J. Infect. Dis.* 145:78–82

87. Leist-Welsh, P., Bjornson, A. B. 1982. Immunoglobulin-independent utilization of the classical complement pathway in opsonophagocytosis of *Escherichia coli* by human peripheral leukocytes. *J. Immunol.* 128:2643–41

88. Levine, R. P., Finn, R., Gross, R. 1983. Interactions between C3b and cell surface molecules. *Ann. NY Acad. Sci.* 421:235–45

89. Levy, M. G., Kakoma, I. 1985. Complement does not facilitate *in vitro* invasion of bovine erythrocytes by *Babesia bovis.* 80:377–78

90. Levy, N. J., Nicholson-Weller, A., Baker, C. J., Kasper, D. L. 1984. Potentiation of virulence by group B streptococcal polysaccharides. *J. Infect. Dis.* 149:851–56

91. Liang-Takasaki, C.-J., Grossman, N., Leive, L. 1983. *Salmonella* activate complement differentially via the alternative pathway depending on the structure of their lipopolysaccharide O-antigen. *J. Immunol.* 130:1867–71

92. Mäkelä, H., Saxèn, H., Valtonen, M., Valtonen, V. 1988. Ability to activate the alternative complement pathway as a virulence determinant in *Salmonella.* See Ref. 19, In press

93. Marikovsky, M., Levi-Schaffer, F., Arnon, R., Fishelson, Z. 1986. *Schistosoma mansoni:* killing of transformed schistosomula by the alternative pathway of human complement. *Exp. Parasitol.* 61:86–94

94. Markham, R. J. F., Nielsen, K. H., Wilkie, B. N. 1979. In vitro activation of complement by *Bacillus subtilis* protease: correlation with the response of guinea pigs to aerosols of the enzyme. *Immunol. Lett.* 1:79–83

95. McGuinness, T. B., Kemp, W. M. 1981. *Schistosoma mansoni:* a complement-dependent receptor on adult male parasites. *Exp. Parsitol.* 51:236–42

96. Medof, M. E., Kinoshita, T., Nussenzweig, V. 1984. Inhibition of complement activation on the surface of cells after incorporation of decay-accelerating factor (DAF) into their membranes. *J. Exp. Med.* 160:1558–78

97. Mogyoros, M., Calef, E., Gitler, C. 1986. Virulence of *Entamoeba histolytica* correlates with the capacity to develop complement resistance. *Isr. J. Med. Sci.* 22:915–17

98. Moll, A., Manning, P. A., Timmis, K. N. 1980. Plasmid-determined resistance to serum activity: a major outer membrane protein, the *traT* gene product, is responsible for plasmid-specified serum resistance in *Escherichia coli. Infect. Immun.* 28:359–67

99. Morrison, D. C., Kline, F. L. 1977. Activation of the classical and properdin pathway of complement by bacterial lipopolysaccharides (LPS). *J. Immunol.* 118:363–68

100. Mosser, D. M., Burke, S. K., Coutavas, E. E., Wedgwood, J. F., Edelson, P. J. 1986. *Leishmania* species: mechanisms of complement activation by five strains of promastigotes. *Exp. Parasitol.* 62: 394–404

101. Mosser, D. M., Edelson, P. J. 1984. Activation of the alternative complement pathway by *Leishmania* promastigotes: parasite lysis and attachment to macrophages. *J. Immunol.* 132:1501–5

102. Mosser, D. M., Edelson, P. J. 1985. The mouse macrophage receptor for C3bi (CR3) is a major mechanism in the phagocytosis of *Leishmania* promastigotes. *J. Immunol.* 135:2785–89

103. Mosser, D. M., Edelson, P. J. 1987. The third component of complement (C3) is responsible for the intracellular survival of *Leishmania major. Nature* 327:329–31

104. Mosser, D. M., Wedgwood, J. F., Edelson, P. J. 1985. *Leishmania* amastigotes: resistance to complement-mediated lysis is not due to a failure to fix C3. *J. Immunol.* 134:4128–31

105. Müller-Eberhard, H. J., Schreiber, R. D. 1980. Molecular biology and chemistry of the alternative pathway of complement. *Adv. Immunol.* 29:1–53

106. Muniz, J., Borriella, A. 1945. Estudo sobre a acão litica de diferentes soros sobre as formas de cultura sanguicolas do *T. cruzi. Rev. Bras. Biol.* 5:563–69

107. Musoke, A. J., Barbet, A. F. 1977. Activation of complement by variant specific antigen of *Trypanosoma brucei. Nature* 270:438–40

108. Myones, B. L., Dalzell, J. G., Hogg, N., Ross, G. D. 1987. CR4 is p150, 95 (CD11c), the third member of the LFA-1/CR3 (CD11a/CD11b) glycoprotein family (CD18). *Complement* 4:199

109. Newman, S. L., Mikus, L. K. 1985. Deposition of C3b and iC3b onto particulate activators of the human complement system. Quantitation with monoclonal antibodies to Human C3. *J. Exp. Med.* 161:1414–31

110. Nicholson-Weller, A., Burge, J., Fearon, D. T., Weller, P. F., Austen, K. F. 1982. Isolation of a human erythrocyte membrane glycoprotein with decay-accelerating activity for C3 convertases of the complement system. *J. Immunol.* 129:184–89

111. Nilsson, T., Carlsson, J., Sundqvist, G. 1985. Inactivation of key factors of the plasma proteinase cascade systems by

Bacteroides gingivalis. Infect. Immun. 50:467–71

112. Nogueira, N., Bianco, C., Cohn, Z. 1975. Studies on the selective lysis and purification of *Trypanosoma cruzi. J. Exp. Med.* 142:224–29

113. O'Connor, S. P., Cleary, P. P. 1986. Localization of the streptococcal C5a peptidase to the surface of group A streptococci. *Infect. Immun.* 53:432–34

114. Ogata, R. T., Levine, R. P. 1980. Characterization of complement resistance in *Escherichia coli* conferred by the antibiotic resistance plasmid R100. *J. Immunol.* 124:1494–98

115. Ouaissi, M. A., Santoro, F., Capron, A. 1980. Interaction between *Schistosoma mansoni* and the complement system: receptors for C3b on cercariae and schistosomula. *Immunol. Lett.* 1:197–210

116. Payne, N. R., Bellinger-Kawahara, C. G., Horwitz, M. A. 1987. Phagocytosis of *Mycobacterium tuberculosis* by human monocytes is mediated by receptors for the third component of complement. *Clin. Res.* 35:617A

117. Payne, N. R., Horwitz, M. A. 1987. Phagocytosis of *Legionella pneumophila* is mediated by human monocyte complement receptors. *J. Exp. Med.* 166: 1377–89

118. Peterson, P. K., Schmeling, D., Cleary, P. P., Wilkinson, B. J., Kim, Y., Quie, P. G. 1979. Inhibition of alternative complement pathway opsonization by group A streptococcal M protein. *J. Infect. Dis.* 139:575–85

119. Pluschke, G., Mayden, J., Achtman, M., Levine, R. P. 1983. Role of the capsule and the O antigen in resistance of O14:K1 *Escherichia coli* to complement-mediated killing. *Infect. Immun.* 42:907–13

120. Porat, R., Johns, M. A., McCabe, W. R. 1987. Selective pressures and lipopolysaccharide subunits as determinants of resistance of clinical isolates of gram-negative bacilli to human serum. *Infect. Immun.* 55:320–28

121. Prellner, K. 1981. C1q binding and complement activation by capsular and cell wall components of *S. pneumoniae* type XIX. *Acta Pathol. Microbiol. Scand.* 89:359–64

122. Puentes, S. M., Bates, P. A., Dwyer, D. M., Joiner, K. A. 1986. C3-binding to *Leishmania donovani* promastigotes (LD). *Fed. Proc.* 36:525

123. Puentes, S. M., Sacks, D., da Silva, R., Joiner, K. A. 1988. Binding of complement by two developmental stages of *Leishmania major* varying in expression

of a cell surface glycolipid. *J. Exp. Med.* 167:887–902

124. Puentes, S. M., Sacks, D. L., Joiner, K. A. 1987. Complement binding to two developmentally distinct stages of *Leishmania major* promastigotes. *Complement* 4:215

125. Quinn, P. H., Crosson, F. J. Jr., Moxon, E. R. 1977. Activation of the alternative complement pathway by *Haemophilus influenzae* type B. *Infect. Immun.* 16:400–2

126. Ramalho-Pinto, F. J., Oliveira-Silva, S., Horta, M. F. M. 1987. Decay accelerating factor (DAF) as the host antigen with protective activity to complement killing of schistosomula. *Mem. Inst. Oswaldo Cruz* In press

127. Ramm, L. E., Whitlow, M. B., Koski, C. L., Shin, M. L., Mayer, M. M. 1983. Elimination of complement channels from the plasma membranes of U937, a nucleated mammalian cell line: temperature dependence of the elimination rate. *J. Immunol.* 131:892–98

128. Rasmussen, K. R., Kemp, W. M. 1987. *Schistosoma mansoni:* interactions of adult parasites with the complement system. *Parasite Immunol.* 9:235–48

129. Reed, S. L., Curd, J. G., Gigli, I., Gillin, F. D., Braude, A. I. 1986. Activation of complement by pathogenic and nonpathogenic *Entamoeba histolytica. J. Immunol.* 136:2265–70

130. Reed, S. L., Gigli, I. 1987. Binding of C9 fails to lyse invasive *Entamoeba histolytica. Clin. Res.* 35:488A

131. Reed, S. L., Sargeaunt, P. G., Braude, A. I. 1983. Resistance to lysis by human serum of pathogenic *Entamoeba histolytica. Trans. R. Soc. Trop. Med. Hyg.* 77:248–53

132. Reid, K. B. M., Bentley, D. R., Campbell, R. D., Chung, L. P., Sim, R. B., et al. 1986. Complement system proteins which interact with C3b or C4b: a superfamily of structurally related proteins. *Immunol. Today* 7:230–34

133. Reynolds, B. L., Rother, J. A., Rother, K. O. 1975. Interaction of complement components with a serum-resistant strain of *Salmonella typhimurium. Infect. Immun.* 1:944

134. Rice, P. A., Vayo, H. E., Tam, M. R., Blake, M. S. 1986. Immunoglobulin G antibodies directed against protein III block killing of serum-resistant *Neisseria gonorrhoeae* by immune serum. *J. Exp. Med.* 164:1735–48

135. Rimoldi, M. T., Sher, A., Heiny, S., Hammer, C. H., Joiner, K. 1988. Developmentally regulated expression by

Trypanosoma cruzi of molecules which accelerate the decay of complement C3 convertases. *Proc. Natl. Acad. Sci. USA* 85:193–97

136. Rimoldi, M. T., Tenner, A., Bobak, D., Joiner, K. A. 1987. C1q enhances *Trypanosoma cruzi* internalization by human mononuclear cells. *Complement* 4:217

137. Ross, G. D., Cain, J. A., Myones, B. L., Newman, S. L., Lachmann, P. J. 1987. Specificity of membrane complement receptor type three (CR3) for beta-glucans. *Complement* 4:61–74

138. Ross, G. D., Medof, M. E. 1985. Membrane complement receptors specific for bound fragments of C3. *Adv. Immunol.* 37:217–67

139. Ruppel, A., McLaren, D. J., Diesfeld, H. J., Rother, U. 1984. *Schistosoma mansoni:* escape from complement-mediated parasiticidal mechanisms following percutaneous primary infection. *Eur. J. Immunol.* 14:702–8

140. Ruppel, A., Rother, U., Diesfeld, H. J. 1984. *Schistosoma mansoni:* loss of the ability of schistosomula to bind mouse complement following intravenous injection into mice. *Tropenmed. Parasitol.* 35:23–28

141. Russell, D. G. 1987. The macrophage-attachment glycoprotein gp63 is the predominant C3-acceptor site on *Leishmania mexicana* promastigotes. *Eur. J. Biochem.* 164:213–21

142. Sacks, D. L., da Silva, R. 1987. Developmentally regulated expression of a surface glycolipid on *L. major* promastigotes. *J. Immunol.* 139:3099–106

143. Samuelson, J. C., Caulfield, J. P. 1986. Cercarial glycocalyx of *Schistosoma mansoni* activates human complement. *Infect. Immun.* 51:181–86

144. Samuelson, J. C., Sher, A., Caulfield, J. P. 1980. Newly transformed schistosomula spontaneously lose surface antigens and C3 acceptor sites during culture. *J. Immunol.* 24:2055–57

145. Santoro, F. 1982. Interaction of complement with parasite surfaces. *Clin. Immunol. Allergy* 2(3):639–53

146. Santoro, F., Lachmann, P. J., Capron, A., Capron, M. 1979. Activation of complement by *Schistosoma mansoni* schistosomula: killing of parasites by the alternative pathway and requirement of IgG for classical pathway activation. *J. Immunol.* 123:1551–57

147. Santoro, F., Ouaissi, M. A., Pestel, J., Capron, A. 1980. Interaction between *Schistosoma mansoni* and the complement system: binding of C1q to schistosomula. *J. Immunol.* 124:2886–91

148. Schenkman, S., Güther, M. L., Yoshida, N. 1986. Mechanism of resistance to lysis by the alternative complement pathway in *Trypanosoma cruzi* trypomastigotes: effect of a specific monoclonal antibody. *J. Immunol.* 137:1623–28

149. Schiller, N. L., Joiner, K. A. 1986. Interaction of complement with serum-sensitive and serum-resistant strains of *Pseudomonas aeruginosa. Infect. Immun.* 54:689–94

150. Schultz, D. R., Miller, K. D. 1974. Elastase of *Pseudomonas aeruginosa:* inactivation of complement components and complement-derived chemotactic and phagocytic factors. *Infect. Immun.* 10:128–35

151. Schweinle, J. E., Hitchcock, P. J., Frank, M. M., Joiner, K. A. 1987. Bactericidal and opsonic activity of monoclonal antibody (McAb 10) for *Neisseria gonorrhoeae* (GC). *Clin. Res.* 35:490A

152. Seinen, W., Stegmann, T., Kuil, H. 1982. Complement does not play a role in promoting *Babesia rodhaini* infections in BALB/c mice. *Z. Parasitenkd.* 68:249–57

153. Shafer, W. M., Joiner, K., Guymon, L. F., Cohen, M. S., Sparling, P. F. 1984. Serum sensitivity of *Neisseria gonorrhoeae:* the role of lipopolysaccharide. *J. Infect. Dis.* 149:1757–83

154. Sher, A., Hieny, S., Joiner, K. 1986. Evasion of the alternative complement pathway by metacyclic trypomastigotes of *Trypanosoma cruzi:* dependence on the developmentally regulated synthesis of surface protein and *N*-linked carbohydrate. *J. Immunol.* 137:2961–67

155. Shin, M. L., Hänsch, G., Hu, V. W., Nicholson-Weller, A. 1986. Membrane factors responsible for homologous species restriction of complement-mediated lysis: evidence for a factor other than DAF operating at the stage of C8 and C9. *J. Immunol.* 136:1777–82

156. Sorvillo, J. M., Pearlstein, E. 1985. C1q, a subunit of the first component of complement, enhances binding of plasma fibronectin to bacteria. *Infect. Immun.* 49:664–69

157. Stanley, H. A., Mayes, J. T., Cooper, N. E., Reese, R. T. 1987. Complement activation by the surface of *Plasmodium falciparum* infected erythrocytes. *Mol. Immunol.* 21:145–50

158. Steele, N. P., Munson, R. S. Jr., Granoff, D. M., Cummins, J. E., Levine, R. P. 1984. Antibody-dependent alternative pathway killing of *Haemophilus influenzae* type B. *Infect. Immun.* 44:452–58

159. Stemmer, F., Loos, M. 1985. Evidence for direct binding of the first component of complement, C1, to outer membrane proteins from *Salmonella minnesota*. *Curr. Top. Microbiol. Immunol.* 121: 73–84

160. Stevens, P., Huang, S. N.-Y., Welch, W. D., Young, L. S. 1978. Restricted complement activation by *Escherichia coli* with the K-1 capsular serotype: A possible role in pathogenicity. *J. Immunol.* 121:1216–17

161. Sturtevant, J. E., Balber, A. E. 1987. *Trypanosoma lewisi:* restriction of alternative complement pathway C3/C5 convertase activity. *Exp. Parasitol.* 63: 260–71

162. Sundqvist, G., Carlsson, J., Herrmann, B., Tärnvik, A. 1985. Degradation of human immunoglobulins G and M and complement factors C3 and C5 by black-pigmented *Bacteroides*. *J. Med. Microbiol.* 19:85–94

163. Tarr, P. I., Hosea, S. W., Brown, E. J., Schneerson, R., Sutton, A., Frank, M. M. 1982. The requirement of specific anticapsular IgG for killing of *Haemophilus influenzae* by the alternative pathway of complement activation. *J. Immunol.* 128:1772–75

164. Taylor, P. W. 1983. Bactericidal and bacteriolytic activity of serum against gram-negative bacteria. *Microbiol. Rev.* 47:46–83

165. Taylor, P. W., Kroll, H. P. 1983. Killing of an encapsulated strain of *Escherichia coli* by human serum. *Infect. Immun.* 39:122–31

166. Taylor, P. W., Kroll, H. P. 1984. Interaction of human complement proteins with serum-sensitive and serum-resistant strains of *Escherichia coli*. *Mol. Immunol.* 21:609–20

167. Taylor, P. W., Kroll, H.-P. 1985. Effect of lethal doses of complement on the functional integrity of target enterobacteria. *Curr. Top. Microbiol. Immunol.* 121:135–58

168. Tenner, A. J., Ziccardi, R. J., Cooper, N. R. 1984. Antibody-independent C1 activation by *E. coli*. *J. Immunol.* 133: 886–91

169. Tertti, R., Eerola, E., Lehtonen, O.-P., Ståhlberg, T. H., Viander, M., Toivanen, A. 1987. Virulence-plasmid is associated with the inhibition of opsonization in *Yersinia enterocolitica* and *Yersinia pseudotubercolis*. *Clin. Exp. Immunol.* 68:266–74

170. Van der Ley, P., de Graaff, P., Tommassen, J. 1986. Shielding of *Escherichia coli* outer membrane proteins as receptors for bacteriophages and colicins by O-antigenic chains of lipopolysaccharide. *J. Bacteriol.* 168:449–51

171. Van der Ley, P., Kuipers, O., Tommassen, J., Lugtenberg, B. 1986. O-antigenic chains of lipopolysaccharide prevent binding of antibody molecules to an outer membrane pore protein in Enterobacteriaceae. *Microb. Pathog.* 1:43–49

172. Van Dijk, W. C., Verbrugh, H. A., van der Tol, M. E., Peters, R., Verhoef, J. 1979. Role of *Escherichia coli* K capsular antigens during complement activation, C3 fixation, and opsonization. *Infect. Immun.* 25:603–9

173. Vik, D. P., Fearon, D. T. 1985. Neutrophils express a receptor for iC3b, C3dg, and C3d that is distinct from CR1, CR2, and CR3. *J. Immunol.* 134:2571–79

174. Vik, D. P., Fearon, D. T. 1987. Cellular distribution of complement receptor type 4 (CR4): Expression on human platelets. *J. Immunol.* 138:312–15

175. Vukajlovich, S. T., Hoffman, J., Morrison, D. C. 1987. Activation of human serum complement by bacterial lipopolysaccharides: structural requirements for antibody independent activation of the classical and alternative pathways. *Mol. Immunol.* 24:319–31

176. Ward, P. A., Chapitis, J., Conroy, M. C., Lepow, I. H. 1973. Generation by bacterial proteinases of leukotactic factors from human serum, and human C3 and C5. *J. Immunol.* 110:1003–9

177. Ward, P. A., Sterzel, R. B., Lucia, H. L., Campbell, G. H., Jack, R. M. 1981. Complement does not facilitate plasmodial infections. *J. Immunol.* 126:1826–28

178. Wexler, D. E., Chenoweth, D. E., Cleary, P. P. 1985. Mechanism of action of the group A streptococcal C5a inactivator. *Proc. Natl. Acad. Sci. USA* 82:8144–48

179. Wexler, D. E., Cleary, P. P. 1985. Purification and characteristics of the streptococcal chemotactic factor inactivator. *Infect. Immun.* 50:757–64

180. Wilkinson, B. J., Sisson, S. P., Kim, Y., Peterson, P. K. 1979. Localization of the third component of complement on the cell wall of encapsulated *Staphylococcus aureus* M: implications for the mechanism of resistance to phagocytosis. *Infect. Immun.* 26:1159–63

181. Winkelstein, J. A., Abramovitz, A. S., Tomasz, A. 1980. Activation of C3 via the alternative complement pathway results in fixation of C3b to the pneumococcal cell wall. *J. Immunol.* 124:2502–6

182. Winkelstein, J. A., Bocchini, J. A., Schiffman, G. 1976. The role of the capsular polysaccharide in the activation of the alternative pathway by the pneumococcus. *J. Immunol.* 116:367–71

183. Wozencraft, A. O., Sayers, G., Blackwell, J. M. 1986. Macrophage type 3 complement receptors mediate serumindependent binding of *Leishmania donovani*. *J. Exp. Med.* 164:1332–37

184. Wright, S. D., Reddy, P. A., Jong, M. T. C., Erickson, B. W. 1987. C3bi receptor (complement receptor type 3) recognizes a region of complement protein C3 containing the sequence Arg-Gly-Asp. *Proc. Natl. Acad. Sci. USA* 84:1965–68

185. Wright, S. D., Silverstein, S. C. 1983. Receptors for C3b and C3bi promote phagocytosis but not the release of toxic oxygen from human phagocytes. *J. Exp. Med.* 158:2016–23

186. Young, J. D.-E., Cohn, Z. A., Podack, E. R. 1986. The ninth component of complement and the pore-forming protein (perforin 1) from cytotoxic T cells: structural, immunological, and functional similarities. *Science* 233:184–90

187. Zalman, L. S., Wood, L. M., Müller-Eberhard, H. J. 1986. Isolation of a human erythrocyte membrane protein capable of inhibiting expression of homologous complement transmembrane channels. *Proc. Natl. Acad. Sci. USA* 83:6975–79

Ann. Rev. Microbiol. 1988. 42:231–61

THE ENZYMES ASSOCIATED WITH DENITRIFICATION[1]

Lawrence I. Hochstein

Ames Research Center, Moffett Field, California 94035

Geraldine A. Tomlinson

Department of Biology, Santa Clara University, Santa Clara, California 95053

CONTENTS

INTRODUCTION

The ability to grow anaerobically by reducing ionic nitrogenous oxides to gaseous products is distributed among a diverse number of eubacteria (72) and in the extremely halophilic branch of the Archaebacteria (57, 101, 150). This respiratory process, in which nitrogen oxides serve as electron acceptors,

[1]The US Government has the right to retain a nonexclusive, royalty-free license in and to any copyright covering this paper.

results in the concomitant generation of ATP (83, 116) and is designated denitrification (126). The enzymes associated with denitrification are synthesized when conditions become anaerobic, although denitrification can occur in the presence of oxygen (17, 56, 113, 130, 131). In some cases, enzyme induction may even require low concentrations of oxygen (3, 12a, 128, 170). The presence of nitrate (or other nitrogenous oxides) is not always necessary; significant levels of enzyme may be present as a consequence of anaerobiosis (44, 108, 116, 129).

A broad overview of denitrification has been published (127). In addition, the ecology (40, 42, 81, 148b), evolution (14), genetics (25, 62–64), and energetics (7, 148a) have been extensively covered. In this review we discuss the enzymes concerned with the reduction of the nitrogenous oxides thought to be the intermediates in denitrification. Because the taxonomy of the denitrifying bacteria is in a state of flux, we retain the nomenclature used by the authors to avoid confusion. Synonyms are listed in Table 1.

NITRATE REDUCTASE

The nitrate reductases associated with denitrification and respiration (i.e. dissimilation) are, with one exception, membrane-bound enzymes that catalyze the reduction of nitrate to nitrite and couple this reduction to the translocation of protons. In the case of the enzyme from *Staphylococcus aureus,* immunologically indistinguishable versions occur as cytoplasmic and membrane-bound enzymes (22). The dissimilatory nitrate reductases can be distinguished from those enzymes associated with assimilative and fermentative reduction by the end product of the reductive pathway (dinitrogen or ammonia) and by how ammonia as well as oxygen regulates enzyme activity. Synthesis of the dissimilatory enzymes is unaffected by ammonium and is

Table 1 Synonyms

Original description	Subsequent description(s)
Micrococcus denitrificans	*Paracoccus denitrificans*
Micrococcus halodenitrificans	*Paracoccus halodenitrificans*
Pseudomonas denitrificans	*Alcaligenes* strain NCIB 11015
	Achromobacter xylosoxidans
	Achromobacter denitrificans subsp. *xylosoxidans*
Pseudomonas perfectomarinus	*Pseudomonas perfectomarina*
	Pseudomonas stutzeri
Rhodopseudomonas sphaeroides f. sp. *denitrificans*	*Rhodobacter sphaeroides* f. sp. *denitrificans*
Rhodopseudomonas capsulata	*Rhodobacter capsulatus*

repressed by oxygen (14). The dissimilatory enzymes share many properties. They contain molybdenum, heme and nonheme iron, and acid-labile sulfur (148). The molybdenum is noncovalently linked to a pterin derivative, where it occurs as the molybdenum cofactor, which is extremely labile when separated from the enzyme (55, 74). The molybdenum cofactor has been identified in the nitrate reductases from *Escherichia coli* (9, 142), *Proteus mirabilis* (30), *Bacillus licheniformis* (156), *Rhodobacter capsulatus* (112), and *Pseudomonas denitrificans* (65).

Nitrate Reductases From Nonphotosynthetic Bacteria

The considerable variation in the reported molecular weight of nitrate reductases (148) reflects the different methods used to release the membrane-bound enzyme (39, 65), the effects of proteolysis during preparation (97), the presence of a modifying enzyme (27), and the tendency of the enzyme to form aggregates (26, 96, 122). The methods used to solubilize the enzyme also account for the disparity in the reported number of subunits. Two subunits (α and β) are detected when the enzyme is removed by incubating membranes at elevated temperatures (26, 65, 99). A third subunit (γ), containing a b-type cytochrome, is sometimes detected when detergents are used (31, 36, 38, 65, 98, 119), although not always (26, 37, 154, 156). The γ subunit is particularly sensitive to heating in the presence of mercaptoethanol and sodium dodecylsulfate (SDS), so even when present it may escape detection by SDS electrophoresis (28, 36).

The α subunit from several respiratory nitrate reductases varies in size from 104 to 150 kd (23, 31, 39, 99, 119, 122, 154, 156); the α subunit of the nitrate reductases from denitrifying bacteria varies from 118 to 150 kd (26, 36, 65). The significance of this variation in the molecular mass of the α subunit is difficult to assess. Limited tryptic proteolysis of the nitrate reductase from *S. aureus* stimulated nitrate reductase activity and reduced the molecular mass from 225 to 200 kd while altering the mass of the α subunit from 140 to 112 kd (23). A two-subunit nitrate reductase from *Pseudomonas denitrificans*, which consists of α and β subunits of 136 and 55 kd, respectively, was also stimulated when incubated with trypsin. Stimulation was associated with an alteration of the α subunit, so that following SDS electrophoresis two subunits of 87 and 47 kd were detected. In addition to this change, a third polypeptide was observed, which has molecular mass of 43 kd and originates from the β subunit (65). The mass of the β subunit of several respiratory nitrate reductases varies from 52 to 63 kd, while the range for the γ subunit is from 19 to 20 kd (38, 99, 119, 154, 156). The values for the β and γ subunits of the nitrate reductases from denitrifying bacteria are 55–64 and 19–21 kd, respectively (36, 65). Several lines of evidence indicate that catalysis takes place at the α subunit: The enzyme can be treated with trypsin so as to alter the β subunit without affecting the ability of the enzyme to reduce nitrate (37, 119); selective iodination of the α subunit inhibits nitrate reductase activity (51); and the isolated α subunit, but not the β subunit, exhibits nitrate reductase activity following reactivation with the molybdenum cofactor (29). The β subunit may be involved in membrane attachment (37, 97). The γ subunit links nitrate reductase to the electron transport chain at the level of ubiquinone (36, 119).

The apparent K_m for nitrate is affected by the reductant. When reduced viologen dyes are used, the values range from about 100 μM to 1.3 mM (148). Nitrate reductases from *E. coli* (119) or *Paracoccus denitrificans* (36) that contain the γ subunit use reduced quinones, such as ubiquinol-1 and duroquinol, in addition to reduced viologen dyes, as electron donors. When quinones are used, the apparent K_m values for nitrate (2–13 μM) approach those observed with intact cells (15, 124).

Dissimilatory nitrate reductases are inhibited by azide, and the inhibition is competitive with respect to nitrate (25, 36, 43, 50, 155, 156). Azide inhibition with methyl or benzyl viologen as the variable substrate is uncompetitive with the enzymes from *Klebsiella aerogenes* (155) and noncompetitive with the enzymes from *B. licheniformis* (156) and *Pseudomonas aeruginosa* (26). The inhibition by azide suggests that benzyl viologen binds first, followed by azide and nitrate (26), and that benzyl viologen and nitrate bind at different sites (26, 155). Nitrate reductase activity is also inhibited by thiocyanate (88, 90) and toluene-3,4-dithiol (65, 90), reagents that chelate molybdenum. Nitrate reductase activity is inhibited by cyanide. In the case of the enzyme from *K. aerogenes,* the inhibition is noncompetitive with respect to nitrate (155). In *E. coli* (1), the enzyme can exist in three oxidation states, of which the dithionite-reduced form is the most cyanide sensitive. Nitrate protects the dithionite-reduced enzyme against cyanide inhibition, even though the inhibition is presumably noncompetitive with respect to nitrate. The other two oxidation forms of the enzyme from *E. coli* (Mo^{5+} and the ferricyanide-oxidized form of the enzyme) are considerably less sensitive to cyanide, and nitrate does not protect against inhibition in either case.

The kinetic behavior of nitrate reductase is markedly affected by the reductants. Morpeth & Boxer (119) compared the properties of the enzyme from *E. coli* using reduced methyl viologen, ubiquinol-1, and duroquinol. The quinols are probably reasonable analogs of the physiological reductant, since ubiquinone is thought to introduce electrons to nitrate reductase via the cytochrome *b* associated with the enzyme (8, 62, 125). Nitrate reduction was inhibited by 2-*n*-heptyl-4-hydroxyquinoline-*N*-oxide (HOQNO) (123), which inhibits nitrate reduction at the quinone (62). Nitrate reduction was not inhibited by myxathiazol or antimycin *a*, which inhibit electron flow at the level of cytochrome b_1c (125). The enzyme from *E. coli* was inhibited by

HOQNO when the quinols, but not reduced methyl viologen, were the reductants (119). Nitrate reduction with reduced methyl viologen results in an ordered mechanism in which nitrate binds prior to the viologen dye. This order of binding differs from that observed in *P. aeruginosa* (26). When quinols are the reductants, the binding of the quinol and the release of the quinone require that nitrate be bound to the enzyme. When reduced methyl viologen is the reductant the release of nitrite is rate limiting, whereas the reduction of the enzyme is rate limiting when the reductant is a quinol. Furthermore, methyl viologen and quinol must bind at different sites, since modification of the enzyme with trypsin, diethyl pyrocarbonate, or HOQNO inhibited nitrate reduction when a quinol but not when methyl viologen was the electron donor. The apparent affinity for nitrate was 0.42 mM in the presence of reduced methyl viologen and about 2 μM with either quinol. On the other hand, V_{max} was 60–700 times greater in the presence of the reduced viologen dye.

The enzyme from *E. coli*, like other so-called type A nitrate reductases (148), reduces chlorate and bromate in addition to nitrate. The turnover numbers follow the order chlorate $>$ nitrate $>$ bromate in the presence of reduced methyl viologen. No significant difference in the turnover number was observed when either duroquinol or ubiquinol-1 was the reductant. Craske & Ferguson (36) examined the kinetic behavior of the cytochrome *b*–containing nitrate reductase from *Paracoccus denitrificans*. They also observed that V_{max} was greater (and the apparent K_m for nitrate was lower) when duroquinol replaced methyl viologen. HOQNO inhibited nitrate reduction only when duroquinol was the electron donor (i.e. when the subunit containing cytochrome *b* was present). These observations iterate the limitations inherent in the use of viologen dyes pointed out by Carlson et al (26).

The active center of the enzyme resides on the inner aspect of the cytoplasmic membrane in *Paracoccus denitrificans* (6, 73, 87) and *E. coli* (76, 84). The location of the enzyme on the cytoplasmic aspect of the membrane raises the question of how nitrate enters the cell. In *Paracoccus denitrificans* nitrate uptake is thought to take place by facilitated diffusion. A uniporter is unlikely because of the unfavorable negative intracellular membrane potential. Two other nitrate uptake systems have been proposed (19, 148b). One operates in symport with protons; the other operates as an NO_3^-/NO_2^- antiporter. The former system initiates nitrate uptake in the absence of nitrite when the antiporter system is inoperative. The latter system also serves to maintain a low intracellular concentration of nitrite and provides a mechanism for export to the location of the nitrite reductase, which appears to be a periplasmic enzyme in *Paracoccus denitrificans*. The argument for a system of active transport derives in part from the high apparent K_m for nitrate observed in cell-free systems when viologen dyes are used as reductants.

Parsonage et al (124) reexamined the question of nitrate transport in sphero-plast preparations from *Parococcus denitrificans*. Spheroplasts would be expected to swell when incubated in the presence of nitrate, nitrite, and nigericin if a NO_3^-/NO_2^- antiporter were operative, owing to the net influx of potassium nitrate. If a NO_3^-/H^+ transport system were operating, incubation of spheroplasts in the presence of ammonium nitrate (in the presence of sufficient azide to inhibit the activity of intracellular nitrate reductase activity) should also result in swelling. No such spheroplast swelling occurred in either case. Swelling took place when carbonyl cyanide *p*-trifluoromethoxyphenyl-hydrazone (FCCP) was added to enhance proton permeability and nitrate entered the cell. However, the rate of entry was the same under aerobic and anaerobic conditions. In these studies the rate of nitrate uptake was linear at concentrations as low as 5 μM nitrate, which implies that the apparent K_m is less than 5 μM. In view of the low apparent K_m, a nitrate-specific pore might suffice for nitrate entry. Craske & Ferguson (36) suggested that in *Pseudomonas denitrificans*, where the apparent K_m for nitrate is less than 5 μM when duroquinol is the reductant, nitrate may enter via a channel provided by nitrate reductase.

Nitrate Reductases From the Rhodospirillaceae

Certain purple nonsulfur photosynthetic bacteria, which ordinarily carry out the assimilatory reduction of nitrate, also reduce nitrate to nitrite. These include *Rhodopseudomonas sphaeroides* f. sp. *denitrificans*, which also re-duces nitrite to dinitrogen (88, 134); several strains of *Rhodopseudomonas palustris* that grow anaerobically in the dark in the presence of nitrate (80); and spontaneous mutants of *Rhodopseudomonas capsulata* that reduce nitrate but are unable to grow anaerobically in the dark using this reductive step (5, 109). These reductases turn out to be a varied lot.

Two nitrate reductases have been found in *R. sphaeroides* var. *de-nitrificans*. One occurs as a periplasmic enzyme, has a molecular mass of 112 kd, and contains molybdenum and cytochrome *c*. The minimum catalytically active form of the enzyme has a molecular mass of 60 kd and contains 1 mol of cytochrome *c* (13 kd), so the native enzyme consists of at least a catalytic subunit (47 kd) and 5 mol of cytochrome *c* (133). The role of cytochrome *c* is not clear. Membranes contain a *b*-type cytochrome, which is reoxidized by nitrate in a reaction inhibited by HOQNO. While it has been proposed that this cytochrome reduces cytochrome *c*, the latter cytochrome does not un-dergo redox changes during nitrate reduction (160). The other nitrate reduc-tase (24) from the same denitrifying strain of *R. sphaeroides* f. sp. *de-nitrificans* is membrane-bound. The purified enzyme, which is detached by incubating membranes at elevated temperatures in an alkaline medium, con-tains molybdenum. There is no evidence for the presence of cytochrome *c*.

The enzyme has a molecular mass of 180 kd and is composed of a catalytic subunit (120 kd) and a structural subunit (60 kd). It appears to be similar to the dissimilatory nitrate reductases from nonphotosynthetic bacteria.

The nitrate reductase from *R. capsulata* AD2 is soluble and contains molybdenum and cytochrome *c* (4). The enzyme has a molecular mass of 185 kd and is a dimer composed of a single subunit (85 kd). This nitrate reductase is relatively insenstive to cyanide, is unaffected by chlorate (although chlorate is a substrate), and is stimulated by azide and thiocyanate. The nitrate reductase from *R. capsulata* BK5 is membrane bound, has a molecular mass of 360 kd, and is inhibited by azide, cyanide, chlorate, and thiocyanate. Unlike the enzyme from AD2, no hemes are associated with this enzyme (157). McEwan et al (109) isolated a respiratory nitrate reductase from a strain of *R. capsulata* (N22DNAR$^+$) by repeatedly subculturing the parent strain in nitrate-containing medium. The enzyme was inhibited by oxygen and light, but not by ammonium ion. This contrasts to the assimilatory nitrate reductase found in the parent strain (71). Although electron flow to nitrate reduction is associated with a proton-translocating electron transport chain, this strain does not grow anaerobically in the dark. The periplasmic dissimilatory nitrate reductase from *R. capsulatus* N22DNAR$^+$ has been extensively purified (112). The enzyme does not reduce chlorate, contains 1 mol of pterin molybdenum cofactor per subunit, and is composed of a single subunit of 90 kd, which is in good agreement with the mass for the subunit from *R. capsulata* AD2 (4). No cytochrome *c* is present in the purified enzyme, although less-purified enzyme fractions contain a *c*-type cytochrome, which is associated with a low–molecular weight polypeptide. Mc-Ewan et al (111) have summarized the characteristics of the two functional nitrate reductases in the rhodopseudomonads. One, a cytoplasmic enzyme, is solely assimilatory. The other is a periplasmic enzyme that may have an assimilatory function, but serves as an alternate electron acceptor for generating a membrane potential and acts to dispose of excess reducing equivalents (32).

NITRITE REDUCTASES

The dissimilatory reduction of nitrite is carried out by two distinct reductases. One is a metalloprotein containing copper (the copper nitrite reductase); the other is a heme protein that contains *c*- and *d*-type cytochromes (the cd_1-cytochrome nitrite reductase). These enzymes may be readily distinguished by their distinct spectral properties and their differential sensitivity to diethyldithiocarbamate, which inhibits the reduction of nitrite to nitric oxide by the copper enzyme but not by the cd_1-cytochrome nitrite reductase (140).

Copper Nitrite Reductases

The copper enzymes show considerable differences in color, mass, number of subunits, and the type of copper present (93). Low-mass blue proteins are implicated as the physiological electron donors. The product of nitrite reduction depends on the reducing system. Nitrous oxide is produced in the presence of dithionite and viologen dyes (77, 117, 166), whereas nitric oxide is the product when the reducing system is phenazine methosulfate (PMS) ascorbate (67, 77, 136, 166). Kakutani et al (77) suggested that nitrous oxide is produced by the chemical reduction of nitric oxide by dithionite. The nitrite reductase from *Alcaligenes* NCIB 11015 reduced nitrite to nitric oxide when N,N,N',N'-tetramethyl-p-phenylene diamine (TMPD) ascorbate was the reductant and to ammonia when methyl viologen dithionite was the reducing system (102). In the latter case, nitrite was first reduced enzymatically to nitric oxide, which was chemically reduced to hydroxylamine. The hydroxylamine was reduced to ammonia by a second enzymatic reaction.

Two nitrite reductases were found in the cytoplasmic fraction of *Alcaligenes faecalis* S-6 (77). The principal one was green, had a molecular mass of 110 kd as determined by gel filtration, and was composed of four subunits (30 kd), each containing an atom of type 1 or 2 copper. The enzyme was inhibited by diethyldithiocarbamate and cyanide, but was unaffected by p-chloromercuribenzoate, N-ethylmaleimide, and iodoacetate. The enzyme was inactivated by high concentrations of mercaptoethanol, but the inactivation could be partially reversed by dialyzing the enzyme against cupric sulfate. The physiological electron donor for the copper nitrite reductase of *A. faecalis* S-6 is a copper protein of 12 kd, which is rapidly oxidized by the nitrite reductase (78). This blue protein also stimulated the inactivation of the enzyme when the nitrite reductase was incubated with either cysteine or ascorbate in the presence of oxygen, presumably owing to the production of hydrogen peroxide.

The nitrite reductase from *Achromobacter cycloclastes,* which is also green, has a molecular mass of 69 kd and is composed of two identical subunits of 37 kd (67, 69, 95). Types 1 and 2 copper were detected by electron paramagnetic resonance (EPR) spectroscopy (69). The electron donor for this nitrite reductase is a blue copper protein (12–14 kd), which is oxidized in a nitrite-dependent reaction (68, 95). Purification of the enzyme in the absence of phenylmethylsulfonyl fluoride resulted in three fractions exhibiting nitrite reductase activity, whereas purification in the presence of this protease inhibitor resulted in a single nitrite reductase fraction (95). This latter preparation differs from others (67, 69) with respect to copper content, spectral properties, and amino acid composition. These differences have been ascribed to the degradation of the native enzyme to less active forms (95).

Several blue nitrite reductases have also been described. The nitrite reductase from *Alcaligenes* strain NCIB 11015 occurs as a soluble enzyme of 70 kd and is composed of two copper-containing subunits (37 kd). The EPR spectrum is consistent with the presence of only type 1 copper (103). The nitrite reductase from *Pseudomonas aureofaciens* has been purified to electrophoretic homogeneity (167). This enzyme has a molecular mass of 85 kd and consists of two identical subunits (40 kd) each containing a gram atom in the form of type 1 copper. This copper is labile and is converted to a type 2 copper. Unlike other copper nitrite reductases (78, 103, 117, 136), the enzyme from *P. aureofaciens* does not exhibit any oxidase activity. It is of interest that the enzymes exhibiting oxidase activity also contain type 2 copper. The mechanism by which type 1 copper–containing nitrite reductases reduce nitrite is not clear. Type 1 copper–containing proteins are supposedly involved in electron transfer reactions and not catalysis (92, 103). A blue copper protein of 16 kd (azurin) serves as the electron donor; the product is nitric oxide, as is the case when PMS ascorbate is the electron donor. The enzyme also reduces nitrite in the presence of either methyl viologen dithionite or hydroxylamine; nitrous oxide is then the sole product.

The other blue nitrite reductase occurs in *R. sphaeroides* f. sp. *denitrificans* as a periplasmic enzyme (135). Two preparations have been described, each resulting in enzymes with different properties. The enzyme purified by Sawada et al (136) had a mass of 80 kd, was composed of two subunits (39 kd), and contained a single atom of copper per subunit. The EPR spectrum was characteristic of types 1 and 2 copper. The purified enzyme catalyzed the disappearance of nitrite when incubated in the presence of duroquinol and its own cytochrome c_2 and cytochrome bc_1 complex. Nitrite and cytochrome c_2 reduction were inhibited by antimycin *a*, and nitrite reduction was inhibited by diethyldithiocarbamate. The pathway of reduction was proposed to proceed from duroquinol through the cytochrome bc_1 complex and cytochrome c_2, which is the immediate electron donor for the enzyme (151). Michalski & Nicholas (117) described the purification of the other nitrite reductase from *R. sphaeroides* f. sp. *denitrificans*. This blue enzyme was composed of two nonidentical subunits of 37.5 and 39.5 kd, with different isoelectric points, which contain type 1 and 2 copper, respectively. Dialysis against EDTA removed half of the copper and inactivated the enzyme. Following EDTA treatment, the subunit with the type 2 copper signal (39.5 kd) was present. The subunit with the type 1 copper was replaced by another subunit that was more acidic and lacked copper. Incubating the dialyzed enzyme with Cu^{2+} restored nitrite reductase activity. The reactivated enzyme contained two atoms of copper per mole of enzyme, was composed of two discrete subunits (37.5 and 39.5 kd), and had an EPR spectrum characteristic of type 1 and 2 copper sites. Changes in the EPR spectrum observed in the presence of nitrite suggest that the type 1 copper is involved in nitrite binding and, for topological considerations, the formation of the N–N bond. The type 2 copper is thought to act as an electron acceptor.

cd_1-Cytochrome Nitrite Reductase

In contrast to the copper nitrite reductases, the heme-containing enzymes show considerable similarity. They are composed of two identical subunits, each containing a c- and d-type cytochrome (54, 61, 89, 100). The d-component is labile (121) and is readily lost during chromatography (149) or electrophoresis (100). The enzyme occurs in a variety of bacterial genera including *Alcaligenes* (105), *Paracoccus* (53, 121), *Thiobacillus* (137), *Azospirillum* (90, 162), and *Pseudomonas* (33, 54, 58, 82, 170). The enzyme exhibits cytochrome oxidase activity and was first described as *Pseudomonas* cytochrome oxidase (58). However, since the apparent affinity of the nitrite reductase for oxygen is considerably less than that for nitrite (54, 149), and since cytochrome oxidases (aa_3 and o) are favored over nitrite reductase in the competition of electrons (7, 92), it is unlikely that the physiological role of the nitrite reductase is that of an oxidase.

The most frequently reported molecular mass values for the nitrite reductases isolated from *Paracoccus denitrificans* (121, 149), *P. aeruginosa* (12, 54), and *Thiobacillus denitrificans* (137) approximate 120 kd. However, values as low as 90 kd for *P. aeruginosa* (58, 121) and *A. faecalis* (66) and as high as 140 kd in *Paracoccus halodenitrificans* (100) have also been reported.

The visible absorption spectrum of the oxidized enzyme consists of maxima located at approximately 407 and 640 nm. Upon reduction, the Soret band shifts to 418 nm, and a maximum located at 460 nm, which corresponds to the Soret band of the d-cytochrome, becomes apparent. A distinctive split absorption band occurs at about 549 and 553 nm and corresponds to the α bands of cytochrome c. The maximum at 640 nm, which is the α band of cytochrome d in the oxidized form, shifted to 655 nm when PMS ascorbate was the reductant, but was located at 625 nm when dithionite was used (12). When the nitrite reductase from *T. denitrificans* was reduced with either PMS ascorbate or freshly prepared solutions of dithionite, the absorption maximum at 640 nm shifted to 650 nm. When the enzyme was reduced with a solution of dithionite that contained sulfite or when sulfite was added to the enzyme after reduction with PMS ascorbate, the maximum was located at 625 nm (61). The location of the absorption maximum of cytochrome d is also affected by the hydrogen ion concentration. When the spectrum of the enzyme from *Paracoccus denitrificans* was determined at pH 6 (121), the major absorption band was at 625 nm with a minor one at 655 nm. When the spectrum was determined at pH 8.2, the absorption maximum at 655 nm was the major one. These observations may well explain the reported spectral differences of the enzyme.

There is considerable controversy concerning the location of the cd_1-cytochrome nitrite reductase activity. The enzyme from *T. denitrificans* (94, 137) is associated with the cytoplasmic membrane. The enzyme from *Paracoccus denitrificans* is variously reported to occur in the cytoplasmic fraction, the periplasmic space, the periplasmic aspect of the cytoplasmic membrane, or the cytoplasmic aspect of the cytoplasmic membrane (6, 14, 18, 87, 114, 121). Nitrite reductase activity is found in the membrane and cytoplasmic fractions in *Pseudomonas perfectomarinus* (169, 170). In *P. halodenitrificans* 97% of the total nitrite reductase activity did not sediment after crude extracts were centrifuged for 1 hr at 175,000 × g (101). The enzyme is located on the inner aspect of the cytoplasmic membrane, since no nitrite reductase or NADH oxidase activity was detected in spheroplasts prepared from *P. halodenitrificans*, whereas both activities are present in membrane vesicles (53). The distribution of the cd_1-cytochrome nitrite reductase from *P. halodenitrificans* between the cytoplasmic and membrane fractions is not due to the fragile binding of the enzyme to the cytoplasmic membrane. Disruption of cells or spheroplasts in a French pressure cell or disruption of spheroplasts by osmotic shock did not affect the distribution of nitrite reductase activity between the membrane and the cytoplasmic fractions (100). In *P. aeruginosa* the nitrite reductase is located in the periplasm along with azurin and cytochrome c_1, which are the physiological electron donors (159). This observation is contrary to those of Saraste & Kuronen (132), who used ferritin-conjugated antibody to demonstrate that the *P. aeruginosa* cd_1-cytochrome nitrite reductase is located on the cytoplasmic aspect of the membrane. They concluded that the enzyme is loosely associated with the membrane and is easily removed during preparation.

Nitric oxide was the product of nitrite reduction by *Paracoccus denitrificans* when the enzyme was assayed with PMS NADH (91). The enzyme from *T. denitrificans* produced a mixture of nitric and nitrous oxide when sulfide, which is the physiological reductant, was employed (137). However, LeGall et al (94) detected only nitric oxide with either sulfide or PMS ascorbate. The purified cd_1-cytochrome nitrite reductase from *P. aeruginosa* produced nitric oxide and small amounts of nitrous oxides when TMPD ascorbate was the electron donor. The ratio of nitric oxide to nitrous oxide was about 56 (79). Nitrite was also reduced to nitric and nitrous oxides with enzymatically reduced *Pseudomonas* cytochrome c_{551} (158) as the electron donor. With the latter donor, a small burst of nitric oxide preceded nitrous oxide production and thereafter the production of nitric oxide declined and attained a steady state while nitrous oxide production continually increased. The nitrite reductase from *P. aeruginosa* reduced hydroxylamine to ammonia when either reduced pyocyanine or reduced methylene blue was the electron donor (143). The enzyme did not reduce hydroxylamine following removal of the heme d, and neither did the isolated heme d. Hyroxylamine reduction was restored upon the reincorporation of the heme d into the aponitrite reductase. Although pyocyanine is an effective electron donor for hydroxylamine or nitrite reduction, cytochrome c_{551}, which is the physiological electron donor for the reduction of nitrite, is ineffective for the reduction of hydroxylamine. A purified cytochrome cd_1-nitrite reductase from A. *faecalis,* which reduced nitrite to nitric oxide (66), also reduced nitric oxide to nitrous oxide in the presence of PMS ascorbate (106). The reduction of nitric oxide was inhibited by cyanide and diethyldithiocarbamate, although the enzyme was less sensitive to these inhibitors than is the membrane-bound enzyme. Membranes prepared from P. *halodenitrificans* only produced nitrous oxide when incubated with PMS ascorbate. When incubated in the presence of formate or succinate the membranes produced nitric and nitrous oxides in about $2:1$ ratio. Nearly equimolar amounts of both gases were produced with NADH as the reductant, but when membranes were incubated with NADH and either flavin mononucleotide (FMN) or flavin adenine dinucleotide (FAD) only nitrous oxide was detected (53).

In addition to the nature of the reducing system, the cellular location of the enzyme also determines the end products of cd_1-cytochrome nitrite reductase activity. The membrane-bound enzyme from P. *halodenitrificans* reduces nitrite to nitrous oxide in the presence of PMS ascorbate, whereas the cytoplasmic nitrite reductase reduces nitrite to nitric oxide. The cytoplasmic enzyme reduces nitrite to ammonia without any gas production with methyl viologen and dithionite as the reducing system, whereas the membrane fraction fails to produce ammonia under these conditions. Following purification, the reduction of nitrite to nitric oxide (in the presence of phenazine methosulfate ascorbate) or ammonia (in the presence of methyl viologen dithionite) was brought about by the same enzyme (100). The membrane-bound cd_1-cytochrome nitrite reductase from *Pseudomonas perfectomarina* also failed to reduce nitrite to nitrous oxide when incubated in the presence of methyl viologen and dithionite; no mention was made of ammonia production (170).

Ferrous ascorbate catalyzes the chemical reduction of nitrite and nitric oxide (165). The nonenzymatic reduction of nitrite to nitric oxide takes place in acidic medium and is virtually nonexistent at pH greater than 6. On the other hand, the nonenzymatic reduction of nitric oxide to nitrous oxide takes place in neutral and alkaline solutions, reaching a maximum at pH 7. This reaction is inhibited by phosphate and EDTA, as well as by other chelating agents. NADH and FMN also reduce nitric oxide nonenzymatically in a reaction enhanced by ferrous iron. These chemical reactions may explain the contradictory observations concerning the nature of the products of nitrite reductase activity.

NITRIC OXIDE REDUCTASES

The reduction of nitric oxide is the least well characterized of the enzymatic steps associated with denitrification, partly because of suspicions about the chemical reactivity of nitric oxide as well as doubts concerning its role as an intermediate. A recent proposal for the formation of nitrous oxide excluded nitric oxide as an intermediate (10). Nitric oxide is rarely detected during denitrification. The products observed with nitrate reduction by *R. sphaeroides* f. sp. *denitrificans*, nitrous oxide and dinitrogen, but not nitric oxide, are typical (88). On the other hand, *Pseudomonas fluorescens*, *Pseudomonas alcaligenes*, and a *Flavobacterium* species accumulate transient quantities of nitric oxide when simultaneously reducing nitrate (or nitrite) and nitrous oxide. The accumulation of nitric oxide (as well as nitrite and nitrous oxide) reflects the relative activities of the reductases (15).

Free nitric oxide does not behave as a free intermediate during the reduction of nitrite to dinitrogen by *P. aeruginosa* (147). Insignificant amounts of ^{15}N were trapped in a pool of [^{14}N]nitric oxide during the reduction of [^{15}N]nitrite; only small quantities of nitric oxide accumulated during nitrite reduction, even though the rate of nitric oxide reduction to dinitrogen was considerably slower than the corresponding rate of nitrite reduction; and the extent of isotopic mixing in dinitrogen when [^{15}N]nitrite was reduced in the presence of [^{14}N]nitric oxide (isotopic scrambling) is inconsistent with the role of nitric oxide as an intermediate. On the other hand, radioactive nitrogen was trapped in the nitric oxide pool when [^{13}N]nitrite was reduced by *P. aureofaciens* and *Pseudomonas chlororaphis*, which implies that free nitric oxide either is an intermediate or is in rapid equilibrium with an unknown intermediate (41). The dissimilarity between these studies could reflect inadequate equilibration between the gas and liquid phases, differences in the concentration of the unlabeled pool constituents, or different rates of nitrite reduction (41). In *Pseudomonas stutzeri*, isotope fractionation is known to be affected by the nitrite and reductant concentrations in whole cells and cell-free extracts (21), and this could also account for observed differences. These disparities may also reflect intrinsic differences among the denitrifiers (46). With *Paracoccus denitrificans*, [^{15}N]nitrite was not trapped in [^{14}N]nitric oxide, and isotope scrambling was observed in dinitrogen. Extensive scrambling was observed with *Pseudomonas denitrificans* and *P. aureofaciens*. However, with these organisms [^{15}N]nitrite was trapped in nitric oxide, but this was due to an enzymatic isotope exchange reaction. No exchange occurred between nitrite and nitrous oxide. Thus nitrite and nitric oxide were proposed to be in equilibrium but reduced to nitrous oxide via separate pathways (46). In the case of *P. stutzeri*, nitrite-nitrogen was not trapped in the nitric oxide pool during nitrite reduction, yet extensive scrambling of nitrite-nitrogen was observed in dinitrogen. In this organism, nitrite and nitric oxide may be reduced via a common mononitrogen intermediate that is not nitric oxide. Thus in *Pseudomonas denitrificans* and *P. aeruginosa* nitrite and nitric oxide are reduced to nitrous oxide by separate pathways. In *Pseudomonas denitrificans* and *P. aureofaciens*, nitrite and nitric oxide are in equilbrium; in the former species both are reduced by separate pathways, whereas in the latter species nitrite reduction via nitric oxide is the major pathway (46). These differences in the pathway of nitrite reduction may be related to the nature of the nitrite reductases, since the enzymes from *P. aeruginosa* (54) and *Paracoccus denitrificans* (121) are cd_1-cytochromes whereas the nitrite reductases of *Pseudomonas denitrificans* (103) and *P. aureofaciens* (167) are copper enzymes.

There are other observations that support the notion that nitric oxide is an intermediate. The accumulation of nitric oxide by purified nitrite reductases suggests that another reductase is present. Diethyldithiocarbamate, which inhibited the production of nitric oxide by the copper-containing nitrite reductases of *A. cycloclastes* and *R. sphaeroides* f. sp. *denitrificans*, did not affect the reduction of nitric oxide to nitrous oxide; thus the reduction of nitric oxide, at least in these organisms, is not an expression of the nitrite reductase (140). Nitric oxide reduction was associated with the translocation of protons and proline transport in *Paracoccus denitrificans*, processes unaffected by acetylene (45). Nitric oxide–dependent proton translocation also occurred in *R. sphaeroides* f. sp. *denitrificans*, *A. cycloclastes*, and *R. japonicum* when they were grown under denitrifying conditions, but not when they were grown aerobically (139). Pichinoty et al (128a) reported that 7 of 15 denitrifying bacilli, isolated by anaerobic enrichment in the presence of nitrite, grew anaerobically in the presence of nitric oxide, which was reduced to nitrous oxide and dinitrogen. In *P. perfectomarina* the level of nitric oxide reductase activity was greater when cells were grown in the presence of nitrate than when cells were grown in the presence of nitrous oxide, which suggests that nitric oxide was used as an electron acceptor (44).

The reaction of nitric oxide with hemoglobin (Hb) can be used to detect nitric oxide. *P. aeruginosa*, *P. stutzeri*, *P. perfectomarina*, *Paracoccus denitrificans*, and *A. cycloclastes* grown under denitrifying conditions produced HbNO when cells were incubated with nitrite (J. Goretski & T. C. Hollocher, manuscript in preparation). Since Hb would not be expected to enter intact cells, these experiments not only confirm that nitric oxide is produced, but also imply that nitric oxide is a free intermediate. This conclusion contradicts the earlier isotopic experiments (46, 48). The basis for the discrepancy with the earlier observations is not understood. Goretski & Hollocher have suggested that a highly active nitric oxide reductase may reduce [^{15}N]nitric oxide produced from [^{15}N]nitrite before the former diffuses into the bulk phase and equilibrates with the nitric oxide pool.

Chromatophore membranes from *R. sphaeroides* f. sp. *denitrificans* grown under denitrifying conditions reduced nitric oxide to nitrous oxide in the presence of PMS ascorbate. The reaction was inhibited by antimycin *a*, which suggests the involvement of the cytochrome bc_1 complex. The cytochrome bc_1 complex, isolated by detergent treatment of membranes, also reduced nitric oxide to nitrous oxide. Since the complex has at least seven components, the agent responsible for nitric oxide reduction is unknown. The cytochrome bc_1 complex per se does not appear to be involved since the complex isolated from photosynthetically grown cells did not reduce nitric oxide (152).

Membranes from *P. perfectomarina* reduced nitric oxide to nitrous oxide with either PMS ascorbic acid or NADH FMN as electron donor (170). The enzyme was inhibited by citrate and diethyldithiocarbamate, but not by EDTA, which inhibited the chemical reaction (165). This nitric oxide reductase, solubilized with Triton X-100, has been purified to electrophoretic homogeneity as a protein-Triton micelle (166). The molecular mass of the micelle is approximately 170 kd as determined by size exclusion chromatography; several components, of 17, 40, and 67 kd, were detected by SDS polyacrylamide gel electrophoresis. The absorption spectrum of the enzyme is similar to the spectrum of a nitric oxide–reducing fraction from *P. halodenitrificans* (52). The cytochrome *b* formed a nitric oxide complex, and when the complex was reduced the heme groups were reoxidized by nitric oxide. This recalls an earlier observation that *b*- and *c*-type cytochomes were involved in the reduction of nitric oxide (118). The reduction of nitric oxide was inhibited by cyanide, 2,2'-dipyridyl, and diethyldithiocarbamate. The membrane-bound enzyme from *A. faecalis* was also inhibited by cyanide and diethyldithiocarbamate (106). In contrast Shapleigh & Payne (139) observed that the chelating agent diethyldithiocarbamate did not inhibit the reduction of nitric oxide by extracts prepared from *P. perfectomarina* or *Paracoccus denitrificans*, although some inhibition was observed with extracts prepared from *A. cycloclastes* and *R. sphaeroides* f. sp. *denitrificans*.

It is possible to resolve the overall reduction of nitrite to nitrous oxide. Incubation of membranes from *P. halodenitrificans* with the detergent 3 - (3-cholamidopropyldimethylammonio) - 1 - (2-hydroxy-1-propanesulfonate) (CHAPSO) abolished their ability to reduce nitrite to nitrous oxide (52). The detergent-soluble material reduced nitrite to a mixture of nitric and nitrous oxides, and nitric oxide to nitrous oxide. Two fractions were separated by ammonium sulfate precipitation. One fraction reduced nitrite to nitric oxide and contained cytochrome cd_1. The other fraction exhibited no nitrite reductase activity but reduced nitric oxide to nitrous oxide. The visible absorption spectrum of the latter fraction had a maximum at 550 nm and a suggestion of other components with maxima at 562 and 565 nm. A different pattern was observed when membranes were extracted with the detergent octyl-β-glucopyranoside. The soluble material catalyzed the reduction of nitrite to nitrous oxide without any nitric oxide accumulation. Furthermore, ammonium sulfate fractionation failed to separate nitrite and nitric oxide reductase activities. Grant et al (52) took these results to suggest that the reduction of nitrite to nitrous oxide is catalyzed by a membrane-bound enzyme complex consisting of nitrite and nitric oxide reductases and that nitric oxide does not occur as a free intermediate. Attempts to purify the nitric oxide reductase from *P. halodenitrificans* have so far been hindered by the tendency of the enzyme to aggregate once solubilized. Shapleigh et al (138) examined the effect of several detergents on the production of nitrous oxide from nitrite by membranes prepared from *A. cycloclastes, Paracoccus denitrificans, P. aeruginosa,* and *P. perfectomarina.* They also concluded that nitrite reduction to nitrous oxide proceeds via nitric oxide and is catalyzed by two discrete enzymes, a nitrite and a nitric oxide reductase.

MECHANISM OF N–N BOND FORMATION

The formation of the N–N bond has been the object of experiments both numerous and varied in approach, yet it still remains a contentious subject. Averill & Tiedje (10) proposed that the reduction of nitrite to nitrous oxide is initiated by the conversion of enzyme-bound nitrite to the nitrosyl derivative $(E \cdot NO^+)$ by protonation and dehydration. The nitrosyl derivative subsequently reacts with a second molecule of nitrite, which after a reductive step forms oxyhyponitrite, also known as trioxodinitrate:

$$E + NO_2^- \rightarrow E \cdot NO_2^- \rightarrow E \cdot NO^+ \rightarrow E \cdot N_2O_3. \qquad 1.$$

Following reduction and dehydration, oxyhyponitrite is converted to enzyme-bound nitrous oxide, which dissociates to yield free nitrous oxide:

$$E \cdot N_2O_3 \rightarrow \rightarrow E \cdot N_2O \rightarrow E + N_2O. \qquad 2.$$

In this construction nitric oxide is not an intermediate, but nitric oxide can react with free enzyme to produce an NO-enzyme complex. Alternatively, the complex may form by a reversible reductive step from the enzyme nitrosyl complex:

$$E + NO \rightarrow E \cdot NO \leftarrow E \cdot NO^+. \qquad 3.$$

Several groups have detected an EPR spectrum of an NO-Fe^{2+} complex (77, 94, 120), although there is no evidence that it is involved in the reduction of nitrite to nitrous oxide. The overall scheme of Averill & Tiedje (10) also accounts for the isotopic exchange of nitrite and nitric oxide and the reduction of nitric oxide to nitrous oxide. Denitrifying bacteria can bring about nitrosyl transfer from nitrite to hydroxylamine (nitrosation), and this reaction provides further evidence for the nitrosyl intermediate. When nitrosation and denitrification occur at the same time in the presence of [^{14}N]hydroxylamine and [^{15}N]nitrite, the enrichment of ^{18}O in nitrous oxide originating from both reactions is similar. This result requires that nitrosation and denitrification proceed through a common intermediate such as an enzyme-nitrosyl complex (48).

While there is consensus that an enzyme-nitrosyl complex is an intermediate in nitrite reduction, there is no agreement about the subsequent fate of the nitrosyl derivative. Hubbard et al (60) examined the reaction of the cytochrome d oxidase from E. coli with oxyhyponitrite to test if reduction of oxyhyponitrite to nitrous oxide could be detected as predicted from Averill & Tiedje's model (10). The choice of the enzyme as an analog of cd_1-cytochrome nitrite reductase does not seem unreasonable. This cytochrome oxidase was formed under conditions of low aeration, which have also been associated with the production of nitrous oxide in E. coli (16). In addition, nitric oxide preferentially bound to the d component of the cd_1-cytochrome nitrite reductase (75, 141). The reduced E. coli cytochrome d oxidase reacted with oxyhyponitrite (as well as nitrite and nitric oxide) to produce a spectral entity suggestive of a cytochrome d–nitrosyl complex rather than nitrous oxide. Oxyhyponitrite was unstable at neutral pH (49), and its decomposition was greater than its rate of reaction with reduced cytochrome d (59, 60). The nitrosyl complex originated from one of the products of oxyhyponitrite decomposition, probably nitroxyl ion (13, 49). Garber et al (49) concluded that oxyhyponitrite is not an intermediate in denitrification on the following grounds: The rate of nitrous oxide production from oxyhyponitrite by whole cells or cell-free extracts was barely greater than that by the controls; and the reduction of oxyhyponitrite to nitrous oxide requires that the N–N bond be preserved, yet the nitrous oxide produced from a mixture of nitrite and oxyhyponitrite (in the presence or absence of cells) was isotopically randomized (i.e. the N–N bond of oxyhyponitrite was not preserved during conversion to nitrous oxide).

If oxyhyponitrite (E·N$_2$O$_3$) arises by a nucleophilic attack of nitrite on enzyme-bound nitrosyl (E·NO$_2^+$), increasing the nitrite concentration should decrease the reverse reaction, whereas the reverse reaction should predominate at low nitrite concentrations:

$$E \cdot NO_2^- \overset{-H_2O}{\underset{+H_2O}{\rightleftharpoons}} E \cdot NO_2^+$$

$$\quad\quad\quad\quad \updownarrow \; +NO_2^-$$

$$E \cdot N_2O_3 \quad\quad\quad\quad\quad\quad 4.$$

In this situation water and nitrite would be expected to compete for the nitrosyl derivative. As a consequence, the exchange of $H_2{}^{18}O$ with the nitrosyl derivative should decrease with increasing concentration of nitrite. Just such a relationship was observed (2).

When crude extracts from *P. stutzeri* were incubated with azide, nitrite, and varying amounts of nitrite, the nitrous oxide produced from azide by nitrosation decreased with increasing nitrite concentration. This reflects the competition between azide and nitrite for the enzyme-nitrosyl complex. Incubation of cell-free extracts in the presence of nitrite, azide, and $H_2{}^{18}O$ allows one to determine the ^{18}O content of the $E \cdot NO^+$ pool. If nitrous oxide originates from nitroxyl, then the nitrous oxide originating by denitrification should have similar ^{18}O content to that arising via nitrosation. If nitrous oxide originates by the sequential addition of two nitrites, then the nitrous oxide from denitrification should have lower ^{18}O content than that originating from nitrosation. Less ^{18}O was indeed found in the nitrous oxide originating from denitrification than in that originating from nitrosation. It was concluded that the formation of nitrous oxide must proceed by the addition of nitrite to $E \cdot NO^+$, since water, azide, and nitrite competed for the same intermediate (E. Weeg-Aerssens, J. M. Tiedje & B. A. Averill, manuscript in preparation).

J. Goretski & T. C. Hollocher (manuscript in preparation) have proposed that nitrous oxide is derived from an enzyme-nitroxyl intermediate (E·NOH) whose dissociation is followed by the dimerization of free nitroxyl to nitrous oxide, as summarized in the reaction:

$$E \cdot NO^+ \rightarrow E \cdot NO \rightarrow E \cdot NOH \rightarrow E + NOH \rightarrow N_2O. \quad\quad 5.$$

Evidence for the nitroxyl intermediate has come from several experiments. *P. stutzeri* produced isotopically mixed nitrous oxide when [^{15}N]nitrite oxide was reduced in the presence of [^{14}N]nitric oxide (46). The abundance of $^{14}N^{15}NO$ was equal to that of $^{15}N^{14}NO$ in [^{14}N, ^{15}N]nitrous oxide, which means that the removal of oxygen occurs with equal probability from both precursors. This requires that nitrite and nitric oxide share a common intermediate such as nitroxyl (47). Additional evidence in favor of nitroxyl has come from experiments in which [^{15}N]nitrite was reduced in the presence of [^{14}N]oxyhyponitrite (49). In this situation the nitrogens of nitrous oxide were isotopically randomized. If nitrite is reduced through oxyhyponitrite, the N–N bond should be preserved and most of the nitrous oxide should be ^{14}N-labeled. For randomization to obtain, oxyhyponitrite and nitrite must pass through a common monomeric intermediate, which, based on the known chemistry of oxyhyponitrite, is nitroxyl. Nitroxyl that could be trapped by methemoglobin (13a) was not trapped when nitrite was reduced by *Paracoccus denitrificans* in the presence of methemoglobin. This result could have been obtained for any of several reasons. Perhaps free nitroxyl was not an intermediate; nitroxyl was enzyme bound; or free nitroxyl did not cross the cytoplasmic membrane because of its charge (J. Goretski & T. C. Hollocher, manuscript in preparation).

NITROUS OXIDE REDUCTASES

The reduction of nitrous oxide to dinitrogen is coupled to ATP formation, since there are organisms that grow anaerobically with nitrous oxide as the sole oxidant (13, 20, 26a, 72, 104, 110) and nitrous oxide respiration is accompanied by proton translocation (18, 152) as well as by the generation of a membrane potential (110).

Nitrous Oxide Reductases From Nonphotosynthetic Bacteria

The reduction of nitrous oxide is inhibited by acetylene (11, 161), carbon monoxide, azide, cyanide (85, 166), and sulfide (146). Several lines of evidence suggest that the enzyme is a copper protein. The purified enzyme from several organisms contains copper as the principal metal (35, 115, 145). The nitrous oxide–dependent anaerobic growth of *A. faecalis* IAM 1015, *Alcaligenes* sp. strain NCIB 11015 (70), and *P. perfectomarina* (104) did not take place in medium extracted with dithiozone. Nitrous oxide–dependent anaerobic growth did take place when dithiozone-treated medium was supplemented with copper, but not when the medium was supplemented with iron, zinc (70), molybdenum, nickel, cobalt, or manganese (104). *P. perfectomarina* produced nitrous oxide reductase when grown anaerobically in the presence of either nitrate or nitrite. Cells grown anaerobically in dithiozone-treated medium in the presence of either nitrate or nitrite lacked nitrous oxide reductase activity. When such cells were subsequently incubated with copper and rifampin they reduced nitrous oxide, which suggests that a copper-deficient apoenzyme is synthesized during growth in copper-depleted medium. Zumft et al (164) isolated three classes of *P. perfectomarina* mutants defective in the ability to grow using nitrous oxide as the terminal electron acceptor. One mutant lacked the structural protein of nitrous oxide reductase. In the second, the structural protein was present, but the protein lacked copper. In the third, low levels of the enzyme were present, which suggested that these were regulatory mutants. No mutants were isolated that failed to grow on nitrous oxide but exhibited nitrous oxide reductase activity, or that reduced nitrous oxide but lacked the copper enzyme. An interesting outcome of these studies is the observation that the amount of nitrous oxide reductase in cells grown on nitrate was high although enzyme activity was low. This contrasts with the amount and activity of the enzyme in cells grown either anaerobically in the presence of nitrous oxide or in oxygen-limited cultures in the absence of any nitrogenous oxide. Cells grown in the absence of nitrate had about 10% the amount of enzyme of cells grown in the presence of nitrate, yet the specific activity of the enzyme in the former cells was three to four times greater.

While the properties of various nitrous oxide reductases differ, it is not always clear to what extent these variations reflect differences in the methods used to prepare the enzyme, particularly with respect to the presence of air during the isolation. The nitrous oxide reductase from *Paracoccus denitrificans* is a cytoplasmic enzyme that rapidly loses activity when exposed to air (85). The partially purified enzyme loses half of its activity in about 30 min. Inactivation is reversed by incubating the inactive enzyme with reduced benzyl viologen. The enzyme is also inactivated by mercaptoethanol or dithionite but not by bisulfite, which suggests that the product of dithionite oxidation is not the inhibitor (86).

The properties of the nitrous oxide reductase from *P. perfectomarina* have recently been summarized (34). A striking feature of the enzyme is its chromogenicity, for it can be pink, blue, or purple (35). The pink form is obtained when purification takes place in the presence of air. The purple form is obtained by maintaining anaerobic conditions during isolation. The purple form of the enzyme is considerably more active than the pink enzyme, particularly after alkaline activation. The blue form, obtained when either the pink or purple form is treated with dithionite, ascorbate, or ferrocyanide, is inactive. When the blue form is treated with ferricyanide it is converted to the pink form with the restoration of enzyme activity. These interconversions are related to a number of poorly understood factors, since the purified purple copper enzyme is not transformed into the pink form upon exposure to air, and cell-free extracts lose activity even when stored under anaerobic conditions. The visible absorption spectrum of the pink form has a shoulder at 480 nm and maxima at 530, 620, and 780 nm. The absorption spectrum of the purple form is dominated by an intense band at 540 nm. An unusual feature of the enzyme is that reduction does not result in the loss of absorbance as with other copper proteins, but results in a shift in the absorption spectrum, which produces the blue form. The purple and pink forms cannot be distinguished using a monospecific antiserum. The enzyme is composed of two identical subunits of 74 kd and contains somewhat less than eight gram atoms of copper per mole of enzyme (120 kd). A small amount of zinc is also associated with the enzyme. The EPR spectra of the purple and pink forms of the enzyme lack a signal characteristic of type 2 copper. This was confirmed by the failure of dimethylglyoxime to remove copper from the enzyme. The enzyme appears to contain an unusal type 1 copper center and a significant amount of EPR "silent" copper. Anaerobic dialysis of the enzyme against cyanide removed virtually all of the copper. The colorless apoenzyme is inactive. Treating the apoenzyme with copper resulted in the restoration of about 90% of the original copper, but did not restore enzyme activity (34). Dialysis must alter the enzyme, since the aponitrous oxide reductase produced by cells grown in copper-free medium was activated upon the incorporation of copper (104).

Nitrous oxide respiration is inhibited by nitric oxide (44). Sensitivity to nitric oxide depends on whether the cells are grown anaerobically in the presence of nitrous oxide or nitrate; calls grown with nitrous oxide are far more sensitive. Nitrous oxide respiration is inhibited in pseudomonads, the photodenitrifier R. sphaeroides f. sp. denitrificans, and Alcaligenes sp. strain NCIB 11015. The basis for nitric oxide inhibition is not clear, although there is reason to suspect the interconversion of the various forms of nitrous reductase. Incubation of the purple form of the enzyme with nitric oxide inhibited nitrous oxide reduction by about 90%. This inhibition was accompanied by the conversion of the enzyme to a nitric oxide–complexed pink form, which, however, differed from the pink form of the enzyme obtained by aerobic isolation. Removal of nitric oxide from the nitric oxide–generated pink form restored the spectral features but not the enzyme activity of the purple form.

Alcaligenes sp. strain NCIB contains a violet protein which has been purified to electrophoretic homogeneity (107). The protein has several properties in common with the nitrous oxide reductase from P. perfectomarina, such as mass (120 kd), absorption spectrum, and color changes. However, the Alcaligenes protein has only five copper atoms per mole of enzyme, and its EPR spectrum is characteristic of type 2 copper, unlike that of the P. perfectomarina enzyme. Since the protein from Alcaligenes strain NCIB 11015 was purified without excluding air, there was no indication of enzyme activity, so the precise function of this protein is not clear.

When crude extracts from Pseudomonas denitrificans, P. perfectomarinus, and Paracoccus denitrificans were passed through a Sephacryl 300 column, nitrous oxide reductase activity and a 120-kd copper protein did not coelute. Furthermore, when a crude extract was passed through an anion exchange column, nitrous oxide reductase activity, but not the copper protein, disappeared. This suggests that the copper protein is not associated with nitrous oxide reductase activity (144). Zumft et al (163) examined the relationship between nitrous oxide reductase activity and the copper protein from P. perfectomarina during gel filtration. When crude extracts were used, enzyme activity and the copper protein coeluted when anaerobic conditions were maintained during gel filtration. When conditions were aerobic, enzyme activity preceded the copper peak. No such separation was observed with either the purified pink or purple form of the enzyme. This displacement probably reflects the occurrence of oxidative inactivation at the leading edge of the migrating chromatographic band in connection with the presence of components in the crude extracts that convert the native enzyme to an active but lower–molecular mass form.

Snyder & Hollocher (145) reexamined the properties of the nitrous oxide reductase from *Paracoccus denitrificans* purified using anaerobic conditions throughout isolation. The resulting enzyme was extremely active. It had a mass of 144 kd and was composed of two identical subunits of 70 kd. There were four gram atoms of copper per subunit; the EPR spectrum suggested an atypical type 1 copper, and much of the copper was silent. The major absorption maxima for the oxidized form of the enzyme was located at 550 nm; that for the reduced enzyme was located at 650 nm. The other absorption bands represented modified low-activity forms of the enzyme. EDTA and EGTA ([ethylenebis(oxyethylenenitrilo)]tetraacetic acid), but not diethyldithiocarbamate, inhibited nitrous oxide reductase activity. The enzyme rapidly lost activity in the presence of oxygen or with reduced benzyl viologen in the presence of oxygen or nitrous oxide. The enzyme was not inactivated when incubated with either reduced benzyl viologen or nitrous oxide. Interestingly, reduced benzyl viologen, but not dithionite, reactivated the enzyme.

Nitrous Oxide Reductases From the Rhodospirillaceae

Nitrous oxide reductase activity is found in members of the Rhodospirillaceae including *Rhodospirillum rubrum*, several strains of *R. capsulata*, and *R. palustris* (110). *R. capsulata* N22 and the dissimilatory nitrate reductase–containing strain N22DNAR$^+$, which do not grow anaerobically in the dark with nitrate (109), grow anaerobically in the dark with nitrous oxide as the electron acceptor. The enzyme from *R. capsulata* N22DNAR$^+$ occurs in the periplasmic fraction and consists of a single component of 76 kd as determined by SDS electrophoresis. The reduction of nitrous oxide was not inhibited by antimycin *a*, which excludes a pathway via ubiquinone–cytochrome c_2. There is some evidence that periplasmic cytochrome *c'* may be the physiological electron donor, since it is rapidly oxidized by nitrous oxide reductase in a reaction that requires the presence of nitrous oxide (110).

The nitrous oxide reductase from *R. sphaeroides* f. sp. *denitrificans* is also a periplasmic enzyme (153). Michalski et al (115) purified the enzyme to homogeneity without taking any special precautions to exclude air. The enzyme was purple and had a mass of 95 kd as determined by gel filtration, although lower values of 89 and 73 kd were obtained when electrophoresis was carried out under dissociating conditions. This discrepency was attributed to changes in tertiary structure induced by the detergent. The visible absorption spectra of the ferricyanide-oxidized and dithionite-reduced enzymes are similar to those of the enzyme from *P. perfectomarina* (168). There were four gram atoms of copper per mole of enzyme, and while type 2 copper was detected, most of the copper associated with the enzyme was EPR silent. In addition to copper, zinc and nickel (2 and 0.76 gram atoms per mole of enzyme, respectively) were also present. Whether zinc has an integral role in enzyme activity is not certain, although it may be significant that the enzyme from *P. perfectomarina* also contains zinc (35). The visible absorption spectrum of the purple enzyme has maxima at 481, 534, 635, and 740 nm, which are atypical for bacterial copper proteins.

CONCLUSIONS

While much is known about the enzymology of denitrification, several areas still remain to be clarified. The mechanism by which the N–N bond of nitrous oxide is formed has still to be resolved in spite of what appear to be definitive experiments that support contrary and apparently mutually exclusive pathways. Connected with this uncertainty is the role of nitric oxide. In one model, nitric oxide is excluded as an intermediate so that the reduction of nitrite to nitrous oxide is brought about by nitrite reductase. There are membrane-bound nitrite reductases, which reduce nitrite to nitrous oxide. It is possible to resolve such activity into distinct activities that reduce nitrite to nitric oxide and nitric oxide to nitrous oxide. Purified nitrite reductases accumulate nitric oxide and produce little, if any, nitrous oxide. A model excluding nitric oxide fails to account for these observations.

Only the nitric oxide reductase from *P. aureofaciens* has been purified and characterized. Since distinct nitrite and nitrous oxide reductases have been described, it would be interesting to compare the enzyme from *P. aureofaciens* with the nitric oxide reductases from other organisms, particularly ones that contain cd_1-cytochrome nitrite reductases.

The Rhodospirillaceae deserve further examination. In particular, it would be of interest to ascertain if denitrification is limited to certain strains of *R. palustris* and *R. sphaeroides* or if these organisms exhibit extreme examples of truncated denitrification pathways that occur in other organisms (72). Some of the reductases from these photosynthetic bacteria appear to differ from the analogous enzymes from nonphotosynthetic bacteria, so additional examples for comparative studies would be appropriate.

ACKNOWLEDGMENTS

We wish to thank Professors B. A. Averill, T. C. Hollocher, and W. G. Zumft for providing us with preprints of their manuscripts. Research from this laboratory is supported by the NASA Program of Exobiology.

Literature Cited

1. Adams, M. W. W., Mortenson, L. E. 1982. The effect of cyanide and ferricyanide on the activity of the dissimilatory nitrate reductase of *Escherichia coli*. *J. Biol. Chem.* 257:1791–99
2. Aerssens, E., Tiedje, J., Averill, B. A. 1986. Isotope labeling studies on the mechanism of N–N bond formation in denitrification. *J. Biol. Chem.* 261: 9652–56
3. Aida, T., Hata, S., Kusunoki, H. 1986. Temporary low oxygen concentrations for the formation of nitrate reductase and nitrous oxide reductase by denitrifying *Pseudomonas* sp. G59. *Can. J. Microbiol.* 32:543–47
4. Alef, K., Klemme, J.-H. 1979. Assimilatory nitrate reductase of *Rhodopseudomonas capsulata* AD2: a molybdo-heme protein. *Z. Naturforsch. Teil C* 34:33–39
5. Alef, K., Klemme, J.-H. 1982. Regulatory aspects of inorganic nitrogen metabolism in the *Rhodospirillaceae*. *Arch. Microbiol.* 133:239–41
6. Alefounder, P. R., Ferguson, S. J. 1980. The location of dissimilatory nitrite reductase and the control of dissimilatory nitrate reductase by oxygen in *Paracoccus denitrificans*. *Biochem. J.* 192:231–40
7. Alefounder, P. R., Greenfield, A. J., McCarthy, J. E. G., Ferguson, S. J. 1983. Selection and organisation of denitrifying electron-transfer pathways in *Paracoccus denitrificans*. *Biochim. Biophys. Acta* 724:20–39
8. Alefounder, P. R., MacCarthy, J. E. G., Ferguson, S. J. 1981. The basis of the control of nitrate reduction by oxygen in *Paracoccus denitrificans*. *FEMS Microbiol. Lett.* 12:321–26
9. Amy, N. K., Rajagopalan, K. V. 1979. Characterization of molybdenum cofactor from *Escherichia coli*. *J. Bacteriol.* 140:114–24
10. Averill, B. A., Tiedje, J. M. 1982. The chemical mechanism of microbial denitrification. *FEBS Lett.* 138:8–12
11. Balderston, W. L., Sherr, B., Payne, W. J. 1976. Blockage by acetylene of nitrous oxide reduction in *Pseudomonas perfectomarinus*. *Appl. Environ. Microbiol.* 31:504–8
12. Barber, D., Parr, S. R., Greenwood, C. 1976. Some spectral and steady-state kinetic properties of *Pseudomonas* cytochrome oxidase. *Biochem. J.* 157:431–38
12a. Bazylinski, D. A., Blakemore, R. P. 1983. Denitrification and assimilatory nitrate reduction in *Aquaspirillum magnetotacticum*. *Appl. Environ. Microbiol.* 46:1118–24
13. Bazylinski, D. A., Hollocher, T. C. 1985. Evidence from the reaction between trioxodinitrate (II) and nitrogen-15-labeled nitric oxide that trioxodinitrate (II) decomposed into nitrosyl hydride and nitrite in neutral soution. *Inorg. Chem.* 24:4285–58
13a. Bazylinski, D. A., Hollocher, T. C. 1985. Metmyoglobin and methemoglobin as efficient traps for nitrosyl hydride (nitroxyl) in neutral aqueous solution. *J. Am. Chem. Soc.* 107:7982–86
14. Betlach, M. R. 1982. Evolution of bacterial denitrification and denitrifier diversity. *Antonie van Leeuwenhoek J. Microbiol. Serol.* 48:585–607
15. Betlach, M. R., Tiedje, J. M. 1981. Kinetic explanation for accumulation of nitrite, nitric oxide, and nitrous oxide during bacterial denitrification. *Appl. Environ. Microbiol.* 42:1074–84
16. Bleakley, B. H., Tiedje, J. M. 1982. Nitrous oxide production by organisms other than nitrifiers and denitrifiers. *Appl. Environ. Microbiol.* 44:1342–48
17. Boody, L. D., Davis, K. J. P. 1987. Persistence of bacterial denitrification capacity under aerobic conditions–the rule rather than the exception. *FEMS Microbiol. Ecol.* 45:185–190
18. Boogerd, F. C., Van Verseveld, H. W., Stouthamer, A. H. 1981. Respiration-driven proton translocation with nitrite and nitrous oxide in *Paracoccus denitrificans*. *Biochim. Biophys. Acta* 638:181–91
19. Boogerd, F. C., Van Verseveld, H. W., Stouthamer, A. H. 1983. Dissimilatory nitrate uptake in *Paracoccus denitrificans* via μ_{H^+}-dependent system and nitrate-nitrite antiport system. *Biochim. Biophys. Acta* 723:415–27
20. Bryan, B. A., Jeter, R. M., Carlson, C. A. 1985. Inability of *Pseudomonas stutzeri* denitrification mutants with the phenotype of *Pseudomonas aeruginosa* to grow in nitrous oxide. *Appl. Environ. Microbiol.* 50:1301–3
21. Bryan, B. A., Shearer, G., Skeeters, J. L., Kohl, D. H. 1983. Variable expression of the nitrogen isotope effect associated with denitrification of nitrite. *J. Biol. Chem.* 258:8613–17
22. Burke, K. A., Brown, A. E., Lascelles, J. 1981. Membrane and cytoplasmic nitrate reductase of *Staphylococcus aureus*

and application of crossed immunoelectrophoresis. *J. Bacteriol.* 148:724–
27
23. Burke, K. A., Lascelles, J. 1979. Partial
purification and some properties of the
Staphylococcus aureus cytoplasmic nitrate reductase. *J. Bacteriol.* 139:120–
25
24. Byrne, M. D., Nicholas, D. J. D. 1987.
A membrane-bound dissimilatory nitrate
reductase from *Rhodobacter sphaeroides* f. sp. *denitrificans. Biochim. Biophys. Acta* 915:120–24
25. Carlson, C. A. 1982. The physiological
genetics of denitrifying bacteria. *Antonie
van Leeuwenhoek J. Microbiol. Serol.*
48:529–607
26. Carlson, C. A., Ferguson, L. P., Ingraham, J. L. 1982. Properties of dissimilatory nitrate reductase purified
from the denitrifier *Pseudomonas aeruginosa. J. Bacteriol.* 151:162–71
26a. Carlson, C. A., Ingraham, J. L. 1983.
Comparison of denitrification of *Pseudomonas stutzeri, Pseudomonas aeruginosa,* and *Paracoccus denitrificans.
Appl. Environ. Microbiol.* 45:1247–
53
27. Chaudry, G. R., Chaiken, I. W., Mac-
Gregor, C. H. 1983. An activity from
Escherichia coli membranes responsible
for the modification of nitrate reductase
to its precursor form. *J. Biol. Chem.*
258:5828–33
28. Chaudry, G. R., MacGregor, C. H.
1983. *Escherichia coli* nitrate reductase.
Its properties and association with the
enzyme complex. *J. Biol. Chem.*
258:5819–27
29. Chaudry, G. R., MacGregor, C. H.
1983. *Escherichia coli* nitrate reductase
subunit A: its role as the catalytic site
and evidence for its modification. *J.
Bacteriol.* 154:387–94
30. Claassen, V. P., Oltmann, L. F., Bus,
S., van't Riet, J., Stouthamer, A. H.
1981. The influence of growth conditions on the synthesis of molybdenum
cofactor in *Proteus mirabilis. Arch.
Microbiol.* 130:44–49
31. Clegg, R. A. 1976. Purification and
some properties of nitrate reductase (EC
1.7.99.4) from *Escherichia coli* K-12.
Biochem. J. 153:533–41
32. Cole, J. A., Brown, C. M. 1980. Nitrite
reduction to ammonia by fermentative
bacteria: a short circuit in the biological
nitrogen cycle. *FEMS Microbiol. Lett.*
7:65–72
33. Cox, C. D., Payne, W. J. 1973. Separation of soluble denitrifying enzymes as
cytochromes from *Pseudomonas per-

fectomarinus. Can. J. Microbiol.* 19:
861–72
34. Coyle, C. L., Zumft, W. G., Kroneck,
P. M. H. 1987. Nitrous oxide reductase
from denitrifying *Pseudomonas perfectomarina,* a novel multicopper enzyme. *Life Chem. Rep.* 5:289–303
35. Coyle, C. L., Zumft, W. G., Kroneck,
P. M. H., Körner, H., Jakob, W. 1985.
Nitrous oxide reductase from *Pseudomonas perfectomarina.* Purification and
properties of a novel multicopper enzyme. *Eur. J. Biochem.* 153:459–67
36. Craske, A., Ferguson, S. J. 1986. The
repiratory nitrate reductase from *Paracoccus denitrificans.* Molecular characterisation and kinetic properties. *Eur. J.
Biochem.* 158:429–36
37. DeMoss, J. A. 1977. Limited proteolysis of nitrate reductase purified
from membranes of *Escherichia coli. J.
Biol. Chem.* 252:1696–701
38. Enoch, H. G., Lester, R. L. 1974. The
role of a novel cytochrome *b*-containing
nitrate reductase and quinone in the *in
vitro* reconstruction of formate–nitrate
reductase activity in *E. coli. Biochem.
Biophys. Res. Commun.* 61:1234–41
39. Enoch, H. G., Lester, R. L. 1975. The
purification and properties of formate
dehydrogenase and nitrate reductase
from *Escherichia coli. J. Biol. Chem.*
250:6693–705
40. Firestone, M. K. 1982. Biological denitrification. In *Nitrogen in Agricultural
Soils, Agron. Monogr. No. 22,* ed. F. J.
Stevenson, pp. 289–326. Madison, Wis:
Am. Soc. Agron. 940 pp.
41. Firestone, M. K., Firestone, R. B.,
Tiedje, J. M. 1979. Nitric oxide as an
intermediate in denitrification: evidence
from nitrogen-13 isotope exchange. *Biochem. Biophys. Res. Commun.* 91:
10–16
42. Focht, D. D., Verstraete, W. 1977.
Biochemical ecology of nitrification and
denitrification. *Adv. Microb. Ecol.* 1:
135–214
43. Forget, P. 1974. The bacterial nitrate
reductases. Solubilization, purification
and properties of the enzyme A from
Escherichia coli K-12. *Eur. J. Biochem.*
42:325–32
44. Frunzke, K., Zumft, W. G. 1986. Inhibition of nitrous oxide respiration by
nitric oxide in the denitrifying bacterium
*Pseudomonas perfectomarina. Biochim.
Biophys. Acta* 852:119–25
45. Garber, E. A. E., Castignetti, D., Hollocher, T. C. 1982. Proton translocation
and proline uptake associated with
reduction of nitric oxide by denitrifying

Paracoccus denitrificans. Biochem. Biophys. Res. Commun. 107:1504–7

46. Garber, E. A. E., Hollocher, T. C. 1981. [15]N tracer studies on the role of NO in denitrification. *J. Biol. Chem.* 256:5459–65

47. Garber, E. A. E., Hollocher, T. C. 1982. Positional isotopic equivalence of nitrogen in N_2O produced by the denitrifying bacterium *Pseudomonas stutzeri*. Indirect evidence for a nitroxyl pathway. *J. Biol. Chem.* 257:4705–8

48. Garber, E. A. E., Hollocher, T. C. 1982. [15]N, [18]O tracer studies on the activation of nitrite by denitrifying bacteria. Nitrite/water-oxygen exchange and nitrosation reactions as indicators of electrophilic catalysis. *J. Biol. Chem.* 257:8091–97

49. Garber, E. A. E., Wehrli, S., Hollocher, T. C. 1983. ^{15}N-tracer and NMR studies on the pathway of denitrification. Evidence against trioxodinitrate but for nitroxyl as an intermediate. *J. Biol. Chem.* 258:3587–91

50. Garland, P. B., Downie, J. A., Haddock, B. A. 1975. Proton translocation and the respiratory nitrate reductase of *Escherichia coli*. *Biochem. J.* 152:547–59

51. Graham, A., Boxer, D. H. 1980. Implication of α-subunit of *Escherichia coli* nitrate reductase in catalytic activity. *Biochem. Soc. Trans.* 8:329–30

52. Grant, M. A., Cronin, S. E., Hochstein, L. I. 1984. Solubilization and resolution of the membrane-bound nitrite reductase from *Paracoccus halodenitrificans* into nitrite and nitric oxide reductases. *Arch. Microbiol.* 140:183–86

53. Grant, M. A., Hochstein, L. I. 1984. A dissimilatory nitrite reductase in *Paracoccus halodenitrificans*. *Arch. Microbiol.* 137:79–84

54. Gudat, J. C., Singh, J., Wharton, D. C. 1973. Cytochrome oxidase from *Pseudomonas aeruginosa*. I. Purification and some properties. *Biochim. Biophys. Acta* 292:376–90

55. Hageman, R. V., Rajagopalan, K. V. 1986. Assay and detection of the molybdenum cofactor. *Methods Enzymol.* 122:399–412

56. Hochstein, L. I., Betlach, M., Kritikos, G. 1984. The effect of oxygen on denitrification during steady-state growth of *Paracoccus halodenitrificans*. *Arch. Microbiol.* 137:74–78

57. Hochstein, L. I., Tomlinson, G. A. 1985. Denitrification by extremely halophilic bacteria. *FEMS Microbiol. Lett.* 27:329–31

58. Horio, T., Higashi, T., Yamanaka, T.,

Matsubara, H., Okunuki, K. 1961. Purification and properties of cytochrome oxidase from *Pseudomonas aeruginosa*. *J. Biol. Chem.* 236:944–51

59. Hubbard, J. A., Hughes, M. N., Poole, R. K. 1983. Nitrate, but not silver, ions induce spectral changes in *Escherichia coli* cytochrome d. *FEBS Lett.* 164:241–43

60. Hubbard, J. A., Hughes, M. N., Poole, R. K. 1985. Reaction of some nitrogen oxyanions and nitric oxide with cytochrome oxidase d from oxygen-limited *Escherichia coli* K-12. In *Microbial Gas Metabolism*, ed. R. K. Poole, C. S. Dow, pp. 231–36. New York: Academic

61. Huynh, B. H., Lui, M. C., Moura, J. J. G., Moura, I., Ljungdahl, P. O., et al. 1982. Mössbauer and EPR studies on nitrite reductase from *Thiobacillus denitrificans*. *J. Biol. Chem.* 257:9576–81

62. Ingledew, W. J., Poole, R. K. 1984. The respiratory chain of *Escherichia coli*. *Microbiol. Rev.* 48:222–71

63. Ingraham, J. L. 1981. Microbiology and genetics of denitrifiers. In *Denitrification, Nitrification, and Atmospheric Nitrous Oxide*, ed. C. C. Delwiche, pp. 45–65. New York: Wiley

64. Ingraham, J. L. 1985. Genetics of denitrification in *Pseudomonas aeruginosa* and *stutzeri*. In *Denitrification in the Nitrogen Cycle*, ed. H. L. Golterman, pp. 67–78. New York: Plenum

65. Ishizuka, M., Toraya, T., Fukui, S. 1984. Purification, properties and limited proteolysis of nitrate reductase from *Pseudomonas denitrificans*. *Biochim. Biophys. Acta* 786:133–43

66. Iwasaki, H., Matsubara, T. 1971. Cytochrome c-577 (551) and cytochrome cd of *Alcaligenes faecalis*. *J. Biochem. Tokyo* 69:847–57

67. Iwasaki, H., Matsubara, T. 1972. A nitrite reductase from *Achromobacter cycloclastes*. *J. Biochem. Tokyo* 71:645–52

68. Iwasaki, H., Matsubara, T. 1973. Purification and some properties of *Achromobacter cycloclastes* Azurin. *J. Biochem. Tokyo* 73:659–61

69. Iwasaki, H., Noji, S., Shidara, S. 1975. *Achromobacter cycloclastes* nitrite reductase. The function of copper, amino acid composition, and ESR spectra. *J. Biochem. Tokyo* 78:355–61

70. Iwasaki, H., Saigo, T., Matsubara, T. 1980. Copper as a controlling factor of anaerobic growth under N_2O and biosynthesis of N_2O reductase in de-

nitrifying bacteria. *Plant Cell Physiol.* 21:1573–84

71. Jackson, M. A., Jackson, J. B., Ferguson, S. J. 1981. Direct observation with an electrode of uncoupler-sensitive assimilatory nitrate uptake by *Rhodopseudomonas capsulata. FEBS Lett.* 136:275–78

72. Jeter, R. M., Ingraham, J. L. 1981. The denitrifying prokaryotes. In *A Handbook on Habitats, Isolation, and Identification of Bacteria,* ed. M. P. Starr, H. Stolp, H. Trüper, A. Balows, H. G. Schlegel, pp. 913–25. New York: Springer-Verlag

73. John, P. 1977. Aerobic and anaerobic bacterial respiration monitored by electrodes. *J. Gen. Microbiol.* 98:231–38

74. Johnson, J. L., Hainline, B. E., Rajagopalan, K. V. 1980. Characterization of the molybdenum cofactor of sulfite oxidase, xanthine oxidase, and nitrate reductase. *J. Biol. Chem.* 255:1783–86

75. Johnson, M. K., Thomson, A. J., Walsh, T. A., Barber, D., Greenwood, C. 1980. Electron paramagnetic resonance studies on *Pseudomonas* nitrosyl nitrite reductase. Evidence for multiple species in the electron paramagnetic resonance spectra of nitrosyl haemoproteins. *Biochem. J.* 189:285–94

76. Jones, R. W., Ingledew, W. J., Graham, A., Garland, P. B. 1978. Topography of nitrate reductase of the cytoplasmic membrane of *Escherichia coli:* the nitrate-reducing site. *Biochem. Soc. Trans.* 6:1287–89

77. Kakutani, T., Watanabe, H., Arima, K., Beppu, T. 1981. Purification and properties of a copper-containing nitrite reductase from a denitrifying bacterium, *Alcaligenes faecalis* strain S-6. *J. Biochem. Tokyo* 89:453–61

78. Kakutani, T., Watanabe, H., Arima, K., Beppu, T. 1981. A blue protein as an inactivating factor for nitrite reductase from *Alcaligenes faecalis* strain S-6. *J. Biochem. Tokyo* 89:463–72

79. Kim, C.-H., Hollocher, T. C. 1983. [15]N tracer studies on the reduction of nitrite by the purified dissimilatory nitrite reductase of *Pseudomonas aeruginosa.* Evidence for direct production of N_2O without free NO as a free intermediate. *J. Biol. Chem.* 258:4861–63

80. Klemme, J.-H., Chyla, I., Preuss, M. 1980. Dissimilatory nitrate reduction by strains of the facultative phototrophic bacterium *Rhodopseudomonas palustris. FEMS Microbiol. Lett.* 9:137–40

81. Knowles, R. 1982. Denitrification. *Microbiol. Rev.* 46:43–70

82. Kodama, T. 1970. Effects of growth conditions on formation of cytochrome system of a denitrifying bacterium, *Pseudomonas stutzeri. Plant Cell Physiol.* 11:231–39

83. Koike, I., Hattori, A. 1975. Energy yield of denitrification: an estimate from growth yield in continuous cultures of *Pseudomonas denitrificans* under nitrate-, nitrite- and nitrous oxide limited conditions. *J. Gen. Microbiol.* 88:11–19

84. Kristjansson, J. K., Hollocher, T. C. 1979. Substrate binding site for nitrate reductase of *Escherichia coli* is on the inner aspect of the membrane. *J. Bacteriol.* 137:1227–33

85. Kristjansson, J. K., Hollocher, T. C. 1980. First practical assay for soluble nitrous oxide reductase of denitrifying bacteria and a partial kinetic characterization. *J. Biol. Chem.* 255:704–7

86. Kristjansson, J. K., Hollocher, T. C. 1981. Partial purification and characterization of nitrous oxide reductase from *Paracoccus denitrificans. Curr. Microbiol.* 6:247–51

87. Kristjansson, J. K., Walter, B., Hollocher, T. C. 1978. Respiration-dependent proton translocation and transport of nitrate and nitrite in *Paracoccus denitrificans* and other denitrifying bacteria. Biochemistry 17:5014–19

88. Kundu, B., Nicholas, D. J. D. 1985. Denitrification in *Rhodopseudomonas sphaeroides* f. sp. *denitrificans. Arch. Microbiol.* 141:57–62

89. Kuronen, T., Saraste, M., Ellfolk, N. 1975. The subunit structure of *Pseudomonas* cytochrome oxidase. *Biochim. Biophys. Acta* 393:48–54

90. Lalande, R., Knowles, R. 1987. Cytoplasmic content of *Azospirillum brasilense* Sp 7 grown under aerobic and denitrifying conditions. *Can. J. Microbiol.* 33:151–56

91. Lam, Y., Nicholas, D. J. D. 1969. A nitrate reductase from *Micrococcus denitrificans. Biochim. Biophys. Acta* 178:225–34

92. Lam, Y., Nicholas, D. J. D. 1969. A nitrite reductase with cytochrome oxidase activity from *Micrococcus denitrificans. Biochim. Biophys. Acta* 180:459–72

93. Lappin, A. G. 1981. Properties of copper "blue" proteins. In *Metal Ions in Biological Systems,* ed. H. Sigel, 13:15–71. New York: Dekker

94. LeGall, J., Payne, W. J., Morgan, T. V., DerVartanian, D. 1979. On the purification of nitrite reductase from *T.*

denitrificans and its reaction with nitrite under reducing conditions. *Biochem. Biophys. Res. Commun.* 87:355–62

95. Liu, M.-Y., Liu, M.-C., Payne, W. J., LeGall, J. 1986. Properties and electron transfer specificity of copper proteins from the denitrifier *"Achromobacter cycloclastes."* *J. Bacteriol.* 166:604–8

96. Lund, K., DeMoss, J. A. 1976. Association-dissociation behavior and subunit structure of heat released nitrate reductase from *Escherichia coli. J. Biol. Chem.* 251:2207–16

97. MacGregor, C. H. 1975. Solubilization of *Escherichia coli* nitrate reductase by a membrane-bound protease. *J. Bacteriol.* 121:1102–10

98. MacGregor, C. H. 1975. Anaerobic cytochrome b_1 in *Escherichia coli:* association with and regulation of nitrate reductase. *J. Bacteriol.* 121:1111–16

99. MacGregor, C. H., Schnaitman, C. A., Normansell, D. E., Hodgins, M. G. 1974. Purification and properties of nitrate reductase from *Escherichia coli* K-12. *J. Biol. Chem.* 249:5321–27

100. Mancinelli, R. L., Cronin, S., Hochstein, L. I. 1986. The purification and properties of a *cd*-cytochrome nitrite reductase from *Paracoccus halodenitrificans. Arch. Microbiol.* 145:202–8

101. Mancinelli, R. L., Hochstein, L. I. 1986. The occurrence of denitrification in extremely halophilic bacteria. *FEMS Microbiol. Lett.* 35:55–58

102. Masuko, M., Iwasaki, H. 1984. Activation of nitrite reductase in a cell-free system from the denitrifier *Alcaligenes* sp. by freeze thawing. *Plant Cell Physiol.* 25:439–46

103. Masuko, M., Iwasaki, H., Sakurai, T., Suzuki, S., Nakahara, A. 1984. Characterization of nitrite reductase from a denitrifier, *Alcaligenes* sp. NCIB 11015. A novel copper protein. *J. Biochem. Tokyo* 96:447–54

104. Matsubara, T., Frunzke, K., Zumft, W. G. 1982. Modulation by copper of the 107 products of nitrite respiration in *Pseudomonas perfectomarinus. J. Bacteriol.* 149:816–23

105. Matsubara, T., Iwasaki, H. 1971. Enzymatic steps in dissimilatory nitrite reduction in *Alcaligenes faecalis. J. Biochem. Tokyo* 69:859–68

106. Matsubara, T., Iwasaki, H. 1972. Nitric oxide reducing activity of *Alcaligenes faecalis* cytochrome *cd. J. Biochem. Tokyo* 72:57–64

107. Matsubara, T., Sano, M. 1985. Isolation and some properties of a novel violet copper protein from a denitrifying bacterium, *Alcaligenes* sp. *Chem. Lett.* 1985 (7):1053–56

108. Matsubara, T., Zumft, W. G. 1982. Identification of a copper protein as part of the nitrous oxide–reducing system in nitrite-respiring (denitrifying) pseudomonads. *Arch. Microbiol.* 132:322–28

109. McEwan, A. G., George, C. L., Ferguson, S. J., Jackson, J. B. 1982. A nitrate reductase activity in *Rhodopseudomonas capsulata* linked to electron transfer and generation of a membrane potential. *FEBS Lett.* 150:277–80

110. McEwan, A. G., Greenfield, A. J., Wetzstein, H. G., Jackson, J. B., Ferguson, S. J. 1985. Nitrous oxide reduction by members of the family *Rhodospirillaceae* and the nitrous oxide reductase of *Rhodopseudomonas capsulata. J. Bacteriol.* 164:823–30

111. McEwan, A. G., Jackson, J. B., Ferguson, S. J. 1984. Rationalization of properties of nitrate reductases in *Rhodopseudomonas capsulata. Arch. Microbiol.* 137:344–49

112. McEwan, A. G., Wetzstein, H. G., Meyer, O., Jackson, J. B., Ferguson, S. J. 1987. The periplasmic nitrate reductase of *Rhodobacter capsulatus:* purification, characterization and distinction from a single reductase from trimethylamine-*N*-oxide, dimethylsulphoxide and chlorate. *Arch. Microbiol.* 147:340–45

113. Meiburg, J. B. M., Bruinenberg, P. M., Harder, W. 1980. Effect of dissolved oxygen tension on the metabolism of methylated amines in *Hypomicrobium* X in the absence and presence of nitrate: evidence for "aerobic" denitrification. *J. Gen. Microbiol.* 120:453–63

114. Meijer, E. M., Van Der Zwaan, J. W., Stouthamer, A. H. 1979. Location of the proton-consuming site in nitrite reduction and stoichiometries for proton pumping in anaerobically grown *Paracoccus denitrificans. FEMS Microbiol. Lett.* 5:369–72

115. Michalski, W. P., Hein, D. H., Nicholas, D. J. D. 1986. Purification and characterization of nitrous oxide reductase from *Rhodopseudomonas sphaeroides* f. sp. *denitrificans. Biochim. Biophys. Acta* 872:50–60

116. Michalski, W. P., Nicholas, D. J. D. 1984. The adaptation of *Rhodopseudomonas sphaeroides* f. sp. *denitrificans* for growth under denitrifying conditions. *J. Gen. Microbiol.* 130:155–65

117. Michalski, W. P., Nicholas, D. J. D. 1985. Molecular characterization of a copper-containing nitrite reductase from

Rhodopseudomonas sphaeroides forma sp. *denitrificans. Biochim. Biophys. Acta* 828:130–37

118. Miyata, M. 1971. Studies on denitrification. XIV. The electron donating system in the reduction of nitric oxide and nitrate. *J. Biochem. Tokyo* 70:205–13

119. Morpeth, F. F., Boxer, D. H. 1985. Kinetic analysis of respiratory nitrate reductase from *Escherichia coli* K-12. *Biochemistry* 24:40–46

120. Muhoberac, B. B., Wharton, D. C. 1980. EPR study of heme·NO complexes of ascorbic acid reduced *Pseudomonas* cytochrome oxidase and corresponding model compounds. *J. Biol. Chem.* 255:8437–42

121. Newton, N. 1969. The two-haem nitrite reductase of *Micrococcus denitrificans. Biochim. Biophys. Acta* 185:316–31

122. Oltmann, L. F., Reijnders, W. N. M., Stouthamer, A. H. 1976. Characterization of purified nitrate reductase A and chlorate reductase C from *Proteus mirabilis. Arch. Microbiol.* 111:25–35

123. Parsonage, D., Ferguson, S. J. 1983. Reassessment of pathways of electron flow to nitrate reductase that are coupled to energy conservation in *Paracoccus denitrificans. FEBS Lett.* 153:108–12

124. Parsonage, D., Greenfield, A. J., Ferguson, S. J. 1985. The high affinity of *Paracoccus denitrificans* cells for nitrate as an electron acceptor. Analysis of possible mechanisms of nitrate and nitrite movement across the plasma membrane and the basis for inhibition by added nitrite of oxidase activity in permeabilised cells. *Biochim. Biophys. Acta* 807:81–95

125. Parsonage, D., Greenfield, A. J., Ferguson, S. J. 1986. Evidence that energy conserving electron transport pathways to nitrate and cytochrome *o* branch at ubiquinone in *Paracoccus denitrificans. Arch. Microbiol.* 145:191–96

126. Payne, W. J. 1973. Reduction of nitrogenous oxides by microorganisms. *Bacteriol. Rev.* 37:409–52

127. Payne, W. J. 1981. *Denitrification*, pp. 79–89. New York: Wiley

128. Payne, W. J., Riley, P. S., Cox, C. D. 1971. Separate nitrite, nitric oxide, and nitrous oxide reducing fractions from *Pseudomonas perfectomarinus. J. Bacteriol.* 106:356–61

128a. Pichinoty, F., Mandel, M., Garcia, J.-L. 1979. The properties of novel mesophilic denitrifying *Bacillus* cultures found in tropical soils. *J. Gen. Microbiol.* 115:419–30

129. Preuss, M., Klemme, J.-H. 1983.

130. Robertson, L. A., Kuenen, J. G. 1984. Aerobic denitrification—a controversy revived. *Arch. Microbiol.* 139:351–54

131. Robertson, L. A., Kuenen, J. G. 1984. Aerobic denitrification—old wine in new bottles? *Antonie van Leeuwenhoek J. Microbiol. Serol.* 50:525–44

132. Saraste, M., Kuronen, T. 1978. Interaction of *Pseudomonas* cytochrome cd_1 with the cytoplasmic membrane. *Biochim. Biophys. Acta* 513:117–31

133. Satoh, T. 1981. Soluble dissimilatory nitrate reductase containing cytochrome *c* from a photodenitrifier, *Rhodopseudomonas sphaeroides* forma sp. *denitrificans. Plant Cell Physiol.* 22:443–52

134. Satoh, T., Hoshino, Y., Kitamura, H. 1976. *Rhodopseudomonas sphaeroides* forma sp. *denitrificans*, a denitrifying strain as a subspecies of *Rhodopseudomonas sphaeroides. Arch. Microbiol.* 108:265–69

135. Sawada, E., Satoh, T. 1980. Periplasmic location of dissilimatory nitrate and nitrite reductases in a denitrifying phototrophic bacterium, *Rhodopseudomonas sphaeroides* forma sp. *denitrificans. Plant Cell Physiol.* 21:205–10

136. Sawada, E., Satoh, T., Kitamura, H. 1978. Purification and properties of a dissimilatory nitrite reductase of a denitrifying phototrophic bacterium. *Plant Cell Physiol.* 19:1339–51

137. Sawhney, V., Nicholas, D. J. D. 1978. Sulphide-linked nitrite reductase from *Thiobacillus denitrificans* with cytochrome oxidase activity: purification and properties. *J. Gen. Microbiol.* 106:119–28

138. Shapleigh, J. P., Davies, K. J. P., Payne, W. J. 1987. Detergent inhibition of nitric-oxide reductase activity. *Biochim. Biophys. Acta* 911:334–40

139. Shapleigh, W. J., Payne, W. J. 1985. Nitric oxide–dependent proton translocation in various denitrifiers. *J. Bacteriol.* 163:837–40

140. Shapleigh, W. J., Payne, W. J. 1985. Differentiation of c,d_1 cytochrome and copper nitrite reductase production in denitrifiers. *FEMS Microbiol. Lett.* 26:275–79

141. Silvestrini, M. C., Colosimo, A., Brunori, M., Walsh, T. A., Barber, D., Greenwood, C. 1979. A re-evaluation of

Purification and characterization of a dissimilatory nitrite reductase from the phototrophic bacterium *Rhodopseudomonas palustris. Z. Naturforsch. Teil C* 38:933–38 some basic structural and functional properties of *Pseudomonas* cytochrome oxidase. *Biochem. J.* 183:701–9

142. Silvestro, A., Pommier, S., Giordano, G. 1986. Molybdenum cofactor—a compound in the *in vitro* activation of both nitrate reductase and trimethyl-amine-*N*-oxide reductase activities in *Escherichia coli* K-12. *Biochim. Biophys. Acta* 872:243–52

143. Singh, J. 1973. Cytochrome oxidase from *Pseudomonas aeruginosa*. III. Reduction of hydroxylamine. *Biochim. Biophys. Acta* 333:28–36

144. Snyder, S. W., Hollocher, T. C. 1984. Nitrous oxide reductase and the 120,000 MW copper protein of N_2-producing de-nitrifying bacteria are different entities. *Biochem. Biophys. Res. Commun.* 119: 588–92

145. Snyder, S. W., Hollocher, T. C. 1987. Purification and some characteristics of nitrous oxide reductase from *Paracoccus denitrificans*. *J. Biol. Chem.* 262:6515–25

146. Sorenson, J., Tiedje, J. M., Firestone, R. B. 1980. Inhibition by sulfide of ni-tric oxide and nitrous oxide reduction by denitrifying *Pseudomonas fluorescens*. *Appl. Environ. Microbiol.* 39:105–8

147. St. John, R. T., Hollocher, T. C. 1977. Nitrogen 15 tracer studies on the path-way of denitrification in *Pseudomonas aeruginosa*. *J. Biol. Chem.* 252:212–18

148. Stouthamer, A. H. 1976. Biochemistry and genetics of nitrate reductase in bac-teria. *Adv. Microb. Physiol.* 14:315–75

148a. Stouthamer, A. H., Boogerd, F. C., van Verseveld, H. W. 1985. The bioenergetics of denitrification. *Antonie van Leeuwenhoek J. Microbiol. Serol.* 48:545–53

148b. Tiedje, J. M., Sexstone, A. J., Myrold, D. D., Robinson, J. A. 1985. Denitrification: ecological niches, com-petition and survival. *Antonie van Leeuwenhoek J. Microbiol. Serol.* 48:569–83

149. Timkovich, R., Dhesi, R., Martinkus, K. J., Robinson, M. K., Rea, T. M. 1982. Isolation of *Paracoccus de-nitrificans* cytochrome cd_1: comparative kinetics with other nitrite reductases. *Arch. Biochem. Biophys.* 215:47–58

150. Tomlinson, G. A., Jahnke, L. L., Hoch-stein, L. I. 1986. *Halobacterium de-nitrificans* sp. nov., an extremely halophilic denitrifying bacterium. *Int. J. Syst. Bacteriol.* 36:66–70

151. Urata, K., Satoh, T. 1984. Evidence for cytochrome bc_1 complex involvement in nitrite reduction in a photodenitrifier,

Rhodopseudomonas sphaeroides forma sp. *denitrificans*. *FEBS Lett.* 172:205–8

152. Urata, K., Satoh, T. 1985. Mechanism of nitrite reduction to nitrous oxide in a photodenitrifier, *Rhodopseudomonas sphaeroides* forma sp. *denitrificans*. *Biochim. Biophys. Acta* 841:201–7

153. Urata, K., Shimada, K., Satoh, T. 1982. Periplasmic location of nitrous ox-ide reductase in a photodenitrifier, *Rho-dopseudomonas sphaeroides* forma sp. *denitrificans*. *Plant Cell Physiol.* 23: 1121–24

154. van't Riet, J., Planta, R. J. 1975. Purification, structure and properties of the respiratory nitrate reductase of *Kleb-siella aerogenes*. *Biochim. Biophys. Acta* 379:81–94

155. van't Riet, J., Van Ee, J. H., Wever, R., Van Gelder, B. F., Planta, R. J. 1975. Characterization of the respiratory nitrate reductase of *Klebsiella aerogenes* as a molybdenum-containing iron-sulfur enzyme. *Biochim. Biophys. Acta* 405: 306–17

156. van't Riet, J., Wientjes, F. B., Van Doorn, J., Planta, R. J. 1979. Purifica-tion and characterization of the respira-tory nitrate reductase of *Bacillus licheni-formis*. *Biochim. Biophys. Acta* 576: 347–60

157. Wesch, R., Klemme, J.-H. 1980. Catalytic and molecular differences be-tween assimilatory nitrate reductases isolated from two strains of *Rhodo-pseudomonas capsulata*. *FEMS Micro-biol. Lett.* 8:37–41

158. Wharton, D. C., Weintraub, S. T. 1980. Identification of nitric oxide and nitrous oxide as products of nitrite reduction by *Pseudomonas* cytochrome oxide (nitrite reductase). *Biochem. Biophys. Res. Commun.* 97:236–42

159. Wood, P. M. 1978. Periplasmic location of the terminal reductase in nitrite respiration. *FEBS Lett.* 92:214–18

160. Yokota, S., Urata, K., Satoh, T. 1984. Redox properties of membrane-bound *b*-type cytochromes and a soluble *c*-type cytochrome of nitrate reductase in a photodenitrifier, *Rhodopseudomonas sphaeroides* forma sp. *denitrificans*. *J. Biochem. Tokyo* 95:1535–41

161. Yoshinari, T., Knowles, R. 1976. Acetylene blockage inhibition of nitrous oxide reduction by denitrifying bacteria. *Biochem. Biophys. Res. Commun.* 69:705–10

162. Zimmer, W., Stephan, M. P., Bothe, H. 1984. Denitrification by *Azospirillum brasilense* Sp 7. *Arch. Microbiol.* 138: 206–11

163. Zumft, W. G., Coyle, C. L., Frunzke, K. 1985. The effect of oxygen on chromatographic behavior and properties of nitrous oxide reductase. *FEBS Lett.* 183:240–44

164. Zumft, W. G., Döhler, K., Körner, H. 1985. Isolation and characterization of transposon Tn5-induced mutants of *Pseudomonas perfectomarina* defective in nitrous oxide respiration. *J. Bacteriol.* 163:918–24

165. Zumft, W. G., Frunzke, K. 1982. Discrimination of ascorbate-dependent nonenzymatic and enzymatic, membrane-bound reduction of nitric oxide in denitrifying *Pseudomonas perfectomarinus*. *Biochim. Biophys. Acta* 681: 459–68

166. Zumft, W. G., Gotzmann, D. J., Frunzke, K., Viebrock, A. 1987. Novel terminal oxidoreductases of anaerobic respiration (denitrification) from *Pseudomonas*. In *Inorganic Nitrogen Metabolism*, ed. W. Ullrich, P. J. Apar-icio, P. J. Syrett, pp. 61–67. Berlin: Springer-Verlag

167. Zumft, W. G., Gotzmann, D. J., Kroneck, P. M. H. 1987. Type 1, blue copper proteins constitute a respiratory nitrite-reducing system in *Pseudomonas aureofaciens*. *Eur. J. Biochem.* 168: 301–7

168. Zumft, W. G., Matsubara, T. 1982. A novel kind of multi-copper protein as terminal oxidoreductase of nitrous oxide respiration in *Pseudomonas perfectomarinus*. *FEBS Lett.* 148:107–12

169. Zumft, W. G., Sherr, B. F., Payne, W. J. 1979. A reappraisal of nitric oxide–binding protein of denitrifying *Pseudomonas*. *Biochem. Biophys. Res. Commun.* 88:1230–36

170. Zumft, W. G., Vega, J. M. 1979. Reduction of nitrite to nitrous oxide by a cytoplasmic membrane fraction from the marine denitrifier *Pseudomonas perfectomarinus*. *Biochim. Biophys. Acta* 584:484–99

Ann. Rev. Microbiol. 1988. 42:263–87

MICROBIAL DEGRADATION OF HALOAROMATICS

Walter Reineke

Bergische Universität—Gesamthochschule Wuppertal, Fachbereich 9, Gauss-Strasse 20, D-5600 Wuppertal 1, Federal Republic of Germany

Hans-Joachim Knackmuss

Institut für Mikrobiologie der Universität Stuttgart, Azenbergstrasse 18, D-7000 Stuttgart 1, Federal Republic of Germany

CONTENTS

INTRODUCTION

Public concern over the possible effects of chemicals on humans and their environment has largely focused on a few classes of compounds. Of these compounds chlorinated aromatics are the most spectacular. Polychlorinated biphenyls (PCBs) and chlorinated benzenes show the entry points of those chemicals into the environment. PCBs were first produced in 1929 in the United States. Total cumulative world production of PCB has been estimated

263

at about 750,000 tonnes. Of this total, approximately 60% have gone into closed electric uses (e.g. transformers, capacitors), 15% to nominally closed uses (hydraulic and heat-transfer fluids), and 25% to dispersive uses (plasticizers, paint and printing ink components, adhesives, and additives for cutting oils, textile auxiliaries, and pesticides). It may be deduced therefore that 300,000 t have entered the environment since 1929 in widely disseminated form and that 450,000 t are either still in service or in landfills.

Most chlorinated benzenes are used mainly as intermediates in the synthesis of fine chemicals, and enter the environment as losses and wastes from production sites. As losses are normally 1–2% of raw materials, their total quantities do not exceed a few hundred tonnes per annum. The exceptionable aspects are the dispersive uses of mono- and dichlorobenzenes and waste disposal.

Two types of pollution may result from releases of chemicals into the environment: In *point source pollution* the concentration of the chemical is high; this may occur in landfills, waste dumps, and industrial effluents or at sites of accidents associated with transportation and application of chemicals. In *dispersed pollution* the concentrations are low and result from losses from production sites via volatilization or from agricultural uses. Different strategies are necessary to stem the impact of low concentrations of dispersed chemicals occurring worldwide and the high concentrations of point source pollutants that occur in smaller areas. If a dispersed chemical such as an agrochemical tends to persist, its use and introduction into the environment can only be regulated legislatively. Technological solutions, however, can be applied for more highly concentrated pollutants, e.g. in industrial wastewater and dumps.

In this paper we discuss metabolic pathways elaborated in single microorganisms that are able to grow with haloaromatics as carbon and energy source. These pathways include the major catabolic routes to amphibolic intermediates. Cometabolism, i.e. the widespread ability of microorganisms to catalyze partial transformation to products that do not support growth, is discussed as a minor aspect. The central role of some simple chlorinated benzene derivatives in the degradation of certain chloroaromatic pollutants is schematized in Figure 1, to show the importance of the pathways in the degradation of some pollutants summarized here. The paper closes with some aspects of the application of chlorinated aromatic-degrading bacteria.

BIODEGRADATION MECHANISMS

The biodegradation of a halogenated arene can be considered complete only when its carbon skeleton is converted into intermediary metabolites and its organic halogen is returned to the mineral state. The crucial point is the

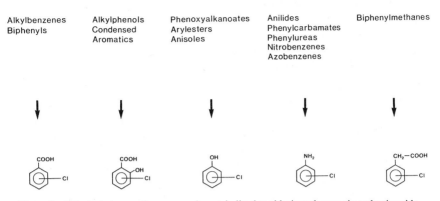

Figure 1 Chlorinated aromatic compounds metabolized to chlorinated aromatic carbonic acids, phenols, and anilines.

removal of halogen substituents from the organic compound. This may occur at an early stage of the degradative pathway with reductive, hydrolytic, or oxygenolytic elimination of the halosubstituent. Alternatively, nonaromatic structures may be generated, which spontaneously lose halide by hydrolysis or hydrogen halide by β-elimination.

Displacement of Halogen Through Hydrogen

Molecular oxygen is required not only as the terminal electron acceptor during respiration, but also for insertion into the aromatic compound during ring-activating hydroxylation and ring cleavage. Microorganisms have of necessity evolved different mechanisms for degradation in the absence of oxygen, including ring fission of aromatic compounds. Although details of the pathways and the enzymes involved are still missing, metabolism of the aromatic ring in the absence of molecular oxygen is now known to proceed in at least five different situations (90): (*a*) through anaerobic photometabolism; (*b*) under nitrate-reducing conditions in mixed cultures and by single strains of *Bacillus* sp., *Pseudomonas* sp., and *Moraxella* sp.; (*c*) with sulfate as electron acceptor; (*d*) in consortia through fermentation coupled to methanogenesis, and (*e*) through fermentation. Anaerobic degradation of aromatic compounds is reviewed by Evans & Fuchs (27a) in this volume.

The first lines of evidence for anaerobic degradation of halogenated aromatics were presented by Horowitz et al (42) and Suflita et al (94). They found that an anaerobic microbial consortium isolated from sewage sludge could degrade a number of *meta*-substituted chlorinated benzoates. The most interesting degradative reaction was the loss of the chloride without the alteration of the aromatic ring. When all the chlorine atoms were successively removed, ring fission led to methane and carbon dioxide. Dechlorination occurred only under methanogenic conditions. Analysis of the kinetics of the dechlorination steps suggested that the dichlorinated parent compound was the preferred substrate and inhibited the dechlorination of the monochlorinated intermediate (95). Shelton & Tiedje (89) recently isolated and characterized organisms in an anaerobic consortium that mineralized 3-chlorobenzoate. Based on the organisms isolated there appears to be a three-tiered food chain (Figure 2). The methanogenic consortium consisted of one dechlorinating bacterium, one benzoate-oxidizing bacterium, two butyrate-oxidizing bacteria, two H_2-consuming methanogens (*Methanospirillum hungatei, Methanobacterium* sp.), and a sulfate-reducing bacterium (*Desulfovibrio* sp.). The dechlorinating bacterium converted 3-chlorobenzoate stoichiometrically to benzoate, which accumulated in the medium. The presence of butyrate-oxidizing bacteria and the sulfidogen in the enrichment is unexplained, since they do not appear to be in the main path of carbon flow. The growth substrate of the dechlorinating bacterium in the enrichment is not clear. Presumably, one or more of the organisms in the enrichment were cross-feeding the dechlorinating bacterium. This may offer an explanation for the presence of the butyrate-oxidizing and sulfate-reducing bacteria.

The dechlorination reaction appeared to be enzymatic, since it occurred after induction and because of the low substrate K_m of 67 μM, loss of activity at temperatures above 39°C, and high degree of substrate specificity (*o*- and *p*-chlorobenzoate were not dechlorinated) (43, 95).

Recently, Dolfing & Tiedje (20) established a defined 3-chlorobenzoate-degrading consortium consisting of the key organisms from the above-

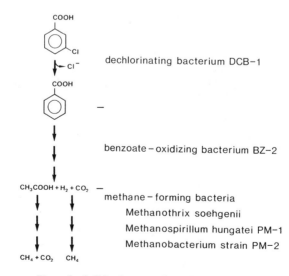

Figure 2   3-Chlorobenzoate-degrading food chain.

mentioned consortium, i.e. the dechlorinating organism (DCB-1), the benzoate degrader (BZ-1), and the lithotrophic methanogen (*Methanospirillum* strain PM-1). The chlorine released from the aromatic ring was recovered in stoichiometric amounts as chloride ion. The reducing power required for reductive dechlorination was obtained from the hydrogen produced in the acetogenic oxidation of benzoate. One third of this hydrogen was consumed via the reductive dechlorination, while two thirds was left to the methanogen.

Reductive dechlorination has also been shown for 2,4,5-trichlorophenoxyacetic acid (2,4,5-T), chlorophenols, and 1,2,4-trichlorobenzene (1,2,4-TCB). A methanogenic consortium grown on 3-chlorobenzoate dechlorinated 2,4,5-T at the *para* position to form 2,5-dichlorophenoxyacetic acid (2,5-D) (96). These microorganisms did not metabolize 2,5-D or several other chlorinated phenoxyacetic acids. However, when 2,4,5-T was incubated in sludge samples the first conversion was to 2,4,5-trichlorophenol (2,4,5-TCP), whereas the initial event observed in pond sediment or methanogenic aquifer samples was dehalogenation (38). The cleavage of the ether bond is consistent with the results of Mikesell & Boyd (63), who found trichlorophenol as the initial metabolic product when 2,4,5-T was incubated in sewage sludge.

Boyd & Shelton and their colleagues (7, 8) investigated the anaerobic degradation of mono- and dichlorophenol isomers by fresh sludge and by sludge acclimated to either 2-, 3-, or 4-chlorophenol. In unacclimated sludge, each of the monochlorophenol isomers was degraded. The rates of disappearance were in the order: *ortho* > *meta* > *para*. For the dichlorophenols, reductive dechlorination of the chlorine-group *ortho* to phenolic OH was observed. The respective monochlorophenol compounds released were subsequently degraded. 3,4-Dichlorophenol (3,4-DCP) and 3,5-DCP were persistent. Specific cross-acclimation patterns were observed for monochlorophenol degradation. Sludge acclimated to 2-chlorophenol (2-CP) cross-acclimated to 4-CP but did not utilize 3-CP. This sludge also degraded 2,4-DCP. Sludge acclimated to 3-CP cross-acclimated to 4-CP but not to 2-CP. This sludge degraded 3,4-DCP and 3,5-DCP but not 2,3-DCP or 2,5-DCP. The data indicated that two unique microbial activities exist that are not present in fresh sludge. The active microbial population in the 4-CP-acclimated sludge appeared to be a mixture of both populations present in the 2-CP- and 3-CP-acclimated sludges, because it was able to degrade all three monochlorophenol isomers and 2,4-DCP and 3,4-DCP. ^{14}C-Labeled 4-CP, 2-CP, and 2,4-DCP were converted to ^{14}CH$_4$ and ^{14}CO$_2$. Even pentachlorophenol (PCP) was completely dechlorinated by a mixture of the 2-CP-, 3-CP-, and 4-CP-acclimated sludges (64). With repeated PCP additions, 3,4,5-trichlorophenol (3,4,5-TCP), 3,5-DCP, and 3-CP accumulated. All chlorinated compounds disappeared after the PCP additions were stopped.

When sludge was incubated with [^{14}C]PCP, 66% of the added ^{14}C was mineralized to ^{14}CO$_2$ and ^{14}CH$_4$.

The proposed PCP degradation pathway, based on the sequential appearance and disappearance of 3,4,5-TCP, 3,5-DCP, and 3-CP (Figure 3), appeared to result from the relatively higher rate of PCP dechlorination by the 2-CP-acclimated sludge. This sludge rapidly removed chlorine from positions 2 and 6 of PCP to give 3,4,5-TCP. The *para*-chlorine was then removed by populations present in the 2-CP- or the 4-CP-acclimated sludge or both, which have been shown to dechlorinate this position (8). 3,5-DCP and 3-CP were probably dechlorinated by the 3-CP-acclimated sludge.

Tsuchiya & Yamaha (102, 103) isolated *Staphylococcus epidermidis* from the intestinal contents of rats. Whole cells converted 1,2,4-TCB to *o*-dichlorobenzene, which was further converted to monochlorobenzene. These conversions, which resulted also from *m*- and *p*-dichlorobenzene, proceeded only in a hydrogen atmosphere. Dried and broken cells also maintained the dechlorinating activity, which was stimulated by the addition of NADPH.

Displacement of Halogen by Hydroxyl

The first lines of evidence implicating replacement of chlorine from the aromatic ring through hydroxyl were presented by Johnston et al (48). They isolated a *Pseudomonas* species capable of utilizing 3-chlorobenzoate as the sole carbon source. Manometric studies showed that cells grown on either benzoate or 3-chlorobenzoate oxidized 3-chloro-, 3-hydroxy-, and 2,5-dihydroxybenzoate without a lag period. During growth with 3-chlorobenzoate, 3-hydroxy- and 2,5-dihydroxybenzoate were excreted into the culture medium. From these data the authors proposed a degradative sequence with chloride elimination in the first step and 3-hydroxybenzoate as the reaction product.

In 1975 Chapman (14) isolated *Micrococcus* spp. able to grow with 4-chlorobenzoate. Degradation apparently proceeded via 4-hydroxybenzoate and protocatechuate (Figure 4). The same degradative sequence was also reported for an *Arthrobacter* sp. (80), a *Nocardia* sp. (52), a *Pseudomonas* sp. (53), and *Arthrobacter globiformis* (106, 107). Chlorinated benzoates other than 4-chlorobenzoate did not support growth. The dechlorinating enzymes of the *Pseudomonas* sp. and *A. globiformis* were inducible by 4-chlorobenzoate but not by 4-hydroxybenzoate. The enzymes from the

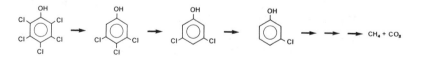

Figure 3 Proposed pentachlorophenol pathway.

Micrococcus sp. and *Pseudomonas* sp. strain CBS3 converted 4-bromobenzoate, whereas 4-fluorobenzoate was not accepted as a substrate. In contrast, an *Arthrobacter* sp. isolated by Marks et al (59) was able to dehalogenate 4-chloro-, 4-fluoro-, and 4-bromobenzoate, although fluorobenzoate did not support growth. Marks et al (59) were able to prepare an active cell extract from this organism, and it converted 4-chlorobenzoate at approximately 5% of the rate observed for whole cells. The dehalogenase activity had an optimum pH of 6.8 and an optimum termperature of 20°C. It was inhibited by dissolved oxygen and stimulated by Mn^{2+}. Unlike the other cell-free aromatic dehalogenase system reported by Klages & Lingens (53), this system was not stimulated by the addition of Fe^{2+}. The apparent Michaelis constant of the cell-free extract for 4-chlorobenzoate was 30 μM. The enzyme also appeared to be highly specific for *para*-substituted monohalobenzoates.

Additional data were recently reported for the dechlorinating enzyme from *Pseudomonas* sp. strain CBS3 (98) partially purified by ammonium sulfate fractionation. The pH optimum was between 7.0 and 7.5. The K_m values for 4-chloro-, 4-bromo-, and 4-iodobenzoate were found to be 0.15, 0.068, and 0.12 mM, respectively. In contrast to the activity reported for the cell-free extract of Klages & Lingens (53), this activity could not be increased through the addition of Fe^{2+}.

The mechanism of the dehalogenation process has recently been clarified by labeling experiments using $^{18}O_2$ and $H_2^{18}O$ (60, 67). The data indicate that the dechlorination reaction utilizes water as the hydroxyl donor and not molecular oxygen. The results showed that the enzymatic conversion of 4-chlorobenzoate to 4-hydroxybenzoate proceeds via a hydrolytic cleavage of the carbon-chlorine bond.

Replacement of a chlorine substituent by a hydroxy group has recently also been shown for the degradation of pentachlorophenol by use of mutants of an aerobic, chlorophenol-utilizing *Flavobacterium* sp. (93). The pathway for PCP degradation was initiated by the conversion of PCP to tetrachloro-*p*-hydroquinone. Labeling experiments using $H_2^{18}O$ and $^{18}O_2$ demonstrated that the initial dechlorination of PCP proceeded by a hydrolytic displacement of chlorine, rather than by an oxygenase-catalyzed mechanism. Two reductive dechlorinations of tetrachloro-*p*-hydroquinone followed, yielding first trichlorohydroquinone and then 2,6-dichlorohydroquinone. These results are

Figure 4 Hydrolytic dechlorination of 4-chlorobenzoate.

in agreement with some pathway intermediates proposed earlier by Suzuki (97) and Reiner et al (79). Chlorine is also hydrolytically displaced from PCP by *Rhodococcus chlorophenolicus* PCP-I (2). However, the novel hydroxy group occurred in position 4 whether or not a substrate had chlorine substituents in this position. This indicates that the enzymes only fortuitously function as a dehalogenating enzyme. The *para*-hydroxylation, although considered a hydrolase reaction, required the presence of molecular oxygen. Further metabolism of the reaction product tetrachloro-*p*-hydroquinone was claimed to proceed under anaerobiosis.

Oxygenolytic Halogen-Carbon Bond Cleavage

Fortuitous dehalogenation by dioxygenases is another mechanism to remove halogen substituents from haloaromatic compounds. Goldman et al (39) reported such oxygenolytic dehalogenation of an arylhalide in investigating 2-fluorobenzoate metabolism by a pseudomonad. This organism metabolized 2-fluorobenzoate by two pathways (65) because nonselective dioxygenation by the benzoate 1,2-dioxygenase generated a mixture of 2- and 6-fluoro-1,2-dihydro-1,2-dihydroxybenzoate (Figure 5). Some 85% of the 2-fluorobenzoate underwent defluorination to produce catechol, which was degraded through the 3-oxoadipate pathway. The other pathway, which accounted for the remaining 15% of the 2-fluorobenzoate utilized by the organism, proceeded through 3-fluorocatechol to 2-fluoro-*cis,cis*-muconate, both of which were isolated from the growth medium. 2-Fluoro-*cis,cis*-muconate was not further metabolized.

The nonenzymatic nature of the elimination of fluorine from 2-fluoro-1,2-dihydro-1,2-dihydroxybenzoate was clearly demonstrated with a mutant of *Alcaligenes eutrophus* that was defective in the dihydrodihydroxybenzoate dehydrogenase (78), but able to use 2-fluorobenzoate as the growth substrate (27). Of the originally covalently bound fluorine, 80% was released as fluoride. The remaining 20% of the substrate utilized was identified as 6-fluoro-1,2-dihydro-1,2-dihydroxybenzoate. As fluorine is a leaving group, it is eliminated spontaneously from 2-fluoro-1,2-dihydro-1,2-dihydroxybenzoate as an anion. Decarboxylation of the resulting β-keto acid gives cate-

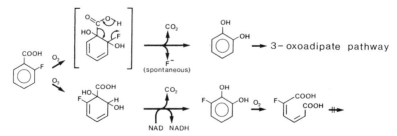

Figure 5   Degradation pathway for 2-fluorobenzoate.

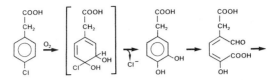

Figure 6   Reaction sequence in the degradation of 4-chlorophenylacetate.

chol. The bacterial defluorination of 2-fluorobenzoate is therefore a fortuitous event. The reaction product of the rearomatization, catechol, can then be degraded by the usual catabolic enzymes of the 3-oxoadipate pathway.

An analogous type of degradation sequence was suggested in the mineralization of 2-chlorobenzoate by *Pseudomonas cepacia* (108).

Oxygenolytic elimination from a *cis*-dihydrodiol produced by dioxygenation probably accounts for the initial dehalogenation of 4-halophenylacetates by *Pseudomonas* sp. strain CBS3 (54). This explanation is in accord with the organism's ability to use phenylacetate, 4-fluoro-, 4-chloro-, and 4-bromophenylacetate as the sole carbon source. Degradation apparently proceeded via homoprotocatechuate, which was further metabolized by a *meta*-cleavage pathway (Figure 6). On fractionation of a crude extract on Sephacryl S-200 the dehalogenating enzyme activity could be separated into two components, which were both necessary for the reaction. The highest activity was obtained in the presence of cofactors Fe^{2+} and NADH (61). One component of this system was purified and characterized (62).

A novel mechanism for expelling halogen by a dioxygenase has been shown recently with a bacterial strain isolated for growth on 5-chlorovanillate (49). Cell suspensions of this strain grown on 5-chlorovanillate released chloride quantitatively and readily oxidized 5-chloroprotocatechuate, a proposed metabolite of 5-chlorovanillate. An inducible protocatechuate 4,5-dioxygenase further converted 5-chloroprotocatechuate. Ring opening is followed by a nucleophilic displacement of chloride from the acylchloride (Figure 7). This reaction does not require an additional cyclizing enzyme to give 2-pyrone-4,6-dicarboxylate. In contrast, this last metabolite arises from the natural compound protocatechuate by the action of the 4,5-dioxygenase

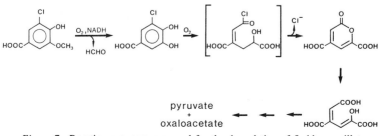

Figure 7   Reaction sequence proposed for the degradation of 5-chlorovanillate.

and further oxidation of the resulting aldehyde by a specific dehydrogenase. Partially purified protocatechuate 4,5-dioxygenase and pure protocatechuate 4,5-dioxygenase from *Pseudomonas testosteroni* showed similar substrate specificities. 5-Chloro-, 5-bromo-, and 5-fluoroprotocatechuate were mainly converted to 2-pyrone-4,6-dicarboxylate.

Chloride Eliminations From Nonaromatic Intermediates

The biochemistry of the biodegradation of chlorinated benzoates, benzenes, anilines, and phenoxyacetates, involving chlorinated intermediates prior to the dechlorination step, has recently been investigated. A common feature of these pathways is the elimination of chloride after *ortho* cleavage of chlorocatechols. Chloride appears to be eliminated spontaneously after the carbon halogen bond has been labilized through isomerases or reductases.

PERIPHERAL ENZYME SEQUENCES GENERATING CHLOROCATECHOLS
Reaction sequences that describe the formation of chlorocatechols from chlorinated aromatics are shown in Figure 8. Various soil bacteria have been reported to cleave the ether linkage of phenoxyacetate and to produce 2,4-dichlorophenol from 2,4-D, 2-methyl-4-chlorophenol from 2-methyl-4-chlorophenoxyacetate (MCPA), and 4-chlorophenol from 4-chlorophenoxyacetate (5, 37, 58). Experiments with resting cells and cell-free extracts from bacteria grown on phenoxyacetate showed that ether bond cleavage requires oxygen (41, 99). Oxygen is incorporated into the side chain so that glyoxylate is released. The enzymatic production of 4-chlorocatechol from 4-chlorophenol and of 3,5-dichlorocatechol from 2,4-dichlorophenol required both oxygen and NADPH (6). 3-Methyl-5-chlorocatechol is the product of hydroxylation of 2-methyl-4-chlorophenol.

Recently, phenol hydroxylases have been purified from 2,4-D–degrading bacteria such as an *Acinetobacter* sp., a *Pseudomonas putida* strain, and *Alcaligenes eutrophus* 335. The 2,4-D plasmid pJP4–encoded hydroxylase is a true 2,4-dichlorophenol hydroxylase because it readily converted 2,4-dichlorophenol, 2,4-dibromophenol, and 2-methyl-4-chlorophenol to the corresponding catechols. 4-Chlorophenol and 2-chlorophenol were hydroxylated at lower rates, and phenol was not hydroxylated (57). The substrate specificity of the pJP4–encoded hydroxylase differed from that of the *Acinetobacter* sp., for which only 2,4-dichlorophenol, 4-chloro-2-methylphenol, and 4-chlorophenol were found to be true substrates. Other substituted phenols evoke the oxidation of ÑAD(P)H and oxygen consumption without undergoing hydroxylation. Instead, the product is hydrogen peroxide, which suggests that electron flow is uncoupled from hydroxylation in the presence of these compounds (4).

The productive degradation of chloroanilines also proceeds via chlor-

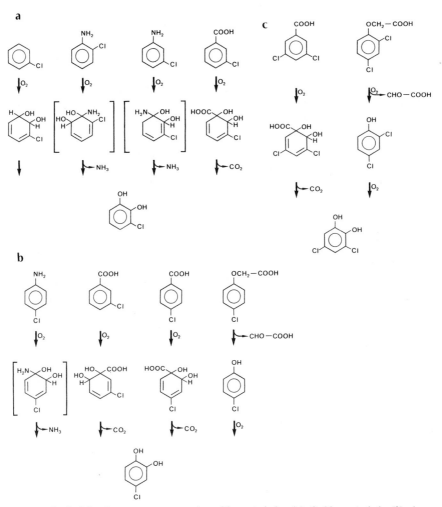

Figure 8 Peripheral sequences generating chlorocatechols; (*a*) 3-chlorocatechol, (*b*) 4-chlorocatechol, (*c*) 3,5-dichlorocatechol.

ocatechols (55, 109, 110). In the catabolism of aniline (3) ammonia is liberated through a dioxygenase reaction. In *Pseudomonas* sp. strain JL4, which uses 2-, 3-, or 4-chloroaniline as the growth substrate, 2-chloroaniline is subject to 1,6-dioxygenation, which yields 3-chlorocatechol (55). This compound is also generated from 3-chloroaniline. Dioxygenation of 4-chloroaniline results in the formation of 4-chlorocatechol. Aniline-utilizing cells of *Rhodococcus* sp. strain An 117, however, hydroxylated 3-chloroaniline in the 1,6 position to give 4-chlorocatechol (85).

Furukawa et al (30–34) have studied the conversion of various isomers of

PCB by *Alcaligenes* sp. and *Acinetobacter* sp. grown on biphenyl. A probable pathway for the degradation is shown in Figure 9. This pathway was proposed based on the detected accumulated metabolites and in analogy to the known pathway for biphenyl (11–13, 101). A *cis*-dihydrodiol is formed and is dehydrogenated to yield a 2',3'-dihydroxybiphenyl; *meta*-cleavage of this intermediate generates chlorinated derivatives of 2-hydroxy-6-oxo-6-phenyl-hexa-2,4-dienoates, which are further metabolized, yielding chlorinated benzoates and 2-hydroxypenta-2,4-dienoate.

The reaction sequences involved in the conversion of chlorinated benzoates into the respective catechols proceed via chlorosubstituted 1,2-dihydro-1,2-dihydroxybenzoates as intermediates (72, 73). This conversion has mostly been studied with the 3-chlorobenzoate-utilizing *Pseudomonas* strain B13 and its derivative strains. Nonselective dioxygenation generated a mixture of 3- and 5-chloro-1,2-dihydro-1,2-dihydroxybenzoate; 67% of the 3-chlorobenzoate was degraded through 3-chlorocatechol and the remaining 33% through 4-chlorocatechol (72). The benzoate 1,2-dioxygenase in *Pseudomonas* strain B13 is rather specific; only benzoate and 3-chlorobenzoate were converted at a considerable rate. This enzyme is unable to oxidize benzoates containing a chlorine substituent in the *ortho* or *para* position. In this respect the B13 dioxygenase is similar to that in benzoate-utilizing bacteria of many different genera (1, 10, 18, 44–46, 92, 104, 105). However, other benzoate 1,2-dioxy-genases with different regioselectivities have been described. Dioxygenation in the 1,6 position was found with an enzyme from an *Arthrobacter* sp. (44), so that only 4-chlorocatechol resulted from 3-chlorobenzoate. In contrast, a *Pseudomonas aeruginosa* strain accumulated solely 3-chlorocatechol from 3-chlorobenzoate (46). Benzoate 1,2-dioxygenases with considerably relaxed substrate specificity function in derivatives of *Pseudomonas* strain B13 containing the TOL plasmid and in *Pseudomonas* sp. strain WR912 (40, 71, 74, 75). Both 3- and 4-chlorobenzoate and even 3,5-dichlorobenzoate were turned over at a considerable rate, yielding 4-chloro- and 3,5-dichloro-1,2-dihydro-1,2-dihydroxybenzoate and finally 4-chloro- and 3,5-dichlorocatechol, respectively.

The metabolism of chlorobenzene to 3-chlorocatechol was described for

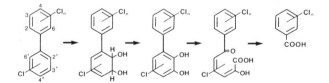

Figure 9 Proposed major metabolic sequences for the degradation of polychlorinated biphenyl (*n* = 1–4).

strain WR1306 (76). It proceeds via 3-chloro-1,2-dihydro-1,2-dihydroxyben-zene as an intermediate. The benzene 1,2-dioxygenase was found to be rather specific, since only benzene, chlorobenzene, and bromobenzene were substrates. Fluorobenzene and dichlorobenzenes were not oxidized. Ana-logous sequences with a dioxygenation *ortho* and *para* to a chloro-substi-tuent were recently observed for the degradation of 1,3- and 1,4-dichloroben-zene, yielding 3,5- and 3,6-dichlorocatechol, respectively (19, 68, 84, 91).

ASSIMILATION OF CHLOROCATECHOLS   In 2,4-D- and 3-chlorobenzoate-degrading bacteria catechol 1,2-dioxygenases were found, which exhibit high activities with mono- and dichlorocatechols (21–23). The products from *ortho*-cleavage of 3- and 4-chlorocatechol were 2- and 3-chloro-*cis,cis*-muconates, respectively (83). For the cycloisomerization of 2-chloro-*cis,cis*-muconate by enzyme preparations from *Pseudomonas* strain B13 grown on 3-chlorobenzoate, Schmidt & Knackmuss (82) proposed 4-carboxychlorome-thylbut-2-en-4-olide as an intermediate; this intermediate spontaneously generated *trans*-4-carboxymethylenebut-2-en-4-olide by *anti* elimination of hydrogen chloride (Figure 10). Identification of *cis*-4-carboxymethylenebut-2-en-4-olide as the enzymatic cycloisomerization product of 3-chloro-*cis,cis*-muconate suggested 4-chloro-4-carboxymethylbut-2-en-4-olide as an in-termediate, from which hydrogen chloride could be released spontaneously by *anti* elimination. Both *cis* and *trans* isomers of 4-carboxymethylene-but-2-en-4-olide were converted into maleylacetate by a 4-carboxymethylene-but-2-en-4-olide hydrolase. Cell-free extracts of B13 grown on 3-chloroben-zoate catalyzed the total degradation of 3,5-dichlorocatechol (86). On the basis of separated enzyme activities, the catabolic pathway of 3,5-dichlorocatechol was proposed to proceed via 2,4-dichloromuconate, *trans*-2-chloro-4-carboxymethylenebut-2-en-4-olide, *cis*-2-chloro-4-carboxymethyl-enebut-2-en-4-olide, β-chloromaleylacetate, and maleylacetate.

Previously Bollag et al (6) used an enzyme preparation that liberated stoichiometric amounts of chloride from 4-chlorocatechol and showed that 4-carboxymethylenebut-2-en-4-olide was the reaction product. Enzymes iso-lated from an *Arthrobacter* sp. catalyzed the conversion of 4-chloro- and 3,5-dichlorocatechol to 3-chloro- and 2,4-dichloro-*cis,cis*-muconate, respectively (100). Sharpee et al (88) prepared 2-chloro-4-carboxymethylenebut-2-en-4-olide from 2,4-dichloro-*cis,cis*-muconate with a partially purified lactonizing enzyme from an *Arthrobacter* sp. grown on 2,4-D. The 2-chloro-4-carboxy-methylenebut-2-en-4-olide and the unsubstituted 4-carboxymethylenebut-2-en-4-olide, the corresponding chlorinated *cis,cis*-muconates, and the chloro-catechols were converted enzymatically and yielded identical products, which were tentatively identified as β-chloromaleylacetate and maleylacetate, respectively (100). On the basis of isolated metabolites and induction experiments, Evans et al (29) reported the same maleylacetate pathway for the degradation of 4-chlorocatechol in a pseudomonad capable of utilizing 4-chlorophenoxyacetate. An *ortho*-fission enzyme that converted 3,5-dichlorocatechol into 2,4-dichloro-*cis,cis*-muconate (28) was reported in cell-free extracts of *Pseudomonas* sp. strain NCIB 9340 grown on 2,4-D. Enzymatic lactonization yielded 2-chloro-4-carboxymethylenebut-2-en-4-olide with release of chloride. Further degradation due to a hydrolyzing enzyme gave β-chloromaleylacetate. Degradation of 3-methyl-5-chlorocatechol, the intermediate in MCPA metabolism, was found to proceed through 2-methyl-4-chloro-*cis,cis*-muconate. Further metabolism to 2-methyl-4-carboxymethylenebut-2-en-4-olide followed by hydrolysis gave β-methylmaleylacetate (Figure 10). Sequences involved in the degradation of 3-chloro-, 4-chloro-, 3,5-dichloro-, and 3-methyl-5-chlorocatechol are summarized in Figure 10.

A maleylacetate reductase and a 3-oxoadipate succinyl-CoA transferase from *Pseudomonas* strain B13 grown on 3-chlorobenzoate and strain WR1306 grown on chlorobenzene were found to metabolize maleylacetate to 3-oxoadipate and 3-oxoadipyl-CoA (Figure 11). Enzymes isolated from the *Arthrobacter* sp. converted maleylacetate and β-chloromaleylacetate to succinate (24) by consumption of two equivalents of NADH or NADPH. Be-

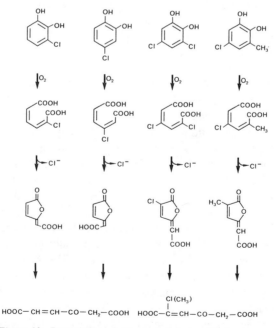

Figure 10 Degradation of chlorocatechols to maleylacetates.

cause succinate was also formed enzymatically from chlorosuccinate, Dux-bury et al (24) hypothesized that β-chloromaleylacetate is degraded via 2-chloro-4-oxoadipate and chlorosuccinate to produce succinate (Figure 11). Succinate was also found as a product of 2,4-D when ring-labeled 2,4-D was metabolized by a soluble enzyme preparation (100).

Chapman (15) proposed an alternative mechanism (Figure 11) for the degradation of β-chloromaleylacetate. He observed that a maleylacetate re-ductase that normally funtions in the degradation of resorcinol (16) was able to convert β-chloromaleylacetate to 3-oxoadipate and chloride. In accordance with this mechanism, two moles of reduced nicotinamide nucleotide were consumed per mole of substrate.

In *Pseudomonas* strain B13 and derivative strains, catechol and chlor-ocatechols were assimilated via two separate *ortho*-cleavage pathways. Cor-respondingly, two types of isofunctional enzymes for ring fission were found. Pyrocatechase type I, highly specific for catechol, was present in cells grown on benzoate. This enzyme, together with the isofunctional enzyme pyro-catechase type II, which exhibited relaxed specificities and high activities for the chlorosubstituted substrates, was induced when 3-chlorobenzoate was the growth substrate (22, 23). As has been shown with derivatives of strain B13 harboring the TOL plasmid, the relative amounts of these two isoenzymes induced are dependent on the mode of chlorine substitution of the aromatic growth substrate (75). Cell-free extracts from cells grown on 4-chloroben-

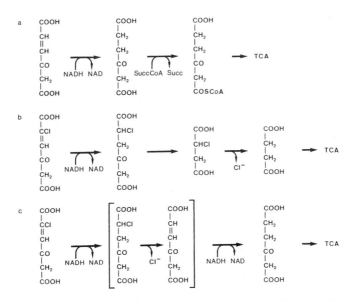

Figure 11 Proposed pathways for the degradation of maleylacetate (*a*) and β-chloromaleylacetate (*b* and *c*).

zoate contained considerably higher amounts of pyrocatechase type II than extracts from cells grown on 3-chloro- or 3,5-dichlorobenzoate. Both types of *ortho*-cleaving enzymes were also induced in chloroaniline-degrading derivatives of *Pseudomonas* strain B13 (55) and the chlorophenol-degrading *Alcaligenes* strain A7-2, which had acquired the genes encoding the maleylacetate pathway from strain B13 (87).

Two types of isofunctional enzymes were also found for cycloisomerization of *cis,cis*-muconate and *cis,cis*-chloromuconates. Cycloisomerase type I, functioning in the 3-oxoadipate pathway, is highly specific for *cis,cis*-muconate, whereas cycloisomerase type II has high activity for *cis,cis*-muconate, 2-chloromuconate, and 3-chloromuconate (82). In contrast to the ring-fission enzymes, cycloisomerase type I is only induced in *Pseudomonas* strain B13 grown on benzoate and cycloisomerase type II is only induced in B13 grown on 3-chlorobenzoate.

Cycloisomerization of *cis,cis*-muconate gave 4-carboxymethylbut-2-en-4-olide, while 4-carboxymethylenebut-2-en-4-olides were formed from chloro-substituted *cis,cis*-muconates. The latter butenolides resulted from chloride elimination that occurred during or after cycloisomerization. The subsequent enzymes in the maleylacetate pathway cannot function in the 3-oxoadipate pathway because of the different reduction states of the respective metabolites (82). The 4-carboxymethylbut-2-en-4-olide was further transformed by an isomerase to 4-carboxymethylbut-3-en-4-olide prior to delactonization. The delactonizing enzymes in both pathways forming the dicarboxylic acids are not isoenzymes because they exhibit no metabolic cross-activities (82). 4-Carboxymethylenebut-2-en-4-olide hydrolase, which was only induced during growth on 3-chlorobenzoate, was highly specific for its substrate and exhibited no activity for 4-carboxymethylbut-3-en-4-olide, the corresponding metabolite of the 3-oxoadipate pathway. Similarly, 4-carboxymethylbut-3-en-4-olide hydrolase did not accept 4-carboxymethylenebut-2-en-4-olide as a substrate.

3-Oxoadipate is a common metabolite of the normal *ortho* pathway, the resorcinol pathway (35, 36), and the modified *ortho* pathway of chlorocatechol and 3,5-dichlorocatechol assimilation. Maleylacetate reductase functions in the convergence of these pathways by introducing reduction equivalents into metabolites of the more highly oxidized substrates, resorcinol and chlorocatechols.

ISOLATION AND APPLICATION OF HALOAROMATIC-DEGRADING BACTERIA

Isolation of Haloaromatic-Degrading Strains

In four billion years microorganisms have evolved an extensive range of enzymes, pathways, and control mechanisms in order to be able to degrade a wide array of naturally occurring aromatic compounds. In contrast, the list of pure, biochemically well characterized cultures able to grow at the expense of haloaromatics is short. Haloaromatic-assimilating microbial strains have been obtained by (*a*) enrichment from nature, (*b*) in vivo genetic manipulation, and (*c*) in vitro genetic engineering.

The theory of enrichment culture is simple. The haloaromatic compound to be degraded is supplied as the growth-limiting and usually sole source of an essential nutrient in a culture medium. Of the many organisms added at the start of the experiment, only those with the necessary degradative ability will grow significantly under these conditions. Enrichment cultures grown on chloroaromatics often require several months for isolation. This indicates that besides selection genetic events might be involved. For the degradation of some chlorinated aromatics natural gene exchange has to occur.

An important prerequisite for the construction of bacterial strains capable of degrading novel chlorinated aromatic compounds is the recognition and acquisition of genes coding an enzyme sequence able to convert chlorinated aromatics to the respective chlorocatechols. Genes allowing total degradation of chlorocatechols are borne on some transmissible plasmids (summarized in 70).

The first report of in vivo construction of a catabolic pathway for the mineralization of chlorinated aromatics using external genetic information for the acquisition of a novel phenotype described work with *Pseudomonas* strain B13 and *Pseudomonas putida* mt-2 and the novel growth substrate 4-chlorobenzoate. Strain B13 was isolated by enrichment culture with 3-chlorobenzoate. It oxidizes 3-chlorobenzoate to 3- and 4-chlorocatechol (see above) and uses the maleylacetate pathway for further breakdown. Strain B13 is unable to utilize 4-chlorobenzoate, since the benzoate 1,2-dioxygenase has a very narrow specificity and will not accept 4-chlorobenzoate as a substrate. However, strain B13 can oxidize 4-chlorocatechol, the expected metabolite in the degradation of 4-chlorobenzoate. The benzoate 1,2-dioxygenase in *Pseudomonas putida* mt-2 determined by the TOL plasmid has a broader specificity than the B13 enzyme and can accept 4-chlorobenzoate as a substrate (72).

The primary transconjugant from the mating between strains mt-2 and B13 grew on 3-chlorobenzoate but not on 4-chlorobenzoate, although it was able to use *m*-toluate, a substrate for the enzyme sequence determined by the TOL plasmid. Transfer of the TOL plasmid from *Pseudomonas putida* mt-2 to *Pseudomonas* sp. strain B13 was not sufficient to produce a strain that could utilize 4-chlorobenzoate for growth, even though this compound is oxidized by the TOL-encoded benzoate 1,2-dioxygenase. However, derivatives of *Pseudomonas* strain WR211 such as WR216 were obtained by spontaneous mutation on plates containing 4-chlorobenzoate. These strains had lost the ability to grow on *m*-toluate because of an insertion found in the catechol 2,3-dioxygenase gene, *xylE* (47).

Chatterjee & Chakrabarty (17) followed the same procedure using the TOL plasmid and the plasmid pAC25, which codes the chlorocatechol-degrading enzyme sequence, to isolate strains that could grow on 4-chlorobenzoate and 3,5-dichlorobenzoate. In addition to the transfer of external genetic information, mutational events have to occur before a novel compound can be used as a growth substrate and misrouting of 4-chlorocatechol can be avoided. Several authors have described similar experiments in which the range of substrates utilized by bacterial isolates has been extended.

Schwien & Schmidt (87) transferred the genes coding the maleylacetate pathway from *Pseudomonas* strain B13 to *Alcaligenes* strain A7, which is able to grow on phenol. The transconjugant *Alcaligenes* strain A7-2 could utilize all three isomeric chlorophenols, which are not attacked by any of the parent strains. A similar transfer to *Pseudomonas* strain WR401, which is capable of growing on methylsalicylate but not on chlorosalicylates, created transconjugants that could use the latter compounds for growth (77). Transfer of the genes coding the maleylacetate pathway to an aniline-degrading *Pseudomonas* strain, JL1, allowed the isolation of chloroaniline-degrading bacteria (55). The manifestation of the same catabolic sequence of strain B13 in *Pseudomonas putida* F1 (degrading benzene) led to strains that could use chlorobenzene as the sole source of carbon and energy (68).

The strategy of combining genes of separate pathways to produce hybrid pathways could also be applied to the use of cloned catabolic genes. This approach has some advantages over conventional crosses for the construction of strains with novel catabolic properties. (*a*) Specific and well-characterized cloned DNA fragments can be used. (*b*) The use of DNA fragments carrying only essential genes avoids the introduction of unproductive enzymes and the consequent need for inactivation of the corresponding genes by mutation. However, this approach has only been shown for the degradation of chlorosalicylate and chlorobenzoate. Lehrbach et al (56) cloned the *xylD* gene alone and the *xylD* and *xylL* genes together (encoding the toluate 1,2-dioxygenase and the dihydrodihydroxytoluate dehydrogenase) into *Pseudomonas* strain B13 to extend the chlorinated benzoates utilized. The introduction of the cloned *xylD* gene alone allowed strain B13 to degrade 4-chlorobenzoate, while the introduction of both the cloned *xylD* and *xylL* genes, followed by spontaneous mutational divergence, resulted in emergence of B13 variants that could also degrade 3,5-dichlorobenzoate. Lehrbach et al (56) also cloned the naphthalene-degradative gene *nahG*, which encodes a broad-specificity salicylate hydroxylase (66), from the plasmid NAH7 and introduced the cloned gene into strain B13. Expression of *nahG* enabled B13 to completely mineralize novel substrates such as chlorosalicylates.

Application of Degradative Strains

Certain xenobiotics, particularly those with polychlorinated aromatic rings, are not known to be growth substrates but are nevertheless subject to co-metabolism. Recent work (9) demonstrated that degradation of PCB in soil was not enhanced by addition of a PCB-cometabolizing, biphenyl-utilizing *Acinetobacter* strain. Enhancement of both substrate disappearance and mineralization was only brought about by the addition of the substrate analog, biphenyl. This substrate-analog enrichment selectively increased the number of biphenyl degraders and thereby the cometabolic activity within the indigenous microflora. Upon depletion of the (co)substrate the number of biphenyl oxidizers declined exponentially.

Xenobiotics present in soil at rather high concentrations may be mineralized by laboratory strains that grow at the expense of the contaminant. Edgehill & Finn (26) have shown that soil was more rapidly cleared of the wood-preserving chemical PCP if inoculated with specialized *Arthrobacter* cells.

In another study, soil contaminated with a high concentration of 2,4,5-T was repeatly inoculated with *Pseudomonas cepacia* strain AC1100, which is capable of utilizing 2,4,5-T as the sole source of carbon and energy (50, 51). The contaminated soil samples showed more than 90% degradation after six treatments with strain AC1100, while the concentration of the contaminant remained unchanged in the uninoculated soil. Although strain AC1100 was effective in removing most of the 2,4,5-T from soil, the residual concentration of the herbicide or its metabolites was still high enough (2–10 μg g^{-1} soil) to impair plant growth. Apparently, some of the 2,4,5-T or its breakdown products became bound to soil particles and was unavailable for microbial degradation. Assessment of the long-term survival of AC1100 in uncontaminated soil indicated that its initial cell titer of 10^7 cells g^{-1} soil fell to an undetectable level within 12 wk of incubation. When 2,4,5-T was added AC1100 became detectable again after an initial lag of 2 wk. By use of a Nal^r mutant it was demonstrated that the 2,4,5-T-degrading capability was not transferred within the indigenous soil population.

Pertsova et al (69) have shown that in soil columns of chernozem soil inoculated with a *Pseudomonas putida* strain able to grow with 3-chlorobenzoate, continuously added 3-chlorobenzoate was totally degraded. The native microflora of the soil, however, was unable to degrade 3-chlorobenzoate, as shown by a breakthrough of 3-chlorobenzoate in the effluent. Taxonomic characteristics of 3-chlorobenzoate-utilizing bacteria isolated from the inoculated soil columns indicated that the 3-chlorobenzoate-degrading plasmid from *Pseudomonas putida* had been transferred in the course of incubation.

Many potentially hazardous pollutants are recalcitrant, highly toxic, or insoluble and thus escape degradation or disturb conventional treatment systems. Conventional biological treatment procedures are often inadequate for the removal of such pollutants from industrial wastewaters. A wide variety of halogenated aromatic compounds can be degraded by adapted microbial strains or defined mixed cultures. The use of such microorganisms for biotreatment of industrial wastewaters containing chlorinated aromatic pollutants requires extrapolation of laboratory studies to real-world treatment systems. Extrapolation is imprecise because only some of the biological and chemical real-world parameters that influence the biodegradation potential of laboratory strains can be assessed under laboratory conditions in well-controlled experiments. A major problem is the retention of a biomass with a specific haloaromatic-degradative capability or the establishment of a specific trait within the autochthonous microflora of a treatment-plant sludge.

The degradation of chlorophenols has been studied in laboratory and scaled-up systems by the addition of laboratory strains. By using a fill-and-draw reactor, domestic activated sludge was adapted to utilize 40 mg liter^{-1} of pentachlorophenol in a synthetic industrial waste. This procedure routinely required 6–7 days. To reduce the lag period another procedure was used, whereby PCP-degrading bacteria were inoculated directly into an operating activated sludge unit in the laboratory. Part of the mixed liquor was removed and replaced by an equivalent volume of a batch culture of *Arthrobacter* strain ATCC 33790 grown with PCP. The level of PCP was reduced from 40 mg liter^{-1} to less than 1 mg liter^{-1} in 1–2 day instead of 6–7. Once acclimated, the activated sludge was stable under laboratory conditions (25).

That degradative sequences for chloroaromatics can be established in activated sludge has been demonstrated with a mixture of the isomeric chlorophenols as model compounds and *Pseudomonas* strain B13 as an organism harboring the capability to assimilate the halosubstituted aromatic ring and to transfer this trait to other bacteria (81). A synthetic sewage containing alkanols, acetone, and phenol was readily degraded by two stably coexisting organisms, a methanol-degrading *Pseudomonas extorquens* strain and *Alcaligenes* strain A7, which can degrade phenol at high concentration in addition to ethanol, isopropanol, and acetone. Unsteady-state transient conditions were observed by increasing loads of chlorophenols. However, a culture augmented with an inoculum of *Pseudomonas* strain B13 grown on 3-chlorobenzoate developed a stable community that totally degraded high loads of chlorophenols. Proof of mineralization was provided by low dissolved organic carbon (DOC) in the cell-free effluent and by the elimination of equimolar amounts of chloride. Population analysis revealed that the establishment of the chlorocatechol-degrading capacity is more important than the number of viable cells of strain B13 for obtaining a functioning mixed culture. Hybrid organisms that emerged in the population, such as *Alcaligenes* strain A7-2, which had acquired the halocatechol-degrading genes from strain B13, were much more competitive with respect to phenol and chlorophenol utilization than strain B13.

To establish the potential for biodegrading chlorinated aromatic compounds as a stable property of the indigineous microflora of a biological treatment system, several requirements must be met. (*a*) The system must productively break down the chemical so that by increasing biomass the system can transiently respond to increasing loads. (*b*) Catabolic traits such as the assimilation of chlorocatechols within the microbial population must be transmissible, to ensure that through the recruitment of existing catabolic enzymes and gene regulators with appropriate effector/substrate specificities new hybrid pathways for haloaromatics can be generated. Thus in addition to physiological flexibility, the adaptability of the system to shifts of substrates such as congenerous compounds would be guaranteed.

Literature Cited

1. Alexander, M., Lustigman, B. K. 1966. Effect of chemical structure on microbial degradation of substituted benzenes. *J. Agric. Food Chem.* 14:410–13

2. Apajalahti, J. H. A., Salkinoja-Salonen, M. S. 1987. Dechlorination and *para*-hydroxylation of polychlorinated phenols by *Rhodococcus chlorophenolicus*. *J. Bacteriol.* 169:675–81

3. Bachofer, R., Lingens, F., Schäfer, W. 1975. Conversion of aniline into pyrocatechol by a *Nocardia* sp. Incorporation of oxygen-18. *FEBS Lett.* 50:288–90

4. Beadle, C. A., Smith, A. R. W. 1982. The purification and properties of 2,4-dichlorophenol hydroxylase from a strain of *Acinetobacter* species. *Eur. J. Biochem.* 123:323–32

5. Bollag, J.-M., Helling, C. S., Alexander, M. 1967. Metabolism of 4-chloro-2-methylphenoxyacetic acid by soil bacteria. *Appl. Microbiol.* 15:1393–98

6. Bollag, J.-M., Helling, C. S., Alexander, M. 1968. 2,4-D metabolism. Enzymatic hydroxylation of chlorinated phenols. *J. Agric. Food Chem.* 16:826–28

7. Boyd, S. A., Shelton, D. R. 1984. Anaerobic biodegradation of chlorophenols in fresh and acclimated sludge. *Appl. Environ. Microbiol.* 47:272–77

8. Boyd, S. A., Shelton, D. R., Berry, D., Tiedje, J. M. 1983. Anaerobic biodegradation of phenolic compounds in digested sludge. *Appl. Environ. Microbiol.* 46:50–54

9. Brunner, W., Sutherland, F. H., Focht, D. D. 1985. Enhanced biodegradation of polychlorinated biphenyls in soil by analog enrichment and bacterial inoculation. *J. Environ. Qual.* 14:324–28

10. Cain, R. B., Tranter, E. K., Darrah, J. A. 1968. The utilization of some halogenated aromatic acids by *Nocardia*. *Biochem. J.* 106:211–27

11. Catelani, D., Colombi, A. 1974. Metabolism of biphenyl. Structure and physicochemical properties of 2-hydroxy-6-oxo-6-phenylhexa-2,4-dienoic acid, the *meta*-cleavage product from 2,3-dihydroxybiphenyl by *Pseudomonas putida*. *Biochem. J.* 143:431–34

12. Catelani, D., Colombi, A., Sorlini, C., Trecanni, V. 1973. Metabolism of biphenyl. 2-Hydroxy-6-oxo-6-phenyl-hexa-2,4-dienoate: the *meta*-cleavage product from 2,3-dihydroxybiphenyl by *Pseudomonas putida*. *Biochem. J.* 134:1063–66

13. Catelani, D., Sorlini, S., Treccani, V. 1971. The metabolism of biphenyl by *Pseudomonas putida*. *Experientia* 27:1173–74

14. Chapman, P. J. 1975. Bacterial metabolism of 4-chlorobenzoic acid. *Abstr. Annu. Meet. Am. Soc. Microbiol.* 1975:192

15. Chapman, P. J. 1979. Degradation mechanisms. In *Microbial Degradation of Pollutants in Marine Environments*. *EPA-600/9-79-012*, ed. A. W. Bourquin, P. H. Pritchard, pp. 28–66. Gulf Breeze, Fla: US Environ. Prot. Agency. 552 pp.

16. Chapman, P. J., Ribbons, D. W. 1976. Metabolism of resorcinylic compounds

by bacteria: alternative pathways of resorcinol catabolism in *Pseudomonas putida*. *J. Bacteriol.* 125:985–98

17. Chatterjee, D. K., Chakrabarty, A. M. 1982. Genetic rearrangements in plasmids specifying total degradation of chlorinated benzoic acids. *Mol. Gen. Genet.* 188:279–85

18. Clarke, K. F., Callely, A. G., Livingstone, A., Fewson, C. A. 1975. Metabolism of monofluorobenzoates by *Acinetobacter calcoaceticus* N.C.I.B. 8250. Formation of monofluorocatechols. *Biochim. Biophys. Acta* 404: 169–79

19. De Bont, J. A. M., Vorage, M. J. A. W., Hartmans, S., van den Tweel, W. J. J. 1986. Microbial degradation of 1,3-dichlorobenzene. *Appl. Environ. Microbiol.* 52:677–80

20. Dolfing, J., Tiedje, J. M. 1986. Hydrogen cycling in a three-tiered food web growing on the methanogenic conversion of 3-chlorobenzoate. *FEMS Microbiol. Ecol.* 38:293–98

21. Dorn, E., Hellwig, M., Reineke, W., Knackmuss, H.-J. 1974. Isolation and characterization of a 3-chlorobenzoate degrading pseudomonad. *Arch. Microbiol.* 99:61–70

22. Dorn, E., Knackmuss, H.-J. 1978. Chemical structure and biodegradability of halogenated compounds. Two catechol 1,2-dioxygenases from a 3-chlorobenzoate–grown pseudomonad. *Biochem. J.* 174:73–84

23. Dorn, E., Knackmuss, H.-J. 1978. Chemical structure and biodegradability of halogenated aromatic compounds. Substituent effects on 1,2-dioxygenation of catechol. *Biochem. J.* 174:85–94

24. Duxbury, J. M., Tiedje, J. M., Alexander, M., Dawson, J. E. 1970. 2,4-D metabolism: enzymatic conversion of chloromaleylacetic acid to succinic acid. *J. Agric. Food Chem.* 18:199–201

25. Edgehill, R. U., Finn, R. K. 1983. Activated sludge treatment of synthetic wastewater containing pentachlorophenol. *Biotechnol. Bioeng.* 25:2165–76

26. Edgehill, R. U., Finn, R. K. 1983. Microbial treatment of soil to remove pentachlorophenol. *Appl. Environ. Microbiol.* 45:1122–25

27. Engesser, K.-H., Schmidt, E., Knackmuss, H.-J. 1980. Adaptation of *Alcaligenes eutrophus* B9 and *Pseudomonas* sp. B13 to 2-fluorobenzoate as growth substrate. *Appl. Environ. Microbiol.* 39: 68–73

27a. Evans, W. C., Fuchs, G. 1988. Anaerobic degradation of aromatic compounds. *Ann. Rev. Microbiol.* 42:289–317

28. Evans, W. C., Smith, B. S. W., Fernley, H. N., Davis, J. I. 1971. Bacterial metabolism of 2,4-dichlorophenoxyacetate. *Biochem. J.* 122:543–51

29. Evans, W. C., Smith, B. S. W., Moss, P., Fernley, H. N. 1971. Bacterial metabolism of 4-chlorophenoxyacetate. *Biochem. J.* 122:509–17

30. Furukawa, K., Matsumura, F., Tonomura, K. 1978. *Alcaligenes* and *Acinetobacter* strains capable of degrading polychlorinated biphenyls. *Agric. Biol. Chem.* 42:543–48

31. Furukawa, K., Tomizuka, N., Kamibayashi, A. 1979. Effect of chlorine substitution on the bacterial metabolism of various polychlorinated biphenyls. *Appl. Environ. Microbiol.* 38:301–10

32. Furukawa, K., Tomizuka, N., Kamibayashi, A. 1983. Metabolic breakdown of Kaneclors (polychlorobiphenyls) and their products by *Acinetobacter* sp. *Appl. Environ. Microbiol.* 46:140–45

33. Furukawa, K., Tonomura, K., Kamibayashi, A. 1978. Effect of chlorine substitution on the biodegradability of polychlorinated biphenyls. *Appl. Environ. Microbiol.* 35:223–27

34. Furukawa, K., Tonomura, K., Kamibayashi, A. 1979. Metabolism of 2, 4,4'-trichlorobiphenyl by *Acinetobacter* sp. P6. *Agric. Biol. Chem.* 43:1577–83

35. Gaal, A., Neujahr, H. Y. 1979. Metabolism of phenol and resorcinol in *Trichosporon cutaneum*. *J. Bacteriol.* 137:13–21

36. Gaal, A. B., Neujahr, H. Y. 1980. Maleylacetate reductase from *Trichosporon cutaneum*. *Biochem. J.* 185:783–86

37. Gaunt, J. K., Evans, W. C. 1961. Metabolism of 4-chloro-2-methyl-phenoxyacetic acid by a soil microorganism. *Biochem. J.* 79:25p–26p

38. Gibson, S. A., Suflita, J. M. 1986. Extrapolation of biodegradation results to groundwater aquifers: reductive dehalogenation of aromatic compounds. *Appl. Environ. Microbiol.* 52:681–88

39. Goldman, P., Milne, G. W. A., Pignataro, M. T. 1967. Fluorine containing metabolites formed from 2-fluorobenzoic acid by *Pseudomonas* species. *Arch. Biochem. Biophys.* 118:178–84

40. Hartmann, J., Reineke, W., Knackmuss, H.-J. 1979. Metabolism of 3-chloro-, 4-chloro-, and 3,5-dichlorobenzoate by a pseudomonad. *Appl. Environ. Microbiol.* 37:421–28

41. Helling, C. S., Bollag, J.-M., Dawson, J. E. 1968. Cleavage of ether-oxygen bond in phenoxyacetic acid by an *Arthrobacter* sp. *J. Agric. Food Chem.* 16:538–39

42. Horowitz, A., Shelton, D. R., Cornell, C. P., Tiedje, J. M. 1982. Anaerobic degradation of aromatic compounds in sediments and digest sludge. *Dev. Ind. Microbiol.* 23:435–44

43. Horowitz, A., Suflita, J. M., Tiedje, J. M. 1983. Reductive dehalogenation of halobenzoates by anaerobic lake sediment microorganisms. *Appl. Environ. Microbiol.* 45:1459–65

44. Horvath, R. S., Alexander, M. 1970. Cometabolism of *m*-chlorobenzoate by an *Arthrobacter*. *Appl. Microbiol.* 20: 254–58

45. Hughes, D. E. 1965. The metabolism of halogen-substituted benzoic acids by *Pseudomonas fluorescens. Biochem. J.* 96:181–88

46. Ichihara, A., Adachi, K., Hosokawa, K., Takeda, Y. 1962. The enzymatic hydroxylation of aromatic carboxylic acids; substrate specificities of anthranilate and benzoate oxidases. *J. Biol. Chem.* 237:2296–302

47. Jeenes, D. J., Reineke, W., Knackmuss, H.-J., Williams, P. A. 1982. TOL plasmid pWWO in constructed halobenzoate-degrading *Pseudomonas* strains: enzyme regulation and DNA structure. *J. Bacteriol.* 150:180–87

48. Johnston, H. W., Briggs, G. G., Alexander, M. 1972. Metabolism of 3-chlorobenzoic acid by a pseudomonad. *Soil Biol. Biochem.* 4:187–90

49. Kersten, P. J., Chapman, P. J., Dagley, S. 1985. Enzymatic release of halogens or methanol from substituted protocatechuic acids. *J. Bacteriol.* 162: 693–97

50. Kilbane, J. J., Chatterjee, D. K., Chakrabarty, A. M. 1983. Detoxification of 2,4,5-trichlorophenoxyacetic acid from contaminated soil by *Pseudomonas cepacia. Appl. Environ. Microbiol.* 45: 1697–700

51. Kilbane, J. J., Chatterjee, D. K., Karns, J. S., Kellog, S. T., Chakrabarty, A. M. 1982. Biodegradation of 2,4,5-trichlorophenoxyacetic acid by a pure culture of *Pseudomonas cepacia. Appl. Environ. Microbiol.* 44:72–78

52. Klages, U., Lingens, F. 1979. Degradation of 4-chlorobenzoic acid by a *Nocardia* species. *FEMS Microbiol. Lett.* 6:201–3

53. Klages, U., Lingens, F. 1980. Degradation of 4-chlorobenzoic acid by a

Pseudomonas sp. *Zentralbl. Bakteriol. Parasitenkd. Infektionskr. Reihe C* 1:215–23

54. Klages, U., Markus, A., Lingens, F. 1981. Degradation of 4-chlorophenylacetic acid by a *Pseudomonas* species. *J. Bacteriol.* 146:64–68

55. Latorre, J., Reineke, W., Knackmuss, H.-J. 1984. Microbial metabolism of chloroanilines: Enhanced evolution by natural genetic exchange. *Arch. Microbiol.* 140:159–65

56. Lehrbach, P. R., Zeyer, J., Reineke, W., Knackmuss, H.-J., Timmis, K. N. 1984. Enzyme recruitment in vitro. Use of cloned genes to extend the range of haloaromatics degraded by *Pseudomonas* sp. strain B13. *J. Bacteriol.* 158: 1025–32

57. Liu, T., Chapman, P. J. 1984. Purification and properties of a plasmid-encoded 2,4-dichlorophenol hydroxylase. *FEBS Lett.* 173:314–18

58. Loos, M. A., Roberts, R. N., Alexander, M. 1967. Phenols as intermediates in the decomposition of phenoxyacetates by an *Arthrobacter* species. *Can. J. Microbiol.* 13:679–90

59. Marks, T. S., Smith, A. R. W., Quirk, A. V. 1984. Degradation of 4-chlorobenzoic acid by *Arthrobacter* sp. *Appl. Environ. Microbiol.* 48:1020–25

60. Marks, T. S., Wait, R., Smith, A. R. W., Quirk, A. V. 1984. The origin of oxygen incorporated during the dehalogenation/hydroxylation of 4-chlorobenzoate by an *Arthrobacter* sp. *Biochem. Biophys. Res. Commun.* 124:669–74

61. Markus, A., Klages, U., Krauss, S., Lingens, F. 1984. Oxidation and dehalogenation of 4-chlorophenylacetate by a two-component enzyme system from *Pseudomonas* sp. strain CBS3. *J. Bacteriol.* 160:618–21

62. Markus, A., Krekel, D., Lingens, F. 1986. Purification and some properties of component A of the 4-chlorophenylacetate 3,4-dioxygenase from *Pseudomonas* species strain CBS. *J. Biol. Chem.* 261:12883–88

63. Mikesell, M. D., Boyd, S. A. 1985. Reductive dechlorination of pesticides 2,4-D, 2,4,5-T, and pentachlorophenol in anaerobic sludge. *J. Environ. Qual.* 14:337–40

64. Mikesell, M. D., Boyd, S. A. 1986. Complete reductive dechlorination and mineralization of pentachlorophenol by anaerobic microorganisms. *Appl. Environ. Microbiol.* 52:861–65

65. Milne, G. W. A., Goldman, P., Holz-

man, J. L. 1968. The metabolism of 2-fluorobenzoic acid. II. Studies with $^{18}O_2$. *J. Biol. Chem.* 243:5374–76

66. Morris, C. M., Barnsley, E. A. 1982. The cometabolism of 1- and 2-chloronaphthalene by pseudomonads. *Can. J. Microbiol.* 28:73–79

67. Müller, R., Thiele, J., Klages, U., Lingens, F. 1984. Incorporation of [^{18}O] water into 4-hydroxybenzoic acid in the reaction of 4-chlorobenzoate dehalogenase from *Pseudomonas* spec. CBS3. *Biochem. Biophys. Res. Commun.* 124:178–82

68. Oltmanns, R. H., Rast, H. G., Reineke, W. 1988. Degradation of 1,4-dichlorobenzene by enriched and constructed bacteria. *Appl. Microbiol. Biotechnol.* 28:609–16

69. Pertsova, R. N., Kunc, F., Golovleva, L. A. 1984. Degradation of 3-chlorobenzoate in soil by pseudomonads carrying biodegradative plasmids. *Folia Microbiol. Prague* 29:242–47

70. Reineke, W. 1986. Construction of bacterial strains with novel degradative capabilities for chloroaromatics. *J. Basic Microbiol.* 26:551–67

71. Reineke, W., Jeenes, D. J., Williams, P. A., Knackmuss, H.-J. 1982. TOL plasmid pWWO in constructed halobenzoate-degrading *Pseudomonas* strains: prevention of *meta* pathway. *J. Bacteriol.* 150:195–201

72. Reineke, W., Knackmuss, H.-J. 1978. Chemical structure and biodegradability of halogenated aromatic compounds. Substituent effects on 1,2-dioxygenation of benzoic acid. *Biochim. Biophys. Acta* 542:412–23

73. Reineke, W., Knackmuss, H.-J. 1978. Chemical structure and biodegradability of halogenated aromatic compounds. Substituent effects on dehydrogenation of 3,5-cyclohexadiene-1,2-diol-1-carboxylic acid. *Biochim. Biophys. Acta* 542:424–29

74. Reineke, W., Knackmuss, H.-J. 1979. Construction of haloaromatics utilising bacteria. *Nature* 277:385–86

75. Reineke, W., Knackmuss, H.-J. 1980. Hybrid pathway for chlorobenzoate metabolism in *Pseudomonas* sp. B13 derivatives. *J. Bacteriol.* 142:467–73

76. Reineke, W., Knackmuss, H.-J. 1984. Microbial metabolism of haloaromatics: Isolation and properties of a chlorobenzene-degrading bacterium. *Appl. Environ. Microbiol.* 47:395–402

77. Reineke, W., Wessels, S. W., Rubio, M. A., Latorre, J., Schwien, U., et al. 1982. Degradation of monochlorinated aromatics following transfer of genes encoding chlorocatechol catabolism. *FEMS Microbiol. Lett.* 14:291–94

78. Reiner, A. M., Hegeman, G. D. 1971. Metabolism of benzoic acid by bacteria. Accumulation of (-)-3,5-cyclohexadiene-1,2-diol-1-carboxylic acid by a mutant strain of *Alcaligenes eutrophus*. *Biochemistry* 10:2530–36

79. Reiner, E. A., Chu, J., Kirsch, E. J. 1978. Microbial metabolism of pentachlorophenol. In *Pentachlorophenol: Chemistry, Pharmacology and Toxicity,* ed. K. R. Rao, pp. 67–81. New York: Plenum

80. Ruisinger, S., Klages, U., Lingens, F. 1976. Abbau der 4-Chlorbenzoesäure durch eine *Arthrobacter*-Species. *Arch. Microbiol.* 10:253–56

81. Schmidt, E., Hellwig, M., Knackmuss, H.-J. 1983. Degradation of chlorophenols by a defined mixed microbial community. *Appl. Environ. Microbiol.* 46:1038–44

82. Schmidt, E., Knackmuss, H.-J. 1980. Chemical structure and biodegradability of halogenated aromatic compounds. Conversion of chlorinated muconic acids into maleoylacetic acid. *Biochem. J.* 192:339–47

83. Schmidt, E., Remberg, G., Knackmuss, H.-J. 1980. Chemical structure and biodegradability of halogenated aromatic compounds. Halogenated muconic acids as intermediates. *Biochem. J.* 192:331–37

84. Schraa, G., Boone, M. L., Jetten, M. S. M., van Neerven, A. R. W., Colberg, P. J., Zehnder, A. J. B. 1986. Degradation of 1,4-dichlorobenzene by *Alcaligenes* sp. strain A175. *Appl. Environ. Microbiol.* 52:1374–81

85. Schukat, B., Janke, D., Krebs, D., Fritsche, W. 1983. Cometabolic degradation of 2- and 3-chloroaniline because of glucose metabolism by *Rhodococcus* sp. An 117. *Curr. Microbiol.* 9:58–67

86. Schwien, U. 1984. *Bakterieller Abbau von Dichlararomaten: Metabolismus von 3,5-Dichlorbrenzcatechin.* PhD thesis. Univ. Göttingen, Fed. Rep. Germany. 101 pp.

87. Schwien, U., Schmidt, E. 1982. Improved degradation of monochlorophenols by a constructed strain. *Appl. Environ. Microbiol.* 44:33–39

88. Sharpee, K. W., Duxbury, J. M., Alexander, M. 1973. 2,4-Dichlorophenoxyacetate metabolism by *Arthrobacter* sp.: Accumulation of a chlorobutenolide. *Appl. Microbiol.* 26:445–47

89. Shelton, D. R., Tiedje, J. M. 1984. Isolation and partial characterization of bacteria in an anaerobic consortium that mineralizes 3-chlorobenzoic acid. *Appl. Environ. Microbiol.* 48:840–48

90. Sleat, R., Robinson, J. P. 1984. The bacteriology of anaerobic degradation of aromatic compounds. *J. Appl. Bacteriol.* 57:381–94

91. Spain, J. C., Nishino, S. F. 1987. Degradation of 1,4-dichlorobenzene by a *Pseudomonas* sp. *Appl. Environ. Microbiol.* 53:1010–19

92. Spokes, J. R., Walker, N. 1974. Chlorophenol and chlorobenzoic acid cometabolism by different genera of soil bacteria. *Arch. Microbiol.* 96:125–34

93. Steiert, J. G., Crawford, R. L. 1986. Catabolism of pentachlorophenol by a *Flavobacterium* sp. *Biochem. Biophys. Res. Commun.* 141:825–30

94. Suflita, J. M., Horowitz, A., Shelton, D. R., Tiedje, J. M. 1982. Dehalogenation: a novel pathway for the anaerobic biodegradation of haloaromatic compounds. *Science* 218:1115–17

95. Suflita, J. M., Robinson, J. A., Tiedje, J. M. 1983. Kinetics of microbial dehalogenation of haloaromatic substrates in methanogenic environments. *Appl. Environ. Microbiol.* 45:1466–73

96. Suflita, J. M., Stout, J., Tiedje, J. M. 1984. Dechlorination of (2,4,5-trichlorophenoxy)acetic acid by anaerobic microorganisms. *J. Agric. Food Chem.* 32:218–21

97. Suzuki, T. 1977. Metabolism of pentachlorophenol by a soil microbe. *J. Environ. Sci. Health Part B* 12:113–27

98. Thiele, J., Müller, R., Lingens, F. 1987. Initial characterisation of 4-chlorobenzoate dehalogenase from *Pseudomonas* sp. CBS3. *FEMS Microbiol. Lett.* 41:115–19

99. Tiedje, J. M., Alexander, M. 1969. Enzymatic cleavage of the ether bond of 2,4-dichlorophenoxyacetate. *J. Agric. Food Chem.* 17:1080–84

100. Tiedje, J. M., Duxbury, J. M., Alexander, M., Dawson, J. E. 1969. 2,4-D metabolism: Pathway of degradation of chlorocatechols by *Arthrobacter* sp. *J. Agric. Food Chem.* 17:1021–26

101. Tittmann, U., Lingens, F. 1980. Degradation of biphenyl by *Arthrobacter simplex* strain BPA. *FEMS Microbiol. Lett.* 8:225–58

102. Tsuchiya, T., Yamaha, T. 1983. Reductive dechlorination of 1,2,4-trichlorobenzene on incubation with intestinal contents of rats. *Agric. Biol. Chem.* 47:1163–65

103. Tsuchiya, T., Yamaha, T. 1984. Reductive dechlorination of 1,2,4-trichlorobenzene by *Staphylococcus epidermidis* isolated from intestinal contents of rats. *Agric. Biol. Chem.* 48:1545–50

104. Walker, N., Harris. D. 1970. Metabolism of 3-chlorobenzoic acid by *Azotobacter* species. *Soil Biol. Biochem.* 2:27–32

105. Yamaguchi, M., Fujisawa, H. 1980. Purification and characterization of an oxygenase component in benzoate 1,2-dioxygenase system from *Pseudomonas arvilla* C-1. *J. Biol. Chem.* 255:5058–63

106. Zaitsev, G. M., Karasevich, Y. N. 1980. Utilization of 4-chlorobenzoic acid by *Arthrobacter globiformis*. *Mikrobiologiya* 50:35–40

107. Zaitsev, G. M., Karasevich, Y. N. 1980. Preparative metabolism of 4-chlorobenzoic acid in *Arthrobacter globiformis*. *Mikrobiologiya* 50:423–28

108. Zaitsev, G. M., Karasevich, Y. N. 1982. Utilization of 2-chlorobenzoic acid by *Pseudomonas cepacia*. *Mikrobiologiya* 53:75–80

109. Zeyer, J., Kearney, P. C. 1982. Microbial degradation of *para*-chloroaniline as sole carbon and nitrogen source. *Pestic. Biochem. Physiol.* 17:215–23

110. Zeyer, J., Wasserfallen, A., Timmis, K. N. 1985. Microbial mineralization of ring-substituted anilines through an *ortho*-cleavage pathway.. *Appl. Environ. Microbiol.* 50:447–53

Ann. Rev. Microbiol. 1988. 42:289–317

ANAEROBIC DEGRADATION OF AROMATIC COMPOUNDS

W. Charles Evans

Department of Biochemistry and Soil Science, University College of North Wales, Bangor, LL 57 2UW Gwynedd, Wales, United Kingdom

Georg Fuchs

Angewandte Mikrobiologie, Universität Ulm, D-7900 Ulm, Federal Republic of Germany

CONTENTS

0066-4227/88/1001-0289$02.00

INTRODUCTION

A large variety of aromatic substances participate in life processes; their biosynthesis and degradation form an important part of the natural carbon cycle. Human activities are an additional source of synthetic organic chemicals in the environment, ultimately to be metabolized. Some xenobiotic benzenoid structures are relatively recalcitrant, requiring the competence of the microbial world for their dissimilation.

The biosphere presents a host of different habitats within which microorganisms can operate provided nutrients are available and physical conditions are appropriate for mineralization to occur. Under aerobic conditions aromatic substrates are metabolized by a variety of bacteria, with ring fission accomplished by mono- and dioxygenases. Molecular oxygen is essential for these enzymes to function since it is incorporated into the reaction products. This field has been extensively reviewed (27, 36; for recent reviews of anaerobic degradation see 13, 37, 39, 52, 95, 116).

Anoxic ecosystems are created when oxygen consumption exceeds its supply, e.g. in soils with impeded drainage, stagnant water, municipal landfills, sewage treatment digesters, industrial plants that produce methane from organic waste, the alimentary tract of all animals, and finally sediments of the oceans and other natural bodies of water. The metabolic fate of organic compounds and their mineralization to CO_2 (and CH_4) depends on the availability of light or of inorganic electron acceptors such as NO_3^-, SO_4^{-2}, or CO_2. Progress in elucidating the pathways of degradation of aromatic acids, phenols, and hydrocarbons has been made through the study of (Table 1) (a) the photosynthetic anaerobic metabolism of benzoate by *Rhodopseudomonas palustris;* (b) the anaerobic metabolism of aromatic acids and phenols by nitrate respiration; (c) the anaerobic dissimilation of aromatic compounds through sulfate respiration; (d) the anaerobic fermentation of many polyphenolic substances; and (e) the methanogenic fermentation of almost all naturally occurring soluble aromatic compounds by undefined consortia of bacteria cooperating to form a food chain. This multiple-species interaction is essential for the methanogenic system to operate. At least three physiological types of bacteria are discernible: fermenters, which convert the initial substrates into organic acids; acetogenic proton-reducing bacteria; and hydrogen- and acetate-consuming methanogens.

In 1934 (102) Tarvin & Buswell provided the first decisive chemical evidence that the aromatic nucleus of common aromatic compounds was completely decomposed when incubated anaerobically with sewage sludge. Virtually all of the ring carbons of the aromatic substrate were accounted for as CO_2, CH_4, and microbial cells. This pioneering quantitative study of the

Table 1 Anaerobic metabolism of aromatic compounds

Energy-yielding process	Organism	Substrates	References
Photosynthetic phosphorylation	*Rhodopseudomonas palustris* *Rhodopseudomonas gelatinosa*	Benzoate *m,p*-Hydroxybenzoate	(82) (32)
		Phloroglucinol	(50, 54, 55, 65, 111)
Denitrification $NO_3^- + 2H^+ + 4H_2 \rightarrow$ $NH_4^+ + 3H_2O$ $\Delta G_0' = -600$ kJ	*Pseudomonas* PN-1 *(Alcaligenes xylosoxidans)* *Moraxella* sp. *(Paracoccus denitrificans)* *Bacillus* sp. *Pseudomonas*	Benzoate Hydroxybenzoates Protocatechuate Vanillate *o,m,p*-Phthalate 2-Aminobenzoate Phenol *o,m,p*-Cresol	(79, 103, 104) (115) (15) (104) (1, 2) (21, 118) (4, 5) (18, 108)
Sulfate reduction $SO_4^{-2} + 2H^+ + 4H_2 \rightarrow$ $H_2S + 4H_2O$ $\Delta G_0' = -152$ kJ	*Desulfovibrio* sp. *Desulfococcus* *Desulfonema* *Desulfosarcina*	Benzoate Hydroxybenzoates Phenylacetate, hippurate Phenol Indol	(112) (113) (112) (27) (2a, 3)
Fermentation	*Coprococcus* sp. *Streptococcus* *Pelobacter acidigallici* *Eubacterium oxidoreducens*	Phloroglucinol Resorcinol/acids Gallate, pyrogallol Polyphenols Quercetin	(105) (106) (89) (71, 72) (107)
Methanogenic fermentation $HCO_3^- + H^+ + 4H_2 \rightarrow$ $CH_4 + 3H_2O$ $\Delta G_0' = -135$ kJ	Microbial consortia: fermentative bacteria + acetogenic and methanogenic bacteria	Lignin (cornstalks) Benzoate Tyrosine Cinnamate, phenylpropionate Phenylacetate, benzoate Phenol Catechol Hydroquinone Ferulate Vanillate Syringate Benzoate Phenylalanine Tyrosine Tryptophan, indole Coniferyl alcohol Benzene, toluene Chlorobenzene Chlorophenols Chlorobenzoates Chlorophenoxyacetates Nitrophenols Chloroguaiacols	(17) (102) (76) (7–9, 11) (10, 25, 41, 38, 68, 78) (56, 57) (24) (12) (56) (67) (94) (40, 93a) (11) (101) (14) (43, 45) (48, 110) (83) (20) (19) (98–100) (98) (77)

methanogenesis of organic compounds established the stoichiometry of the reaction as follows:

$$C_nH_aO_b + (n - \tfrac{a}{4} - \tfrac{b}{2})H_2O = (\tfrac{n}{2} - \tfrac{a}{8} + \tfrac{b}{4})CO_2 + (\tfrac{n}{2} + \tfrac{a}{8} - \tfrac{b}{4})CH_4. \quad 1.$$

The fact that the phenomenon only occurs in a mixed culture probably inhibited further progress for some time; the doctrine of single substrate–single organism had become entrenched. With the advent of radioactive tracers, Clark & Fina (24) showed in 1952 that in a methanogenic culture adapted to metabolize benzoate, [ring-^{14}C]benzoate was completely degraded, and 50% of the radioactivity was recovered as CH_4. With specifically labeled carbon atoms, Fina & Fiskin (41) demonstrated that the carboxyl group and C-4 on the ring were largely converted to CO_2, while C-1 was mainly reduced to CH_4. Nottingham & Hungate (78), using uniformly labeled [ring-^{14}C]benzoate, confirmed that the composition of the fermentation gases agrees with the stoichiometry of the equation

$$4\,^{14}C_6H_5CO_2H + 18H_2O = 15\,^{14}CH_4 + 9\,^{14}CO_2 + CO_2. \quad 2.$$

Further progress in understanding the anaerobic dissimilation of benzoate was made through detailed study of its photometabolism by *R. palustris*.

ANAEROBIC PHOTOMETABOLISM OF AROMATIC COMPOUNDS

Benzoate Metabolism by Athiorhodaceae

Several species of the purple nonsulfur Rhodospirillaceae family, which can obtain energy from light or from aerobic respiration, grow anaerobically in the light at the expense of simple aromatic compounds as sole carbon source.

EARLY STUDIES   Early attempts to elucidate the anaerobic breakdown of aromatic acids such as benzoate were based on a literal interpretation of van Niel's general scheme for photosynthesis in bacteria and plants (109). One of the products of the light reaction was a strong oxidant (OH) which, in plants, was converted into molecular oxygen; in bacteria this proposedly light-induced "bound oxygen" was used to oxidize substrates. Thus Proctor & Scher (82, 88) surmised in 1960 that the light-induced oxidant and molecular oxygen were equivalent and that the anaerobic photosynthetic metabolic pathway of benzoate was similar to the pathways of well-known aerobic processes; however, Leadbetter & Hawk (73) and Dutton & Evans (29) could not reproduce the results of Proctor & Scher. *R. palustris* cells were grown photosynthetically on benzoate or 3-hydroxybenzoate; neither of these growth substrates promoted any respiratory response in the cells under aerobic conditions. Hence oxygen and any light-generated oxidant were not equivalent. Return to anaerobic conditions in the light permitted benzoate utilization to proceed again promptly and normally.

In 1968 Dutton & Evans (31) determined the key intermediates of the anaerobic photosynthetic metabolism of benzoate in *R. palustris* by incubating unlabeled test intermediates with whole-cell suspensions actively photometabolizing ^{14}C-labeled benzoate. Five products incorporated the isotope: cyclohex-1-enecarboxylate, cyclohexanecarboxylate, 2-hydroxycyclohexanecarboxylate, 2-oxocyclohexanecarboxylate, and pimelate; thus all seven carbon atoms of benzoate remain together as far as pimelate. Interpreting these results at the simplest level, we might propose that the aromatic ring may become fully reduced with the incorporation of six hydrogen equivalents to form cyclohexanecarboxylate. The subsequent reactions, starting from a fully saturated alicyclic acid, would be analogous to a classical β-oxidation of a fatty acid (Figure 1). Of the two other reduced benzoate compounds tested by Dutton & Evans (32), cyclohex-2-enecarboxylate showed some labeling after incubation, while cyclohex-3-enecarboxylate did not. This may indicate that C-3 and C-4 are obligatorily reduced first, irrespective of what happens next. If this is true, the first intermediate of the breakdown pathway would be cyclohex-1,5-dienecarboxylate. [In an investigation of benzoate metabolism by nitrate respiration, Williams & Evans (115) detected the presence of a diene, which could well be cyclohex-1,5-dienecarboxylate, but unfortunately an authentic sample was not available for direct comparison.]

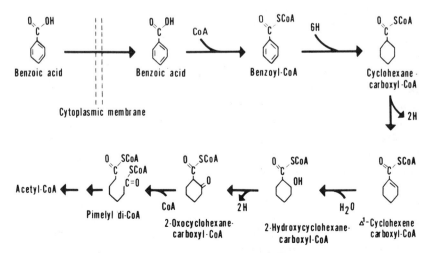

Figure 1 Proposed pathway of benzoate metabolism in *Rhodopseudomonas palustris*.

Guyer & Hegeman (50) provided a valuable confirmation of the essential correctness of the reductive pathway in the photometabolism of benzoate using a different approach. They isolated mutant strains of *R. palustris* and tested their ability to grow at the expense of the proposed intermediates (benzoate, cyclohexanecarboxylate, cyclohex-l-enecarboxylate, 2-hydroxy-cyclohexanecarboxylate, pimelate, and acetate) in the light. Each strain was able to grow on all proposed intermediates later in the pathway than the compound for which it was selected, but not on the preceding intermediates. [*ring*-^{14}C]benzoate or [*carboxy*-^{14}C]benzoate also gave rise to radioactivity in cyclohex-1-enecarboxylate.

CELL-FREE PREPARATIONS AND THE ROLE OF COENZYME A   Cell-free extracts of photosynthetically grown cells severely inhibited benzoate photometabolism by whole-cell suspensions. The inhibitory material was mainly octadeca-11-enoic acid (vaccenic acid) associated with the chromatophore fraction. Dutton & Evans (33) subsequently found that in the 1–10 mM range all of the monocarboxylic acid homologs (but not their methyl esters), even propionate, can inhibit the rate of benzoate utilization by whole cells 80%. The original rate was later recovered, which suggests that the fatty acids compete as substrates for enzymes or cofactors, such as coenzyme A, involved in benzoate photoassimilation (30). Whittle et al (111) verified these ideas experimentally; cell-free extracts from cells grown photosynthetically on benzoate were capable of catalyzing the thioesterification of benzoate to benzoyl-CoA in the presence of ATP, Mg^{2+}, and CoA. Furthermore, this cell-free system converted cyclohex-1-enecarboxylate to pimelate in the obligatory presence of coenzyme A, ATP, Mg^{2+}, and NAD^+.

Lunt (cited in 34) found that cell-free extract (sonicated, 15,000 × *g* supernatant containing chromatophores) of *R. palustris* grown photosynthetically on benzoate, when incubated anaerobically in the light with benzoyl-phosphate and coenzyme A, lost its aromatic absorption; neither benzoate nor benzoyl-CoA was metabolized in this system. Unfortunately, these results could not be repeated with consistency. In 1983, Hutber & Ribbons (65) continued investigations on this phenomenon; they assayed the enzymes involved in β-oxidation of cyclohexanecarboxyl-CoA by *R. palustris* cells grown anaerobically and aerobically on benzoate, succinate, and cyclohexanecarboxylate. All the enzymes necessary for the β-oxidation of short-chain fatty acids were constitutively present in cell extracts at comparable levels; in addition, alicyclic CoA esters could also act as substrates. Growth on either benzoate or cyclohexanecarboxylate did, however, induce a very active acyl-CoA synthetase; this might be taken as evidence that the substrate for the reduction of the aromatic ring is benzoyl-CoA. Attempts to demonstrate this reduction in vitro have not, so far, been successful. Interestingly, the appearance of benzoate in culture fluids of cells grown photosynthetically on p-hydroxybenzoate indicated that dehydroxylation was involved, as did the lack of acyl-CoA synthetase activity toward p-hydroxybenzoate. Benzoyl-CoA synthetase activity, however, was elevated in these cells (64).

Recently, Harwood & Gibson (54) have made a careful study of the kinetics of uptake of [7-^{14}C]benzoate into induced cells of R. palustris grown photosynthetically on benzoate. This process is extraordinarily efficient; it is energy dependent but tolerant of a wide pH range. There was no evidence of an active transport system. Intracellularly, the benzoate was immediately converted into benzoyl-CoA by an inducible benzoyl-CoA synthetase. Autoradiograms also revealed some reduction products of benzoyl-CoA, but no trace of benzoyl phosphate could be seen. Geissler et al (42) have purified an O_2-insensitive benzoyl-CoA ligase (60 kd) to homogeneity. It catalyzes the Mg^{2+}- and ATP-dependent formation of acyl-CoA from carboxylate and coenzyme A with high specificity for benzoate and 2-fluorobenzoate. The kinetic properties of the enzyme match the kinetics of benzoate uptake by whole cells and confirm its role in catalyzing the first degradation step (Figure 1). Harwood & Gibson (55) also maintained that auxanography indicated that R. palustris is one of the most versatile of the anaerobic bacteria described to date with respect to aromatic degradation; they suggested that two major routes operate, one that flows through benzoate and another that flows through 4-hydroxybenzoate prior to cleavage of the aromatic ring. However, the R. palustris strain of Hutber et al (64) grown photosynthetically on p-hydroxybenzoate produced benzoate as an intermediate, which indicates that a dehydroxylation occurs.

Phloroglucinol Metabolism by Rhodopseudomonas

Whittle et al (111) showed that Rhodopseudomonas gelatinosa grows anaerobically in the light on phloroglucinol (1,3,5-trihydroxybenzene); dihydrophloroglucinol and aliphatic acids (one of them 2-oxo-4-hydroxyadipate) appeared in the culture fluid. Extracts of R. gelatinosa reduced phloroglucinol to dihydrophloroglucinol in the dark in the presence of cysteine, NADPH, and EDTA. The enzymatic reactions beyond this point remain to be revealed, including the ring-cleavage mechanism.

METABOLISM OF AROMATIC COMPOUNDS BY NITRATE-REDUCING BACTERIA

Metabolism of Benzoate

Nitrate-reducing bacteria couple the oxidation of organic compounds with water to the exergonic reduction of nitrate via nitrite to N_2 or, less often, to NH_3. Energy is derived mainly from electron transport phosphorylation during nitrate respiration, and cell carbon is derived from breakdown products of the organic compound. In the early studies importance was attached to the fact that the bacteria in question were photosynthetic, and therefore it was thought that some product derived from the light reaction might be involved. Even after the pathway had been determined to be reductive, there was still a tendency to overvalue the possibility that the forces for benzoate reduction might be particular to photosynthetic bacteria or that nitrate might serve as oxygen source. However, in 1970 Taylor et al (103) isolated a gram-negative bacterium, designated *Pseudomonas* strain PN-1, which aerobically metabolized *p*-hydroxybenzoate through protocatechuate using the *meta* pathway of ring cleavage; anaerobically, it grew on benzoate in the obligatory presence of nitrate as electron acceptor. [This organism has recently been reclassified as *Alcaligenes xylosoxidans* subsp. *denitrificans* (15).]

Taylor and his collaborators (103, 104) showed that anaerobically grown nitrate-respiring cells of PN-1 were devoid of the oxygenase enzymes that operate aerobically; using [^{14}C]benzoate they also confirmed its dissimilation to CO_2. A nonreductive mechanism was proposed for this degradation, involving the addition of three molecules of H_2O to the benzene nucleus in preparation for ring fission.

Williams & Evans (114, 115) also isolated an organism capable of anaerobic utilization of benzoate by nitrate respiration. It was initially thought to be a strain of *Pseudomonas stutzeri* or a *Moraxella* sp., but it is now believed to be a strain of *Paracoccus denitrificans* [National Collection of Industrial Bacteria (Torry Institute, Aberdeen, Scotland), personal communication]. Aerobically, it oxidatively decarboxylated benzoate to catechol and subsequently metabolized it by the *ortho* pathway. When cells were incubated anaerobically with [*ring*-^{14}C]benzoate in nitrate-phosphate buffer, labeled cyclohexanecarboxylate, cyclohex-1-enecarboxylate, 2-hydroxycyclohexanecarboxylate, and adipate were formed in the reaction mixture. With [*carboxy*-^{14}C]benzoate, the above intermediates were again radioactive with the exception of adipate; the production of adipate instead of pimelate in these cultures is explained by a divergence at the 2-oxocyclohexanecarboxylate stage; this intermediate is decarboxylated to cyclohexanone and further metabolized to adipate. In similar experiments, 2-oxocyclohexanecarboxylate gave rise to cyclohexanone, 2-hydroxycyclohexanone, and adipate. Williams & Evans (115) were not successful in preparing cell-free extracts capable of accomplishing the reductive sequence of transformations.

Recently, Blake & Hegeman, using *A. xylosoxidans* subsp. *denitrificans* PN-1, the strain originally isolated by Taylor et al (103), have made spectacular progress in understanding the anaerobic metabolism of benzoate by nitrate respiration (C. K. Blake & G. D. Hegeman, personal communication). Extracts of PN-1 cells grown on benzoate by nitrate respiration converted

[^{14}C]benzoate to two radioactive compounds: The first was formed quickly and was identified as benzoyl-CoA; the second was formed more slowly and appeared to be attached to a small protein. Both products yielded back benzoate on mild alkaline hydrolysis.

Since the resonance stability of the aromatic ring is great (163 kJ mol^{-1}), an extremely low-potential reductant would be required for its reduction, e.g. a flavodoxin or ferredoxin. The antimicrobial agent metronidazole, which affects only anaerobes whose metabolism requires ferredoxin or an equivalent low-potential reductant, inhibited anaerobic growth at a 50-fold lower concentration with benzoate than when succinate was the substrate under denitrifying conditions. This observation supports the view that a low-potential reductant operates in the reductive phase of anaerobic benzoate metabolism.

ROLE OF PLASMIDS   Blake & Hegeman (15), using Tn5 transposon mutagenesis, obtained mutants of PN-1 that were blocked in anaerobic benzoate catabolism and that were cured of a resident plasmid owing to apparent incompatibility with the incoming, selected plasmid-carrying Tn5. This 17.4-kbp resident plasmid, pCBI, can be transferred from PN-1 to strains of *P. aeruginosa* and *P. stutzeri*. *Pseudomonas* strains to which pCBI is transferred by conjugation including the *P. stutzeri* strains, which are unable to use benzoate aerobically, concomittantly acquire the ability to grow anaerobically on benzoate. A plasmid the size of pCBI could encode 7–15 average-sized proteins (50–100 kd) required for anaerobic benzoate catabolism. In vitro mutagenesis of the plasmid should provide valuable blocked mutants unavailable by mutagenesis of cells.

HALOGENATED BENZOATES   Aftring and coworkers (1, 2) described a *Bacillus* species that could grow anaerobically on *o*-phthalate-nitrate medium. This substrate was decarboxylated to benzoate before subsequent metabolism. *Pseudomonas* strain PN-1 would not grow on halogenated benzoates alone, but cells grown anaerobically on *p*-hydroxybenzoate degraded *o*- and *p*-fluorobenzoate through benzoate with the release of fluoride ions. Schennen et al (87) studied the anaerobic degradation of 2-fluorobenzoate by a benzoate-utilizing denitrifying bacterium. The fluoride ion was eliminated, probably by a reductive process, and the benzoate and 2-fluorobenzoate were thioesterified by an induced benzoyl-CoA synthetase before being metabolized.

Metabolism of Anthranilate

Using a denitrifying *Pseudomonas* species isolated by Braun & Gibson (21), Ziegler et al (118) have studied anaerobic 2-aminobenzoate (anthranilate) metabolism. Aerobically, this organism grew on 2-aminobenzoate and on benzoate. It failed to grow on catechol and protocatechuate, and therefore possibly metabolized benzoate via gentisate. In the presence of nitrate, benzoate and 2-aminobenzoate were oxidized to CO_2 with the concurrent reduction of NO_3^- to NO_2^-; only after complete NO_3^- consumption was NO_2^- reduced to N_2 gas. Extracts catalyzed the activation of benzoate to benzoyl-CoA, with ATP being cleaved to AMP and PP_i; two synthetase activities were present. Extracts from cells grown on 2-aminobenzoate catalyzed a NADH- or NADPH-dependent reduction of 2-aminobenzoyl-CoA, but not reduction of benzoyl-CoA. Purified 2-aminobenzoyl-CoA reductase converted 2-[carboxy-^{14}C]aminobenzoyl-CoA utilizing 2–3 mol NADH and produced at least two labeled CoA esters of undetermined constitution and 1 mol NH_3. One of the reaction products was unstable, undergoing decarboxylation; a tentative explanation is that reduction of the ring is followed by a hydrolytic deamination and the eventual formation of the CoA ester of a 2-oxocyclohexanecarboxylate derivative, which would be easily decarboxylated. The purified reductase, which was oxygen stable, was a flavoenzyme (R. Buder & G. Fuchs, unpublished results).

Metabolism of Phenolic Compounds

Bakker (4, 5) demonstrated that an enrichment culture (mainly Pseudomonas sp. and Spirillum sp.) degraded phenol and the cresols under anaerobic conditions in a nitrate–mineral salts medium. [ring-U-^{14}C]phenol was converted to $^{14}CO_2$; radioactive cell material and labeled n-caproate and acetate were detected in the culture fluid. Bakker suggested that phenol is reduced to cyclohexanol, which is dehydrogenated to cyclohexanone, and that hydrolytic fission of the alicyclic ring then produces n-caproate. Under denitrifying conditions CH_4 was not produced. Chmielowski et al (23) had already described a methanogenic fermentation of phenol and catechol.

Recently, Tschech & Fuchs (108) have isolated several strains of Pseudomonas-type bacteria capable of anaerobic growth on phenol with nitrate. While benzoate and p-hydroxybenzoate were degraded both anaerobically and aerobically, phenol was only metabolized under anaerobic conditions; reduced alicyclic compounds were not degraded. Cells grown on phenol catalyzed the exchange of $^{14}CO_2$ and the carboxyl of 4-hydroxybenzoate; this reaction was not catalyzed by cells grown on 4-hydroxybenzoate (108). The exchange reaction was catalyzed by a "phenol carboxylase," a Mn^{2+} enzyme that is presently being studied by these authors. Further metabolism of 4-hydroxybenzoate was not investigated. p-Cresol was metabolized by the Pseudomonas strain used to study phenol degradation via anaerobic oxidation of the methyl group through the alcohol and aldehyde to the carboxyl, yielding 4-hydroxybenzoate (Figure 2). Hopper (60, 61) had previously observed these transformations during the aerobic metabolism of this substrate and had elucidated the enzymology of the reactions. Metabolism of *m*-cresol seemed to be different.

Bakker's results were obtained with a mixed culture. Balba & Evans (10) used a methanogenic consortium to investigate catechol degradation via transformation to the diol followed by dehydroxylation to phenol prior to reduction of the ring. The question of the most common pathway of anaerobic phenol utilization through nitrate respiration in a pure culture of a competent organism is now at issue.

METABOLISM OF AROMATIC COMPOUNDS BY SULFATE-REDUCING BACTERIA

Sulfate-reducing bacteria couple the oxidation of organic compounds with water to the exergonic reduction of sulfate via sulfite to sulfide. Energy is

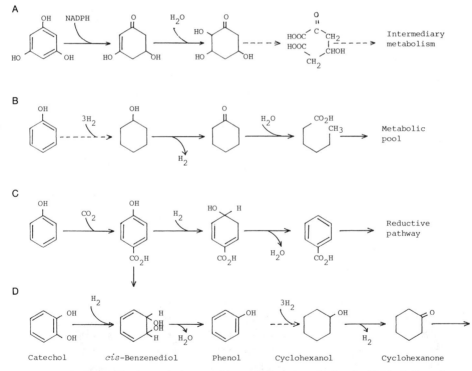

Figure 2   Proposed pathways in the anaerobic metabolism of certain phenols: (*A*) Metabolism of phloroglucinol by *Rhodopseudomonas gelatinosa* (111). (*B*) Metabolism of phenol through *n*-caproate by nitrate-reducing enrichment cultures. Initial reduction was accomplished by an unidentified hydrogen donor (5). (*C*) Metabolism of phenol by a denitrifying *Pseudomonas* sp. and by methanogenic enrichment cultures (108). (*D*) Catechol metabolism by a methanogenic consortium (10, 101).

derived mainly from electron transport phosphorylation during sulfite reduction. Cell carbon is derived from breakdown products of the organic compound. Sulfate reducers are mainly responsible for degradation of organic matter in anaerobic marine environments that contain approximately 27 mM sulfate. The well studied *Desulfovibrio vulgaris* is not capable of sulfate-dependent anaerobic growth on benzoate. Accidential contamination of a *D. vulgaris* culture by a strain of *Pseudomonas aeruginosa* produced a syntrophic association that resulted in both benzoate utilization and sulfate reduction (12). Mountfort & Bryant (76) also characterized an anaerobic benzoate degrader isolated in coculture with a sulfate reducer.

Widdel (112) provided conclusive evidence that axenic cultures of sulfate reducers that are able to use aromatic compounds as the sole source of carbon and electrons exist in nature. Four distinct genera, *Desulfovibrio, Desulfococcus, Desulfonema,* and *Desulfosarcina,* were isolated from anaerobic mud of freshwater, brackish, and marine sediments with benzoate as substrate. Some of these species required selenite and molybdate as trace elements and three vitamins. Some strains had an unusually high requirement for Ca^{2+} ions, while others were sensitive to light. They were all versatile in regard to aromatic substrates, utilizing benzoate, cyclohexanecarboxylate, phenylacetate, 3-phenylpropionate, and 2-, 3-, and 4-hydroxybenzoates. Recently, Cord-Ruwisch & Garcia (26) have isolated a spore-forming anaerobe, *Desulfotomaculum sapomandens,* that reduced sulfate, sulfite, thiosulfate, and elemental sulfur to H_2S using a variety of electron donors including aromatic acids; nitrate was not reduced. Bak & Widdel (3) recently isolated the first obligate anaerobic bacterium, *Desulfobacterium phenolicum,* that completely oxidized phenol with sulfate as terminal electron acceptor. They also isolated the indol-oxidizing *Desulfobacterium indolicum* (2a). No pathway of the anaerobic metabolism of aromatic substrates by sulfate respiration has yet been published.

FERMENTATION OF HYDROXYLATED AROMATIC COMPOUNDS

In fermentation, microorganisms derive their energy from substrate-level phosphorylation reactions; organic compounds serve as electron donors and acceptors. Patel et al (81) and Tsai & Jones (105) were the first to isolate from the rumen species of *Coprococcus* and *Streptococcus* that fermented phloroglucinol via dihydrophloroglucinol to acetate [a pathway analogous to that of photometabolism of phloroglucinol by *R. gelatinosa* (111)]. Schink & Pfennig (89) isolated five strains of a strictly anaerobic bacterium from marine mud with gallic acid, pyrogallol, phloroglucinol, or 2,4,6-trihydroxybenzoate as substrate; the substrates were fermented stoichiometrically to 3 mol acetate and 1 mol CO_2. Neither sulfate nor sulfur nor nitrate was reduced. The bacterium was named *Pelobacter acidigallici* gen. nov. sp. nov. and was classified in the family Bacteroidaceae. Of great interest is the fact that when *P. acidigallici* was grown in coculture with *Acetobacterium woodii*, syringic acid was also completely converted to acetate. A demethylation by the latter microbe provided a metabolizable substrate for the aromatic ring-splitter; this syntrophic cooperation is familiar in nature (see below). Mixed cultures consisting of *Pelobacter* sp., *A. woodii*, and *Methanosarcina barkeri* or *Methanothrix soehngenii* transform methoxylated gallic acid derivatives into methane and CO_2; this was one of the first examples described of the reconstitution of an ecological food chain. Samain et al (84b) have studied the initial steps of catabolism of trihydroxybenzenes in *P. acidigallici*.

Tschech & Schink (106) studied anaerobic fermentative degradation of resorcinol (1,3-dihydroxybenzene) and α,β,γ-resorcylic acids (3,5-, 2,4-, and 2,6-dihydroxybenzoates) in enrichment cultures. α-Resorcylate was rapidly degraded to acetate and methane; part of the acetate may have been derived from homoacetogenic H_2 oxidation. With resorcinol and β- and γ-resorcylate, different strict anaerobes, gram-positive, spore-forming *Clostridium* sp., were isolated. From freshwater enrichments, *Clostridium* strains were isolated in defined coculture with *Campylobacter* sp. They fermented resorcinol and β- and γ-resorcylate stoichiometrically to acetate and butyrate (Figure 3). In contrast, isolates from marine sediments formed acetate and hydrogen from resorcylic compounds, therefore depending on H_2-scavenging partners.

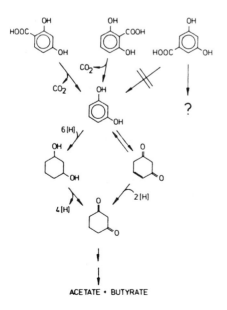

Figure 3 Proposed pathway of anaerobic degradation of resorcinol and resorcylates. β-Resorcylate and γ-resorcylate *(upper middle)* are decarboxylated to resorcinol; α-resorcylate *(upper right)* takes a different, so far unknown pathway.

Tschech & Schink (107) studied fermentative degradation of monohydroxybenzoates by defined syntrophic cocultures. From anaerobic freshwater enrichment cultures with 3-hydroxybenzoate, they isolated a bacterium in coculture with either *D. vulgaris,* a methanogen, or *A. woodii* as hydrogen scavenger; the hydroxybenzoate was degraded via reductive dehydroxylation through benzoate to acetate and methane. With 2-hydroxybenzoate (salicylate), short coccoid rods were enriched from anaerobic freshwater mud and were isolated in defined coculture in the presence of sulfate with *D. vulgaris.* Alternatively, *Methanospirillum hungatei* or *A. woodii* was used when sulfate was omitted. Dehydroxylation to benzoate could not be demonstrated; the 2-hydroxy group is in the correct position for subsequent β-oxidation of the ring after its reduction, and is thought to be conserved. In enrichment cultures with 4-hydroxybenzoate, decarboxylation to phenol was the initial step in degradation, which finally led to acetate, methane, and CO_2.

A defined mixed culture of *A. woodii, P. acidigallici,* and *Desulfobacter postgatei* anaerobically degraded 3,4,5-trimethoxybenzoate to acetate (70). Complete mineralization followed.

In a recent investigation Krumholz & Bryant (71) isolated from the rumen an anaerobic chemoorganotroph, *Syntrophococcus sucromutans,* that cleaved methyl-ether linkages of substituted monobenzenoids, e.g. syringate, caffeate, and vanillin. It also grew well with the H_2-consuming *Methanobrevibacter smithii* in coculture. Krumholz & Bryant (72) also isolated from the rumen a strictly anaerobic chemoorganotroph, *Eubacterium oxidoreducens,* that degrades gallate, pyrogallol, and phloroglucinol to acetate, butyrate, and occasionally CO_2; it requires either formate or hydrogen as an electron donor to catabolize these aromatic substrates. Gallate was decarboxylated; pyrogallol was converted to phloroglucinol, reduced to dihydrophloroglucinol, and finally hydrolyzed to 3-hydroxy-5-oxohexanoate (72a). A facultative anaerobe, *Enterobacter cloacae,* fermentatively O-demethylated and dehydroxylated ferulate (3-CH$_3$ O-4-OH-cinnamate) and reduced the side chain to phenylpropionate; the latter was not metabolized further (47).

METABOLISM OF AROMATIC COMPOUNDS BY UNDEFINED METHANOGENIC CONSORTIA

Bacterial methanogenesis occurs on a vast scale in anoxic nonmarine environments and is mainly responsible for the degradation of organic matter in these ecosystems. Pure cultures of methanogenic bacteria are able to use only a few simple substrates for growth, e.g. acetate, formate, methanol, and CO_2; hydrogen is used as the electron donor for CO_2 reduction. All methanogenic consortia must therefore rely on syntrophic associations with fermenters that degrade complex organic compounds into usable products for the methanogens. Since these obligate anaerobes require an oxidation-reduction potential of less than -300 mV for growth, progress in handling these fastidious methanogens had to await the development of Hungate's techniques (62).

Metabolism of Benzoate

The pioneering work of Boruff & Buswell (17) and Tarvin & Buswell (102) showed that organic acids, including aromatic acids, are quantitatively decomposed into CO_2 and CH_4 in sewage consortia at equilibrium. Further progress in understanding the fate of aromatic substrates in these complex systems had to await the availability of radioactive tracers coupled with various forms of chromatography and mass spectrometry.

Clark & Fina (24) and Fina & Fiskin (41) worked with [^{14}C]benzoate as substrate and an adapted methanogenic consortium converting it into CH_4 and CO_2. They observed that [$carboxy$-^{14}C]benzoate behaved like $^{14}CO_2$ in that it was not primarily reduced to methane; [$ring$-1-^{14}C]benzoate appeared mainly as $^{14}CH_4$; and [$ring$-4-^{14}C]benzoate was converted largely into $^{14}CO_2$. Propionate, acetate, and formate were detected; the propionate was labeled in the carboxyl carbon when [$ring$-4-^{14}C]benzoate was used, but not when [$ring$-1-^{14}C]benzoate or [$carboxy$-^{14}C]benzoate were substrates. Nottingham & Hungate (78) confirmed that the composition of the fermentation gases agreed with the initially mentioned stoichiometry.

Isotopic trapping experiments using ^{14}C-labeled benzoate afforded cyclohexanecarboxylate, cyclohex-1-enecarboxylate, heptanoate, valerate, butyrate, and acetate as fermentation products (40, 68, 69). The presence of heptanoate among these products was explained by a reductive fission of cyclohexanecarboxylate at the C-1–C-2 bond; it is hazardous to make such a definitive inference from the results presented. Balba & Evans (7), using similar techniques, also detected adipate, cyclohexanecarboxylate, cyclohex-1-enecarboxylate, propionate, and acetate production from radioactive benzoate. Although there were about five other unidentified labeled acidic components present none corresponded with heptanoate, n-caproate, valerate, or butyrate. Shlomi et al (93a), in a [^{14}C]benzoate-degrading methanogenic consortium, identified $trans$-2-hydroxycyclohexanecarboxylate, 2-oxocyclohexanecarboxylate, pimelate, n-caproate, butyrate, acetate, and H_2 as products.

All the results, even for different consortia made up of multiple species, support the view that benzoate is first reduced, probably as the CoA ester; a β-oxidation sequence follows, leading to ring cleavage and aliphatic acids. At this stage the diversity of bacteria result in a plethora of products on the way to acetate, formate, and CO_2, the substrates for methanogenesis. Figure 4 illustrates these reactions.

Ferry & Wolfe (38) found that o-chlorobenzoate inhibited benzoate utiliza-

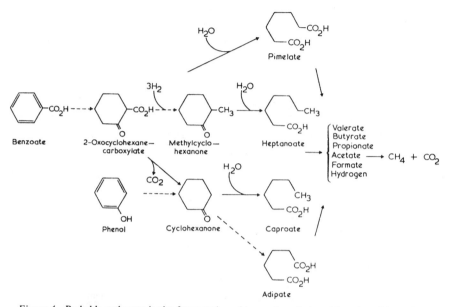

Figure 4 Probable pathways in the fermentation of benzoate and phenol by adapted bacterial consortia from a variety of methanogenic ecosystems. 1-Methylcyclohexanone is a hypothetical intermediate, but is one of a few possible precursors of heptanoate; others involve reduction of C-2 after a cleavage or have a biosynthetic origin.

tion without affecting the production of methane from acetate in a stable consortium containing three dominant organisms. The methanogens isolated were unable to metabolize benzoate, and the anaerobic benzoate utilizer could not be obtained in pure culture. However, uptake studies afforded tentative evidence that cyclohexanecarboxylate, 2-hydroxycyclohexanecarboxylate, and pimelate could be intermediates. Grbic-Galic & Young (49) were also able to uncouple benzoate degradation from methane production using bromoethanesulfonate (a structural analog of coenzyme M) as a specific inhibitor of methanogenesis; benzoate utilization was complete and acetate accumulated. In this case acetogenic bacteria may have acted as electron acceptors. Shelton & Tiedje (93) and Dolfing & Tiedje (28) studied another interesting consortium degrading 3-chlorobenzoic acid.

Metabolism of Other Aromatic Compounds

Healy & Young (56) studied the conversion of phenol and catechol to methane and CO_2 by adapted consortia from sewage. The conversion required 3–4 wk, but once the cultures were primed, further added substrates disappeared within a matter of days with the production of stoichiometrical quantities of fermentation gases. Knoll & Winter (69a) found that benzoate formed from phenol in sewage sludge in the presence of H_2 and CO_2; the metabolism may have proceeded via 4-hydroxybenzoate. Smolinski & Suflita (96) studied cresol metabolism in anoxic water. Mikesell & Boyd (75b) even described the complete mineralization of pentachlorophenol.

Balba & Evans (10) examined the reactions involved in the degradation of phenol and catechol. An enrichment culture actively metabolizing catechol converted cis-1,2-[U-^{14}C]benzenediol into radioactive phenol, cyclohexanol, cyclohexanone, 2-hydroxycyclohexanone, adipate, succinate, propionate, acetate, CH_4, and CO_2; cis-benzenediol was actually identified in the culture fluid of enrichment cultures. Balba et al (9) and Balba & Evans (10, 11) examined the methanogenesis of several naturally occurring aromatic compounds using a consortium initially adapted to benzoate, with the following results. (a) The methoxybenzoates immediately gave rise to methane and the hydroxybenzoates; a lag occurred before the culture could metabolize the hydroxybenzoates by dehydroxylation to benzoate and dissimilation via the reductive pathway. In the case of p-hydroxybenzoate a small amount of phenol was detected, which indicated that some decarboxylation also took place (9). (b) After adaptation, protocatechuate was mainly decarboxylated to catechol, which was then dehydroxylated to phenol, which was metabolized by the reductive pathway; a small quantity of m-hydroxybenzoate and benzoate were also detected (10). (c) A benzoate-adapted but benzoate-depleted methanogenic consortium immediately started to utilize β-phenylpropionate; benzoate, cyclohexanecarboxylate, and adipate were detected as products. This culture was not adapted to ferment β-cyclohexylpropionate, nor was it detected among the intermediates of β-phenylpropionate degradation. Cinnamate, however, was immediately utilized. Phenylacetate was metabolized only after about 18 days' adaptation; cyclohexylacetate, cyclohexylideneacetate, and adipate were identified in the acid ether extracts of the culture fluid, but benzoate and cyclohexanecarboxylate were not. Cyclohexanone and 2-hydroxycyclohexanone were present in the neutral ether extract. These results support the view that the methanogenic degradation of phenylpropionate occurs via β-oxidation of its CoA-ester to benzoyl-CoA. Phenylacetate is initially dealt with quite differently. The carbon atom in the position β to the carboxyl group is part of the aromatic ring, and its β-oxidation is blocked. α-Oxidation is prohibited since it normally requires molecular oxygen. This situation could be circumvented by reduction of the aromatic ring to give cyclohexylacetate, which is amenable to a CoA-mediated β-oxidation sequence similar to the aerobic metabolism reported by Ougham & Trudgill (80). Alternatively, a radical mechanism of α-oxidation may be envisaged. (d) The methanogenesis of the naturally occurring amino acids in this consortium gave the following results (11). Phenylalanine gave rise to phenylpyruvate and phenylacetate, followed by its reductive degradation as described above. Tyrosine afforded *p*-hydroxyphenylacetate, phenylacetate, and cyclohexylacetate along with small amounts of *p*-hydroxyphenylpyruvate, adipate, propionate, and acetate in the acid ether fraction of the culture fluid; the neutral ether extract contained *p*-cresol and phenol as major components and cyclohexanol and cyclohexanone as minor constituents. Tryptophan was deaminated to indol-3-ylacetate, then metabolized through indole; anthranilate, salicylate, and benzoate appeared in acid ether extracts of the culture, while indole and cyclohexanone were detected in the neutral ether extract.

From these results, tentative metabolic pathways for the methanogenesis of the aromatic amino acids have been proposed (Figure 5). Anaerobic transformations of the alanine side chain are similar to those observed in the rumen (74a, 92) and in pure cultures of clostridia (35); a unique characteristic of the process is the dismemberment of the aromatic residues for cell growth and methanogenesis. Determination of the exact sequence of enzymatic reactions is always difficult in a mixed culture; different bacteria possess different types of biochemical competence, and each intermediate is liable to lead to more than one product. Different metabolic processes involve dehydroxylation or use of a molecule of water for hydroxylation, the addition and abstraction of hydrogen, carboxylation and decarboxylation, and the elimination of substituent groups, e.g. halogens by a reductive process or amino groups by a hydroxyl. Some bacteria have evolved special enzymes that can operate anaerobically. For example, *p*-cresol methylhydroxylase converts the methyl group to carboxyl (60, 61), provided that a positive electron acceptor is available, and a β elimination enzyme system produces phenol from tyrosine directly in *Escherichia coli phenologenes* (66) and *Clostridium tetanomorphum* (22).

Grbic-Galic & Young (49) and Healy et al (57) investigated the methanogenesis of ferulate and other monomers of lignin degradation. Bromoethanesulfonate was used to inhibit the production of methane from acetate and thus facilitate the accumulation of intermediates. Ferulate methanogenesis cultures gave cinnamate, phenylpropionate, phenylacetate, cyclohexanecarboxylate, benzoate, pimelate, adipate, butyrate, propionate, and acetate. Grbic-Galic (44, 45) augmented this list to include caffeate, *p*-hydroxycinnamate, cyclohexane, methylcyclohexane, cyclohexanone, methylcyclohexanone, and heptanoate. When the consortium was deprived of hydrogen-scavenging bacteria, the fermentative bacteria in the consortium produced different, more reduced products, including toluene, benzene, ethylbenzene, *p*-cresol, phenol, and catechol (46). A facultative anaerobe, *Enterobacter cloacae,* was isolated in pure culture from the ferulate methanogenic consortium; it could demethoxylate and dehydroxylate the ring substituents and reduce the side chain under anaerobic conditions, producing

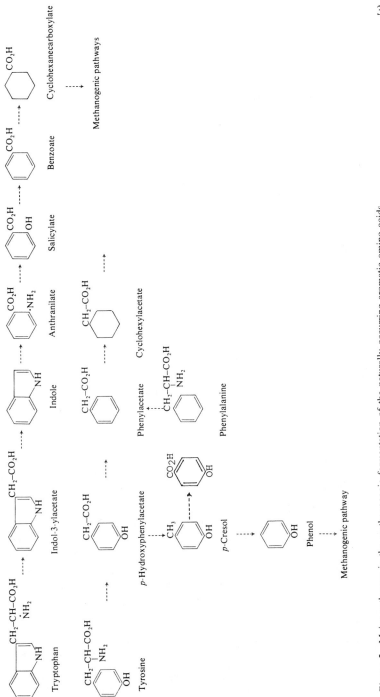

Figure 5 Major pathways in the methanogenic fermentation of the naturally occurring aromatic amino acids.

phenylpropionate and phenylacetate. However, it was only capable of metabolizing these products further by ring fission after an extended incubation period (46, 47).

Metabolism of Aromatic Hydrocarbons

Anaerobic degradation of aromatic hydrocarbons (benzene, toluene, and xylenes) under denitrifying conditions has been reported (12a, 72b). Grbic-Galic & Young (49) detected toluene in ferulate methanogenic consortia partially inhibited with 2-bromoethanesulfonate (to disrupt interspecies hydrogen transfer). Later, Grbic-Galic & Vogel (48, 110) reported that subcultures derived from these cultures had a remarkable capability to transform rapidly relatively high amounts of benzene or toluene as sole carbon and energy source into hydroxylated intermediates under strictly anaerobic methanogenic conditions. Using gas chromatography–mass spectrometry, ^{18}O-labeled water, and ^{14}C-labeled compounds, they established the following: (a) The oxygen incorporated into the hydroxylated products comes from water. (b) A high percentage of $^{14}CO_2$ was recovered from [methyl-^{14}C]toluene, which suggests that toluene is converted to carboxyl and not to methane. A low percentage of $^{14}CO_2$ was produced from ring-labeled toluene or benzene, which indicates incomplete conversion of the ring carbon to CO_2. (c) Aromatic intermediate products detected in the toluene-fed cultures were p-cresol, o-cresol (smaller amount), benzyl alcohol, benzaldehyde, benzoate, 2-hydroxybenzoate, benzene, and phenol; alicyclic products were 2- and 4-methylcyclohexanol, methylcyclohexane, cyclohexanone, and cyclohexene; and aliphatic acid products were heptanoate, 4-methylhexanoate, hexanoate, 4-methylpentanoate, pentanoate, 2-methylbutanoate, butanoate, propanoate, acetate, and formate. (d) Benzene gave phenol as the major aromatic metabolite, followed by cyclohexanone and aliphatic acids, including propanoate. Figure 6 illustrates these transformations. Catechol bodies, which are produced in aerobic aromatic ring metabolism, were never encountered as intermediates in any anaerobic dissimilation of benzenoid structures.

The contribution of intestinal microflora to the fate of chemical components of the diet and to the metabolism of drugs in humans has been extensively studied. Since this environment is largely anaerobic, it is not surprising that some of the transformations of substituent aromatic groups described in this review occur (reviewed in 85, 86). Rarely, however, is the aromatic ring disrupted.

POSSIBLE MECHANISMS AND ENZYMES INVOLVED

Anaerobic Aromatic Ring Reduction

That some microorganisms have evolved enzymatic systems that degrade benzenoid structures under anaerobic conditions, through reduction of the

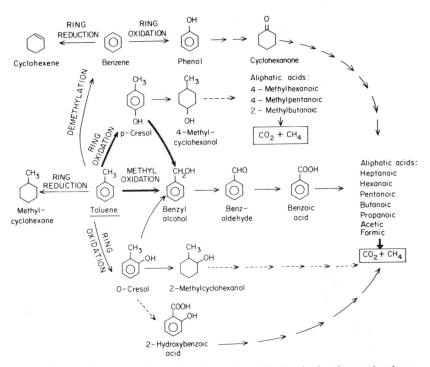

Figure 6 Tentative sequence of anaerobic toluene degradation by mixed methanogenic cultures. All the compounds shown were detected and identified by gas chromatography and mass spectroscopy.

ring followed by cleavage, is a remarkable biochemical feat. The resonance stabilization of the aromatic ring is great (~ 160 kJ mol^{-1}); chemical considerations require that for additions to an aromatic nucleus to occur, the system of bonding characterized by very extensive delocalization of the π-electrons must be converted to one in which little of the delocalization energy remains. Accordingly, only a few substituents, e.g. chlorine and hydrogen, can be added to the benzene ring under easily attainable conditions. The critical step in the reduction of benzene [$\Delta G_0(6H \triangleq 3H_2) \sim -100$ kJ mol^{-1}] is the addition of the first two hydrogen atoms, since the reaction is endergonic [$\Delta G_0(2H \triangleq H_2) \sim +60$ kJ mol^{-1}]. Nature probably overcomes this problem by binding the reaction product and reducing it further. The reductions of the second and third double bonds are exergonic reactions [$\Delta G_0(2H \triangleq H_2) \sim -90$ and ~ -70 kJ mol^{-1}, respectively] and are therefore thermodynamically favored.

The first step in the biochemical reduction of benzoate is its activation through conversion into benzoyl-CoA by an inducible ligase. The thioesterification of α-β unsaturated acids increases their reactivity toward reducing agents owing to the resonance effect between the double bond and the S–CO group (74). This may apply to the aromatic nucleus. Another contribution of coenzyme A may be that the adenine in the CoA thioester interacts with the aromatic nucleus, producing an intramolecular complex (75a, 110a). Activation may be required for covalent binding of the aromatic acid to the active site of a protein.

The reductive system has not yet been obtained cell free with benzoyl-CoA as the substrate, and the system remains unclear. Apparently the benzoyl-CoA is first transferred to a peptide molecule, whose precise role is not yet known. There is preliminary evidence for the participation of a low-potential reductant (58).

The pathways operating after the formation of cyclohexanoyl-CoA have been demonstrated in cell-free systems (16, 84, 111); they conform with the known pathways of the β-oxidation of fatty acids (16, 84). There has been no progress in understanding the enzymology of the anaerobic reductive degradation of phenol; none of the steps have been demonstrated in cell-free systems.

Anaerobic Hydroxylation of Aromatic Hydrocarbons

Benzene and toluene degradation in acclimatized methanogenic consortia, which occurs by way of a hydroxylation reaction to phenol and cresol, is a remarkable process. The methyl carbons of m-cresol and p-cresol appear to be treated differently (84a). The source of the hydroxyl must be water (48, 110). Although there is no information regarding the initial oxidation reaction, parallels may be drawn from the hydroxylation of heterocyclic substrates, e.g. nicotinate fermentation by *Clostridium barkeri* (53, 63, 97). The first step is the formation of 6-hydroxynicotinate by an $NADP^+$-dependent nicotinate dehydrogenase (hydroxylase) (59), whose formation is selenite dependent. This is followed by a ring reduction and a hydrolytic ring-cleavage sequence. Hopper (60, 61) discovered another type of hydroxylase in *Pseudomonas putida*, p-cresol methylhydroxylase [4-cresol: (acceptor) oxidoreductase (methyl-hydroxylating) EC 1.17.99.1]. This anaerobic dehydrogenase is a flavocytochrome c. It reacts with p-cresol, probably forming a quinone methide intermediate by dehydrogenation, which then adds on water to give the alcohol (75) (Figure 7). Note that in the first example cited, hydroxylation of the substrate precedes dehydrogenation of the product, while in the latter case dehydrogenation takes place prior to hydration. The reactions involved in the anaerobic oxidation of the aromatic hydrocarbons are both endergonic with most electron acceptors:

toluene + H_2O = p-cresol + H_2; ΔG_0 = +71 kJ mol^{-1},   3.

benzene + H_2O = phenol + H_2; ΔG_0 = +73 kJ mol^{-1}.   4.

Energy or a positive electron acceptor is therefore required. Perhaps the molecules must be activated in some way for the reactions to proceed. Benzene does have very weak basicity, so attack by a proton would produce a positive charge on the ring for subsequent addition of a negatively charged hydroxyl moiety. More often, reductive dehydroxylation or dehalogenation seems to be used as an energy-yielding reaction. Carboxylation/decarboxylation reactions (Kolbe-Schmidt carboxylations), α-oxidation of acetyl side groups, and many reactions of heterocyclic compounds are interesting. The enzymatic mechanisms and energetics of all these reactions remain a challenge.

ANAEROBIC DEGRADATION OF LIGNIN

Lignocelluloses make up about 95% of the earth's renewable, land-produced biomass; about one quarter of this is made up of the aromatic polymer lignin. Lignin's relative inertness makes it a valuable structural component of plants; however, it has one of the slowest turnover rates of all natural products (37). In contrast to most other natural polymers, lignin is produced randomly by radical coupling of coniferyl alcohols; its structural disorder does not lend itself to the intimate stereochemical relationship of enzyme and substrate.

A white-rot fungus, *Phanerochaete chrysosporium*, produces an extracellular enzyme that degrades lignin in the presence of hydrogen peroxide (91). This peroxidase reacts with H_2O_2 to produce an oxo-Fe(IV)-porphyrin cation radical; this highly reactive species removes an electron from one of lignin's many oxygenated aryl rings, forming an aromatic cation radical.

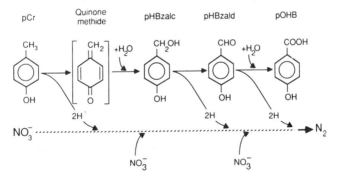

Figure 7 Proposed model pathway for the anaerobic oxidation of *p*-cresol to *p*-hydroxybenzoate under denitrifying conditions. Initial incorporation of water-derived oxygen is mediated by a proposed methyl hydroxylase. Further oxidation can be mediated by alkyl dehydrogenase. Nitrate serves as the terminal electron acceptor for each oxidation step, so the reducing equivalents (6H) generated in the pathway reduce NO_3^- to N_2. *pHBzalc*, *p*-hydroxybenzyl alcohol; *pHBzald*, *p*-hydroxybenzaldehyde.

These radicals cause side-chain fragmentation (C–C cleavage), which leads to breakdown of the polymer.

Dehydrogenation polymers of *aryl*-^{14}C-labeled coniferyl alcohols have been used as synthetic lignins to study lignin degradation. When these polymers were incubated anaerobically with stacked steer bedding or freshwater lake sediment in the presence or absence of nitrate, no label release was detected. When they were aerobically incubated $^{14}CO_2$ formed. Hackett et al (51) therefore concluded that anaerobic lignin degradation does not occur.

Thermochemical pretreatment, using a high temperature and alkaline pH, partially solubilized lignin, producing lower–molecular weight fractions (14,000, 700, and 300). Anaerobic degradation occurred when the solubilized lignin fractions were incubated with an inoculum from an anaerobic, mesophilic sludge digester; after 30 days the fraction with the highest molecular weight was much reduced. The reduction was attributed to cleavage of the intermonomer bonds during anaerobic attack, which produced lignin monomers readily degraded to methane and CO_2 (25). In similar experiments, Zeikus and collaborators (117) found that lignin moieties with molecular weight greater than 850 were not degraded by anaerobic methanogenic consortia during 110 days' incubation.

Thus, the fervent dream of workers in this field, that of developing microbial methods for the utilization of native lignin, has not yet been realized.

EPILOG

Results obtained from laboratory investigations allow anaerobic pathways of individual aromatic substances to be determined. However, caution must be exercised in the attempt to predict the environmental fate of organic compounds based on laboratory results alone. These results should be critically correlated with results obtained in field studies; natural environments differ in physical, chemical, and microbiological aspects. For instance, an inoculum from rumen liquor can be adapted to accomplish the methanogenesis of the aromatic amino acids; yet, as Balba & Evans (6) showed, phenylalanine and tyrosine do not undergo ring cleavage in the rumen. Adaptation reveals potentialities, while natural situations present actualities. Conversely, modern analytical methods allow us to detect compounds in the parts-per-million range in field studies; more critical assessment of their concentrations and kinetics would be desirable.

The outlines of the subject have now been delineated, but much remains to be accomplished on the enzymology, energetics, regulation, genetics, and ecological significance of these degradative processes.

ACKNOWLEDGMENT

W. Charles Evans wishes to thank the Royal Society for a grant to defray his laboratory expenses.

Literature Cited

1. Aftring, P. R., Chalker, B. E., Taylor, B. F. 1981. Degradation of phthalic acids by a denitrifying mixed culture of bacteria. *Appl. Environ. Microbiol.* 41: 1117–83
2. Aftring, P. R., Taylor, B. F. 1981. Aerobic and anaerobic catabolism of phthalic acid by a nitrate-respiring bacterium. *Arch. Microbiol.* 130:101–4
2a. Bak, F., Widdel, F. 1986. Anaerobic degradation of indolic compounds by sulfate-reducing enrichment cultures, and description of *Desulfobacterium indolicum* gen. nov., sp. nov. *Arch. Microbiol.* 146:170–76
3. Bak, F., Widdel, F. 1986. Anaerobic degradation of phenol and phenolic derivatives by *Desulfobacter phenolicum* sp. nov. *Arch. Microbiol.* 146:177–80
4. Bakker, G. 1977. *Degradation of aromatic compounds by microorganisms in dissimilatory nitrate reduction.* PhD thesis. Technical Univ., Delft, the Netherlands
5. Bakker, G. 1977. Anaerobic degradation of aromatic compounds in the presence of nitrate. *FEMS Microbiol. Lett.* 1:103–8
6. Balba, M. T., Evans, W. C. 1977. The origin of hexahydrohippurate (cyclohexanoylglycine) in the urine of herbivores. *Biochem. Soc. Trans.* 5:300–2
7. Balba, M. T., Evans, W. C. 1977. The methanogenic fermentation of aromatic compounds. *Biochem. Soc. Trans.* 5: 302–4
8. Balba, M. T., Evans, W. C. 1979. The methanogenic fermentation of ω-phenylalkane carboxylic acids. *Biochem. Soc. Trans.* 7:403–5
9. Balba, M. T., Clarke, N. A., Evans, W. C. 1979. The methanogenic fermentation of plant phenolics. *Biochem. Soc. Trans.* 7:1115–16
10. Balba, M. T., Evans, W. C. 1980. The methanogenic biodegradation of catechol by a microbial consortium: evidence for the production of phenol through *cis*-benzenediol. *Biochem. Soc. Trans.* 8:452–54
11. Balba, M. T., Evans, W. C. 1980. Methanogenic fermentation of the naturally occurring aromatic amino acids by a microbial consortium. *Biochem. Soc. Trans.* 8:625–27
12. Balba, M. T., Evans, W. C. 1980. The anaerobic dissimilation of benzoate by *Pseudomonas aeruginosa* coupled with *Desulfovibrio vulgaris*, with sulphate as terminal electron acceptor. *Biochem. Soc. Trans.* 8:632–35

12a. Battermann, G., Werner, P. 1984. Beseitigung einer Untergrundkontamination mit Kohlenwasserstoffen durch mikrobiellen Abbau. *Wasser Abwasser* 125:366–73
13. Berry, D. F., Francis, A. J., Bollag, J. M. 1987. Microbial metabolism of homocyclic and heterocyclic aromatic compounds under anaerobic conditions. *Microbiol. Rev.* 51:43–59
14. Berry, D. F., Madsen, E. L., Bollag, J. M. 1987. Conversion of indole to oxindole under methanogenic conditions. *Appl. Environ. Microbiol.* 53:180–82
15. Blake, C. K., Hegeman, G. D. 1987. Plasmid pCB1 carries genes for anaerobic benzoate catabolism in *Alcaligenes xylosoxidans* subsp. *denitrificans* PN-1. *J. Bacteriol.* 169:4878–83
16. Blakley, E. R. 1978. The microbial degradation of cyclohexane-carboxylic acid by a β-oxidation pathway with simultaneous adaptation to the utilization of benzoate. *Can. J. Microbiol.* 24:847–55
17. Boruff, C. S., Buswell, A. M. 1934. The anaerobic fermentation of lignin. *J. Am. Chem. Soc.* 56:886–88
18. Bossert, I., Young, L. Y. 1986. Anaerobic oxidation of *p*-cresol by a denitrifying bacterium. *Appl. Environ. Microbiol.* 52:1117–22
19. Boyd, S. A., Shelton, D. R. 1984. Anaerobic degradation of chlorophenols in fresh and acclimated sludge. *Appl. Environ. Microbiol.* 47:272–77
20. Boyd, S. A., Shelton, D. R., Berry, D., Tiedje, J. M. 1983. Anaerobic biodegradation of phenolic compounds in digested sludge. *Appl. Environ. Microbiol.* 46:50–54
21. Braun, K., Gibson, D. T. 1984. Anaerobic degradation of 2-aminobenzoate (anthranilate) by denitrifying bacteria. *Appl. Environ. Microbiol.* 48:102–7
22. Brot, N., Smit, Z., Weissbach, H. 1965. A β-elimination enzyme from *Clostridium tetanomorphum* producing phenol from tyrosine. *Arch. Biochem.* 112:1–6
23. Chmielowski, J., Grossman, A., Wegnzynowska, I. 1964. Biochemical degradation of some phenols during the methane fermentation. *Zesz. Nauk. Politech. Slask. Inz. Sanit.* 8:97–122
24. Clark, F. M., Fina, L. R. 1952. The anaerobic decomposition of benzoic acids during methane fermentation. *Arch. Biochem. Biophys.* 36:26–32
25. Colberg, P. J., Young, L. Y. 1982. Biodegradation of lignin-derived mole-

cules under anaerobic conditions. *Can. J. Microbiol.* 28:886–89

26. Cord-Ruwisch, R., Garcia, J. L. 1985. Isolation and characterization of an anaerobic benzoate-degrading, spore-forming, sulphate-reducing bacterium, *Desulfotomaculum sapomandens* sp. nov. *FEMS Microbiol. Lett.* 29:325–30

27. Dagley, S. 1971. Catabolism of aromatic compounds by microorganisms. *Adv. Microb. Physiol.* 6:1–46

28. Dolfing, J., Tiedje, J. M. 1986. Hydrogen cycling in a three-tiered food web growing on the methanogenic conversion of 3-chlorobenzoate. *FEMS Microbiol. Ecol.* 38:293–98

29. Dutton, P. L., Evans, W. C. 1967. Dissimilation of aromatic substrates by *Rhodopseudomonas palustris. Biochem. J.* 104:30–31P

30. Dutton, P. L., Evans, W. C. 1968. The photometabolism of benzoate by *Rhodopseudomonas palustris;* inhibition by chromatophores and fatty acids. *Biochem. J.* 107:28P

31. Dutton, P. L., Evans, W. C. 1968. The photometabolism of benzoic acid by *Rhodopseudomonas palustris;* a new reductive pathway. *Biochem. J.* 109:5P

32. Dutton, P. L., Evans, W. C. 1969. The metabolism of aromatic compounds by *Rhodopseudomonas palustris. Biochem. J.* 113:525–35

33. Dutton, P. L., Evans, W. C. 1970. Inhibition of aromatic photometabolism in *Rhodopseudomonas palustris* by fatty acids. *Arch. Biochem. Biophys.* 136:228–32

34. Dutton, P. L., Evans, W. C. 1978. The metabolism of aromatic compounds by Rhodospirillaceae. In *The Photosynthetic Bacteria,* ed. R. K. Clayton, W. R. Sistrom, pp. 719–26. New York: Plenum

35. Elsden, S. R., Hilton, M. G., Waller, J. M. 1976. Metabolic transformations of the alanine side chain of the aromatic amino acids by pure cultures of clostridia. *Arch. Microbiol.* 107:283–88

36. Evans, W. C. 1969. Microbial transformations of aromatic compounds. In *Fermentation Advances,* ed. D. Perlman, pp. 649–87. New York: Academic

37. Evans, W. C. 1977. Biochemistry of the bacterial catabolism of aromatic compounds in anaerobic environments. *Nature* 270:17–22

38. Ferry, J. G., Wolfe, R. S. 1976. Anaerobic degradation of benzoate to methane by a microbial consortium. *Arch. Microbiol.* 107:33–40

39. Fewson, C. A. 1981. Biodegradation of aromatics with industrial relevance. In *Microbial Degradation of Xenobiotics and Recalcitrant Compounds,* ed. T. Leisinger, A. M. Cook, R. Hütter, J. Nüesch, pp. 141–79. London: Academic

40. Fina, L. R., Bridges, R. L., Coblentz, T. H., Roberts, F. F. 1978. The anaerobic decomposition of benzoic acid during methane fermentation. III. The fate of carbon four and the identification of propanoic acid. *Arch. Microbiol.* 118:169–72

41. Fina, L. R., Fiskin, A. M. 1960. The anaerobic decomposition of benzoic acid during methane fermentation. II. Fate of carbon atoms one and seven. *Arch. Biochem. Biophys.* 91:163–65

42. Geissler, J. F., Harwood, C. S., Gibson, J. 1988. Purification and properties of benzoyl-CoA ligase, a *Rhodopseudomonas palustris* enzyme involved in the anaerobic degradation of benzoate. *J. Bacteriol.* 170:1709–14

43. Grbic-Galic, D. 1983. Anaerobic degradation of coniferyl alcohol by methanogenic consortia. *Appl. Environ. Microbiol.* 46:1442–46

44. Grbic-Galic, D. 1985. Fermentative and oxidative transformation of ferulate by a facultative anaerobic bacterium isolated from sewage sludge. *Appl. Environ. Microbiol.* 50:1052–57

45. Grbic-Galic, D. 1986. Anaerobic production and transformation of aromatic hydrocarbons and substituted phenols by ferulic acid–degrading BESA-inhibited methanogenic consortia. *FEMS Microbiol. Ecol.* 38:161–69

46. Grbic-Galic, D. 1986. O-Demethylation, dehydroxylation, ring-reduction and cleavage of aromatic substrates by Enterobacteriaceae under anaerobic conditions. *J. Appl. Bacteriol.* 62:702–8

47. Grbic-Galic, D., Pat-Polasko, L. 1985. *Enterobacter cloacae* DG-6: a strain that transforms methoxylated aromatics under aerobic and anaerobic conditions. *Curr. Microbiol.* 12:321–24

48. Grbic-Galic, D., Vogel, T. M. 1987. Transformation of toluene and benzene by mixed methanogenic cultures. *Appl. Environ. Microbiol.* 53:254–60

49. Grbic-Galic, D., Young, L. Y. 1985. Methane fermentation of ferulate and benzoate: anaerobic degradation pathways. *Appl. Environ. Microbiol.* 50:292–97

50. Guyer, M., Hegeman, G. D. 1969. Evidence for a reductive pathway for the anaerobic metabolism of benzoate. *J. Bacteriol.* 99:906–7

51. Hackett, W. F., Connors, W. J., Kirk, T. K., Zeikus, J. G. 1977. Microbial

decomposition of synthetic ^{14}C-labeled lignins in nature: lignin biodegradation in a variety of natural materials. *Appl. Environ. Microbiol.* 33:43–51

52. Hanselmann, K. W. 1982. Lignochemicals. *Experientia* 38:176–89

53. Harary, I. 1957. Bacterial fermentation of nicotinic acid. *J. Biol. Chem.* 226: 815–31

54. Harwood, C. S., Gibson, J. 1986. Uptake of benzoate by *Rhodopseudomonas palustris* grown anaerobically in light. *J. Bacteriol.* 165:504–9

55. Harwood, C. S., Gibson, J. 1988. Anaerobic and aerobic metabolism of diverse aromatic compounds by the photosynthetic bacterium *Rhodopseudomonas palustris*. *Appl. Environ. Microbiol.* 54:712–17

56. Healy, J. B., Young, L. Y. 1978. Catechol and phenol degradation by a methanogenic population of bacteria. *Appl. Environ. Microbiol.* 35:216–18

57. Healy, J. B., Young, L. Y., Reinhard, M. 1980. Methanogenic decomposition of ferulic acid, a model lignin derivative. *Appl. Environ. Microbiol.* 39: 436–44

58. Hegeman, G. D., 1988. Anaerobic growth of bacteria on benzoate under denitrifying conditions. *J. Bacteriol.* In press

59. Holcenberg, J., Stadtman, E. R. 1969. Nicotinic acid metabolism. III. Purification and properties of a nicotinic acid hydroxylase. *J. Biol. Chem.* 244:1194–203

60. Hopper, D. J. 1976. The hydroxylation of *p*-cresol and its conversion to *p*-hydroxybenzyl alcohol in *Pseudomonas putida*. *Biochem. Biophys. Res. Commun.* 69:162–68

61. Hopper, D. J. 1978. Incorporation of [^{18}O]water in the formation of *p*-hydroxybenzyl alcohol by the *p*-cresolmethylhydroxylase from *Pseudomonas putida*. *Biochem. J.* 175:345–47

62. Hungate, R. E. 1969. A roll tube method for cultivation of strict anaerobes. In *Methods in Microbiology*, ed. J. R. Norris, D. W. Ribbons, 3B: 117–32. London: Academic

63. Hunt, A. L., Hughes, D. E., Lowenstein, J. M. 1958. The hydroxylation of nicotinic acid by *Pseudomonas fluorescens*. *Biochem. J.* 69:170–73

64. Hutber, G., Evans, W. C., Ribbons, D. W. 1988. Evidence for anaerobic dehydroxylation of *p*-hydroxybenzoate by *Rhodopseudomonas palustris*. Submitted

65. Hutber, G., Ribbons, D. W. 1983. Involvement of coenzyme A esters in the

metabolism of cyclohexanecarboxylate by *Rhodopseudomonas palustris*. *J. Gen. Microbiol.* 129:2413–20

66. Ichihara, K., Yoshimatsu, H., Sakamoto, Y. 1956. Conversion of tyrosine to phenol by *E. coli phenologenes*. *J. Biochem. Tokyo* 43:803

67. Kaiser, J. P., Hanselmann, K. W. 1982. Aromatic chemicals through microbial conversion of lignin monomers. *Experientia* 38:176

68. Keith, C. L. 1972. The methanogenic fermentation of benzoate by a consortium. *Diss. Abstr. Int. B* 33:3214–15

69. Keith, C. L., Bridges, R. L., Fina, L. R., Iverson, K. L., Cloran, J. A. 1978. The anaerobic decomposition of benzoic acid in methane fermentation. 1V. Decomposition of the ring and volatile fatty acids forming on ring rupture. *Arch. Microbiol.* 118:173–76

69a. Knoll, G., Winter, J. 1987. Anaerobic degradation of phenol in sewage sludge. Benzoate formation from phenol and CO_2 in the presence of hydrogen. *Appl. Microbiol. Biotechnol.* 25:384–91

70. Kreikenbohm, R., Pfennig, N. 1985. Anaerobic degradation of 3,4,5-trimethoxybenzoate by defined mixed culture of *Acetobacterium woodii*, *Pelobacter acidigallici*, and *Desulfobacter postgatei*. *FEMS Microbiol. Ecol.* 31:29–38

71. Krumholz, L. R., Bryant, M. P. 1986. *Syntrophococcus sucromutans* sp. nov. gen. nov. uses carbohydrates as electron donors and formate, methoxymonobenzoids or *Methanobrevibacter* as electron acceptor system. *Arch. Microbiol.* 143:313–18

72. Krumholz, L. R. Bryant, M. P. 1986. *Eubacterium oxidoreducens* sp. nov. requiring H_2 or formate to degrade gallate, pyrogallol, phloroglucinol, and quercetin. *Arch. Microbiol.* 144:8–14

72a. Krumholz, L. R. Crawford, R. L., Hemling, M. E., Bryant, M. P. 1987. Metabolism of gallate and phloroglucinol in *Eubacterium oxidoreducens* via 3-hydroxy-5-oxohexanoate, *J. Bacteriol.* 169:1886–90

72b. Kuhn, E. P., Colberg, P. J., Schnoor, J. L., Wanner, O., Zehnder, A. J. B., Schwarzenbach, R. P. 1985. Microbial transformations of substituted benzenes. *Environ. Sci. Technol.* 19:961–68

73. Leadbetter, F. R., Hawk, A. 1965. Aromatic acid utilization by Athiorhodaceae. *J. Appl. Bacteriol.* 27:448

74. Lynen, F. 1953, Functional group of coenzyme A and its metabolic relations, especially in the fatty acid cycle. *Fed. Proc.* 12:683–91

74a. Martin, A. K. 1982. The origin of urinary aromatic compounds excreted by ruminants. 3. The metabolism of phenolic compounds to simple phenols. *Br. J. Nutr.* 48:497–507

75. McIntire, W., Hopper, D. J., Singer, T. P. 1985. *p*-Cresol-methylhydroxylase—assay and general properties. *Biochem. J.* 228:325–35

75a. Mieyal, J. J., Webster, L. T., Siddiqui, U. A. 1974. Benzoyl and hydroxybenzoyl esters of coenzyme A. *J. Biol. Chem.* 249:2633–40

75b. Mikesell, M. D., Boyd, S. A. 1986. Complete reductive dechlorination and mineralization of pentachlorophenol by anaerobic microorganisms. *Appl. Environ. Microbiol.* 52:861–65

76. Mountfort, D. O., Bryant, M. P. 1982. Isolation and characterization of an anaerobic syntrophic benzoate degrading bacterium from sewage sludge. *Arch. Microbiol.* 133:249–56

77. Neilson, A. H., Allard, A. S., Lindgren, C., Remberger, M. 1987. Transformations of chloroguaiacols, chloroveratroles and chlorocatechols by stable consortia of anaerobic bacteria. *Appl. Environ. Microbiol.* 53:2511–19

78. Nottingham, P. M., Hungate, R. E. 1969. Methanogenic fermentation of benzoate. *J. Bacteriol.* 98:1170–72

79. Oshima, T. 1965. On the anaerobic metabolism of aromatic compounds in the presence of nitrate by soil microorganisms. *Z. Allg. Mikrobiol.* 5:386–94

80. Ougham, H. J., Trudgill, P. W. 1978. The aerobic microbial metabolism of cyclohexylacetate by *Nocardia*. *Biochem. Soc. Trans.* 6:1324–26

81. Patel, T. R., Jure, K. G., Jones, G. A. 1981. Catabolism of phloroglucinol by the rumen anaerobe *Coprococcus*. *Appl. Environ. Microbiol.* 42:1010–17

82. Proctor, M. H., Scher, S. 1960. Decomposition of benzoate by a photosynthetic bacterium. *Biochem. J.* 76:33P

83. Reineke, W., Knackmuss, H. J. 1984. Microbial metabolism of haloaromatics: isolation and properties of a chlorobenzene-degrading bacterium. *Appl. Environ. Microbiol.* 47:395–402

84. Rho, E. M., Evans, W. C. 1975. The aerobic metabolism of cyclohexanecarboxylate by *Acinetobacter anitratum*. *Biochem. J.* 148:11–15

84a. Roberts, D. J., Fedorak, P. M. 1986. Comparison of the fates of the methyl carbons of *m*-cresol and *p*-cresol in methanogenic consortia. *Can. J. Microbiol.* 33:335–38

84b. Samain, E., Albagnac, G., Dubourguier, H. C. 1986. Initial steps of catabolism of trihydroxybenzenes in *Pelobacter acidigallici*. *Arch. Microbiol.* 144:242–44

85. Scheline, R. R. 1973. Metabolism of foreign compounds by gastro-intestinal microorganisms. *Microbiol. Rev.* 25:451–523

86. Scheline, R. R. 1978. *Mammalian Metabolism of Plant Xenobiotics.* London: Academic

87. Schennen, U., Braun, K., Knackmuss, H. J. 1985. Anaerobic degradation of 2-fluorobenzoate by benzoate-degrading, denitrifying bacteria. *J. Bacteriol.* 161:321–25

88. Scher, S. 1960. Photometabolism of benzoic acid by Athiorhodaceae. In *Progress in Photobiology. Proc. 3rd Int. Congr. Photobiol.*, ed. B. Christensen, B. Buckman, p. 583. London: Elsevier

89. Schink, B., Pfennig, N. 1982. Fermentation of trihydroxybenzenes by *Pelobacter acidigallici* gen. nov. sp. nov., a new strictly anaerobic non-spore-forming bacterium. *Arch. Microbiol.* 133:195–201

90. Deleted in proof

91. Schoemaker, H. E., Harvey, P. J., Bowen, R. M., Palmer, J. M. 1985. On the mechanism of enzymatic lignin breakdown. *FEBS Lett.* 183:7–12

92. Scott, T. W., Ward, P. F., Dawson, R. M. C. 1964. Transformations of the aromatic amino acids in rumen liquor. *Biochem. J.* 90:12–23

93. Shelton, D. R., Tiedje, J. M. 1984. Isolation and partial characterization of bacteria in an anaerobic consortium that mineralizes 3-chlorobenzoic acid. *Appl. Environ. Microbiol.* 48:840–48

93a. Shlomi, E. R., Lankhorst, A., Prins, R. A. 1978. Methanogenic fermentation of benzoate in an enrichment culture. *Microb. Ecol.* 4:249–61

94. Sleat, R., Robinson, J. P. 1983. Methanogenic degradation of sodium benzoate in a profundal sediment from a small eutrophic lake. *J. Microbiol.* 129:141–52

95. Sleat, R., Robinson, J. P. 1984. A review. The bacteriology of anaerobic degradation of aromatic compounds. *J. Appl. Bacteriol.* 57:381–94

96. Smolinski, W. J., Suflita, J. M. 1987. Biodegradation of cresol isomers in anoxic aquifers. *Appl. Environ. Microbiol.* 53:710–16

97. Stadtman, E. R., Stadtman, T. C., Pastan, I., Smith, L. D. 1972. *Clostridium barkeri* sp. n. *J. Bacteriol.* 110:758–60

98. Suflita, J. M., Gibson, S. A. 1985. Biodegradation of haloaromatic substrates in a shallow anoxic ground water

aquifer. *Proc. 2nd Int. Conf. Ground Water Qual. Res.*, ed. N. N. Durham, A. E. Redelfs, pp. 30–32. Stillwater, Okla: Univ. Cent. Water Res. Okla. State Univ.

99. Suflita, J. M., Horowitz, A., Schelton, D. R., Tiedje, J. M. 1982. Dehalogenation: a novel pathway for the anaerobic biodegradation of haloaromatic compounds. *Science* 218:1115–17

100. Suflita, J. M., Miller, G. D. 1985. The microbial metabolism of chlorophenolic compounds in ground water aquifers. *Environ. Toxicol. Chem.* 4:751–58

101. Szewzyk, U., Szewzyk, R., Schink, B. 1985. Methanogenic degradation of hydroquinone and catechol via reductive dehydroxylation to phenol. *FEMS Microbiol. Ecol.* 31:79–87

102. Tarvin, D., Buswell, A. M. 1934. The methane fermentation of organic acids and carbohydrates. *J. Am. Chem. Soc.* 56:1751–55

103. Taylor, B. F., Campbell, W. L., Chinoy, I. 1970. Anaerobic degradation of the benzene nucleus by a facultative anaerobic microorganism. *J. Bacteriol.* 102:430–37

104. Taylor, B. F., Heeb, M. J. 1972. The anaerobic degradation of aromatic compounds by a denitrifying bacterium. *Arch. Microbiol.* 83:165–71

105. Tsai, D. G., Jones, G. A. 1975. Isolation and identification of rumen bacteria capable of anaerobic phloroglucinol degradation. *Can. J. Microbiol.* 21:794–801

106. Tschech, A., Schink, B. 1985. Fermentative degradation of resorcinol and resorcylic acids. *Arch. Microbiol.* 143:52–59

107. Tschech, A., Schink, B. 1986. Fermentative degradation of monohydroxybenzoates by defined syntrophic coculture. *Arch. Microbiol.* 145:396–402

108. Tschech, A., Fuchs, G. 1987. Anaerobic degradation of phenol by pure cultures of newly isolated denitrifying *Pseudomonas. Arch. Microbiol.* 148:213–17

109. van Niel, C. B. 1941. The bacterial photosynthesis. *Adv. Enzymol.* 1:263–328

110. Vogel, T. M., Grbic-Galic, D. 1986. Incorporation of oxygen from water into toluene and benzene during anaerobic fermentation. *Appl. Environ. Microbiol.* 52:200–2

110a. Webster, L. T., Mieyal, J. J., Siddiqui, U. A. 1974. Benzoyl and hydroxybenzoyl esters of coenzyme A. *J. Biol. Chem.* 249:2641–45

111. Whittle, P. J., Lunt, D. O., Evans, W. C. 1976. Anaerobic photometabolism of aromatic compounds by *Rhodopseudomonas* sp. *Biochem. Soc. Trans.* 4:490–91

112. Widdel, F. 1980. *Anaerober Abbau von Fettsäuren und Benzoesäure durch neu isolierte Arten sulfat-reduzierender Bakterien.* PhD thesis. Georg-August-Univ., Göttingen, Fed. Rep. Germany

113. Widdel, F., Pfennig, N. 1984. Dissimilatory sulphate-sulphur reducing bacteria. In *Bergey's Manual of Systematic Bacteriology*, ed. N. R. Krieg, J. C. Holt, pp. 663–79. Baltimore, Md: Williams & Wilkins

114. Williams, R. J., Evans, W. C. 1973. Anaerobic metabolism of aromatic substrates by certain microorganisms. *Biochem. Soc. Trans.* 1:186–87

115. Williams, R. J., Evans, W. C. 1975. The metabolism of benzoate by *Moraxella* sp. through anaerobic nitrate respiration. *Biochem. J.* 143:1–10

116. Young, L. Y. 1984. Anaerobic degradation of aromatic compounds. In *Microbial Degradation of Organic Compounds*, ed. D. T. Gibson, pp. 487–523. New York: Dekker

117. Zeikus, J. G., Wellstein, A. L., Kirk, T. K. 1982. Molecular basis for the biodegradative recalcitrance of lignin in anaerobic evironments. *FEMS Microbiol. Lett.* 15:193–97

118. Ziegler, K., Braun, K., Böckler, A., Fuchs, G. 1987. Studies on the anaerobic degradation of benzoic acid and 2-aminobenzoic acid by a denitrifying *Pseudomonas* strain. *Arch. Microbiol.* 149:62–69

Ann. Rev. Microbiol. 1988. 42:319–38

SMALL, ACID-SOLUBLE SPORE PROTEINS OF *BACILLUS* SPECIES: Structure, Synthesis, Genetics, Function, and Degradation

Peter Setlow

Department of Biochemistry, University of Connecticut Health Center, Farmington, Connecticut 06032

CONTENTS

INTRODUCTION

It was first noted more than 30 years ago that germination of spores of *Bacillus* species is accompanied by generation of significant amounts of free amino acids, most of which are produced by degradation of dormant spore protein (15, 37, 47). We now know that during the first 20–40 min of

319

germination of spores of *Bacillus* species up to 20% of total dormant spore protein is degraded to free amino acids (43, 44). Similar observations have been made with spores of several *Clostridium* species (19, 51) and spores of one species of *Thermoactinomyces* (27). Detailed studies with *Bacillus megaterium* spores have shown that the amino acids produced in this process are reutilized during spore germination for synthesis of small molecules such as nucleotides, and most importantly for synthesis of new protein (43, 47). Production of amino acids by proteolysis is essential for rapid protein synthesis during spore germination in an amino acid–deficient environment, since dormant spores lack many free amino acids as well as a number of amino acid–biosynthetic enzymes (37, 43). These amino acid–biosynthetic enzymes are absent even from dormant spores of amino acid prototrophs, as they are degraded during spore formation. They are only synthesized at defined times in spore germination, using amino acids provided by degradation of dormant spore protein (43, 44). Consequently, a major function of the proteolysis in the first minutes of bacterial spore germination is to generate the precursors necessary for protein synthesis during this period of development.

The magnitude of the proteolysis during spore germination (up to 20% of total spore protein in *B. megaterium*) allowed identification of the proteins degraded as a group of acid-soluble proteins (43, 44). Since these acid-soluble proteins were later found to have low molecular weights, they have been given the acronym SASP, for small, acid-soluble spore protein(s); this term is used here to designate both the singular and plural. Detailed studies in *B. megaterium* showed that the SASP contribute more than 90% of the protein degraded during spore germination. The SASP are also unique to the spore stage of the life cycle; they are synthesized only late in sporulation and degraded early in spore germination (43, 44). In this review I focus on work of the past ten years on the structure, synthesis, genetics, and function of SASP in various gram-positive organisms, as well as on SASP-degradative enzymes, since most of these topics have either never been reviewed or were reviewed some time ago. Earlier reviews discuss the proteolysis during germination in detail (10, 43, 44), and this process is not covered further in this review.

SASP PURIFICATION, CHARACTERIZATION, AND NOMENCLATURE

Purification of SASP from spores of various species is greatly facilitated by their acid solubility (44). However, SASP cannot be extracted from intact spores directly, but only after spore disruption (44). The procedure for spore disruption that has consistently given the highest yield of SASP unmodified by proteolysis during breakage is mechanical disruption of dry spores followed by extraction with 0.5-M acetic acid (25, 43, 44). Disruption of wet spores has often allowed significant proteolysis during breakage (25, 44). Acid rupture of spores in high concentrations (0.5–2 N) of mineral acids has also given good yields of SASP, but these strong acids may cause some SASP modification (25, 42, 57).

The amount of SASP varies between 8 and 20% of total spore protein depending on the species examined (25, 27, 44, 51, 57). A major reason for this apparent difference among species is the amount of protein in spore coats (25, 44, 57). Spores of species that have a high percentage of protein in spore coats have lower percentages of total protein as SASP than do those spores with smaller amounts of protein in spore coats (25, 44). In general, neither the SASP level nor the type of SASP present is affected significantly by changes in the sporulation medium (25, 44). However, the effect of varying a parameter such as spore genome number (52) on the SASP level has not yet been determined. In view of the close association between spore DNA and several major SASP (35, 40), this is a measurement that would be well worth having.

Analysis of acetic acid extracts of dry ruptured spores of a variety of species has shown that three SASP (termed α, β, and γ in *Bacillus subtilis*) comprise 75–90% of the protein in this fraction (25, 27, 39, 44, 51, 57). These major SASP have been purified to homogeneity using column chromatography (25, 39, 44, 57) and preparative high-performance liquid chromatography (2, 19). The latter technique may prove of great utility in SASP purification because of the similarity of several of the major SASP from a single species (10, 25, 44). To date one or more major SASP have been purified and/or partially characterized from spores of *Bacillus cereus, B. megaterium, Bacillus sphaericus, Bacillus stearothermophilus, B. subtilis, Clostridium bifermentans, Clostridium perfringens,* and *"Thermoactinomyces thalpophilus"* (2, 19, 25, 27, 39, 44, 51, 57). In all of these organisms except the clostridia two distinct types of major SASP have been identified. These two types are named for the SASP of these types in *B. subtilis,* an α/β type and a γ type (25). In spores of any given organism there are two major SASP of the α/β type, as well as many minor α/β type SASP, each coded for by a unique gene (10, 25, 27, 44, 57). The α/β type SASP have extremely similar amino acid sequences, are closely related immunologically, have molecular weights of 5–7000, and have a significant percentage of hydrophobic amino acids (2, 10, 11, 27) (Table 1). In contrast, all organisms so far examined contain only a single γ type SASP (and a single γ type SASP gene) (21, 22, 53), which has a very different amino acid sequence from α/β type SASP (44, 53), does not cross-react immunologically with antisera to α/β type SASP (2, 22), has a molecular weight of 8–11,000 (22, 50, 53), and is extremely low in large hydrophobic amino acids (53) (Table 1). The γ type SASP also have higher isoelectric points than the α/β type SASP from the

Table 1 Comparison of various properties of α/β and γ type SASP[a]

Property	α/β type SASP	γ type SASP
Number of amino acid residues	61–72	81–96
Net charge[b]	−3 to +2	+1 to +6
Percentage of large hydrophobic residues[c]	21–29	9–11
Number of genes in any one species	≥ 7	1

[a]Data are for all *Bacillus* species and the one *Thermoactinomyces* species tested (2, 3, 11–14, 21, 22, 27, 48, 49, 50, and 53). All values are for proteins without the amino-terminal methionine, since this appears to be removed posttranslationally.

[b]Determined from the amino acid sequence by assigning histidine, lysine, and arginine a charge of +1 and glutamic acid and aspartic acid a charge of −1.

[c]Residues include isoleucine, leucine, methionine, phenylalanine, proline, tyrosine, and valine.

same species (25, 44, 53). However, the isoelectric points of γ type SASP from different species range from below pH 7.0 to greater than pH 9.5 (25, 44) (Table 1).

The major SASP contain 18 of the 20 common amino acids, but lack cysteine and tryptophan (10, 44, 53). While no SASP containing cysteine has yet been identified, several minor SASP contain tryptophan (42, 44). Where detailed analyses have been carried out, all three major SASP have been shown to lack both carbohydrate and phosphate (25, 44). However, a recent report suggests that SASP from *C. perfringens* may contain carbohydrate (19).

In general, the two major α/β type SASP comprise 40–50% of the total SASP pool in any one species, with the γ type SASP comprising 30–40% (25, 39, 44). The remaining SASP are minor α/β type SASP as well as a number of other minor SASP that are unrelated immunologically to the major SASP (25, 42, 44). A number of the latter minor SASP have been purified from *B. megaterium* spores and are known to be synthesized in parallel with the major SASP during sporulation and to be degraded during germination (42). However, these minor SASP have not been studied further.

LOCALIZATION OF SASP WITHIN THE SPORE

Early work on SASP localization within the spore demonstrated that little if any SASP was in the spore coats or cortex and strongly suggested a location in the spore core (44). This localization has been confirmed with *B. subtilis* spores using immunoelectronmicroscopy with affinity-purified antibodies specific for α/β or γ type SASP and a gold-labeled second antibody (S. Francesconi, T. J. McAlister & P. Setlow, unpublished results). These studies showed that SASP antigens are located only in the spore core, where the γ type SASP antigen is rather evenly distributed. However, the α/β type SASP antigen is found only in those regions of the spore core that also contain the spore DNA. This association of α/β type SASP with spore DNA in vivo is consistent with studies in which *B. megaterium* spores were given high doses of UV light to cross-link proteins to DNA (40). In these spores the α/β type SASP were readily cross-linked to spore DNA, which suggests that these molecules were present in a complex in vivo. However, thorough in vitro studies of SASP-DNA binding have not yet been carried out. As discussed below, the association of α/β type but not γ type SASP with spore DNA in vivo is consistent with a role for α/β type SASP in the resistance of dormant spores to UV irradiation (35).

CLONING OF SASP GENES AND SASP AMINO ACID SEQUENCES

A major advance that has facilitated the detailed study of SASP synthesis, structure, and function has been the cloning of a number of SASP genes. The first SASP gene cloned was that for one of the two major α/β type SASP of *B. megaterium* (4). The gene was cloned using a plasmid expression vector and immunological screening for SASP production (4). This SASP gene was then used as a hybridization probe to facilitate the cloning of the gene that codes for the other major α/β type SASP of *B. megaterium* (11). Strikingly, in addition to the two genes coding for the major α/β type SASP, *B. megaterium* also has at least five other genes that code for α/β type SASP (5, 11, 13, 14). Where tested, these other five genes were all expressed. They code for SASP that are synthesized in parallel with the major α/β type SASP, but at much lower levels (10, 11, 13, 14). All seven of the *B. megaterium* α/β type SASP genes code for proteins of very similar, although not identical, amino acid sequence (Figure 1) (10–14). Consequently, these genes represent a prokaryotic divergent multi-gene family.

The *B. megaterium* α/β type SASP genes were also used as hybridization probes to facilitate the cloning of the analogous genes from *B. subtilis*. SASP genes in *B. subtilis* have been named *ssp,* for spore-specific protein (2). To date four α/β type SASP genes have been cloned from *B. subtilis*. Two of these genes, termed *sspA* and *sspB,* code for the two major α/β type SASP (SASP-α and SASP-β, respectively) of this organism (2, 10). The other two genes, termed *sspC* and *sspD,* code for related SASP expressed at a much lower level (2, 3, 36). There are also at least two α/β type SASP genes that have not yet been cloned from *B. subtilis* (2). α/β type SASP genes have also been cloned from *B. cereus, B. stearothermophilus,* and "*T. thalpophilus*" (27). These organisms also appear to have more than five α/β type SASP genes (27).

SASP gene

B. cereus 1: MGKNNSGSRNEVLVRGAEQALDQMKYEIAQEFGVQLGADTTARSNGSVGGEITKRLVAMAEQQLGGRANR

2: MSRSTNKLAVPGAESALDQMKYEIAQEFGVQLGADATARANGSVGGEITKRLVSLAEQQLGGYQK

B. megaterium A: MANTNKLVAPGSAAAIDQMKYEIASEFGVNLGPEATARANGSVGGEITKRLVQMAEQQLGGK

C: MANYQNASNRNSSNKLVAPGAQAAIDQMKFEIASEFGVNLGPDATARANGSVGGEITKRLVQLAEQNLGGKY

C-1: MANNNSSNNNELLVYGAEQAIDQMKYEIASEFGVNLGADTTARANGSVGGEITKRLVQLAEQQLGGGRF

C-2: MANNKSSNNNELLVYGAEQAIDQMKYEIASEFGVNLGADTTARANGSVGGEITKRLVQLAEQQLGGGRSKTTL

C-3: MARTNKLLTPGVEQFLDQYKYEIAQEFGVTLGSDTAARSNGSVGGEITKRLVQQAQHLSGSTQK

C-4: MANNKSSNNHELLVYGAEQAIDQMKYEIASEFGVNLGADTTARANGSVGGEITKRLVQLAEQQLGGGRF

C-5: MANSRHKSSNELAVHGAQAIDQMKYEIASEFGVTLGPDTTARANGSVGGEITKRLVQMAEQQLGGGRSKSLS

B. stearothermophilus: MPNQSGSMSSNQLLVPGAQVIDQMKFEIASEFGVNLGAETTSRANGSVGGEITKRLVSFAQQQMGGGVQ

B. subtilis A: MANNNSGNSNNLLVPGAAQAIDQMKLEIASEFGVNLGADTTSRANGSVGGEITKRLVSFAQQNMGGGQF

B: MANQNSSNDLLVPGAAQAIDQMKLEIASEFGVNLGADTTSRANGSVGGEITKRLVSFAQQOMGGRVQ

C: MAQQSRSRSNNNNDLLIPQAASAIEQMKLEIASEFGVQLGAETTSRANGSVGGEITKRLVRLAQQNMGGQFH

D: MASRNKLVVPGVEQALDQFKLEVAQEFGVNLGSDTVARANGSVGGEMTKRLVQQAQSQLNGTTK

"T. thalpophilus": MAQQGRNRSSNQLLVAGAAQAIDQMKFEIAQEFGVTLGADTTSRANGSVGGEITKRLVSLAQQQLGGGTSF

C. perfringens: SQHLVPEAKNGLSKFKNEVAAELGVPFSDYNGDLQSGSVGGEMVKRIVEQYEQSMK

Figure 1 Comparison of amino acid sequences of α/β type SASP from various species (2, 3, 11–14, 19, 27, 48, 49). The arrows above the sequences denote the site of cleavage by the SASP-specific protease. The arrows below the "*T. thalpophilus*" sequence denote residues conserved in all *Bacillus* SASP as well as the one "*T. thalpophilus*" SASP analyzed. The arrows below the *C. perfringens* sequence denote residues also conserved in this SASP.

Determination of the nucleotide sequence of all 15 of the α/β type SASP genes that have been cloned to date has shown that they have a strong ribosome-binding site immediately preceding the coding sequence (2, 3, 10–14, 27). Where the amino acid sequence of the mature SASP is known it has been found identical to that coded for by the gene, with the exception of the amino-terminal methionine, which is presumably removed posttranslationally (2, 11, 12, 27, 48, 49, 57). In all α/β type SASP genes examined the translation termination codon is UAA; shortly after the termination codon there is a region of dyad symmetry followed by a T-rich region, which is presumably a transcription stop signal (2, 10, 11, 27).

Comparison of the amino acid sequences of all α/β type SASP determined to date has shown a remarkable sequence conservation within as well as across species, as maximal alignment of the sequences requires introduction of no gaps (2, 3, 10, 11, 27) (Figure 1). In all seven *B. megaterium* α/β type SASP there are 57 residues between and including the first and last residues conserved (10, 11). Of these 57 amino acids, 34 residues are conserved in all seven proteins; an additional 15 residues are either conserved in all but one protein or are one of two analogous amino acids (11) (Figure 1). When all 15 sequences from different *Bacillus* species as well as "*T. thalpophilus*" are compared, 26 residues are found to be conserved exactly (27) (Figure 1). Of these 26 conserved residues, 15 are also conserved in an α/β type SASP of *C. perfringens* (19) (Figure 1). Given that *B. megaterium* and *B. subtilis* diverged ~300 million years ago, that the thermoactinomyces diverged from the bacilli ~550 million years ago, and that the clostridia diverged from the bacilli ~1.2 billion years ago (16, 56), this degree of sequence conservation is extremely striking and suggests that these proteins have some sequence-specific function. The highest degree of sequence conservation both within and across species is seen in two regions of these proteins. One is the site recognized and cleaved by the SASP-specific protease, which acts during germination (regions around the arrows above the sequences, Figure 1) (see section on SASP-degradative enzymes) (2, 11, 27); the second is a more carboxyl-terminal region whose function is not known. The sequences of the most amino-terminal and most carboxyl-terminal residues in these proteins do not appear to be conserved (Figure 1).

Like the first α/β type SASP gene cloned, the first γ type SASP gene was also cloned from *B. megaterium* using an expression vector, but in this case a λ expression vector (21). This gene was then used as a hybridization probe to clone the homologous genes from *B. cereus, B. stearothermophilus, B. subtilis,* and "*T. thalpophilus*" (22, 53). In these organisms only a single γ type SASP gene was found (21, 22, 53). This gene has been designated *sspE* in *B. subtilis* (22). Like α/β type SASP genes, γ type SASP genes contain a strong ribosome-binding site immediately preceding the coding sequence, which codes for a protein identical to the mature SASP with the exception of the amino-terminal methionine, which is missing in the mature protein (21, 22, 53). The translation termination codon is again UAA, and this stop codon is followed by a region of dyad symmetry and then a T-rich region, which is probably a transcription termination signal (21, 22, 53). The amino acid sequences of the γ type SASP are not as well conserved across species as those of α/β type SASP, as maximal alignment of γ type SASP sequences requires introduction of gaps (22, 53) (Figure 2). Even with gaps there are only 19 residues conserved in these five larger SASP, compared to the 26 residues conserved in all 15 smaller α/β type SASP from the same species. One somewhat bizarre feature of γ type SASP sequences is the presence of regions of sequence repeated two or three times (50, 53). However, the function of these sequence repeats is not known.

REGULATION OF SASP SYNTHESIS DURING SPORULATION

SASP Synthesis During Sporulation

SASP are not found in vegetative or young sporulating cells, as their synthesis only begins 3–4 hr into sporulation, when all three major SASP are synthesized in parallel (25, 44). Minor SASP are synthesized at this time as well (10, 33, 42). In a population of sporulating cells, synthesis of 75% of final SASP levels requires about 2 hr, and after this period the rate of SASP synthesis falls back to zero (14, 33, 44). SASP synthesis occurs essentially at the same time as synthesis of glucose dehydrogenase and the product of the spoVA gene and 1–2 hr before synthesis of dipicolinic acid (32, 33, 44). Where the site of synthesis has been examined all newly synthesized SASP have been found only in the developing forespore, which is also the location of glucose dehydrogenase and the spoVA gene product (7, 32, 33, 44). The developing forespore appears to be the site of SASP synthesis, since this is the location of those SASP mRNAs that have been analyzed (2, 7, 10). β-Galactosidase synthesized from various SASP gene–lacZ gene fusions was also found only in the developing forespore (33).

Effect of Asporogenous Mutations on SASP Synthesis

Analysis of SASP synthesis in asporogenous B. subtilis mutants (34), as well as measurement of expression of β-galactosidase from sspA-lacZ, sspB-lacZ, and sspE-lacZ gene fusions in these mutants, has allowed the determination of the dependence of SASP gene expression on other sporulation genes (33). Strikingly, all SASP genes examined responded identically to all asporogenous mutations tested. SASP genes were not expressed in spo0 and spoII mutants; the only exception was the spoIID298 mutation, which is a

Species

B. cereus: MSKKQQGYMKATSGASIQ——STNAS————YGTEFSTETDVQAVKQANAQSEAKKAQASGAQSANASYGTEFATETDVHSVKKQNAKSAAKQSQSSSNQ

B. megaterium: MAKQTNKTASGTSTQHVKQQNAQASKNN——FGTEFGSETNVQEVKQQNAQAANKSQNAQASKNN——FGTEFASETSAQEVRQQNAQAKKNQNSGKYRG

B. stearothermophilus: MANSNNFSK——TNAQQVRKQNQQSAAGQGQ FGTEFASETNVQQVRKQNQQSAGQQGQ————FGTEFASETDAQQVRQQNQSAEQNKQ

B. subtilis: MANSNNK————TNAQQVRKQNQQSASGQGQ FGTEFASETNVQQVRKQNQQSAAGQGQ————FGTEFASETDAQQVRQQNQSAEQNKQ

"T. thalpophilus": MNTKNFTPQESRTNAQQVRQQNQQSAQGTSSGFATEFASETNAQQVRQQNQQSAQANRMSGATAGG——FNTEFASETNVQQVRQQNQQSEAKKRNNQQ

Figure 2 Comparison of amino acid sequences of γ type SASP from various species (21, 23, 50, 53). The horizontal lines between residues denote gaps that must be introduced to maximally align the sequences. Arrows above the sequences give the sites of cleavage by the SASP-specific protease. Arrows below the "*T. thalpophilus*" sequence denote residues conserved in all five γ type SASP.

leaky mutation in the *spoIID* locus (31, 33). Mutations in those *spoIII* genes that only affect gene expression in the mother-cell compartment [*spoIIIB*, *spoIIIC*, and *spoIIID* (32)] had no effect on SASP gene expression, while mutations in *spoIII* genes that block gene expression only in the forespore compartment [*spoIIIA* and *spoIIIE* (32)] abolished SASP gene expression (33). Mutants blocked beyond stage III of sporulation (i.e. *spoIV* and *spoV* mutants) had essentially no effect on SASP gene expression (33, 34). These asporogenous mutations had identical effects on expression of two other genes expressed only in the forespore, the glucose dehydrogenase and the *spoVA* genes (32, 33; J. Errington, personal communication). This suggests that all of these forespore-specific genes may share some common regulatory features.

Regulation of SASP Gene Expression

There are extensive data demonstrating that the major site for regulation of expression of both α/β and γ type SASP genes is at the transcriptional level. Measurements of SASP mRNA levels by in vitro translation (7, 24, 26) and Northern blot analysis (2, 3, 11–14, 21, 22) of RNA isolated throughout growth and sporulation have shown that these mRNAs are absent during vegetative growth and early sporulation. The mRNAs for all SASP appear in parallel only midway in sporulation; their levels rise to a peak and then fall back to zero. The changes in rates of SASP synthesis during sporulation can be explained completely by changes in these mRNA levels (7). The Northern blot analyses have further shown that all SASP genes are transcribed as monocistronic mRNAs whose sizes do not change at different times in sporulation (2, 3, 11–14, 21, 22). To date stability has been measured only for α/β type SASP mRNAs, with different results from different organisms. *B. megaterium* α/β type SASP mRNAs had half-lives of ~4 min, a value comparable to that found for other mRNAs in this organism (7). However, *B. subtilis* α/β type SASP mRNAs had half-lives of ~11 min, and they were much more stable than other mRNAs (26). Since mRNA stability has a significant influence on the amount of gene product obtained from a given amount of transcript synthesized, the stability of SASP mRNAs seems worth investigating in detail.

While mRNAs for both SASP classes are synthesized in parallel during sporulation, there is significant evidence that there are differences in the regulation of the α/β and γ type SASP genes. In *B. subtilis* increasing the *sspA* gene copy number about 10-fold resulted in an increase of only ~1.5-fold in *sspA* mRNA levels and a corresponding 3–5-fold decrease in *sspB* mRNA levels, with parallel changes in the levels of SASP-α and SASP-β accumulated in spores. Similarly, increasing the *sspB* gene copy number approximately 10-fold elevated *sspB* mRNA levels about 2-fold and decreased *sspA* mRNA levels 2–3-fold, with parallel changes in the levels of SASP-β and SASP-α accumulated (33, 35; J. M. Mason, P. Fajardo-Cavazos & P. Setlow, unpublished results). These changes in *sspA* or *sspB* gene dosage had no effect on levels of *sspE* mRNA or SASP-γ (33, 45). However, a 10-fold increase in the *sspE* gene copy number was accompanied by a 10-fold increase in the level of *sspE* mRNA (J. M. Mason, P. Fajardo-Cavazos & P. Setlow, unpublished results), although spore levels of SASP-γ were not increased (45). These data not only indicate that the α/β and γ type SASP genes have differences in their transcriptional regulation, but also suggest that both types of genes exhibit some features indicative of feedback regulation. This feedback regulation might involve the intact SASP, with α/β type SASP genes regulated at the transcriptional level and the γ type SASP gene regulated at the posttranscriptional level. However, as yet there is no understanding of how this feedback regulation is achieved.

Feedback regulation was also shown from analysis of SASP levels in mutants lacking various intact SASP genes. A *B. subtilis* mutant carrying an *sspE* gene with the promoter, ribosome-binding site, and first 59 codons of (SASP-γ^-) deleted made no SASP-γ, but had wild-type levels of SASP-α and SASP-β in spores (23). Similarly, an analogous SASP-α mutant had normal spore levels of SASP-γ, but the SASP-β level was increased about 2-fold. SASP-α levels increased by about 50% in an analogous SASP-β mutant (35).

Strikingly, this feedback regulation was not seen when *ssp-lacZ* fusion gene expression was analyzed (33). Increasing the gene dosage of *sspA-*, *sspB-*, *sspD-*, or *sspE-lacZ* gene fusions in the *B. subtilis* chromosome up to 17-fold resulted in an equivalent increase in both *lacZ* mRNA levels and β-galactosidase production, with no effect on SASP production (33, 45; J. M. Mason, P. Fajardo-Cavazos & P. Setlow, unpublished results). Similarly, 10-fold increases in the gene dosage of the *sspA, sspB,* or *sspE* gene had no effect on the expression of any *ssp-lacZ* gene fusion tested (33). Since the *ssp-lacZ* fusions contained the SASP gene's promoter and ribosome-binding site, but only the first 22–31 amino acids of the SASP (33), it is possible that some more downstream regions of the SASP genes, SASP mRNAs, or the SASP themselves are involved in the feedback regulation. However, these putative regulatory regions have not yet been identified.

As noted above, a number of *B. subtilis* strains have been constructed with an increased gene dosage of either intact *ssp* genes or *ssp-lacZ* fusions. While some of these strains showed alterations in the amounts of SASP or β-galactosidase produced, in none of these strains was the timing of SASP synthesis altered from that seen in the single-copy strains (17, 33, 36). The timing was even preserved in strains that carried intact *sspA, sspB,* or *sspE* genes cloned in plasmid pUB110, which has a copy number of 50–100 (33). The fact that these greatly elevated SASP gene copy numbers did not alter the timing of SASP gene expression (i.e. there was no expression in vegetative or early sporulating cells) suggests that SASP gene expression is regulated, at least in part, by positive control of transcription. There is abundant evidence for the regulation of other sporulation-specific genes by positive control of transcription (8, 30, 31).

Transcription of SASP Genes

In vitro transcription studies with cloned SASP genes are consistent with a positive control mechanism for regulation of SASP gene expression; purified RNA polymerase from vegetative cells of *B. subtilis* would not transcribe cloned SASP genes in vitro. However, RNA polymerase purified from sporulating cells that were actively transcribing SASP genes in vivo transcribed both the *sspA* and *sspE* genes in vitro, although *sspB* transcription in vitro was weak at best (45; D. Sun, W. Nicholson & P. Setlow, unpublished results). The in vitro transcription activity of sporulating-cell RNA polymerase on the *sspE* gene was due to one or more polymerase-associated factors, i.e. σ factors, of 35–40 kd (D. Sun & P. Setlow, unpublished results). This suggests that, as found previously with other sporulation genes (8, 30, 31), SASP gene transcription requires synthesis during sporulation of one or more new σ factors to modulate RNA polymerase specificity.

The start sites for transcription in vivo of a number of SASP genes of *B. megaterium* and *B. subtilis* have now been determined by nuclease mapping techniques, allowing localization of regions upstream of the transcription start sites (10, 12, 45; B. Setlow & P. Setlow, unpublished results). The in vivo transcription start sites tested are identical to those utilized in vitro (D. Sun & P. Setlow, unpublished results). The *sspA, sspB,* and *sspE* genes of these two organisms show significant sequence homology in the 50 bases upstream from the transcription start sites (2, 11, 13, 21, 22). However, these sequences, in particular those located ~10 and ~35 bases upstream from the transcription start point in the regions thought to be most important in determining promoter specificity (8, 30, 31), are not identical in the *sspA, sspB,* and *sspE* genes (2, 11, 13, 21, 22). Consequently, there may be different regulatory proteins, i.e. σ factors, involved in the regulation of the expression of these different genes. It has been shown that efficient in vitro transcription of the *sspE* genes requires at most only the 75 nucleotides upstream from the transcription start site (P. Fajardo-Cavazos, D. Sun & P. Setlow, unpublished results), although the exact importance of various nucleotide sequences in this region has not yet been determined. However, for the *sspA* and *sspB* genes we do not yet have any knowledge of the regions of nucleotide sequence determining the regulation of gene expression.

CHROMOSOMAL LOCATIONS OF SASP GENES

The chromosomal locations of four α/β type SASP genes have been determined in *B. subtilis* (1). These genes are scattered about the chromosome, mapping at ~70° *(sspB)*, 115° *(sspD)*, 180° *(sspC)*, and 260° *(sspA)* (1). It has been suggested that this dispersal of SASP genes on the chromosome may reflect ancestral chromosomal duplications (31). However, since there are at least six α/β SASP genes (2), this dispersal would require more than two chromosome duplications. Nevertheless, it could be informative to learn the map position of those α/β type SASP genes not yet cloned from *B. subtilis*. The gene coding for SASP-γ of *B. subtilis*, termed *sspE*, maps at ~65° (22). Interestingly, several SASP genes are found in regions devoid of other known sporulation loci (1, 22). At least one SASP gene *(sspC)* is absent from a strain carrying a deletion in the *sspC* region of the chromosome; the deletion has no obvious effects on growth, sporulation, or spore properties (1).

To date one α/β type SASP gene has been mapped on the *B. megaterium* chromosome (M. Sussman, P. S. Vary, & P. Setlow, unpublished results). This gene is extremely homologous to the *sspA* gene of *B. subtilis* in both its coding and flanking sequences, and it is undoubtedly the *sspA* gene of *B. megaterium*. Strikingly, the *B. megaterium sspA* gene lies in the same region of the chromosome (between *argA* and *hisA*) as the *B. subtilis sspA* gene. This suggests that at least this SASP gene has not moved on the chromosome since the divergence of these two species.

SASP FUNCTION

As mentioned in the introduction, it was clear over 10 years ago that a major function of SASP is to supply, by their degradation, amino acids for protein synthesis during spore germination. However, a number of early observations suggested that SASP might have another function as well. Thus, the findings that SASP are associated with spore DNA in vivo and appear at the time spores become resistant to UV light led to the suggestion that SASP might have some direct role in spore UV resistance (43, 44). That this early suggestion was in fact correct, at least for α/β type SASP, has been demonstrated using strains with deletion mutations in one or more of the genes coding for the three major *B. subtilis* SASP (35, 36). These mutations involved deletions of the SASP gene's promoter, ribosome-binding site, and amino-terminal coding sequence. Mutants were constructed in vitro using the cloned genes (23, 35). The mutant gene was then transferred to the *B. subtilis* chromosome by homologous recombination, and strains in which this process had been accompanied by gene conversion with loss of the wild-type gene were identified (22, 35). Using this approach, strains with deletion mutations in one, two, or three of the genes coding for major SASP were isolated (23, 35, 36). All SASP deletion mutants grew and sporulated with kinetics indistinguishable from those of the wild-type strain (23, 35). However, the spores lacked that SASP whose gene had been deleted, and in spores of the SASP-$\alpha^- \beta^- \gamma^-$ strain no other SASP was more abundant than in wild-type spores (23, 35). All spores lacking one or more SASP also initiated germination normally, but spores lacking SASP were slowed in spore outgrowth (23, 35). This decrease in the rate of spore outgrowth was particularly noticeable in amino acid–free media (23, 35), which is consistent with a role for SASP as amino acid storage proteins.

Other than this outgrowth defect, spores lacking SASP-γ exhibited no other phenotype, as their resistance to heat and UV or γ-irradiation was identical to that of wild-type spores (23). However, spores lacking SASP-α and SASP-β had UV resistance below that of vegetative cells and well below that of wild-type spores (35). This UV-sensitive phenotype of SASP-$\alpha^- \beta^-$ spores could be cured by synthesis of sufficient levels of either SASP-α or SASP-β when the sspA or sspB genes were supplied either on plasmids or by integration in the chromosome (36). Even elevated levels of a normally minor α/β type B. subtilis SASP (the product of the sspD gene) restored much of the UV resistance of the SASP-$\alpha^- \beta^-$ spores (36). These SASP-$\alpha^- \beta^-$ spores had no increased sensitivity to γ-irradiation, but were somewhat more heat sensitive than wild-type spores (23, 35). This heat-sensitive phenotype was also cured by synthesis of sufficient spore levels of either SASP-α or SASP-β or the sspD gene product (36).

The resistance of dormant bacterial spores to UV radiation is due to the production of a spore photoproduct (SP) in spore DNA upon UV irradiation. The SP is a 5-thyminyl-5,6-dihydrothymine adduct, which is efficiently repaired during spore germination (9, 18, 55). UV irradiation of dormant spores produces no detectable cyclobutane-type pyrimidine dimers (PP) in DNA, although these are the major lethal photoproducts formed in cell DNA (9, 18, 41). Analysis of the UV photoproducts formed in DNA in SASP-$\alpha^- \beta^-$ spores showed that SP formation was reduced at least twofold. More importantly, PP formation took place with a quantum yield about one half that found in vegetative cells (41). These data suggest that α/β type SASP have some key role in spore DNA photochemistry, which in turn determines spore UV resistance. However, it is not clear how α/β type SASP are involved, although their contribution may involve a direct effect on DNA, since these SASP are associated with spore DNA in vivo. The extremely high degree of amino acid sequence conservation among α/β type SASP from different species is certainly consistent with an important structural role for these proteins. However, there is essentially no information on structural interaction(s) between DNA and α/β type SASP.

SASP-DEGRADATIVE ENZYMES

Since a major function of SASP is to be degraded in the first minutes of spore germination, some work has gone into the identification and characterization of the enzyme(s) involved in this process. One SASP-specific protease has been identified in spores of various *Bacillus* species. This enzyme has been purified to homogeneity from germinated spores of *B. megaterium* (28, 29). As purified, this protease is a tetramer of 40-kd subunits (P_{40}); only the tetramer is enzymatically active (28). The protease is synthesized during sporulation, in parallel with synthesis of its SASP substrates, as a 46-kd polypeptide (P_{46}), which associates into tetramers but is enzymatically inactive (20, 29). After 1–2 hr of further sporulation P_{46} is processed, probably by proteolysis, to a 41-kd polypeptide (P_{41}), which can form tetramers and is enzymatically active in vitro, although not in vivo (20, 29, 44). During spore germination P_{41} is converted to P_{40} by proteolytic removal of 13 amino acids from the amino terminus of P_{41} (29; M. D. Sussman & P. Setlow, unpublished results); the function of this processing is not known. Later in germination P_{40} is degraded completely in an energy-dependent process (28, 29). At present essentially nothing is known about the enzymes involved in the processing of this protease or the regulation of protease activity (20, 29).

This SASP-specific protease cleaves within a peptide sequence that is well conserved in all major SASP (Figure 3) (46). This sequence is similar in both α/β and γ type SASP, although there are some differences. However, in all cases cleavage is adjacent to a glutamyl residue (arrow, Figure 3) (46, 57). A heptapeptide containing this sequence (Figure 3) is also cleaved in vitro by the *B. megaterium* protease, with a high V_{max} and an extremely high K_m (6). All major SASP contain either one (α/β type SASP) or two (γ type SASP) of these protease cleavage sites (46, 57). Consequently, spore protease cleavage generates large (20–35-residue) oligopeptides, which in vitro, and presumably in vivo, are rapidly degraded to amino acids by peptidases (44).

Several mutants of *B. megaterium* with decreased spore protease activity have been obtained (38). In the best of these mutants spore germination exhibits normal kinetics, but degradation of the γ type SASP is greatly retarded (38). However, degradation of the α/β type SASP is only marginally slowed (38). It is not known if these mutants are only leaky or are tight but subjected to a second SASP-specific protease acting rapidly on α/β type SASP. The gene for the SASP-specific protease described above (termed the *gpr* gene) has recently been cloned from *B. megaterium* (M. D. Sussman & P.

α/β type SASP:

M (13)	K (16)	Y (7)	E (16)↓	I (14)	A (15)	S (10)	E (16)	F (15)	G (16)	V (16)	N (9)
F (2)	F (4)			V (2)	S (1)	Q (5)		L (1)			Q (3)
Y (1)	L (4)				A (1)						T (3)
	N (1)										L (1)

γ type SASP:

F (9)	G (8)	T (10)	E (10)↓	F (10)	A (9)	S (8)	E (10)	T (10)	N (5)	V (7)	Q (9)
Y (1)	A (1)				G (1)	T (2)			D (4)	A (3)	H (1)
	N (1)								S (1)		

Synthetic heptapeptide substrate: N-Acetyl-T   E↓   F   A   S   E   F-COOH

Figure 3   Comparison of the amino acid sequences around the cleavage sites of the SASP-specific protease (2, 3, 11–14, 19, 21, 22, 27, 46, 48–50, 53, 57). Positions of the protease cleavage site were either determined directly or inferred from an SASP sequence. The value below each residue is the number of times that residue is found in that position at the cleavage site. The arrow indicates the site of protease cleavage.

Setlow, unpublished results). Its availability should allow the cloning and subsequent deletion of the analogous gene from *B. subtilis*. This will allow definitive assessment of this protease's complete function during germination, and it could allow identification of a second SASP-specific protease if it exists.

UNANSWERED QUESTIONS AND FUTURE DIRECTIONS

While much has been learned about the SASP system in the gram-positive spore formers, there are a number of key unanswered questions.

What is the reason for the large number of α/β type SASP genes? The presence of several α/β type SASP genes expressed at high levels can be explained based on their function in spore UV resistance and on some advantage of redundant genetic information for this system. However, the need for at least five other α/β type SASP genes expressed at much lower levels is not clear, nor are the selective mechanisms that maintain the sequences of these minor SASP and eliminate any mutations giving rise to pseudogenes.

How did these different α/β type SASP genes arise and become scattered on the chromosome? An obvious possibility is some duplicative transposition–like function. However, analysis of the sequences of many α/β type SASP genes has revealed no transposon–like associated sequences. Furthermore, the *sspA* gene is found in the same place on the chromosome in both *B. megaterium* and *B. subtilis*. Since *Clostridium* spores also appear to contain multiple α/β type SASP genes, they probably appeared before the *Clostridium-Bacillus* divergence over one billion years ago and may have stopped moving on the chromosome prior to the *B. subtilis–B. megaterium* divergence ~300 million years ago (16, 56). It may then be impossible to answer this question definitively.

What are the details of the regulation of SASP gene expression? It seems most likely that a major regulatory mechanism is positive control of initiation of transcription by a combination of SASP gene–specific σ factors and novel promoters for SASP genes. However, much remains to be done to identify and characterize the new σ factor(s) and to characterize their overall role in sporulation. Also, the mechanism of feedback regulation of the different SASP genes is not at all understood. The findings that α/β type SASP are involved in DNA binding in vivo suggest one model of feedback regulation that can readily be tested.

What is the role of SASP-γ in spores? SASP-γ is the most predominant SASP in *B. subtilis* spores. While it has a clear role in supplying amino acids during spore germination, no other role has been ascribed to this predominant protein.

What are the functions of minor SASP unrelated to α/β SASP? These minor proteins make up 10–20% of the SASP pool and are composed of more than four different proteins (42). At present little characterization of these proteins has been achieved, and nothing is known of their function.

How do the α/β type SASP affect spore DNA photochemistry? Their involvement in spore DNA photochemistry suggests that these proteins have some key structural role in the spore. At present essentially nothing is known about SASP structure in solution, and absolutely nothing is known about how SASP might affect DNA structure. Since large amounts of purified SASP can now be produced by recombinant DNA techniques, this problem seems readily approachable in the future.

ACKNOWLEDGMENTS

The work in the author's laboratory has been supported by grants from the Army Research Office and the National Institutes of Health (GM-19698). It is a great pleasure to acknowledge the many recent contributors to the work in the author's laboratory: Michael Connors, Everardo Curiel-Quesada, Susan Dignam, Patricia Fajardo-Cavazos, Edward Fliss, Susan Goldrick, Rebecca Hackett, Charles Loshon, James Mason, Wayne Nicholson, Cynthia Postemsky, Barbara Setlow, Dongxu Sun, Michael Sussman, and Bonnie Swerdlow.

Literature Cited

1. Connors, M. J., Howard, S., Hoch, J. A., Setlow, P. 1986. Determination of the chromosomal map location of four *Bacillus subtilis* genes which code for a family of small, acid-soluble spore proteins. *J. Bacteriol.* 166:412–16
2. Connors, M. J., Mason, J. M., Setlow, P. 1986. Cloning and nucleotide sequence of genes for three small, acid-soluble proteins of *Bacillus subtilis* spores. *J. Bacteriol.* 166:417–25
3. Connors, M. J., Setlow, P. 1985. Cloning of a gene for a small, acid-soluble spore protein from *Bacillus subtilis,* and determination of its complete nucleotide sequence. *J. Bacteriol.* 161:333–39
4. Curiel-Quesada, E., Setlow, B., Setlow, P. 1983. The cloning of the gene for C-protein, a low molecular weight spore-specific protein from *Bacillus megaterium. Proc. Natl. Acad. Sci. USA* 80:3250–54
5. Curiel-Quesada, E., Setlow, P. 1984. Cloning of a new low–molecular weight

spore-specific protein gene from *Bacillus megaterium. J. Bacteriol.* 157:751–57
6. Dignam, S. S., Setlow, P. 1980. *Bacillus megterium* spore protease: action of the enzyme on peptides containing the amino acid sequence cleaved *in vivo. J. Biol. Chem.* 255:8408–12
7. Dignam, S. S., Setlow, P. 1980. *In vivo* and *in vitro* synthesis of the spore specific proteins A and C of *Bacillus megaterium. J. Biol. Chem.* 255:8417–23
8. Doi, R. H., Wang, L.-F. 1986. Multiple procaryotic ribonucleic acid polymerase sigma factors. *Microbiol. Rev.* 50:227–43
9. Donnellan, J. E. Jr., Setlow, R. B. 1963. Thymine photoproducts but not thymine dimers found in ultraviolet irradiated bacterial spores. *Science* 149:308–10
10. Fliss, E. R., Connors, M. J., Loshon, C. A., Curiel-Quesada, E., Setlow, B., Setlow, P. 1985. Small, acid-soluble spore proteins of *Bacillus:* products of a

sporulation-specific, multigene family. In *Molecular Biology of Microbial Development,* ed. J. Hoch, P. Setlow, pp. 60–66. Washington, DC: Am. Soc. Microbiol.

11. Fliss, E. R., Loshon, C. A., Setlow, P. 1986. Genes for *Bacillus megaterium* small acid soluble proteins: cloning and nucleotide sequence of three additional genes from this multi-gene family. *J. Bacteriol.* 165:467–73

12. Fliss, E. R., Setlow, P. 1984. The complete nucleotide sequence and the translational and transcriptional start sites for the protein C gene of *Bacillus megaterium. J. Bacteriol.* 158:809–13

13. Fliss, E. R., Setlow, P. 1984. *Bacillus megaterium* spore protein C-3: nucleotide sequence of its gene and the amino acid sequence at its spore protease cleavage site. *Gene* 30:167–70

14. Fliss, E. R., Setlow, P. 1985. Genes for *Bacillus megaterium* small, acid soluble, spore proteins: nucleotide sequence of two genes and their expression during sporulation. *Gene* 35:151–57

15. Foster, J. 1957. Discussion of paper by J. F. Powell. Chemical changes occurring during spore germination. In *Spores,* ed. H. O. Halvorson, pp. 76–82. Washington, DC: Am. Inst. Biol. Sci.

16. Fox, G. E., Stackebrandt, E., Hespell, R. B., Gibson, H., Maniloff, J., et al. 1980. The phylogeny of prokaryotes. *Science* 209:451–63

17. Goldrick, S., Setlow, P. 1983. Expression of a *Bacillus megaterium* sporulation specific gene in *Bacillus subtilis. J. Bacteriol.* 155:1459–62

18. Gould, G. W. 1983. Mechanisms of resistance and dormancy. In *The Bacterial Spore,* ed. A. Hurst, G. W. Gould, 2:173–209. London/New York: Academic

19. Granum, P. E., Richardson, M., Blom, H. 1987. Isolation and amino acid sequence of an acid soluble protein from *Clostridium perfringens* spores. *FEMS Lett.* 42:225–30

20. Hackett, R. H., Setlow, P. 1983. Determination of the enzymatic activity of the precursor forms of the *Bacillus megaterium* spore protease. *J. Bacteriol.* 153:375–78

21. Hackett, R. H., Setlow, B., Setlow, P. 1986. Cloning and nucleotide sequence of the *Bacillus megaterium* gene coding for small, acid-soluble, spore protein B. *J. Bacteriol.* 168:1023–25

22. Hackett, R. H., Setlow, P. 1987. Cloning, nucleotide sequencing and genetic mapping of the gene for small, acid-soluble spore protein-γ of *Bacillus subtilis. J. Bacteriol.* 169:1985–92

23. Hackett, R. H., Setlow, P. 1988. Studies on the properties of spores of *Bacillus subtilis* strains which lack the major small, acid-soluble protein. *J. Bacteriol.* 170:1403–4

24. Johnson, W. C., Mahler, I., Phillips, K., Tipper, D. J. 1985. Transcriptional control of synthesis of acid-soluble proteins in sporulating *Bacillus subtilis. J. Bacteriol.* 163:543–51

25. Johnson, W. C., Tipper, D. J. 1981. Acid-soluble spore proteins of *Bacillus subtilis. J. Bacteriol.* 146:972–82

26. Leventhal, J. M., Chambliss, G. H. 1982. Synthesis of acid-soluble spore proteins by *Bacillus subtilis. J. Bacteriol.* 152:1117–25

27. Loshon, C. A., Fliss, E. R., Setlow, B., Foerster, H. F., Setlow, P. 1986. Cloning and nucleotide sequence of genes for small, acid-soluble, spore proteins of *Bacillus cereus, Bacillus stearothermophilus,* and *"Thermoactinomyces thalpophilus." J. Bacteriol.* 167:417–25

28. Loshon, C. A., Setlow, P. 1982. *Bacillus megaterium* spore protease: purification, radioimmunoassay, and analysis of antigen level and localization during growth, sporulation and spore germination. *J. Bacteriol.* 150:303–11

29. Loshon, C. A., Swerdlow, B. M., Setlow, P. 1982. *Bacillus megaterium* spore protease: synthesis and processing of precursor forms during sporulation and germination. *J. Biol. Chem.* 257:10838–45

30. Losick, R., Pero, J. 1981. Cascades of sigma factors. *Cell* 25:582–84

31. Losick, R., Youngman, P., Piggot, P. J. 1986. Genetics of endospore formation in *Bacillus subtilis. Ann. Rev. Genet.* 20:625–69

32. Mandelstam, J., Errington, J. 1987. Dependent sequences of gene expression controlling spore formation in *Bacillus subtilis. Microbiol. Sci.* 4:238–44

33. Mason, J. M., Hackett, R. H., Setlow, P. 1988. Studies on the regulation of expression of genes coding for small, acid-soluble proteins of *Bacillus subtilis* spores using *lacZ* gene fusions. *J. Bacteriol.* 170:239–44

34. Mason, J. M., Setlow, P. 1984. Expression of *Bacillus megaterium* and *Bacillus subtilis* small acid-soluble spore protein genes during stationary phase growth of asporogenous mutants of *Bacillus subtilis. J. Bacteriol.* 157:931–33

35. Mason, J. M., Setlow, P. 1986. Evidence for an essential role for small,

acid-soluble, spore proteins in the resistance of *Bacillus subtilis* spores to ultraviolet light. *J. Bacteriol.* 167:174–78

36. Mason, J. M., Setlow, P. 1987. Different small, acid-soluble proteins of the α/β type have interchangeable roles in the heat and ultraviolet radiation resistance of *Bacillus subtilis* spores. *J. Bacteriol.* 169:3633–37

37. Nelson, D. L., Kornberg, A. 1970. Biochemical studies of bacterial sporulation and germination. XVIII. Free amino acids in spores. *J. Biol. Chem.* 245:1101–7

38. Postemsky, C. J., Dignam, S. S., Setlow, P. 1978. Isolation and characterization of *Bacillus megaterium* mutants containing decreased levels of spore protease. *J. Bacteriol.* 135:841–50

39. Setlow, B., Hackett, R. H., Setlow, P. 1982. Non-involvement of the spore cortex in the acquisition of low molecular weight basic proteins and ultraviolet light resistance during sporulation of *Bacillus sphaericus*. *J. Bacteriol.* 149:494–98

40. Setlow, B., Setlow, P. 1979. Localization of low molecular weight basic proteins in *Bacillus megaterium* spores by irradiation with ultraviolet light. *J. Bacteriol.* 139:486–94

41. Setlow, B., Setlow, P. 1987. Thymine containing dimers as well as spore photoproducts are found in ultraviolet-irradiated *Bacillus subtilis* spores that lack small acid-soluble proteins. *Proc. Natl. Acad. Sci. USA* 84:421–23

42. Setlow, P. 1978. Purification and characterization of additional low–molecular weight basic proteins degraded during germination of *Bacillus megaterium* spores. *J. Bacteriol.* 136:331–40

43. Setlow, P. 1978. Degradation of dormant spore protein during germination of *Bacillus megaterium* spores. In *Limited Proteolysis in Microorganisms*, ed. G. N. Cohen, H. Holzer, pp. 109–13. Washington, DC: US Dep. Health Educ. Welfare

44. Setlow, P. 1985. Protein degradation during bacterial spore germination. In *Fundamental and Applied Aspects of Bacterial Spores*, ed. D. J. Ellar, pp. 285–96. London: Academic

45. Setlow, P., Fajardo, P., Hackett, R. H., Mason, J. M., Setlow, B. C., Sun, D. 1987. Regulation of expression of the *sspE* gene which codes for small, acid soluble protein-γ of *Bacillus subtilis* spores. In *Proc. 4th Int. Conf. Genet. Biotechnol. Bacilli, San Diego, Calif.* New York: Academic. In press

46. Setlow, P., Gerard, C., Ozols, J. 1980. The amino acid sequence specificity of a protease from spores of *Bacillus megaterium*. *J. Biol. Chem.* 255:3624–28

47. Setlow, P., Kornberg, A. 1970. Biochemical studies of bacterial sporulation and germination. XXIII. Nucleotide metabolism during spore germination. *J. Biol. Chem.* 245:3645–52

48. Setlow, P., Ozols, J. 1979. Covalent structure of protein A: a low molecular weight protein degraded during germination of *Bacillus megaterium* spores. *J. Biol. Chem.* 254:11938–42

49. Setlow, P., Ozols, J. 1980. Covalent structure of protein C: a second major low molecular weight protein degraded during germination of *Bacillus megaterium* spores. *J. Biol. Chem.* 255:8413–16

50. Setlow, P., Ozols, J. 1980. The complete covalent structure of protein B: the third major protein degraded during germination of *Bacillus megaterium* spores. *J. Biol. Chem.* 255:10445–50

51. Setlow, P., Waites, W. M. 1976. Identification of several unique low molecular weight basic proteins in dormant spores of *Clostridium bifermentans* and their degradation during spore germination. *J. Bacteriol.* 127:1015–17

52. Slee, A. M., Slepecky, R. A. 1977. The formation in media affording different growth rates of spores of *Bacillus megaterium* containing varying amounts of deoxyribonucleic acid. In *Spore Research 1976*, ed. H. N. Barker, J. Wolf, D. J. Ellar, G. J. Dring, G. W. Gould, pp. 183–94. London/New York: Academic

53. Sun, D., Setlow, P. 1987. Cloning and nucleotide sequencing of genes for the second type of small, acid-soluble spore proteins of *Bacillus cereus*, *Bacillus stearothermophilus*, and "*Thermoactinomyces thalpophilus.*" *J. Bacteriol.* 169:3088–93

54. Deleted in proof

55. Wang, T.-Z. V., Rupert, C. S. 1977. Evidence for the monomerization of spore photoproduct to two thymines by the light independent "spore repair" process in *Bacillus subtilis*. *Photochem. Photobiol.* 25:123–27

56. Wilson, A. C., Ochman, H., Prager, E. M. 1987. Molecular time scale for evolution. *Trends Genet.* 3:241–47

57. Yuan, K., Johnson, W. C., Tipper, D. J., Setlow, P. 1981. Comparison of various properties of low–molecular weight proteins from dormant spores of various *Bacillus* species. *J. Bacteriol.* 146:965–71

Ann. Rev. Microbiol. 1988. 42:339–58
Copyright © 1988 by Annual Reviews Inc. All rights reserved

REPLICATION OF KINETOPLAST DNA IN TRYPANOSOMES

Kathleen A. Ryan, Theresa A. Shapiro, Carol A. Rauch, and Paul T. Englund

Department of Biological Chemistry, Johns Hopkins School of Medicine, Baltimore, MD 21205

CONTENTS

INTRODUCTION

Kinetoplast DNA is the mitochondrial DNA of parasitic protozoa such as those of the genera *Trypanosoma, Leishmania,* and *Crithidia.* This DNA is in the form of a network of thousands of topologically interlocked DNA circles, which is a structure unique in nature. Each cell contains only one network, which resides in the matrix of its single mitochondrion. Two kinds of DNA circles constitute the network. Most abundant are the minicircles, present in several thousand copies per network, which vary in size from 0.5 to 2.8 kilobases (kb) in different species. Their function is not yet known. Less abundant are the maxicircles. Networks contain 25–50 maxicircles, and their

0066-4227/88/1001-0339$02.00

sizes vary from about 20 to 37 kb in different species. The maxicircles resemble mitochondrial DNAs in other eukaryotes in both their structure and genetic function. See References 3, 9, 24, 25, 57, 68, 70, 71, 75, and 76 for previous reviews on kinetoplast DNA.

Because of the network structure of kinetoplast DNA, its replication presents unique problems. The number of minicircles and maxicircles must be doubled, and the progeny circles must be distributed into two daughter networks. These daughter networks, whose composition and size are indistinguishable from those of the parent, must then segregate into the daughter cells during the cell division process.

In this review we first describe kinetoplast DNA structure and sequence organization, with emphasis on properties relevant to replication. We then discuss current knowledge of the replication mechanisms.

PROPERTIES OF KINETOPLAST DNA

Kinetoplast Network Structure

The electron micrograph in Figure 1 shows the minicircles and a maxicircle from the edge of an isolated kinetoplast DNA network. It is possible to distinguish maxicircles in electron micrographs only when they extend past the network edge. Because the kinetoplast DNA circles are topologically interlocked, topoisomerase II will efficiently decatenate a network in vitro, generating individual minicircles and maxicircles (50). Networks are two-dimensional sheets of DNA on the order of 5 μm in diameter, and they have molecular weights in the range of 10^{10} (22). Despite the large size of the networks, electron microscopy and sedimentation velocity experiments indicate that they are remarkably uniform in size and shape, except when undergoing replication.

Minicircles in nonreplicating networks are covalently closed, and individual covalently closed circles released from such networks by sonication have a very low superhelical density (C. A. Rauch, unpublished observations). As discussed below, shortly after replication network minicircles contain nicks and gaps in their newly synthesized strand.

A network resembling that in Figure 1 is found in most kinetoplastid parasites, including *Trypanosoma*, *Leishmania*, and *Crithidia* species. It is likely that kinetoplast DNA replication mechanisms are similar in all of these organisms. Some unconventional kinetoplast DNAs are found in distantly related organisms. For example, *Herpetomonas ingenoplastis* has networks of large and heterogeneous circles (16–23 kb) (S. L. Hajduk, cited in 25), and *Bodo caudatis* has DNA resembling kinetoplast DNA, in which the circles (10–12 kb) are not topologically interlocked (30). Nothing is known about the replication mechanisms of these DNAs.

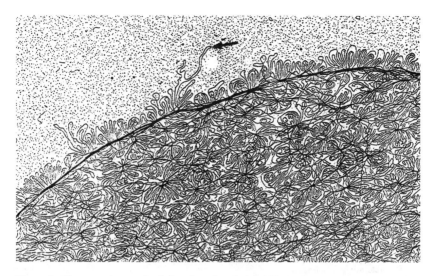

Figure 1 Electron micrograph of the edge of an isolated kinetoplast network from *Crithidia fasciculata*. Small loops are minicircles (2.5 kb) and the large loop *(arrow)* is part of a maxicircle (37 kb). The intact network is a two-dimensional sheet of DNA about 4 × 6 μm.

Properties of Minicircles

For most species, all the minicircles within a network are about the same size, which ranges from about 0.5 (8) to 2.8 kb (60). Nevertheless, minicircles within a single network are often heterogeneous in sequence, to varying degrees. There are about 250 different minicircle sequence classes in a *Trypanosoma brucei* network (74), whereas in a *Leishmania tarentolae* network, minicircle heterogeneity is less extensive (five sequence classes constitute about 50% of the total kinetoplast DNA) (36). In contrast, *Trypanosoma equiperdum* minicircles are homogeneous in sequence (2), and those from *Crithidia fasciculata* are nearly so (77). There is little similarity among minicircle sequences from different species, and even different isolates of the same species may diverge widely in minicircle sequence. It appears, therefore, that there is little selective pressure to maintain specific minicircle sequences.

Some features, however, are conserved in minicircles from all species. Sequencing of cloned minicircles has revealed characteristic patterns of sequence organization. Although minicircles within a network are heterogeneous in sequence, they always have conserved regions of approximately 100–200 base pairs (bp) (2, 13, 36, 46, 56, 77). For example, in the *T. brucei* network all of the 1-kb minicircles have a ~125-bp conserved region and a ~875-bp variable region (13). *L. tarentolae* minicircles have a similar sequence organization (Figure 2A; see also 36). In contrast, minicircles from *C.*

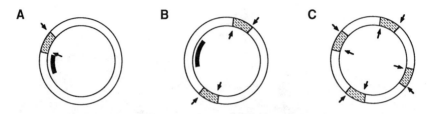

Figure 2 Sequence organization of kinetoplast minicircles. *Stippled region* represents conserved sequence (approximately 100–200 bp). *Outward-pointing arrows* indicate the sequence ACGCCC on the L strand, and *inward-pointing arrows* represent the sequence GGGGTTGGTGTA on the H strand. The possible role of these sequences in replication is discussed in the text. (*A*) Minicircles with one conserved sequence from *T. brucei* (1 kb), *T. equiperdum* (1 kb), or *L. tarentolae* (0.9 kb). (*B*) Minicircles with two conserved sequences from *C. fasciculata* (2.5 kb) or *T. lewisi* (1 kb). (*C*) Minicircles with four conserved sequences from *T. cruzi* (1.4 kb). *Black bar* indicates the position of the major bent helix in minicircles from *T. brucei, T. equiperdum* (54), *L. tarentolae* (36), and *C. fasciculata* (58). *T. lewisi* and *T. cruzi* minicircles do not have a region of strongly bent helix, although they do have isolated A runs that could dictate slight bending.

fasciculata (2.5 kb) (77) and *Trypanosoma lewisi* (1 kb) (56) each have two conserved regions which are direct repeats 180° apart (Figure 2*B*). Finally, *Trypanosoma cruzi* minicircles (1.4 kb) have four conserved regions which are direct repeats 90° apart (Figure 2*C*; see also 46).

Overall, there are few similarities among conserved sequences of different species. Two short sequences, however, are present in the conserved regions of every minicircle that has been studied: GGGGTTGGTGTA in the heavy (H) strand and ACGCCC in the light (L) strand.[1] [The only exception is one *L. tarentolae* minicircle, which contains GAGCCC at the latter position (36).] The positions of these sequences, which are separated by about 70 or 95 nucleotides, are indicated by arrows in Figure 2. We discuss below the possible role of these short sequences in minicircle replication.

Minicircles in most species have a single major region of bent helix (48, 54; see Figure 2 for location). DNA bending was first discovered in minicircles (48), but it has since been found in many eukaryotic and prokaryotic DNAs, especially in control regions such as promoters and replication origins. The helix bends because one of its strands has runs of four to six adenine residues periodically positioned within every ten base pairs (e.g. 5' AAAAAANN-NNNNAAAAANNNNNAAAAA 3') (79). Small bends are believed to be associated with each A run, and their periodic positioning, approximately in phase with the 10.5-bp helical repeat, results in a systematic curvature of the

---

[1] "Light" and "heavy" refer to the relative density of *C. fasciculata* minicircle strands in an alkaline CsCl gradient; the same nomenclature is used for strands from other species that contain the same conserved sequences.

helix. One ~200-bp restriction fragment from a *C. fasciculata* minicircle has 18 A runs (41). This fragment is the most extremely bent DNA molecule ever found in nature; as visualized by electron microscopy it bends about 360° (27). Bending causes anomalously slow electrophoretic migration of DNA fragments through polyacrylamide gels (12, 48, 69), and it also affects other physical behavior such as rotational diffusion (48). The biological role of the minicircle bent helix is not yet known, but we mention two possible functions below.

The function of minicircles is unclear. The wide divergence of minicircle sequences among different species and the absence of conserved open reading frames suggest that minicircles do not code for proteins. Nevertheless, indirect evidence has been presented for a minicircle protein product in *C. fasciculata* (64). Also, a minicircle transcript of 240 nucleotides was recently discovered in *T. brucei* (61). The function of this transcript is not known.

Properties of Maxicircles

Depending on the species, maxicircles range in size from 20 to 37 kb. All of the maxicircles within a network seem to be identical in sequence. They apparently have no sequences in common with minicircles. Maxicircles from *T. brucei* and *L. tarentolae* have been almost completely sequenced (see 17, 31, 72 for complete set of references); they contain genes for rRNA, cytochrome *b*, and subunits of cytochrome oxidase and NADH dehydrogenase. There are several unidentified open reading frames, and there is a divergent or variable region of several kilobases which contains a variety of repetitive sequences (52). Recent reviews describe in detail maxicircle gene organization, transcription, evolution, and developmental regulation of gene expression (71, 76). Of special interest is a recently described and novel mechanism of processing maxicircle transcripts (U insertion) (4, 26).

Packing of the Network Within the Mitochondrion

The network is condensed in vivo into a small volume within the mitochondrial matrix. Electron micrographs of thin sections indicate that in most species the network is packed in a disk, with the DNA fibers aligned parallel to its axis (Figure 3). The thickness of the kinetoplast disk varies among different species and is about half the circumference of a minicircle (see 68 for a tabulation of these dimensions from several species). These observations led to the suggestion nearly 20 years ago that minicircles in networks are organized in vivo as shown schematically in Figure 4 (18).

The forces that stabilize the kinetoplast disk in vivo are not known, but it is possible that helix bending could help the DNA form a structure like that in Figure 4 (49). Proteins or polyamines could also facilitate the organized packing of the network in vivo. Even networks compacted in vitro by

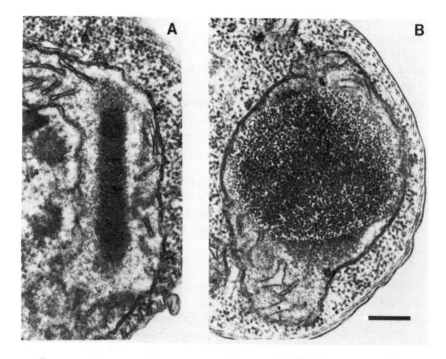

Figure 3 Electron micrographs of thin sections through the kinetoplast disk of *Phytomonas davidi*. Sections are parallel (*A*) and perpendicular (*B*) to the disk axis. The thickness of the disk is about 0.2 μm, which is about half the circumference of the 1.1-kb minicircle (14). The *bar* represents 0.24 μm. Micrographs were provided by Drs. Ronald A. Walkosz and William B. Cosgrove.

spermidine may have some organized structure; under these conditions restriction sites in some regions of the minicircle are inaccessible to cleavage (66).

Nothing is known about how maxicircles pack into the condensed network. They could be tucked completely inside the network disk, or they could be fully extended with only a small segment interlocked with some minicircles. Since maxicircles are 15–30 times the size of the minicircles, if they are fully extended they could reach far from the kinetoplast disk into the mitochondrial matrix.

REPLICATION OF KINETOPLAST DNA

Network Replication

EXPERIMENTAL DATA   Our current view of the molecular mechanism of kinetoplast replication is based on the following observations. (*a*) Kinetoplast DNA synthesis occurs only during a discrete S phase of the cell cycle, with

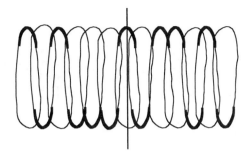

Figure 4 Schematic diagram of kinetoplast DNA packed within the mitochondrial matrix. The diagram is a cross section through the kinetoplast disk, and the *vertical line* represents the disk axis. The diagram is comparable to Figure 3A. *Heavy lines* indicate suggested locations of minicircle bent helices at either face of the kinetoplast disk.

timing similar to that of nuclear DNA synthesis (16). (*b*) Based on sedimentation velocity analysis, molecular weight measurements, and microscopy of isolated kinetoplast DNA, networks of replicating cells can be up to twice as large as those of nonreplicating cells; some double-size networks appear to be splitting in two (22, 32, 73). (*c*) Electron microscopy of thin sections has indicated that during replication the *C. fasciculata* kinetoplast disk enlarges radially but not in thickness; subsequently the enlarged disk appears to split parallel to its axis so that progeny networks can segregate into daughter cells (1). (*d*) Density shift experiments have suggested that minicircles replicate exactly once per generation (47, 73). (*e*) During replication a network grows from a structure of about 5000 covalently closed minicircles to a structure of about 10,000 nicked minicircles (22). (*f*) In vivo pulse-labeling experiments with [3H]thymidine have revealed that newly synthesized minicircles in networks are localized on the network periphery (73). (*g*) Electron microscopy of replicating networks in the presence of ethidium bromide has indicated that minicircles on the periphery are nicked or gapped, whereas those near the center are covalently closed (22). (*h*) Pulse-chase studies have indicated that free minicircles replicate after release from the network; after replication the progeny are reattached to the network (23, 39). (*i*) Electron microscopy has suggested that free minicircles replicate as theta structures (11, 23).

These experiments have led to the kinetoplast network replication scheme presented in Figure 5. This scheme focuses on minicircle replication; maxicircle synthesis is discussed later.

NETWORK REPLICATION SCHEME   Prior to replication, in the G1 phase of the cell cycle, the network contains about 5000 minicircles which are all covalently closed. This type of network is designated Form I (22). When the S phase begins, individual minicircles are released from the network by a

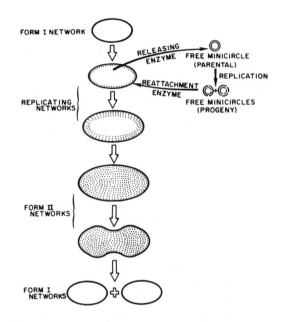

Figure 5   Scheme for network replication. *Elliptical structures* represents the outline of an entire network. *Dots* indicate the location of nicked or gapped minicircles that have undergone replication. See text for details. (From 25.)

releasing enzyme, presumably a topoisomerase, to form covalently closed free minicircles (23). These are replicated as theta structures; each forms two progeny minicircles, which are nicked or gapped (11, 23). At any time during the S phase there are several hundred free minicircles that have been released from the network for the purpose of replication; the detailed mechanism of free minicircle replication is presented below. After replication, a reattachment enzyme, also presumably a topoisomerase, returns the progeny to the network (23). Reattachment occurs exclusively at the network periphery (22, 73); therefore a replicating network has two zones. The central zone contains still unreplicated covalently closed minicircles, and the peripheral zone contains nicked or gapped progeny minicircles that have already been reattached (22). As replication proceeds, the central zone shrinks, the peripheral zone enlarges, and, because two minicircles are reattached for every one removed, the whole network grows in size (22, 73). Finally, when the S phase ends, all minicircles in the network have undergone replication and are nicked or gapped. The network at this stage, designated Form II, has doubled in size. By electron microscopy it often appears narrowed in the middle, as if it were in the process of dividing in two (22, 32). The final steps in this pathway, which occur during the G2 phase, include both the physical splitting of the double-size network and the covalent closure of all of the minicircles (22). The ultimate products of replication are two Form I networks, each indistinguishable from the parent, which segregate into the progeny cells during cell division.

Figure 6 shows how network replication may occur in vivo within the mitochondrial matrix. Minicircles are released from the central region of the network disk, and after replication they reattach to the periphery. Although topologically separate from the network, the free minicircles may be constrained in some way to remain in the region of the mitochondrial matrix containing the network.

OTHER FEATURES OF NETWORK REPLICATION   The replication of kinetoplast networks, as presented above, contrasts with that of mitochondrial DNA in higher eukaryotes. Replication of mitochondrial DNA circles in higher organisms is not restricted to the nuclear S phase, but occurs throughout the cell cycle (15); also, in any generation some circles replicate several times and some not at all (15). Because of its complex structure, a kinetoplast DNA network probably requires a replication mechanism that is more organized than that of mitochondrial DNA in higher cells. Perhaps nicking or gapping of minicircles is necessary to insure that each minicircle replicates once per generation. The releasing and reattachment enzymes may distinguish between the nicked or gapped (replicated) and covalently closed (still unreplicated) minicircles.

Although the releasing and reattachment enzymes may be specific for covalently closed and nicked or gapped minicircles, respectively, the choice

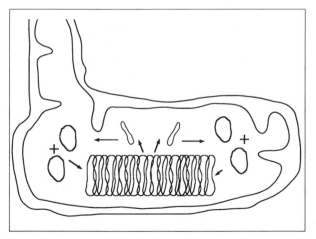

Figure 6   Scheme for network replication in vivo. The diagram shows a condensed network within a mitochondrial double membrane.

of minicircles for release and the site of reattachment could be random. Randomness could account for the fact that minicircles of identical sequence do not appear to be clustered within a network. [Clustering seems unlikely because minor sequence classes of minicircles can be released from a network by restriction enzyme cleavage without apparently creating holes in the structure (78).]

Randomness in release and reattachment of minicircles could be a powerful evolutionary force driving toward homogeneity of minicircles in a network (12). Upon replication of a minicircle, its two progeny could be distributed by chance into different daughter networks or into the same network. If they were distributed into the same network, the two daughter networks would differ in their copy number of this minicircle: One network would be enriched whereas the other would be depleted. If there were only one copy of the parent minicircle, the second network would lose that minicircle sequence forever. Many rounds of replication and stochastic distribution of minicircles could lead, by random walk, to the eventual fixation of a given minicircle class as the dominant or exclusive one in a network (19). This postulated mechanism may be a factor in explaining how two species of *T. equiperdum* evolved different homogeneous networks of minicircles (2, 66).

Why then do networks in most species have heterogeneous minicircles? There are several possible reasons. First, the rate of homogenization of a network of thousands of minicircles by random genetic drift would be extremely slow. This slowness could allow weak selection forces to become more important. Perhaps there is even a selective advantage of minicircle heterogeneity in a network (19). Secondly, new sequences are continually introduced by mutation and possibly by recombination. Many point mutations are detectable in *C. fasciculata* minicircles (77), and slight differences in restriction digests of kinetoplast DNA are detectable in cells passaged in laboratory culture for only a few years (10). Finally, minicircles from two different strains of a parasite could mix during mating to form hybrid networks (7). Mating has been described for *T. brucei* (a parasite with roughly 250 minicircle sequence classes in its network) during the insect vector stage of its life cycle (35). Since *T. equiperdum* has lost the ability to passage through the insect vector, it may never have this opportunity for minicircle exchange.

There are important unanswered questions regarding the replication of kinetoplast networks. One concerns the precise site where minicircles are reattached to the network after their replication. Autoradiography of networks isolated from cells that had been pulse-labeled for a short time with [³H]thymidine revealed radioactivity only at two peripheral sites 180° apart (67). This result suggests that minicircle reattachment occurs only at these sites. On the other hand, longer pulse-labeling led to a distribution of radioactivity around the entire network periphery (73). These results could be explained if the reattachment sites move around the network. Alternatively, minicircles could somehow be translocated from their initial reattachment site to locations along the rest of the network periphery. Another intriguing question concerns the mechanism by which the double-size Form II network splits in two. This process could involve a topoisomerase that unlinks neighboring minicircles, but the mechanism by which division is restricted to the center of the double-size network is completely unknown.

Free Minicircle Replication

The replication of free minicircles has been investigated most extensively in *C. fasciculata* and *T. equiperdum,* and studies so far have focused on the characterization of replication intermediates. Electron microscopy of isolated free minicircles has revealed molecules that resemble theta structures (11, 23). Agarose gel electrophoresis has allowed resolution of free minicircles that are covalently closed, nicked, gapped, linearized, and interlocked as dimers (5, 6, 39, 40, 55). Replication intermediates can be radioactively labeled either in vivo or in an isolated kinetoplast system in which the kinetoplast DNA and its replication enzymes are enclosed within the mitochondrial membrane (5).

Free minicircles that are covalently closed and relaxed appear to be the substrates for replication (39), which involves theta intermediates (11, 23). Synthesis of the L strand begins by RNA priming at a unique site and proceeds continuously and unidirectionally (5, 6, 39, 53, 55); synthesis of the H strand is probably discontinuous (5, 6, 39, 40). Pulse-chase experiments have indicated that the two progeny molecules, one with a newly synthesized L strand and the other with a newly synthesized H strand, are eventually reattached to the network (39). In the following paragraphs we discuss some aspects of these processes in more detail.

In *T. equiperdum* the newly synthesized L strand has a single gap of about 10 nucleotides (Figure 7; see also 53, 55). The gap is opposite the GGGGTTGGTGTA sequence common to the H strand of minicircles from all species examined (see Figure 2), which serves as the origin of unidirectional L-strand synthesis. One or two ribonucleotides, remnants of a replication primer, are detectable at unique sites on the 5' end of the newly synthesized L strand. The 10-nucleotide gap may arise from partial excision of a longer primer; alternatively, it may form if synthesis of the L strand stops a few bases before the 5' end of the primer. Both the gap and the primer are detectable on free minicircles and on minicircles that have been reattached to the network. A gap of almost identical structure is found opposite either of the two GGGGTTGGTGTA sequences in the newly synthesized L strand of *C. fasciculata* minicircles (see Figure 2B for location of these sequences; see also

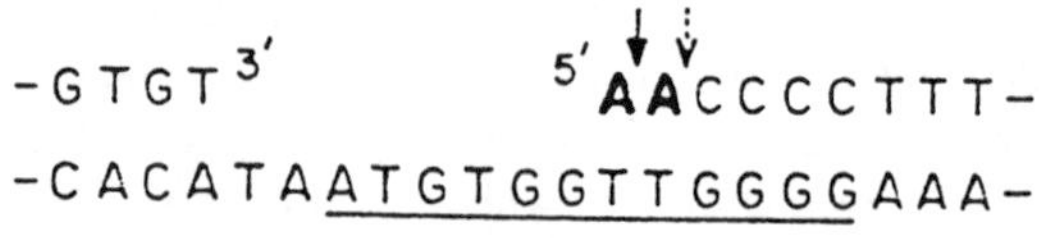

-GTGT $^{3'}$ $5'$ **AA**CCCCTTT-

-CACATA<u>ATGTGGTT</u>GGGGAAA-

Figure 7 Structure of the L-strand gap in newly replicated minicircles from *T. equiperdum*. **Bold** letters indicate ribonucleotides. The *underlined* sequence is found in the H strand of all minicircles examined. (From 55.)

6). Since there is only one gap per molecule, replication can be initiated at either site but not at both sites in the same molecule (6). In *T. cruzi* minicircles, L-strand synthesis might be initiated at any of the four GGGGTTGGTGTA sequences in the minicircle (see Figure 2*C*).

The mechanism of initiation of replication is not known. If the L strand initiated first, at the GGGGTTGGTGTA sequence, it would displace a parental L strand, which could then serve as a template for H-strand synthesis. A strong possibility for the initiation site of the first discontinuously synthesized H-strand fragment is the ACGCCC sequence, which is present, with minor variation, on the L strand of minicircles from all species examined. In *T. equiperdum* this sequence is located about 70 bp from the GGGGTTGGTGTA sequence (2). In free minicircles from this parasite, an abundant H-strand fragment of 73 nucleotides was detectable by denaturing gel electrophoresis (K. A. Ryan, unpublished observation). This fragment is present not only in progeny molecules, but also in theta structures. The 5' end of this fragment is complementary to the ACGCCC sequence, and the 3' end is within the GGGGTTGGTGTA sequence. This 73-nucleotide fragment has no detectable ribonucleotides at its 5' end. Nevertheless, it maps to the position expected for the first discontinuously synthesized H-strand fragment. These results suggest ACGCCC as a prominent initiation site for H-strand synthesis. A schematic drawing of an early theta structure that incorporates these features is shown in Figure 8*A*. The ACGCCC sequence is discussed further below.

Denaturing gel electrophoresis of newly synthesized H-strand fragments from *T. equiperdum* free minicircles indicated that they were of discrete size (ranging from about 40 nucleotides to nearly 1 kb). Mapping studies showed that their 5' ends were at unique locations, although there were no obvious similarities in their 5' terminal sequences (K. A. Ryan, unpublished observations). These fragments did not contain detectable ribonucleotides at their 5' ends, but since most derived from more mature molecules rather than theta structures, they may have undergone processing (e.g. primer removal or ligation) after their synthesis. Newly synthesized H-strand fragments from *C.*

A   **B**

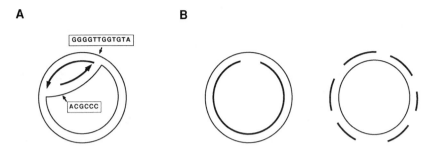

Figure 8 (*A*) Possible structure of early theta forms in *T. equiperdum*. L-strand synthesis, which is initiated at the GGGGTTGGTGTA sequence, is continuous. Synthesis of a predominant H strand fragment starts at ACGCCC. This H-strand fragment is 73 nucleotides. The fork moves unidirectionally and counterclockwise. (*B*) Possible structures of progeny minicircles prior to reattachment to the network. *Left*, molecule with continuously synthesized L strand. *Right*, molecule with discontinuously synthesized H strand.

fasciculata were smaller (30–120 nucleotides) and were separated by small gaps (40). Mapping studies have not been reported for the *C. fasciculata* H-strand fragments.

The diagram in Figure 8*B* shows possible structures for the progeny minicircles with newly synthesized H and L strands. After partial repair of nicks or gaps and possibly even the introduction of additional nicks (39), the progeny minicircles are reattached to the network. In *C. fasciculata* the progeny with a newly synthesized L strand are reattached faster than those with a newly synthesized H strand, enriching the free minicircle pool in those with a newly synthesized H strand (39). The final removal of L-strand primers and the repair of the nicks and gaps occur after the molecules are reattached to the network. Some nicks and gaps are repaired quickly, but others persist. As discussed above, molecules with a newly synthesized L strand have a nick or gap that remains opposite a GGGGTTGGTGTA sequence (6, 53). *C. fasciculata* minicircles with a newly synthesized H strand have interruptions that persist opposite both of the conserved ACGCCC sequences (6); therefore, after initial processing the H strands in these molecules are half minicircle length (6). These persistent interruptions may function to distinguish minicircles that have already undergone replication. The mechanism by which they are maintained at the conserved sequences has not been established.

Maxicircle Replication

Maxicircle synthesis occurs at the same time during the cell cycle as that of minicircles, and each maxicircle probably replicates once per generation (29). However, the mechanism of maxicircle replication differs from that of minicircles in that maxicircles replicate as network-bound rolling circles (29). These rolling circles are easily seen by electron microscopy; in some cases the tail of the rolling circle appears to be single-stranded (29). The presence of a single strand suggests that the initiation site for lagging-strand synthesis may be distant from that of leading-strand synthesis.

Our current view of maxicircle replication is shown in Figure 9. Rolling-circle replication is initiated at a specific site on a network-bound maxicircle, and the replication fork then proceeds around the circle. The direction of fork movement is not yet known. The tail, initially a single strand, is converted to a double strand when lagging-strand synthesis begins. Once the fork has moved completely around the circle and then a few kilobases past the original initiation site, the tail is cut off to form a linearized free maxicircle. This structure, which can be resolved by gel electrophoresis, is a few kilobases longer than a maxicircle genome and has terminally repetitive sequences (29). Mapping studies have revealed that these molecules have unique termini, of which one must correspond to the site where rolling-circle replication was initiated. The linearized free maxicircle presumably circularizes by homologous recombination within its terminally repetitive sequences. The linearized molecule may be reattached to the network by threading through the network prior to recircularization; alternatively, reattachment could be a topoisomerase-mediated process occurring after recircularization. Completion of this replication process results in double the number of maxicircles associ-

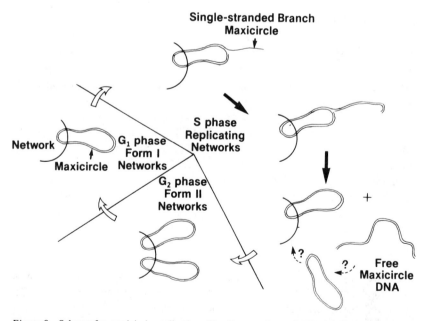

Figure 9 Scheme for maxicircle replication. The diagram shows the maxicircle attached to the network. See text for details. (From 28.)

ated with a double-size Form II network. Splitting of this Form II network then restores the normal maxicircle number.

The nucleotide sequence of the initiation site for rolling-circle replication has not yet been determined, although it probably resides in the divergent or variable region. A 189-bp *C. fasciculata* maxicircle restriction fragment behaves as an autonomously replicating sequence in yeast (37, 42). Because of this activity, and because similar sequences are found in maxicircles from *L. tarentolae* (34) and other trypanosomatid species (38), it is a possible maxicircle replication origin. Since this sequence is not near one of the termini of the linearized free maxicircles, it is unlikely to be the site of initiation of leading-strand synthesis (29), but it could be the site of initiation of lagging-strand synthesis.

There are additional complexities of maxicircle replication. Electron microscopy of double-size *T. brucei* Form II networks has revealed that most of the maxicircles are clustered near the central site where network division takes place (32). This clustering could reflect a mechanism by which the maxicircles are distributed evenly between the two progeny networks, but the molecular details are completely unknown.

Enzymology of Kinetoplast DNA Replication

There has been little progress in understanding the enzymology of kinetoplast DNA replication. Several DNA-metabolizing enzymes have been purified from kinetoplastid parasites, but none has been shown to have a mitochondrial localization.

Topoisomerases must have a central role in kinetoplast DNA replication. They undoubtedly function in the replication of free minicircles, and they are probably involved in release of minicircles from the network and reattachment of their progeny. Topoisomerases may also be involved in reattachment of progeny maxicircles. Furthermore, networks undergo considerable remodeling during replication, including division of double-size Form II networks and a loosening of network structure. (Form II networks appear to have more interlocks between neighboring minicircles than do Form I networks.) These processes could be mediated by topoisomerases.

A type II topoisomerase, unusual for its ATP independence, has been partially purified from *T. cruzi* (20). This enzyme, which in its native form has a molecular mass of 200 kd, was purified about 80-fold from a nuclear fraction. It can catenate, decatenate, and unknot double-stranded DNA, and it changes the linking number of its substrate in steps of two.

A type II toposiomerase has also been isolated from *C. fasciculata* (63, 65). This enzyme was purified 6000-fold, and it migrates as a single 60-kd polypeptide on SDS-polyacrylamide gel. It is ATP and Mg^{2+} dependent, and it catenates, decatenates, and relaxes supercoiled DNA. Of considerable interest in view of the discussion above is the fact that this enzyme decatenates networks of covalently closed minicircles but not minicircles nicked by the enzyme described in the following paragraph (63).

The *C. fasciculata* nicking enzyme, which has been purified to homogeneity and which contains two 60-kd subunits (62), attacks negatively or positively supercoiled DNA (44). In addition, it also attacks relaxed DNA circles that contain the minicircle bent helix (44). Binding assays have indicated that the enzyme binds to the minicircle bent sequence (43) and also to bent sequences associated with replication origins or promoters from several prokaryotic and eukaryotic DNAs (J. Shlomai, personal communication). Although this enzyme seems to recognize and bind to a bend in the DNA induced either by a specific sequence or by supercoiling, nicking can occur at a different site. There are many possible sites, and they seem to be localized in AT-rich regions of low melting temperature (45). The existence of this enzyme suggests another possible function for the minicircle bent helix. Although it has not been proven that kinetoplast DNA is a substrate for this enzyme in vivo, the specificity for bent DNA and the effect on topoisomerase II–mediated decatenation suggest that the enzyme may affect the topological interconversion of kinetoplast DNA networks and minicircles (44, 62).

Two other types of DNA-metabolizing enzymes have been purified from kinetoplastid parasites. Topoisomerase I activities have been purified from both *T. cruzi* (59) and *C. fasciculata* (51), and both resemble eukaryotic type I topoisomerases. Based on immunofluorescence, the *C. fasciculata* enzyme has been localized to the nucleus; the intracellular localization of the *T. cruzi* enzyme is unknown. DNA polymerases have been partially purified from both *T. brucei* (21) and *C. fasciculata* (33). *C. fasciculata* has two types of DNA polymerases, A and B, which have some resemblance to the mammalian alpha and beta polymerases, respectively (33). No enzyme has been detected in these parasites that resembles the gamma DNA polymerase in mammalian mitochondria.

CONCLUSIONS

We now have substantial knowledge of the kinetoplast DNA replication processes. However, there is still much work to be done, and the prospects are exciting. Investigation of this amazing DNA may continue to reveal basic principles of DNA structure and metabolism. Furthermore, study of the enzymes that replicate and maintain this unique structure could lead to novel modes of chemotherapy that may be useful in conquering the diseases caused by kinetoplastid parasites.

ACKNOWLEDGMENTS

We thank Tamara Doering for valuable comments on the manuscript and Viiu Klein and Shirley Metzger for expert help. We also thank Drs. Ronald Walkosz and William Cosgrove for permitting us to use unpublished electron micrographs. Research in the authors' laboratory was supported by grants from the NIH (GM 27608) and the MacArthur Foundation. TAS is supported by the Rockefeller Foundation and CAR by a Medical Scientist Training Program grant (2T32G-M07309).

Literature Cited

1. Anderson, W., Hill, G. C. 1969. Division and DNA synthesis in the kinetoplast of *Crithidia fasciculata. J. Cell Sci.* 4:611–20
2. Barrois, M., Riou, G., Galibert, F. 1981. Complete nucleotide sequence of minicircle kinetoplast DNA from *Trypanosoma equiperdum. Proc. Natl. Acad. Sci. USA* 78:3323–27
3. Benne, R. 1985. Mitochondrial genes in trypanosomes. *Trends Genet.* 1:117–21
4. Benne, R., Van den Burg, J., Brakenhoff, J. P., Sloof, P., Van Boom, J. H., Tromp, M. C. 1986. Major transcript of the frameshifted *coxII* gene from trypanosome mitochondria contains four nucleotides that are not encoded in the DNA. *Cell* 46:819–26
5. Birkenmeyer, L., Ray, D. S. 1986. Replication of kinetoplast DNA in isolated kinetoplasts from *Crithidia fasciculata:* identification of minicircle DNA replication intermediates. *J. Biol. Chem.* 261:2362–68
6. Birkenmeyer, L., Sugisaki, H., Ray, D. S. 1987. Structural characterization of site-specific discontinuities associated with replication origins of minicircle DNA from *Crithidia fasciculata. J. Biol. Chem.* 262:2384–92
7. Borst, P., Fase-Fowler, F., Gibson, W. C. 1987. Kinetoplast DNA of *Trypanosoma evansi. Mol. Biochem. Parasitol.* 23:31–38
8. Borst, P., Fase-Fowler, F., Weijers, P. J., Barry, J. D., Tetley, L., Vickerman, K. 1985. Kinetoplast DNA from *Trypanosoma vivax* and *T. congolense. Mol. Biochem. Parasitol.* 15:129–42
9. Borst, P., Hoeijmakers, J. H. 1979. Kinetoplast DNA. *Plasmid* 2:20–40
10. Borst, P., Hoeijmakers, J. H. J. 1979. Structure and function of kinetoplast DNA of the African trypanosomes. *Mol. Cell. Biol.* 15:515–31
11. Brack, C., Delain, E., Riou, G. 1972. Replicating, covalently closed, circular DNA from kinetoplasts of *Trypanosoma cruzi. Proc. Natl. Acad. Sci. USA* 69:1642–46
12. Challberg, S. S., Englund, P. T. 1980. Heterogeneity of minicircles in kinetoplast DNA of *Leishmania tarentolae. J. Mol. Biol.* 138:447–72
13. Chen, K. K., Donelson, J. E. 1980. Sequences of two kinetoplast DNA minicircles of *Trypanosoma brucei. Proc. Natl. Acad. Sci. USA* 77:2445–49
14. Cheng, D., Simpson, L. 1978. Isolation and characterization of kinetoplast DNA and RNA of *Phytomonas davidi. Plasmid* 1:297–315
15. Clayton, D. A. 1982. Replication of animal mitochondrial DNA. *Cell* 28:693–705
16. Cosgrove, W. B., Skeen, M. J. 1970. The cell cycle in *Crithidia fasciculata:* temporal relationships between synthesis of deoxyribonucleic acid in the nucleus and in the kinetoplast. *J. Protozool.* 17:172–77
17. de la Cruz, V. F., Neckelmann, N., Simpson, L. 1984. Sequences of six genes and several open reading frames in the kinetoplast maxicircle DNA of *Leishmania tarentolae. J. Biol. Chem.* 259:15136–47
18. Delain, E., Riou, G. 1969. DNA ultrastructure of the kinetoplast of *Trypanosoma cruzi* cultivated in vitro. *C. R. Acad. Sci. Ser. D* 268:1225–27
19. Dobzhansky, T. 1970. *Genetics of the Evolutionary Process*, pp. 230–66. New York: Columbia Univ. Press. 505 pp.
20. Douc-Rasy, S., Kayser, A., Riou, J. F., Riou, G. 1986. ATP-independent type II topoisomerase from trypanosomes. *Proc. Natl. Acad. Sci. USA* 83:7152–56
21. Dube, D. K., Williams, R. O., Seal, G., Williams, S. C. 1979. Detection and

characterization of DNA polymerase from *Trypanosoma brucei*. *Biochim. Biophys. Acta* 561:10–16

22. Englund, P. T. 1978. The replication of kinetoplast DNA networks in *Crithidia fasciculata*. *Cell* 14:157–68

23. Englund, P. T. 1979. Free minicircles of kinetoplast DNA in *Crithidia fasciculata*. *J. Biol. Chem.* 254:4895–900

24. Englund, P. T. 1981. Kinetoplast DNA. In *Biochemistry and Physiology of Protozoa*, ed. M. Levandowsky, S. H. Hutner, 4:333–83. New York: Academic

25. Englund, P. T., Hajduk, S. L., Marini, J. C. 1982. The molecular biology of trypanosomes. *Ann. Rev. Biochem.* 51:695–726

26. Feagin, J. E., Jasmer, D. P., Stuart, K. 1987. Developmentally regulated addition of nucleotides within apocytochrome *b* transcripts in *Trypanosoma brucei*. *Cell* 49:337–45

27. Griffith, J., Bleyman, M., Rauch, C. A., Kitchin, P. A., Englund, P. T. 1986. Visualization of the bent helix in kinetoplast DNA by electron microscopy. *Cell* 46:717–24

28. Hajduk, S. L., Englund, P. T. 1984. The replication of kinetoplast DNA. *3rd John Jacob Abel Symp. Mol. Parasitol., Baltimore, Md., 1983*, ed. J. T. August, pp. 53–62. New York: Academic

29. Hajduk, S. L., Klein, V. A., Englund, P. T. 1984. Replication of kinetoplast DNA maxicircles. *Cell* 36:483–92

30. Hajduk, S. L., Siqueira, A. M., Vickerman, K. 1986. Kinetoplast DNA of *Bodo caudatus:* a noncatenated structure. *Mol. Cell. Biol.* 6:4372–78

31. Hensgens, L. A., Brakenhoff, J., De Vries, B. F., Sloof, P., Tromp, M. C., et al. 1984. The sequence of the gene for cytochrome *c* oxidase subunit I, a frameshift containing gene for cytochrome *c* oxidase subunit II and seven unassigned reading frames in *Trypanosoma brucei* mitochondrial maxicircle DNA. *Nucleic Acids Res.* 12:7327–44

32. Hoeijmakers, J. H., Weijers, P. J. 1980. The segregation of kinetoplast DNA networks in *Trypanosoma brucei*. *Plasmid* 4:97–116

33. Holmes, A. M., Cheriathundam, E., Kalinski, A., Chang, L. M. S. 1984. Isolation and partial characterization of DNA polymerases from *Crithidia fasciculata*. *Mol. Biochem. Parasitol.* 10:195–205

34. Hughes, D., Simpson, L., Kayne, P. S., Neckelmann, N. 1984. Autonomous replication sequences in the maxicircle kinetoplast DNA of *Leishmania tar-*

entolae. *Mol. Biochem. Parasitol.* 13: 263–75

35. Jenni, L., Marti, S., Schweizer, J., Betschart, B., Le Page, R. W. F., et al. 1986. Hybrid formation between African trypanosomes during cyclical transmission. *Nature* 322:173–75

36. Kidane, G. Z., Hughes, D., Simpson, L. 1984. Sequence heterogeneity and anomalous electrophoretic mobility of kinetoplast minicircle DNA from *Leishmania tarentolae*. *Gene* 27:265–77

37. Kim, R., Ray, D. S. 1984. A 189-bp fragment of *Crithidia fasciculata* maxicircle DNA confers autonomous replication in *Saccharomyces cerevisiae*. *Gene* 29:103–12

38. Kim, R., Ray, D. S. 1985. Conservation of a 29-base-pair sequence within maxicircle ARS fragments from six species of trypanosomes. *Gene* 40:291–99

39. Kitchin, P. A., Klein, V. A., Englund, P. T. 1985. Intermediates in the replication of kinetoplast DNA minicircles. *J. Biol. Chem.* 260:3844–51

40. Kitchin, P. A., Klein, V. A., Fein, B. I., Englund, P. T. 1984. Gapped minicircles: a novel replication intermediate of kinetoplast DNA. *J. Biol. Chem.* 259:15532–39

41. Kitchin, P. A., Klein, V. A., Ryan, K. A., Gann, K. L., Rauch, C. A., et al. 1986. A highly bent fragment of *Crithidia fasciculata* kinetoplast DNA. *J. Biol. Chem.* 261:11302–9

42. Koduri, R., Ray, D. S. 1984. Identification of autonomous replication sequences in genomic and mitochondrial DNA of *Crithidia fasciculata*. *Mol. Biochem. Parasitol.* 10:151–60

43. Linial, M., Shlomai, J. 1987. Sequence-directed bent DNA helix is the specific binding site for *C. fasciculata* nicking enzyme. *Proc. Natl. Acad. Sci. USA* 84:8205–9

44. Linial, M., Shlomai, J. 1987. The sequence-directed bent structure in kinetoplast DNA is recognized by an enzyme from *Crithidia fasciculata*. *J. Biol. Chem.* 262:15194–201

45. Linial, M., Shlomai, J. 1988. A unique endonuclease from *Crithidia fasciculata* which recognizes a bend in the DNA helix: specificity of the cleavage reaction. *J. Biol. Chem.* 263:290–97

46. Macina, R. A., Sanchez, D. O., Gluschankof, D. A., Burrone, O. R., Frasch, A. C. C. 1986. Sequence diversity in the kinetoplast DNA minicircles of *Trypanosoma cruzi*. *Mol. Biochem. Parasitol.* 21:25–32

47. Manning, J. E., Wolstenholme, D. R.

1976. Replication of kinetoplast DNA of *Crithidia acanthocephali*. I. Density shift experiments using deuterium oxide. *J. Cell Biol.* 70:406–18

48. Marini, J. C., Levene, S. D., Crothers, D. M., Englund, P. T. 1982. Bent helical structure in kinetoplast DNA. *Proc. Natl. Acad. Sci. USA* 79:7664–68

49. Marini, J. C., Levene, S. D., Crothers, D. M., Englund, P. T. 1983. A bent helix in kinetoplast DNA. *Cold Spring Harbor Symp. Quant. Biol.* 47:279–83

50. Marini, J. C., Miller, K. G., Englund, P. T. 1980. Decatenation of kinetoplast DNA by topoisomerases. *J. Biol. Chem.* 255:4976–79

51. Melendy, T., Ray, D. S. 1987. Purification and nuclear localization of a type I topoisomerase from *Crithidia fasciculata*. *Mol. Biochem. Parasitol.* 24:215–25

52. Muhich, M. L., Neckelmann, N., Simpson, L. 1985. The divergent region of the *Leishmania tarentolae* kinetoplast maxicircle DNA contains a diverse set of repetitive sequences. *Nucleic Acids Res.* 13:3241–60

53. Ntambi, J. M., Englund, P. T. 1985. A gap at a unique location in newly replicated kinetoplast DNA minicircles from *Trypanosoma equiperdum*. *J. Biol. Chem.* 260:5574–79

54. Ntambi, J. M., Marini, J. C., Bangs, J. D., Hajduk, S. L., Jimenez, H. E., et al. 1984. Presence of a bent helix in fragments of kinetoplast DNA minicircles from several trypanosomatid species. *Mol. Biochem. Parasitol.* 12:273–86

55. Ntambi, J. M., Shapiro, T. A., Ryan, K. A., Englund, P. T. 1986. Ribonucleotides associated with a gap in newly replicated kinetoplast DNA minicircles from *Trypanosoma equiperdum*. *J. Biol. Chem.* 261:11890–95

56. Ponzi, M., Birago, C., Battaglia, P. A. 1984. Two identical symmetrical regions in the minicircle structure of *Trypanosoma lewisi* kinetoplast DNA. *Mol. Biochem. Parasitol.* 13:111–19

57. Ray, D. S. 1987. Kinetoplast DNA minicircles: high-copy-number mitochondrial plasmids. *Plasmid* 17:177–90

58. Ray, D. S., Hines, J. C., Sugisaki, H., Sheline, C. 1986. kDNA minicircles of the major sequence class of *C. fasciculata* contain a single region of bent helix widely separated from the two origins of replication. *Nucleic Acids Res.* 14:7953–65

59. Riou, G. F., Gabillot, M., Douc-Rasy, S., Kayser, A., Barrois, M. 1983. A type I DNA topoisomerase from *Trypanosoma cruzi*. *Eur. J. Biochem.* 134:479–84

60. Riou, J. F., Dollet, M., Ahomadegbe, J. C., Coulaud, D., Riou, G. 1987. Characterization of *Phytomonas* sp. kinetoplast DNA. *FEBS Lett.* 213:304–8

61. Rohrer, S. P., Michelotti, E. F., Torri, A. F., Hajduk, S. L. 1987. Transcription of kinetoplast DNA minicircles. *Cell* 49:625–32

62. Shlomai, J., Linial, M. 1986. A nicking enzyme from trypanosomatids which specifically affects the topological linking of duplex DNA circles: purification and characterization. *J. Biol. Chem.* 261:16219–25

63. Shlomai, J., Zadok, A. 1983. Reversible decatenation of kinetoplast DNA by a DNA topoisomerase from trypanosomatids. *Nucleic Acids Res.* 11:4019–34

64. Shlomai, J., Zadok, A. 1984. Kinetoplast DNA minicircles of trypanosomatids encode for a protein product. *Nucleic Acids Res.* 12:8017–28

65. Shlomai, J., Zadok, A., Frank, D. 1984. A unique ATP-dependent DNA topoisomerase from trypanosomatids. *Adv. Exp. Med. Biol.* 179:409–22

66. Silver, L. E., Torri, A. F., Hajduk, S. L. 1986. Organized packaging of kinetoplast DNA networks. *Cell* 47:537–43

67. Simpson, A. M., Simpson, L. 1976. Pulse-labeling of kinetoplast DNA: localization of 2 sites of synthesis within the networks and kinetics of labeling of closed minicircles. *J. Protozool.* 23:583–87

68. Simpson, L. 1972. The kinetoplast of the hemoflagellates. *Int. Rev. Cytol.* 32:139–207

69. Simpson, L. 1979. Isolation of maxicircle component of kinetoplast DNA from hemoflagellate protozoa. *Proc. Natl. Acad. Sci. USA* 76:1585–88

70. Simpson, L. 1986. Kinetoplast DNA in trypanosomid flagellates. *Int. Rev. Cytol.* 99:119–79

71. Simpson, L. 1987. The mitochondrial genome of kinetoplastid protozoa: genomic organization, transcription, replication, and evolution. *Ann. Rev. Microbiol.* 41:363–82

72. Simpson, L., Neckelmann, N., de la Cruz, V. F., Simpson, A. M., Feagin, J. E., et al. 1987. Comparison of the maxicircle (mitochondrial) genomes of *Leishmania tarentolae* and *Trypanosoma brucei* at the level of nucleotide sequence. *J. Biol. Chem.* 262:6182–96

73. Simpson, L., Simpson, A. M., Wesley, R. D. 1974. Replication of the kinetoplast DNA of *Leishmania tarentolae* and

Crithidia fasciculata. Biochim. Biophys. Acta 349:161–72
74. Steinert, M., Van Assel, S. 1980. Sequence heterogeneity in kinetoplast DNA: reassociation kinetics. *Plasmid* 3:7–17
75. Stuart, K. 1983. Kinetoplast DNA, mitochondrial DNA with a difference. *Mol. Biochem. Parasitol.* 9:93–104
76. Stuart, K. D. 1987. Regulation of mitochondrial gene expression in *T. brucei. Bioessays* 6:178–81
77. Sugisaki, H., Ray, D. S. 1987. DNA

sequence of *Crithidia fasciculata* kinetoplast minicircles. *Mol. Biochem. Parasitol.* 23:253–63
78. Weislogel, P. O., Hoeijmakers, J. H., Fairlamb, A. H., Kleisen, C. M., Borst, P. 1977. Characterization of kinetoplast DNA networks from the insect trypanosome *Crithidia luciliae. Biochim. Biophys. Acta* 478:167–79
79. Wu, H. M., Crothers, D. M. 1984. The locus of sequence-directed and protein-induced DNA bending. *Nature* 308:509–13

Ann. Rev. Microbiol. 1988. 42:359–93
Copyright © 1988 by Annual Reviews Inc. All rights reserved

STRUCTURE AND FUNCTION OF PORINS FROM GRAM-NEGATIVE BACTERIA

Roland Benz

Lehrstuhl für Biotechnologie der Universität Würzburg, Röntgenring 11, D-8700 Würzburg, Federal Republic of Germany

CONTENTS

INTRODUCTION

The outer membrane of gram-negative bacteria has an important role in the physiology of these organisms. Although it appears as a normal membrane

359

0066-4227/88/1001-0359$02.00

from its structure in the electron microscope (21), its properties are quite different from those of other biomembranes. All nutrients or antibiotics, whether hydrophilic or hydrophobic, have to cross this permeability barrier. Only hydrophilic solutes smaller than the exclusion limit specific for a given gram-negative bacterium can pass the outer membrane. The special sieving properties of this barrier are due to the presence of a few major proteins called "porins" (82). General porins form transmembrane channels across the membrane that sort according to the molecular weight of the solutes. Specific porins have also been characterized, which contain binding sites for substrates.

Porin channels are instructive as models of transmembrane diffusion of hydrophilic solutes. Furthermore, it is possible to isolate porin channels in large quantities and to study their function in reconstituted systems. The stability of the pore-forming oligomers in the presence of denaturating detergents and the possibility of their chemical modification gave the first insight into porin structure (32, 51, 114, 124). The genetics of porins of *Escherichia coli* and other enteric bacteria is under full control at present. Thus the groups responsible for special porin function are easily investigated via the construction of chimeric proteins (77, 128, 138) and site-directed mutagenesis (1). Both types of investigation will facilitate further study of the pores' structure-function relationship and will help to identify the amino acids in the primary structure responsible for pore characteristics.

Excellent reviews have recently appeared on the architecture (21, 71), function (84, 90, 94), and biogenesis (34) of the outer membrane of gram-negative bacteria. Further reviews have treated specific components of the outer membrane such as porins (10, 46, 47), lipopolysaccharides (43), and lipoprotein (23). This review describes the structure and function of the general diffusion porins in only a general way; emphasis is given to the description of the specific porins.

STRUCTURE AND COMPOSITION OF THE BACTERIAL OUTER MEMBRANE

The outer membrane is described here only to the extent required for the understanding of its function as a permeability barrier for hydrophilic solutes. The cytoplasmic membrane of gram-negative and most other bacteria is supported on the outside by the peptidoglycan layer, which gives the bacteria its characteristic form and protects it from osmotic lysis (21). The peptidoglycan consists of a network of amino sugars and amino acids. The amino sugars (N-acetylglucosaminyl-N-actylmuramyl dimers) form long linear strands which are covalently linked together between two muramyl residues by short tetrapeptides (21, 138). The components of the outer membrane are either covalently bound to the peptidoglycan or involved with it through ionic bridges in such a way that a tight network is produced (71).

The periplasmic space between the inner and outer membranes represents an additional cellular compartment and occupies between 5 and 20% of the total cell volume according to different estimations (94, 120). It has an important role in the physiology of gram-negative bacteria (84) because binding proteins for solutes such as maltose and phosphate and the β-lactamase activity are located there. The periplasmic space is almost isoosmotic with the cytoplasm (around 300 mOsm); thus the osmotic pressure, which is normally around 3.5 bars and at maximum about 7–8 bars (21), is maintained across the outer membrane and not across the inner membrane. The periplasmic space is strongly anionic compared to the external medium (120) because of anionic groups attached to the outer membrane and because of anionic membrane-derived oligosaccharides (MDO) present in the periplasmic space to maintain part of the osmolarity. These oligosaccharides also contribute to the Donnan potential across the outer membrane, which can be as large as 80–100 mV (inside negative) in media of low ionic strength (117a, 120).

Lipids

The outer membrane of gram-negative bacteria contains lipids, lipopolysaccharides (LPS), and proteins as major components. Its lipid composition is very similar to that of the inner membrane (30). Electron microscopic studies have provided some evidence of contacts between the cytoplasmic membrane and the outer membrane (7). These adhesion or fusion sites could be structures whose inner and outer membranes are fused; thus they allow diffusion of newly synthesized membrane components, including lipids and lipopolysaccharides, from the inner to the outer membrane. Similar contact sites have also been found between mitochondrial outer and inner membranes, and they probably have a major role in the physiology and protein import of mitochondria (62). The contact sites may not be as important in bacteria, but their presence would result in a similar lipid composition of the bacterial inner and outer membranes. This would explain why the outer membrane of cyanobacteria contains carotenoids (108).

The major lipid component of the enteric bacteria *E. coli* and *Salmonella typhimurium* is the zwitterionic phosphatidylethanolamine (30). Besides this neutral lipid, the outer membranes of both enteric organisms also contain small amounts of the negatively charged phospholipids phosphatidylglycerol and cardiolipin. In enteric bacteria the phospholipids are exclusively located in the inner monolayer of the outer membrane, where they cover about 50% of the surface (131). The rest is covered by proteins. Nonenteric bacteria such as *Neisseria gonorrhoeae*, the cyanobacteria, and rough mutants of enteric bacteria may also contain phospholipids in the outer monolayer of the outer membrane (90, 108).

Lipopolysaccharides

The outer surface of enteric bacteria is covered by about 40% LPS and 60% protein (131). This means that it is extremely hydrophilic. LPS are amphipathic molecules that exhibit a structural similarity to lipids. The hydrophobic lipid A moiety (endotoxin) is common to all LPS, while the hydrophilic polysaccharide core (O-antigen) may vary within a single species (37, 43, 136). The basic structure of the lipid A is a D-glucosaminyl-β-D-glucosamine backbone to which five to seven side chains are linked via ester and amide bonds. The saturated fatty acid residues are 14 and 16 carbon atoms long. They often contain hydroxyl groups in the C-3 position, which can be used for the binding of additional fatty acid residues via ester bonds (43).

In enteric bacteria the lipopolysaccharides form a strong barrier to the diffusion of hydrophobic molecules through the outer membrane (71, 84, 90). This barrier is partly formed by the long oligosaccharide side chain attached to the lipid A and to the ionic bridges between charged groups in the polysaccharide moiety (71). Mutants with drastically shorter oligosaccharides ("rough" mutants) often have an increased permeability of the outer membrane for hydrophobic compounds (10, 90). LPS also shows a considerable affinity to the pore-forming complexes (34, 45, 71) and other proteins (8). LPS is, however, not essential for pore function (102), although the opposite has been suggested from other reconstitution experiments (113).

Proteins

The protein composition of the outer membrane of gram-negative bacteria is relatively simple. Sodium dodecyl sulfate (SDS)-polyacrylamide gel electrophoretograms show only a limited number of bands that correspond to major proteins (40, 69, 76, 86, 95, 110, 115). Because of the peculiar aggregation behavior of the proteins, a clear and reproducible polypeptide pattern in the SDS-PAGE of the outer membrane proteins may only be obtained if the samples are boiled for at least several minutes in 1% SDS (49, 69, 86, 115). There is also considerable confusion in the nomenclature of the outer membrane proteins of E. coli and S. typhimurium because of the use of several different definitions by different laboratories (25, 54, 127). A uniform nomenclature system for proteins of E. coli and S. typhimurium is now widely accepted. In this system the protein or gene product is named after its structural gene, i.e. OmpF is the product of the ompF gene. (Omp stands for outer membrane protein.)

The outer membrane of gram-negative bacteria contains, in addition to the porins, murein-lipoprotein [Braun's lipoprotein (23, 34), MW 7,200 in E. coli], OmpA [MW 35, 160 (26, 29)], and minor proteins. OmpA and lipoprotein have no pore function. The minor proteins have important roles, e.g. in the uptake of iron (53) and vitamins (59). Under special starvation conditons some of these proteins are induced and become major bands on SDS-PAGE (see below).

Porins

Porins are also called "peptidoglycan-associated proteins" (90), "peptidogly-can-associated general diffusion pore proteins" (71), or "matrix proteins" (109). These proteins are very important for gram-negative bacteria because all hydrophilic nutrients must pass through them in the outer membrane (10, 47, 71, 84, 94). Their role as diffusion channels led to the name "porin," which was first proposed by Nakae (83). In most cases they appear to be closely associated with the peptidoglycan layer (54, 71). The name "matrix protein" is somewhat misleading because the real matrix proteins (i.e. the proteins that are important for the integrity of the outer membrane) are presumably the lipoproteins and OmpA (34, 118). However, the porins may have a role in the stability of the outer membrane (96, 118). Furthermore, the hexagonal structure that resulted in this definition (109, 119) is only observed in *E. coli* B under artificial conditions (this strain has only one porin). The name "peptidoglycan-associated protein" does not give any information on the function of this major class of outer membrane proteins. Throughout this paper the name porin is used, especially because some of the porins are not tightly associated with the peptidoglycan layer (49, 72). The outer membrane of gram-negative organisms contains up to 10^5 copies of the different porins per cell (109).

Porins from *E. coli* (17, 66, 111), *S. typhimurium* (13, 110, 123), and a number of other gram-negative bacteria such as *Pseudomonas aeruginosa* (49–51, 139), *Haemophilus influenzae* (129), *Neisseria gonorrhoeae* (35, 140), *Proteus mirabilis* (95, 110), *Rhodobacter capsulatus* (20, 41), *Paracoccus denitrificans* (141), and others (3, 33, 36, 60, 65) have been isolated and characterized. The proteins have molecular mass ranging from 30 to 50 kd. Most but not all porins are organized in the outer membrane as trimers of three identical subunits (2,100,101). [The porin of *Paracoccus denitrificans* is a dimer (141).] Whereas the outer membrane of *E. coli* B contains only one porin (OmpF) under normal growth conditions (109), two porins, OmpF and OmpC (61, 70), are present in the outer membrane of *E. coli* K-12. Their expression is influenced by the osmolarity of the culture medium and the nature of the carbon source. OmpF is preferentially expressed if the cells grow in media of low osmolarity or high cAMP levels (61, 116). On the other hand, OmpC is preferentially expressed in media of high osmolarity. The genes involved in the regulation of the osmolarity-sensitive expression of both porins are *ompR* (coding for a cytoplasmic regulator), and *envZ* (coding for a sensor and located in the cell envelope) (78). However, it was shown that

OmpC and OmpF directly affect the expression of one another, which means that there is also some feedback between both proteins (117). It is not clear how the cells sense the osmolarity of their environment. In vitro experiments have shown that the effective diameter of OmpF is a little larger than that of OmpC (17, 92). It has to be noted, however, that under different conditions the synthesis of one or both porins may not be critical for cell growth in the laboratory (73).

Many porins of gram-negative bacteria have properties similar to those of the *E. coli* porins. Most porins form wide water-filled channels in the outer membrane which sort solutes of different structures, mostly according to their molecular weights. These general diffusion porins have only minor interaction between the solutes and the pore interior. The largest known exclusion limit for hydrophilic solutes is that of the outer membrane of *P. aeruginosa*, which is permeable for substrates of up to about 5000 daltons (50). This large exclusion limit is due to the presence of the protein F, which forms trimers in the outer membrane (2, 50). On the other hand, *P. aeruginosa* is quite resistant to antibiotics (89). From this it has been concluded that only a limited number of pores are open at a given time. This conclusion has been confirmed by vesicle permeability measurements and artificial bilayer experiments in the presence of protein F (11, 139). On the other hand, in vivo studies have recently suggested that the high resistance of this organism to certain antibiotics is caused by the small exlusion limit (350 daltons) of the outer membrane (24a, 138a).

A number of porin pores are inducible in the outer membranes if the gram-negative bacteria are grown under special conditions. In the Enterobacteriaceae and other gram-negative bacteria, phosphate limitation leads to the induction of outer membrane proteins that form anion-selective channels (5, 104, 133). These PhoE porins are part of the *pho* regulon (98, 133) and are coregulated with the expression of several other P_i starvation–inducible proteins located in the periplasmic space and the inner membrane. An additional outer membrane protein, protein P, is induced in *P. aeruginosa* under similar conditions. However, whereas the different PhoE porins form general diffusion pores, protein P is a specific porin with a small selectivity filter for anions (12, 48, 51).

Other inducible specific porins are the LamB proteins of *E. coli* (122) and *S. typhimurium* (99), which are part of the maltose-inducible *mal* regulon of Enterobacteriaceae. This regulon consists of two regions, *malA* and *malB* (57, 105). Both regions contain a number of genes coding for proteins involved in the uptake of maltose and maltodextrins into the cell and their degradation. The LamB protein is involved in the permeation of sugars across the outer membrane. Similarly, Protein D1 is induced in *P. aeruginosa* when glucose is added to the growth medium (49, 76). LamB and the proteins D1 and P of *P.*

aeruginosa [but not PhoE (6)] are substrate specific because they contain binding sites for sugar or phosphate inside the channel (12, 18, 19, 31, 47, 48, 67, 68; H. Nikaido, personal communication). There are also specific porins in the outer membrane of gram-negative bacteria under normal growth conditions. *E. coli* (64) and other members of the Enterobacteriaceae (E. Bremer, personal communication) contain outer membrane proteins (Tsx proteins) that are important in the uptake of nucleosides into the cells (64, 80). Recently, it was shown that Tsx of *E. coli* is also a specific porin with a binding site for nucleosides inside the pore (72). This protein is constitutive, with about 10^4 copies per cell (72).

INVESTIGATION OF PORIN FUNCTION IN VIVO

The permeability of the outer membrane of gram-negative bacteria for hydrophilic solutes has been studied in two different ways. First, radioactively labeled substrates were added to the external media, and the uptake of the radioactivity into the cell was measured (55, 63, 118, 119). Secondly, the β-lactamase activity in the periplasmic space was used to study the uptake of lactamase-sensitive β-lactam antibiotics through the outer membrane; this method was developed separately by Zimmermann & Rosselet (142) and Sawai et al (111). The β-lactamase activity was used to study the effect of hydrophobicity on the permeation of different β-lactam antibiotics through the outer membrane (93). The results clearly showed that increasing the hydrophobicity reduced the flux of the β-lactams through the outer membrane. This was the first clear indication that the porins form hydrophilic pores, which represent a considerable diffusion barrier for the permeation of hydrophobic solutes.

The β-lactamase activity was also used to study the function of different porin species, such as OmpF, OmpC, and PhoE, in the *E. coli* outer membrane (63, 93). The results clearly indicated that PhoE had a preference for negatively charged solutes, because the rate of hydrolysis of a negatively charged β-lactam was much larger than that of a positively charged analog. The opposite was shown for OmpC. The function of PhoE (63), PhoE mutants (1), and PhoE-OmpC hybrids (128, 134) was studied in detail using hydrolysis of the neutral antibiotic cephaloridine and the negatively charged cefsulodin. Although the absolute rates of hydrolysis in the different organisms were not meaningful because of different expression of the porins, a comparison could be made on the basis of the uptake ratio of the two antibiotics in a given system (128). This ratio clearly demonstrated that certain amino acids, especially lysines, along the PhoE sequence were responsible for the anion selectivity of PhoE. Another study revealed an influence of additional positively or negatively charged groups inserted into the PhoE pore in other regions of the sequence that were probably exposed to the surface of the cell (1).

The uptake of radioactively labeled solutes into the cells can also be used to study outer membrane permeability. The solutes must be actively taken up into the cytoplasm, which limits the use of this procedure to the uptake of sugars, amino acids, and other nutrients (137). Using this method the selectivity of the PhoE pore for anionic substrates (63) and the selectivity of LamB and LamB mutants for sugars (55) were studied in detail.

Both methods only give precise information on outer membrane permeability if the flux through the membrane is rate limiting (93, 137). Thus with the first method the β-lactamase–mediated rate of hydrolysis has to be much larger than the flux of β-lactam antibiotics across the outer membrane. With the second method the uptake of the substrates across the inner membrane must be much faster than that across the outer membrane. Under both conditions the gradient is established across the outer membrane and the measurements reflect the flux of substrates through the outer membrane, but only if the number of pores is drastically reduced or if the substrate concentration is very low (137). The properties of specific porins such as LamB have occasionally been studied in vivo using methods that are not generally applicable (39, 42).

ISOLATION AND PURIFICATION OF PORINS

Many [but not all (49, 72)] porins isolated to date have been tightly associated with the peptidoglycan layer (71, 81, 82). Their isolation is relatively simple because only a few outer membrane proteins are associated with the murein, whereas the others are lost during the washing of the cell envelope fraction with SDS-containing solutions. The SDS-insoluble material contains the peptidoglycan with the covalently bound lipoprotein and the peptidoglycan-associated proteins, including OmpA and some porins. The porin can be separated from the peptidoglycan either by digestion of the peptidoglycan layer using lysozyme and trypsin (82, 83) or in a more elegant way by the salt extraction method (86, 88, 125). Boiling of the porin-peptidoglycan complex for 5 min in SDS also leads to dissociation, but it denatures the protein.

The salt extraction method for the isolation of functionally active porin trimers was first proposed by Nakamura & Mizushima (86). The method was extended by Tokunaga et al (125), and an excellent description has been given by Nikaido (88). The basic procedure is the treatment of the peptidoglycan porin complex with a buffer that contains as essential components 1% SDS, 0.4 M NaCl, and 5 mM EDTA at 37°C. The supernatant of a subsequent centrifugation step is applied to a Sepharose 4B or a Sephacryl S-200 superfine column. The porin trimers are eluted from the column with the same buffer. Ion-exchange columns (e.g. DEAE-Sephacel) can also be used for the purification of porins because the pore-forming complexes may absorb to the column material according to their net surface charge (72). The high ionic strength of the salt extraction must first be dialyzed against a low–ionic strength buffer. The column is first washed with low–ionic strength detergent buffer and then eluted with linear salt gradients ranging from 50 mM to 500 mM at a given pH. The salt and detergent content of the eluted protein can be decreased by another dialysis procedure, but this is not essential for the pore-forming activity, which remains constant for several months at 4°C or below. The protein solution can also be lyophilized in many cases without any loss of the pore-forming activity. In lyophilized form the protein remains active for at least 1 yr when stored in a freezer at −20°C.

Another method for the isolation of porin is saline extraction of whole cells. This method utilizes the susceptibility of the outer membranes of photosynthetic bacteria to 150-mM NaCl solution (136), but it is not generally applicable. Freshly harvested cells are shaken for 2 hr in 150 mM NaCl at 37°C. The supernatant of a centrifugation step is dialyzed against distilled water and lyophilized. The obtained material is composed of porin, lipid, LPS, and occasionally other outer membrane proteins (136). The porin may be purified as described above. Its pore function is identical to that of the porin obtained by the salt extraction method (72).

For the isolation of porins that are not closely associated with the peptidoglycan layer the outer membrane has to be isolated first. The procedure basically follows that first proposed by Miura & Mizushima (74). The cell envelope fraction is layered on top of a two-step sucrose gradient [70 and 54% (72)]. After centrifugation in an ultracentrifuge for about 12 hr at 80,000 × g the outer membrane lies at the interface between the sucrose density steps.

RECONSTITUTION OF PORIN PORES

The pore-forming properties of purified porins cannot be studied in detergent solutions. Reconstitution is necessary to reestablish the pore function and to show the integrity of the pore-forming complex after the isolation process. Furthermore, the in vitro systems allow detailed study of the pores. They contain only a few components and allow good control of the growth conditions. The main disadvantages of these systems are that artifacts are possible and that components are missing that may be necessary for the full activity of the porin pores. Many of the porin pores of gram-negative bacteria have successfully been reconstituted into liposomes, vesicles, or lipid bilayer membranes. Short descriptions of the three different methods and their possible artifacts are given below.

The vesicle permeability method was the first that was used to identify the pore-forming proteins of the outer membrane of *E. coli* and *S. typhimurium* (81–83). The basic principle is as follows. Liposomes are formed in a buffer solution containing two radiolabeled solutes (e.g. [^{14}C]sucrose and [^{3}H]dextran) from bacterial or other lipids in the presence of porin. During liposome formation both radiolabeled substances are entrapped. Subsequently, the liposomes are passed through a Sepharose 4B column, and they elute just after the void volume. The impermeable [^{3}H]dextran is retained in the liposomes, whereas the low–molecular weight [^{14}C]sucrose may leave the liposomes through the pores during the elution process and be retained on the column. Similar experiments performed with ^{14}C-labeled solutes of different molecular weights allow the evaluation of the exclusion molecular weight and thus of the effective diameter of the pore (82, 83, 130).

The vesicle permeability method has been used to measure the exclusion molecular weight of porins of *E. coli*, *S. typhimurium*, *P. aeruginosa*, and *H. influenzae* (50, 81–83, 130). The exclusion limits of the porins of *E. coli* and *S. typhimurium* were between 550 and 800, which would be consistent with diameters around 1.1–1.3 nm. The diameter of the protein F pore from the *P. aeruginosa* outer membrane appears to be much larger because solutes with molecular weights of more than 5000 could permeate the pore (50). The vesicle permeability assay does not allow the measurement of the relative kinetics of the permeation of different substrates. So much time is needed for the elution of the column that it is not possible to see any difference between the rates of permeation of different solutes or any specificity of the pore for one class of solutes. Artificial results may be obtained if the permeability of only one solute is measured or if more than one porin is present in the protein preparation. In the latter case only the larger pores can be detected even if they account for only a small fraction of the total number of pores.

Nikaido and coworkers (67, 68, 88, 91, 92, 139) introduced the liposome swelling method for the study of the permeability properties of porin pores following a method established by Bangham et al (4). Porin-containing liposomes are formed from lipid and protein in a buffer containing a large–molecular weight dextran or stachyose to maintain a certain osmolarity (around 40 mOsm). The stachyose or large–molecular weight dextran is not rendered permeable by the porin and is entrapped inside the liposomes. The liposomes are added under rapid mixing to an isotonic solution of a test solute. If this solute can penetrate the pores, the total concentration of solutes inside the liposomes increases because stachyose or dextran is retained. The influx of the test solute is followed by influx of water. The subsequent swelling of the liposomes is revealed by a decrease of the average refractive index of the liposome suspension, which can be detected by measurement of the optical density (see Figure 1). The initial swelling rate may be used as a measure of the rate of penetration of the test solute through the porin pore. Using the same liposome preparation the penetration rate of different solutes can be compared (Figure 1). For a general diffusion pore, i.e. a water-filled cylinder, the logarithm of the relative rate of permeation is a linear function of the molecular weight of the solutes, which allows an estimation of the effective diameter of the pore according to the theory of Renkin (107).

The liposome swelling assay allowed a meaningful comparison of different porin pores from the *E. coli* outer membrane (92). Furthermore, this method was used to demonstrate the greater specificity of the LamB channel for maltose and maltodextrins than for other saccharides (see Table 1) (67, 68) and to characterize the properties of different LamB mutants (85). The properties of a variety of porin pores from different gram-negative bacteria were also investigated using this method (35, 36, 41, 130, 139, 141). The pore diameter varied between 1.1 and 2.0 nm. Although the liposome swelling assay is far more precise than the vesicle permeability method, it can have a number of complications. The Renkin equation used to calculate the pore diameter is based on a number of simplifications (107), which may influence the results. In fact, disaccharides of identical molecular weight caused a substantial variability of the swelling rates (3–7 fold) (92). The large–molecular weight solute necessary to maintain the osmotic pressure (dextran or stachyose) may also create problems. The osmolarity of the large–molecular weight dextran was difficult to adjust in the experiments with LamB (67). Stachyose, on the other hand, which was also used in these

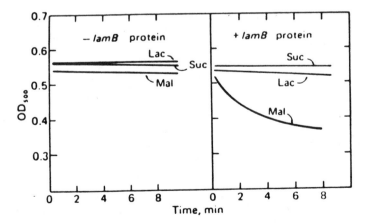

Figure 1 Liposome swelling in the presence of different 342-dalton disaccharides. *Left:* Liposome swelling in the absence of LamB. *Right:* Liposome swelling in the presence of LamB. The test solute had a concentration of 40 mM. Note that the initial swelling rate is higher in maltose than in sucrose or lactose. Taken from Reference 67 by permission.

experiments (85), binds to the binding site inside the LamB channel (19). Thus the relative rate of permeation of different sugars through LamB mutants may be influenced by the binding of stachyose.

The swelling experiments with reconstituted liposomes provide excellent information about the size of porin pores. They provide less detailed information on the selectivity of the pores because membrane potentials are created by the charged test solutes (92). Experiments with lipid bilayer membranes allow much better measurement of porin selectivity. For this type of measurement a salt gradient is established under zero current conditions across membranes in which at least 100 channels have been reconstituted (15, 66, 129, 140). Ions move through the pores according to their permeability until the chemical concentration gradient is balanced by the membrane potential. From the measured zero current membrane potential V_m and the concentration gradient c''/c' across the membrane, the ratio of the permeabilities for cations and anions, P_c/P_a, may be calculated according to the Goldman-Hodgkin-Katz equation (15).

For most porins of gram-negative bacteria the potential was found to be positive on the more dilute side of the membrane. This may be explained by the preferential movement of cations through the pores, i.e. the pores were cation selective (17). Other porins, e.g. NmpC and PhoE of *E. coli,* had a negative potential and were anion selective. The ionic selectivity of the general diffusion pores was found to be dependent on the mobility of the permeant ions in the aqueous phase (17). This means that their selectivity was not absolute as for nerve channels or the highly anion-selective protein P pore of the *P. aeruginosa* outer membrane (12, 17).

Porins have been successfully reconstituted into lipid bilayer membranes by two different methods. The simplest method consists of the addition of detergent-solubilized porin, in very small concentrations ($1-100$ ng ml^{-1}), to the aqueous phase bathing a painted lipid bilayer membrane (14, 66). After an initial lag of some minutes, presumably caused by the diffusion of the protein through the unstirred layers, the conductance (i.e. membrane current per unit voltage) of the membrane increases by many orders of magnitude within about 20–30 min (10, 13, 66). Only a slight additional increase (compared with the initial one) occurs after that time. Experiments with many channels have usually been performed with surface areas of $1-2$ mm^2, whereas the membrane areas have been below 0.1 mm^2 in single-channel recordings (14, 15). A maximum of 10^6-10^8 pores cm^{-2} could be incorporated into lipid bilayer membranes using this method. This means that the reconstituted porin pores are not rare, although they do not reach the channel density of the outer membrane [10^{12} channels cm^{-2} (13)].

Schindler & Rosenbusch (112, 113) have used an alternative method for the reconstitution of porin pores. Vesicles reconstituted from lipid and protein are spread on the surface of the aqueous phase on both sides of a thin piece of Teflon foil that has a small circular hole (about 100 μm diameter) above the initial water levels. The surfaces of both aqueous compartments are covered by monolayers. The water levels on both sides of the membrane are then raised, and a folded lipid bilayer membrane is formed across the small hole. It is not clear how the porin pores are incorporated into the lipid bilayers. In the early investigations the pores were always activated as multiples of the single conductance unit at large voltages, presumably because of the fusion of porin-containing vesicles with the membrane (112). More recently, especially with LamB (31), the pores have been activated as single conductive units. The diameter of the folded membranes is limited to about 100–200 μm (area about 0.07 mm^2).

Both lipid bilayer methods allow the resolution of single channels (14, 112). Besides the feasibility of selectivity measurements, this is the main advantage of the lipid bilayer approach. Small amounts of porin (1–10 ng ml^{-1}) are added to the aqueous phase bathing a black membrane of small surface area. Subsequently, a stepwise increase of the current through the membrane is observed. Figure 2 shows an experiment of this type. All conductance steps were directed upward. Closing events are in general only rarely observed, which means that the lifetime of porin pores from gram-negative bacteria usually exceeds 1 min.

The single-channel conductance of many but not all porin pores was found to be a linear function of the bulk aqueous conductivity of the bulk aqueous phase (15, 20, 66). This means that despite a large variation of the average single-channel conductance, the ratio between pore conductance (Λ) and the

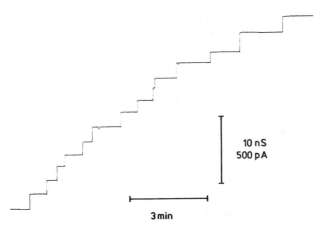

10 nS
500 pA

3 min

Figure 2   Stepwise increase of the membrane conductance (membrane current per unit voltage) in the presence of 1 ng ml^{-1} OmpF from *E. coli* K-12 in a 1 M KCl solution. The membrane was formed from phosphatidylcholine/*n*-decane; $T = 25°C$, $V_m = 50$ mV.

bulk aqueous conductivity (σ) varies only a little (15). The single-channel conductance of these general diffusion pores can be used to calculate the effective diameter of the porin pores. Assuming that the pores are filled with a solution of the same specific conductivity as the external solution and assuming that the pore is cylindrical with a length l of 6 nm, the effective pore diameter can be calculated from the equation

$$\Lambda = \sigma\pi r^2/l. \qquad\qquad\qquad 1.$$

According to this method the diameters of the pores vary between 1.0 and 2.3 nm (46).

Like the other methods, the lipid bilayer technique has its advantages and disadvantages. (See Reference 47 for a recent critique of this method.) The natural substrates of porin pores are sugars, amino acids, and only to a smaller extent ions. In lipid bilayer experiments the movement of ions is measured as the consequence of an applied voltage. Therefore this method provides information about ion transport only. This information is appropriate for protein P of the *P. aeruginosa* outer membrane, which is actually an ion channel similar to those in nerve and muscle membranes (12, 48, 51). For the general diffusion pores, however, this is only part of the information required to build a picture of the pore.

The basic advantage of the lipid bilayer method is in the measurement of molecular events. Every step of Figure 2 reflects the conductance due to the insertion of one conductive unit, i.e. one protein oligomer, into the membrane. Thus it is easy to check the purity of the porin preparation and the integrity of the pore-forming units after the isolation and purification of the porin (18). The interpretation of single-channel studies requires a statistically significant number of measurements. Hundreds of pores must be studied to avoid identifying random fluctuations that may occur in a limited number of single-channel events as characteristic properties of the pore (87).

FUNCTION OF GENERAL DIFFUSION PORINS

The porin pores of the outer membrane of gram-negative bacteria can be divided into two classes according to their permeability toward hydrophilic solutes. All gram-negative bacteria contain at least one general diffusion pore in the outer membrane which simply filters the solutes. Molecules with molecular weight larger than the exclusion limit are excluded from the pores and cannot enter the periplasmic space, whereas smaller solutes seem to permeate freely. However, the pore selectivity may modulate the permeation of charged molecules through the pore, as has been shown for the anion-selective PhoE porins of enteric bacteria (17, 63, 92). The effective radii of the general diffusion pores of different gram-negative organisms could be evaluated by the different methods described in the previous section by assuming that the pores form cylinders with effective diameter d and length l. In the liposome swelling assay the relative rate of permeation of different solutes is an exponential function of their molecular weights (91, 107). In the lipid bilayer experiments the single-channel conductance in the presence of different salts is a linear function of their specific aqueous conductances (11, 13, 20). The ion selectivity of the pores in different salt solutions follows the mobility sequence of the ions in the aqueous phase (15, 17, 20).

Many general diffusion pores have been characterized to date. (See Reference 46 for a complete list of general diffusion pores and their diameters calculated by the different methods.) The diameters calculated according to the different methods show satisfactory agreement except for those of the major outer membrane protein of *H. influenzae* (129). The reason for the inconsistency is not completely clear. It could be caused by structural differences of the pore in different environments, i.e. by oxidized cholesterol in the lipid bilayer membrane and by asolectin in the lipid vesicles. On the other hand, the reason proposed by the authors, i.e. the branching of the channel, may be valid in this case (47, 129).

FUNCTION OF SPECIFIC PORINS

The outer membrane of certain gram-negative bacteria contains, besides one or more general diffusion pores, channels that are specific for one class of solutes (9, 12, 28, 19, 49, 68, 72). These specific porins have a very different effect on the permeation of molecules because they contain binding sites with a defined half-saturation concentration for the molecules. The specific porins LamB and Tsx of *E. coli* and proteins P and D1 of *P. aeruginosa* are discussed in detail here.

Theoretical Considerations

It is assumed that the transport of the substrates through these channels can be explained by a simple two-barrier, one-site model (12, 18, 19, 72). This model assumes a binding site for a substrate in the center of the channel. The rate constant k_1 describes the jump of the substrates from the aqueous phase (concentration c) across the barriers to the central binding site, whereas the inverse movement is described by the rate constant k_2. So far experiments with the different specific porins have provided no indication of channel asymmetry; therefore symmetry of the channel with respect to the binding site is assumed.

The stability constant of the binding between a substrate molecule and the binding site is $K = k_1/k_2$. Furthermore, it is assumed that only one molecule can bind to the binding site at a given time and that no solute or ion (12, 19) can pass through the channel if the binding site is occupied. This means that a solute or ion can enter the channel only when the binding site is free. The probability, p, that the binding site is occupied by a sugar (LamB), nucleoside (Tsx), or anion (protein P) (identical concentrations on both sides) and does not conduct ions is given by

$$p = Kc/(1 + Kc).$$
  2.

The probability that the channel is free and conducts ions is given by

$$1 - p = 1/(1 + Kc).$$
  3.

The net flux of substrate molecules, Φ, through the channel under stationary conditions as the result of a concentration gradient $c'' - c'$ across the membrane is given by the net movement of substrates across one of the two barriers:

$$\Phi = k_1 c''/(1 + K') - k_2 K'/(1 + K'),$$
  4.

where K' is given by

$$K' = K(c' + c'')/2.$$
  5.

In Equation 4 the rate constants k_1 and k_2 are multiplied by the probabilities that the binding site is free and occupied, respectively. This is justified by the assumption that the substrate molecules or ions cannot pass one another in the pore.

When $c'' = c$ and $c' = 0$ [the conditions for the initial rate of permeation in the liposome swelling assay (67) or the in vivo experiments (42)] Equation 4 has the form

$$\Phi = k_1 c /(2 + Kc).$$
  6.

This equation may be used for the quantitative description of the relative rate of permeation as given by Reference 67 for the permeation of sugars through the LamB channel.

LamB of Escherichia coli

The ion conductance through a LamB-containing membrane can be titrated by increasing the concentration of sugars (18, 19, 31). Figure 3 shows an experiment of this type, in which increasing concentrations of maltopentaose were added to the aqueous phase bathing a LamB-containing membrane. The membrane conductance decreased as a function of the maltopentaose concentration. The data of Figure 3 (and of similar experiments with Tsx or protein P) can be analyzed using the following equations (see section on theoretical consideration, above). The conductance $\lambda(c)$ of a LamB-containing membrane in the presence of a sugar with the stability constant K and a sugar concentration c is given by the probability that the binding site is free:

$$\lambda(c) = \lambda_{max}/(1 + Kc). \qquad\qquad 7.$$

Here λ_{max} is the membrane conductance before sugar is added to the aqueous phase. Equation 7 may also be written

$$[\lambda_{max} - \lambda\,(c)]/\lambda_{max} = Kc\,/(Kc + 1), \qquad\qquad 8.$$

which means that the titration curves can be analyzed using a Lineweaver-Burke plot as shown in Figure 4 for the data of Figure 3. The straight line in Figure 4 corresponds to a stability constant K of 17×10^3 M^{-1} (half-saturation constant $K_8 = 5.9 \times 10^{-5}$ M). The results are identical if the experiment is repeated with 0.1 M KCl instead of 1 M KCl. Figure 4 also

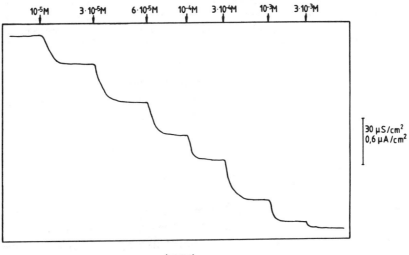

Figure 3   Titration of LamB-induced membrane conductance with maltopentaose. The membrane was formed from diphytanoyl phosphatidylcholine/n-decane. The aqueous phase contained 50 ng ml^{-1} LamB, 1 M KCl, and maltopentaose at the concentrations shown at the top of the figure. The temperature was 25°C and the applied voltage was 20 mV. Taken from Reference 19 by permission.

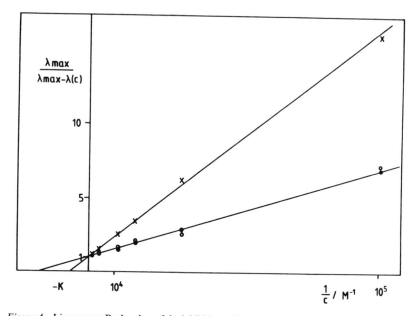

Figure 4 Lineweaver-Burke plots of the inhibition of LamB-induced membrane conductance by maltopentaose. Three different experiments are shown. *Full circles:* Experiment of Figure 2 (1 M KCl); $K = 18 \times 10^3$ M^{-1}. *Open circles:* Similar experiment to that in Figure 2, but with 0.1 M KCl; $K = 17 \times 10^3$ M^{-1}. *Crosses:* Experiment similar to that in Figure 2, but maltopentaose was added to only one side of the membrane; $K = 8 \times 10^3$ M^{-1}. Taken from Reference 19 by permission.

shows the reduced data of an experiment in which the maltopentaose was only added to one side of the LamB-containing membrane. In this case K was about 8×10^3 M^{-1}, which is about half the value obtained when the sugar was added to both sides of the membrane. This result indicates that the binding site inside the LamB channel is accessible from both sides of the membrane and that sugar crosses the membrane, because the binding site sees only half of the sugar concentration (i.e. the half-saturation constant is doubled).

The stability constants for the binding of a large variety of sugars to the binding site inside the LamB channel have been derived from measurements similar to those described above (19). The results are given in Table 1. In the series glucose to maltose to maltopentaose the binding constant increases about 1800-fold, whereas there is no further increase when there are five to seven glucose residues in the maltodextrins. All disaccharides, such as maltose, lactose, and sucrose, have stability constants between 7 and 250 M^{-1} for the binding to LamB. Sucrose and maltose have approximately the same

Table 1 Binding and permeation properties of the LamB channel for different sugars

Sugar	K $(M^{-1})^a$	K_s $(mM)^b$	P $(s^{-1})^c$	k_1 $(Ms^{-1})^d$	k_2 (s^{-1})
Maltose	100	10	100	15000	150
Maltotriose	2500	0.40	66	170000	68
Maltotetraose	10000	0.10	19	190000	19
Maltopentaose	17000	0.059	—	—	—
Maltohexaose	15000	0.067	—	—	—
Maltoheptaose	15000	0.067	2.5	38000	2.5
Trehalose	46	22	76	7300	160
Lactose	18	56	9	610	34
Sucrose	67	15	2.5	290	4.3
Gentibiose	250	4.0	42	13000	52
Melibiose	180	5.5	33	7600	42
Cellobiose	6.7	150	13	740	110
D-Glucose	9.5	110	290	17000	1800
L-Glucose	22	46	—	—	—
D-Galactose	24	42	225	17000	710
D-Fructose	1.7	600	135	7000	4100
D-Mannose	6.3	160	160	34000	5300
Stachyose	20	50	<1	—	—
Raffinose	46	22	—	—	—

[a]Stability constant calculated from titration experiments similar to those in Figures 3 and 4.
[b]Half-saturation constant (= K^{-1}).
[c]Rate of permeation relative to that of maltose. Data (adjusted to 100 s^{-1} for maltose) from Reference 67. The LamB-containing liposomes were added to buffer solutions containing 40 mM of the corresponding test sugars.
[d]Rate constants of permeation k_1 and k_2 were calculated assuming that the flux of maltose through the LamB channel under the conditions of Reference 67 (compare Equation 9) is 100 s^{-1} and that the flux of the other sugars relative to that of maltose is given by the relative rate of permeation.

affinity to LamB. This result seems to contrast with data from the liposome swelling assay (67), in which substantial differences have been found.

The discrepancy may be explained by differences in the transport kinetics of these sugars, which are predicted by the equations given above. Substituting the experimental conditions of Reference 67 ($c'' = 40$ mM, $c' = 0$), Equation 6 gives

$$\Phi = k_1(40 \text{ mM})/[2 + K(40 \text{ mM})], \qquad\qquad 9.$$

so that k_1 is given by

$$k_1 = \Phi(50 \text{ liter mol}^{-1} + K). \qquad\qquad 10.$$

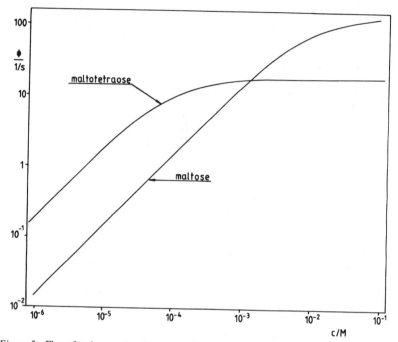

Figure 5   Flux of maltose and maltotetraose through LamB as a function of the concentration of the corresponding sugar on one side of the channel. The concentration on the other side was set at zero. The flux was calculated from Equation 6 assuming that the flux of maltose under the conditions of Reference 67 is 100 s^{-1} and using the stability constants and k_1 in Table 1. Taken from Reference 19 by permission.

Equation 10 can be used for a more quantitative description of the kinetics of sugar transport through LamB. If it is assumed that the flux of maltose under the conditions of Luckey & Nikaido (67) is 100 s^{-1}, k_1 is calculated to be 15 × 10^3 Ms^{-1} for maltose, whereas it is only 290 Ms^{-1} for sucrose. The rate constant of the off process, k_2, differs greatly between the two sugars (150 s^{-1} and 4.3 s^{-1}, respectively). Similar considerations apply to the transport of the other carbohydrates. The transport rate constants relative to that of maltose are summarized in Table 1. The sugars of the maltodextrin series have the largest association constants, whereas the dissociation constants, k_2, are largest for the monosaccharides.

The analysis also shows that the LamB channel is saturated by 40 mM concentration of those sugars with a large affinity for the binding site (i.e. the conditions of Reference 67). A further increase of the sugar concentration on one side of the channel cannot increase the flux. In this case the maximum flux is given by k_2, i.e. the maximum turnover number of a one-site, two-barrier channel. The maximum permeability of such a channel is given by $k_1/2$. This can also be seen from Figure 5, which shows the dependence of the flux of maltose and maltotetraose as a function of the sugar concentration on one side of the LamB channel while the concentration on the other side is zero (compare Equation 5). The maximum flux for the two sugars is given by the individual rate constants k_2, whereas the flux in the linear range (at small c) is given by $k_1c/2$. It is interesting that at small substrate concentrations (which corresponds to the in vivo situation) the flux of maltotetraose through the LamB channel is much higher than the flux of maltose. This result demonstrates how transport via a binding site inside a channel is faster than transport through a general diffusion pore, which always shows a linear dependence between driving force (gradient) and substrate flow (18).

Tsx of Escherichia coli

The Tsx protein is part of the nucleoside uptake system in E. coli, since mutants lacking this protein are impaired in the uptake of several nucleosides (64, 80) including adenosine and thymidine, whereas the uptake of cytidine is similar in tsx^+ and tsx^- strains. As with LamB and maltose uptake, transport by Tsx protein becomes rate limiting at nucleoside concentrations below about 1 μM (64, 80). Reconstitution experiments with purified Tsx showed that it is able to increase the conductance of lipid bilayer membranes by many orders of magnitude (72). As with LamB, the conductance was inhibited by the addition of nucleosides to the membranes, and half-saturation constants were calculated for the binding of the nucleoside to the membrane from Lineweaver-Burke plots (Table 2). The results are consistent with the assumption that the Tsx channel contains a binding site for nucleosides. However, the

Table 2 Stability constants for the binding of nucleobases, nucleosides, and deoxynucleosides to the Txs channel[a]

Nucleoside	K (M^{-1})[b]	K_s (mM)[c]
Adenine	500	2.0
Adenosine	2000	0.50
Deoxyadenosine	7100	0.14
Cytosine	No binding detectable	
Cytidine	45	22
Deoxycytidine	100	10

[a]Data were taken from Reference 72 and from R. Benz, A. Schmid, C. Maier, & E. Bremer, unpublished results.
[b]Calculated from titration experiments similar to those in Figures 3 and 4 for LamB.
[c]Half-saturation constant $K_s = K^{-1}$.

single-channel conductance of Tsx in 1 M KCl was only 10 pS, much smaller than that of other outer-membrane pores (160 pS for LamB and 1.9 nS for OmpF of *E. coli* K-12) (72).

As can be seen from Table 2, the stability constant for binding adenosine was much higher (corresponding to a smaller half-saturation constant) than that for binding cytidine. This result is consistent with the in vivo observation that Tsx facilitated the uptake of adenosine but not that of cytidine (64). Accordingly, a saturation of the flux of adenosine through the Tsx channel may be expected to increase concentrations of this nucleoside (72). It has been reported that Tsx also allows the flux of amino acids (56). Titration experiments with lipid bilayer membranes showed that there is no binding site for amino acids or sugars inside the channel (R. Benz, A. Schmid, C. Maier & E. Bremer, unpublished results). This means that if Tsx acts as a pore for these substrates it is a general diffusion pore for them.

Porin P of Pseudomonas aeruginosa

Under conditions of phosphate limitation, a porin (protein P) is induced in the outer membrane together with an alkaline phosphatase and a phosphate-binding protein (48, 52, 103). The induced system therefore has some similarities to the *pho* system of the Enterobacteriaceae, and protein P and PhoE are immunologically cross-reactive (104). Protein P forms highly anion-selective channels in lipid bilayer membranes which are the only in vitro system for the study of this channel. Zero-current membrane potentials have a Nernstian slope (corresponding to $P_c/P_a \ll 1$) and indicate that the protein P channel is at least 1000 times more permeable for chloride than for potassium (12, 48, 52). Similarly, the type of cation in the aqueous salt solutions had no influence on the single-channel conductance. In 100 mM chloride solutions this conductance was always about 160 pS, irrespective of the size and charge of the cations. The type of anion had a more substantial influence on the single-channel conductance, and anions of different size could be used to probe the diameter and the selectivity filter of the protein P channel (12). The binding of anions to this filter accounts for the concentration dependence of the single-channel conductance (48; R. Benz & R. E. W. Hancock, unpublished). The saturation of the single-channel conductance can easily be fitted to the following equation, which has been derived from the formalism described above (12):

$$\Lambda(c) = \Lambda_{max} K c / (1 + K c), \qquad\qquad 11.$$

where Λ_{max} is the maximum single-channel conductance given by

$$\Lambda_{max} = e_0^2 k_2 / (2kT), \qquad\qquad 12.$$

e_0 is the elementary charge, k is the Boltzmann constant, and T is the absolute temperature. The binding constant K (corresponding to the inverse half-saturation) and the maximum single-channel conductance can be used to calculate the kinetics for the movement of the different ions through the protein P channel. This analysis has shown that the channel is not efficient at high phosphate concentrations; its single-channel conductance was only 9 pS in 1 M KH_2PO_4, as compared with 280 pS in 1 M KCl (12, 48). However, under phosphate starvation (i.e. phosphate concentrations below 1 mM) the channel conducts phosphate much better than chloride because of its smaller half-saturation constant for phosphate (150 μM) than for chloride (20 mM) (R. Benz & R. E. W. Hancock, unpublished results). This is another clear demonstration of the advantage of having a binding site inside a specific porin.

Protein D1 of Pseudomonas aeruginosa

The addition of glucose to the growth medium of *P. aeruginosa* resulted in the synthesis of the 46-kd outer membrane protein D1 (49, 76). Protein D1 can be considered a glucose-specific channel because its presence results in a release of glucose from liposomes (49). The rate of penetration of different sugars through protein D1 was studied using the liposome swelling assay (H. Nikaido, personal communication). Comparison between protein F–containing liposomes and protein D1–containing liposomes clearly indicated that D1 is sugar specific (H. Nikaido, personal communication). The use of different glucose analogs allowed some identification of the groups responsible for the interaction with the pore interior (H. Nikaido, personal communication). As with LamB (19), it seems that the sugars interact with carbonyl groups in the pore interior by means of hydrogen bonds.

STRUCTURE OF PORIN PORES

The primary structure is now known for a large number of porins. In one case it was derived directly from the amino acid sequence (25) and in all the others from the DNA sequence (22, 28, 37a, 45a, 58, 75, 97). The best studied organism in this respect is *E. coli;* the primary sequences of most of the outer membrane porins, including that of the sugar-specific LamB channel, are known. The general diffusion porins of *E. coli* are acidic proteins. OmpF has an excess of 11 negatively charged amino acids, OmpC 14, and PhoE 9 (75). Their pIs are close to pH 5, which is somewhat surprising because of the difference in their ionic selectivities. OmpF and OmpC are both cation selective (17, 92), whereas PhoE is anion selective (17, 92). The extremely high degree of homology between OmpC and PhoE is demonstrated in Figure

```
           -21                                              -1
pro-OmpC   M K V K V L S L L V P A L L V A G A A N A
pro-PhoE       K S T     A     V     M G I V A S A S V Q

           +1                  10                  20                  30
OmpC       A E V Y N K D G N K L D L Y G K V D G L H Y F S D N K D V D G D G T
PhoE           I                     V           K A M       M       A S K         Q S

                        40                  50                  60
           Y M R L G F K G E T Q V T D Q L T G Y G Q W E Y Q I Q G N S A E N E
               I   F                 I N               R       A E F A       K       S D

           70                  80                  90                      100
           N N S - W T R V A F A G L K F Q D V G S F D Y G R N Y G V V Y D V T
           T A Q Q K       L                 Y K   L                 L   A L         E

                        110                 120                 130
           S W T D V L P E F G G D T Y G   S D N F M Q Q R G N G F A T Y R N T
           A       M F                 S S A Q T           T K     A S ' L

                        140                 150                 160
           D F F G L V D G L N F A V Q Y Q G K N G N P S G E G F T S G V T N N
                   V I           L T L             E   - - - - - - - - - - - - -

           170                 180                 190
           G R D A L R Q N G D G V G G S I T Y D Y E G - - F G I G G A I S S
           -       V K K             F   T   L         F G   S D   A   S       Y T N

           200                 210                 220                     230
           S K R T D A Q N T A A Y I G N G D R A E T Y T G G L K Y D A N N I
           D       N E   E L Q S R -   T   K       A W A T

                        240                 250                 260
           Y L A A Q Y T Q T Y N A T R V G S L G W A N K A Q N F E A V A Q Y
               T F   S E   R K M     P I T G -   F           T

           270                 280                 290
           Q F D F G L R P S L A Y L Q S K G K N L G R G Y D D E D I L K Y V
                       G   V L           D I E -   I G           L V N     I

           300                 310                 320                     330
           D V G A T Y Y P N K N M S T Y V D Y K I N L L D - D N Q F T R D A
                       F               A F             N Q     S       K L N - - -

                        340             346
           G I N T D N I V A L G L V Y Q F
           -       N   D       V   M T
```

Figure 6 Primary structures of OmpC (75) and PhoE (97) from *E. coli* K-12. The one-letter codes are those of the IUPAC-UIB Commission on Biological Nomenclature.

6, which shows the primary structure of both porins. The primary sequences of PhoE porins from *E. coli*, *Enterobacter cloacae*, and *Klebsiella pneumoniae* were also found to be highly homologous (133). This suggests that all these sequences may have evolved from a common ancestral gene.

Figure 6 also shows that the primary structures of the general diffusion porins OmpC and PhoE of *E. coli* are not as hydrophobic as would be expected for membrane proteins. The polarity indices of both proteins are approximately 48% (10). Such a high polarity is in principle more typical for soluble proteins than for membrane proteins. The unusual high polarity may be related to the pathway of assembly of the outer membrane proteins during which they may pass through the periplasmic space (see Reference 106 for a recent review on protein export). It is interesting that mitochondrial porins in eukaryotic cells also have an unusually high polarity (10). Only a few of the many charged amino acids in both bacterial and mitochondrial porins are exposed to the surface of the pore-forming complex or to the inside of the pore (10, 32, 48, 52, 114, 124). The rest most likely form internal ionic bridges that stabilize the pore-forming complex, because these amino acids are not accessible for chemical modification such as acetylation, succinylation, and amidation (10, 124).

Ionic selectivity is influenced by the charged amino acids at the surface and in the interior of the pores. The charges exposed to the inside of the channel seem to have a greater influence (1, 6, 32, 51). In general diffusion pores such as PhoE, anion selectivity is produced by a net positive charge due to lysine residues, rather than by a selectivity filter (32). However, the highly anion-selective protein P channel contains a selectivity filter of several lysines. This was shown by chemical modification of PhoE and protein P (32, 48). In fact, comparison of the two sequences of Figure 6 shows that several lysines are located in the PhoE sequence at places at which OmpC contains either an acidic or a neutral amino acid. The exchange of a short N-terminal piece of PhoE with the corresponding part of OmpC led to a dramatic change in selectivity; the hybrid channel became almost unselective (128, 134). This indicates that Lys18 is important for the anion selectivity of the PhoE pore. Further exchange of lysines along the primary sequence of PhoE led to additional jumps of the ion selectivity until it became identical to that of OmpC. The direct study of the role of these lysines by site-directed mutagenesis may give further insight into the ionic selectivity of the PhoE pore (K. Bauer, D. Bosch, R. Benz & J. Tommassen, unpublished results).

The primary structures of many outer membrane proteins such as OmpA, LamB, and the porins of *E. coli* indicate that there are no α-helical regions in the secondary structures (27). Instead, the sequences suggest that the pore-forming complexes of LamB and the porins have a β-pleated sheet structure. This suggestion agrees with circular dichroism measurements (86, 109, 125) of a variety of porin pores. If it is correct, 9–11 amino acids are sufficient to span the hydrocarbon core of a membrane 5–6 nm thick. This would mean that the polypeptide chain of the porins could cross the membrane more than 25 times (about 340 amino acids total). A more detailed study of the structure of the PhoE pore by J. Tommassen and his colleagues (1, 126, 128, 132, 133, 135) has shown, however, that the polypeptide chain of PhoE spans the

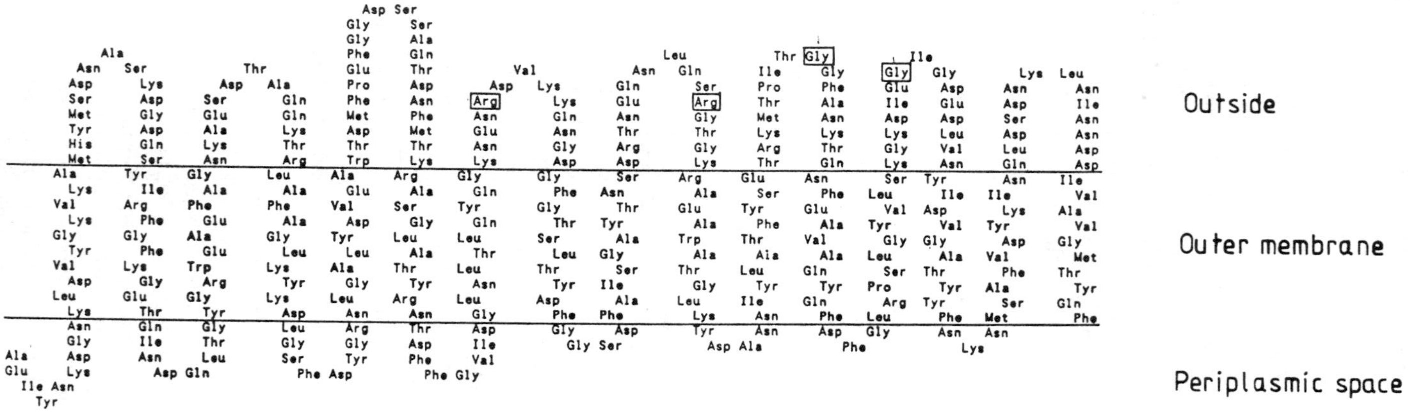

Outside

Outer membrane

Periplasmic space

Figure 7 Tentative model for the arrangement of the primary sequence of PhoE in the outer membrane of *E. coli* K-12. The topology of the protein was deduced from the binding of monoclonal antibodies and phages to PhoE, PhoE mutants, and PhoE-OmpC chimeric proteins (132, 135). From Reference 126 with permission.

membrane only 16 times as shown in Figure 7. The membrane-spanning segments show a high degree of homology to OmpF and OmpC. The N-terminus and the C-terminus of the proteins are located at the periplasmic side of the outer membrane (135). A detailed analysis of the structure of PhoE was performed by using site-directed mutagenesis of the protein (1) and binding of monoclonal antibodies and phages to the pore-forming complex (126, 132, 135). The existence of only eight loops and the binding of antibodies and phages to the surface of the pore-forming complex indicate that a considerable part of the sequence is hydrophilic and exposed to the outer surface. The hydrophilic parts of the sequence vary more than the rest (see above). The change of amino acids in the conserved regions creates problems and could lead to a block of the translation process (24; J. Tommassen, personal communication).

The structure of the pores of the bacterial outer membrane has been studied using electron microscopy and X-ray diffraction. Early electron-microscopic studies have shown that the matrix protein forms a hexagonal lattice in the outer membrane (119). The ordered arrays suggest that the pore-forming unit contains trimeric pores at the surface. More recent studies using image reconstruction of reconstituted OmpF trimers have shown that the pore is basically one channel with three openings facing the external surface (38). The three openings merge approximately in the center of the membrane, and the channel has only one outlet to the periplasmic side. The central constriction has a diameter of about 1 nm (38). This value is a rough approximation because of the rather low resolution of the image reconstruction. Better resolution is expected from the X-ray diffraction of OmpF trimers, but the analysis has not been completed (R. M. Garavito, personal communication). The central constriction of the pore should be rate limiting for both the movement of ions and the diffusion of uncharged solutes. Therefore the effective diameter calculated from the liposome swelling assay or the lipid bilayer experiments is only a rough approximation. On the other hand, the diffusion of molecules through a pore of any shape could be described by their diffusion through a cylinder with a given effective diameter.

OmpF trimers have been crystallized using a polyethylene glycol or salt-generated two-phase system (44). The crystals were analyzed by X-ray diffraction. The dimensions of the OmpF trimers could be evaluated, but not those of the pore (44). The thickness of the trimers perpendicular to the membrane was about 5.5 nm, whereas their diameter was about 8 nm. These results agree with earlier studies using outer membrane sheets. Furthermore, the X-ray diffraction analysis was found to be consistent with the assumption of β-pleated sheets oriented perpendicular to the membrane plane (44). Similar complete studies have not been published for other general diffusion porins or LamB, although the latter seems to have a similar structure (66a). The existence of three channels in a trimer that may or may not be separated by small walls could also explain why *E. coli* cells with OmpF and OmpC mutations grow on maltose and maltodextrins (9). These mutants have only minor changes in their primary sequence (73a), which may be sufficient for a large enough change in the pore structure that maltodextrins can pass through.

CONCLUSIONS

The outer membrane of gram-negative bacteria has very special properties compared with other membranes. Its hydrophilic and rigid surface reduces its permeability toward hydrophobic substances. On the other hand, small hydrophilic solutes are highly permeable, but they are poorly discriminated because of the large number of unselective hydrophilic pores (up to 10^{12} pores cm^{-2}). Similar special molecular sieving properties are only found in the mitochondrial outer membrane (10, 16) and the outer membrane of the chloroplast envelope (10), although any evolutionary relationship between these membranes is questionable.

The function of the general diffusion pores has been well studied on the basis of pore characteristics, chemical modification, and genetic mutations. Our understanding of the structure of the pore-forming complex is less complete because of the lack of good X-ray data. Knowledge of the translation and assembly of outer membrane proteins into the outer membrane is also poor, and a lot of work remains to be done. The molecular basis of the selectivity of the specific porins is completely open at present, and further studies are needed to give insight into the structure of the binding site inside the specific porin channels.

ACKNOWLEDGMENTS

I would like to thank Jan Tommassen for helpful discussions, W. Michael Arnold for a critical reading of the manuscript, and many colleagues for providing me with reprints and unpublished results. My own research was supported by grants of the Deutsche Forschungsgemeinschaft.

Literature Cited

1. Agterberg, M., Benz, R., Tommassen, J. 1988. Insertion mutagenesis on a cell surface exposed region of outer membrane protein PhoE of *Escherichia coli* K-12, *Eur. J. Biochem.* 169:65–71
2. Angus, B. L., Hancock, R. E. W. 1983. Outer membrane porin proteins F, P, and D1 of *Pseudomonas aeruginosa* and PhoE of *Escherichia coli:* chemical cross-linking to reveal native oligomers. *J. Bacteriol.* 155:1042–51
3. Armstrong, S. K., Parr, T. R. Jr., Parker, D. C., Hancock, R. E. W. 1986.

Bordetella pertussis major outer membrane porin protein forms small, anion-selective channels in lipid bilayer membranes. *J. Bacteriol.* 166:212–16
4. Bangham, A. D., Hill, M. W., Miller, N. G. A. 1974. Preparation and use of liposomes as models for biological membranes. *Methods Membrane Biol.* 1:1–68
5. Bauer, K., Schmid, A., Boos, W., Benz, R., Tommassen, J. 1988. Pore formation by *pho*-controlled outer membrane porins of various Enterobacter-

iaceae in lipid bilayers. *Eur. J. Biochem.* In press

6. Bauer, K., Van der Ley, P., Benz, R., Tommassen, J. 1988. The *pho*-controlled outer membrane porin PhoE does not contain specific binding-sites for phosphate or polyphosphate. *J. Biol. Chem.* In press

7. Bayer, M. E. 1979. The fusion sites between outer membrane and cytoplasmic membrane of bacteria: their role in membrane assembly and virus infection. In *Bacterial Outer Membranes,* ed. M. Inouye, pp. 167–202. New York: Wiley

8. Beher, M., Pugsley, A., Schnaitman, C. 1980. Correlation between the expression of an *Escherichia coli* cell surface protein and the ability of the protein to bind to lipopolysaccharide. *J. Bacteriol.* 143:403–10

9. Benson, S. A., Decloux, A. 1985. Isolation and characterization of outer membrane permeability mutants in *Escherichia coli* K-12. *J. Bacteriol.* 161: 36–67

10. Benz, R. 1985. Porins from bacterial and mitochondrial outer membranes. *CRC Crit. Rev. Biochem.* 19:145–90

11. Benz, R., Hancock, R. E. W. 1981. Properties of the large ion-permeable pores formed from protein F of *Pseudomonas aeruginosa* in lipid bilayer membranes. *Biochim. Biophys. Acta* 646: 298–308

12. Benz, R., Hancock, R. E. W. 1987. Mechanism of ion transport through the anion-selective channel of *Pseudomonas aeruginosa* outer membrane. *J. Gen. Physiol.* 89:275–95

13. Benz, R., Ishii, J., Nakae, T. 1980. Determination of ion permeability through the channels made of porins from the outer membrane of *Salmonella typhimurium* in lipid bilayer membranes. *J. Membrane Biol.* 56:19–29

14. Benz, R., Janko, K., Boos, W., Läuger, P. 1978. Formation of large, ion-permeable membrane channels by the matrix protein (porin) of *Escherichia coli. Biochim. Biophys. Acta* 511:305–19

15. Benz, R., Janko, K., Läuger, P. 1979. Ionic selectivity of pores formed by the matrix protein (porin) of *Escherichia coli. Biochim. Biophys. Acta* 551:238–47

16. Benz, R., Ludwig, O., De Pinto, V., Palmieri, F. 1985. Permeability properties of mitochondrial porins of different eukaryotic cells. In *Achievements and Perspectives of Mitochondrial Research,* ed. E. Quagliarello, E. C. Slater, F. Palmieri, C. Saccone, A. M. Kroon, 1: 317–27. Amsterdam: Elsevier

17. Benz, R., Schmid, A., Hancock, R. E. W. 1985. Ion selectivity of gram-negative bacterial porins. *J. Bacteriol.* 162:722–27

18. Benz, R., Schmid, A., Nakae, T., Vos-Scheperkeuter, G. H. 1986. Pore formation by LamB of *Escherichia coli* in lipid bilayer membranes. *J. Bacteriol.* 165: 978–86

19. Benz, R., Schmid, A., Vos-Scheperkeuter, G. H. 1988. Mechanism of sugar transport through the sugar-specific LamB channel of *Escherichia coli* outer membrane. *J. Membrane Biol.* 100:12–29

20. Benz, R., Woitzik, D., Flammann, H. T., Weckesser, J. 1987. Pore forming activity of the major outer membrane protein of *Rhodobacter capsulatus* in lipid bilayer membranes. *Arch. Microbiol.* 148:226–30

21. Beveridge, T. J. 1981. Ultrastructure, chemistry and function of the bacterial wall. *Int. Rev. Cytol.* 72:229–317

22. Blasband, A. J., Marcotte Jr., W. R., Schnaitman, C. A. 1986. Structure of the *lc* and *nmpC* outer membrane porin protein genes of lambdoid bacteriophage. *J. Biol. Chem.* 261:12723–32

23. Braun, V. 1975. Covalent lipoprotein from the outer membrane of *Escherichia coli. Biochim. Biophys. Acta* 415:335–77

24. Catron, K. M., Schnaitman, C. A. 1987. Export of protein in *Escherichia coli:* a novel mutation in *ompC* affects expression of other outer membrane proteins. *J. Bacteriol.* 169:4327–34

24a. Caulcott, C. A., Brown, M. R. W., Gonda, I. 1984. Evidence for small pores in the outer membrane of *Pseudomonas aeruginosa. FEMS Microbiol. Lett.* 21:119–23

25. Chen, R., Krämer, C., Schmidmayr, W., Chen-Schmeisser, U., Henning, U. 1982. Primary structure of outer-membrane protein I (*ompF* protein, porin) of *Escherichia coli* B/r. *Biochem. J.* 203:33–43

26. Chen, R., Schmidmayr, W., Krämer, C., Chen-Schmeisser, U., Henning, U. 1980. Primary structure of major outer membrane protein II (*ompA* protein) of *Escherichia coli* K-12. *Proc. Natl. Acad. Sci. USA* 77:4592–96

27. Chou, P. Y., Fasman, G. D. 1978. Prediction of secondary structure of proteins from their amino acid sequence. *Adv. Enzymol.* 47:45–148

28. Clement, J. M., Hofnung, M. 1981. Gene sequence of the λ-receptor, an outer membrane protein of *Escherichia coli* K-12. *Cell* 27:507–14

29. Cole, S. T., Sonntag, I., Henning, U. 1982. Cloning and expression in *Escherichia coli* K-12 of the genes for major outer membrane protein OmpA from *Shigella dysenteriae, Enterobacter aerogenes* and *Serratia marcescens. J. Bacteriol.* 149:145–50

30. Cronan, J. E., Vagelos, P. R. 1972. Metabolism and function of the membrane phospholipids of *Escherichia coli. Biochim. Biophys. Acta* 265:25–60

31. Dargent, B., Rosenbusch, J., Pattus, F. 1987. Selectivity for maltose and maltodextrins of maltoporin, a pore-forming protein of *E. coli* outer membrane. *FEBS Lett.* 220:136–42

32. Darveau, R. P., Hancock, R. E. W., Benz, R. 1984. Chemical modification of the anion selectivity of the PhoE porin from the *Escherichia coli* outer membrane. *Biochim. Biophys. Acta* 774:69–74

33. Darveau, R. P., MacIntyre, S., Buckley, J. T., Hancock, R. E. W. 1983. Purification and reconstitution in lipid bilayer membranes of an outer membrane, pore-forming protein of *Aeromonas salmonicida. J. Bacteriol.* 156:1006–11

34. Di Rienzo, J. M., Nakamura, K., Inouye, M. 1978. The outer membrane proteins of gram-negative bacteria: biosynthesis, assembly and functions. *Ann. Rev. Biochem.* 47:481–532

35. Douglas, J. T., Lee, M. D., Nikaido, H. 1981. Protein I of *Neisseria gonorrhoeae* outer membrane is a porin. *FEMS Microbiol. Lett.* 12:305–9

36. Douglas, J. T., Rosenberg, E. Y., Nikaido, H., Verstreate, D. R., Winter, A. J. 1984. Porins of *Brucella* species. *Infect. Immun.* 44:16–21

37. Drews, G., Weckesser, J., Mayer, H. 1978. Cell envelopes. In *The Photosynthetic Bacteria,* ed. R. Y. Clayton, W. R. Sistrom, pp. 61–78. New York/London: Plenum

37a. Duchene, M., Schweizer, A., Lottspeich, F., Krauss, G., Marget, M., et al. 1988. Sequence and transcriptional start site of the *Pseudomonas aeruginosa* outer membrane porin protein F gene. *J. Bacteriol.* 170:155–62

38. Engel, A., Massalski, A., Schindler, M., Dorset, D. L., Rosenbusch, J. P. 1985. Porin channels merge into single outlets in *Escherichia coli* outer membranes. *Nature* 317:643–45

39. Ferenci, T., Schwentorat, M., Ullrich, S., Vilmart, J. 1980. Lambda receptor in the outer membrane of *Escherichia coli* as a binding protein for maltodextrins and starch polysaccharides. *J. Bacteriol.* 142:521–26

40. Flammann, H. T., Weckesser, J. 1984. Characterization of the cell wall and outer membrane of *Rhodopseudomonas capsulata. J. Bacteriol.* 159:191–98

41. Flammann, H. T., Weckesser, J. 1984. Porin isolated from the cell envelope of *Rhodopseudomonas capsulata. J. Bacteriol.* 159:410–12

42. Freundlieb, S., Ehmann, U., Boos, W. 1988. Facilitated diffusion of *p*-nitrophenyl-α-D-maltohexaoxide through the outer membrane of *Escherichia coli:* characterization of LamB as a specific and saturable channel for maltooligosaccharides. *J. Biol. Chem.* 263:314–20

43. Galanos, C., Lüderitz, O., Rietschel, E. T., Westphal, O. 1977. Newer aspects of the chemistry and biology of bacterial lipopolysaccharides with special reference to their lipid A component. In *International Review of Biochemistry,* Vol. 14, *Biochemistry of Lipids II,* ed. T. W. Goodwin, pp. 239–335 Baltimore: Univ. Park Press

44. Garavito, R. M., Jenkins, J. A., Jansonius, J. M., Karlsson, R., Rosenbusch, J. P. 1983. X-ray diffraction analysis of matrix porin, an integral membrane protein A from *Escherichia coli* outer membranes. *J. Mol. Biol.* 164:313–27

45. Gmeiner, J., Schlecht, S. 1980. Molecular composition of the outer membrane of *Escherichia coli* and the importance of protein-lipopolysaccharide interactions. *Arch. Microbiol.* 127:81–86

45a. Gotschlich, E. C., Seiff, M. E., Blake, M., Koomey, M. 1987. Porin protein of *Neisseria gonorrhoeae:* cloning and gene structure. *Proc. Natl. Acad. Sci. USA* 84:8135–39

46. Hancock, R. E. W. 1986. Model membrane studies of porin function. In *Bacterial Outer Membranes as Model Systems* ed. M. Inouye, pp. 187–225. New York: Wiley

47. Hancock, R. E. W. 1987. Role of porins in outer membrane permeability. *J. Bacteriol.* 169:929–33

48. Hancock, R. E. W., Benz, R. 1986. Demonstration and chemical modification of a specific phosphate binding site in the phosphate-starvation-inducible outer membrane porin protein P of *Pseudomonas aeruginosa. Biochim. Biophys. Acta* 860:699–707

49. Hancock, R. E. W., Carey, A. M. 1980. Protein D1—a glucose-inducilbe, pore-forming protein from the outer membrane of *Pseudomonas aeruginosa. FEMS Microbiol. Lett.* 8:105–9

50. Hancock, R. E. W., Decad, G. M., Nikaido, H. 1979. Identification of the protein producing transmembrane diffusion pores in the outer membrane of *Pseudomonas aeruginosa* PAO1. *Biochim. Biophys. Acta* 554:323–31

51. Hancock, R. E. W., Poole, K., Benz, R. 1982. Outer membrane protein P of *Pseudomonas aeruginosa:* regulation by phosphate deficiency and formation of small anion-specific channels in lipid bilayer membranes. *J. Bacteriol.* 150:730–38

52. Hancock, R. E. W., Schmid, A., Bauer, K., Benz, R. 1986. Role of lysines in ion selectivity of bacterial outer membrane porins. *Biochim. Biophys. Acta* 860:263–67

53. Hantke, K. 1980. Regulation of ferric ion transport in *Escherichia coli* K-12: isolation of a constitutive mutant. *Mol. Gen. Genet.* 182:288–94

54. Hasegawa, Y., Yamada, H., Mizushima, S. 1976. Interactions of outer membrane proteins O-8 and O-9 with peptidoglycan sacculus of *Escherichia coli* K-12. *J. Biochem.* 80:1401–9

55. Heine, H.-G., Kyngdon, J., Ferenci, T. 1987. Sequence determinants in the *lamB* gene of *Escherichia coli* influencing the binding and pore selectivity of maltoporin. *Gene* 53:287–92

56. Heuzenroeder, W. M., Reeves, P. 1981. The *tsx* protein of *Escherichia coli* can act as a pore for amino acids. *J. Bacteriol.* 147:1113–16

57. Hofnung, M., Lepouce, E., Braun-Breton, C. 1981. General method for fine mapping of *Escherichia coli* K-12 *lamB* gene: localization of missense mutations affecting bacteriophage lambda absorptions. *J. Bacteriol.* 148:853–60

58. Inokuchi, K., Mutah, N., Matsuyama, S., Mizushima, S. 1982. Primary structure of the *ompF* gene that codes for a major outer membrane protein of *Escherichia coli* K-12. *Nucleic Acids Res.* 10:6957–68

59. Kadner, R. J. 1978. Repression of the synthesis of the vitamin B_{12} receptor in *Escherichia coli*. *J. Bacteriol.* 1367:1050–57

60. Kaneko, M., Yamaguchi, A., Sawai, T. 1984. Purification and characterization of two kinds of porins from *Enterobacter cloacae* outer membrane. *J. Bacteriol.* 158:1179–81

61. Kawaiji, H., Mizuno, T., Mizushima, S. 1979. Influence of molecular size and osmolarity of sugars and dextrans of the synthesis of outer membrane proteins O-

8 and O-9 of *Escherichia coli* K-12. *J. Bacteriol.* 140:843–47

62. Knoll, G., Brdiczka, D. 1983. Changes in freeze-fractured mitochondrial membranes correlated to their energetic state: dynamic interaction of the bond membranes. *Biochim. Biophys. Acta* 733:102–10

63. Korteland, J., De Graff, P., Lugtenberg, B. 1984. PhoE protein pores in the outer membrane of *Escherichia coli* K-12 not only have a preference for P_i and P_i-containing solutes but are general anion-preferring channels. *Biochim. Biophys. Acta* 778:311–16

64. Krieger-Brauer, H. J., Braun, V. 1980. Functions related to the receptor protein specified by the the the *tsx* gene of *Escherichia coli*. *Arch. Microbiol.* 124:233–42

65. Kropinski, A. M., Parr, T. R. Jr., Angus, B. L., Hancock, R. E. W., Ghiorse, W. C., Greenberg, E. P. 1987. Isolation of the outer membrane and characterization of the major outer membrane protein from *Spirochaeta aurantia*. *J. Bacteriol.* 169:172–79

66. Lakey, J. H., Watts, J. P., Lea, E. J. A. 1985. Characterisation of channels induced in planar bilayer membranes by detergent solubilised *Escherichia coli* porins. *Biochim. Biophys. Acta* 208–16

66a. Lepault, J., Dargent, B., Tichelaar, W., Rosenbusch, J. P., Leonard, K., Pattus, F. 1988. Three-dimensional reconstitution of maltoporin from electron microscopy and image processing. *EMBO J.* 7:261–68

67. Luckey, M., Nikaido, H. 1980. Specificity of diffusion channels produced by λ-phage receptor protein of *Escherichia coli*. *Proc. Natl. Acad. Sci. USA* 77:165–71

68. Luckey, M., Nikaido, H. 1980. Diffusion of solutes through channels produced by phage lambda receptor protein of *Escherichia coli:* Inhibition of glucose transport by higher oligosaccharides of maltose series. *Biochem. Biophys. Res. Commun.* 93:166–71

69. Lugtenberg, B., Meyers, J., Peters, R., van der Hoek, P., van Alphen, L. 1975. Electrophoretic resolution of the "major outer membrane protein" of *Escherichia coli* K-12 into four bands. *FEBS. Lett.* 58:254–58

70. Lugtenberg, B., Peters, R., Bernheimer, H., Berendsen, W. 1976. Influence of cultural conditions and mutations on the composition of the outer membrane proteins of *Escherichia coli*. *Mol. Gen. Genet.* 147:251–62

71. Lugtenberg, B., Van Alphen, L. 1983. Molecular architecture and functioning of the outer membrane of *Escherichia coli* and other gram-negative bacteria. *Biochim. Biophys. Acta* 737:51–115

72. Maier, C., Bremer, E., Schmid, A., Benz, R. 1987. Pore-forming activity of the Tsx protein from the outer membrane of *Escherichia coli*. Demonstration of a nucleoside-specific binding site. *J. Biol. Chem.* 263:2493–99

73. Matsuyama, S., Inokuchi, K., Mizushima, S. 1984. Promoter exchange between *ompF* and *ompC*, genes for osmoregulated major outer membrane proteins of *Escherichia coli* K-12. *J. Bacteriol.* 158:1041–47

73a. Misra, R., Benson, S. A. 1988. Isolation and characterization of OmpC porin mutants with altered pore properties. *J. Bacteriol.* 170;528–33

74. Miura, T., Mizushima, S. 1968. Separation by density gradient centrifugation of two types of membranes from spheroplast membrane of *Escherichia coli*. *Biochim. Biophys. Acta* 150:159–64

75. Mizuno, T., Chou, M.-Y., Inouye, M. 1983. A comparative study on the genes for three porins of the *Escherichia coli* outer membrane: DNA sequence of the osmoregulated *ompC* gene. *J. Biol. Chem.* 258:6932–40

76. Mizuno, T., Kageyama, M. 1978. Separation and characterization of the outer membrane of *Pseudomonas aeruginosa*. *J. Biochem. Tokyo* 84:179–91

77. Mizuno, T., Kasai, H., Mizushima, S. 1987. Construction of a series of *ompC-ompF* chimeric genes by *in vivo* homologous recombination in *Escherichia coli* and characterization of their translational products. *Mol. Gen. Genet.* 207:217–23

78. Mizuno, T., Wurtzel, E., Inouye, M. 1982. Cloning of the regulatory genes (*ompR* and *envZ*) for the matrix proteins of *Escherichia coli* outer membrane. *J. Bacteriol.* 150:1462–66

79. Mühlradt, P. F., Golecki, J. R. 1975. Asymmetrical distribution and artificial reorientation of lipopolysaccharide in the outer membrane bilayer of *Salmonella typhimurium*. *Eur. J. Biochem.* 51:343–52

80. Munch-Petersen, A., Mygind, B., Nicolaisen, A., Pihl, N. J. 1979. Nucleoside transport in cells and membrane vesicles from *Escherichia coli* K-12. *J. Biol. Chem.* 254:3730–37

81. Nakae, T. 1975. Outer membrane of *Salmonella typhimurium:* reconstitution of sucrose-permeable membrane vesicles. *Biochem. Biophys. Res. Commun.* 64:1224–30

82. Nakae, T. 1976. Identification of the major outer membrane protein of *Escherichia coli* that produces transmembrane channels in reconstituted vesicle membranes. *Biochem. Biophys. Res. Commun.* 71:877–84

83. Nakae, T. 1976. Outer membrane of *Salmonella*. Isolation of protein complex that produces transmembrane channels. *J. Biol. Chem.* 251:2176–78

84. Nakae, T. 1986. Outer-membrane permeability in bacteria. *CRC Crit. Rev. Microbiol.* 13:1–62

85. Nakae, T., Ishii, J., Ferenci, T. 1986. The role of maltodextrin binding site in determining the transport properties of maltoporin. *J. Biol. Chem.* 261:622–26

86. Nakamura, K., Mizushima, S. 1976. Effects of heating in dodecyl sulfate solution on the conformation and electrophoretic mobility of isolated major outer membrane proteins from *Escherichia coli* K-12. *J. Biochem.* 80:1411–22

87. Neuhaus, J. M., Schindler, H., Rosenbusch, J. P. 1983. The periplasmic maltose-binding protein modifies the channel-forming characteristics of maltoporin. *Embo J.* 2:1987–91

88. Nikaido, H. 1983. Proteins forming large channels from bacterial and mitochondrial outer membranes: porins and phage lambda receptor protein. *Methods Enzymol* 97:85–100

89. Nikaido, H., Hancock, R. E. W. 1985. Outer membrane permeability of *Pseudomonas aeruginosa*, In *The Bacteria: A Treatise on Structure and Function*, ed. J. R. Sokatch, 10:145–93. New York: Academic

90. Nikaido, H., Nakae, T. 1979. The outer membrane of gram-negative bacteria. *Adv. Microb. Physiol.* 20:163–250

91. Nikaido, H., Rosenberg, E. Y. 1981. Effect of solute size on diffusion rates through the transmembrane pores of the outer membrane of *Escherichia coli*. *J. Gen. Physiol.* 77:121–35

92. Nikaido, H., Rosenberg, E. Y. 1983. Porin channels in *Escherichia coli:* studies with liposomes reconstituted from purified proteins. *J. Bacteriol.* 153:241–52

93. Nikaido, H., Rosenberg, E. Y., Foulds, J. 1983. Porin channels in *Escherichia coli:* studies with β-lactams in intact cells. *J. Bacteriol.* 153:232–40

94. Nikaido, H., Vaara, M. 1985. Molecular basis of bacterial outer membrane permeability. *Microbiol. Rev.* 49:1–32

95. Nixdorf, K., Fitzer, H., Gmeiner, J., Martin, H. H. 1977. Reconstitution of model membranes from phospholipid and outer membrane proteins from *Proteus mirabilis:* role of porteins in the formation of hydrophilic pores and protection of membranes against detergents. *Eur. J. Biochem.* 81:63–68

96. Nogami, T., Mizushima, S. 1983. Outer membrane porins are important in maintenance of the surface structure of *Escherichia coli. J. Bacteriol.* 156:402–8

97. Overbeeke, N., Bergmans, H., van Mansfield, F., Lugtenberg, B. 1983. Complete nucleotide sequences of *phoE,* the structural gene for the phosphate limitation-inducible outer membrane pore protein of *Escherichia coli* K-12. *J. Mol. Biol.* 163:513–32

98. Overbeeke, N., Lugtenberg, B. 1980. Expression of outer membrane protein e of *Escherichia coli* K-12 by phosphate limitation. *FEBS. Lett.* 112:229–32

99. Palva, E. T. 1978. Major outer membrane protein in *Salmonella typhimurium* induced by maltose. *J. Bacteriol.* 136:286–94

100. Palva, E. T., Randall, L. L. 1978. Arrangement of protein I in *Escherichia coli* outer membrane: cross-linking study. *J. Bacteriol.* 133:279–86

101. Palva, E. T., Westermann, P. 1979. Arrangement of the maltose-inducible major outer membrane proteins, the bacteriophage receptor in *Escherichia coli* and the 44K protein in *Salmonella typhimurium. FEBS Lett.* 99:77–80

102. Parr, T. R. Jr., Poole, K., Crockford, G. W. K., Hancock, R. E. W. 1986. Lipopolysaccharide-free *Escherichia coli* OmpF and *Pseudomonas aeruginosa* protein P porins are functionally active in lipid bilayer membranes. *J. Bacteriol.* 165:523–26

103. Poole, K., Hancock, R. E. W. 1984. Phosphate transport in *Pseudomonas aeruginosa:* involvement of a phosphate binding protein. *Eur. J. Biochem.* 144:607–12

104. Poole, K., Hancock, R. E. W. 1986. Phosphate-starvation-induced outer membrane proteins of members of the families *Enterobacteriaceae* and *Pseudomonodaceae:* demonstration of immunological cross-reactivity with an antiserum specific for porin protein P of *Pseudomonas aeruginosa. J. Bacteriol.* 165:987–93

105. Raibaud, O., Roa, M., Braun-Breton, C., Schwartz, M. 1979. Structure of the *malB* region in *Escherichia coli* K-12.

Genetic map of the *malK-lamB* operon. *Mol. Gen. Genet.* 174:241–48

106. Randall, L. L., Hardy, S. J. S., Thom, J. R. 1987. Export of protein: a biochemical view. *Ann. Rev. Microbiol.* 41:507–41

107. Renkin, E. M. 1954. Filtration, diffusion and molecular sieving through porous cellulose membranes. *J. Gen. Physiol.* 38:225–53

108. Resch, C. M., Gibson, J. 1983. Isolation of the carotenoid-containing cell wall of three unicellular cyanobacteria. *J. Bacteriol.* 155:345–50

109. Rosenbusch, J. P. 1974. Characterization of the major envelope protein from *Escherichia coli.* Regular arrangement on the peptidoglycan and unusual dodecylsulfate binding. *J. Biol. Chem.* 249:8019–29

110. Sawai, T., Hiruma, R., Kawana, N., Kaneko, M., Taniyasu, F., Inami, A. 1982. Outer membrane permeation of β-lactam antibiotics in *Escherichia coli, Proteus mirabilis,* and *Enterobacter cloacae. Antimicrob. Agents Chemother.* 22:585–92

111. Sawai, T., Matsuba, K., Yamagishi, S. 1977. A method for measuring outer membrane-permeability of β-lactam antibiotics in gram-negative bacteria. *J. Antibiot.* 30:1134–36

112. Schindler, H., Rosenbusch, J. P. 1978. Matrix protein from *Escherichia coli* outer membranes forms voltage-controlled channels in lipid bilayers. *Proc. Natl. Acad. Sci. USA* 75:3751–55

113. Schindler, H., Rosenbusch, J. P. 1981. Matrix protein in planar membranes; clusters of channels in a native environment and their functional reassembly. *Proc. Natl. Acad. Sci. USA* 78:2302–6

114. Schlaeppi, J. M., Ichihara, S., Nikaido, H. 1985. Accessibility of lysyl residues of *Escherichia coli* B/r porin (OmpF) to covalent coupling reagents of different sizes. *J. Biol. Chem.* 260:9775–83

115. Schnaitman, C. A. 1973. Outer membrane proteins of *Escherichia coli.* I. Effect of preparation conditions on the migration of protein in polyacrylamide gels. *Arch. Biochem. Biophys.* 157:541–52

116. Schnaitman, C. A. 1974. Outer membrane proteins of *Escherichia coli.* IV. Differences in outer membrane proteins due to strain and cultural differences. *J. Bacteriol.* 118:454–64

117. Schnaitman, C. A., McDonald, G. A. 1984. Regulation of outer membrane protein synthesis in *Escherichia coli* K-

12: deletion of *ompC* affects expression of the OmpF protein. *J. Bacteriol.* 159:555–63

117a. Sen, K., Hellman, J., Nikaido, H. 1988. Porin channels in intact cells of *Escherichia coli* are not affected by Donnan potentials across the outer membrane. *J. Biol. Chem.* 263:1182–87

118. Sonntag, I., Schwarz, H., Hirota, Y., Henning, U. 1978. Cell envelope and shape of *Escherichia coli:* multiple mutants missing the outer membrane lipoprotein and other major outer membrane proteins. *J. Bacteriol.* 136:280–85

119. Steven, A. C., ten Heggeler, B., Müller, R., Kistler, J., Rosenbusch, J. P. 1977. Ultrastructure of a periodic protein layer in the outer membrane of *Escherichia coli. J. Cell Biol.* 72:292–301

120. Stock, J. B., Rauch, B., Roseman, S. 1977. Periplasmic space in *Salmonella typhimurium* and *Escherichia coli. J. Biol. Chem.* 252:7850–61

121. Szmelcman, S., Hofnung, M. 1975. Maltose transport in *Escherichia coli* K-12: involvement of the bacteriophage lambda receptor. *J. Bacteriol.* 124:112–18

122. Szmelcman, S., Schwartz, M., Silhavy, T. J., Boos, W. 1976. Maltose transport in *Escherichia coli* K-12: a comparison of transport kinetics in wild-type and λ-resistant mutants with the dissociation constant of the maltose-binding protein as measured by flourescence quenching. *Eur. J. Biochem.* 65:13–19

123. Tokunaga, H., Tokunaga, M., Nakae, T. 1979. Characterization of porins from the outer membrane of *Salmonella typhimurium. Eur. J. Biochem.* 95:433–47

124. Tokunaga, H., Tokunaga, M., Nakae, T. 1981. Permeability properties of chemically modified porin trimers from *Escherichia coli* B. *J. Biol. Chem.* 256:8024–29

125. Tokunaga, M., Tokunaga, H., Okajima, Y., Nakae, T. 1979. Characterization of porins from the outer membrane of *Salmonella typhimurium.* 2. Physical properties of the functional oligomeric aggregates. *Eur. J. Biochem.* 95:441–48

126. Tommassen, J. 1988. Biogenesis and membrane topology of outer membrane proteins in *Escherichia coli.* In *Membrane Biogenesis, NATO ASI Ser. Vol. H16,* ed. J. A. F. Op den Kamp, pp. 351–73. New York: Springer-Verlag

127. Tommassen, J., Lugtenberg, B. 1980. Outer membrane protein e of *Escherichia coli* K-12 is coregulated with alkaline phosphatase. *J. Bacteriol.* 143:151–57

128. Tommassen, J., Van der Ley, P., Van Zeijl, M., Agterberg, M. 1985. Localization of functional domains in *E. coli* K-12 outer membrane porins. *EMBO J.* 4:1583–87

129. Vachon, V., Laprade, R., Coulton, J. W. 1986. Properties of the porin of *Haemophilus influenzae* type b in planar lipid bilayer membranes. *Biochim. Biophys. Acta* 861:74–82

130. Vachon, V., Lyew, D. J., Coulton, J. W. 1985. Transmembrane permeability channels across the outer membrane of *Haemophilus influenzae* type b. *J. Bacteriol.* 162:918–24

131. Van Alphen, L., Lugtenberg, B., Van Boxtel, R., Verhoef, K. 1977. Architecture of the outer membrane of *Escherichia coli* K-12: action of phospholipase A_2 and C on wild-type strains and outer membrane mutants. *Biochim. Biophys. Acta* 466:257–68

132. Van der Ley, P., Amesz, H., Tommassen, J., Lugtenberg, B. 1985. Monoclonal antibodies directed against the cell-surface-exposed part of PhoE pore protein of the *Escherichia coli* K-12 outer membrane. *Eur. J. Biochem.* 147:401–7

133. Van der Ley, P., Bekkers, A., Van Meersbergen, J., Tommassen, J. 1987. A comparative study on the *phoE* genes of three enterobacterial species. Implications for structure-function relationships in a pore-forming protein of the outer membrane. *Eur. J. Biochem.* 164:469–75

134. Van der Ley, P., Burm, P., Agterberg, M., Van Meersbergen, J., Tommassen, J. 1987. Analysis of structure-function relationships in *Escherichia coli* K-12 outer membrane porins with the aid of *ompC-phoE* and *phoE-ompC* hybrid genes. *Mol. Gen. Genet.* 209:585–91

135. Van der Ley, P., Struyvé, M., Tommassen, J. 1986. Topology of outer membrane protein PhoE of *Escherichia coli.* Identification of cell surface–exposed amino acids with the aid of monoclonal antibodies. *J. Biol. Chem.* 261:12222–25

136. Weckesser, J., Drews, G., Ladwig, R. 1972. Localization and biological and physicochemical properties of the cell wall lipopolysaccharide of *Rhodopseudomonas capsulata. J. Bacteriol.* 110:346–53

137. West, I. C., Page, M. G. P. 1984. When is the outer membrane of *Escherichia coli* rate-limiting for uptake of galactosides? *J. Theor. Biol.* 110:11–19

138. Yamada, H., Oshima, N., Mizuno, T.,

Matsui, H., Kai, Y., et al. 1987. Use of a series of OmpF-OmpC chimeric proteins for locating antigenic determinants recognized by monoclonal antibodies against the OmpC and OmpF proteins of the *Escherichia coli* outer membrane. *J. Biochem. Tokyo* 102:In press

138a. Yoneyama, H., Nakae, T. 1986. A small diffusion pore in the outer membrane of *Pseudomonas aeruginosa*. *Eur. J. Biochem.* 157:33–38

139. Yoshimura, F., Zalman, L. S., Nikaido, H. 1983. Purification and properties of *Pseudomonas aeruginosa* porin. *J. Biol. Chem.* 258:2308–14

140. Young, J. D.-E., Blake, M., Mauro, A., Cohn, Z. A. 1983. Properties of the major outer membrane protein from *Neisseria gonorrhoeae* incorporated into model lipid membranes. *Proc. Natl. Acad. Sci. USA* 80:3831–35

141. Zalman, L., Nikaido, H. 1985. Dimeric porin from *Paracoccus denitrificans*. *J. Bacteriol.* 162:430–33

142. Zimmermann, W., Rosselet, A. 1977. Function of the outer membrane of *Escherichia coli* as a permeability barrier to beta-lactam antibiotics. *Antimicrob. Agents Chemother.* 12:368–72

Ann. Rev. Microbiol. 1988. 42:395–419
Copyright © 1988 by Annual Reviews Inc. All rights reserved

AEROMONAS AND PLESIOMONAS AS ETIOLOGICAL AGENTS

Nancy Khardori and Victor Fainstein

Section of Infectious Diseases, Department of Medical Specialties, The University of Texas M. D. Anderson Hospital and Tumor Institute, Houston, Texas 77030

CONTENTS

BACTERIOLOGY

The genus *Aeromonas* (Kluyver and Van Niel, 1936) belongs to the family of Vibrionaceae. The organisms in this family are straight or curved gram-

395

0066-4227/88/1001-0395$02.00

negative rods and are usually motile by polar flagella. Some cells may produce lateral flagella under certain growth conditions. These organisms are chemoorganotrophic with both fermentative and respiratory metabolism. The oxidase reaction is positive. Several species produce butylene glycol from glucose. Some are proteolytic, and some produce indole. They are facultative anaerobes without fastidious nutritional requirements. The G+C ratio of their DNA ranges from 39 to 63 mol%. The members of this family are usually found in fresh water or seawater and occasionally in fish or humans.

The word *Aeromonas* is derived from the Greek words *aer,* meaning air or gas, and *monas,* meaning unit or monad, i.e. *Aeromonas*=gas-producing monad. The cells range from straight rods with rounded ends to cocci. They are 1.0–4.4 μm long, occasionally forming filaments up to 8 μm long. They occur singly, in pairs, or in chains. They are gram-negative, motile by polar flagella, and generally monotrichous. Some species are nonmotile. No resting stages are known. *Aeromonas* species break down carbohydrates to acid or acid and gas (CO_2 and H_2). They ferment glucose, fructose, maltose, and trehalose, but not adonitol, dulcitol, inositol, inulin, melezitose, sorbose, and xylose. Starch, dextrin, and glycerol are hydrolyzed. Nitrate is reduced to nitrite. Cytochrome oxidase, oxidase, and catalase reactions are positive. Aeromonads are active producers of protease, diastase, lipase, DNase, and lecithinase. The organisms are not halophilic and do not grow in 7.5% NaCl. Minimum growth temperature for the members of the genus is 0–5°C, and maximum temperature is 38–41°C. Some species do not grow at 37°C. Most biochemical tests are performed at the optimal temperature of 30°C, although some are best performed at 20°C. The organisms grow at pH 5.5–9.0. After 24 hr of incubation on nutrient agar colonies are 1–3 mm in diameter, convex, smooth, whitish, and transluscent; they become light beige on further incubation. There is a wide zone of β-hemolysis on blood agar after 1 day. The growth becomes dark green after 2–3 days. The G+C content of the DNA is 57–63 mol%. Aeromonads are not susceptible to the vibriostatic agent 2,4-diamino-6,7-di-isopropylteridine (0/129). The proposed type species is *Aeromonas hydrophila* (Chester) Stainer, 1943. *Aeromonas* species share some properties with the members of the family Enterobacteriaceae and with other genera of the family Vibrionaceae, namely *Plesiomonas* and *Vibrio* (104, 127).

The various differentiating characteristics for the species, subspecies, and biotypes of the genus *Aeromonas* are given in Table 1. The most important species is *Aeromonas hydrophila* (formerly called *Bacillus hydrophilus fuscus, Bacillus hydrophilus, Bacterium hydrophilum, Pseudomonas hydrophila, Pseudomonas ichthyosmia, Flavobacterium fermentans,* and *Vibrio jamaicensis*). *A. hydrophila* subsp. *hydrophila* causes red leg disease in frogs, septicemia and stomatitis in snakes, and infections of freshwater fish. It is

Table 1 Differentiation between *Aeromonas hydrophila*, *Aeromonas caviae*, *Aeromonas sobria*, *Aeromonas salmonicida*, and *Plesiomonas shigelloides*[a]

Characteristics	*A. hydrophila*	*A. caviae*	*A. sobria*	*A. salmonicida* subsp.			*P. shigelloides*
				salmonicida	*achromogenes*	*masoucida*	
Motility	+	+	+	–	–	–	+
Monotrichous flagellation in liquid medium	+	+	+	–	–	–	–
Lophotrichous flagellation in liquid medium	–	–	–	–	–	–	+
Coccobacilli in pairs, chains and clumps	–	–	–	+	+	+	–
Rods in singles and pairs	+	+	+	+	+	–	+
Brown water-soluble pigment	–	–	–	+	–	–	–
Growth in nutrient broth at 37°C	+	+	+	–	–	+	+
Indole production in 1% peptone water	+	+	+	–	+	+	+
Esculin hydrolysis	+	+	–	+	–	+	–
Growth in KCN broth (Møller technique)	+	+	–	–	–	–	–
L-Histidine and L-arginine utilization	+	+	–	+	–	–	–
L-Arabinose utilization	+	+	–	+	–	+	–
Fermentation of salicin	+	+	–	d	d	d	–
Fermentation of sucrose	+	+	+	–	+	+	–
Fermentation of mannitol	+	+	+	+	–	+	–
Breakdown of inositol	–	–	–	–	–	–	+
Acetoin from glucose (Voges-Proskauer)	+	–	d	–	–	+	–
Gas from glucose	+	–	+	+	–	+	–
H_2S from cysteine	+	–	+	–	–	+	–

[a]From Reference 112, with permission. Symbols: +, typically positive; –, typically negative; d, differs among strains.

found in fresh, uncontaminated water. *A. hydrophila* subsp. *anaerogenes* does not cause disease in frogs. It is found in sewage and contaminated fresh water. *Aeromonas hydrophila* subsp. *proteolytica* is named for its proteolytic activity. It was originally isolated from the gastrointestinal tract of the marine borer *Limnoria tripunctata*.

Aeromonas punctata was formerly named *Bacillus punctatus* and *Pseudomonas punctata*. *A. punctata* subsp. *punctata* can produce experimental red leg disease in frogs. It is found mostly in fresh uncontaminated brooks and rivers. *A. punctata* subsp. *caviae* derives its name from its original isolation from guinea pigs (genus *Cavia*). It is found in sewage and contaminated water.

The nonmotile species *Aeromonas salmonicida* causes furunculosis in salmonid fish and may also cause serious infections in other fish. It is not found in surface waters. Subspecies include *A. salmonicida* subsp. *salmonicida, A. salmonicida* subsp. *achromogenes,* and *A. salmonicida* subsp. *masoucida.* The last of these subspecies, which is isolated from Japanese salmon, is aerogenic.

Pivnick & Sabina (111) proposed the division of *Aeromonas hydrophila* into two distinct species, *Aeromonas formicans* (anaerogenic) and *Aeromonas liquefaciens* (aerogenic). In addition to gas production, consistent acid reaction in mannose, sucrose, and glycerol, production of indole, hemolysis, and variability in utilizing citrate separate the aerogenic strains from anaerogenic strains.

The ninth edition of *Bergey's Manual of Determinative Bacteriology* (112) lists four species and separates them into two main groups on the basis of motility. The nonmotile aeromonads are clustered into one species *(A. salmonicida)* with four subspecies and the motile aeromonads are divided into three species *(A. hydrophila, A. caviae,* and *A. sobria).* Most human *Aeromonas* infections involve the motile aeromonads and have been attributed to *A. hydrophila* in the past because many clinical laboratories do not speciate motile aeromonads.

Plesiomonas (Habs and Schubert, 1962) is also in the family Vibrionaceae. The name is derived from the Greek words *plesios,* meaning neighbor, and *monad,* meaning unit, for neighbor to *Aeromonas.* The cells are gram negative. They are straight or rod-shaped with rounded ends and measure 0.8–1.0 by 3.0 μm. They grow singly, in pairs, or in chains, are motile by polar flagella, and are generally lophotrichous. Resting stages are not known. The cells break down carbohydrates with production of acid but no gas. Cytochrome oxidase, oxidase, and catalase reactions are positive. The bacteria do not produce diastase, lipase, deoxyribonuclease, or proteinases. Optimal growth temperature is 30°C; however, good growth occurs at 37°C. Maximum growth temperature ranges betwen 39 and 41°C. There is no growth in nutrient broth containing 7.5% sodium chloride. The G+C content of the DNA is 51 mol%. Most strains are sensitive to the vibriostatic agent 0/129. The type species is *Plesiomonas shigelloides* (Bader) Habs and Schubert 1962.

This monotypic genus differs from the other genera of the family Vibrionaceae, namely *Aeromonas* and *Vibrio,* in being lophotrichous (with two to seven flagella), lacking exoenzymes, fermenting inosital, and having a restricted range of carbohydrate fermentation. Some strains of *Plesiomonas shigelloides* share a common O antigen with *Shigella sonnei* (122).

EPIDEMIOLOGY

Aeromonas hydrophila has been recognized as a pathogen of amphibians (33, 130), reptiles (86, 130), fish (39, 54), snails (91), cows (147), and humans (30). The ectothermic vertebrate hosts develop an ulcerative stomatitis that often progresses to septicemia and death. Studies have suggested that densities of *A. hydrophila* in natural bodies of water may be an important contributing factor to episodes in fish (30, 39, 54). Commercial and sport fishery losses to *A. hydrophila* may be extensive (95).

Studies have also suggested (130) that *A. hydrophila* is cosmopolitan in distribution. Hazen et al (57) studied the prevalence and distribution of *A. hydrophila* in the United States. The organism was isolated at all but 12 of the 147 lotic and lentic habitats sampled in 30 states and Puerto Rico. The density was higher in lotic than in lentic systems and higher in saline systems than in freshwater systems. *A. hydrophila* was not isolated from extremely saline, thermal, or polluted waters even though it was found over wide ranges of temperature, pH, conductivity, salinity, and turbidity. Of these parameters, only conductivity was significantly regressed with density of *A. hydrophila.* Isolates have been recovered from rivers, tap water, hospital water sources, swimming pools, and lakes (89, 92, 132, 142).

The majority of clinical isolates are recovered during the spring and summer months (142). The organism is not part of the normal human flora, and the rare *Aeromonas* isolates from stools of healthy persons are usually transient. However, von Graevenitz & Zinterhofer (143) found a 3.2% isolation rate on DNase agar from stools of hospital inpatients with no gastrointestinal symptoms. A study done in India (21) using alkaline peptone water enrichment reported an isolation rate of 8%. Using better quality enrichment and selective media, Millership et al (96) documented a 4.2% isolation rate from unselected routine fecal specimens from 815 patients. The very low isolation rates reported earlier could have been due to geographical variation or more probably to poor media selection.

Aeromonas species in numbers comparable to those in raw surface water were isolated from a metropolitan water supply in Australia (14) that conformed to international standards for drinking water. The coliform count did not correlate with the *Aeromonas* count in this and two other studies from North America (24, 77). Burke et al (14) showed that cases of *Aeromonas*-associated diarrhea in Perth, Western Australia paralleled the isolation of *Aeromonas* spp. from the chlorinated domestic water supply, both being greatest during summer. In contrast, studies on the unchlorinated domestic water supply of another Western Australian city with population of 21,000 revealed that both clinical and environmental isolations continued in the winter (13). These observations suggest that clinical incidence is related to bacterial counts in drinking water, rather than to some other variable, e.g. environmental temperature.

Aeromonas spp. in domestic water supplies may also be an important source of nongastrointestinal infection in patients with immunological abnormalities (148). Such infections after exposure to contaminated water have been reported (49, 55, 70). Patients with hepatobiliary disease may be another group at risk from exposure to water-borne *Aeromonas* spp. The potential for *Aeromonas* spp. to cause enteric and nonenteric infections in humans suggests that domestic water supply should be free of these organisms.

The infection can sometimes be hospital acquired. *A. hydrophila* was isolated from 19 patients during a 5-wk outbreak of hospital-acquired infection in Sheffield, United Kingdom (93). Fourteen isolates were from the respiratory tract; Mellersh et al (93) believed that 11 of these represented postantibiotic colonization and three represented pneumonia. Three strains were isolated from wounds, two from high vaginal swabs, and one from urine. None of the patients with infections were immunosuppressed. Although the authors were not able to culture the organism from an environmental source and were unable to cut the chain of infection, the incident was brought to a close by increased microbiological surveillance activity, probably combined with a decrease in ambient temperature.

During a 4-yr surveillance for *Aeromonas*, the Regional Public Health Laboratory in Tilburg, the Netherlands (75) detected the organism in 0.67% of 28,981 fecal samples. Most isolations were in the summer months. The investigators used a semiselective blood agar plate with 10 mg ampicillin liter^{-1} for isolation, and the identification tests were done at 36°C with a well-defined control. Of 156 *Aeromonas* isolates 15% were identified as *A. hydrophila*, 17% as *A. sobria*, and 68% as *A. caviae*. There was a good correlation between the isolation of *A. hydrophila* and *A. sobria* and clinical symptoms of gastroenteritis or enterocolitis.

Arai et al (4) conducted a survey during the period 1974–1976 to determine the distribution of *Plesiomonas shigelloides* in human and animal feces and aquatic environments in Japan. *P. shigelloides* was isolated from 0.0078% of

38,545 healthy Tokyo residents, 3.8% of 967 dogs, 10.3% of 389 cats, 10.2% of 246 freshwater fish, 12.8% of 497 river water samples, and 10.5% of 19 sludge samples. Many of the cultures from dogs and cats were of the same serovars as those from diarrhoeic patients, suggesting that these animals are important carriers and are involved in human infections. The strains isolated from river water and freshwater fish were also of the same serovars as those from diarrhoeic patients. Most of the infectious gastroenteritis outbreaks in humans caused by this organism occurred during the summer, which was the time of environmental contamination as shown by ecological surveys.

CULTURAL CHARACTERISTICS

Aeromonads grow well on blood agar in colonies that have a ground glass appearance and a fruity odor. They also grow on enteric media (eosin-methylene blue, *Salmonella-Shigella,* and MacConkey agar) and triple-sugar iron. Several media have been evaluated for their usefulness for detecting *Aeromonas* spp. from fecal specimens. The optimal media include alkaline peptone water, trypticase soy broth with ampicillin, dextrin-fuchsin-sulfite agar, inositol–brilliant green bile salts agar, xylose-sodium deoxycholate-citrate agar, and Pril-xylose-ampicillin agar (141). The colonies on these media vary from colorless (the last three) to dark red (dextrin-fuchsin-sulfite agar).

Aeromonas is most commonly confused with the members of the Entero-bacteriaceae. The oxidase test is a major distinguishing test and should be routinely performed on all suspected isolates. McGrath et al (90) described oxidase-variable strains that are oxidase positive when grown on nonselective media but oxidase-negative when grown on gram-negative or differential media. A pH of 5.1 or less and the presence of lactose in the medium also correlate with oxidase negativity. The organic end products of lactose fermentation are thought to cause inhibition of the oxidase reaction (90). A platinum wire loop should be used for oxidase testing, as an iron-containing loop can give false positive results.

NATURAL INFECTIONS IN ANIMALS

Aeromonas species mainly affect cold-blooded animals, but infections have been described in dogs and in turkeys.

Red leg disease of frogs is a septicemic illness that can be fulminant. More commonly, however, the animals develop muscle flaccidity followed by cutaneous hemorrhages and ulcerations. Hemoptysis and convulsions are terminal events (15, 90).

Fish develop furunculosis or hemorrhagic septicemia due to *Aeromonas.*

Furunculosis is a septicemic illness caused by *A. salmonicida* and is manifest as either a rapidly fatal infection or a slow infection with lethargy and focal necrotic lesions in various muscles. Hemorrhagic septicemia is a disease of warm-water fish and is usually caused by *A. liquefaciens*. It causes red mouth of rainbow trout, red sore of pike, and abdominal dropsy in several species of fish.

Snakes and lizards can get three types of infections (15, 87). (*a*) Acute septicemia is characterized by lethargy, weakness, and convulsions. Pathology reveals pulmonary and pericardial hemorrhages and hemorrhagic enteritis. (*b*) Pneumonia caused by *Aeromonas* has been responsible for considerable mortality in snake colonies. The animals exhibit nasal discharge, anorexia, and respiratory insufficiency and die within 5 days. The snake mite *ophionyssus natricis* transmits the disease. (*c*) The third presentation in snakes is that of ulcerative stomatitis or mouth rot, characterized by frothy, fibrinous exudate around the mouth and inability to eat. Caseous masses develop in the buccal cavity within a week and may progress to involve teeth and bones.

Aeromonas also causes sepsis in dogs (108), pneumonia and dermatitis in dolphins (28), abortion in cattle (147), black rot in hen eggs (94), and diarrhea in piglets (35).

HUMAN INFECTION

Gastrointestinal Infections

Sporadic isolation of *Aeromonas* from stool specimens was described as early as 1937. In 1961, Martinez-Silva et al (88) described an outbreak of enteritis in which eight newborns and one 7-day-old baby were affected. Six of the nine patients had *Aeromonas* in their stools. The only patient with an apparently pure growth of *Aeromonas* from the stool died.

Lautrop (76) isolated the organism from stool cultures of 8 of 4500 patients, 7 of whom had gastrointestinal symptoms (76). *A. hydrophila* was grown from four stool cultures from a child with severe gastroenteritis (119). Several workers have subsequently reported the isolation of *Aeromonas* from patients with diarrhea (5, 10, 51–53, 59, 66, 67, 97).

Several investigators have determined the asymptomatic carrier rate for the organism to assess its role in diarrheal disease. The different carriage rates reported have been 0.7% (18), 3.2% (143), and 2% (52). The carrier rate might be as high as 16–27% in certain districts in Thailand (109). *A. hydrophila* was isolated with the same frequency in native Thais with and without diarrhea. Among American Peace Corps volunters in the same area, *A. hydrophila* was isolated more frequently from those with diarrhea.

Chatterjee & Neogy (21) isolated *Aeromonas* and *Plesiomonas* species in 8% (32 of 3877) of cases of choleric diarrhea in Calcutta, India. The organisms were isolated on *Salmonella-Shigella* agar and appeared as oxidase positive non-lactose-fermenting colonies. Colonies did not grow on fresh thiosulphate citrate bile salt sucrose agar. *Aeromonas* and *Plesiomonas* were the only potential pathogens in 5.6% of the patients. In 2.5% (11 cases) aeromonads were found in association with *Vibrio* species. The close association of *Vibrio* and *Aeromonas* species in these cases suggests a common vehicle of infection; infection is presumably waterborne in nature.

Agger et al (2) reported the prevalence of *A. hydrophila* in stool specimens from patients with diarrhea during an 18-mo period. The organsim was found in 1.1% of the patients with diarrhea and in none of the 533 control patients. Isolation of this organism was 1.5 times higher during the summer months than during the winter months. Most incidence of *A. hydrophila* occurred in children less than two years of age. Clinical features included fever, abdominal cramps, and vomiting. The illness lasted more than 10 days. Detection of *A. hydrophila* in stools was facilitated by the use of sheep blood agar with 15 μg ampicillin ml^{-1} and flooding with oxidase reagent. A cytotoxin was demonstrated in 62% of the isolates, and the cytotoxic strains showed positive results in a hemolysin assay and a lysine decarboxylase reaction.

Species of *Aeromonas* other than *A. hydrophila* have also been reported to cause diarrhea. Champsaur et al (20) reported the isolation of *A. sobria* as the only pathogen in the rice-water stool of a Thai woman. The strain produced enterotoxin, cytolysin, proteolysin, hemolysin, and a cell-rounding factor. Microbiological studies revealed 30 episodes of traveler's diarrhea in 20 of 35 Peace Corps volunteers during their first 6 wk in Thailand (138). Ninety percent of these episodes were associated with enteric pathogens, predominantly enterotoxigenic *Escherichia coli* and various *Salmonella* serotypes. Infections with nine other enteric pathogens were also identified [*Campylobacter jejuni* (17%), *P. shigelloides* (13%), *A. hydrophila* (10%), *Blastocystis hominis* (7%), Norwalk virus (7%), *Vibrio parahaemolyticuus* (5%), non-O1 *Vibrio cholerae* (3%), *Vibrio fluvialis* (3%), and rotavirus (3%).

Two epidemics of water-borne diarrheal disease involving 1000 persons in which no enteropathogens other than *P. shigelloides* were isolated were reported from Japan (140). *P. shigelloides* was isolated from 132 of 342 samples of water and mud collected from ponds, rivers, and shallow streams. This organism was also isolated from several indigenous animals including fish, shellfish, and newts (140).

An oyster-associated outbreak of diarrheal disease in North Carolina may have been caused by *P. shigelloides* (120). *P. shigelloides* was isolated as the predominant organism from the only acute stool specimen obtained. The organism was found in 1 of 10 oyster samples collected from the oyster-harvesting area. The authors implicated *P. shigelloides* in this oyster-associated outbreak, as only 0.008% of the healthy population carries *P. shigelloides* in the intestines (139). Although the oysters were roasted on a grate over an open flame for 6–7 min, many of the consumers described them as wet or undercooked. Sanyal et al (123a) have reported that *P. shigelloides* is enterotoxigenic.

P. shigelloides was thought to be the cause of diarrhea in six patients seen in Los Angeles (117) during a 3-yr period.

Holmberg et al (62, 62a) evaluated the clinical and epidemiological aspects of *Aeromonas* and *Plesiomonas* intestinal infections in the United States. They studied the cases of 34 persons nationwide from whom *A. hydrophila* had been isolated in large numbers from stools in 1984. In comparison with the 68 control subjects, these patients were more likely to have drunk untreated water, usually from private wells. No clear correlation was observed between the type of illness and the tested genotypic or phenotypic characteristics of the organisms. Symptoms tended to be chronic in adults and acute and severe in children. Patients who took antibiotics to which their *Aeromonas* strains were susceptible experienced alleviation or resolution of their gastrointestinal symptoms. These observations indicate that some *Aeromonas* strains are enteropathogenic for the normal host and that the organisms are acquired by drinking untreated water. Holmberg et al (62a) studied 32 persons nationwide from whom *P. shigelloides* had been isolated in large numbers from stools in 1984 and compared them with 62 matched controls. The factors that strongly correlated with infections were consumption of uncooked shellfish, usually raw oysters, in the 48 hr before the onset of illness and foreign travel, usually in Mexico. The illness in most patients was self-limited. Blood and mucus in the stool and other clinical findings suggested enteroinvasiveness of the organism. Organisms recovered from two patients who developed the illness after taking ampicillin for unrelated reasons were resistant to ampicillin. These isolates were sensitive to the antimicrobial agents that caused alleviation of symptoms in seven other persons. These findings suggest that *P. shigelloides* may cause enteric disease in the normal host, may be acquired from eating uncooked shellfish, and may cause traveler's diarrhea.

Gastrointestinal infections are usually self-limited in normal individuals and remain localized (68, 119, 123, 124). However, dissemination is more likely to occur in compromised patients and may result in intraabdominal infection or septicemia (121). *Aeromonas* species have been cultured from several intraabdominal sites: bile in biliary obstruction and cholecystitis (9, 34, 142, 144), intraabdominal abscesses (144), intestinal sinus tracts (142), and surgical wound infections (89, 144). These organisms have also been implicated in spontaneous peritonitis in cirrhotics (26) and in peritonitis secondary to perforated bowel (142, 144). The predominant species isolated from these sources is *A. hydrophila*. Several small liver abscesses were reported at autopsy of a 16-year-old leukemic girl who died from overwhelming *A. hydrophila* sepsis.

Extraintestinal Infections

Skin and soft tissue infections are the second most common site for *A. hydrophila* isolation after the gastrointestinal tract. Most of these infections are directly associated with traumatic exposure to water or soil (89, 106, 118, 142, 144). The lower extremity is involved in 75% of patients. There is rapid onset of cellulitis and purulent discharge. Fever and leukocytosis are present in some patients. Most wound infections remain localized in healthy hosts. Dissemination from localized wound infections has been observed in compromised hosts (89).

Ecthyma gangrenosum due to *A. hydrophila* infection has been reported in three children with leukemia (72, 98, 128) and five adults with *A. hydrophila* sepsis (6, 72, 137, 148). Rosenthal et al (118) described severe *Aeromonas* infection of an alligator bite. *A. hydrophila* infection associated with gas formation in the subcutaneous tissues was reported in four patients (58, 78, 114, 136), one of whom was injured while scuba diving in fresh water (78). Although most isolates in the literature have been *A. hydrophila,* a small number have been *P. shigelloides* (38, 89, 142).

Aeromonas species are especially pathogenic for muscle (9, 60, 72, 129). The organisms cause intense local muscle inflammation and necrosis (89, 106, 118, 135, 142). *Aeromonas* sepsis was associated with a fulminant necrotizing myositis in three of the earliest reports of human infection (9, 61, 129). The necrotizing myositis appears to be a metastitic localization of hematogenous infection with *Aeromonas*. *A. hydrophila* is the species that causes virtually all of these infections.

Smith (133) reported the growth of *A. hydrophila* and *Clostridium perfringens* from gelatinous greenish-brown leg muscle of a 9-year-old girl following cut injury by a rusty knife. Gas was seen on X ray, and the patient required a below-the-knee amputation. Deepe & Coonrod (32) reported a similar course in a 16-year-old boy. Bulger & Sherris (9) described a 5-year-old girl with acute lymphocytic leukemia with *Aeromonas* sepsis who developed widespread foul-smelling necrosis of muscles. There have been several reports of *Aeromonas* myositis in adults (42, 58, 61, 72, 78, 129, 133, 136).

Meningitis

Aeromonas meningitis has been reported three times in the literature. One case was *A. hydrophila* meningitis and sepsis that developed as a complication of a craniotomy in a healthy male with a subdural hematoma (113). Direct inoculation at the time of craniotomy was thought to be the portal of entry. The patient responded to treatment with chloramphenicol and gentamicin. The second case was that of a 23-month-old boy with sickle cell anemia who developed fever, diarrhea, and seizures and died within a few hours of onset. *A. hydrophila* was grown from postmortem blood and cerebrospinal fluid cultures (149). The third was in a neonate born after a pregnancy complicated by fever and premature rupture of membranes. The newborn was started on ampicillin and gentamicin. All cultures were sterile, so the antibiotics were discontinued after 5 days and the patient was discharged. He was readmitted with upper respiratory tract symptoms, irritability, and lethargy at 13 days of age. Cerebrospinal fluid showed polymorphonuclear leukocytosis, and cultures grew *A. hydrophila*. The strain was sensitive in vitro to gentamicin but resistant to ampicillin. The patient developed and continued to have seizures while on therapy and died 24 hr after admission (47).

Osteomyelitis

Lopez et al (84) described *A. hydrophila* osteomyelitis and sepsis in a child with leukemia. Davis et al (30) reported localized *A. hydrophila* infection at an amputation site following traumatic neurovascular damage. The injury had been heavily contaminated with soil. Revision of the stump and therapy with sulfamethoxazole trimethoprim resulted in rapid improvement. Four other adult patients with *Aeromonas* osteomyelitis have been described (17, 71, 146).

Septic Arthritis

Dean & Post (31) described a leukemic child who developed suppurative arthritis of her second metacarpophalangeal joint during a severe sepsis with *A. hydrophila*. The organism was isolated from the infected joint at autopsy. Chmel & Armstrong (22) described two leukemic adults with *A. hydrophila* sepsis complicated by knee arthritis, which could not be eradicated in spite of appropriate antimicrobial therapy. The patients eventually succumbed to their underlying illnesses.

Endocarditis

Davis et al (30) described a case of *A. hydrophila* endocarditis in a patient with cirrhosis and renal failure. The patient presented with chills, dyspnea, and tachycardia during hemodialysis. She was noted to have a II/VI systolic ejection murmur at the lower left sternal border. She was started on clindamycin and gentamicin, which brought about defervescence. Blood cultures taken on admission grew *Aeromonas hydrophila*. The patient was continued on gentamicin for 10 days. On a thorough search to reveal the source of the sepsis, it was found that the patient had bailed water from her flooded basement shortly after having had percutaneous vascular punctures for hemodialysis, and she had let the water come into contact with her fresh puncture wounds. Three weeks after discharge, the patient was readmitted following an incident of dizziness and falling. She was noted to be febrile on admission and was started on nafcillin and gentamicin. Two of the blood cultures taken on admission again grew *A. hydrophila,* and a new blowing VI/VI diastolic murmur was noted on the third hospital day. The patient was treated for infective endocarditis. She refused surgical therapy for aortic insufficiency, and expired after a cardiorespiratory arrest on the forty-third hospital day. Autopsy revealed two large perforations near the base of the left anterior cusp of the aortic valve. Postmorten blood cultures and cultures of the associated vegetative lesions were negative. Another case of endocarditis caused by *A. hydrophila* was reported by Ben-Chetrit et al (7).

Peritonitis

Peritonitis caused by *Aeromonas* sp., as indicated by growth of *Aeromonas* sp. from peritoneal fluid, has been reported in at least 15 adults (23, 26, 66, 99, 100, 118, 121, 142), three of whom had a ruptured appendix (142). It has also been reported in a 5-year-old child with ruptured appendix. Other pathogens were also grown from peritoneal fluid of patients with ruptured appendices.

Eye Infections

Cohen et al (25) reported on an 8-year-old boy who sustained a corneal laceration following a penetrating injury by a fish hook that led to a rapidly developing endophthalmitis. Anterior chamber cultures following enucleation of the eye grew *A. hydrophila* and *P. shigelloides*. Six cases of ocular infection with *Aeromonas* have been reported in adults. Two patients had conjunctivitis (133, 134), two had corneal ulcers (45), and two had penetrating injuries to the eye (45, 144, 145). In contrast, Smith (133, 134) reported conjunctival colonization with *Aeromonas* without any evidence of infection.

Urinary Tract Infections

McCracken & Barkley (89) reported a pure growth of *A. hydrophila* from the urine of a 5-month-old child with diarrhea and dehydration. Ten urinary isolates have been reported in adults, five of which were mixed cultures. Underlying disorders included chronic urinary tract infection (132), hydronephrosis (144), and paraplegia with suprapubic cystostomy (144).

Miscellaneous Isolates

Cultures positive for *Aeromonas* isolates have been reported from urine (70, 74, 75, 95), pyometra (70), decubitus ulcers (75), stress ulcers (74), sputum (70, 74, 95), chronic otitis media (74), eye wound (74), localized inflammation without trauma (74, 75), and throat and nose (70, 74). Most of the isolates have been *A. hydrophila* alone or in association with other organisms. The clinical significance of these isolates is difficult to evaluate. However, some of these isolates have been associated with significant clinical symptoms. The source of infection is also difficult to ascertain in many of these reports. Pelvic and urinary isolates may have originated in the gastrointestinal tract. Some of the other infections may have been related to water exposure.

Unusual Sources of Infection

Other sources of infection include contamination of hospital or home hemodialysis systems (66, 115), alligator bite (107, 118), exposure of a burn wound by jumping in a contaminated pond (3), tornado-associated wound contamination (50), and contamination of blood or blood products (110, 116).

Infections Caused by Other Species of Aeromonas

Significant rates of isolation of *A. sobria* and *A. caviae* from stool specimens from patients with gastroenteritis have been reported recently (66, 97). One case of *A. sobria* bacteremia (66) and one case of a cholera-like illness caused by enterotoxigenic *A. sobria* (20) have been described. *A. punctata* has been isolated from stools of patients with gastroenteritis (10, 48) and from a 1-year-old child with enteric fever syndrome (135). Some workers (29) have suggested that *A. sobria* may be a more important human pathogen than *A. hydrophila* and that *A. hydrophila* is associated more often with environmental isolates. However, these observations are not supported by general experience or, more importantly, strong clinical and laboratory data.

Bacteremia

Aeromonas bacteremia is reported relatively infrequently. Freij (47) reported three instances of bacteremia and cited 15 additional cases of *Aeromonas* sepsis in children (1, 8, 9, 31, 64, 72, 98, 105, 125, 132, 148–150). Ten of these patients had an underlying malignancy; most were leukemia patients 4–17 years of age. The patients were severely neutropenic, and nine of them had either never achieved remission or were in relapse. One patient had gone fishing 2 days before the onset of sepsis, and the natural habitat of *Aeromonas* spp. was thought to be the source of infection (148); in all the other immunocompromised patients the gastrointestinal tract was the presumed source of the pathogen. Sirinavin et al (131) reported 20 infants and children from Thailand who had *Aeromonas* septicemia. Eighteen of them had underlying diseases; five had leukemia, four had aplastic anemia, two had cirrhosis, three had thalassemia/hemoglobinopathy, and one each had renal failure, ileal perforation, marasmus, and cavernous hemangioma with thrombocytopenia.

Four of the 18 patients had polymicrobial bacteremia. Fifteen episodes were community acquired and five were hospital acquired. Clinical manifestations in addition to the usual signs of gram-negative sepsis included ecthyma gangrenosum, necrotizing fascitis, and meningitis in two patients each. The overall mortality rate was 50%. The same review (47) reported 65 cases of sepsis in adults (6, 23, 30, 40, 49, 51, 66, 72, 85, 89, 92, 93, 100, 107, 132, 135, 137, 142, 144, 148), with a cumulative case fatality rate of more than 50% (30). Several of these adult patients did not have an underlying malignancy.

Several reports have recognized an association between hepatobiliary disease and *Aeromonas* sepsis in adults (9, 26, 30, 34, 66, 74, 89, 92, 93, 129, 132, 142). A lack of hepatic filtration in cirrhosis is thought to favor bacteremia due to pathogenic bacteria from the gastrointestinal tract.

Harris et al (56) reported on 17 hospitalized patients with *Aeromonas* bacteremia, all of whom had an underlying malignancy. Two patients had carcinoma of the lung, and leukemia against which therapy was considered to be failing was the underlying malignancy in all the others. All of the patients were febrile with various signs of sepsis; only four had hemodynamic evidence of shock. Ecthyma gangrenosum occurred in two of the 17 patients. The mortality rate in this report, defined as death within 2 wks of onset of bacteremia, was only 28%. The most likely source of infection was thought to be the patient's own gastrointestinal tract. There was no clustering of cases suggesting a common source. Six patients had bacteremia documented within 48 hr of admission. One of these had recently been active outdoors, but had not been exposed to fish or to fresh or salt water. A seasonal variation was noted, with 14 of 17 cases occurring between May and September. In this report 82% of the patients were male. Similarly, from 27 other reports of *Aeromonas* sepsis in cancer patients, 70% of the patients were male.

IMMUNOLOGY

No studies have been made on the immune reponse to *A. hydrophila* infections; reports on serologic observations of individual patients are available. Agglutinating, precipitating, and antihemolysin antibodies to *A. hydrophila* have been detected in patients with deep tissue infections (16), but not in those with superficial infections (126). Hemolysin titers in the range of $1:320$ to $1:1280$ and agglutinin titers of $1:160$ to $1:640$ have been demonstrated (16). Neutralizing antibodies to *Aeromonas* hemolysin were not detected in normal adult sera. Studies on a small number of patients (72) demonstrated that normal serum promotes phagocytosis and intracellular killing of the organisms by white blood cells. However, no bacterial activity was demonstrated in pooled serum from four donors. Ketover et al (72) also demonstrated a rise in serum opsonic titers in one patient, from less than 1:5 in the acute phase to 1:5120 in the convalescent stage. The studies suggest that both a specific opsonizing antibody present in normal serum and the activity of neutrophils are required to prevent invasive *A. hydrophila* infection. Sera from two patients with fatal infections were not active in the bactericidal assay.

PATHOGENESIS AND VIRULENCE FACTORS

Extracellular Enzymes

A. hydrophila produces several extracellular enzymes with presumed toxic activity, namely diastase, lipase, DNase, lecithinase, and elastase. *A. sobria* produces similar extracellular enzymes, but elastase production is rare (82). *A. salmonicida* produces a protease and a hemolysin (47).

Two hemolysins, alpha- and beta-hemolysin, are produced by *A. hydrophila*. Most clinical isolates are beta-hemolytic, and hemorrhage is a conspicuous feature of *Aeromonas* infections (see section on natural infection in animals, above), implicating hemolysins in the pathogenesis of *A. hydrophila* infections. In high doses both hemolysins cause hemorrhagic enteritis in the rabbit ileal loop test, and antibodies to either hemolysin neutralize both toxins (47).

Alpha-hemlysin is released from cells during the stationary growth phase. It has a molecular mass of 50–65 kd. It is stable at pH 3.5–9.5 at room temperature. It is inactivated after 10 min at 56°C. Zinc stimulates its in vitro production, and iron has an inhibitory effect (82, 83). Increased hemolytic activity has been seen in the presence of oxygen (83). The hemolysin is cytotoxic to HeLa cells and human embryonic lung fibroblasts, but the effect is reversed when the cells are resuspended in fresh medium. Alpha-hemolysin is lethal when injected intraperitoneally in mice and rabbits and causes dermonecrosis on injection into rabbit skin.

Beta-hemolysin (aerolysin, cytolytic factor) has a molecular mass of 49–53 kd and is released toward the end of the logarithmic phase of growth. It is heat labile, with its activity abolished by heating for 1 hr at 50°C and pH 7 or at 37°C and pH 8.2. Beta-hemolysin is not destroyed by the proteolytic enzymes trypsin and pronase (83). It is lethal to rats, mice, and rabbits and causes dermonecrosis in rabbit skin. It is cytotoxic to several cell systems, including human diploid lung fibroblasts and HeLa cells. This effect is irreversible, in contrast to that of alpha-hemolysin.

Enterotoxins

There is considerable controversy regarding the assays most appropriate for detecting *Aeromonas* enterotoxins. Sanyal et al (124) first demonstrated that

A. hydrophila caused an accumulation of fluid in ligated rabbit ileal loops. Because they injected viable cells, it was not clear if the effect was cytotoxic, cytolytic, or inflammatory rather than truly enterotoxic. Later investigators showed the same response (81) using culture filtrates. However, there is still skepticism about the reliability of this test. The suckling mouse assay, which uses a combined score of the ratio of intestinal weight to body weight and the presence of diarrhea, has recently been advocated by some investigators (12).

The nature of the *Aeromonas* enterotoxin is also controversial. Keusch & Donta (73) have divided enterotoxins into two types: cytotonic and cytotoxic. The cytotonic enterotoxins, e.g. the heat-labile enterotoxins of *Vibrio cholerae* and *E. coli*, stimulate the cAMP-mediated sequence of events in cells. The cytotoxic enterotoxins, e.g. *Shigella dysenteriae* type 1 and *Clostridium perfringens* enterotoxins, give rise to cell damage or cell death. There is debate as to whether the *Aeromonas* enterotoxin is cytotonic, causing typical rounding, production of cAMP, and steroidogenesis in mouse adrenal Y1 cells, or cytotoxic, causing cell death in Y1 or other tissue culture lines, e.g. Vero or Chinese hamster ovary cells. Ljungh et al (79) classified *Aeromonas* enterotoxin as a cytotonic enterotoxin (79), as it induced steroid secretion in adrenal Y1 cells and increased the intracellular cAMP content of Y1 as well as rabbit intestinal epithelial cells. It was not injurious to mucosa (81). Chlorpromazine, a cAMP inhibitor, caused a 60% decrease in intestinal loop secretion, whereas prostaglandin inhibitors caused no reduction in the diarrheal response to *Aeromonas* enterotoxin (81). The enterotoxin-induced fluid had an albumin and electrolyte content similar to that induced by cholera toxin. Ljungh & Kronevi (80) and Ljungh & Wadstrom (82) reported both cytotoxic and cytotonic effects. However, true Y1 cell rounding and steroidogenesis were not observed by Donta & Haddow (36) or by Cumberbatch et al (27), who ascribed early rounding of some cells to the transient effect of a proteolytic enzyme or cytotoxin. The timing of the appearance of exotoxins during bacterial growth, effects of heating, isoelectric focusing, and dialysis have suggested that the cytotoxin and the enterotoxin are separate toxins (11, 69).

The relationship of *Aeromonas* toxins to cholera toxin and to heat-labile *E. coli* enterotoxins is also controversial. James et al (65) suggested that there are both heat-stable and heat-labile *Aeromonas* toxins, the latter being cross-reactive with cholera antitoxin.

Ljungh & Kronevi (80), using more than 70 biochemical reactors, could not correlate any single test with enterotoxigenicity in the strains of *A. hydrophila* examined. All strains were found to belong to the ideal phenotype of *A. hydrophila*, but each strain had its own biochemical profile. Cumberbatch et al (27) reported that cytotoxin production correlated with a positive lysine decarboxylase phenotype (98%) and a positive Voges-Proskauer phenotype (94%), compared to 27% lysine decarboxylase positivity and 23% Voges-Proskauer positivity among cytotoxin-negative strains. In fecal samples cytotoxin production correlated with diarrheal disease; 80% of the diarrheal isolates were toxigenic, compared to 41% of the nondiarrheal isolates. Cytotoxin production and capability of producing diarrheal disease also correlated with cytotoxicity to HeLa cells (67). Positivity for lysine decarboxylase, the Voges-Prauskauer test, and beta-hemolysis also correctly predicted the outcome of 97% of cultures in the suckling mouse assay (53). In contrast, Pitarangsi et al (109) reported that biochemical characteristics, production of cytotoxin, and ability to distend suckling mouse intestine were similar among *A. hydrophila* isolates from individuals with and without diarrhea.

Chakraborty et al (19), using gene cloning techniques, demonstrated that the determinants of enterotoxic, cytotoxic, and hemolytic activities are located at three different segments of the *A. hydrophila* chromosome. Two heated culture filtrates of *A. hydrophila* and nonheated filtrates of an *E. coli* clone containing the *A. hydrophila* enterotoxin gene provoked fluid accumulation in the rabbit ileal loop and suckling mouse models and caused elongation of Chinese hamster ovary cells. The authors concluded that *A. hydrophila* produces a cytotonic enterotoxin that is distinct from the *A. hydrophila* cytotoxin and hemolysin and from known *E. coli* enterotoxins.

Relatively little is known about *Plesiomonas* toxins. Several studies using a variety of tests that have been used to evaluate other toxins have shown no effects with *P. shigelloides* (63, 69, 81, 109). This could indicate that *P. shigelloides* does not produce an enterotoxin or that the enterotoxin cannot be demonstrated by the systems used to study *Aeromonas*, *Vibrio*, or *E. coli* toxins. In a volunteer study performed in 1959, Sakazaki et al (122) were not able to induce human disease by either oral or rectal administration of live cells.

Olsvik et al (101) tested 11 strains of *P. shigelloides* for heat-stable and heat-labile enterotoxins resembling those of *E. coli* by genetic probing, infant mouse assay, and ELISA; all were negative. One strain had cytotoxic activity in HeLa cell cultures. The strains were negative in the Sereny test for invasiveness, but 2 of 11 and 3 of 11 were invasive for HeLa cells in two different laboratories. Plasmid profiles showed the absence of a *Shigella*-like invasiveness plasmid, but all strains possessed a 200-Md plasmid of unknown function. These isolates were not from a common source, as indicated by otherwise diverse plasmid profiles. The authors concluded that *P. shigelloides* may be associated with a dysentery-like illness but the mechanism of pathogenicity is different from that of *Shigella* species.

In Vitro Antimicrobial Susceptibilities

A. hydrophila isolates are usually susceptible to chloramphenicol, colistin, gentamicin, kanamycin, nitrofurantoin, tetracycline, tobramycin, and trimethoprim-sulfamethoxazole (103). Overman (102) compared minimal inhibitory concentrations (MIC), disk diffusion, and Autobac 1 susceptibility tests for 22 strains of A. hydrophila. Eleven of the strains had discrepancies between Autobac and disk diffusion or MIC results. Autobac 1 results with strains of A. hydrophila may falsely indicate susceptibilities to ampicillin, carbenicillin, or cephalothin. Fass & Barnishan (44) investigated the MICs of 32 antimicrobial agents for 20 strains of A. hydrophila by a micro dilution method. Moxalactam was the most active drug tested. All strains were also susceptible to clinically achievable concentrations of mecillinam, cefomandole, cefuroxime, cefotaxime, aminoglycosides (except streptomycin), tetracycline, chloramphenicol, and trimethoprim-sulfamethoxazole. On the basis of in vitro susceptibility tests reported in the literature (37, 41, 43, 44, 102), Aeromonas species are almost always resistant to ampicillin, carbenicillin, ticarcillin, cefazolin, and cephalothin. Susceptibility to azlocillin, mezlocillin, and piperacillin is variable. The new β-lactam antibiotics, especially the third-generation cephalosporins, have been found to be very active in vitro (41). The most efficacious therapy for bacteremia due to A. hydrophila is not totally clear. However, at present the combination of an aminoglycoside and a broad-spectrum cephalosporin constitutes appropriate therapy. Trimethoprim-sulfamethoxazole constitutes an alternate therapy.

Of the 11 strains of P. shigelloides tested by Olsvik et al (101), all were resistant to streptomycin, ten were resistant to ampicillin, and eight were resistant to gentamicin.

Prophylaxis

Echeverria et al (37) reported on a randomized double-blind study to determine the efficacy of a 3-wk course of doxycycline (100 mg daily) for prevention of traveler's diarrhea among 63 United States Peace Corps volunteers during their first 5 wk in Thailand. Doxycycline-resistant enterotoxigenic E. coli are known to be common in this area. Three of 30 patients taking doxycycline (10%) and 8 of 33 patients in the placebo group (24%) developed diarrhea. The calculated protection was 59%, but this was not statistically significant ($p = 0.12$). A. hydrophila was isolated from 8 of 19 volunteers in the placebo group, but from only 1 of 12 in the doxycycline group ($p = <0.05$). Doxycycline significantly prevented colonization of the gastronintestinal tract with A. hydrophila during therapy ($p = <0.01$). These data further support the role of A. hydrophila as an enteric pathogen, even though doxycycline was not shown to be significantly protective against traveler's diarrhea.

Literature Cited

1. Abrams, E., Zierdt, C. H., Brown, J. A. 1971. Observations on *Aeromonas hydrophila* septicemia in a patient with leukemia. *J. Clin. Pathol.* 24:491–92
2. Agger, W. A., McCormick, J. D., Gurwith, M. J. 1985. Clinical and microbiological features of *Aeromonas hydrophila*–associated diarrhea. *J. Clin. Microbiol.* 21:909–13
3. Ampel, N., Peter, G. 1981. *Aeromonas* bacteraemia in a burn patient. *Lancet* 2:987
4. Arai, T., Ikejima, N., Itoh, T., Sakai, S., Shimada, T., et al. 1980. A survey of *Plesiomonas shigelloides* from aquatic enviroments, domestic animals, pets and humans. *J. Hyg.* 84:203–11
5. Baman, S. I. 1980. *Aeromonas hydrophila* as the etiologic agent in severe gastroenteritis: report of a case. *Am. J. Med. Technol.* 46:179–81
6. Beaune, J., Llorca, G., Gonin, A., Brun, Y., Fleurette, J., et al. 1978. Phlegmon nécrotique de la main, point de départ d'une septicémie due à *Aeromonas hydrophila:* Démonstration de l'origine ichthyologique de l'infection. *Nouv. Presse Med.* 7:1206–7
7. Ben-Chetrit , E., Nashif, M., Levo, Y. 1983. Infective endocarditis caused by uncommon bacteria. *Scand. J. Infect. Dis.* 15:179–83
8. Blatz, D. J. 1979. Open fracture of the tibia and fibula complicated by infection with *Aeromonas hydrophila:* a case report. *J. Bone Jt. Surg. Am. Vol.* 61:790–91
9. Bulger, R., Sherris, J. 1966. The clinical significance of *Aeromonas hydrophila. Arch. Intern. Med.* 118:562–64
10. Burke, V., Gracey, M., Robinson, J., Peck, D., Beauman, J., et al. 1983. The microbiology of childhood gastroenteritis: *Aeromonas* species and other infective agents. *J. Infect. Dis.* 148:68–74
11. Burke, V., Robinson, J., Atkinson, H. M., Dibley, M., Berry, R. J., et al. 1981. Exotoxins of *Aeromonas hydrophila. Aust. J. Exp. Biol. Med. Sci.* 59:753–61
12. Burke, V., Robinson, J., Berry, R. J., Gracey, M. 1981. Detection of enterotoxins of *Aeromonas hydrophila* by a suckling-mouse test. *J. Med. Microbiol.* 14:401–8
13. Burke, V., Robinson, J., Gracey, M., Peterson, D., Meyer, N., et al. 1984. Isolation of *Aeromonas* spp. from an unchlorinated domestic water supply. *Appl. Environ. Microbiol.* 48:367–70
14. Burke, V., Robinson, J., Gracey, M., Peterson, D., Partridge, K. 1984. Isolation of *Aeromonas hydrophila* from a metropolitan water supply: seasonal correlation with clinical isolates. *Appl. Environ. Microbiol.* 48:361–66
15. Carlton, W. W., Hunt, R. D. 1978. Bacterial diseases. In *Pathology of Laboratory Animals*, ed. K. Benirschke, F. M. Garner, T. C. Jones, 2:1373–77. New York: Springer-Verlag
16. Caselitz, F. H., Freitag, V., Jannasch, G. 1975. Demonstration of specific antibodies in sera of patients with infections caused by *Aeromonas hydrophila. Zentralbl. Bakteriol. Parasitenkd. Infektionskr. Hyg. Abt. 1 Orig. Reihe A* 233:347–54
17. Caselitz, F. H., Hofmann, A., Martinez-Silva, R. 1958. Unbeschriebener keim der familie *Pseudomonadaceae* als infecktionserreger. *Zentralbl. Bakteriol. Parasitenkd. Infektionskr. Hyg. Abt. 1 Orig.* 170:564–70
18. Catsaras, M., Buttiaux, R. 1965. Les *Aeromonas* dans les matieres fecales humaines. *Ann. Inst. Pasteur Lille* 16:85–88
19. Chakraborty, T., Montenegro, M. A., Sanyal, S. C., Helmuth, R., Bulling, E., et al. 1984. Cloning of enterotoxin gene from *Aeromonas hydrophila* provides conclusive evidence of production of a cytotonic enterotoxin. *Infect. Immun.* 46:435–41
20. Champsaur, H., Andremont, A., Mathieu, D., Rottman, E., Auzepy, P. 1982. Cholera-like illness due to *Aeromonas sobria. J. Infect. Dis.* 145:248–54
21. Chatterjee, B. D., Neogy, K. N. 1972. Studies on *Aeromonas* and *Plesiomonas* species isolated from cases of choleraic diarrhoea. *Indian J. Med. Res.* 60:520–24
22. Chmel, H., Armstrong, D. 1976. Acute arthritis caused by *Aeromonas hydrophila:* clinical and therapeutic aspects. *Arthritis Rheum.* 19:169–72
23. Chong, Y., Yi, K. N., Lee, S. Y. 1980. Cultural and biochemical characteristics of clinical isolates of *Aeromonas hydrophila. Yonsei Med. J.* 21:52–57
24. Clark, J. A., Burger, G. A., Sabatinos, L. E. 1982. Characterization of indicator bacteria in municipal raw water, drinking water and new main water samples. *Can. J. Microbiol.* 28:1002–13
25. Cohen, K. L., Holyk, P. R., McCarthy, L. R., Peiffer, R. L. 1983. *Aeromonas*

hydrophila and *Plesiomonas shigelloides* endophthalmitis. *Am. J. Ophthalmol.* 96:403–4

26. Conn, H. O. 1964. Spontaneous peritonitis and bactermia in Laennec's cirrhosis caused by enteric organisms: a relatively common but rarely recognized syndrome. *Ann. Intern. Med.* 60:568–80

27. Cumberbatch, N., Gurwith, M. J., Langston, C., Sack, R. B., Brunton, J. I. 1979. Cytotoxic enterotoxin produced by *Aeromonas hydrophila:* relationship of toxigenic isolates to diarrheal disease. *Infect. Immun.* 23:829–37

28. Cusick, P. K., Bullock, B. C. 1973. Ulcerative stomatitis and pneumonia associated with *Aeromonas hydrophila* infection in the bottle-nosed dolphin. *J. Am. Vet. Med. Assoc.* 163:578–79

29. Daily, O. P., Joseph, S. W., Coolbaugh, J. C., Walker, R. I., Merrill, B. R., et al. 1981. Association of *Aeromonas sobria* with human infection. *J. Clin. Microbiol.* 13:769–77

30. Davis, W. A. II, Kane, J. G., Garagusi, V. F. 1978. Human *Aeromonas* infections: a review of the literature and a case report of endocarditis. *Medicine Baltimore* 57:267–77

31. Dean, H. M., Post, R. M. 1967. Fatal infection with *Aeromonas hydrophila* in a patient with acute myelogenous leukemia. *Ann. Intern. Med.* 66:1177–79

32. Deepe, G. S., Coonrod, J. D. 1980. Fulminant wound infection with *Aeromonas hydrophila. South. Med. J.* 73:1546–47

33. DeFigueiredo, J., Plumb, J. A. 1977. Virulence of different isolates of *Aeromonas hydrophila* in channel catfish. *Aquaculture* 11:349–54

34. DeFronzo, R. A., Murray, G. F., Maddrey, W. C. 1973. *Aeromonas* septicemia from hepatobiliary disease. *Am. J. Dig. Dis.* 18:323–31

35. Dobrescu, L. 1978. Enterotoxigenic *Aeromonas hydrophila* from a case of piglet diarrhea. *Zentralbl. Veterinaermed. Reihe B* 25:713–18

36. Donta, S. T., Haddow, A. D. 1978. Cytotoxic activity of *Aeromonas hydrophila. Infect. Immun.* 21:989–93

37. Echeverria, P., Sack, R. B., Blacklow, N. R., Bodhidatta, P., Rowe, B., McFarland, A. 1984. Prophylactic doxycycline for traveler's diarrhea in Thailand: Further supportive evidence of *Aeromonas hydrophila* as an enteric pathogen. *Am. J. Epidemiol.* 120:912–21

38. Ellner, P. D., McCarthy, L. R. 1973. *Plesiomonas shigelloides* bacteremia: a case report. *Am. J. Clin. Pathol.* 59:216–18

39. Esch, G. W., Hazen, T. C. 1978. Thermal ecology and stress: a case history for red-sore disease in large-mouth bass *(Micropterus salmoides).* In *Energy and Environmental Stress in Aquatic Systems. Dep. Energy Symp. Ser. No. CONF-771114,* ed. J. H. Thorpe, J. W. Gibbons. Springfield, Va: Nat. Tech. Inf. Serv.

40. Ewing, W. H., Hugh, R., Johnson, J. G. 1961. *Studies on the Aeromonas Group.* Atlanta, Ga: US Dep. Health Educ. Welfare, Public Health Serv., Cent. Dis. Control

41. Fainstein, V., Weaver, S., Bodey, G. P. 1982. In vitro susceptibilities of *Aeromonas hydrophila* against new antibiotics. *Antimicrob. Agents Chemother.* 22:513–14

42. Farrington, M., Gray, H. H. 1983. A fishy tale: trout borne *Aeromonas hydrophila* septicaemia. *Br. Med. J.* 287:1184

43. Fass, R. J. 1980. In vitro activity of cefoperazone against nonfermenters and *Aeromonas hydrophila. Antimicrob. Agents Chemother.* 18:483–86

44. Fass, R. J., Barnishan, J. 1981. In vitro susceptibilities of *Aeromonas hydrophila* to 32 antimicrobial agents. *Antimicrob. Agents Chemother.* 19:357–58

45. Feaster, F. T., Nisbet, R. M., Barber, J. C. 1978. *Aeromonas hydrophila* corneal ulcer. *Am. J. Ophthalmol.* 85:114–17

46. Deleted in proof

47. Freij, B. J. 1984. *Aeromonas:* biology of the organism and diseases in children. *Pediatr. Infect. Dis.* 3:164–75

48. Fritsche, D., Dahn, R., Hoffman, G. 1975. *Aeromonas punctata* subsp. *caviae* as the causative agent of acute gastroenteritis. *Zentralbl. Bakteriol. Parasitenkd. Infektionskr. Hyg. Abt. 1 Orig. Reihe A* 233:232–35

49. Fulghum, D. D., Linton, W. R., Taplin, D. 1978. Fatal *Aeromonas hydrophila* infection of the skin. *South. Med. J.* 71:739–41

50. Gilbert, D. N., Sanford, J. P., Kutscher, E., Sanders, C. V. Jr., Luby, J. P., et al. 1973. Microbiology study of wound infections in tornado casualties. *Arch. Environ. Health* 26:125–30

51. Goodwin, C. S., Harper, W. E. S., Stewart, J. K., Gracey, M., Burke, V., et al. 1983. Enterotoxigenic *Aeromonas hydrophila* and diarrhoea in adults. *Med. J. Aust.* 1:25–26

52. Gracey, M., Burke, V., Robinson, J. 1982. *Aeromonas*-associated gastroenteritis. *Lancet* 2:1304–6

53. Gracey, M., Burke, V., Rockhill, R. C., Suharyono, S. 1982. *Aeromonas* species as enteric pathogens (Letter). *Lancet* 1:223–24

54. Haley, R., Davis, S. P., Hyde, J. M. 1967. Environmental stress and *Aeromonas liquefaciens* in American and threadfin shad mortalities. *Prog. Fish Cult.* 29:193

55. Hanson, P. G., Standridge, J., Jarrett, F., Maki, D. G. 1977. Freshwater wound infection due to *Aeromonas hydrophila*. *J. Am. Med. Assoc.* 238:1053–54

56. Harris, R. L., Fainstein, V., Elting, L., Hopfer, R. L., Bodey, G. P. 1985. Bacteremia caused by *Aeromonas* species in hospitalized cancer patients. *Rev. Infect. Dis.* 7:314–20

57. Hazen, T. C., Fliermans, C. B., Hirsch, R. P., Esch, G. W. 1978. Prevalence and distribution of *Aeromonas hydrophila* in the United States. *Appl. Environ. Microbiol.* 36:731–38

58. Heckerling, P. S., Stine, T. M., Pottage, J. C. Jr., Levin, S., Harris, A. A. 1983. *Aeromonas hydrophila* myonecrosis and gas gangrene in a nonimmunocompromised host. *Arch. Intern. Med.* 143:2005–7

59. Helm, E. B, Stille, W. 1970. Akute enteritis durch *Aeromonas hydrophila*. *Dtsch. Med. Wochenschr.* 94:18–24

60. Heywood, R. 1968. Aeromonas infection in snakes. *Cornell Vet.* 88:236–41

61. Hill, K. R., Caselitz, F. H., Moody, L. M. 1954. Case of acute metastatic myositis caused by new organism of the family: Pseudomonadaceae. *West Indian Med. J.* 3:9–11

62. Holmberg, S. D., Schell, W. L., Fanning, G. R., Wachsmuth, I. K., Hickman-Brenner, F. W., et al. 1986. *Aeromonas* intestinal infections in the United States. *Ann. Intern. Med.* 105:683–89

62a. Holmberg, S. D., Wachsmuth, I. K., Hickman-Brenner, F. W., Blake, P. A., Farmer, J. J. 1986. *Plesiomonas* enteric infections in the United States. *Ann. Intern. Med.* 105:690–94

63. Hostacka, A., Ciznar, I., Korych, B., Karolcek, J. 1982. Toxic factors of *Aeromonas hydrophila* and *Plesiomonas shigelloides*. *Zentralbl. Bakteriol. Mikrobiol. Hyg. Ser. A* 252:525–34

64. Hunter, W. F., Atkinson, H. M. 1968. Infection due to *Aeromonas hydrophila*. *Med. J. Aust.* 1:565

65. James, C., Dibley, M., Burke, V., Robinson, J., Gracey, M. 1982. Immunological cross-reactivity of enterotoxins of *Aeromonas hydrophila* and

cholera toxin. *Clin. Exp. Immunol.* 47:34–42

66. Janda, J. M., Bottone, E. J., Reitano, M. 1983. *Aeromonas* species in clinical microbiology: significance, epidemiology and speciation. *Diagn. Microbiol. Infect. Dis.* 1:221–28

67. Janda, J. M., Bottone, E. J., Skinner, C. V., Calcaterra, D. 1983. Phenotypic markers associated with gastrointestinal *Aeromonas hydrophila* isolates from symptomatic children. *J. Clin. Microbiol.* 17:588–91

68. Jandl, G., Linke, K. 1976. Bericht uber zwei Fälle von akuter gastroenteritis durch *Plesiomonas shigelloides*. *Zentralbl. Bakteriol. Parasitenkd. Infektionskr. Hyg. Abt. 1 Orig. Reihe A* 236:136–40

69. Johnson, W. M., Lior, H. 1981. Cytotoxicity and suckling mouse reactivity of *Aeromonas hydrophila* isolated from human sources. *Can. J. Microbiol.* 27:1019–27

70. Joseph, S. W., Daily, O. P., Hunt, W. S., Seidler, R. J., Allen, D. A., et al. 1979. *Aeromonas* primary wound infection of a diver in polluted waters. *J. Clin. Microbiol.* 10:46–49

71. Karam, G. H., Ackley, A. M., Dismukes, W. E. 1983. Posttraumatic *Aeromonas hydrophila* osteomyelitis. *Arch. Intern. Med.* 143:2073–74

72. Ketover, B. P., Young, L. S., Armstrong, D. 1973. Septicemia due to *Aeromonas hydrophila*: clinical and immunologic aspects. *J. Infect. Dis.* 127:284–90

73. Keusch, G. T., Donta, S. T. 1975. Classification of enterotoxins on the basis of activity in cell culture. *J. Infect. Dis.* 131:58–63

74. Kjems, E. 1955. Studies on five bacterial strains of the genus *Pseudomonas*. *Acta. Pathol. Microbiol. Scand.* 51:531–36

75. Kuijper, E. J., Zanen, H. C., Peeters, M. F. 1987. *Aeromonas*-associated diarrhea in the Netherlands. *Ann. Intern. Med.* 106:640–41

76. Lautrop, H. 1961. *Aeromonas hydrophila* isolated from human faeces and its possible pathological significance. *Acta. Pathol. Microbiol. Scand.* 51(Suppl. 144):299–301

77. Le Chevallier, M. W., Evans, T. M., Seidler, R. J., Daily, O. P., Merrell, B. R., et al. 1982. *Aeromonas sobria* in chlorinated drinking water supplies. *Microb. Ecol.* 8:325–33

78. Levin, M. L. 1973. Gas-forming *Aeromonas hydrophila* infection in a diabetic. *Postgrad. Med.* 54:127–29

79. Ljungh, A., Eneroth, P., Wadstrom, T. 1982. Cytotonic enterotoxin from *Aeromonas hydrophila*. *Toxicon* 20:787–94

80. Ljungh, A., Kronevi, T. 1982. *Aeromonas hydrophila* toxins: intestinal fluid accumulation and mucosal injury in animal models. *Toxicon* 20:397–407

81. Ljungh, A., Popoff, N., Wadstrom, T. 1977. *Aeromonas hydrophila* in acute diarrheal disease: detection of enterotoxin and biotyping of strains. *J. Clin. Microbiol.* 6:96–100

82. Ljungh, A., Wadstrom, T. 1982. *Aeromonas* toxins. *Pharmacol. Ther.* 15:339–54

83. Ljungh, A., Wadstrom, T. 1983. Toxins of *Vibrio parahaemolyticus* and *Aeromonas hydrophila*. *J. Toxicol.—Toxin Rev.* 1:257–307

84. Lopez, J. F., Quesada, J., Saied, A. 1968. Bacteremia and osteomyelitis due to *Aeromonas hydrophila*: a complication during the treatment of acute leukemia. *Am. J. Clin. Pathol.* 50:587

85. Lynch, J. M., Tilson, W. R., Hodges, G. R., Barnes, W. G., Bopp, W. J., et al. 1981. Nosocomial *Aeromonas hydrophila* cellulitis and bacteremia in a nonimmunocompromised patient. *South. Med. J.* 74:901–2

86. Marcus, L. C. 1971. Infectious diseases of reptiles. *J. Am. Vet. Med. Assoc.* 159:1629–31

87. Marcus, L. C. 1981. *Veterinary Biology and Medicine of Captive Amphibians and Reptiles*, pp. 83–95. Philadelphia: Lea & Febiger

88. Martinez-Silva, V. R., Guzmann-Urrego, M., Caselitz, F. H. 1961. Zur frage der bedeutung von Aeromonasasstammen bei Sauglingsen-teritis. *Z. Tropenmed. Parasitol.* 12:445–51

89. McCracken, A. W., Barkley, R. 1972. Isolation of *Aeromonas* species from clinical sources. *J. Clin. Pathol.* 25:970–75

90. McGrath, V. A., Overman, S. B., Overman, T. L. 1977. Media-dependent oxidase reaction in a strain of *Aeromonas hydrophila*. *J. Clin. Microbiol.* 5:112–13

91. Mead, A. R. 1969. *Aeromonas liquefaciens* in the leukodermia syndrome of *Achatina fulica*. *Malacol. Int. J. Malacol.* 9:43

92. Meeks, M. 1963. The genus *Aeromonas*: methods for identification. *Am. J. Med. Technol.* 29:361–78

93. Mellersh, A. R., Norman, P., Smith, G. H. 1984. *Aeromonas hydrophila*: an outbreak of hospital infection. *J. Hosp. Infect.* 5:425–30

94. Miles, A. A., Halnan, E. T. 1937. A new species of micro-organism *(Proteus melanovogenes)* causing black rot in eggs. *J. Hyg.* 37:79–97

95. Miller, R. M., Chapman, W. R. 1976. *Epistytis* sp. and *Aeromonas hydrophila* infections in fishes from North Carolina reservoirs. *Prog. Fish Cult.* 38:165–68

96. Millership, S. E., Curnow, S. R., Chattopadhyay, B. 1983. Faecal carriage rate of *Aeromonas hydrophila*. *J. Clin. Pathol.* 36:920–23

97. Motyl, M. R., Janda, J. M. 1983. *Aeromonas* gastroenteritis: a two year survey. *23rd Intersci. Conf. Antimicrob. Agents Chemother. Las Vegas*, p. 268. Washington, DC: Am. Soc. Microbiol. (Abstr.)

98. Moyes, C. D., Sykes, P. A., Rayner, J. M. 1977. *Aeromonas hydrophila* septicaemia producing ecthyma gangrenosum in a child with leukemia. *Scand. J. Infect. Dis.* 9:151–53

99. Naumann, G., Gartner, L., Schade, R. 1969. *Aeromonas hydrophila* as eiterer-reger. *Med. Welt* 20:1711–12

100. Nygard, G. S., Bissett, M. L., Wood, R. M. 1970. Laboratory identification of aeromonads from man and other animals. *Appl. Microbiol.* 19:618–20

101. Olsvik, O., Wachsmuth, K., Bradford, A., Thomas, R., Bradley, R. 1985. Pathogenicity studies of clinical isolates of *Plesiomonas shigelloides*. *Abstr. Annu. Meet. Am. Soc. Microbiol.* 1985:46

102. Overman, T. L. 1980. Antimicrobial susceptibility of *Aeromonas hydrophila*. *Antimicrob. Agents. Chemother.* 17:612–14

103. Overman, T. L., Seabolt, J. P. 1983. Minimal inhibitory concentrations of antimicrobial agents against *Aeromonas hydrophila* determined with the Autobac MTS. *J. Clin. Microbiol.* 17:1175–76

104. Parker, M. T., Smith, G. 1984. *Topley and Wilson's Principles of Bacteriology, Virology and Immunology*, Vol. 2. Baltimore, Md: Williams & Wilkins. 7th ed.

105. Pearson, T. A., Mitchell, C. A., Hughes, W. T. 1972. *Aeromonas hydrophila* septicemia. *Am. J. Dis. Child.* 123:579–82

106. Phillips, J. A., Bernhardt, H. E., Rosenthal, S. G. 1974. *Aeromonas hydrophila* infections. *Pediatrics* 53:110–12

107. Picard, B., Arlet, G., Goullet, P. 1983. Origine hydrique d'infections hospitalières à *Aeromonas hydrophila*. *Presse Med.* 12:700

108. Pierce, R. L., Daley, C. A., Gates, C. E., Wohlgemuth, K., Brookings, V.

M., et al. 1973. *Aeromonas hydrophila* septicemia in a dog. *J. Am. Vet. Med. Assoc.* 162:469

109. Pitarangsi, C., Echeverria, P., Whitmire, R., Tarapai, C., Formal, S., et al. 1982. Enteropathogenicity of *Aeromonas hydrophila* and *Plesiomonas shigelloides:* prevalence among individuals with and without diarrhea in Thailand. *Infect. Immun.* 35:666–73

110. Pittman, M. 1953. A study of bacteria implicated in transfusion reactions and of bacteria isolated from blood products. *J. Lab. Clin. Med.* 42:273–88

111. Pivnick, H., Sabina, L. R. 1957. Studies of *Aeromonas formicans. J. Bacteriol.* 73:247–52

112. Popoff, M. 1984. *Aeromonas*. In *Bergey's Manual of Determinative Bacteriology*, Vol. 1, ed. N. R. Krieg, J. G. Holt, pp. 545– 48. Baltimore, Md: Williams & Wilkins. 9th ed.

113. Quadri, S. M., Gordan, L. P., Wende, R. D., Williams, R. P. 1976. Meningitis due to *Aeromonas hydrophila. J. Clin. Microbiol.* 3:102

114. Quinot, J. F., Delatte, P., Flye Saint Marie, F., Richard, C. 1982. Gangrène gazeuse à *Aeromonas hydrophila* un piège théapeutique. *Nouv. Presse Med.* 11:2783–84

115. Ramsey, A. M., Rosenbaum, B. J., Yarbrough, C. L. 1978. *Aeromonas hydrophila* sepsis in a patient undergoing hemodialysis therapy. *J. Am. Med. Assoc.* 239:128–29

116. Raszeja, S., Krynski, S., Krueger, A., et al. 1973. Blood contamination with *Aeromonas hydrophilus* as a cause of lethal post-transfusion complications. *Pol. Tyg. Lek.* 28:1159–62

117. Reinhardt, J. F., George, W. L. 1985. *Plesiomonas shigelloides–*associated diarrhea. *J. Am. Med. Assoc.* 253:3294–95

118. Rosenthal, S. G., Bernhardt, H. E., Phillips, J. A. 1974. *Aeromonas hydrophila* wound infection. *Plast. Reconstr. Surg.* 53:77–79

119. Rosner, R. 1964. *Aeromonas hydrophila* as the etiologic agent in a case of severe gastroenteritis. *Am. J. Clin. Pathol.* 42:402–4

120. Rutala, W. A., Sarubbi, F. A., Finch, C. S., MacCormack, J. N., Steinkraus, G. E. 1982. Oyster-associated outbreak of diarrhoeal disease possibly caused by *Plesiomonas shigelloides* (Letter). *Lancet* 1:739

121. Saito, R., Schick, S. 1973. *Aeromonas hydrophila* peritonitis. *Cancer Chemother. Rep.* 57:489–91

122. Sakazaki, R., Namioka, R., Nakaya, R., Fukumi, H. 1959. Studies on so-called paracolon C27 (Ferguson). *Jpn J. Med. Sci. Biol.* 12:355–63

123. Sanyal, S. C., Gaur, S. D., Shrivastava, D. L., Sen, P. C., Marwah, S. M., et al. 1972. Enteric infections in Sunderpur slum area. *Indian J. Med. Res.* 60:979

123a. Sanyal, S. C., Saraswathi, B., Sharma, P. 1980. Enteropathogenicity of *Plesiomonas shigelloides. J. Med. Microbiol.* 13:401–9

124. Sanyal, S. C., Singh, S. J., Sen, P. C. 1975. Enterpathogenicity of *Aeromonas hydrophila* and *Plesiomonas shigelloides. J. Med. Microbiol.* 8:195–98

125. Sasu, D., Apostica, E. 1967. On a strain of *Aeromonas liquefaciens* isolated from blood. *Microbiologia* 12:437–41

126. Schubert, R. H. W. 1967. Die pathogenitaet der Aeromonaden fuer Mensch und Tier. *Arch. Hyg. Bakteriol.* 150:709–16

127. Schubert, R. H. W. 1974. *Aeromonas*. In *Bergey's Manual of Determinative Bacteriology*, ed. R. E. Buchanan, N. E. Gibbons, pp. 345–48. Baltimore, Md: Williams & Wilkins. 8th ed.

128. Shackelford, P. G., Ratzan, S. A., Shearer, W. T. 1973. Ecthyma gangrenosum produced by *Aeromonas hydrophila. J. Pediatr.* 83:100–1

129. Shilkin, K. B., Annear, D. I., Rowett, L. R., Laurence, B. H. 1968. Infection due to *Aeromonas hydrophila. Med. J. Aust.* 1:351–53

130. Shotts, E. B. Jr., Gaines, J. L., Martin, C., Prestwood, A. K. 1972. *Aeromonas-*induced deaths among fish and reptiles in an eutrophic inland lake. *J. Am. Vet. Med. Assoc.* 161:603–7

131. Sirinavin, S., Likitnukul, S., Lolekha, S. 1984. *Aeromonas* septicemia in infants and children. *Pediatr. Infect. Dis.* 3:122–25

132. Slotnick, I. J. 1970. *Aeromonas* species isolates. *Ann. NY Acad. Sci.* 174:503–10

133. Smith, J. A. 1980. *Aeromonas hydrophila:* analysis of 11 cases. *Can. Med. Assoc. J.* 122:1270–72

134. Smith, J. A. 1980. Ocular *Aeromonas hydrophila. Am. J. Ophthalmol.* 89:449–51

135. Stephen, S., Rao, K. N. A., Kumar, M. S., Indurani, R. 1975. Human infection with *Aeromonas* species: varied clinical manifestations. *Ann. Intern. Med.* 83:368–69

136. Suthipintawongs, C., Wanvaree, S. 1982. Gas gangrene: an unusual manifestation of *Aeromonas* infection. *J. Med. Assoc. Thailand* 65:678–81

137. Tapper, M. L., McCarthy, L. R., Mayo, J. B., Armstrong, D. 1975. Recurrent

Aeromonas sepsis in a patient with leukemia. *Am. J. Clin. Pathol.* 64:525–30

138. Taylor, D. N., Echeverria, P., Blaser, M. J., Pitarangsi, C., Blacklow, N., et al. 1985. Polymicrobial aetiology of travellers' diarrhoea. *Lancet* 1:381–83

139. Teruyoski, A., Nobuyuki, I., Itoh, T., Sakai, S., Shimada, T., et al. 1980. A survey of *Plesiomonas shigelloides* from aquatic environments, domestic animals, pets and humans. *J. Hyg.* 84:203–11

140. Tsukamoto, T., Kinoshita, Y., Shimada, T., Sakazaki, R. 1978. Two epidemics of diarrhoeal disease possibly caused by *Plesiomonas shigelloides*. *J. Hyg.* 80:275–80

141. von Graevenitz, A., Bucher, C. 1983. Evaluation of differential and selective media for isolation of *Aeromonas* and *Plesiomonas* spp. from human feces. *J. Clin. Microbiol.* 17:16–21

142. von Graevenitz, A., Mensch, A. H. 1968. The genus *Aeromonas* in human bacteriology: Report of 30 cases and review of the literature. *N. Engl. J. Med.* 278:245–49

143. von Graevenitz, A., Zinterhofer, L. 1970. The detection of *Aeromonas hydrophila* in stool specimens. *Health Lab. Sci.* 7:124–27

144. Washington, J. A. 1972. *Aeromonas hydrophila* in clinical bacteriologic specimens. *Ann. Intern. Med.* 76:611–14

145. Washington, J. A. II. 1973. The role of *Aeromonas hydrophila* in clinical infection. *Infect. Dis. Rev.* 2:75–86

146. Weinstock, R. E., Bass, S. J., Lauf, E., Sorkin, B. A. 1982. *Aeromonas hydrophila:* a rare and potentially life-threatening pathogen to humans. *J. Foot Surg.* 21:45–53

147. Wohlegemuth, D., Pierce, R. L., Kirkbride, C. A. 1972. Bovine abortion association with *Aeromonas hydrophila*. *J. Am. Vet. Med. Assoc.* 160:1001–2

148. Wolff, R. L., Wiseman, S. L., Kitchens, C. G. 1980. *Aeromonas hydrophila* bacteremia in ambulatory immunocompromised hosts. *Am. J. Med.* 68:238–42

149. Yadava, R., Seeler, R. A., Kalelkar, M., Royal, J. E. 1979. Fatal *Aeromonas hydrophila* sepsis and meningitis in a child with sickle cell anemia. *Am. J. Dis. Child.* 133:753–54

150. Zajc-Satler, J. 1972. Morphological and biochemical studies of 27 correct strains belonging to the genus *Aeromonas* isolated from clinical sources. *J. Med. Microbiol.* 5:263–65

Ann. Rev. Microbiol. 1988. 42:421–40

HOST RANGE DETERMINANTS IN PLANT PATHOGENS AND SYMBIONTS

N. T. Keen

Department of Plant Pathology, University of California, Riverside, California 92521

B. Staskawicz

Department of Plant Pathology, University of California, Berkeley, California 94720

CONTENTS

INTRODUCTION

This paper reviews recent research into the mechanisms underlying plant-microorganism interactions. A major impetus for the study of these interactions has been the economic impact of microorganisms, either as beneficial plant symbionts in the case of *Rhizobium* spp. or as detrimental pathogens such as *Agrobacterium* spp. and members of the *Pseudomonas*

0066-4227/88/1001-0421$02.00

syringae group. In addition to providing an understanding of the mechanisms involved in symbiosis or pathogenicity, research has also provided important new insights into host range determinants, i.e. those factors determining which plant(s) a particular microorganism can colonize. The host range may include a large number of plant species at one extreme or only a single genotype of a single plant species at the other. We summarize the research progress to date, outline future directions, and finally speculate on how the emerging information will be used to manipulate plant-microbe interactions for the improvement of crop plant production and quality. Since space is limited and several recent reviews have appeared in the area (e.g. 9, 18, 19, 31, 35, 58), we do not attempt to be comprehensive. Instead, we focus on recent findings directly germane to host range determination.

Microorganisms require unique genetic information to colonize plant tissues and establish symbiotic or parasitic relationships. These genes confer what has been called the "basic compatibility" of the microorganism with its plant partner (29). They enable the microorganism to infect and grow in the host, obtain nutrients, and avoid plant defense mechanisms. Superimposed on these basic required genes are other genes that may extend the host range of the microorganism to include additional plants (e.g. the *Rhizobium* host range genes to be discussed later) or restrict the host range to fewer plant species or genotypes (e.g. avirulence genes).

AGROBACTERIUM TUMEFACIENS, THE CAUSAL AGENT OF CROWN GALL

Agrobacterium tumefaciens and the related *Agrobacterium rhizogenes,* causal agent of hairy root, generally have wide host ranges, including most dicotyledenous plants. Individual strains, however, exhibit relatively narrow host ranges (2). Pathogenicity and host range are determined by several positive bacterial functions and by the general failure of plants to recognize the bacteria and invoke defense reactions. In at least some cases, host range is restricted by bacterial functions that operate negatively in that they lead to plant recognition and invocation of an active plant defense mechanism.

A. tumefaciens is a soil-borne bacterium that may colonize plant surfaces without causing disease. However, local wounding typically leads to the characteristic gall or neoplastic symptoms. Following initiation, tumor development no longer requires the presence of the bacteria. The oncogenic principle was identified as a large extrachromosomal element, the Ti plasmid (94). A 13–26-kb T-DNA region, but not the remainder of the Ti plasmid, is transferred to the plant cell. There it expresses three genes with eukaryotic signals that lead to overproduction of the plant hormones zeatin (a cytokinin) and indole acetic acid (an auxin), which in turn cause disorganized plant overgrowths (for review, see 97). The T-DNA also contains one or more genes that cause plant cells to produce novel metabolites, called opines, which are utilized by the infecting bacteria. The unique transfer of T-DNA from the bacterium to plant cells requires the gene products of bacterial virulence *(vir)* genes, which occur both on the bacterial chromosome (e.g. 23) and on the Ti plasmid, but are not transferred to the plant cell.

Tumor formation requires attachment of *Agrobacterium* cells to wounded plant cells (58). While the mechanism of attachment is not totally understood, two different genes, one of them involved in extracellular polysaccharide (EPS) synthesis, may be required through their role in the synthesis of a cyclic β-1,2-linked glucan (8, 70).

The Ti plasmid-borne *vir* genes are normally repressed, but activation occurs after bacterial contact with certain plant substances (81). Several monocotyledonous plants contain only insignificant amounts of these inducers (92), which may be related to the general lack of *Agrobacterium* pathogenicity on these plants. Elegant work on the nature and function of the *Agrobacterium vir* genes in several laboratories has shown that transfer of the T-DNA to plant cells is accomplished by a conjugative process that is regulated by plant products (see 82 for a recent review). The T-DNA is excised from the Ti plasmid by gene products of the *virD* locus as a single-stranded DNA prior to transfer to plant cells (1, 68, 96), which is in turn facilitated by the product of the *virC* locus (39). Several other *vir* proteins are located in the bacterial envelope, which suggests that they may also be involved in T-DNA transfer (30). Activation of *vir* gene expression involves two regulatory genes, *virA* and *virG*. The membrane-associated *virA* protein appears to interact with plant phenolics and then covalently modify the intracellular *virG* protein product such that it binds and activates other *vir* gene promoters (53, 72, 99).

Genes That Broaden the Host Range

The T-DNA genes encoding phytohormone production also clearly function as pathogenicity genes, since their deletion eliminates the oncogenic phenotype. However, more specific host range effects have also been associated with certain of the *vir* genes. For example, Hirooka & Kado (36) showed that deletions at the 3' end of *virE* progressively reduced the number of plant species on which tumors formed. Similarly, Yanofsky & Nester (102) observed that mutations in either of two open reading frames constituting the *virC* locus reduced the number of susceptible host plants. Jin et al (42) have further shown that a segment of the *vir* region of the supervirulent *A. tumefaciens* A281 increases virulence and extends the host range of certain other *Agrobacterium* strains. The A281 DNA was found to include the *virG* locus as well as the 3' end of the *virB* operon. The supervirulent phenotype may therefore involve the increased expression of *vir* genes in response to the altered *virG* gene.

Bacterial Factors That Function Negatively to Narrow the Host Range

Unlike most *Agrobacterium* strains, those isolated from grapevine plants form tumors on only a small number of plant species. The Ti plasmid carried by these strains contains a defective cytokinin biosynthetic gene, and introduction of the cytokinin gene from a wide–host range Ti plasmid expands the host range of grapevine strains (37, 56). However, two virulence region genes, *virA* and *virC*, are also involved in the narrow–host range phenotype. For example, recent work by Leroux et al (53) showed that the grapevine strain has a *virA* gene that is divergent from that in other wide–host range strains. Since the *virA* protein is believed to recognize plant phenolics, the *virA* gene of grapevine strains may respond to a metabolite(s) unique to grapevines. In its association with determination of host range *virA* is therefore analogous to the *Rhizobium nodD* gene, which is discussed below.

Wide–host range Ti plasmids are ineffective in causing galls on grapevine plants because of a plant defense reaction, possibly the hypersensitive reaction (HR) discussed below (101). Significantly, mutations within the *virC* locus of a wide–host range plasmid prevented the HR, and normal tumors formed on grapevines. These results suggest that the *virC* gene product may be an elicitor of the grapevine HR or may result in the formation of an elicitor. Thus, *virC* may be analogous to the avirulence genes recently cloned from several bacterial pathogens, as discussed below.

RHIZOBIUM-LEGUME SYMBIOSIS

Rhizobium spp. are soil-inhabiting bacteria that may infect root hairs of certain legume plants and initiate a complex series of events resulting in the formation of specialized structures called nodules. Nodules are composed of both plant and bacterial elements and carry out the conversion of atmospheric nitrogen to reduced forms, which are translocated to other plant parts, where they decrease or eliminate the need for fixed nitrogen from the soil. *Rhizobium* strains have well-defined and generally narrow host ranges involving one or a few plant genera, usually but not always legumes. A picture is emerging of the initial processes involved in root nodulation as well as of the factors involved in host range determination. Most of these factors are positive, bacteria-encoded functions that are necessary for nodule formation, but some also operate negatively to restrict the host range.

Rhizobia harbor large symbiosis (Sym) plasmids, which contain several genes essential for host nodulation (*nod* genes) as well as genes required for nitrogen fixation in the nodules. The so-called common *nod* genes occur as a cluster in most strains and share homology among strains (for reviews see 19, 26, 35). Transfer of certain cloned *nod* genes from one *Rhizobium* strain to another [or indeed into *A. tumefaciens* (25, 38)] renders the recipient strain able to infect the specific legume host of the donor *Rhizobium* strain. Thus, in addition to having basic nodulation functions, the *nod* genes are major determinants of host specificity. However, rhizobia also contain genes similar to the chromosomal virulence *(chv)* genes of *A. tumefaciens,* which are required for nodule development (28, 95). These genes, *chvA* and *chvB,* are associated with production of a novel cyclic 1,2-linked β-glucan (34, 69). The role of the glucan in nodulation is not yet clear.

The Sym plasmids of various rhizobia contain a highly conserved operon of three genes, *nodABC,* which are required for initial root infection. Their expression is regulated by the *nodD* gene, which is usually closely linked to the *nodABC* genes. While little is known about how *nod* gene expression leads to host infection, it appears that some *nod* genes may affect bacterial production of cell-surface carbohydrates, which are in turn recognized by plant root hair lectins as signals to induce the events leading to nodulation (see 14). The precise biochemical functions of the *nodABC* genes are not known, but their regulation by the *nodD* gene is controlled by a ~40–base pair (bp) sequence called a *nod* box. The *nod* box occurs upstream from the *nodABC* genes and perhaps also in front of *nodD* and other *nod* genes described below (e.g. 76). The *nod* box sequence appears to be a *cis*-acting regulatory element.

Recently, it was discovered that plant phenolic substances called flavones induce the expression of the *nodABC* genes and other operons located downstream from a *nod* box (67, 71, 103). Isoflavones, however, appear to be the *nod* inducers in *Bradyrhizobium japonicum* (49). Certain structurally related coumarins and other compounds have also been shown to function as inhibitors (21). The inducers as well as several of the inhibitors function at very low concentrations and are therefore assumed to interact with the *nodD* protein to regulate expression of the *nod* genes. Indeed, Burn et al (7) recently produced several mutants of *Rhizobium leguminosarum* that exhibited altered *nodD* regulatory properties. One of these mutants involved a single amino acid change in the *nodD* protein, resulting from a one-base alteration. The results were interpreted as support for the idea that the *nodD* protein interacts directly with the flavone inducers. Thus, the activated *nodD* protein may act as a positive regulator that interacts with the *nod* box elements. Since only the *nodD* protein is involved, *nod* regulation differs from the two-component plant-sensing system of *Agrobacterium* discussed above.

Positive Host-Range Determinants

As with *A. tumefaciens,* several mechanisms determine whether a certain *Rhizobium* strain will nodulate a plant, and some may involve the earliest nodulation events. For example, Horvath et al (40) found that *nodD* genes from two *Rhizobium meliloti* strains differed considerably in structure as well as in the plant factors that activated them. The *nodD* gene from a narrow–host range strain was activated only by the flavone luteolin, but the *nodD* gene from a wide–host range strain, MPIK3030, was also activated by other unknown plant metabolites. The MPIK3030 *nodD* gene complemented narrow–host range *nodD* strains to extend their host range, but the *nodD* gene of narrow–host range strains did not affect the host range of MPIK3030. Thus, *nodD* appears to govern host range. In similar work, Spaink et al (80) also observed that *nodD* genes differ in their response to plant inducers and concluded that they determine host range. These results are similar to those with the *Agrobacterium virA* gene and establish that *nodD* genes constitute one level of host range determination.

Rhizobium symbiotic plasmids carry other genes that are involved with host range determination. For instance, Lewin et al (54) recently showed that three regions of the Sym plasmid of a promiscuous *Rhizobium* strain were involved in nodulation of various legume hosts. These results were similar to those obtained with *Rhizobium trifolii* by Djordevic et al (20). Some of these host range *nod* genes have been characterized. One of them, *nodF*, from *R. leguminosarum*, encodes a gene product that is related to acyl-carrier proteins from other organisms (78). This gene product may have a role in lipopolysaccharide or exopolysaccharide biosynthesis; when it is activated, an altered cell-surface carbohydrate structure may result. The bacterium with its new coat is then specifically recognized, presumably by a carbohydrate-specific lectin on the plant root, and the biochemical events involved in root infection and nodule formation are initiated. Four host-specific nodulation genes have been cloned from *R. meliloti* (41). One of them determines the earliest interaction between bacterium and root hair, the so-called "root hair curling." These results have also led to the hypothesis that the host range genes of *Rhizobium* sp. may be involved in the adornment of the bacterial cell surface with specific ligands that are recognized by the correct legume host (33, 69).

Several recent reports have noted that *Rhizobium* mutants deficient in EPS production have altered nodulation efficiency. Some are able to form nodules, but these nodules do not fix nitrogen (e.g. 52). Borthakur et al (4) showed that mutation of a gene involved in EPS synthesis did not affect the ability of *Rhizobium phaseoli* to induce normal nitrogen-fixing nodules on its host plant, bean. However, introduction of the mutated gene into the *R. leguminosarum* genome by a gene replacement experiment destroyed the ability of this bacterium to nodulate its normal host, pea. Significantly, the mutation in *R. leguminosarum* was corrected and nodulation was restored by the introduction of a cloned EPS gene obtained from *R. phaseoli* or from the plant pathogen *Xanthomonas campestris* pv. *campestris*. This finding demonstrates that these two bacteria contain functionally conserved genes in their EPS pathways. The data also establish that the EPS gene confers a positive function required for nodulation of peas by *R. leguminosarum* but not for nodulation of beans by *R. phaseoli*.

Negative Factors

Djordjevic et al (22) discovered that certain genes in *R. trifolii* act negatively to restrict the host range. In this sense, they are functionally the equivalent of the pathogen avirulence genes discussed below. Werner et al (98) observed that soybean roots reacted in a defensive manner to infection by an incompatible *Rhizobium* strain. The defense reaction included production of the soybean phytoalexins, which have been associated with active defense against several plant pathogens.

FACTORS DETERMINING THE HOST RANGE OF OTHER PLANT PATHOGENS

Plant pathogens are a large group, including viruses, nematodes, fungi, and bacteria. It is therefore impossible in this paper to discuss adequately the pathogenic strategies and host range determinants of all of them. However, in a now familiar pattern, their host ranges are clearly determined by positive factors that extend virulence or host range as well as by negative factors that reduce it. The former class includes agents such as toxins or enzymes that debilitate or degrade the host plant so that the pathogen can utilize the resulting food base. Some pathogens such as *Pseudomonas syringae* pv. *savastanoi* (e.g. 100) produce plant growth hormones that alter the development of plant tissue, causing a tumorous growth that provides a suitable environment for the pathogen. Several pathogens are equipped with enzymatic and other machinery for avoiding or combatting chemical and structural plant defense mechanisms (e.g. 15, 65, 93). In addition to these positive factors, pathogens also frequently contain avirulence genes, which lead to triggering of the plant hypersensitive reaction, thus restricting pathogen host range.

Pathogenicity and Virulence Factors

The major pathogenicity factors that have been studied are (*a*) enzymes that degrade plant cell walls or other structures, thus facilitating pathogen entry or dispersion through the host, and (*b*) toxins that injure or kill plant cells, permitting the pathogen to colonize the disabled cells. Considerable evidence suggests that these mechanisms are either requisites for basic pathogenicity or virulence factors that increase disease severity.

Cutinases produced by certain plant-pathogenic fungi are required for penetration of the plant cuticle. Kolattukudy and coworkers (see 48 for review) first showed that an anticutinase antibody severely reduced the infection frequency of *Fusarium solani* f. sp. *pisi* on pea epicotyls, presumably because the fungus could not breach the cuticle. Dickman & Patil (17) obtained similar results for the cutinase of *Colletotrichum gloeosporioides*. They isolated mutants of the fungus that were deficient in cutinase and showed that all of the nonleaky mutants had lost pathogenicity. However, when the cuticle of the host plant was artificially wounded prior to inoculation, the mutants were fully pathogenic. The results therefore establish that the *C. gloeosporioides* cutinase is required for initial penetration of the host but is not necessary for pathogenicity once penetration has occurred.

Among the best understood pathogenicity or virulence mechanisms are the pectate lyases produced by soft-rotting *Erwinia* spp. The *pel* genes encoding these proteins have been cloned in several laboratories, and sequence data are available for some of them (for recent reviews see 10, 50). Marker exchange mutagenesis has proven the role of several of the *Erwinia pel* genes in pathogenicity (50). Significantly, however, other bacterial proteins associated with secretion of the pectate lyases into the external medium are also required for pathogenicity. Expression of the *pel* genes is catabolite repressed by glucose and other sugars but inducible by pectic substances. Thus, like the *nod* and *vir* genes of *Rhizobium* spp. and *Agrobacterium* spp., respectively, the *Erwinia pel* genes are regulated by plant substances.

Several bacterial pathogens produce toxins that may increase virulence (e.g. 3), and progress has been made in the cloning of genes responsible for their synthesis (e.g. 66). Recent genetic studies with bacterial pathogens have also identified several pathogenicity genes whose functions are not yet known (6, 55, 57, 63, 79, 91). In two cases, large clusters of closely linked pathogenicity genes [called *hrp* (hypersensitive reaction, pathogenicity) genes] have been cloned (55, 91). Recent work with *P. syringae* pv. *phaseolicola* and *P. syringae* pv. *glycinea* has shown that some of the *hrp* genes are induced when the bacteria are inoculated into plants (P. B. Lindgren, N. J. Panopoulos & B. J. Staskawicz, unpublished data). Surprisingly, the *hrp* genes from *Xanthomonas campestris* pv. *campestris* appear to have high homology with those in *Pseudomonas solanacearum* but not with those of several *P. syringae* pathovars (6). Dow et al (24) observed that many of these genes from *X. campestris* pv. *campestris* are required for the secretion of three pectate lyases through the bacterial outer membrane. While it is not known if the analogous *P. syringae* or *P. solanacearum hrp* genes have similar functions, their mutation abolishes pathogenicity as well as the ability to elicit a plant hypersensitive reaction (6, 55).

The wilt-inducing bacterial pathogens *P. solanacearum* and *Erwinia stewartii* also possess many genes required for pathogenicity. In the former species, production of several extracellular enzymes as well as EPS seems to be involved in pathogenicity. Many of the genes for these factors have recently been shown to occur on a large plasmid, and the deletion of a large segment of this plasmid results in loss of pathogenicity (5). With *E. stewartii,* genes leading to the production of EPS have also been shown to be required for pathogenicity; some of these genes have been cloned (11).

The factors discussed above are among the minimum requirements for pathogenicity and virulence. In addition, there are a few positive functions that specifically extend the host range of a pathogen. For example, Mellano & Cooksey (59) recently isolated mutants of *X. campestris* pv. *translucens* with a narrowed host range. In nature various strains of this bacterium attack a variable number of grass hosts. Mutagenesis of a wild-type strain that normally attacks rye, barley, wheat, and triticale yielded mutants that were pathogenic on some but not all of the host plants. Thus, none of the mutants appeared to be affected in basic pathogenicity genes. The observed loss of pathogenicity on one or more grass hosts suggests that specific bacterial genes are required for the bacterium to attack each host. These genes thus confer positive pathogenicity functions that dictate the host range of the bacterium, as with certain of the *Rhizobium nod* genes.

Host-selective toxins broaden the host range of certain fungal pathogens (for reviews see 27, 64). For example, several *Helminthosporium* spp. produce host-selective toxins that cause their pathogenicity on a certain plant. Scheffer et al (74) performed a sexual cross of *Helminthosporium carbonum* (which attacks only a certain genotype of corn and produces a toxin that only affects that genotype) with *Helminthosporium victoriae* (which attacks only a certain genotype of oats owing to a specific toxin). The researchers recovered equal numbers of progeny that attacked both hosts, corn but not oats, oats but not corn, and neither host. This $1:1:1:1$ segregation ratio suggests that single loci were segregating, a surprising result since the structures of the two toxins are very different. While it is not clear whether this result was due to clustering of biosynthetic genes in the two fungi, it demonstrates that the toxins can expand host range. Turgeon et al (90) have recently developed a DNA transformation system for *Helminthosporium* spp., and advances in our knowledge of toxin production should be forthcoming.

Avirulence Genes

The inducible hypersensitive reaction of plants is a major factor that restricts the host range of pathogens. The HR may confer resistance to entire pathogen species (general resistance) or to only certain genotypes of a single pathogen species (specific resistance). The molecular expression of the hypersensitive response modulated by plant disease resistance genes is only partially understood. It involves relatively rapid effects on plant plasma membranes followed by the rapid necrosis of plant cells in the vicinity of an invading pathogen. Inducibly synthesized enzymatic, structural, and chemical (phytoalexin) barriers are deposited in and around these plant cells (for more details see 9, 18, 44). These last events are a consequence of the rapid derepression of several plant genes, some of which encode enzymes in secondary biosynthetic pathways that produce phytoalexins (e.g. 51).

In specific resistance, the pathogen contains dominant avirulence genes that are genetically complementary to disease resistance genes in the plant host (29). If the pathogen harbors a dominant avirulence gene and the plant host contains a complementary disease resistance gene, the infected plant responds with a hypersensitive reaction, and the pathogen is unsuccessful. Avirulence genes act negatively from the point of view of the pathogen, since they reduce its host range. The protein products of avirulence genes lead, in as yet unknown ways, to the occurrence in the pathogen of specific recognition elements, called elicitors. These elicitors are believed to be recognized by plant receptors containing the protein products of dominant plant disease resistance alleles. Elicitor-receptor binding then initiates the HR (18, 44). Because pathogens can survive, at least in the short run, without certain dominant avirulence alleles, why have these genes not all been lost? Clearly, avirulence genes must confer survival advantages of which we are not aware. Perhaps at least some avirulence genes are also required for high virulence or for survival outside the host.

VIRUSES   Several recent papers have provided insight on factors that determine viral host range and have raised the possibility that some of these factors may behave as avirulence genes. An example is the work of Schoelz et al (75) in which two strains of cauliflower mosaic virus were compared. One strain gave a compatible, systemic infection on *Datura stramonium,* while the other caused a hypersensitive reaction, characterized by small local lesions and the prevention of virus spread through the plant. A 496-bp DNA segment from the first half of the open reading frame of viral gene VI accounted for the difference in host response, since interchanging this region made the virulent virus strain avirulent on *Datura.* Unfortunately, it is not yet known which gene is genetically dominant, i.e. whether the gene VI protein product of the virulent strain suppresses a plant HR caused by some other viral product, or whether the gene VI protein product of the avirulent strain is itself an HR elicitor.

Dawson et al (12) and Meshi et al (60) obtained full-length infectious cDNA clones of tobacco mosaic virus (TMV) and used them to study the basis of host range specificity. For example, Saito et al (73) replaced the coat protein gene of a *tomato* strain of TMV with that of a tobacco strain and showed that the recombinant virus exhibited the host range of the tobacco strain rather than that of the *tomato* strain on tobacco plants carrying the N' resistance gene. Thus the coat gene determined host range. Knorr & Dawson (46) investigated mutants of the common tobacco strain of TMV that caused a hypersensitive, local lesion response on N' tobacco plants rather than the normal susceptible response. Six of seven independent avirulent mutants that elicited the HR were shown to have the same cytosine-to-uracil point mutation at position 6157 of the viral genome, a site near the 3' end of the coat protein gene. As in the work with cauliflower mosaic virus, it is not yet clear whether the mutant coat protein of TMV is an HR elicitor per se in N' tobacco. If it is, then the mutant coat protein gene would constitute an avirulence gene. A dominance test in which both wild-type and mutant coat protein genes are cloned into the virus should test this possibility.

Beachy and collaborators (89 and references therein) demonstrated that transformation of the wild-type coat protein gene of TMV or alfalfa mosaic virus into the tobacco genome resulted in a degree of resistance to these as well as several other plant viruses. Dawson et al (13) have further shown that various insertions and deletions in the coat protein gene of TMV resulted in altered viral symptoms on tobacco, ranging from the absence of symptoms to yellowing or necrosis. Interestingly, two mutants with internal deletions yielded hypersensitive reactions on tobacco and other plants rather than normal systemic virus spread. While the mechanisms underlying these results are not yet known, the narrowing of viral host range in these studies has implications for practical plant disease control.

FUNGI   Many fungal pathogens contain avirulence genes complementary to various plant disease resistance genes. As yet, however, none of them have been cloned and characterized. Fungal pathogens frequently contain carbohydrate or glycoprotein elicitors that initiate the plant hypersensitive reaction in the absence of the pathogen (18). Some of them are nonspecific elicitors of general resistance because they uniformly affect all cultivars of a plant species. For example, Sharp et al (77) elucidated the structure of a branched, β-linked heptaglucan elicitor from *Phytophthora megasperma* f. sp. *glycinea*. In contrast to such nonspecific elicitors, race-specific elicitors have been detected from several pathogens that exist in gene-for-gene relationships with their hosts (18, 44). For example, Tepper & Anderson (88) and DeWit et al (16) have recently isolated carbohydrate and peptide compounds, respectively, that function as race-specific elicitors from two fungal pathogens. The precise role of avirulence genes in dictating the structures of these elicitors, however, is not currently known.

BACTERIA   Avirulence *(avr)* genes that modulate hypersensitive reactions have recently been cloned and characterized from several bacterial pathogens. It is not yet known, however, what their functions are in the bacteria and how the *avr* genes lead to plant recognition of the bacteria. The first avirulence gene was cloned from race 6 of *Pseudomonas syringae* pv. *glycinea* by Staskawicz et al (84). This gene, called *avrA,* has been sequenced (62) and encodes a ~100-kd protein whose function in the bacterium is unknown. The protein does not contain a leader peptide secretion sequence and does not have the significant hydrophobic domains that would be expected if it were membrane associated. Like other cloned *avr* genes, the *avrA* gene behaves strictly as a dominant genetic character. Two other avirulence genes were cloned from a single isolate of race 0 of *P. syringae* pv. *glycinea* (85), and one of them *(avrB)* turned out to be identical to a single *avr* gene present in race 1 of the pathogen. The *avrB* and *avrC* genes from race 0 were sequenced (87) and found to encode single protein products of 36 and 39 kd, respectively. Like *avrA,* the *avrB* and *avrC* proteins contain neither signal secretion sequences nor significant hydrophobic domains. The *avrB* and *avrC* genes have been highly expressed in *Escherichia coli* cells, and polyclonal antisera have been raised against the native and β-galactosidase fusion proteins (T. Huynh & B. Staskawicz, in preparation; S. Tamaki & N. T. Keen, in preparation). Use of the antibodies in Western blotting experiments showed that these *avr* genes were not expressed in *P. syringae* pv. *glycinea* cells grown on complex culture media, but were highly expressed when the bacteria were inoculated into soybean leaves. However, plant induction of *avrB* does not seem to result from specific plant substances as in *Agrobacterium* spp. or *Rhizobium* spp. Instead, results indicate that the *avrB* gene is expressed in minimal culture media but repressed in complex media (T. Huynh & B. Staskawicz, in preparation).

Although the functions of the *avrB* and *avrC* proteins are not known, DNA sequence data have disclosed considerable homology between the protein products (87). Although these genes condition markedly different plant specificities, their protein products may have identical or related functions in the bacteria. Recombinant genes have recently been constructed to deduce which domains of the *avrB* and *avrC* proteins generate the respective plant reaction phenotypes.

Following the initial cloning of *avr* genes from *P. syringae* pv. *glycinea,* several avirulence genes were cloned from other phytopathogenic bacteria. For example, Gabriel et al (32) cloned five different avirulence genes from a single race of *X. campestris* pv. *malvacearum.* Another race of the bacterium appeared to contain recessive (virulence) alleles with homology to two of the cloned *avr* genes. In other recent work, a cosmid clone has been obtained from race 1 of *P. syringae* pv. *phaseolicola* that appears to contain an avirulence gene (M. Shintaku, D. Klupfel, C. K. L. Too & S. Patil, personal communication). In similar work, an avirulence gene has been cloned from race 3 of *P. syringae* pv. *phaseolicola* (F. Hitchin, S. Harper, C. D. Jenner, J. W. Mansfield & M. Daniels, personal communication). S. Hutcheson & A.

Collmer (personal communication) recently obtained cosmid clones from *P. syringae* pv. *syringae* that elicited an HR in tobacco plants and therefore may contain one or more avr genes. Significantly, one of the cosmid clones elicited the tobacco HR when introduced into the tobacco pathogen *P. syringae* pv. *tabaci* as well as the plant saprophytes *Pseudomonas putida* and *E. coli*. The last result is of considerable interest, since along with the *P. syringae* pv. *tomato avrD* gene discussed below, this is the only avr gene that has permitted *E. coli* to elicit the plant HR.

Swanson et al (86) cloned an avirulence gene from race 2 of *X. campestris* pv. *vesicatoria* that had previously been shown to reside on a 200-kb self-transmissible copper resistance plasmid (83). This *avr* gene was of special interest because bacteria lose the *avr* phenotype at $\sim 10^2$ times the frequency expected via mutation. Swanson et al (86) observed that spontaneous race change mutants involved genomic rearrangements at the *avr* gene locus, which explained the high mutation frequency. Recent observations (43) have further shown that a 1225-bp bacterial insertion element is responsible for the high frequency of inactivation of the *avr* gene. Further, if the insertion element is able to excise precisely at low frequency to restore gene function, the pathogenic bacterium may be able to switch *avr* gene function depending on host plant or other selection pressures. In this regard, it would be of interest to know whether loss of the *avr* gene affects the virulence or long-term fitness of the pathogen.

Recent Findings From Our Own Laboratories

Kobayashi & Keen (47) isolated several cosmid clones from a DNA cosmid library of *Pseudomonas syringae* pv. *tomato*, which upon introduction into *P. syringae* pv. *glycinea* caused a HR in certain but not all soybean cultivars. Thus, *P. syringae* pv. *tomato* contains avirulence genes that elicit the soybean HR in a race-specific manner when they are transferred to a related bacterium. Surprisingly, one of the *P. syringae* pv. *tomato* avirulence genes was indistinguishable from the *avrA* gene previously cloned and sequenced from *P. syringae* pv. *glycinea* (62). Furthermore, the *avrA* gene was present in all of 14 different isolates of *P. syringae* pv. *tomato*. This result demonstrates that avirulence genes are not unique to a single pathovar of *P. syringae*, but may be cosmopolitan. In view of this finding, might a single avirulence gene elicit the hypersensitive reaction in more than one host plant? And might avirulence genes determine pathogen specificities above the race level? Kobayashi & Keen (47) addressed the first question by introducing the cloned *avrA* gene from *P. syringae* pv. *tomato* into several other *P. syringae* pathovars and inoculating them into their normal host plants. Introduction of *avrA* did not affect the pathogenicity of *P. syringae* pv. *lachrymans*, *P. syringae* pv. *phaseolicola*, or *P. syringae* pv. *pisi* in their host plants, but *P. syringae* pv. *tabaci* cells carrying *avrA* elicited a hypersensitive response on three different tobacco species rather than the normal pathogenic reaction. In connection with the second question, P. Minsavage & B. Staskawicz (in preparation) cloned an avirulence gene from a race of *X. campestris* pv. *vesicatoria* that causes disease on tomato but not on any known pepper cultivar. The cloned gene resulted in an HR on pepper leaves. Significantly, inactivation of this *avr* gene in the tomato pathogen by a gene replacement experiment extended its host range to include pepper. This provides the first proof that *avr* genes may be responsible for host range specificities above the race-cultivar level.

One of the avirulence genes from *P. syringae* pv. *tomato*, *avrD*, was sequenced and found to encode a 34-kd protein. *P. syringae* pv. *glycinea* also contains a gene with considerable homology to *avrD*, but it does not elicit the soybean HR (D. K. Kobayashi & N. T. Keen, unpublished). Thus, *P. syringae* pv. *glycinea* produces a variant *avrD* gene product that is not recognized by soybean plants. Surprisingly, infiltration of *E. coli* cells expressing the *P. syringae* pv. *tomato avrD* gene into soybean leaves caused a hypersensitive reaction in exactly the same differential cultivars as *P. syringae* pv. *glycinea* race 4 cells carrying the same gene. It is not yet clear, however, if the *avrD* protein produced by *E. coli* is an elicitor per se of the soybean HR. Noteworthy, however, is the recent report (41a) that a 60-kd protein isolated from *Pseudomonas solanacearum* elicited a necrotic response, presumed to be the HR, in tobacco cell cultures. It will be of interest to investigate further whether this protein as well as the protein products of *avrD* and other well characterized avirulence genes are physiologically important elicitors of plant hypersensitive reactions.

Results similar to those with *P. syringae* pv. *tomato* were obtained when a genomic library of *X. campestris* pv. *vesicatoria* was mobilized into the bean pathogen *X. campestris* pv. *phaseoli* (M. Whalen & B. Staskawicz, in preparation). A closmid clone was identified that caused *X. campestris* pv. *phaseoli* to induce a HR on the bean cultivar Sprite, but not on the cultivar Bush Blue Lake. Again, an avirulence gene has been shown to function in a different bacterial pathovar and to behave as a race-specific *avr* gene, affecting one but not another host cultivar.

SUMMARY, FUTURE DIRECTIONS, AND ANTICIPATED PAYOFFS

The study of plant-microbe and plant-virus interactions has produced recent information that promises considerable practical benefit to agriculture. For example, the results of Tumer et al (89) in which defined viral sequences were cloned into plant hosts encourage optimism that virus diseases may be controlled by such approaches. Similarly, a control measure has been devised based on the use of strain K84 of *Agrobacterium radiobacter*, which produces an antibiotic that affects only pathogenic strains containing the Ti plasmid

(45, 61). The exciting recent results on mechanisms conferring the host range of *Rhizobium* spp. raise the possibility of engineering new strains that will nodulate heretofore refractory plants. The significant advances in elucidating mechanisms that determine microbial and viral host ranges have exposed pathogen vulnerabilities that present opportunities for systematic new approaches to disease control.

The recent research has shown that host range determination is considerably more complex than was imagined ten years ago. As we learn more, fascinating parallels between *Agrobacterium*, *Rhizobium*, and other plant pathogen systems are emerging. A beginning has been made toward elucidating the basic mechanisms required for pathogenicity and for broadening the host range; we are also beginning to understand the function of negatively acting avirulence genes, which narrow the pathogen host range. Surprising results have emerged indicating that, in addition to conferring race specificity, *avr* genes may also determine host range at the pathovar and other levels. Despite these advances, the study of pathogen avirulence genes is in its infancy, and many questions remain to be answered about what appears to be a complex dialog between plant and microbe. In the future lies an understanding of the biochemical functions of *avr* gene products in pathogens and their precise role in elicitation of the plant HR. Do avirulence gene proteins directly interact with plant resistance gene protein products to trigger the HR? Or do *avr* gene products require certain basic pathogenicity genes to initiate the host defense response? With *Rhizobium* spp., we do not yet understand the functions of the common *nod* genes; with *Agrobacterium* spp., little is known about *vir* gene function. The future should see continued advances on these questions. In all of these systems knowledge on the plant side is also limited, and additional research is needed. Concerted steps need to be taken to clone and characterize plant disease resistance genes, since their transformation into foreign plants may lead to revolutionary disease control possibilities. Much of the available information from plants has not yet been integrated with that from the microorganisms; this process will require a broad approach. Most of the genes involved in host range determination that have been cloned and sequenced produce proteins whose function we do not know. Thus, while molecular genetic analysis will continue to be a major tool for further advances, a fuller understanding of host range determination will require its integration with cellular and biochemical techniques.

ACKNOWLEDGMENTS

The literature review was concluded in December, 1987. We thank Hans Thordall-Christensen and Donald Cooksey for commenting on the manuscript and acknowledge grants from the USDA and NSF (to NTK) and the USDA, NSF, and McKnight Foundation (to BS), which support our research. Several colleagues generously supplied preprints of their work before publication.

Literature Cited

1. Albright, L. M., Yanofsky, M. F., Leroux, B., Deqin, M. A., Nester, E. W. 1987. Processing of the T-DNA of *Agrobacterium tumefaciens* generates border nicks and linear, single-stranded T-DNA. *J. Bacteriol.* 169:1046–55

2. Anderson, A. R., Moore, L. W. 1979. Host specificity in the genus *Agrobacterium*. *Phytopathology* 69:320–23

3. Bender, C. L., Stone, H. E., Sims, J. J., Cooksey, D. A. 1987. Reduced pathogen fitness of *Pseudomonas syringae* pv. *tomato* Tn5 mutants defective in coronatine production. *Physiol. Mol. Plant Pathol.* 30:273–84

4. Borthakur, D., Barber, C. E., Lamb, J. W., Daniels, M. J., Downie, J. A., Johnston, A. W. B. 1986. A mutation that blocks extracellular polysaccharide synthesis prevents nodulation of peas by *Rhizobium leguminosarum* but not of beans by *R. phaseoli* and is corrected by cloned DNA from *Rhizobium* or the phytopathogen, *Xanthomonas*. *Mol. Gen. Genet.* 203:320–23

5. Boucher, C., Martinel, A., Barberis, P., Alloing, G., Zischek, C. 1986. Virulence genes are carried by a megaplasmid of the plant pathogen *Pseudomonas solanacearum*. *Mol. Gen. Genet.* 205:270–75

6. Boucher, C. A., Van Gijsegem, F., Barberis, P. A., Arlat, M., Zeschek, C. 1987. *Pseudomonas solanacearum* genes controlling both pathogenicity on tomato and hypersensitivity on tobacco are clustered. *J. Bacteriol.* 1619:5626–32

7. Burn, J., Rossen, L., Johnston, A. W. B. 1987. Four classes of mutations in the *nodD* gene of *Rhizobium leguminosarum* biovar. *viciae* that affect its ability to autoregulate and/or activate other *nod* genes in the presence of flavonoid inducers. *Genes Dev.* 1:456–64

8. Cangelosi, G. A., Hung, L., Puvanesarajah, B., Stacey, G., Ozga, D. A., et al. 1987. Common loci for *Agrobacterium tumefaciens* and *Rhizobium meliloti* exopolysaccharide synthesis and their roles in plant interactions. *J. Bacteriol.* 169:2086–91

9. Collinge, D. B., Slusarenko, A. J. 1987. Plant gene expression in response to pathogens. *Plant Mol. Biol.* 9:389–410

10. Collmer, A., Keen, N. T. 1986. The role of pectic enzymes in plant pathogenesis. *Ann. Rev. Phytopathol.* 24:383–409

11. Coplin, D. L., Frederick, R. D., Ma-jerczak, D. R., Haas, E. S. 1986. Molecular cloning of virulence genes from *Erwinia stewartii*. *J. Bacteriol.* 168:619–23

12. Dawson, W. O., Beck, D. L., Knorr, D. A., Grantham, G. L. 1986. cDNA cloning of the complete genome of tobacco mosaic virus and production of infectious transcripts. *Proc. Natl. Acad. Sci. USA* 83:1832–36

13. Dawson, W. O., Bubrick, P., Grantham, G. L. 1988. The effects of modification of the coat protein gene of TMV upon replication, movement and symptomology. *Phytopathology* 78:In press

14. Dazzo, F. B., Truchet, G. L., Sherwood, J. E., Hrabek, E. M., Abe, M., Pankratz, S. H. 1984. Specific phases of root hair attachment in the *Rhizobium trifolii*–clover symbiosis. *Appl. Environ. Microbiol.* 48:1140–50

15. Defago, G., Kern, H., Sedlar, L. 1983. Genetic analysis of tomatine insensitivity, sterol content and pathogenicity for green tomato fruits in mutants of *Fusarium solani*. *Physiol. Plant Pathol.* 22:39–43

16. Dewit, P. J. G. M., Hofman, A. E., Velthuis,, G. C. M., Kuc, J. 1985. Isolation and characterization of an elicitor of necrosis isolated from intercellular fluids of compatible interactions of *Cladosporium fulvum* (syn. *Fulvia fulva*) and *tomato*. *Plant Physiol.* 717:642–47

17. Dickman, M. B., Patil, S. S. 1986. Cutinase deficient mutants of *Colletotrichum gloeosporioides* are nonpathogenic to papaya fruit. *Physiol. Mol. Plant Pathol.* 28:235–42

18. Dixon, R. A. 1986. The phytoalexin response: elicitation, signalling and control of host gene expression. *Biol. Rev.* 61:239–91

19. Djordjevic, M. A., Gabriel, D. W., Rolfe, B. G. 1987. *Rhizobium*—the refined parasite of legumes. *Ann. Rev. Phytopathol.* 25:145–68

20. Djordjevic, M. A., Innes, R. W., Wijffelman, C. A., Schofield, P. R., Rolfe, B. G. 1986. Nodulation of specific legumes is controlled by several distinct loci in *Rhizobium trifolii*. *Plant Mol. Biol.* 6:389–401

21. Djordjevic, M. A., Redmond, J. W., Batley, M., Rolfe, B. G. 1987. Clovers secrete specific phenolic compounds which either stimulate or repress *nod* gene expression in *Rhizobium trifolii*. *EMBO J.* 6:1173–79

22. Djordjevic, M. A., Schofield, P. R.,

Rolfe, B. G. 1985. Tn5 mutagenesis of *Rhizobium trifolii* host-specific nodulation genes results in mutants with altered host-range ability. *Mol. Gen. Genet.* 200:463–71

23. Douglas, C. J., Staneloni, R. J., Rubin, R. A., Nester, E. W. 1985. Identification and genetic analysis of an *Agrobacterium tumefaciens* chromosomal virulence region. *J. Bacteriol.* 161:850–60

24. Dow, J. M., Scofield, G., Trafford, K., Turner, P. C., Daniels, M. J. 1987. A gene cluster in *Xanthomonas campestris* pv. *campestris* required for pathogenicity controls the excretion of polygalacturonate lyase and other enzymes. *Physiol. Mol. Plant Pathol.* 31:261–71

25. Downie, J. A., Hombrecher, G., Ma, Q.-S., Knight, C. D., Wells, B., Johnston, A. W. B. 1983. Cloned nodulation genes of *Rhizobium leguminosarum* determine host-range specificity. *Mol. Gen. Genet.* 190:359–65

26. Downie, J. A., Johnston, A. W. B. 1986. Nodulation of legumes by *Rhizobium:* the recognized root? *Cell* 47:153–54

27. Durbin, R. D. 1981. *Toxins in Plant Disease.* New York: Academic

28. Dylan, T., Ielpi, L., Stanfield, S., Kashyap, L., Douglas, C., et al. 1986. *Rhizobium meliloti* genes required for nodule development are related to chromosomal virulence genes in *Agrobacterium tumefaciens. Proc. Natl. Acad. Sci. USA* 83:4403–7

29. Ellingboe, A. H. 1981. Changing concepts in host-pathogen genetics. *Ann. Rev. Phytopathol.* 19:125–43

30. Engstrom, P., Zambryski, P., Van Montagu, M., Stachel, S. 1987. Characterization of *Agrobacterium tumefaciens* virulence proteins induced by the plant factor acetosyringone. *J. Mol. Biol.* 197:635–45

31. Gabriel, D. W. 1986. Specificity and gene function in plant-pathogen interactions. *ASM News* 52:19–25

32. Gabriel, D. W., Burges, A., Lazo, G. R. 1986. Gene-for-gene interactions of five cloned avirulence genes from *Xanthomonas campestris* pv. *malvacearum* with specific resistance genes in cotton. *Proc. Natl. Acad. Sci. USA* 83:6415–19

33. Gardiol, A. E., Hollingsworth, R. I., Dazzo, F. B. 1987. Alteration of surface properties in a Tn5 mutant strain of *Rhizobium trifolii* 0403. *J. Bacteriol.* 169:1161–67

34. Geremia, R. A., Cavaignac, S., Zorreguieta, A., Toro, N., Olivares, J., Ugalde, R. A. 1987. A *Rhizobium meli-*

loti mutant that forms ineffective pseudonodules in alfalfa produces exopolysaccharide but fails to form β-(1-2) glucan. *J. Bacteriol.* 169:880–84

35. Halverson, L. J., Stacey, G. 1986. Signal exchange in plant-microbe interactions. *Microbiol. Rev.* 50:193–225

36. Hirooka, T., Kado, C. I. 1986. Location of the right boundary of the virulence region on *Agrobacterium tumefaciens* plasmid pTiC58 and a host-specifying gene next to the boundary. *J. Bacteriol.* 168:237–43

37. Hoekema, A., dePater, B. S., Fellinger, A. J., Hooykaas, P. J. J., Schilperoort, R. A. 1984. The limited host range of an *Agrobacterium tumefaciens* strain extended by a cytokinin gene from a wide host range T-region. *EMBO J.* 3:3043–47

38. Hooykass, P. J. J., Den Dulk-Ras, H., Regensburg-Tuink, A. J. G., van Brussel, A. A. N., Schilperoort, R. A. 1985. Expression of a *Rhizobium phaseoli* Sym plasmid in *R. trifolii* and *Agrobacterium tumefaciens:* incompatibility with a *R. trifolii* Sym plasmid. *Plasmid* 14:47–52

39. Horsch, R. B., Klee, H. J., Stachel, S., Winans, S. C., Nester, E. W., et al. 1986. Analysis of *Agrobacterium tumefaciens* virulence mutants in leaf discs. *Proc. Natl. Acad. Sci. USA* 83:2571–75

40. Horvath, B., Bachem, C. W. B., Schell, J., Kondorosi, A. 1987. Host-specific regulation of nodulation genes in *Rhizobium* is mediated by a plant signal, interacting with the *nodD* gene product. *EMBO J.* 6:841–48

41. Horvath, B., Kondorosi, E., John, M., Schmidt, J., Torok, I., et al. 1986. Organization, structure and symbiosis function of *Rhizobium meliloti* nodulation genes determining host specificity for alfalfa. *Cell* 46:335–43

41a. Huang, Y., Helgeson, J., Sequeira, L. 1987. A necrosis-inducing factor produced by *Pseudomonas solanacearum* in callus tissues or extracts of incompatible clones of *Solanum phureja. Phytopathology* 77:1736 (Abstr.)

42. Jin, S., Komari, T., Gordon, M. P., Nester, E. W. 1987. Genes responsible for the supervirulence phenotype of *Agrobacterium tumefaciens* A281. *J. Bacteriol.* 169:4417–25

43. Kearney, B., Ronald, P. C., Dahlbeck, D., Staskawicz, B. J. 1988. Molecular basis for evasion of plant host defence in bacterial spot disease of pepper. *Nature* 332:541–43

44. Keen, N. T. 1986. Phytoalexins and

their involvement in plant disease resistance. *Iowa State J. Sci.* 60:477–99

45. Kerr, A. 1972. Biological control of crown gall: seed inoculation. *J. Appl. Bacteriol.* 35:493–97

46. Knorr, D. A., Dawson, W. O. 1987. A point mutation in the tobacco mosaic virus capsid protein gene induces hypersensitivity in *Nicotiana sylvestris*. *Proc. Natl. Acad. Sci. USA* 85:170–74

47. Kobayashi, D. K., Keen, N. T. 1988. The cloning of avirulence genes from *Pseudomonas syringae* pv. *tomato* which function in *Pseudomonas syringae* pv. *glycinea* to elicit a hypersensitive reaction in soybean. *Proc. Natl. Acad. Sci. USA* In press

48. Kolattukudy, P. E. 1985. Enzymatic penetration of the plant cuticle by fungal pathogens. *Ann. Rev. Phytopathol.* 23:223–50

49. Kosslak, R. M., Bookland, R., Barkei, J., Paaren, H. E., Appelbaum, E. R. 1987. Induction of *Bradyrhizobium japonicum* common *nod* genes by isoflavones isolated from *Glycine max*. *Proc. Natl. Acad. Sci. USA* 84:7428–32

50. Kotoujansky, A. 1987. Molecular genetics of pathogenesis by soft-rot erwinias. *Ann. Rev. Phytopathol.* 25:405–30

51. Lawton, M. A., Lamb, C. J. 1987. Transcriptional activation of plant defense genes by fungal elicitor, wounding and infection. *Mol. Cell. Biol.* 7:335–41

52. Leigh, J. A., Signer, E. R., Walker, G. C. 1985. Exopolysaccharide-deficient mutants of *Rhizobium meliloti* that form ineffective nodules. *Proc. Natl. Acad. Sci. USA* 82:6231–35

53. Leroux, B., Yanofsky, M. F., Winans, S. C., Ward, J. E., Ziegler, S. F., Nester, E. W. 1987. Characterization of the *virA* locus of *Agrobacterium tumefaciens*: a transcriptional regulator and host range determinant. *EMBO J.* 6:849–56

54. Lewin, A., Rosenberg, C., Meyer, H., Wong, C. H., et al. 1987. Multiple host-specificity loci of the broad host-range *Rhizobium* sp. NGR234 selected using the widely compatible legume *Vigna unguiculata*. *Plant Mol. Biol.* 8:447–59

55. Lindgren, P. B., Peet, R. C., Panopoulos, N. J. 1986. Gene cluster of *Pseudomonas syringae* pv. "*phaseolicola*" controls pathogenicity of bean plants and hypersensitivity on nonhost plants. *J. Bacteriol.* 168:512–22

56. Loper, J. E., Kado, C. I. 1979. Host range conferred by the virulence-specifying plasmid of *Agrobacterium tumefaciens*. *J. Bacteriol.* 139:591–96

57. Malik, A. N., Vivian, A., Taylor, J. D.

1987. Isolation and partial characterization of three classes of mutant in *Pseudomonas syringae* pathovar *pisi* with altered behavior towards their host, *Pisum sativum*. *J. Gen. Microbiol.* 133:2393–99

58. Matthysse, A. G. 1987. Surface interactions between gram negative bacteria and plants. In *Molecular Determinants of Plant Diseases*, ed. S. Nishimura, C. Vance, N. Doke, pp. 97–115. Berlin: Springer-Verlag

59. Mellano, V. J., Cooksey, D. A. 1988. Development of host range mutants of *Xanthomonas campestris* pv. *translucens*. *Appl. Environ. Microbiol.* 54: 884–89

60. Meshi, T., Ishikawa, M., Motoyoshi, F., Semba, K., Okada, Y. 1986. *In vitro* transcription of infectious RNAs from full length cDNAs of tobacco mosaic virus. *Proc. Natl. Acad. Sci. USA* 83:5043–47

61. Moore, L. W., Warren, G. 1979. *Agrobacterium radiobacter* strain 84 and biological control of crown gall. *Ann. Rev. Phytopathol.* 17:163–79

62. Napoli, C., Staskawicz, B. 1987. Molecular characterization and nucleic acid sequence of an avirulence gene from race 6 of *Pseudomonas syringae* pv. *glycinea*. *J. Bacteriol.* 169:572–78

63. Niepold, F., Anderson, D., Mills, D. 1985. Cloning determinants of pathogenesis from *Pseudomonas syringae* pathovar *syringae*. *Proc. Natl. Acad. Sci. USA* 82:406–10

64. Nishimura, S., Kohmoto, K. 1983. Host-specific toxins and chemical structures from *Alternaria* species. *Ann. Rev. Phytopathol.* 21:87–116

65. Oku, H., Shiraishi, T., Ouchi, S., Ishiura, M. 1980. A new determinant of pathogenicity in plant disease. *Naturwissenschaften* 67:310–11

66. Peet, R. C., Lindgren, P. B., Willis, D. K., Panopoulos, N. J. 1986. Identification and cloning of genes involved in phaseolotoxin production by *Pseudomonas syringae* pv. "*phaseolicola*." *J. Bacteriol.* 166:1096–105

67. Peters, N. K., Frost, J. W., Long, S. R. 1986. A plant flavone, luteolin, induces expression of *Rhizobium meliloti* nodulation genes. *Science* 233:977–80

68. Porter, S. G., Yanofsky, M. F., Nester, E. W. 1987. Molecular characterization of the *virD* operon from *Agrobacterium tumefaciens*. *Nucleic Acids Res.* 15: 7503–18

69. Puvanesarajah, V., Schell, F., Gerhold, D., Stacey, G. 1987. Surface polysaccharides from *Bradyrhizobium japoni-*

cum and a nonnodulating mutant. *J. Bacteriol.* 169:137–41

70. Puvanesarajah, V., Schell, F. M., Stacey, G., Douglas, C. J., Nester, E. W. 1985. Role for 2-linked β-D-glucan in the virulence of *Agrobacterium tumefaciens*. *J. Bacteriol.* 164:102–6

71. Redmond, J. W., Batley, M., Djordjevic, M. A., Innes, R. W., Kuempel, P. L., Rolfe, B. G. 1986. Flavones induce expression of nodulation genes in *Rhizobium*. *Nature* 323:632–35

72. Ronson, C. W., Nixon, B. T., Ausubel, F. M. 1987. Conserved domains in bacterial regulatory proteins that respond to environmental stress. *Cell* 49:579–81

73. Saito, T., Meshi, T., Takamatsu, N., Okado, Y. 1987. Coat protein gene sequence of tobacco mosaic virus encodes a host response determinant. *Proc. Natl. Acad. Sci. USA* 84:6074–77

74. Scheffer, R. P., Nelson, R. R., Ullstrup, A. J. 1967. Inheritance of toxin production and pathogenicity in *Cochliobolus carbonum* and *Cochliobolus victoriae*. *Phytopathology* 57:1288–91

75. Schoelz, J., Shepherd, R. J., Daubert, S. 1986. Region VI of cauliflower mosaic virus encodes a host range determinant. *Mol. Cell. Biol.* 6:2632–37

76. Schofield, P. R., Watson, J. M. 1986. DNA sequence of *Rhizobium trifolii* nodulation genes reveals a reiterated and potentially regulatory sequence preceding *nodABC* and *nodFE*. *Nucleic Acids Res.* 14:2891–903

77. Sharp, J. K., McNeil, M., Albersheim, P. 1984. The primary structures of one elicitor-active and seven elicitor-inactive hexa(β-D-glucopyranosyl)-D-glucitols isolated from the mycelial walls of *Phytophthora megasperma* f. sp. *glycinea*. *J. Biol. Chem.* 259:11321–36

78. Shearman, C. A., Rossen, L., Johnston, A. W. B., Downie, J. A. 1986. The *Rhizobium leguminosarum* nodulation gene *nodF* encodes a polypeptide similar to acyl-carrier protein and is regulated by *nodD* plus a factor in pea root exudate. *EMBO J.* 5:647–52

79. Somlyai, G., Hevesi, M., Banfalvi, Z., Klement, Z., Kondorosi, A. 1986. Isolation and characterization of nonpathogenic and reduced virulence mutants of *Pseudomonas syringae* pv. *phaseolicola* induced by Tn5 transposon insertions. *Physiol. Mol. Plant Pathol.* 29:369–80

80. Spaink, H. P., Wijffelman, C. A., Pees, E., Okker, R. J. H., Lugtenberg, B. J. J. 1987. *Rhizobium* nodulation gene *nodD* as a determinant of host specificity. *Nature* 328:337–40

81. Stachel, S. E., Messens, E., Van Montagu, M., Zambryski, P. 1985. Identification of the signal molecules produced by wounded plant cells that activate T-DNA transfer in *Agrobacterium tumefaciens*. *Nature* 318:624–29

82. Stachel, S. E., Zambryski, P. C. 1986. *Agrobacterium tumefaciens* and the susceptible plant cell: a novel adaptation of extracellular recognition and DNA conjugation. *Cell* 47:155–57

83. Stall, R. E., Loschke, D. C., Jones, J. B. 1986. Linkage of copper resistance and avirulence loci on a self-transmissible plasmid in *Xanthomonas campestris* pv. *vesicatoria*. *Phytopathology* 76:240–43

84. Staskawicz, B. J., Dahlbeck, D., Keen, N. T. 1984. Cloned avirulence gene of *Pseudomonas syringae* pv. *glycinea* determines race-specific incompatibility on *Glycine max* (L.) Merr. *Proc. Natl. Acad. Sci. USA* 81:6024–28

85. Staskawicz, B., Dahlbeck, D., Keen, N., Napoli, C. 1987. Molecular characterization of cloned avirulence genes from race 0 and race 1 of *Pseudomonas syringae* pv. *glycinea*. *J. Bacteriol.* 169:5789–94

86. Swanson, J., Kearney, B., Dahlbeck, D., Staskawicz, B. 1988. Cloned avirulence gene of *Xanthomonas campestris* pv. *vesicatoria* complements spontaneous race-change mutants. *Mol. Plant-Microbe Interact.* 1:5–9

87. Tamaki, S., Dahlbeck, D., Staskawicz, B., Keen, N. T. 1988. Characterization and expression of two avirulence genes cloned from *Pseudomonas syringae* pv. *glycinea*. *J. Bacteriol.* In press

88. Tepper, C. S., Anderson, A. J. 1986. Two cultivars of bean display a differential response to extracellular components from *Colletotrichum lindemuthianum*. *Physiol. Mol. Plant Pathol.* 29:411–20

89. Tumer, N. E., O'Connell, K. M., Nelson, R. S., Sanders, P. R., Beachy, R. N., et al. 1987. Expression of alfalfa mosaic virus coat protein gene confers cross-protection in transgenic tobacco and tomato plants. *EMBO J.* 6:1181–88

90. Turgeon, B. G., Garber, R. C., Yoder, O. C. 1987. Development of a fungal transformation system based on selection of sequences with promoter activity. *Mol. Cell. Biol.* 7:3297–305

91. Turner, P., Barber, C., Daniels, M. 1985. Evidence for clustered pathogenicity genes in *Xanthomonas campestris* pv. *campestris*. *Mol. Gen. Genet.* 199:338–43

92. Usami, S., Morikawa, S., Takebe, I., Machida, Y. 1987. Absence in mono-

cotyledonous plants of the diffusible plant factors inducing T-DNA circularization and *vir* gene expression in *Agrobacterium. Mol. Gen. Genet.* 209: 221–26

93. VanEtten, H. D., Berg, W. 1981. Expression of pisatin demethylating ability in *Nectria haematococca. Arch. Microbiol.* 129:56–60

94. Van Larebeke, N., Engler, G., Holsters, M., Van den Elsacker, S., Zaenen, I., et al. 1974. Large plasmid in *Agrobacterium tumefaciens* essential for crown gall–inducing ability. *Nature* 252:169–70

95. van Veen, R. J. M., Den Dulk-Ras, H., Schilperoort, R. A., Hooykaas, P. J. J. 1987. Chromosomal nodulation genes: sym-plasmid containing *Agrobacterium* strains need chromosomal virulence genes (*chvA* and *chvB*) for nodulation. *Plant. Mol. Biol.* 8:105–8

96. Veluthambi, K., Jayaswal, R. K., Gelvin, S. B. 1987. Virulence genes A, G, and D mediate the double-stranded border cleavage of T-DNA from the *Agrobacterium* Ti plasmid. *Proc. Natl. Acad. Sci. USA* 84:1881–85

97. Weiler, E. W., Schroder, J. 1987. Hormone genes and crown gall disease. *Trends Biochem. Sci.* 12:271–76

98. Werner, D., Mellor, R. B., Hahn, M. G., Grisebach, H. 1985. Soybean root response to symbiotic infection. Glyceollin I accumulation in an in-effective type of soybean nodules with an early loss of the peribacteroid membrane. *Z. Naturforsch. Teil C* 40:179–81

99. Winans, S. C., Ebert, P. R., Stachel, S. E., Gordon, M. P., Nester, E. W. 1986. A gene essential for *Agrobacterium* virulence is homologous to a family of positive regulatory loci. *Proc. Natl. Acad. Sci. USA* 83:8278–82

100. Yamada, T., Palm, C. J., Brooks, B., Kosuge, T. 1985. Nucleotide sequences of the *Pseudomonas savastanoi* indoleacetic acid genes show homology with *Agrobacterium tumefaciens* T-DNA. *Proc. Natl. Acad. Sci. USA* 82: 6522–26

101. Yanofsky, M., Lowe, B., Montoya, A., Rubin, R., Krul, W., et al. 1985. Molecular and genetic analysis of factors controlling host range in *Agrobacterium tumefaciens. Mol. Gen. Genet.* 201: 237–46

102. Yanofsky, M. F., Nester, E. W. 1986. Molecular characterization of a host-range-determining locus from *Agrobacterium tumefaciens. J. Bacteriol.* 168: 244–50

103. Zaat, S. A. J., Wijffelman, C. A., Spaink, H. P., Van Brusse, A. A. N., Okker, R. J. H., Lugtenberg, S. B. J. J. 1987. Induction of the *nodA* promoter of *Rhizobium leguminosarum* sym plasmid pRL1JI by plant flavanones and flavones. *J. Bacteriol.* 169:198–204

Ann. Rev. Microbiol. 1988. 42:441–64

MICROBIAL ECOLOGY OF THE SKIN[1],[2]

Rudolf R. Roth and William D. James

Dermatology Service, Department of Medicine, Walter Reed Army Medical Center, Washington, DC 20307-5001

CONTENTS

INTRODUCTION

Normal human skin is colonized by large numbers of organisms that live harmlessly as commensals on its surface. Of the many different species of microorganisms found in nature, only a few species are found repeatedly on the skin of groups of individuals. This is surprising considering the large variety of organisms found in other areas such as the gastrointestinal tract. These frequently found organisms constitute the resident flora of the skin.

Price (89), in his original discussion of skin bacteria, classified the microflora into two groups. Those organisms that grew on the skin and were relatively stable in number and composition he called resident flora. He proposed that these organisms were attached to the skin and that their hardiness might be related to this attachment. Increases in the number of resident flora were believed to be due to multiplication of organisms already present and not due to additions from exogenous sources. A second group of organisms, proposed to lie free on the skin surface, he called transient flora. They were believed to be derived from exogenous sources and were found primarily on exposed skin.

The resident organisms live as small microcolonies on the surface of the stratum corneum and within the outermost layers of the epidermis. Flora from any one area of the skin cannot be taken as representative of the entire flora; samples must be taken from multiple sites to determine the floral composition. Different methods of harvesting these organisms recover them from varying depths and therefore yield different quantitative results (47). The type of media used (81), as well as the area of the body tested (99), can cause differences in quantitative tests. In spite of these differences, most studies have revealed similar if not identical qualitative results.

CLASSIFICATION OF ORGANISMS

Micrococcaceae

Micrococci and staphylococci are gram-positive, catalase-positive cocci in the family Micrococcaceae. Baird-Parker (12) made the initial subclassification of these organisms based on the ability of staphylococci and inability of micrococci to produce acid from glucose under anaerobic conditions. DNA homology and cell wall analysis of these organisms revealed discrepancies with this classification scheme, so most microbiologists presently subclassify these organisms according to criteria established by Schleifer & Kloos (96). This separation is based on the ability of staphylococci to produce acid anaerobically from glycerol in the presence of erythromycin (0.4 μg ml^{-1}) and on their susceptibility to lysostaphin and nitrofuran. Staphylococci are divided into the coagulase-positive *Staphylococcus aureus* and coagulase-negative species. Humans have a high degree of natural resistance to skin colonization by *S. aureus,* although the organism can be found in the perineum of 20% of individuals (14). It is most often found as part of the transient flora of normal skin or the nasopharynx. Persistent nasal carriage of the organism is present in 20–40% of normal adults (38) and can lead to persistent skin colonization and recurrent infection. Some groups are more susceptible to *S. aureus* colonization of the skin, including hospital workers (110), patients on hemodialysis, diabetics, and intravenous drug abusers (16). In patients with dermatitis such as psoriasis (8) and atopic dermatitis (113), *S. aureus* may be found widely over the skin surface, over both diseased and normal skin, often constituting up to 80% of the normal flora.

Coagulase-negative staphylococci are the most frequently found organisms of the normal flora. At least 18 different species have been isolated from normal skin (46). *Staphylococcus epidermidis* and *Staphylococcus hominis* are the species recovered most frequently. *S. epidermidis* colonizes the upper part of the body preferentially (82) and constitutes over 50% of the resident staphylococci (66). Following these in prevalence are several species with nearly identical DNA base pairings: *Staphylococcus haemolyticus, Staphylococcus capitis,* and *Staphylococcus warneri* (61). *Staphylococcus saprophyticus* is often found as a resident in the perineum.

Peptococcus saccharolyticus, also called *Staphylococcus saccharolyticus,* is a strict anaerobic staphylococcus and a member of the normal flora in 20% of individuals. It may be present in large numbers on the forehead and antecubital fossa, and it is even the primary organism of the flora of these locations in some individuals (28).

Although less frequently seen than the staphylococci, at least eight different *Micrococcus* species have been identified from human skin (60). *Micrococcus luteus* is by far the most common of the micrococci seen, followed by *Micrococcus varians* (81). *Micrococcus lylae* and *Micrococcus kristinae* may be more significant in children (81), and *M. lylae* is more frequently seen in the cold months (61).

Coryneform Organisms

Coryneform organisms are gram-positive pleomorphic rods that were originally all thought to be *Corynebacterium* species. These organisms were poorly classified in the past; they were simply divided into lipophilic organisms, which require lipid supplements for growth in artificial media, and nonlipophilic organisms. More recently, attempts have been made to classify the organisms on the basis of cell wall analysis (86). Based on the presence and type of amino acids, mycolic acid, and neutral sugars, only 60% of cutaneous coryneform organisms were found to belong to the classical *Corynebacterium* genus. Another 20% were classified as *Brevibacterium* species, organisms that had previously been identified in milk. The other 20% of organisms fell into various other groups according to cell wall analysis; most are transients since they are found primarily on exposed areas.

Classical *Corynebacterium* species compose a significant part of the normal flora, particularly in moist intertriginous areas. Most of these lipophilic organisms are distinct from known coryneform species according to cell wall analysis (74); McGinley et al (74) have suggested the name *C. lipophilicus* for them. These organisms were found in over 50% of the toe-web spaces of Danish military recruits, with the incidence greater in those with hyperhidrosis (107). *C. minutissimum,* once thought to be a single organism distinguished by the ability to produce porphyrin, now appears to be a complex of as many as eight different species (105).

Group JK coryneforms are organisms that have acquired resistance to nearly all antibiotics except vancomycin (115). Studies of their cell walls have shown that they are nearly identical to normal cutaneous lipophilic coryneforms in all parameters except antibiotic resistance (75). They are found on the the skin of normal individuals, but are found more frequently in immunosuppressed hosts (115). They are thought to colonize the intertriginous areas preferentially, and can be found in up to 35% of patients in hospitals (6). There is an inverse correlation between the number of antibiotic-sensitive lipophilic coryneforms that colonize the skin and the number of antibiotic-resistant JK bacteria (65). Although the latter are sensitive to vancomycin in vitro, systemic use of vancomycin neither prevents nor controls their growth on the skin.

Brevibacterium species produce proteolytic enzymes, are penicillin resistant, and are probably the most rapidly growing of the coryneforms (87). They are frequently isolated from the toe webs of patients, especially in patients with tinea pedis, and they are implicated in foot odor (51).

Propionibacteria

Propionibacterium species are non-spore-forming, anaerobic, gram-positive bacteria that are normal inhabitants of hair follicles and sebaceous glands. They are the most prevalent anaerobes of the normal flora. Based on colony morphology and susceptibility to lysis by bacteriophages, they have been divided into three species. *P. acnes* is most numerous on the skin of the scalp, forehead, and back, and is the most predominant species by far (82). The follicular density of *P. acnes* reaches its peak level around puberty and remains relatively constant through adult life. *P. granulosum,* next in prevalence and composing nearly 20% of the propionibacteria, is found in small numbers at all tested sites. *P. avidum* strains are found most often in intertriginous areas, especially the axilla (82).

Gram-Negative Rods

Gram-negative rods are rarely found on the skin, probably because of the skin's desiccation (71). Transient organisms are frequently found as contaminants from the gastrointestinal system. These occasionally become resident flora in moist intertriginous areas and mucosal surfaces such as the axilla, toe webs, or nose of some individuals (81).

Acinetobacter species are nonfermentative aerobic gram-negative rods that are widely dispersed in nature and are found in up to 25% of individuals as normal flora (91). They include species that were formerly referred to as members of the genera *Mima* and *Herellea*. Males are more frequently colonized than females, and there is a significant increase in colonization during the summer (61). These trends are probably secondary to amounts of perspiration, since a high moisture content is probably necessary for survival of these bacteria.

Mycoflora

In the normal human flora, fungi are regularly present. Surveys of the resident cutaneous mycoflora have revealed a predominance of yeast organisms, both in temperate (76) and tropical (77) environments.

Pityrosporum species are lipophilic yeasts that require lipids for growth. Their in vitro growth must be accomplished on olive oil–enriched media. *Pityrosporum ovale* and *Pityrosporum orbiculare* are probably identical organisms, which are prominent in sebaceous areas. In the normal flora they exist in the blastospore form. In quantitative testing, the organisms are most numerous on the back and chest, with highest numbers paralleling areas of highest sebum excretion (31).

Candida species, normally found on up to 40% of mucous membranes (102), seldom colonize the normal skin. When it is present, *C. albicans* is the most common species, existing in the blastospore form. Increased colonization of skin by *Candida* species is seen in immunosuppressed patients, diabetics, and patients with psoriasis or atopic dermatitis (81).

Transient Flora

In addition to some of the species listed above that may be present as transient flora, numerous other organisms may be found temporarily on the skin surface. Any organism that is found in nature or that belongs to the resident flora in noncutaneous areas of the body may be found on the skin transiently. Perhaps the most important of these clinically are group A streptococci, which only rarely colonize the skin because they die rapidly when placed on normal intact skin (69). Streptococci have been recovered from clinically normal skin in children for up to 10 days prior to the development of impetigo, however (39). Recovery of these bacteria on normal skin was associated with a high risk (76%) of subsequent development of lesions due to the same serotype.

FACTORS MODIFYING THE NORMAL FLORA

Although the resident flora remains relatively constant, a number of factors can change the quantity of organisms and the relative percentages of each organism in the normal flora. These factors can be either endogenous to the individual or secondary to environmental or bacterial influences.

Climate

The resident flora can be influenced by their microenvironment. Increased temperature and humidity increase the density of bacterial colonization and alter the relative ratios of organisms. Artificially applied organisms survived longer on wetted skin than on dry skin (90). These environmental changes have been experimentally reproduced by applying occlusive materials on forearm skin. After 24 hr of occlusion, bacterial counts had increased 10,000-fold and the relative numbers of gram-negative rods and coryneform bacteria had increased over coccal forms (7). Once the occlusive material was removed, the numbers of bacteria decreased toward normal amounts, but only very slowly (48). The effect of experimental occlusion is similar to the localized proliferation of the resident flora in areas of the body normally under occlusion, such as the toe webs. Studies have shown that the increases in both temperature and humidity are necessary to duplicate these results, and that neither alone will cause as significant a change in the normal flora (24).

Any decrease in ambient temperature or humidity would be expected to cause decreased colonization. People tend to wear more clothing as it gets colder, however, thus keeping the microenvironment warm and humid and therefore negating any inhibitory effect of the external environment. Nonpathogenic staphylococci have a definite ecological advantage over pathogenic staphylococci when the temperature decreases (110), so virulent infections decrease when the microenvironmental temperature decreases.

Minor differences in the resident organisms in different geographical locations have been reported. The species of coagulase-negative staphylococci seen in patients living in North Carolina differed from those of patients in London (81) and New Jersey (61). Whether these differences are due solely to variables in climate or to other local differences is unknown.

Body Location

The composition of the normal flora varies depending on body location. The face, neck, and hands are exposed areas of the body and as a result may have a higher proportion of transient organisms and a higher bacterial density. The head and upper trunk have a higher number of sebaceous glands and a greater number of lipophilic organisms. The axilla, perineum, and toe webs are partially occluded areas with an increased temperature and moisture level. These areas are colonized more heavily with all organisms, but particularly with those organisms, such as gram-negative rods or coryneforms, that need moisture for survival. The upper arms and legs are relatively dry and often have lower bacterial counts. When the ecological conditions of an area are changed, as with occlusion of a dry skin surface, the flora change to adapt to the new environment.

Hospitalization

Various studies have shown significant differences between the flora of healthy individuals and of hospitalized patients. One of these studies showed that hospitalized patients have increased colonization with pathogenic and antibiotic-resistant organisms (64). The changes in the resident flora of these individuals help to explain the patients' propensity to develop nosocomial infections with the more aggressive organisms.

Age

The age of the individual has a profound influence on the microflora (104). The flora is most varied in young children, who carry micrococci, coryneform bacteria, and gram-negative organisms more frequently and in larger numbers than older children and adults. Infants also carry a higher proportion of pathogens or potential pathogens on their skin.

Pityrosporum and *Propionibacterium* species, on the other hand, are present at much lower levels before puberty. These organisms require higher skin lipid levels, and their appearance parallels age-related changes in sebum production. *Pityrosporum orbiculare,* for example, is rare in children under 5 yr of age and becomes increasingly established in individuals over the next 10 yr, achieving adult levels by age 15 (32).

NEONATAL MICROFLORA   Colonization of the skin begins at birth. The skin of babies born by Caesarean section is sterile, becoming colonized on first contact with the outside world (93), whereas babies born by vaginal delivery are already colonized by organisms encountered in the birth canal prior to delivery. The organisms that are found at birth are usually only present in small numbers, except for *Staphylococcus epidermidis,* which is the predominant organism of the vaginal flora just prior to birth. Coryneform bacteria are also prominent in the resident flora of the newborn, but unlike *S. epidermidis,* they take several hours after delivery to establish themselves.

Before these organisms become fully established, however, other organisms may readily colonize the skin surface, such as gram-negative rods and streptococcal organisms. *Staphylococcus aureus* infection is much more common in the newborn, colonizing the nasopharynx and umbilicus of many infants. Despite the readiness of these pathogenic organisms to colonize the skin, the infant's flora begins to resemble that of the adult after the first few weeks (71).

Sex

There is evidence that males carry higher absolute numbers of organisms, as well as more biotypes (72, 84). This may be due to higher sweat production in men, as well as to the tendency of men to wear more occlusive clothing.

Race

Various reports have distinguished differences between the races in some areas, such as a higher nasal carriage of *S. aureus* in whites (80), an earlier appearance of *P. acnes* in blacks (73), and fewer cutaneous streptococcal infections (5) or neonatal streptococcal infections (13) in blacks. Differences in carriage rates of organisms may be due to differences in HLA antigen expression (59), adhesion, or environmental conditions.

Occupation

Just as other factors in the environment can influence the pattern of the resident flora, occupation can also have an effect. Those who work in environments that have high temperature and humidity, for example, might develop microflora patterns similar to those seen in occlusion studies. Hospital workers have been shown to harbor more pathogenic organisms as transient organisms, which may become established as resident flora if the individuals are consistently exposed (89).

Soaps and Detergents

Repeated washing with soap makes the skin more alkaline than washing with medicated detergents (62). Neither of these products significantly altered the count of coagulase-negative staphylococci, but propionibacteria were markedly increased when soap was used and depressed in the presence of medicated detergents. Preoperative disinfection by shower bath with chlorhexidine soap effectively decreased the incidence of postoperative infections caused by *S. aureus* from 8 to 2% (18). The medicated soap significantly reduced numbers of aerobic bacteria, while the control, a nonmedicated soap, gave an increase in measured organisms. In a related observation, Price (89) found that scrubbing the hands without soap decreased the numbers of bacteria faster than washing with nonmedicated soap.

Medications

Of all exogenous influences, drugs cause the most rapid and radical changes in the normal flora. Antibiotics may suppress the normal flora and increase colonization by other organisms. In addition to destroying the flora, antibiotics may impair bacterial adherence to epithelial cells (26) and allow for the natural selection of other organisms such as gram-negative rods (54).

Oral retinoids, such as 13-*cis*-retinoic acid, cause a decrease in sebum excretion and a significant decrease in the numbers of *P. acnes* (67), which persists even after treatment is discontinued and sebum excretion returns to pretreatment levels (68). Retinoids cause drying of the mucous membranes, which results in a decrease in the total number of organisms. The number of gram-negative rods decreases significantly, although *S. aureus* recovery from the anterior nares is increased (54).

Oral steroids and hormones are also associated with changes in the normal flora. Corticosteroids have a suppressive effect at each stage of the immune response and can increase susceptibility to a variety of bacterial, fungal, viral, and parasitic infections. Females on estrogen therapy experience an increase in vaginal as well as cutaneous candidiasis.

Topical medications as well as systemic drugs can change the composition of the normal flora. Treatment of the axilla with topical neomycin resulted in a decrease in the total microflora (99). After continued treatment there was an increase in the number of gram-negative organisms, which became the dominant flora, and there were marked decreases in the coryneform population. Even after the topical antibiotic was stopped and the resident flora returned, the gram-negative organisms were slightly more numerous than before treatment (99). Topical steroids, on the other hand, have little effect on the numbers or types of cutaneous microflora (20). On eczematous skin, topical steroids actually decrease bacterial counts, probably by healing the dermatitis (79).

Ultraviolet Light

Although no statistically significant change was seen in the normal flora after psoralen plus ultraviolet A (PUVA) therapy for psoriasis (112), ultraviolet B (UVB) has inhibited bacterial growth (36). *Pityrosporum* and *Candida* species are more sensitive than staphylococci.

Bacterial Adherence

For organisms to colonize the skin, they must first become attached to the epithelial surface (58). The capacity to colonize is proportional to the ability of the organism to adhere to the surface (41). Adherence involves the attraction of a specific molecular structure on the cell wall of the bacterium, called an adhesin, and a specific receptor on the host cell surface. Adhesins are microbial surface antigens, or lectins, often in the form of filamentous projections. The adhesin functions as a bridge between the microbe and the host cell. The binding between adhesin and receptor is virtually irreversible. Epithelial cells from different anatomic systems, such as the skin and the gastrointestinal tract, have striking variability in adherence of bacteria, which helps to explain the differences in resident microflora in different locations on the body (41). Group A streptococci isolated from skin adhere much better to skin epithelium than to buccal mucosa, whereas streptococci isolated from the oral cavity bind better to the buccal mucosa (2). The inability of certain organisms, such as viridans streptococci and C. albicans (11), to attach to normal unbroken skin may explain why they are so rarely found as resident organisms.

Teichoic acids are cell wall components of staphylococci and streptococci that are thought to be skin epithelium adhesins for those organisms. Epithelial cells treated with teichoic acid competitively bind with epithelial receptors and inhibit the binding of S. aureus (11). The receptor on the epithelial cell for these adhesins is thought to be fibronectin (63), which binds the adhesin at at least two separate binding sites. Soluble fibronectin inhibits the adherence of streptococci to epithelial cells (37). The binding of S. aureus to fibronectin is thought to be time dependent and irreversible, and it occurs on both live and heat-killed cells (92).

The inability of streptococci and S. aureus to colonize the unbroken epithelium may be due to the absence of available fibronectin on the skin surface. In dermatitis, microscopic breaks of the skin surface may uncover fibronectin receptors and increase the adherence of S. aureus to both normal and abnormal skin (21).

Pathogenic bacteria have higher potential for adherence to the host, giving them greater virulence (41). The enhanced potential may be due to multiple adhesins, which would increase the chance for adherence and give the organism a selective advantage over other organisms. Studies with S. aureus have found that protein A in the cell wall is another possible adhesin (21). Additional adhesins might allow the organism to bind to additional receptors. S. aureus has been found to bind to fibrinogen as well as fibronectin (58). Patients who tend to have increased colonization by pathogens might also have a greater proclivity for adherence of those organisms. Skin cells from atopic patients, for example, have greater adherence for S. aureus than skin cells from either normal individuals or psoriatic patients (17). The propensity for bacterial adhesion in atopic patients may be due to additional receptors for S. aureus or more numerous primary receptors available.

Pityrosporum species have been found to selectively adhere to stratum corneum cells (30). The number of organisms that can attach per cell is lower for these yeasts than for bacteria because the former are larger. Propulsive forces between yeast cells on the stratum corneum surface and the availability of binding sites may also influence adherence. Adherence of these organisms increases with increased binding time and elevated skin temperature.

Viral infection causes changes on the cell surface, altering existing cell receptors and forming new potential receptors for bacteria. Acutely ill patients may have increased adherence of unusual organisms such as *Pseudomonas*, perhaps because of changes on the epithelial surface (1).

Knowledge of the importance of bacterial adherence might affect the treatment of bacterial infection. Application of purified bacterial adhesin or membrane receptor might selectively block the adherence of specific organisms by competitive inhibition. Vaccines could also be developed against specific microbial adhesins to prevent adherence.

NATURAL RESISTANCE OF THE SKIN

Host Defense

Normal skin is resistant to colonization and invasion by most bacteria. The presence of bacteria on the surface does not make infection inevitable. Many factors prevent colonization and invasion of pathogenic organisms.

INTACT STRATUM CORNEUM The most important feature offering protection of the skin from invasion is an intact stratum corneum. Overlapping cells, joined by modified desmosomes, function as an armor against organisms. In addition, the relative dryness of intact skin limits the growth of organisms that require moisture, such as *Candida* species. Cracks and crevices in the armor still exist, however, and appendages may offer a route for infection by some organisms, such as *S. aureus*.

Experimental production of localized infection with pathogenic organisms is very difficult if the skin is intact. Leyden (69) found that a break in the stratum corneum was an absolute requirement for inducing a streptococcal infection. The presence of a single silk suture through the skin increased the infectivity of *S. aureus* by a factor of 10,000 (27). *S. aureus* colonization of patients with dermatitis was also directly related to the degree of epidermal change (79).

RAPID CELL TURNOVER Another feature of the stratum corneum that offers protection from invasion is its rapid turnover; the transit time is only 14 days. If pathogenic organisms adhere, they have a limited time to invade. The resident flora are better suited for reattachment after desquamation than the weakly adhering transient organisms (58).

LIPID LAYER   Many components of the resident microflora have lipase activity and liberate fatty acids, such as oleic, stearic, or palmitic acid, from the triglycerides of sebum. These acids create an acid mantle on the surface of the skin, which has a potent antimicrobial effect on *S. aureus* and streptococcal organisms and a stimulating effect on some other organisms such as *Propionibacterium* species. Propionibacteria also produce propionic acid, which has an antimicrobial effect on many organisms. At the mean pH value of normal skin, 5.5, this antimicrobial effect is much more selective against transient organisms than against resident flora. Occluded areas of the body, such as intertriginous areas, have a neutral or slightly alkaline pH, owing to the diluting effect of the skin's secretions, and become densely populated with microorganisms.

Factors that change the composition of the lipids or the pH of skin may affect the normal flora. Occlusion of forearm skin not only increased the growth of resident organisms, but also increased the pH from 4.9 to a relatively alkaline 7.1 (48). When the skin surface layer was stripped with acetone, pathogenic organisms were able to colonize (10); antimicrobial activity was restored when the skin surface lipids were replaced.

IMMUNE SYSTEM   The skin has both humoral and cellular immune systems which can influence the composition of the microbial communities. In normal skin, the humoral mechanisms involve the local production and secretion of immunoglobulins. Antibacterial antibodies occur in the secretions of the normal patient. On the skin, these are primarily IgA antibodies which are brought to the surface through the eccrine system. The secretory IgA in sweat may prevent infection by a number of possible mechanisms, and its presence may explain the absence of colonization in the eccrine duct.

Elements of cellular immunity in the epidermis include antigen presentation by Langerhans cells and T-cell activation by epidermal thymocyte-activating factor, a product of epidermal cells. The skin may also have a role in T-cell differentiation. These mechanisms are vital in preventing cutaneous infection. Patients with defects in cellular immunity are more susceptible to infection, whether they have diseases that cause limited defects, such as chronic mucocutaneous candidiasis, or diseases that cause more widespread defects, such as the acquired immunodeficiency syndrome.

Role of the Organism

ANTIBACTERIAL SUBSTANCES   Many microorganisms produce protein or protein-complex antibiotics that have an antagonistic effect on other organisms, but not on the producer bacterium. Substances produced by gram-negative bacteria generally have a wide range of antibacterial activity, while those produced by gram-positive organisms are usually effective only against strains of the same or closely related species. These latter substances, those with a narrow spectrum of activity, are called bacteriocins.

In the resident flora, cyclic peptide bacteriocins are produced by the coagulase-negative staphylococci and, to a lesser extent, by coryneform bacteria (82). Their activity is lethal for closely related organisms and is initiated by adsorption to specific outer membrane receptors (44). Bacteriocin producers may be found in up to 20–25% of individuals (98), but bacteriocins, if present, are usually only produced by less than 5% of the organisms (82). When skin becomes diseased, however, these bacteriocin producers increase in number and become the predominant members of the flora (98). Dermatitic patients that have organisms producing bacteriocins have a significantly decreased incidence of secondary infection.

Dermatophytes on the skin can also produce antibiotics, including penicillin and other substances with antibacterial and antifungal activity. Penicillin production can suppress the bacterial flora in fungal lesions, but also tends to select a penicillin-resistant flora (82).

BACTERIAL INTERFERENCE   A resident bacterium sometimes prevents colonization by another strain of a similar species. This most likely occurs by competitive inhibition of binding sites. The best example is the inhibition of colonization by other strains of *S. aureus* when a nonvirulent strain, *S. aureus* 502A, colonizes the skin or anterior nares. This strain has been implanted artificially to prevent recurrent furunculosis, to eradicate persistent nasal carriage, and to prevent life-threatening nursery epidemics. Similarly, commensal staphylococci inhibit colonization with *S. aureus,* and only when they are removed can *S. aureus* colonize forearm skin (103). The protective role of the resident staphylococcal flora is also demonstrated by its absence in the newborn; during the first week of life pathogenic organisms such as *S. aureus* frequently colonize the skin because of the absence of an established resident microflora.

OTHER INTERACTIONS   Certain resident organisms also exercise antagonistic inhibition over other residents that are of dissimilar species. The presence of gram-positive organisms in the axilla, for example, exerts a restraining influence on the number of gram-negative organisms there. Suppression of the gram-positive organisms by topical antibiotics causes a corresponding proliferation of gram-negative bacteria (99). A reciprocal relationship between carriage of *S. aureus* and gram-negative organisms in the nose has also been seen (54).

The large numbers of *S. aureus* in patients with atopic dermatitis seem to eliminate the lipophilic coryneform bacteria from the flora (9, 79). Similarly,

Jackman & Noble (52) showed that large numbers of staphylococci in the axilla tended to decrease the numbers of coryneform bacteria, and vice versa. Staphylococci also displace coryneform bacteria on the plaques of psoriasis (8).

The development of candidiasis after suppression of the normal flora suggests that the inhibitory function of the normal flora has an influence on fungal organisms as well. The presence of normal oral microflora was found to block the adherence of *C. albicans* to host epithelium (70). When the normal flora was suppressed with oral antibiotics, on the other hand, *C. albicans* quickly colonized (49).

RESIDENT FLORA AS PATHOGENS

Micrococci

Micrococcus species are usually considered contaminants when isolated from clinical specimens. *M. luteus,* the most prevalent species of the genus, has occasionally been implicated in bacteremia, pneumonia, septic arthritis, and meningitis. Other micrococci have rarely been implicated in infections.

Staphylococci

Coagulase-negative staphylococci are generally of low pathogenicity, but under certain circumstances the organisms can cause severe and even life-threatening infection. Most infections are hospital acquired; they are transmitted by hospital staff colonized with antibiotic-resistant organisms. The most common infections are associated with indwelling foreign devices such as artificial heart valves, indwelling catheters, CNS shunts, prosthetic joints, breast implants, and vascular grafts. *S. epidermidis* is the most common cause of these infections, and *S. haemolyticus* is the second most common pathogen (42). The organisms most likely gain access to the device by direct inoculation at the time of insertion. The ability of these organisms to adhere and multiply on foreign bodies is mediated by the production of a mucopolysaccharide-like substance, called slime, which enhances adherence to the smooth surface (46). Over 90% of indwelling catheters in place for over 100 days are colonized by slime-producing *S. epidermidis* (109), although the bacteria only rarely cause septicemia. The presence of slime appears to contribute to the virulence of these organisms in the presence of foreign bodies, drastically reducing the lymphoproliferative response of mononuclear cells (45), interfering with granulocyte chemotaxis and phagocytosis (56), and decreasing bactericidal activity of antibiotics, perhaps by sequestering the organism from its action (22). Suppression of the host response may not only allow the proliferation of *S. epidermidis,* but may also increase the possibility for other opportunistic infections.

Infection of native cardiac valves with coagulase-negative staphylococci can also occur, accounting for 5% of all bacterial endocarditis. Infection occurs by seeding of the valves with the organism following transient bacteremia.

Coagulase-negative staphylococci can also cause urinary tract infections. *S. saprophyticus,* occasionally found as a resident in the perineum, can cause symptomatic infection in outpatient females, usually in the sexually active years. *S. epidermidis,* on the other hand, is more frequently isolated from older patients with hospital-acquired infections secondary to an indwelling urinary catheter.

S. epidermidis is also the most frequent cause of bacteremia in immunosuppressed patients (116). The risk of bacteremia in a hospitalized patient is also increased in the presence of neutropenia, multiple trauma, malnutrition, or prior abdominal surgery (46). Septicemia in neonates is much more common in premature infants of low birth weight (40).

STAPHYLOCOCCUS AUREUS INFECTIONS   Individuals that carry *S. aureus* on their skin or anterior nares are particularly vulnerable to infection if their normal protective skin barrier is broken. Individuals with persistent nasal carriage, for example, often note recurrent small furuncles on the upper lip. Recurrent staphylococcal furunculosis is usually associated with nasal carriage of the same phage type (14, 38). Nasal carriers also have more postoperative infections than controls (110). Individuals with increased skin colonization of *S. aureus,* such as atopic patients, also have an increased infection rate (113). The presence of *S. aureus* may both aggravate the eczema and prevent its resolution (23). Other conditions that increase *S. aureus* carriage, such as isotretinoin therapy, are also associated with increased staphylococcal infections (53). *S. aureus* binds to the fibronectin on epithelial cells, and when the skin is broken, invasion is facilitated by the even larger quantities of fibronectin in deeper tissues. Additional possible receptors, such as laminin and type IV collagen, also increase infectivity (111).

S. aureus infection can result in a number of clinical appearances, including folliculitis, furuncles and carbuncles, impetigo, wound infections, mastitis, hidradenitis suppurativa, staphylococcal scalded skin syndrome, and toxic shock syndrome. Septicemia and deep organ involvement can also occur, usually following local infection. The degree of infection depends on the specific extracellular enzymes and toxins produced by the organism.

Corynebacteria

Coryneform organisms have been implicated in several types of infections. An axillary flora in which coryneform bacteria dominate over cocci has been associated with an increase in conditions such as axillary odor (51, 52). Treatment for these conditions should be aimed toward discouraging growth of corynebacteria by keeping the area clean and dry.

ERYTHRASMA   Erythrasma is characterized by reddish-brown, slightly scaling, and slowly spreading patches, most often in intertriginous areas. It may be associated with mild pruritus, but is usually asymptomatic. Examination with an ultraviolet light (Wood's light) reveals a characteristic coral-red fluorescence, secondary to the production of porphyrin by the implicated *C. minutissimum*. The incidence of infection is 19% in normal young adults (105), but 43% in institutionalized patients (106). Diabetics also have an increased incidence of extensive erythrasma.

TRICHOMYCOSIS   Trichomycosis is a condition that affects axillary or pubic hairs; coryneform organisms colonize the outside of the hair shaft and form concretions of bacterial colonies. The condition is asymptomatic and has been associated with poor hygiene. The colonies appear clinically as adherent yellow, red, or black nodules along the hair shaft. Although trychomycosis was originally thought to be due to an organism described as *C. tenuis,* Savin et al (94) isolated at least three distinct species from each infected individual tested. In studies with scanning EM, however, although three distinct species types were entrapped in an adhesive cement substance around the hair, only one species was able to adhere to the hair and colonize on it (100).

PITTED KERATOLYSIS   Pitted keratolysis is characterized by usually asymptomatic pitted erosions in the stratum corneum involving the soles of the feet or rarely the palms. The suspected causative organism has been isolated from pits and found to be a coryneform species by cell wall studies (117). This species is usually accompanied by filamentous bacteria (108). The implicated coryneform organisms are believed to have proteolytic enzymes that digest the keratin and leave the crateriform pits (101). Infection has been reproduced with the organism in healthy volunteers (108). More recently, *Micrococcus sedentarius* was also implicated in the disease, having been found along with coryneform bacteria in 100% of cases (85). In that study, pitted keratolysis was reproduced experimentally by applying *M. sedentarius* to the heels of human volunteers under occlusion. This organism may act synergistically with coryneform bacteria to produce the disease.

DERMATOPHYTOSIS COMPLEX   Dermatophytosis complex presents as white maceration and soggy interdigital scaling accompanied by bad odor. It is caused by the proliferation of the web-space flora, particularly coryneform bacteria, in anatomically occluded tinea pedis. With time it becomes symptomatic, with tenderness and fissuring, and dermatophytes become increasingly difficult to isolate.

GROUP JK INFECTIONS   Infections with group JK coryneform bacteria are remarkably similar to those that can occur with coagulase-negative staphylococci; the JK bacteria can cause bacteremia associated with indwelling foreign devices in hospitalized patients (97). First reported in patients with lymphoreticular malignancies, the organisms are much more common in patients at high risk due to neutropenia or immunosuppressive therapy (65). Other risk factors include prolonged hospitalization, prolonged multiple antibiotic therapy, and violated mucocutaneous barriers (97). The organism first causes a localized skin infection at the site of skin breakage, characterized by an erythematous appearance (55). Bacteremia can be associated with an erythematous maculopapular eruption scattered over the trunk and extremities.

Propionibacteria

P. acnes is the predominant organism colonizing acne lesions, and adolescent patients with acne have more surface *P. acnes* than matched controls (50). *P. acnes* lipases act on sebaceous triglycerides in the follicle to release free fatty acids, which may contribute to comedogenesis (57). The use of antibiotics in the treatment of acne has focused on decreasing the density of these organisms, with resultant clinical improvement. *P. acnes* may also initiate or contribute to the inflammatory stage of acne. Extracellular factors elaborated by *P. acnes* include chemotactic factors for granulocytes and macrophages (4), prostaglandin-like substances that have a direct irritant effect (3), and enzymes that may contribute directly to the degenerative changes in the follicular wall. *P. granulosum* shows greater lipolytic activity than *P. acnes* (114), and may also have significance in the inflammatory stage of acne.

There have also been reports of *P. acnes* and other *Propionibacterium* species causing complications in severely ill hospitalized patients that have indwelling foreign devices or that have recently had surgery. These infections can resemble those caused by *S. epidermidis* in similarly immunocompromised patients (83).

Gram-Negative Rods

ACINETOBACTER   Although infections with *A. calcoaceticus* may be acquired in the community, they occur much more commonly in hospitals and in immunosuppressed patients. Postoperative wound infections and burn wound infections account for the majority of skin infections (43). Infections with *A. calcoaceticus* var. *anitratus* peak in the summer, especially in young males

(91). *Acinetobacter* can cause infections of a wide variety of organs, including bacteremia, endocarditis, meningitis, and diseases of the genitourinary tract and, most commonly, the respiratory tract (43). Demonstration of colonization on the hands of hospital personnel during outbreaks has been demonstrated (91).

GRAM-NEGATIVE FOLLICULITIS   Gram-negative folliculitis can present as small superficial pustules around the nose, most often due to *E. coli, Klebsiella* spp., or *Enterobacter* spp., or as deep nodular and cystic lesions, most often due to *Proteus* spp. The source of infection is usually the anterior nares of acne patients on chronic antibiotic therapy, in which the normal ecological balance has been disrupted due to suppression of the normal flora.

GRAM-NEGATIVE TOE-WEB INFECTIONS   These infections present as toe-web denudation with serous or pustular discharge and marked edema and erythema of the surrounding tissues. Predisposing factors include recurrent tinea pedis and tight interdigital spaces, which result in ecological factors favoring the proliferation of the normal gram-negative flora (25). *Pseudomonas* and *Proteus* spp. are the most common organisms isolated.

Pityrosporum

P. orbiculare may, under the proper environmental conditions, change from the resident blastospore stage to its pathogenic mycelial stage (33). Conditions that may predispose the organism to this transformation include heat and humidity, hyperhidrosis, greasy skin, a genetic predisposition, systemic steroids, depressed cellular immunity, and diabetes. This transformation can be duplicated in vitro when *P. orbiculare* are incubated with stratum corneum cells, and may be secondary to the release of substances by the stratum corneum cells after adherence. *Malassezia furfur* is the name given the mycelial form.

TINEA VERSICOLOR   Tinea versicolor is a superficial, scaling yeast infection of the stratum corneum. It is characterized by hypo- and hyperpigmented macules, usually on the trunk and proximal extremities, where *P. orbiculare* is most prevalent. Less common areas of involvement include the face and scalp. The condition has been duplicated by the inoculation of organisms onto the skin of normal subjects (19, 33). The organisms contain lipoperoxides, which are thought to damage melanocytes as well as to inhibit tyrosinase, which results in the dyspigmentation (78).

PITYROSPORUM FOLLICULITIS   *Pityrosporum* folliculitis is characterized by erythematous, pruritic papules or pustules that typically appear on the trunk and upper extremities. *Pityrosporum* spp. can be found within the dilated hair follicles. Diabetics and patients on systemic steroids or antibiotics have an increased incidence of infection (15). Infection may be due to a predisposition of the host to the lipolytic properties of the organism. In both tinea versicolor and pityrosporum folliculitis, an increased number of circulating antibodies to blastospores has been reported (35).

OTHER CONDITIONS   *Pityrosporum* spp. have been implicated but not proven as a causative factor in seborrheic dermatitis (29) and confluent and reticulate papillomatosis (34). *Malesezzia furfur* has also been implicated as a cause of sepsis and pulmonary vessel infection in infants receiving long-term alimentation with fat emulsions (88).

Candida

Candidiasis is usually caused by *C. albicans* and, less commonly, by other *Candida* species such as *C. tropicalis* and *C. parapsilosis*. *Candida* can rapidly colonize on damaged skin. Active invasion of the stratum corneum can occur when the organism transforms to the mycelial form (95). Just as environmental factors such as heat and humidity may predispose an individual to infection with *Pityrosporum* spp., these same factors can predispose the individual to disease with candidal organisms. Clinical infection is often secondary to immune deficiency, diabetes, or suppression of the resident flora by antibiotics (14).

   Although candidal organisms only rarely colonize normal skin, their prevalence on mucous membrane surfaces results in an endogenous source of infection when cutaneous host defense mechanisms break down. Localized infection occurs most commonly on mucous membranes, but can also occur cutaneously, especially on moist macerated skin or skin in close contact with the mucous membranes. Oral colonization with *Candida* spp., for example, can lead to thrush as well as to atrophic or pseudomembranous oral lesions. Angular cheilitis, an inflammatory condition at the corners of the mouth, can also develop. Other cutaneous manifestations of *Candida* infection include generalized cutaneous candidiasis, candida balanitis, erosio interdigitalis blastomycetica, intertrigo, paronychia, onycholysis, candidal diaper rash, perianal candidiasis, *Candida* folliculitis, cutaneous lesions of disseminated candidiasis, and chronic mucocutaneous candidiasis. More extensive mucocutaneous and deep organ involvement can also occur.

SUMMARY

Humans exist in an environment replete with microorganisms, yet only a few of these microorganisms become residents on the skin surface. These resident flora and the skin constitute a complex ecosystem in which organisms adapt to changes in the microenvironment and to coactions among microorganisms. The skin possesses an assortment of protective mechanisms to limit colonization, and the survival of organisms on the surface lies in part in the ability of the organisms to resist these mechanisms. Microbial colonization on the skin adds to the skin's defense against potentially pathogenic organisms. Although microbes normally live in synergy with their hosts, at times colonization can lead to clinical infection. Common infections consist of superficial infections of the stratum corneum or appendages, which can respond dramatically to therapy but commonly relapse. In rare circumstances these infections can be severe, particularly in immunocompromised patients or hospitalized patients with indwelling foreign devices.

Literature Cited

1. Abraham, S. N., Beachey, E. H., Simpson, W. A. 1983. Adherence of *Streptococcus pyogenes, Escherichia coli,* and *Pseudomonas aeruginosa* to fibronectin-coated and uncoated epithelial cells. *Infect. Immun.* 41:1261–68
2. Alkan, M., Ofek, I., Beachey, E. 1977. Adherence of pharyngeal and skin strains of group A streptococci to human skin and oral epithelial cells. *Infect. Immun.* 18:555–57
3. Allaker, R. P., Greenman, J., Osborne, R. H. 1987. The production of inflammatory compounds by *Propionibacterium acnes* and other skin organisms. *Br. J. Dermatol.* 117:175–83
4. Allaker, R. P., Greenman, J., Osborne, R. H., Gowers, J. I. 1985. Cytotoxic activity of *Propionibacterium acnes* and other skin organisms. *Br. J. Dermatol.* 113:229–35
5. Allen, A. M., Taplin, D., Twigg, L. 1971. Cutaneous streptococcal infections in Vietnam. *Arch. Dermatol.* 104:271–80
6. Allen, K. D., Green, H. T. 1986. Infections due to a group JK corynebacterium. *J. Infect.* 13:41–44
7. Aly, R. 1982. Effect of occlusion of microbial population and physical skin conditions. *Semin. Dermatol.* 1:137–42
8. Aly, R., Maibach, H. I., Mandel, A. 1976. Bacterial flora in psoriasis. *Br. J. Dermatol.* 95:603–6
9. Aly, R., Maibach, H. I., Shinefield, H. R. 1977. Microbial flora of atopic dermatitis. *Arch. Dermatol.* 113:780–82
10. Aly, R., Maibach, H. I., Strauss, W. G., Shinefield, H. R. 1972. Survival of pathogenic microorganisms on human skin. *J. Invest. Dermatol.* 58:205–10

11. Aly, R., Shinefield, H. R., Maibach, H. I. 1981. *Staphylococcus aureus* adherence to nasal epithelial cells: studies of some parameters. In *Skin Microbiology: Relevance to Clinical Infection,* ed. H. Maibach, R. Aly, pp. 171–79. New York: Springer-Verlag. 354 pp.
12. Baird-Parker, A. C. 1963. A classification of micrococci and staphylococci based on physiological and biochemical tests. *J. Gen. Microbiol.* 30:409–27
13. Baker, C. J., Barrett, F. F., Gordon, R. C., Yow, M. D. 1973. Suppurative meningitis due to streptococcal meningitis of Lancefield group B: a study of 33 infants. *J. Pediatr.* 82:724–29
14. Barth, J. H. 1987. Nasal carriage of staphylococci and streptococci. *Int. J. Dermatol.* 26:24–26
15. Beretty, P. J. M., Neumann, H. A. M., Hulsebosch, H. J. 1980. *Pityrosporum* folliculitis: is it a real entity? *Br. J. Dermatol.* 103:565–66
16. Berman, D. S., Schaefler, S., Simberkoff, M. S., Rahal, J. J. 1987. *Staphylococcus aureus* colonization in intravenous drug abusers, dialysis patients, and diabetics. *J. Infect. Dis.* 155:829–31
17. Bibel, D. J., Aly, R., Shinefield, H. R., Maibach, H. I., Strauss, W. G. 1982. Importance of the keratinized epithelial cell in bacterial adherence. *J. Invest. Dermatol.* 79:250–53
18. Brandberg, A., Andersson, I. 1981. Preoperative whole body disinfection by shower bath with chlorhexidine soap: effect on transmission of bacteria from skin flora. See Ref. 11, pp. 92–97
19. Burke, R. C. 1951. Tinea versicolor: susceptibility factors and experimental

infection in human beings. *J. Invest. Dermatol.* 17:267–72

20. Chan, H. L., Aly, R., Maibach, H. I. 1982. Effect of topical corticosteroid on the microbial flora of human skin. *J. Am. Acad. Dermatol.* 7:346–48

21. Cole, G. W., Silverberg, N. L. 1986. The adherence of *Staphylococcus aureus* to human corneocytes. *Arch. Dermatol.* 122:166–69

22. Davenport, D. S., Massanari, R. M., Pfaller, M. A., Bale, M. J., Streed, S. A., et al. 1986. Usefulness of a test for slime production as a marker for clinically significant infections with coagulase-negative staphylococci. *J. Infect. Dis.* 153:332–39

23. David, T. J., Cambridge, G. C. 1986. Bacterial infection and atopic eczema. *Arch. Dis. Child.* 61:20–23

24. Duncan, W. C., McBride, M. E., Knox, J. M. 1969. Bacterial flora. The role of environmental factors. *J. Invest. Dermatol.* 52:479–84

25. Eaglstein, N. F., Marley, W. M., Marley, N. F., Rosenberg, E. W., Hernandez, A. D. 1983. Gram negative toe web infection: successful treatment with a new third generation cephalosporin. *J. Am. Acad. Dermatol.* 8:225–28

26. Eisenstein, B. I., Ofek, I., Beachey, E. H. 1979. Interference with the mannose binding and epithelial cell adherence of *Escherichia coli* by sublethal concentrations of streptomycin. *J. Clin. Invest.* 63:1219–1228

27. Elek, S. D. 1956. Experimental staphylococcal infections in the skin of man. *Ann. NY Acad. Sci.* 65:85–90

28. Evans, C. A., Mattern, K. L., Hallam, S. L. 1978. Isolation and identification of *Peptococcus saccharolyticus* from human skin. *J. Clin. Microbiol.* 7:261–64

29. Faergemann, J. 1986. Seborrhoeic dermatitis and *Pityrosporum orbiculare:* treatment of seborrhoeic dermatitis of the scalp with miconazole-hydrocortisone (Daktacort), miconazole and hydrocortisone. *Br. J. Dermatol.* 114:695–700

30. Faergemann, J., Aly, R., Maibach, H. I. 1983. Adherence of *Pityrosporum orbiculare* to human stratum corneum cells. *Arch. Dermatol. Res.* 275:246–50

31. Faergemann, J., Aly, R., Maibach, H. I. 1983. Quantitative variations in distribution of *Pityrosporum orbiculare* on clinically normal skin. *Acta. Derm. Venereol.* 63:346–48

32. Faergemann, J., Fredriksson, T. 1980. Age incidence of *Pityrosporum orbiculare* on human skin. *Acta. Derm. Venereol.* 60:531–33

33. Faergemann, J., Fredriksson, T. 1981. Experimental infections in rabbits and humans with *Pityrosporum orbiculare* and *P. ovale. J. Invest. Dermatol.* 77:314–19

34. Faergemann, J., Fredriksson, T., Nathorst-Windahl, G. 1980. One case of confluent and reticulate papillomatosis (Gougerot-Carteaud). *Acta. Derm. Venereol.* 60:269–71

35. Faergemann, J., Johansson, S., Back, O., Scheynius, A. 1986. An immunologic and cultural study of *Pityrosporum* folliculitis. *J. Am. Acad. Dermatol.* 14:429–33

36. Faergemann, J., Larko, O. 1987. The effect of UV-light on human skin microorganisms. *Acta. Derm. Venereol.* 67:69–72

37. Feingold, D. S. 1986. Bacterial adherence, colonization, and pathogenicity. *Arch. Dermatol.* 122:161–63

38. Fekety, F. R. Jr. 1964. The epidemiology and prevention of staphylococcal infection. *Medicine* 43:593–613

39. Ferrieri, P., Dajani, A. S., Wannamaker, L. W., Chapman, S. S. 1972. Natural history of impetigo I: site sequence of acquisition and familial patterns of spread of cutaneous streptococci. *J. Clin. Invest.* 51:2851–62

40. Fleer, A., Verhoef, J. 1986. Coagulase-negative staphylococci as nosocomial pathogens in neonates. *Am. J. Med.* 80:161–65

41. Gibbons, R. J., van Houte, J. 1975. Bacterial adherence in oral microbial ecology. *Ann. Rev. Microbiol.* 29:19–44

42. Gill, V. J., Selepak, S. T., Williams, E. C. 1983. Species identification and antibiotic susceptibilities of coagulase-negative staphylococci isolated from clinical specimens. *J. Clin. Microbiol.* 18:1314–19

43. Glew, R. H., Moellering, R. C., Kunz, L. J. 1977. Infections with *Acinetobacter calcoaceticus (Herellea vaginicola):* clinical and laboratory studies. *Medicine* 56:79–97

44. Govan, J. R. W. 1986. In vivo significance of bacteriocins and bacteriocin receptors. *Scand. J. Infect. Dis.* 49:31–37

45. Gray, E. D., Peters, G., Verstegen, M., Regelmann, W. E. 1984. Effect of extracellular slime substance from *Staphylococcus epidermidis* on the human cellular immune response. *Lancet* 1:365–67

46. Hamory, B. H., Parisi, J. T. 1987. *Staphylococcus epidermidis:* a significant nosocomial pathogen. *J. Infect. Control* 15:59–74

47. Hartmann, A. A. 1982. A comparative investigation of methods for sampling skin flora. *Arch. Dermatol. Res.* 274:381–85

48. Hartmann, A. A. 1983. Effect of occlusion on resident flora, skin-moisture and skin-pH. *Arch. Dermatol. Res.* 275:251–54

49. Helstrom, P. B., Balish, E. 1979. Effect of oral tetracycline, the microbial flora, and the athymic state on gastrointestinal colonization and infection of BALB/c mice with *Candida albicans. Infect. Immun.* 23:764–74

50. Holland, K. T., Cunliffe, W. J., Roberts, C. D. 1977. Acne vulgaris: an investigation into the number of anaerobic diphtheroids and members of the Micrococcaceae in normal and acne skin. *Br. J. Dermatol.* 96:623–26

51. Jackman, P. J. H. 1982. Body odor—the role of skin bacteria. *Semin. Dermatol.* 1:143–48

52. Jackman, P. J. H., Noble, W. C. 1983. Normal axillary skin microflora in various populations. *Clin. Exp. Dermatol.* 8:259–68

53. James, W. D. 1984. Using isotretinoin wisely. *J. Assoc. Mil. Dermatol.* 10:63

54. James, W. D., Leyden, J. J. 1985. Treatment of gram-negative folliculitis with isotretinoin: positive clinical and microbiologic response. *J. Am. Acad. Dermatol.* 12:319–24

55. Jerdan, M. S., Shapiro, R. S., Smith, N. B., Virshup, D. M., Hood, A. F. 1987. Cutaneous manifestations of *Corynebacterium* group JK sepsis. *J. Am. Acad. Dermatol.* 16:444–47

56. Johnson, G. M., Lee, D. A., Regelmann, W. E., Gray, E. D., Peters, G., Quie, P. G. 1986. Interference with granulocyte function by *Staphylococcus epidermidis* slime. *Infect. Immun.* 54:13–20

57. Kellum, R. E. 1979. Free fatty acid hypothesis. In *Acne: Update for the Practitioner*, ed. S. B. Frank, pp. 65–73. New York: Yorke Med.

58. Kinsman, O. S. 1982. Attachment to the host as a preliminary to infection. *Semin. Dermatol.* 1:127–36

59. Kinsman, O. S., McKenna, R., Noble, W. C. 1981. Host genetics (HLA antigens) and carriage of *Staphylococcus aureus. Soc. Gen. Microbiol. Q.* 8:117

60. Kloos, W. E. 1981. The identification of *Staphylococcus* and *Micrococcus* species isolated from human skin. See Ref. 11, pp. 3–12

61. Kloos, W. E., Musselwhite, M. S. 1975. Distribution and persistence of *Staphylococcus* and *Micrococcus* species and other aerobic bacteria on human skin. *Appl. Microbiol.* 30:381–95

62. Korting, H. C., Kober, M., Mueller, M., Braun-Falco, O. 1987. Influence of repeated washings with soap and synthetic detergents on pH and resident flora of the skin of forehead and forearm. *Acta. Derm. Venereol.* 67:41–47

63. Kuusela, P., Vartio, T., Vuento, M., Myhre, E. B. 1984. Binding sites for streptococci and staphylococci in fibronectin. *Infect. Immun.* 45:433–36

64. Larson, E. L., McGinley, K. J., Foglia, A. R., Talbot, G. H., Leyden, J. J. 1985. Composition and antimicrobic resistance of skin flora in hospitalized and healthy adults. *J. Clin. Microbiol.* 23:604–8

65. Larson, E. L., McGinley, K. J., Leyden, J. J., Cooley, M. E., Talbot, G. H. 1986. Skin colonization with antibiotic-resistant (JK group) and antibiotic-sensitive lipophilic diphtheroids in hospitalized and normal adults. *J. Infect. Dis.* 153:701–6

66. Leeming, J. P., Holland, K. T., Cunliffe, W. J. 1984. The microbial ecology of pilosebaceous units isolated from human skin. *J. Gen. Microbiol.* 130:803–7

67. Leyden, J. J., McGinley, K. J. 1982. Effect of 13-*cis*-retinoic acid on sebum production and *Propionibacterium acnes* in severe nodulocystic acne. *Arch. Dermatol. Res.* 272:331–37

68. Leyden, J. J., McGinley, K. J., Foglia, A. N. 1986. Qualitative and quantitative changes in cutaneous bacteria associated with systemic isotretinoin therapy for acne conglobata. *J. Invest. Dermatol.* 86:390–93

69. Leyden, J. J., Stewart, R., Kligman, A. M. 1980. Experimental infections with group A streptococci in humans. *J. Invest. Dermatol.* 75:196–201

70. Liljemark, W. F., Gibbons, R. J. 1973. Suppression of *Candida albicans* by human oral streptococci in gnotobiotic mice. *Infect. Immun.* 8:846–49

71. Mackowiak, P. A. 1982. The normal microbial flora. *N. Engl. J. Med.* 307:83–93

72. Marples, R. R. 1982. Sex, constancy, and skin bacteria. *Arch. Dermatol. Res.* 272:317–20

73. Matta, M. 1974. Carriage of *Corynebacterium acnes* in school children in relation to age and race. *Br. J. Dermatol.* 91:557–61

74. McGinley, K. J., Labows, J. N., Zechman, J. M., Nordstrom, K. M., Webster, G. F., Leyden, J. J. 1985. Analysis of cellular components, biochemical reactions, and habitat of human

cutaneous lipophilic diphtheroids. *J. Invest. Dermatol.* 85:374–77

75. McGinley, K. J., Labows, J. N., Zeckman, J. M., Nordstrom, K. M., Webster, G. F., Leyden, J. J. 1985. Pathogenic JK group corynebacteria and their similarity to human cutaneous lipophilic diphtheroids. *J. Infect. Dis.* 152:801–6

76. McGinnis, M. R., Rinaldi, M. G., Halde, C., Hilger, A. E. 1975. Mycotic flora of the interdigital spaces of the human foot: a preliminary investigation. *Mycopathologia* 55:47–52

77. Mok, W. Y., Barreto da Silva, M. S. 1984. Mycoflora of the human dermal surfaces. *Can. J. Microbiol.* 30:1205–9

78. Nazzaro-Porro, M., Passi, S., Picardo, M., Mercantini, R., Breathnach, A. S. 1986. Lipoxygenase activity of *Pityrosporum* in vitro and in vivo. *J. Invest. Dermatol.* 87:108–12

79. Nilsson, E., Henning, C., Hjorleifsson, M. L. 1986. Density of the microflora in hand exzema before and after topical treatment with a potent corticosteroid. *J. Am. Acad. Dermatol.* 15:192–97

80. Noble, W. C. 1974. Carriage of *Staphylococcus aureus* and beta haemolytic streptococci in relation to race. *Acta. Derm. Venereol.* 54:403–5

81. Noble, W. C. 1981. *Microbiology of Human Skin*. London: Lloyd-Luke. 433 pp.

82. Noble, W. C. 1984. Skin microbiology: coming of age. *J. Med. Microbiol.* 17:1–12

83. Noble, R. C., Overman, S. B. 1987. *Propionibacterium acnes* osteomyelitis: case report and review of the literature. *J. Clin. Microbiol.* 25:251–54

84. Noble, W. C., Pitcher, D. G. 1978. Microbial ecology of the human skin. *Adv. Microb. Ecol.* 2:245–89

85. Nordstrom, K. M., McGinley, K. J., Cappiello, L., Zechman, J. M., Leyden, J. J. 1987. Pitted keratolysis: the role of *Micrococcus sedentarius*. *Arch. Dermatol.* 123:1320–25

86. Pitcher, D. G. 1977. Rapid identification of cell wall components as a guide to the classification of aerobic coryneform bacteria from human skin. *J. Med. Microbiol.* 10:439–44

87. Pitcher, D. G. 1978. Aerobic cutaneous coryneforms: recent taxonomic findings. *Br. J. Dermatol.* 98:363–70

88. Powell, D. A., Aungst, J., Snedden, S., Hansen, N., Brady, M. 1984. Broviac catheter–related *Malassezia furfur* sepsis in five infants receiving intravenous fat emulsions. *J. Pediatr.* 105:987–90

89. Price, P. B. 1938. The bacteriology of normal skin; a new quantitative test applied to a study of the bacterial flora and the disinfectant action of mechanical cleansing. *J. Infect. Dis.* 63:301–18

90. Rebel, G., Pillsbury, D. M., Phalle, G., de Saint, M., Ginsberg, D. 1950. Factors affecting the rapid disappearance of bacteria placed on the normal skin. *J. Invest. Dermatol.* 14:247–63

91. Retailliau, H. F., Hightower, A. W., Dixon, R. E., Allen, J. R. 1977. *Acinetobacter calcoaceticus:* a nosocomial pathogen with an unusual seasonal pattern. *J. Infect. Dis.* 139:371–75

92. Ryden, C., Rubin, K., Speziale, P., Hook, M., Lindberg, M., Wadstrom, T. 1982. Fibronectin receptors from *Staphylococcus aureus*. *J. Biol. Chem.* 258:3396–401

93. Sarkany, I., Gaylarde, C. C. 1968. Bacterial colonisation of the skin of the newborn. *J. Pathol. Bacteriol.* 95:115–22

94. Savin, J. A., Somerville, D. A., Noble, W. C. 1970. The bacterial flora of trichomycosis axillaris. *J. Med. Microbiol.* 3:352–56

95. Scherwitz, C. 1982. Ultrastructure of human cutaneous candidosis. *J. Invest. Dermatol.* 78:200–5

96. Schleifer, K. H., Kloos, W. E. 1975. A simple test system for the separation of staphylococci from micrococci. *J. Clin. Microbiol.* 1:337–38

97. Schoch, P. E., Cunha, B. A. 1986. The JK diphtheroids. *Infect. Control* 7:466–69

98. Selwyn, S. 1975. Natural antibiosis among skin bacteria as a primary defence against infection. *Br. J. Dermatol.* 93:487–93

99. Shehadeh, N. H., Kligman, A. M. 1963. The effect of topical antibacterial agents on the bacterial flora of the axilla. *J. Invest. Dermatol.* 40:61–71

100. Shelley, W. B., Miller, M. A. 1984. Electron microscopy, histochemistry, and microbiology of bacterial adhesion in trichomycosis axillaris. *J. Am. Acad. Dermatol.* 10:1005–14

101. Shelley, W. B., Shelley, E. D. 1982. Coexistent erythrasma, trichomycosis axillaris, and pitted keratolysis: an overlooked corynebacterial triad? *J. Am. Acad. Dermatol.* 7:752–57

102. Shepherd, M. G., Poulter, R. T. M., Sullivan, P. A. 1985. *Candida albicans:* biology, genetics, and pathogenicity. *Ann. Rev. Microbiol.* 39:579–614

103. Singh, G., Marples, R. R., Kligman, A. M. 1971. Experimental *Staphylococcus aureus* infections in humans. *J. Invest. Dermatol.* 57:149–62

104. Somerville, D. A. 1969. The normal flora of the skin in different age groups. *Br. J. Dermatol.* 81:248–58
105. Somerville, D. A. 1970. Erythrasma in normal young adults. *J. Med. Microbiol.* 3:352–56
106. Somerville, D. A., Seville, R. H., Cunningham, R. C., Noble, W. C., Savin, J. A. 1970. Erythrasma in a hospital for the mentally subnormal. *Br. J. Dermatol.* 82:355–59
107. Svejgaard, E., Christophersen, J., Jelsdorf, H. 1986. Tinea pedis and erythrasma in Danish recruits. *J. Am. Acad. Dermatol.* 14:993–99
108. Taplin, D., Zaias, N. 1968. The etiology of pitted keratolysis. *13th Int. Congr. Dermatol., Munich,* 1:593–95. Berlin: Springer
109. Tenney, J. H., Moody, M. R., Newman, K. A., Schimpff, S. C., Wade, J. C., et al. 1986. Adherent microorganisms on lumenal surfaces of long-term intravenous catheters. *Arch. Intern. Med.* 146:1949–54
110. Tuazon, C. U. 1984. Skin and skin infections in the patient at risk: carrier state of *Staphylococcus aureus*. *Am. J. Med.* 76(5A):166–71
111. Vercellati, G. M., Lussenhop, D., Peterson, P. K., Furcht, L. T., McCarthy, J. B., et al. 1984. Bacterial adherence to fibronectin and endothelial cells: a possible mechanism for bacterial tissue tropism. *J. Lab. Clin. Med.* 103:34–43
112. Weissmann, A., Noble, W. C. 1980. Photochemotherapy of psoriasis: effects on bacteria and surface lipids in uninvolved skin. *Br. J. Dermatol.* 102:185–93
113. White, M. I., Noble, W. C. 1986. Consequences of colonization and infection by *Staphylococcus aureus* in atopic dermatitis. *Clin. Exp. Dermatol.* 11:34–40
114. Whiteside, J. A., Voss, J. G. 1973. Incidence and lipolytic activity of *Propionibacterium acnes* (*Corynebacterium acnes* group I) and *P. granulosum* (*C. acnes* group II) in acne and in normal skin. *J. Invest. Dermatol.* 60:94–97
115. Wichmann, S., Wirsing von Koenig, C. H., Becker-Boost, E., Finger, H. 1985. Group JK corynebacteria in skin flora of healthy persons and patients. *Eur. J. Clin. Microbiol.* 4:502–4
116. Winston, D. J., Dudnick, D. V., Chapin, M., Ho, W. G., Gale, R. P., et al. 1983. Coagulase-negative staphylococcal bacteremia in patients receiving immunosuppresive therapy. *Arch. Intern. Med.* 143:32–36
117. Zaias, N. 1982. Pitted and ringed keratolysis. *J. Am. Acad. Dermatol.* 7:787–91

Ann. Rev. Microbiol. 1988. 42:465–88

ENERGY METABOLISM OF PROTOZOA WITHOUT MITOCHONDRIA

Miklós Müller

The Rockefeller University, New York, New York 10021

CONTENTS

INTRODUCTION

Pathways of energy metabolism in prokaryotic microorganisms are highly diverse. A great variety of inorganic and organic reduced compounds are utilized as electron donors, and in addition to O_2, a number of oxidized compounds serve as terminal electron acceptors. In contrast, metabolic capabilities of eukaryotic organisms are more uniform and restricted, with

465

0066-4227/88/1001-0465$02.00

only a limited number of organic nutrients serving as substrates for energy generation. The predominant portion of the chemical energy of these compounds is conserved by mitochondrial oxidative phosphorylation, with O_2 as the terminal electron acceptor.

Among protists, however, we find interesting exceptions. Several major protozoan groups consist of or contain species that lack mitochondria. The peculiar metabolic properties of these organisms could be due to the adaptation of mitochondrion-containing organisms to an anaerobic mode of life (29, 85). Alternatively, however, the absence of mitochondria might be a primary feature that has been conserved from times when life was still anaerobic (7) and mitochondria were unknown. However daring such a proposal might have been a few years ago, recent data on molecular properties of ribosomes and rRNA (6, 27, 34, 126) reveal that at least some of these amitochondrial protozoa are related so distantly to other eukaryotes that they must have separated from the main trunk of eukaryotic evolution very early and might have descended from primarily amitochondrial ancestors (10, 12).

Energy metabolism is as vital to any organism as protein synthesis. Characteristics of energy-generating processes could contain important clues to the evolution of amitochondrial organisms. Few species, all parasitic and symbiotic, have been studied so far (Table 1). These belong to different major systematic groups, however, and exhibit markedly different biochemical characteristics. Behind this diversity there is a common core of fundamental properties, reflected primarily in pyruvate oxidation, a key reaction of energy metabolism. With the use of selected data, I delineate the metabolic nature of anaerobic protozoa and draw tentative conclusions concerning their evolutionary history.

Much of the earlier information has been reviewed and is not cited explicitly [metabolism in several species (23, 35, 125), *Entamoeba histolytica* (78, 79, 105), *Giardia lamblia* (57), *Trichomonas vaginalis* (41, 89, 111), *Tritrichomonas foetus* (40, 84, 111), and rumen ciliates (43, 134); hydrogenosomes (13, 85, 88); interaction of 5-nitroimidazole drugs with anaerobic protozoa (30, 86, 87)].

GENERAL CONSIDERATIONS

The energy metabolism of all aerobic protozoa considered is based on exogenous and endogenous carbohydrates. These organisms contain abundant carbohydrate reserves in the form of glycogen, with the exception of rumen ciliates, which contain amylopectin. The metabolism is fermentative under anaerobic and (when tolerated) aerobic conditions [*E. histolytica* (80, 130), *G. lamblia* (56), *T. vaginalis* (76, 90), *T. foetus* (110), *Dasytricha ruminantium* (124, 135), *Isotricha* species (101)]. Carbohydrates are converted to

Table 1 Anaerobic protozoa (Kingdom Protista; Subkingdom Protozoa)

Species[a]	Habitat	Hydrogenosome	Major end products	
			Anaerobic	Aerobic
Phylum Sarcomastigophora, Subphylum Sarcodina, Class Lobosea				
Entamoeba histolytica	Human large intestine	No	Ethanol, CO_2	Ethanol, acetate, CO_2
Entamoeba invadens	Reptile intestine	No	no data	no data
Subphylum Mastigophora, Class Zoomastigophora				
Giardia lamblia	Human small intestine	No	Ethanol, acetate, CO_2	Ethanol, acetate, CO_2
Trichomonas vaginalis	Human genitourinary tract	Yes	Glycerol, lactate, acetate, H_2, CO_2	Glycerol, lactate, acetate, CO_2
Tritrichomonas foetus	Bovine genitourinary tract	Yes	Glycerol, succinate, acetate, H_2, CO_2	Glycerol, succinate, acetate, CO_2
Phylum Ciliophora, Subphylum Rhabdophora				
Dasytricha ruminantium	Ovine rumen	Yes	Lactate, acetate, butyrate, H_2, CO_2	not viable
Isotricha species	Ovine and bovine rumen	Yes	Lactate, acetate, butyrate, H_2, CO_2	not viable

[a]Systematic positions from Reference 53.

various organic compounds (Table 1) without complete oxidation to CO_2 and H_2O; thus the energy metabolism is inefficient in utilization of nutrients. Some of the organisms are able to remove reducing equivalents as H_2, and all are able to reduce O_2.

Pyruvate formed by glycolysis is a key intermediate of energy metabolism and is the precursor of most metabolic end products released into the medium. Glycolysis proceeds by the ubiquitous Embden-Meyerhof pathway, which has been well treated in earlier reviews (23, 35, 57, 105, 125). The pathways of pyruvate metabolism show significant differences in various anaerobic protozoa and are discussed in some detail. ATP is produced by substrate-level phosphorylation only, both in the glycolytic pathway proper and in the further metabolism of pyruvate.

The organisms considered can be divided into two groups on the basis of compartmentation of their metabolism (Table 1). The cytosol is the site of all or most steps of energy metabolism [*Entamoeba invadens* (54), *G. lamblia* (56), *T. vaginalis* (2), *D. ruminantium* (142)]. Trichomonad flagellates and rumen ciliates, however, are exceptional, since their conversion of pyruvate to acetate, a key metabolic sequence, is localized in a separate organelle, the hydrogenosome (85). This organelle is discussed in a separate section. No evidence for glycosomes has been found in trichomonad flagellates (118).

GLYCOLYSIS

Glucose or reserve carbohydrates are converted to phosphoenolpyruvate (PEP) and subsequently to pyruvate via a classical Embden-Meyerhof pathway. As in other organisms, several steps of this pathway are accompanied by substrate-level phosphorylation. The presence of most of the enzymes of the pathway has been demonstrated in all species, but few of these enzymes have been studied in any detail. No information is available on the regulatory properties of the glycolytic enzymes, which in many other parasitic protozoa lack such properties.

E. histolytica represents an exception, however. In this species inorganic pyrophosphate (PP_i) replaces adenine nucleotides as a high-energy compound in several reactions. This is regarded as an ancient metabolic feature (139). In the glycolytic pathway proper, phosphofructokinase is PP_i linked (108). This organism also lacks pyruvate kinase. PEP is converted to pyruvate by pyruvate phosphate dikinase:

$$PEP + AMP + PP_i \rightarrow pyruvate + ATP + P_i. \qquad 1.$$

Alternatively, pyruvate is formed by sequential action of phosphopyruvate carboxylase (104), malate dehydrogenase (NAD), and malate dehydrogenase (decarboxylating, NADP). Together with the dikinase these enzymes constitute a branched pathway (loop): Phosphopyruvate carboxylase generates PP_i in the carboxylation of PEP,

$$PEP + CO_2 + P_i \rightarrow oxaloacetate + PP_i, \qquad 2.$$

and the PP_i is utilized in the conversion of PEP to pyruvate by the dikinase (105). At the same time, the two dehydrogenases provide the functional equivalent of an NADH:NADP transhydrogenase activity (105), which probably corresponds to the transhydrogenase described from *E. histolytica* (37).

Some of the species form end products from glycolytic intermediates. *T. vaginalis* and *T. foetus* produce glycerol (21, 113) from dihydroxyacetone phosphate by the action of cytosolic glycerol 3-phosphate dehydrogenase and glycerol 3-phosphatase (113). *T. foetus* produces succinate as a major end product (110). The enzymes that form succinate from PEP are a cytosolic PEP carboxykinase (ADP), malate dehydrogenase (NAD), fumarate hydratase, and fumarate reductase (84, 91). This sequence corresponds to the C_4 part of a reverse tricarboxylic acid cycle. The fumarate reductase is not membrane bound, and its action is probably not linked to phosphorylation. Although its natural electron donor is unknown, it accepts electrons from free flavins and methyl viologen.

Lactate is produced by *T. vaginalis* and *D. ruminantium* via NADH-linked reduction of pyruvate by cytosolic lactate dehydrogenase (2, 142). In *D. ruminantium* part of this activity might be hydrogenosomal (142).

METABOLISM OF PYRUVATE BY DECARBOXYLATION

Oxidation of Pyruvate to Acetyl Coenzyme A

The enzyme catalyzing the oxidative decarboxylation of pyruvate, a central step in the energy metabolism of almost all organisms (47), is a key characteristic of the anaerobic amitochondrial protozoa. Of the two distinct enzymes that catalyze this reaction, the operative one in these eukaryotes is pyruvate:ferredoxin oxidoreductase (47), regarded as the ancestral one, and not the more advanced pyruvate dehydrogenase complex (103). Pyruvate:ferredoxin oxidoreductase (47) is a dimeric or tetrameric protein of about 240 kd, while pyruvate dehydrogenase complex is a multienzyme system with an aggregate mass of over 10^3 kd (103). Pyruvate decarboxylation by pyruvate:ferredoxin oxidoreductase is reversible; thus another name of this enzyme is pyruvate synthase. The pyruvate dehydrogenase reaction is irreversible. Pyruvate:ferredoxin oxidoreductase utilizes an Fe-S protein, ferredoxin, as acceptor, which usually has a more negative midpoint potential than pyridine nucleotides. Thus the high reducing power of pyruvate makes it useful for reactions requiring stronger reductants than NADH, e.g. evolution of H_2 or N_2 fixation (47). Pyruvate dehydrogenase complex, in contrast, transfers electrons to NAD. Pyruvate:ferredoxin oxidoreductase and pyruvate dehydrogenase complex are regarded as mutually exclusive in any given organism (47). The former is restricted to strictly anaerobic (47) and nitrogen-fixing (127) organisms and halophilic archaebacteria (45).

Pyruvate:ferredoxin oxidoreductase has been purified from only two protozoan species [E. histolytica (109) and T. vaginalis (136)], but data on cell extracts have permitted its identification in other anaerobic protozoa [E. invadens (54), G. lamblia (56), T. foetus (59), Monocercomonas sp. (61), D. ruminantium (142), and other rumen ciliates (140, 143)]. These activities are O_2 labile. An acetyl-CoA–forming pyruvate oxidase reported from E. histolytica (115) is probably the same enzyme (105). Reversibility of the reaction has been observed in E. histolytica (109) and T. foetus (59). Arsenate is not inhibitory, which shows that lipoate does not participate in the reaction [E. histolytica (109), T. foetus (59), D. ruminantium (144)].

Purified pyruvate:ferredoxin oxidoreductase of T. vaginalis has a mass of approximately 240 kd and consists of two 120-kd subunits (136). It contains thiamine pyrophosphate and two [4Fe-4S]centers. This enzyme is similar to prokaryotic pyruvate:ferredoxin oxidoreductases (47). In addition to pyruvate, the enzyme decarboxylates 2-oxobutyrate and 2-oxoglutarate. The purified E. histolytica pyruvate:ferredoxin oxidoreductase has not been characterized in detail (109). Both purified enzymes were shown to reduce homologous ferredoxins.

T. foetus pyruvate:ferredoxin oxidoreductase in hydrogenosomal preparations has been studied with electron paramagnetic resonance (EPR) spectroscopy (25). The reaction proceeds by a free-radical mechanism. A hydroxethyl radical arises from oxidation of the thiamine pyrophosphate–bound C_2 moiety, and a thilyl radical is formed from CoA. The latter accepts the bound acetyl group, whereby acetyl-CoA is formed. The Fe-S centers of the enzyme undergo redox cycling and transfer electrons to ferredoxin or other external electron acceptors. This mechanism is similar to that proposed for pyruvate:ferredoxin oxidoreductase in halobacteria; however, the thilyl radical was not detected in those organisms (46).

Acetate and Butyrate Formation

Acetate is formed from acetyl-CoA by all of the protozoa under consideration, although by E. histolytica only under aerobic conditions (109). The mechanism of acetate formation differs from species to species. The energy of the thioester bond of acetyl-CoA is always conserved by substrate-level phosphorylation; thus acetate formation represents an important contribution to energy metabolism.

E. histolytica (109) and *G. lamblia* (56) contain an acetate thiokinase, which catalyzes the reaction

$$\text{acetyl-CoA} + \text{ADP} + P_i \rightarrow \text{acetate} + \text{CoA} + \text{ATP}. \qquad 3.$$

E. histolytica also contains an acetate kinase (pyrophosphate) of unknown function (106):

$$\text{acetate} + PP_i \rightarrow \text{acetyl phosphate} + P_i. \qquad 4.$$

In *T. foetus* and probably in all other trichomonad flagellates acetate is formed by the sucessive action of two enzymes (55); acetate:succinate CoA-transferase catalyzes the reaction

$$\text{acetyl-CoA} + \text{succinate} \rightarrow \text{acetate} + \text{succinyl-CoA}, \qquad 5.$$

and then succinate thiokinase catalyzes the reaction

$$\text{succinyl-CoA} + \text{ADP} + P_i \rightarrow \text{succinate} + \text{ATP}. \qquad 6.$$

The presence of acetate thiokinase in *T. foetus* (55) is doubtful. Still another mechanism has been reported for *D. ruminantium,* in which acetyl phosphate is an intermediate (144). The first step is a phosphotransacetylase reaction,

$$\text{acetyl-CoA} + P_i \rightarrow \text{acetyl phosphate} + \text{CoA}, \qquad 7.$$

and the second is an acetate kinase reaction,

$$\text{acetyl phosphate} + \text{ADP} \rightarrow \text{acetate} + \text{ATP}. \qquad 8.$$

This mechanism was previously known to occur only in prokaryotes (119); thus its presence in a eukaryotic organism was unexpected.

D. ruminantium also produces butyrate from acetyl-CoA (135). The corresponding enzymes have been detected: acetyl-CoA:acetyl-CoA acetyltransferase, 3-hydroxybutyryl-CoA dehydrogenase, hydroxyacyl-CoA hydrolyase, 3-hydroxyacyl-CoA reductase, phosphate butyryltransferase, and butyrate kinase. The last step is energy conserving. This sequence is cytosolic; thus its substrate, acetyl-CoA, is probably exported from the hydrogenosomes, its only site of formation.

Ethanol Formation

Ethanol is produced by several anaerobic protozoa. *E. histolytica* forms ethanol from acetyl-CoA, since it does not contain a pyruvate decarboxylase (72). Acetyl-CoA is first converted to an enzyme-bound thiohemiacetal by acetaldehyde dehydrogenase (acylating); the thiohemiacetal is then reduced to ethanol by alcohol dehydrogenase (72). Part of the enzyme-bound thiohemiacetal hydrolyzes to free acetaldehyde, which is reduced by a distinct alcohol dehydrogenase (71, 74). The mechanism of ethanol formation by *G. lamblia* is not known, but participation of an alcohol dehydrogenase (NADP) is likely (56). Wild-type *T. foetus* forms only small amounts of ethanol. If the organism loses its functional hydrogenosomes, however, ethanol becomes the only organic end product, formed by cytosolic pyruvate decarboxylase and alcohol dehydrogenase (17).

ELECTRON TRANSFER

General Comments and Electron Carriers

Electrons generated during carbohydrate catabolism in the amitochondrial anaerobic protozoa are transferred to various carriers and disposed of by the reduction of several different glycolytic intermediates. In hydrogenosome-containing organisms they are also disposed of by the reduction of protons. These protozoa respire, too, which shows that they are able to transfer electrons to O_2. The main carriers are Fe-S proteins, pyridine nucleotides, and flavins. Notable is the lack of cytochromes.

Of special interest is the dominance of Fe-S proteins. The low–midpoint potential Fe-S protein ferredoxin is regarded as the ancestral electron transport compound (9, 26, 36). Ferredoxins have been purified from *E. histolytica* (107), *T. vaginalis* (33), and *T. foetus* (77). The other species are also likely to contain a ferredoxin, in view of the specificity of the pyruvate:ferredoxin oxidoreductase. Low-temperature EPR spectra have a number of features attributed to Fe-S centers [*E. histolytica* (131), *G. lamblia* (129), *T. vaginalis* (20, 32), *T. foetus* (97)], some of which represent ferredoxin signals. Other proteins, such as pyruvate:ferredoxin oxidoreductase and probably hydrogenase, also contain Fe-S centers.

Ferredoxins of *E. histolytica* and trichomonads differ considerably. The former has an approximate mass of 6 kd, and its chemical analysis indicates two [4Fe-4S]centers (107). These properties and its optical spectrum are similar to those of clostridial [8Fe-8S]ferredoxins. The ferredoxins of *T. vaginalis* (33) and *T. foetus* (77), in contrast, have a mass of approximately 12 kd and one [2Fe-2S]center. The EPR spectrum shows axial symmetry. These properties suggest that the trichomonad proteins are related to bacterial and mitochondrial [2Fe-2S]ferredoxins (9).

Pyridine nucleotides have not been detected directly in any of the species studied. Several oxidoreductases are NAD or NADP specific, so these compounds have to be present. Flavins are also present in large quantities, as shown by direct determinations [*E. histolytica* (73), *G. lamblia* (129), *T. foetus* (58)] and EPR spectroscopy [*G. lamblia* (129), *T. vaginalis* (20), *T. foetus* (97)].

No cytochromes have been detected in anaerobic protozoa with spectroscopic methods [*E. histolytica* (131), *G. lamblia* (129), *T. vaginalis* (110), *T. foetus* (65, 110), *D. ruminantium* (144)]. No heme iron was found by chemical analysis of *E. histolytica* (132) and *G. lamblia* (129). However, *T. foetus* contains catalase (83).

Transfer of Electrons to Organic Acceptors

Pyruvate:ferredoxin oxidoreductase is functional in anaerobic *E. histolytica* and *G. lamblia*, as shown by the production of ethanol from acetyl-CoA (72) in the former and the production of acetate and ethanol (56), possibly also from acetyl-CoA, in the latter. The mechanisms of reoxidation of ferredoxin, which is reduced in these processes, remain unknown, however. It is likely that ferredoxin provides reducing equivalents for ethanol production. Since the corresponding alcohol dehydrogenases are pyridine nucleotide dependent, the participation of a ferredoxin:NAD(P) oxidoreductase is likely. These putative processes might also function under aerobic conditions, when O_2 also accepts electrons.

Isolated hydrogenosomes of *T. vaginalis* and *T. foetus* produce some malate from pyruvate by the action of a decarboxylating, pyridine nucleotide–specific malate dehydrogenase (114). The electrons from pyruvate:ferredoxin oxidoreductase pass through ferredoxin and are transferred to NAD or NADP by ferredoxin:NAD(P) oxidoreductase (114).

Hydrogen Production

In hydrogenosome-containing organisms a major means of reoxidation of ferredoxin is by production of H_2, a process not available to organisms without this organelle. This reaction is catalyzed by hydrogenase (H_2:ferredoxin oxidoreductase) through the transfer of electrons to protons.

Hydrogenase activity has been detected in all hydrogenosome-containing organisms discussed [*T. vaginalis* (63), *T. foetus* (59), *Monocercomonas* sp. (61), *D. ruminantium* (142), and other rumen ciliates (140, 143)]. The electron-donor specificity of these enzymes, when H_2 is used as reductant, is similar to that of pyruvate:ferredoxin oxidoreductase. No further details are known concerning these hydrogenases, which are likely to be Fe-S proteins as in other organisms (1).

The enzymatic pathway of H_2 production in these organisms consists of only three components: pyruvate:ferredoxin oxidoreductase, ferredoxin, and hydrogenase (114). This pathway exists in a number of prokaryotes (119), primarily anaerobic bacteria, but it is exceptional for eukaryotes. In trichomonad flagellates, moreover, this pathway contains a [2Fe-2S]ferredoxin, which does not participate in prokaryotic H_2 production.

When O_2 acts as alternative electron acceptor, no H_2 is produced (102). The apparent inhibitor constant of O_2 for H_2 production by *T. vaginalis* (68) and *D. ruminantium* (147) is in the micromolar range. H_2 production is also inhibited by other electron acceptors. These include various nitro-derivatives, such as 2,4-dinitrophenol (94) and the antianaerobic 5-nitroimidazoles discussed below. The inhibition by alternative electron acceptors is reversible, and H_2 evolution recommences after anaerobiosis is restored or all added acceptors have been reduced. Diversion of electrons from H_2 generation to alternative acceptors by itself is not deleterious for trichomonads (94). The nature of the electron sink seems to be of little consequence for these organisms unless the product of its reduction is a toxic compound, such as an O_2 derivative or the product of a 5-nitroimidazole drug.

Axenic *E. histolytica* does not produce H_2, since it has no hydrogenosomes (109, 117). However, amebae that have been cultivated with growth-suppressed *Bacteroides symbiosus* will produce H_2 after removal of the bacteria (80). Reeves (105) has suggested that bacterial hydrogenase is "picked up and retained" by the amebae, where it remains functional for several hours. This observation might indicate an unusual symbiotic metabolic interaction.

Respiration

All protozoa discussed here are described as anaerobic organisms. Nonetheless they take up O_2 at rates comparable to those of aerobic protozoa, showing the transfer of reducing equivalents from metabolism to O_2 as terminal acceptor [*E. histolytica* (130), *G. lamblia* (56, 129), *T. vaginalis* (76, 90), *T. foetus* (69, 110), *D. ruminantium* (69, 144)].

Aerobiosis, which is accompanied by O_2 uptake in anaerobic protozoa, does not affect the energy metabolism. All data indicate that reducing equivalents are diverted from anaerobic acceptors to O_2 without providing major energetic advantage to the organism. The rates of overall metabolism are similar under anaerobic and aerobic conditions [*E. histolytica* (80), *G. lamblia* (56), *T. vaginalis* (76, 90), *T. foetus* (15, 110)], and the metabolism remains fermentative. Some shifts, but not major ones, occur in the overall carbon flow, and the proportion of more oxidized end products increases. No Pasteur effect is seen, i.e. the availability of O_2 does not have a substrate-sparing effect; this implies that there is little, if any, energetic advantage to aerobiosis. The rate of O_2 utilization is comparable to that of the excretion of organic end products. This indicates that only a fraction of the total reducing power generated in metabolism is transferred to O_2; much of the reoxidation of reduced carriers still occurs via reduction of organic compounds.

Respiratory systems of intact anaerobic protozoa are enigmatic. The organisms have a high O_2 affinity, as indicated by straight uptake curves in O_2-electrode experiments with *E. histolytica* (130) and *G. lamblia* (129). Apparent K_m values for O_2 of about 1 μM have been found for *T. foetus* and *D. ruminantium* with the use of bacterial bioluminescence at low pO_2 values (69). The ultimate product of O_2 reduction by living anaerobic protozoa is most probably H_2O and not H_2O_2. Respiration rates are not affected by exogeneous catalase [*E. histolytica* (130), *G. lamblia* (129), *T. vaginalis* (110)]. All organisms considered, with the exception of the rumen ciliates, tolerate aerobiosis relatively well, while they are highly susceptible to exogenous H_2O_2 (110). Affinity for O_2 and reduction of O_2 to H_2O are properties similar to those found in organisms that contain cytochrome oxidase (69).

Respiration in anaerobic protozoa is fundamentally different from that in mitochondrion-containing cells, however. O_2 uptake in anaerobic species is not affected by the inhibitors of mitochondrial electron transport, cytochrome oxidase and alternative mitochondrial oxidases, or by uncouplers of oxidative phosphorylation. Only high concentrations of flavin antagonists and iron-chelating agents are inhibitory [*E. histolytica* (130, 131), *G. lamblia* (56, 129), *T. vaginalis* (121, 137), *T. foetus* (67, 110), *D. ruminantium* (144)].

The enzymatic activities able to reduce O_2 are poorly known. Oxidases acting on reduced pyridine nucleotides are present in the cytosol [*E. histolytica* (71), *G. lamblia* (56), *T. vaginalis* (64, 90, 116), *T. foetus* (16, 83), *D. ruminantium* (142), *Isotricha* sp. (100)]. An unusual NADH oxidase of approximately 200 kd has been purified from *T. vaginalis* (64, 116). This enzyme contains bound flavin and is apparently able to reduce O_2 to H_2O. In *Isotricha* sp., NADH is oxidized by a coupled oxidase-peroxidase reaction (100). Activities oxidizing NADPH require added free flavins as electron carrier to O_2; thus their natural acceptor remains unknown [*E. histolytica* (74), *T. vaginalis* (64)].

Intact hydrogenosomes utilize O_2 in the presence of pyruvate and some other substrates [*T. vaginalis* (148), *T. foetus* (14, 93), *D. ruminantium* (144), ophryoscolecid ciliates (112)]. This process shows high O_2 affinity (about 10 μM) in *T. vaginalis* (148) and, like the respiration of whole cells (67, 102, 144), is insensitive to metabolic inhibitors and uncouplers. The terminal oxidase is not known. Autooxidation of pyruvate:ferredoxin oxidoreductase or ferredoxin has been suggested as the mechanism of hydrogenosomal respiration in *T. foetus* (81), but the presence of a distinct terminal oxidase has not been disproved.

Reduction of Nitro-Derivatives Active Against Anaerobic Microorganisms

Of the alternative acceptors for ferredoxin-mediated reactions, an important group consists of 5-nitroimidazole derivatives. These compounds, including metronidazole (1-hydroxyethyl,2-methyl,5-nitroimidazole), are used as selective drugs against anaerobic protozoan and bacterial infections.

5-Nitroimidazole derivatives become cytotoxic only when activated by reduction of the nitro-group (87). Since the electron affinity of the nitro-group is low (141), corresponding electron donors are necessary. Ferredoxins of anaerobic organisms are appropriate donors, whereas aerobes and facultative anaerobes lack suitable electron donors (28, 87).

In anaerobic protozoa, ferredoxin reduced by pyruvate:ferredoxin oxidoreductase is the electron donor to the nitro-group. 5-Nitroimidazoles are rapidly reduced by reduced ferredoxins (33, 62, 77, 141). Although ferrodoxin has a low midpoint potential, other properties might be of importance, since ferredoxin-linked oxidoreductases are still poor electron donors for 5-nitroimidazoles (33, 77, 141). 5-Nitroimidazoles are reduced poorly or not at all by other carriers such as pyridine nucleotides or flavins (22, 98).

5-Nitroimidazoles inhibit H_2 evolution in living $T.$ $vaginalis$ (28) because they act as alternative acceptors for reduced ferredoxin. On the other hand, reductive activation of 5-nitroimidazoles can be inhibited if other acceptors compete for electrons (94) or if the electron flow to ferredoxin is diminished (96). O_2 is an effective competitor for nitro-reduction and thus inhibits antimicrobial activity of 5-nitroimidazoles (92). Reductive activation of 5-nitroimidazoles by $T.$ $vaginalis$ isolates from treatment-refractory patients (75), although normal under anaerobic conditions, reveals an increased sensitivity to inhibition by O_2 (68, 90). The mechanism of this effect is unknown, but this enhanced inhibition by O_2 is likely to account for the observed clinical resistance. Trichomonad flagellates without pyruvate:ferredoxin oxidoreductase activity are not susceptible to 5-nitroimidazoles [$T.$ $vaginalis$ (52) and $T.$ $foetus$ (49)].

The process of 5-nitroimidazole reduction is poorly understood (87). On the average, four electrons are transferred stepwise to the nitro-group. On complete reduction, the compounds fragment to molecules of little or no biological activity (4, 30). It is assumed that the products responsible for antimicrobial action are short-lived intermediates of the reaction. Of these, a one–electron adduct nitro-free radical has been detected by EPR spectroscopy in intact cells and in isolated hydrogenosomes [$T.$ $vaginalis$ (19, 68), $T.$ $foetus$ (81, 82)]. DNA is assumed to be the primary target of the intermediate reduction products (31, 44, 149).

HYDROGENOSOMES

Hydrogenosomes constitute a separate compartment of energy metabolism in trichomonad flagellates and rumen ciliates, as mitochondria do in most other eukaryotes. They harbor a key metabolic sequence discussed above, the ATP-producing pyruvate-to-acetate pathway linked to H_2 production (Figure 1) (59, 85). Hydrogenosomal functions confer an energetic advantage to the cell (145). However, at least in trichomonads, hydrogenosomal metabolism is not absolutely indispensible.

Hydrogenosomes have been detected in trichomonad flagellates [*T. vaginalis* (63), *T. foetus* (59), *Monocercomonas* sp. (61)], in rumen ciliates [*D. ruminantium* (142), *Isotricha* spp. (143), entodiniomorphid ciliates (112, 140)], and also in the rumen fungus *Neocallimastix patriciarum* (146); these organisms all have membrane-bounded organelles in which pyruvate:ferredoxin oxidoreductase and hydrogenase coexist. Morphological and physiological data indicate the presence of these organelles in other flagellates (38) and ciliates (29, 123), but their definitive identification must await biochemical characterization.

Hydrogenosomes are predominantly spherical structures, approximately 0.5–1 μm in diameter, and are surrounded by an envelope. They contain a granular matrix and often an electron-dense core. In trichomonad flagellates

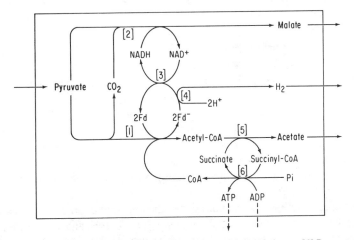

Figure 1 Map of hydrogenosomal metabolism in *T. vaginalis* and *T. foetus*. [*1*] Pyruvate:ferredoxin oxidoreductase, [*2*] malate dehydrogenase (decarboxylating, NAD), [*3*] NAD:ferredoxin oxidoreductase, [*4*] H_2:ferredoxin oxidoreductase, [*5*] acetate:succinate CoA-transferase, [*6*] succinate thiokinase. *Fd:* ferredoxin. *Arrows* indicate the assumed physiological directions in vivo. *Dashed arrows* indicate a postulated adenyl nucleotide transfer. (Reproduced with permission from Reference 114.)

the envelope consists of two closely apposed membranes (5, 42, 51), but the inner membrane does not form cristae. The hydrogenosomes of rumen ciliates might also have a double envelope, but morphological data (142) are insufficient. Arguments have also been made for a single envelope in rumen ciliates (12). A membrane-bounded, flattened vesicle is seen below the envelope in many organelles. Morphological data indicate that hydrogenosomes multiply by division (51, 95). Based on morphology, hydrogenosomes differ from both mitochondria and microbodies.

Our biochemical understanding of hydrogenosomes is based on metabolic studies of isolated organelles [anaerobic: *T. vaginalis* (114) and *T. foetus* (114); aerobic: *T. vaginalis* (148), *T. foetus* (14, 93), *D. ruminantium* (144)] and on the localization of various constituents by centrifugation techniques [*T. vaginalis* (8, 24, 63), *T. foetus* (55, 59, 60, 83), *D. ruminantium* (142)]. The results demonstrate that the major role of these organelles is in the conversion of pyruvate to acetate. This conversion is accompanied by substrate-level phosphorylation and the transfer of electrons either to protons or to O_2. Hydrogenosomes possibly also participate in the metabolism of malate and *sn*-glycerol 3-phosphate of cytoplasmic origin [*T. foetus* (14, 83)].

Pyruvate metabolism in hydrogenosomes shows striking similarities to that in *Clostridium* species and other anaerobic bacteria (119), which are regarded as primitive, possibly ancestral prokaryotes (138). These metabolic similarities are as compelling as those noted between mitochondria and certain aerobic bacteria (12, 133). In brief, hydrogenosomes can be regarded as the anaerobic equivalents of mitochondria. The two organelles differ in so many fundamental properties, however, that independent origins are indicated. Both organelles use pyruvate as a major substrate and form acetyl-CoA. The mechanisms involved and the fate of the electrons generated are different, however. Anaerobic protozoa, and hence their hydrogenosomes, lack a number of major mitochondrial constituents and functions, including a complete tricarboxylic acid cycle, cytochromes, cytochrome oxidase, and electron transport–linked phosphorylation. The absence of some of these components was confirmed in isolated hydrogenosomes (14, 65, 66). Unlike mitochondria, hydrogenosomes contain no cardiolipin (99) and most probably no DNA (122, 128).

Enzymatic activities accounting for the metabolic processes mentioned above have been detected in *T. vaginalis, T. foetus,* and *D. ruminantium* hydrogenosomes. *sn*-Glycerol 3-phosphate dehydrogenase has been detected in *T. foetus* (83). Superoxide dismutase has been found in *T. foetus* and *Monocercomonas* sp. (60), but not in *D. ruminantium* (144). Adenylate kinase has also been demonstrated in *T. vaginalis* (24) and *T. foetus* (93). Data on respiration indicate the presence of an NADH oxidase activity (16) and possibly a terminal oxidase of unknown nature (14). Besides hydrogenosomal ferredoxins (33, 77), only pyruvate:ferredoxin oxidoreductase (136) and adenylate kinase of *T. vaginalis* have been purified and characterized (24).

The metabolism of isolated hydrogenosomes has been little studied. They utilize pyruvate with generation of H_2 [*T. vaginalis* (114), *T. foetus* (114)] or respiration [*T. foetus* (14, 93), *D. ruminantium* (144)]. Trichomonad hydrogenosomes convert pyruvate quantitatively to acetate, malate, CO_2, and H_2 (Figure 1) (114), with acetate as the major product. In intact trichomonad organelles this process is accompanied by the phosphorylation of added ADP to ATP. In *D. ruminantium,* respiration requires not ADP, but P_i. This requirement agrees with the role of acetyl phosphate as a high-energy intermediate (144). Intact trichomonad hydrogenosomes require succinate for the acetate:succinate CoA-tranferase reaction (14, 93, 114).

Acetate, the principal product of hydrogenosomes, is a major metabolic end product of living cells under both anaerobic and aerobic conditions [*T. vaginalis* (65, 90), *T. foetus* (110), *D. ruminantium* (135)]. This shows that these organelles are functionally significant. The high level of anaerobic H_2 production is in agreement with this conclusion. In the presence of O_2 no H_2 is formed (70), and data on isolated organelles suggest that acetate production is coupled to O_2 reduction (93). The contribution of hydrogenosomes to the total respiration, however, remains unknown.

Pyruvate metabolism and possibly the other hydrogenosomal functions are, however, not vital for trichomonads. Organisms without hydrogenosomal functions [*T. vaginalis* (18), *T. foetus* (17)] can be obtained by long-term in vitro cultivation in the presence of increasing concentrations of metronidazole [*T. vaginalis* (52), *T. foetus* (49)], although the procedure is arduous. These organisms produce no H_2, acetate, or succinate, and their metabolism becomes entirely cytosolic [lactate fermentation in *T. vaginalis* (18) and ethanol fermentation in *T. foetus* (17)]. They multiply more slowly than wild-type cells, probably because of their less efficient energy metabolism.

EVOLUTIONARY CONSIDERATIONS

The energy metabolism of anaerobic protozoa separates them from other eukaryotes. They contain enzymes and functions not present in most eukaryotic cells, notably pyruvate:ferredoxin oxidoreductase and ferredoxin-mediated electron transport. It is just as significant, however, that they lack two major evolutionary advances of energy metabolism, electron transport–linked phosphorylation and cytochrome oxidase–mediated utilization of O_2. These organisms contain no morphologically recognizable mitochondria.

The data reemphasize two significant but often disregarded points: (*a*)

Mitochondria, regarded as a hallmark of eukaryotic cells, are by no means universally present. (*b*) Substrate-level phosphorylation can sustain eukaryotic life without mitochondrial energy generation.

The organisms discussed represent only a few of the protozoan groups that contain species without mitochondria. Additional groups include the ubiquitous parasitic Microsporidia (10), the highly evolved hypermastiginid flagellates of termites and roaches (38), and amebae (123, 133) and ciliates (29, 123) of anaerobic aquatic sediments. Amitochondrial species of other higher taxons might still be discovered, although the ultrastructure of most major protist groups is already well known.

The lack of mitochondria is thus a striking common property of a number of eukaryotic groups. A new subkingdom, Archezoa, was proposed to distinguish these organisms from mitochondrion-containing protozoa, the Mitozoa (11). This classification implies that amitochondrial animals are primitive. The four phyla of Archezoa are Archamebae, which contains the genus *Entamoeba;* Metamonadea, containing the genus *Giardia;* Parabasalia, in which trichomonads belong; and Microsporidia. Energy metabolism in the last phylum has not been studied yet. Although rumen ciliates share an impressive array of biochemical characteristics with trichomonad flagellates, they belong to the ciliates, a mitozoan group. Archezoa are probably not monophyletic (10), since the phylum Parabasalia differs markedly from the other phyla; thus the inclusion of this phylum seems to be artificial.

The origin of amitochondrial eukaryotes is a challenging problem. All extant examples live in anaerobic or microaerobic environments. The absence of mitochondrial functions corresponds adequately to anaerobic life. However, this condition could be either a primary feature or a secondary adaptation. If it is the former, the extant amitochondrial organisms descended from ancestors that never contained mitochondria (10). If the second case is correct, the organisms lost the mitochondria at a later stage of evolution. A number of multicellular animals, among them parasitic worms, also have an anaerobic fermentative metabolism, but they contain mitochondria and have various facets of the mitochondrial machinery incorporated into their energy metabolism (3, 48).

Organisms without hydrogenosomes probably represent a primitive stage of evolution (10) and not a secondary adaptation to anaerobiosis. In these organisms pyruvate:ferredoxin oxidoreductase and ferredoxin are localized in the cytosol, the primary cellular compartment. Parsimony suggests that this is their original location and that they were not acquired secondarily. The cellular organization of these organisms is simple, and several common organelles are absent (10, 34, 50). Their primitive nature is also indicated by the reliance of *Entamoeba* species on inorganic pyrophosphate as a high-energy compound (105). On the basis of such information *E. histolytica* was regarded as a "living fossil" for some time (105). The ancestral nature of one archezoan group, Microsporidia, received recent support from sequence comparisons of ribosomal 18S RNA (126). Although such information is lacking for the organisms discussed here, preliminary small-subunit rRNA sequence data affirm the primitiveness of the *Giardia* genus (34). Other molecular data on ribosomes support this contention (6, 34, 126).

Assessment of the position of organisms containing hydrogenosomes is more difficult. In this group pyruvate:ferredoxin oxidoreductase and ferredoxin are found in an organelle and not in the cytosol. The organelle clearly reflects ancestral metabolic properties. If we disregard hydrogenosomes, however, trichomonad flagellates and rumen ciliates are not markedly different from other eukaryotes, and their complex morphology suggests that they represent an advanced level of evolution (40, 41, 43, 51). These considerations force us to separate the questions of the origin of the hydrogenosome and of its host cell.

Hydrogenosomes contain a characteristic pathway of pyruvate oxidation otherwise restricted to a limited group of anaerobic prokaryotes and eukaryotes. As proposed earlier (85), I favor the hypothesis of their endosymbiotic origin from anaerobic prokaryotes that contained this pathway. A mitochondrial origin of hydrogenosomes (12) seems to me less likely. Transformation of mitochondria into hydrogenosomes would have required the insertion of the complete primitive pathway of pyruvate oxidation into the organelle to replace the universal mitochondrial system.

The absence of DNA does not argue against this hypothesis, since hydrogenosomes do not contain those polypeptides that are coded by genes of other organelles of endosymbiotic origin (120).

The trichomonad ferredoxin poses a special problem, however. It is different from ferredoxins of H_2-producing anaerobic bacteria and more closely resembles the [2Fe-2S]ferredoxins involved in hydroxylation reactions in aerobic bacteria and mitochondria. Thus no real counterpart of the hydrogenosomal H_2-generating system has been detected in any anaerobic bacterium. Hydrogenosomal energy conservation also shows marked differences in trichomonads and rumen ciliates. In rumen ciliates, but not in trichomonads, it is likely to depend on acetyl phosphate formation, known only from prokaryotes. This difference, if corroborated, would indicate that trichomonad and rumen ciliate hydrogenosomes had separate precursors or evolved independently.

Hydrogenosome-containing protozoa belong to systematic groups that cannot be linked to a single evolutionary lineage (53). The origin of trichomonad flagellates (Parabasalia) is enigmatic, and they show no obvious links to free-living forms (39). Their hydrogenosomes could have arisen at any time during evolution. The rumen ciliates, however, are closely related to a number of mitochondrion-containing free-living ciliates. The ancestors of hydrogenosome-containing ciliates were probably free-living organisms with mitochondria in which one type of organelle became supplanted by another. These considerations also suggest that hydrogenosomes could have arisen several times in independent events. The acquisition of these organelles seems to be related to anaerobic environments.

CONCLUSIONS

Certain protozoa of various taxonomic groups have no mitochondria, organelles almost universally present in eukaryotic organisms. The energy metabolism of amitochondrial parasitic and symbiotic protozoa reveals a number of key features, probably ancestral and primitive, which set them apart from protozoa with mitochondria. A key difference is in pyruvate oxidation, which is catalyzed by pyruvate:ferredoxin oxidoreductase and not by the pyruvate dehydrogenase complex. Electron transport is primarily via Fe-S proteins, and in several organisms it leads to H_2 production. Energy is conserved only by substrate-level phosphorylation. Reduction of O_2 is not connected with energy conservation. Two major aspects of advanced energy metabolism are absent: electron transport–linked phosphorylation and utilization of O_2 as terminal acceptor for the electrons involved in this process.

The amitochondrial protozoa show a striking dichotomy in the subcellular localization of the ancestral type of pyruvate oxidation. In certain species this occurs in the cytosol and hence is a characteristic of the cell itself. In other species, pyruvate oxidation proceeds in a membrane-bounded hydrogenosome, and its primitive features are characteristic of this organelle, and not of the host cell. This dichotomy suggests differences in the evolutionary history of the two groups of organisms. I suggest that the cytosolic location of the enzymes in the first group reflects a primary property of organisms that diverged from the main trunk of eukaryotic evolution before the appearance of mitochondria. Nucleic acid sequence information supports this notion. In contrast, hydrogenosome-containing organisms probably descended from more evolved eukaryotes and acquired an organelle with ancestral metabolic features at a later stage. It is possible that several groups of organisms acquired hydrogenosomes independently.

As molecular data on key constituents of anaerobic eukaryotes and their hydrogenosomes accumulate, we will gain further insights into the evolution of both, which is a most challenging and intriguing problem.

Literature Cited

1. Adams, M. W. W., Mortenson, L. E., Chen, J.-S. 1981. Hydrogenase. *Biochim. Biophys. Acta* 594:105–76
2. Arese, P., Cappuccinelli, P. 1974. Glycolysis and pentose phosphate cycle in *Trichomonas vaginalis*: enzyme activity pattern and the constant proportion quintet. *Int. J. Biochem.* 5:859–65
3. Barrett, J. 1981. *Biochemistry of Parasitic Helminths*. Baltimore, Md: Univ. Park. 308 pp.
4. Beaulieu, B. B. Jr., McLafferty, M. A., Koch, R. L., Goldman, P. 1981. Metronidazole metabolism in cultures of *Entamoeba histolytica* and *Trichomonas vaginalis*. *Antimicrob. Agents Chemother.* 20:410–14
5. Benchimol, M., de Souza, W. 1983. Fine structure and cytochemistry of the hydrogenosome of *Tritrichomonas foetus*. *J. Protozool.* 30:422–25
6. Boothroyd, J. C., Wang, A., Campbell, D. A., Wang, C. C. 1987. An unusually compact ribosomal DNA repeat in the protozoan *Giardia lamblia*. *Nucleic Acids Res.* 15:4065–84
7. Broda, E. 1975. *The Evolution of the Bioenergetic Process*. Oxford, UK: Pergamon. 211 pp.
8. Brugerolle, G., Méténier, G. 1973. Localisation intracellulaire et caracterisation de deux types de malate deshydrogenase chez *Trichomonas vaginalis* Donné. *J. Protozool.* 20:320–27
9. Cammack, R. 1983. Evolution and diversity in the iron-sulphur proteins. *Chem. Scr.* 21:87–95
10. Cavalier-Smith, T. 1987. Eukaryotes with no mitochondria. *Nature* 326:332–33
11. Cavalier-Smith, T. 1987. The origin of eukaryotes and archaebacterial cells. *Ann. NY Acad. Sci.* 503:17–54
12. Cavalier-Smith, T. 1987. The simultaneous symbiotic origin of mitochondria, chloroplasts, and microbodies. *Ann. NY Acad. Sci.* 503:55–72
13. Čerkasov, J., Čerkasovová, A., Kulda, J. 1980. Carbohydrate metabolism of *Tritrichomonas foetus* with particular reference to enzyme reactions occurring in hydrogenosomes. In *Trends in Enzymology*, Vol. 2, *Industrial and Clinical Enzymology. FEBS Spec. Meet. Enzymes, Dubrovnik, Yugoslavia, 1979*, ed. L. Vitale, V. Simeon, pp. 257–75. Oxford, UK: Pergamon
14. Čerkasov, J., Čerkasovová, A., Kulda, J., Vilhelmová, D. 1978. Respiration of hydrogenosomes of *Tritrichomonas foetus*. I. ADP-dependent oxidation of malate and pyruvate. *J. Biol. Chem.* 253:1207–14
15. Čerkasovová, A. 1970. Energy-producing metabolism of *Tritrichomonas foetus*. I. Evidence for the control of the intensity and the contribution of aerobiosis to the total energy production. *Exp. Parasitol.* 27:165–78
16. Čerkasovová, A., Čerkasov, J. 1974. Location of NADH-oxidase activity in fractions of *Tritrichomonas foetus* homogenate. *Folia Parasitol. (Prague)* 21:193–203
17. Čerkasovová, A., Čerkasov, J., Kulda, J. 1984. Metabolic differences between metronidazole resistant and susceptible strains of *Tritrichomonas foetus*. *Mol. Biochem. Parasitol.* 11:105–18
18. Čerkasovová, A., Novak, J., Čerkasov, J., Kulda, J., Tachezy, J. 1988. Metabolic properties of *Trichomonas vaginalis* resistant to metronidazole under anaerobic conditions. *Acta Univ. Carol. Biol.* In press
19. Chapman, A., Cammack, R., Linstead, D., Lloyd, D. 1985. The generation of metronidazole radicals in hydrogenosomes isolated from *Trichomonas vaginalis*. *J. Gen. Microbiol.* 131:2141–44
20. Chapman, A., Cammack, R., Linstead, D. J., Lloyd, D. 1986. Respiration of *Trichomonas vaginalis*. Components detected by electron paramagnetic resonance spectroscopy. *Eur. J. Biochem.* 156:193–98
21. Chapman, A., Linstead, D. J., Lloyd, D., Williams, J. 1985. ^{13}C-NMR reveals glycerol as an unexpected major metabolite of the protozoan parasite *Trichomonas vaginalis*. *FEBS. Lett.* 191:287–92
22. Clarke, E. D., Wardman, P., Goulding, K. H. 1980. Anaerobic reduction of nitroimidazoles by reduced flavin mononucleotide and by xanthine oxidase. *Biochem. Pharmacol.* 29:2684–87
23. Danforth, W. F. 1967. Respiratory metabolism. In *Research in Protozoology*, ed. T.-T. Chen, 1:201–306. Oxford, UK: Pergamon
24. Declerck, P. J., Müller, M. 1987. Hydrogenosomal ATP:AMP phosphotransferase of *Trichomonas vaginalis*. *Comp. Biochem. Physiol. B* 88b:575–80
25. Docampo, R., Moreno, S. N. J., Mason, R. P. 1987. Free radical intermediates in the reaction of pyruvate:ferredoxin oxidoreductase in *Tri-*

trichomonas foetus hydrogenosomes. *J. Biol. Chem.* 262:12417–20

26. Eck, R., Dayhoff, M. O. 1966. Evolution of the structure of ferredoxin based on living relics of primitive amino acid sequences. *Science* 152:363–66

27. Edlind, T. D., Chakraborty, P. R. 1987. Unusual ribosomal RNA of the intestinal parasite *Giardia lamblia*. *Nucleic Acids Res.* 15:7889–901

28. Edwards, D. I., Dye, M., Carne, H. 1973. The selective toxicity of antimicrobial nitroheterocyclic drugs. *J. Gen. Microbiol.* 76:135–45

29. Fenchel, T., Perry, T., Thane, A. 1977. Anaerobiosis and symbiosis with bacteria in free-living ciliates. *J. Protozool.* 24:154–63

30. Goldman, P., Koch, R. L., Yeung, T.-C., Chrystal, E. J. T., Beaulieu, B. B. Jr., et al. 1986. Comparing the reduction of nitroimidazoles in bacteria and mammalian tissues and relating it to biological activity. *Biochem. Pharmacol.* 35:43–51

31. Goldstein, B. P., Nielsen, E., Berti, M., Bolzoni, G., Silvestri, L. G. 1977. The mechanism of action of nitro-heterocyclic antimicrobial drugs. Primary target of 1-methyl-2-nitro-5-vinylimidazole is DNA. *J. Gen. Microbiol.* 100:271–81

32. Gorrell, T. E. 1985. Effect of culture medium iron content on the biochemical composition and metabolism of *Trichomonas vaginalis*. *J. Bacteriol.* 161:1228–30

33. Gorrell, T. E., Yarlett, N., Müller, M. 1984. Isolation and characterization of *Trichomonas vaginalis* ferredoxin. *Carlsberg Res. Commun.* 49:259–68

34. Gunderson, J. H., Elwood, H. J., Sogin, M. L. 1987. Evolutionary relationships of protistan groups inferred from small-subunit rRNA sequences. *Program Abstr. 40th Annu. Meet. Soc. Protozool., Champaign-Urbana, Ill.*, p. 32 (Abstr.)

35. Gutteridge, W. E., Coombs, G. H. 1977. *Biochemistry of Parasitic Protozoa.* Baltimore, Md: Univ. Park. 172 pp.

36. Hall, D. O., Cammack, R., Rao, K. K. 1971. Role of ferredoxins in the origin of life and evolution. *Nature* 133:136–38

37. Harlow, D. R., Weinbach, E. C., Diamond, L. S. 1976. Nicotinamide nucleotide transhydrogenase in *Entamoeba histolytica*. *Comp. Biochem. Physiol. B* 53:141–44

38. Hollande, A., Valentin, J. 1987. Appareil de Golgi, pinocytose, lyso-

somes, mitochondries, bactéries symbiontiques, atractophores et pleuromitose chez les hypermastigines du genre *Joenia*. Affinités entre joeniides et trichomonadines. *Protistologica* 5:39–86

39. Honigberg, B. M. 1963. Evolutionary and systematic relationships in the flagellate order Trichomonadida Kirby. *J. Protozool.* 10:20–63

40. Honigberg, B. M. 1978. Trichomonads of veterinary importance. In *Parasitic Protozoa*, Vol. 2, *Intestinal Flagellates, Histomonads, Trichomonads, Amoeba, Opalinids, and Ciliates*, ed. J. P. Kreier, pp. 163–273. New York: Academic

41. Honigberg, B. M. 1978. Trichomonads of importance in human medicine. See Ref. 40, pp. 275–454

42. Honigberg, B. M., Volkmann, D., Entzeroth, R., Scholtyseck, E. 1984. A freeze-fracture electron microscope study of *Trichomonas vaginalis* Donné and *Tritrichomonas foetus* (Riedmüller). *J. Protozool.* 31:116–31

43. Hungate, R. E. 1978. The rumen protozoa. See Ref. 40, pp. 655–95

44. Ings, R. M. J., McFadzean, J. A., Ormerod, W. E. 1974. The mode of action of metronidazole in *Trichomonas vaginalis*. *Biochem. Pharmacol.* 23:1421–29

45. Kerscher, L., Oesterhelt, D. 1981. Purification and properties of two 2-oxoacid:ferredoxin oxidoreductases from *Halobacterium halobium*. *Eur. J. Biochem.* 116:587–94

46. Kerscher, L., Oesterhelt, D. 1981. The catalytic mechanism of 2-oxoacid:ferredoxin oxidoreductases from *Halobacterium halobium*. One-electron transfer at two distinct steps of the catalytic cycle. *Eur. J. Biochem.* 116:595–600

47. Kerscher, L., Oesterhelt, D. 1982. Pyruvate:ferredoxin oxidoreductase—new findings on an ancient enzyme. *Trends Biochem. Sci.* 7:371–74

48. Köhler, P. 1985. The strategies of energy conservation in helminths. *Mol. Biochem. Parasitol.* 17:1–18

49. Kulda, J., Čerkasov, J., Demeš, P., Čerkasovová, A. 1984. *Tritrichomonas foetus*: stable anaerobic resistance to metronidazole in vitro. *Exp. Parasitol.* 57:93–103

50. Kulda, J., Nohynková, E. 1978. Flagellates of the human intestine and of intestines of other species. See Ref. 40, pp. 1–138

51. Kulda, J., Nohynková, E., Ludvik, J. 1986. Basic structure and function of the trichomonad cell. *Acta Univ. Carol. Biol.* 30:181–98

52. Kulda, J., Tachezy, J., Čerkasovová, A., Čerkasov, J. 1988. *Trichomonas vaginalis:* anaerobic resistance to metronidazole induced in vitro. *Acta Univ. Carol. Biol.* In press

53. Lee, J. J., Hutner, S. H., Bovee, E. C. 1985. *An Illustrated Guide to Protozoa.* Lawrence, Kans: Soc. Protozool. 629 pp.

54. Lindmark, D. G. 1976. Certain enzymes of the energy metabolism of *Entamoeba invadens* and their subcellular localization. In *Proc. Int. Conf. Amebiasis, Mexico City,* ed. B. Sepulveda, L. S. Diamond, 185–89. Mexico City: Inst. Mex. Siguro Soc.

55. Lindmark, D. G. 1976. Acetate production in *Tritrichomonas foetus.* In *Biochemistry of Parasites and Host-Parasite Relationships,* ed. H. Van den Bossche, pp. 16–21. Amsterdam: North-Holland

56. Lindmark, D. G. 1980. Energy metabolism of the anaerobic protozoon *Giardia lamblia. Mol. Biochem. Parasitol.* 1:1–12

57. Lindmark, D. G., Jarroll, E. L. 1984. Metabolism of trophozoites. In *Giardia and Giardiasis,* ed. E. A. Meyer, S. L. Erlandsen, pp. 65–80. New York: Plenum

58. Lindmark, D. G., Müller, M. 1973. Subcellular distribution of flavins in two trichomonad species. *J. Protozool.* 20:500 (Abstr.)

59. Lindmark, D. G., Müller, M. 1973. Hydrogenosome, a cytoplasmic organelle of the anaerobic flagellate, *Tritrichomonas foetus,* and its role in pyruvate metabolism. *J. Biol. Chem.* 248:7724–28

60. Lindmark, D. G., Müller, M. 1974. Superoxide dismutase in the anaerobic flagellates, *Tritrichomonas foetus* and *Monocercomonas* sp. *J. Biol. Chem.* 249:4634–37

61. Lindmark, D. G., Müller, M. 1974. Biochemical cytology of trichomonad flagellates. II. Subcellular distribution of oxidoreductases and hydrolases in *Monocercomonas* sp. *J. Protozool.* 21: 374–78

62. Lindmark, D. G., Müller, M. 1976. Antitrichomonad action, mutagenicity, and reduction of metronidazole and other nitroimidazoles. *Antimicrob. Agents Chemother.* 10:476–82

63. Lindmark, D. G., Müller, M., Shio, H. 1975. Hydrogenosomes in *Trichomonas vaginalis. J. Parasitol.* 61:552–54

64. Linstead, D. J., Bradley, S. 1988. The purification and properties of two soluble reduced nicotinamide:acceptor oxidoreductases from *Trichomonas vaginalis. Mol. Biochem. Parasitol.* 27:125–34

65. Lloyd, D., Lindmark, D. G., Müller, M. 1979. Respiration of *Tritrichomonas foetus:* Absence of detectable cytochromes. *J. Parasitol.* 65:466–69

66. Lloyd, D., Lindmark, D. G., Müller, M. 1979. Adenosine triphosphatase activity of *Tritrichomonas foetus. J. Gen. Microbiol.* 115:301–7

67. Lloyd, D., Ohnishi, T., Lindmark, D. G., Müller, M. 1983. Respiration of *Tritrichomonas foetus. Wiad. Parazytol.* 29:37–39

68. Lloyd, D., Pedersen, J. Z. 1985. Metronidazole radical anion generation in vivo in *Trichomonas vaginalis.* Oxygen quenching is enhanced in a drug-resistant strain. *J. Gen. Microbiol.* 131:87–92

69. Lloyd, D., Williams, J., Yarlett, N., Williams, A. G. 1982. Oxygen affinities of the hydrogenosome-containing protozoa *Tritrichomonas foetus* and *Dasytricha ruminantium,* and two aerobic protozoa, determined by bacterial bioluminescence. *J. Gen. Microbiol.* 128: 1019–22

70. Lloyd, D., Yarlett, N., Yarlett, N. C. 1986. Inhibition of hydrogen production in drug-resistant and susceptible *Trichomonas vaginalis* strains by a range of nitroimidazole derivatives. *Biochem. Pharmacol.* 35:61–64

71. Lo, H.-S., Chang, C.-J. 1982. Purification and properties of NADP-linked alcohol dehydrogenase from *Entamoeba histolytica. J. Parasitol.* 68:372–77

72. Lo, H.-S., Reeves, R. E. 1978. Pyruvate to ethanol pathway in *Entamoeba histolytica. Biochem. J.* 171:225–30

73. Lo, H.-S., Reeves, R. E. 1979. *Entamoeba histolytica:* Flavins in axenic organisms. *Exp. Parasitol.* 47:180–84

74. Lo, H.-S., Reeves, R. E. 1980. Purification and properties of NADPH: flavin oxidoreductase from *Entamoeba histolytica. Mol. Biochem. Parasitol.* 2:23–30

75. Lossick, J. G., Müller, M., Gorrell, T. E. 1986. In vitro drug susceptibility and doses of metronidazole required for cure in cases of refractory vaginal trichomoniasis. *J. Infect. Dis.* 153:948–55

76. Mack, S. R., Müller, M. 1980. End products of carbohydrate metabolism in *Trichomonas vaginalis. Comp. Biochem. Physiol. B* 67:213–16

77. Marczak, R., Gorrell, T. E., Müller, M. 1983. Hydrogenosomal ferredoxin of the anaerobic protozoon *Tritrichomonas foetus. J. Biol. Chem.* 258:12427–33

78. Martinez-Palomo, A. 1982. *The Biology of* Entamoeba histolytica. Chichester, UK: Res. Stud. 161 pp.
79. McLaughlin, J., Aley, S. 1985. The biochemistry and functional morphology of *Entamoeba. J. Protozool.* 32:221–40
80. Montalvo, F. E., Reeves, R. E., Warren, L. G. 1971. Aerobic and anaerobic metabolism in *Entamoeba histolytica. Exp. Parasitol.* 30:249–56
81. Moreno, S. N. J., Mason, R. P., Docampo, R. 1984. Distinct reduction of nitrofurans and metronidazole to free radical metabolites by *Tritrichomonas foetus* hydrogenosomal and cytosolic enzymes. *J. Biol. Chem.* 259:8252–59
82. Moreno, S. N. J., Mason, R. P., Muniz, R. P. A., Cruz, F. S., Docampo, R. 1983. Generation of free radicals from metronidazole and other nitroimidazoles by *Tritrichomonas foetus. J. Biol. Chem.* 258:4051–54
83. Müller, M. 1973. Biochemical cytology of trichomonad flagellates. I. Subcellular localization of hydrolases, dehydrogenases, and catalase in *Tritrichomonas foetus. J. Cell Biol.* 57:453–74
84. Müller, M. 1976. Carbohydrate and energy metabolism of *Tritrichomonas foetus.* See Ref. 55, pp. 3–14
85. Müller, M. 1980. The hydrogenosome. *Symp. Soc. Gen. Microbiol.* 30:127–42
86. Müller, M. 1981. Action of clinically utilized 5-nitroimidazoles on microorganisms. *Scand. J. Infect. Dis. Suppl.* 26:31–41
87. Müller, M. 1986. Reductive activation of nitroimidazoles in anaerobic microorganisms. *Biochem. Pharmacol.* 35:37–41
88. Müller, M. 1987. Hydrogenosomes of trichomonad flagellates. *Acta Univ. Carol. Biol.* 30:249–60
89. Müller, M. 1988. Biochemistry of *Trichomonas vaginalis.* In *Trichomonads Parasitic in Humans,* ed. B. M. Honigberg. New York: Springer. In press
90. Müller, M., Gorrell, T. E. 1983. Metabolism and metronidazole uptake in *Trichomonas vaginalis* isolates with different metronidazole susceptibilities. *Antimicrob. Agents Chemother.* 24:667–73
91. Müller, M., Lindmark, D. G. 1974. Enzymes involved in succinate formation in *Tritrichomonas foetus. Program Abstr. 49th Annu. Meet. Am. Soc. Parasitol., Kansas City, Mo.,* p. 43 (Abstr.)
92. Müller, M., Lindmark, D. G. 1976. Uptake of metronidazole and its effect on viability in trichomonads and *Entamoeba invadens* under anaerobic and aerobic conditions. *Antimicrob. Agents Chemother.* 9:696–700
93. Müller, M., Lindmark, D. G. 1978. Respiration of hydrogenosomes of *Tritrichomonas foetus.* II. Effect of CoA on pyruvate oxidation. *J. Biol. Chem.* 253:1215–18
94. Müller, M., Nseka, V., Mack, S. R., Lindmark, D. G. 1979. Effects of 2,4-dinitrophenol on trichomonads and *Entamoeba invadens. Comp. Biochem. Physiol. B* 64:97–100
95. Nielsen, M. H., Diemer, N. H. 1976. The size, density, and relative area of chromatic granules ("hydrogenosomes") in *Trichomonas vaginalis* Donné from cultures in logarithmic and stationary growth. *Cell Tissue Res.* 167:461–65
96. Nseka, K., Müller, M. 1979. L'action des inhibiteurs de la glycolyse sur l'absorption du metronidazole par les protozoaires *Tritrichomonas foetus* et *Entamoeba invadens. C. R. Soc. Biol.* 172:1094–98
97. Ohnishi, T., Lloyd, D., Lindmark, D. G., Müller, M. 1980. Respiration of *Tritrichomonas foetus.* Components detected in hydrogenosomes and in intact cells by electron paramagnetic resonance spectrometry. *Mol. Biochem. Parasitol.* 2:39–50
98. Olive, P. L. 1979. Correlation between metabolic rates and electron affinity of nitroheterocycles. *Cancer Res.* 39:4512–15
99. Paltauf, F., Meingassner, J. G. 1982. The absence of cardiolipin in hydrogenosomes of *Trichomonas vaginalis* and *Tritrichomonas foetus. J. Parasitol.* 68:949–50
100. Prins, R. A., Prast, E. R. 1973. Oxidation of NADH in a coupled oxidase-peroxidase reaction and its significance for the fermentation in the rumen protozoa of the genus *Isotricha. J. Protozool.* 20:471–77
101. Prins, R. A., van Hoven, W. 1977. Carbohydrate fermentation by the rumen ciliate *Isotricha prostoma. Protistologica* 13:549–56
102. Read, C. P., Rothman, A. H. 1955. Preliminary notes on the metabolism of *Trichomonas vaginalis. Am. J. Hyg.* 61:249–60
103. Reed, L. J. 1974. Multienzyme complexes. *Acc. Chem. Res.* 7:40–46
104. Reeves, R. E. 1970. Phosphopyruvate carboxylase from *Entamoeba histolytica. Biochim. Biophys. Acta* 220:346–49
105. Reeves, R. E. 1984. Metabolism of *Entamoeba histolytica* Schaudinn, 1903. *Adv. Parasitol.* 23:105–42
106. Reeves, R. E., Guthrie, J. D. 1975.

Acetate kinase (pyrophosphate). A fourth pyrophosphate-dependent kinase from *Entamoeba histolytica*. *Biochem. Biophys. Res. Commun.* 66:1389–95

107. Reeves, R. E., Guthrie, J. D., Lobelle-Rich, P. 1980. *Entamoeba histolytica:* isolation of ferredoxin. *Exp. Parasitol.* 49:83–88

108. Reeves, R. E., South, D. J., Blytt, H. J., Warren, L. G. 1974. Pyrophosphate:D-fructose 6-phosphate 1-phosphotransferase. A new enzyme with the glycolytic function of 6-phosphofructokinase. *J. Biol. Chem.* 249:7737–41

109. Reeves, R. E., Warren, L. G., Susskind, B., Lo, H.-S. 1977. An energy-conserving pyruvate-to-acetate pathway in *Entamoeba histolytica*. Pyruvate synthase and a new acetate thiokinase. *J. Biol. Chem.* 252:726–31

110. Ryley, J. F. 1955. Studies on the metabolism of the protozoa. 5. Metabolism of the parasitic flagellate *Trichomonas foetus*. *Biochem. J.* 59:361–69

111. Shorb, M. S. 1964. The physiology of trichomonads. In *Biochemistry and Physiology of Protozoa*, ed. S. H. Hutner, 3:383–457. New York: Academic

112. Snyers, L., Hellings, P., Bovy-Kesler, C., Thines-Sempoux, D. 1982. Occurrence of hydrogenosomes in the rumen ciliates Ophryoscolecidae. *FEBS. Lett.* 137:35–39

113. Steinbüchel, A., Müller, M. 1986. Glycerol, a metabolic end product of *Trichomonas vaginalis* and *Tritrichomonas foetus*. *Mol. Biochem. Parasitol.* 20:45–55

114. Steinbüchel, A., Müller, M. 1986. Anaerobic pyruvate metabolism of *Tritrichomonas foetus* and *Trichomonas vaginalis* hydrogenosomes. *Mol. Biochem. Parasitol.* 20:57–65

115. Takeuchi, T., Weinbach, E. C., Diamond, L. S. 1975. Pyruvate oxidase (CoA acetylating) in *Entamoeba histolytica*. *Biochem. Biophys. Res. Commun.* 65:591–96

116. Tanabe, M. 1979. *Trichomonas vaginalis:* NADH oxidase activity. *Exp. Parasitol.* 48:135–43

117. Tanowitz, H. B., Wittner, M., Rosenbaum, R. M., Kress, Y. 1975. In vitro studies on the differential toxicity of metronidazole in protozoa and mammalian cells. *Ann. Trop. Med. Parasitol.* 69:19–28

118. Taylor, M. B., Berghausen, H., Heyworth, P., Messenger, N., Rees, L. J., Gutteridge, W. E. 1980. Subcellular localization of some glycolytic enzymes in parasitic flagellated protozoa. *Int. J. Biochem.* 11:117–20

119. Thauer, R. K., Jungermann, K., Decker, K. 1977. Energy conservation in chemotrophic anaerobic bacteria. *Bacteriol. Rev.* 41:100–80

120. Tsagoloff, A., Myers, A. M. 1986. Genetics of mitochondrial biogenesis. *Ann. Rev. Biochem.* 55:249–85

121. Tsukahara, T. 1961. Respiratoy metabolism of *Trichomonas vaginalis*. *Jpn. J. Microbiol.* 5:157–69

122. Turner, G., Müller, M. 1983. Failure to detect extranuclear DNA in *Trichomonas vaginalis* and *Tritrichomonas foetus*. *J. Parasitol.* 69:234–36

123. van Bruggen, J. J. A., Stumm, C. K., Vogels, G. D. 1983. Symbiosis of methanogenic bacteria and sapropelic protozoa. *Arch. Microbiol.* 136:89–95

124. van Hoven, W., Prins, R. A. 1977. Carbohydrate fermentation by the rumen ciliate *Dasytricha ruminantium*. *Protistologica* 13:599–606

125. von Brand, T. 1973. *Biochemistry of Parasites*. New York: Academic. 499 pp. 2nd ed.

126. Vossbrinck, C. R., Maddox, J. V., Friedman, S., Debrunner-Vossbrinck, B. A., Woese, C. R. 1987. Ribosomal RNA sequence suggests microsporidia are extremely ancient eukaryotes. *Nature* 326:411–14

127. Wahl, R. C., Orme-Johnson, W. H. 1987. Clostridial pyruvate oxidoreductase and the pyruvate-oxidizing enzyme specific for nitrogen fixation in *Klebsiella pneumoniae* are similar enzymes. *J. Biol. Chem.* 262:10489–96

128. Wang, A. L., Wang, C. C. 1985. Isolation and characterization of DNA from *Tritrichomonas foetus* and *Trichomonas vaginalis*. *Mol. Biochem. Parasitol.* 14:323–35

129. Weinbach, E. C., Claggett, C. E., Keister, D. B., Diamond, L. S., Kon, H. 1980. Respiratory metabolism of *Giardia lamblia*. *J. Parasitol.* 66:347–50

130. Weinbach, E. C., Diamond, L. S. 1974. *Entamoeba histolytica:* I. Aerobic metabolism. *Exp. Parasitol.* 35:232–43

131. Weinbach, E. C., Diamond, L. S., Claggett, C. E., Kon, H. 1976. Iron-sulfur proteins of *Entamoeba histolytica*. *J. Parasitol.* 62:127–28

132. Weinbach, E. C., Takeuchi, T., Claggett, C. E., Inohue, F., Kon, H., Diamond, L. S. 1980. Role of iron-sulfur proteins in the electron transport system of *Entamoeba histolytica*. *Arch. Invest. Med.* 11(Supl. 1):75–81

133. Whatley, J. M., John, P., Whatley, F. R. 1979. From extracellular to in-

tracellular: the establishment of mitochondria and chloroplasts. *Proc. R. Soc. London Ser. B* 204:165–87

134. Williams, A. G. 1986. Rumen holotrich ciliate protozoa. *Microbiol. Rev.* 50:25–49

135. Williams, A. G., Harfoot, C. G. 1987. Factors affecting the uptake and metabolism of soluble carbohydrates by the rumen ciliate *Dasytricha ruminantium* isolated from ovine rumen contents by filtration. *J. Gen. Microbiol.* 96:125–36

136. Williams, K., Lowe, P. N., Leadlay, P. F. 1987. Purification and characterization of pyruvate:ferredoxin oxidoreductase from the anaerobic protozoon *Trichomonas vaginalis*. *Biochem. J.* 246:529–36

137. Wirtschafter, S., Saltman, P., Jahn, T. L. 1956. The metabolism of *Trichomonas vaginalis:* the oxidative pathway. *J. Protozool.* 3:86–88

138. Woese, C. R. 1987. Bacterial evolution. *Microbiol. Rev.* 51:221–71

139. Wood, H. G. 1977. Some reactions in which inorganic pyrophosphate replaces ATP and serves as a source of energy. *Fed. Proc.* 36:2197–205

140. Yarlett, N., Coleman, G. S., Williams, A. G., Lloyd, D. 1984. Hydrogenosomes in known species of rumen entodiniomorphid protozoa. *FEMS. Microbiol. Lett.* 21:15–19

141. Yarlett, N., Gorrell, T. E., Marczak, R., Müller, M. 1985. Reduction of nitroimidazole derivatives by hydrogenosomal extracts of *Trichomonas vaginalis*. *Mol. Biochem. Parasitol.* 14:29–40

142. Yarlett, N., Hann, A. C., Lloyd, D.,

Williams, A. 1981. Hydrogenosomes in the rumen protozoon *Dasytricha ruminantium* Schuberg. *Biochem. J.* 200:365–72

143. Yarlett, N., Hann, A. C., Lloyd, D., Williams, A. G. 1983. Hydrogenosomes in a mixed isolate of *Isotricha prostoma* and *Isotricha intestinalis* from ovine rumen contents. *Comp. Biochem. Physiol. B* 74:357–64

144. Yarlett, N., Lloyd, D., Williams, A. G. 1982. Respiration of the rumen ciliate *Dasytricha ruminantium* Schuberg. *Biochem. J.* 206:259–66

145. Yarlett, N., Lloyd, D., Williams, A. G. 1985. Butyrate formation from glucose by the rumen protozoon *Dasytricha ruminantium*. *Biochem. J.* 228:187–92

146. Yarlett, N., Orpin, C. G., Munn, E. A., Yarlett, N. C., Greenwood, C. A. 1986. Hydrogenosomes in the rumen fungus *Neocallimastix patriciarum*. *Biochem. J.* 236:729–39

147. Yarlett, N., Scott, R. I., Williams, A. G., Lloyd, D. 1983. A note on the effects of oxygen on hydrogen production by the rumen protozoon *Dasytricha ruminantium* Schuberg. *J. Appl. Bacteriol.* 55:359–61

148. Yarlett, N., Yarlett, N. C., Lloyd, D. 1986. Metronidazole resistant clinical isolates of *Trichomonas vaginalis* have lowered oxygen affinities. *Mol. Biochem. Parasitol.* 19:111–16

149. Yeung, T.-C., Beaulieu, B. B. Jr., McLafferty, M. A., Goldman, P. 1984. Interaction of metronidazole with DNA repair mutants of *Escherichia coli*. *Antimicrob. Agents Chemother.* 25:65–70

Ann. Rev. Microbiol. 1988. 42:489–516

ASSEMBLY OF ANIMAL VIRUSES AT CELLULAR MEMBRANES

Edward B. Stephens

Department of Infectious Diseases, College of Veterinary Medicine, University of Florida, Gainesville, Florida 32610

Richard W. Compans

Department of Microbiology, University of Alabama at Birmingham, Birmingham, Alabama 35294

CONTENTS

INTRODUCTION

Many animal viruses possess a lipid-containing envelope acquired as the virion is assembled by a process of budding at a cellular membrane. One or more virus-coded glycoproteins are present on the external surface of these viruses; these glycoproteins have important functions including attachment to

489

0066-4227/88/1001-0489$02.00

specific receptors and a membrane fusion activity involved in virus penetration. The internal components of the virus include a nucleoprotein core and, in many cases, other nonglycosylated proteins. Several viruses with helical nucleocapsids (orthomyxoviruses, paramyxoviruses, and rhabdoviruses) possess a nonglycosylated matrix (M) protein, which appears to be localized under the lipid bilayer and which has an important role in virus assembly. The structure and biosynthesis of enveloped viruses, as well as many aspects of the budding process, have been described in detail in previous reviews (17, 28, 66, 67, 121, 142). The viral glycoproteins are transmembrane proteins that are synthesized on membrane-bound ribosomes in the endoplasmic reticulum and are transported to the site of virus assembly by the same process used by cellular membrane proteins. In contrast, nonglycosylated viral proteins are synthesized in the cytoplasm and associate with the cytoplasmic surface of the membranes where budding takes place. Although the precise interactions that lead to virus budding are not understood, it is likely that interactions between the glycoproteins and internal proteins are involved. The assembly process is specific in that host-cell membrane proteins are excluded from the completed virus particles, whereas the lipids closely reflect those of the host-cell membrane.

In many families of enveloped viruses assembly occurs at the cell surface, and the completed virions are immediately released from the cell. Alpha viruses, arena viruses, orthomyxoviruses, paramyxoviruses, retroviruses, and rhabdoviruses are all assembled by budding at the cell surface. In contrast, virions of several other families are assembled at intracellular membranes. Coronaviruses are assembled by budding at membranes of the rough endoplasmic reticulum or the Golgi complex, and bunyaviruses form by budding into cisternae of the Golgi complex. In each case, viral envelope proteins appear to accumulate at the site of virus maturation and probably serve as sites for the alignment of the internal components of the virion. The herpesviruses are a family of DNA-containing enveloped viruses with icosahedral nucleocapsids that are assembled within the nucleoplasm and acquire an envelope by budding at the inner nuclear envelope. In each of the families of viruses that form at intracellular membranes, completed virions are transported to the cell surface within the cisternal spaces of intracytoplasmic membranes and are released by exocytosis.

Recent studies of glycoproteins of enveloped viruses and their involvement in the assembly process have emphasized several topics. The role of specific protein domains and structural features of glycoproteins in their transport and assembly into virions has been investigated using molecular genetic approaches. Some of the viral molecules that are being studied most actively are depicted schematically in Figure 1. The role of protein folding and oligomerization of subunits in the transport and functional properties of viral glycoproteins is also being actively studied. Since viral glycoproteins are targeted to specific cellular locations, they provide an excellent system for analysis of the structural features of membrane proteins involved in targeting. In this regard, the finding that virus maturation at cell surfaces is restricted to specific plasma membrane domains of polarized epithelial cells has stimulated interest in the use of enveloped viruses in studies of epithelial polarity. In this review we emphasize these aspects of research on enveloped virus assembly. We include examples of families of viruses that are assembled at various cellular membranes. A summary of the structure and protein composition of the virions of some of these families is presented in Table 1.

VIRUSES THAT ASSEMBLE AT THE ENDOPLASMIC RETICULUM: ROTAVIRUSES

The rotaviruses are nonenveloped viruses that have the unique property of maturing intracellularly by budding through the rough endoplasmic reticulum (RER), in a process that involves the transient acquisition of a lipid envelope.

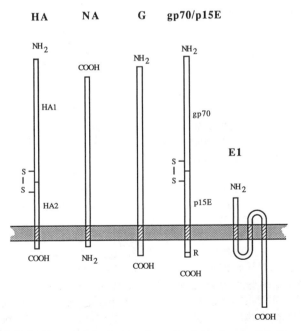

Figure 1   Schematic diagram of the transmembrane topology of some viral glycoproteins. The molecules depicted *(left to right)* include the HA and NA glycoproteins of influenza virus, the G protein of VSV, the envelope [gp70/pl5(E)] protein of murine leukemia virus, and the E1 glycoprotein of coronaviruses. The *shaded* layer represents a lipid bilayer, with the external surface above.

Table 1 Major proteins of some enveloped viruses[a]

Virus (family, example)	Proteins (Designation: MW)		
	Glycoprotein(s)	Nucleocapsid protein	Other major proteins
Bunyavirus, snowshoe hare	G1: 115,000	N: 24,000	
	G2: 38,000		
Coronavirus, mouse hepatitis	E1: 23,000	N: 50,000	
	E2: 180,000		
Orthomyxovirus, influenza A	HA: 77,000[b]	NP: 56,100	M_1: 27,800
	NA: 50,000		
Retrovirus, murine leukemia	gp70/p15(E): 80–85,000	p30: 30,000	p15: 15,000
			p12: 12,000
			p10: 10,000
Rhabdovirus, vesicular stomatitis	G: 57,416	N: 47,355	L: 240,707
			M: 26,064
			NS: 20,989

[a]For references, see text.
[b]Cleaved into two subunits designated HA_1 (~50,000) and HA_2 (~27,000).

Mature rotavirus particles have 11 segments of double-stranded RNA contained within a core particle surrounded by an inner and outer capsid typical of the Reoviridae (3). The simian rotavirus SA11 replicates well in cell culture and has been used to study rotavirus morphogenesis. Virions contain up to eight structural polypeptides (VP), four of which are derived from proteolytic cleavage of precursor polypeptides (34). The VP3 protein is responsible for the viral hemagglutinating activity, and proteolytic cleavage of VP3 into VP5 and VP8 enhances the infectivity of rotaviruses in cell culture (30, 33, 58). The other protein of the outer capsid, VP7, is a glycoprotein and the type-specific neutralization antigen (46, 59). In addition to the virus structural proteins, at least four nonstructural viral proteins (variously designated as NCVP or NS) have been demonstrated in virus-infected cells (32, 70). One of these nonstructural proteins, designated NCVP5 or NS29, is a virally encloded glycoprotein (32).

The intracellular localization of viral polypeptides in SA11 virus–infected cells and their proposed function in the process of rotavirus morphogenesis have recently been summarized (91). The core particles assemble within characteristic perinuclear viroplasmic inclusions, which have been postulated to be the site of RNA replication. Following their assembly, the core particles appear to acquire the inner shell protein (VP6) and one of the outer capsid proteins (VP3) in transit from the viroplasmic inclusions to the RER. The final step in maturation involves budding through ribosome-free areas of the RER. In the process of budding, the particles appear to transiently acquire a lipid envelope, which then disintegrates, releasing the virus particles within the lumen of the RER. As the envelope disintegrates, the VP7 glycoprotein is retained in the outer capsid of the virus. There is no apparent involvement of the Golgi apparatus in viral maturation, and no particles have been observed budding from the plasma membrane.

The use of inhibitors has demonstrated the importance of posttranslational glycosylation for the assembly and infectivity of rotaviruses (89, 115). The presence of tunicamycin (TM) during replication of bovine rotavirus was shown to reduce the size of the glycoproteins and reduce the yield of virus owing to the inhibition of attachment of N-linked oligosaccharides (115). Similarly, the presence of TM during growth of wild-type SA11 virus resulted in a decrease in the titer of infectious virus (89). Using a mutant with an unglycosylated VP7 protein, Petrie et al (89) determined that glycosylation of VP7 is not essential for the maturation and infectivity of the SA11 virus. However, they observed that in the presence of TM, 80–100% of the particles within the lumen of the RER remained enveloped, compared with 10% for virus grown in the absence of TM. These results suggest that glycosylation of the virus-coded nonstructural glycoprotein is essential for proper maturation of the SA11 rotavirus.

Both the VP7 and NCVP5 glycoproteins have primary amino acid sequences that are atypical of most membrane-bound proteins (9, 10). Sequence analysis of VP7 has revealed the presence of two in-phase initiation codons, each of which is followed by hydrophobic domains at the amino terminus; no hydrophobic domain is present in the carboxyl terminus (9). These N-terminal hydrophobic sequences are thought to act as signal sequences. Recently, Chan et al (12) presented evidence that the RNA segment 9, which codes for the VP7 glycoprotein, is functionally bicistronic, since the initiation of translation occurred at either ATG codon. However, these authors determined that if initiation occurred at the first ATG then the hydrophobic signal sequence was removed upon translocation across the membrane of the RER, whereas if the second ATG codon was utilized the signal sequence was not cleaved upon translocation. Both of these VP7s were found to be synthesized in infected cells at similar times, and both were found in mature virions (12). It remains to be determined whether these VP7s are functionally different.

The amino acid sequence of NCVP5 deduced from DNA sequence analysis has revealed that like VP7, the NCVP5 glycoprotein of SA11 rotavirus has two hydrophobic domains at the amino terminus (10). Since the only two potential N-linked glycosylation sites are within the first hydrophobic domain, the N-terminal hydrophobic region is probably not cleaved after translocation across the RER membrane (10).

The VP7 and NCVP5 glycoproteins remain sensitive to endoglycosidase H (endo-H), an enzyme specific for high-mannose oligosaccharides, during the course of an SA11 virus infection and can be localized within the RER by immune electron microscopy techniques (13, 56, 90). Thus, VP7 and NCVP5 glycoproteins may provide useful models for examining the molecular features that are required for retention of membrane proteins within the RER. Mutagenesis techniques were used to create deletions within the coding regions for the first or second hydrophobic domain of the VP7 glycoprotein (96). When mutants with deletions within the first hydrophobic domain were inserted into SV40-based vectors and used to transfect susceptible cells, the expressed proteins were localized within the RER and remained endo-H sensitive. In contrast, glycoproteins with deletions in the second hydrophobic domain were found to be secreted from the cell and became endo-H resistant, which indicates that the secreted forms passed through the Golgi complex in transit to the cell surface. These results indicate that the second hydrophobic domain may contain signals for the retention of VP7 in the RER and is responsible for anchoring this glycoprotein within the RER membrane (96).

VIRUSES THAT MATURE AT THE ENDOPLASMIC RETICULUM AND/OR GOLGI APPARATUS: CORONAVIRUSES AND BUNYAVIRUSES

The coronaviruses and bunyaviruses are assembled at intracellular membranes (RER and/or Golgi apparatus) and appear to utilize the exocytic pathway for virus release. The most widely studied of the coronaviruses are mouse hepatitis virus (MHV) and avian infectious bronchitis virus (IBV). Coronavirus assembly may occur at the RER and/or Golgi apparatus, depending on the cell line (130–132). The bunyaviruses are usually assembled at the Golgi apparatus (77), although budding sometimes occurs at the plasma membrane (4).

The mature coronavirus particles contain two surface glycoproteins (E1 and E2) embedded within the cell-derived lipid bilayer, and one nucleocapsid protein (N), which makes up the protein component of the ribonucleoprotein core. The E1 glycoprotein contains oligosaccharides added to the polypeptide chain via O-glycosidic linkages (49, 79). These oligosaccharides are added by a posttranslational mechanism in the transcisternae of the Golgi apparatus (79). As shown by sequence analysis and protease digestion experiments, the E1 glycoprotein has several unusual structural properties, which include (a) integration into membranes without a cleavable N-terminal sequence, (b) three areas of hydrophobicity within the N-terminal half, which could allow spanning of the membrane multiple times, (c) a long cytoplasmic domain of 124 amino acids, and (d) a small glycosylated region (5 kd) protruding from the envelope (5, 111, 113). Thus, E1 has propeties of both a transmembrane glycoprotein and, because of its long cytoplasmic tail, a matrix protein that interacts with the nucleocapsid on the cytoplasmic face of the membrane during viral maturation. In addition, this glycoprotein has the unusual property of being retained within the Golgi apparatus.

The E2 glycoprotein forms the large peplomeric structures on the envelope and is responsible for binding to cell-surface receptors and virus-mediated cell fusion. The E2 glycoprotein is typical of many cellular membrane proteins in that it is synthesized as a precursor molecule, which is translocated across the RER and transported through the Golgi apparatus, where the high-mannose oligosaccharides are modified to produce complex structures. However, whereas the E1 glycoprotein is retained within the Golgi apparatus, the E2 glycoprotein is readily transported to the plasma membrane. The E2 glycoprotein of MHV is synthesized as a 180-kd molecule, which can be cleaved by trypsin into two subunits of approximately 90 kd each (129). Recent studies have shown that proteolytic cleavage of the gp180 precursor is essential for viral cell-fusion activity (130). In this respect, the E2 glycoprotein appears to be similar to other cleavage-activated viral fusion glycoproteins (influenza HA, parainfluenza F), which require proteolytic activation to mediate fusion of susceptible cells as well as the fusion process involved in virus entry.

Because the E1 glycoprotein is localized within the Golgi apparatus of infected cells, it is a useful model for defining the molecular features that target proteins to the Golgi apparatus. When cDNAs encoding the E1 glycoprotein of MHV or IBV were expressed in cells using SV40-based vectors or vaccinia vectors, both products were found to be localized at the Golgi regions, which indicates that other viral proteins are not required for Golgi localization (68, 112). Using site-directed mutagenesis techniques, Machamer & Rose (68) constructed mutant IBV E1 glycoproteins that lacked two of the three hydrophobic domains. The mutant proteins were glycosylated, which suggests that the first or third transmembrane domain can act as an internal signal sequence. Like the wild-type protein, a deletion mutant lacking the second and third transmembrane domains was apparently retained at intracellular membranes. In contrast, a mutant lacking the first and second domains was transported to the cell surface, which indicates that the first transmembrane domain of the IBV E1 glycoprotein contains a signal for its retention within the Golgi apparatus (68). The structural features of the first transmembrane domain of E1 that permit it to act as a signal for retention are unknown. However, they clearly differ from the Lys-Asp-Glu-Leu (KDEL) sequence, which was recently found to be a signal for retention of proteins within the lumen of the RER (76).

Virions of the family Bunyaviridae consist of a nucleocapsid containing a tripartite single-stranded RNA genome (with segments designated L, M, and S) of negative polarity associated with a major nucleoprotein (N), a minor virus RNA polymerase (L) protein, and two glycoprotein species (G1 and G2) embedded within a host-derived lipid bilayer (7). The bunyaviruses are similar to the coronaviruses in that they lack a true matrix protein. One of their glycoproteins may interact with the nucleocapsid during virus maturation, like the E1 protein of coronaviruses. Most bunyavirus glycoproteins accumulate within the Golgi region of infected cells, although a small fraction of G1 and G2 is reported to be transported to the cell surface (64, 69).

A single mRNA species is transcribed from the M RNA segment, which presumably encodes a polyprotein precursor to the viral glycoproteins (15, 31, 50). The putative polyprotein precursor has never been detected in virus-infected cells, but was produced in an in vitro system (133). Sequence analysis has revealed that the mRNA from the M segment will encode for a polyprotein precursor with or without a nonstructural protein (designated NS_M) fused to the amino terminus, depending on the genus within the Bunyaviridae (31, 50). To date, no information has been obtained about the role of the NS_M protein or about whether it is involved in the localization of viral glycoproteins within the Golgi apparatus.

Treatment of MHV-infected cells with TM resulted in the generation of noninfectious particles lacking the characteristic peplomeric structures (49, 99). The synthesis and incorporation into virions of the E1 glycoprotein was unaffected by the presence of TM, since E1 contains O-linked oligosaccharides. However, the E2 glycoprotein was not detected in TM-treated cells; this accounts for the lack of peplomers on these viral particles. It is unknown whether the absence of E2 was due to an inhibition of E2 synthesis or to the synthesis of a molecule that was readily degraded intracellularly. However, based on these results and on immunofluorescence studies that have localized E1 to the RER and Golgi apparatus, it has been concluded that the E1 glycoprotein determines the site of virus maturation (49, 79, 130). Treatment of bunyavirus-infected cells with TM yielded different results (11). Treatment of snowshoe hare virus–infected cells with TM did not decrease the synthesis of G1 and G2. However, the presence of the drug did prevent glycosylation of the two viral glycoproteins. With snowshoe hare virus, in contrast to the coronaviruses, treatment with TM blocked the assembly and release of virus particles. These results suggest that glycosylated forms of G1 and G2 are necessary for the assembly of this bunyavirus at the Golgi apparatus.

VIRUS ASSEMBLY AT THE PLASMA MEMBRANE

As indicated above, many viruses assemble at the plasma membrane. We have selected three virus families for inclusion in this review because of the large amount of information that has been obtained on each of these systems.

Orthomyxoviruses (Influenza Viruses)

The structure of the influenza viruses has been reviewed in detail elsewhere (16, 78, 81). They contain a ribonucleoprotein core composed of a nucleoprotein (NP) and three polymerase polypeptides (PA, PB1, PB2) in association with eight segments of single-stranded RNA of negative polarity: The envelope consists of an internal matrix (M) protein and two surface glycoproteins, the hemugglutinin (HA) and neuraminidase (NA), which span a host-derived lipid bilayer. The influenza HA glycoprotein has been used as a model membrane glycoprotein to study the mechanism of virus-mediated cell fusion, the process of intracellular glycoprotein transport, and the function of various protein domains in transport and in determining biological properties. The HA and NA glycoproteins have been crystallized, and their three-dimensional structures have been determined by X-ray crystallography (134, 139, 141). The HA glycoprotein is a trimeric molecule, whereas NA is a tetramer. The NA glycoprotein is anchored to membranes by an uncleaved signal-anchor sequence near its N-terminus, whereas the HA glycoprotein is synthesized with a cleaved N-terminal signal peptide and is anchored by a hydrophobic domain near the C-terminus (8, 36, 47). Influenza A virus also codes for three additional nonstructural proteins; one of these, the M2 protein, is a nonglycosylated membrane protein found on the surface of infected cells (143). Influenza B virus encodes a protein of similar size, which has been designated NB (119, 140).

The role of the HA proteins in virus assembly has been investigated using temperature-sensitive *(ts)* mutants defective in HA transport (84). At nonpermissive temperature, cells infected with these mutants produced noninfectious virus particles containing NA but not HA glycoproteins. Thus, the presence of HA on cell surfaces is not required for virus budding. Whether NA is required for the formation of such particles remains to be determined.

The influenza HA glycoprotein is the prototype of a cleavage-activated fusion glycoprotein that has an essential function in virus penetration. The uncleaved protein (HA) is cleaved into two subunits (HA_1 and HA_2) by trypsin-like enzymes. Influenza virus enters susceptible cells via receptor-mediated endocytosis and becomes part of the endosomal compartment, whose contents are acidified (138). The cleaved HA undergoes a conformational change in the acidic environment, after which a hydrophobic domain at the N-terminal end of the HA_2 molecule is exposed; this peptide probably mediates the fusion event (21).

The ability of the fusion peptide to serve as a hydrophobic anchor sequence was examined with the F protein of the paramyxovirus SV5, another cleavage-activated fusion protein (83). Hydrophobic fusion peptides with and without five adjacent arginine residues that serve as a cleavage site were introduced at the carboxyl end of influenza HA molecules in place of the normal transmembrane and cytoplasmic domains. When the sequences were expressed in cells using SV40 vectors, those chimeras with only the fusion peptide sequence were found to be anchored in cell membranes. In contrast, chimeras with the fusion peptide and five adjacent arginines were no longer anchored, but were transported into the lumen of microsomal vesicles. Thus, cleavage of the adjacent charged residues confers an ability on the hydrophobic domain to interact stably with the lipid bilayer. These results indicate that an intricate mechanism permits certain hydrophobic domains to serve as membrane anchor sequences, whereas other hydrophobic domains may be translocated and serve other functions. The position as well as the flanking sequences of the hydrophobic domain may determine whether the domain is translocated or anchored in the membrane.

The role of the cytoplasmic domain of the influenza HA glycoprotein in intracellular transport has been examined using deletion mutants (25, 26). Some changes made within the cytoplasmic domain affected the intracellular transport of HA from the RER to the cell surface (25). However, when the cytoplasmic domain was deleted entirely the resulting molecules were transported to the cell surface and remained anchored in the membrane (26). These results demonstrate that the cytoplasmic domain is not essential for transport and that the hydrophobic transmembrane domain possesses the stop-translocation function for the glycoprotein. Truncation of the hydrophobic domain from 27 to 17 residues also yielded molecules that were anchored in membranes, whereas further truncation was found to result in a loss of membrane anchorage.

Recent studies have demonstrated that correct folding and oligomerization of the HA glycoprotein is essential for its intracellular transport (20, 41). The initial folding and assembly of trimeric HA structure occurs within minutes after synthesis and before the protein leaves the RER. The initial interaction is followed by a stabilization event, which probably occurs after the protein has reached the Golgi complex (20). There was also evidence of an increase in hydrophobicity upon trimer formation, which may stabilize the interaction of the glycoprotein with the lipid bilayer. Several mutants of HA defective in transport from the RER were unable to form native trimers or did so inefficiently (41). The resulting HA monomers were found to be associated with a cellular protein in the RER, the heavy-chain binding protein (BiP). This protein possesses a specific signal (KDEL) for retention in the RER (76). Gething et al (41) suggested that the BiP may have a role in retaining incompletely folded HA proteins in the RER. However, it is also possible that oligomerization is a positive requirement for the mechanism of exit of HA from the RER. Oligomers, but not monomers, have the ability to form multiple, identical contacts with neighboring molecules, either by self-association or by interaction with a specific cellular protein. Such interactions may be necessary for incorporation of HA into a transport vesicle, which may involve lateral redistribution and clustering in the plane of the membrane.

The hydrophobic domain of NA, which serves as a signal-anchor sequence, is about 30 amino acids in length. The effects of introducing charged residues at various sites in this hydrophobic domain have recently been investigated (122). The presence of a single charged residue at position 11 or 17 had no effect on glycosylation or surface expression of the resulting molecule. However, introduction of an arginine residue at position 26 completely prevented surface expression. Double mutants containing two charged residues were also blocked in surface expression, irrespective of the position of the insertions examined. Most of the mutants, however, were able to undergo translocation into the RER and produced glycosylated proteins. These results indicate that the functional properties of the NA signal-anchor sequence may be less sensitive to the presence of charged residues than those of C-terminal hydrophobic anchor sequences (see below), possibly because the former sequence consists of a significantly longer stretch of hydrophobic residues.

As noted above, influenza A and B viruses encode nonstructural membrane proteins that are expressed on the surface of infected cells. Both M2 and NB have a similar size (97 and 100 amino acids), a short extracellular amino terminus and a longer cytoplasmic domain, and an uncleaved hydrophobic sequence that may serve as a signal-anchor (65, 140, 143). However, the transmembrane topology of M2 and NB is opposite that of influenza NA or most other membrane proteins with an uncleaved signal-anchor sequence. The major difference between M2 and NB is that the latter is glycosylated, whereas the former lacks any carbohydrate addition. Both proteins are found on surfaces of virus-infected cells, and the M2 protein, when expressed from cloned DNA, is transport competent (143). The functions of the M2 and NB proteins in influenza virus replication are unknown, but their presence in large numbers on surfaces of infected cells suggests a possible role in virus assembly.

Rhabdoviruses

Vesicular stomatitis virus (VSV), the prototype of the rhabdoviruses, has been widely used to study the assembly of eukaryotic viruses at the plasma membrane. The structure of the VSV virion has been described in detail in previous reviews (28, 135). The mature VSV particle is bullet-shaped and consists of a nucleocapsid containing a single-stranded RNA genome of negative polarity complexed with nucleocapsid protein (N) and smaller amounts of the viral transcriptase proteins (NS and L), a matrix (M) protein, and a membrane glycoprotein (G), which spans the lipid bilayer. The nucleotide sequence of the gene that encodes the G protein has been determined (107). It predicts (*a*) a protein of 511 amino acids with two N-linked glycosylation sites, (*b*) a cleavable hydrophobic signal sequence at the amino terminus, (*c*) a hydrophobic domain of 20 amino acids at the carboxyl terminus, which anchors the protein in the lipid bilayer, and (*d*) a hydrophilic cytoplasmic domain of 29 amino acids at the carboxyl terminus.

Recent studies have indicated that the interaction of the M protein with the nucleocapsid is essential for the budding process. Using *ts* mutants in the M protein and immunofluoresence techniques, Ono et al (80) observed that infection of cells at the permissive temperature resulted in a diffuse localization of M protein within the cytoplasm. Later in infection the M protein could be observed associated with the cell plasma membrane. At permissive temperature the viral nucleocapsids appeared to be coiled, and virus was observed budding from the cell. However, at the nonpermissive temperature the M protein synthesized by these mutants aggregated at the outer nuclear membrane and other vesicular structures, and the virus was defective in nucleocapsid coiling and maturation. From these results the authors concluded that localization of the M protein at the cell plasma membrane is essential for nucleocapsid coiling and the initiation of virus budding.

A *ts* mutant with a lesion in the glycoprotein (ts045) has been used to investigate the role of the G protein in virus assembly (37). This mutant has a reversible defect in transport of the G glycoprotein from the RER to the Golgi complex (6). Sequence analysis of cDNA clones and recombination experiments with wild-type virus and ts045 mutants have demonstrated that the substitution of a serine residue for a phenylalanine within the hydrophobic transmembrane domain is responsible for the defect in transport (40). The ts045 mutant forms spikeless virions at the nonpermissive temperature, and the mechanism of their formation is of interest, since it was proposed that the localization of the G protein in the cell plasma membrane may be essential for initiating the process of virus maturation (121). It appears that a small fraction of the G protein of ts045 is transported to the plasma membrane at the nonpermissive temperature and may be involved in the budding event. Following release of the virus from the cell, the G protein is cleaved by a cell-derived protease. This results in the generation of a soluble component that is released from the virus and a membrane anchor component that remains associated with the spikeless particles (75).

The importance of the cytoplasmic domain in the transport of G protein to the cell surface has been investigated using truncated G proteins or chimeric glycoproteins containing the ectodomain of VSV G protein and the cytoplasmic domains from other viral or cellular membrane glycoproteins. Removal of either the cytoplasmic domain or both the cytoplasmic and transmembrane domains resulted in truncated glycoproteins that were transported at a decreased rate from the RER to the Golgi complex (105, 106). The cytoplasmic domain of the VSV G protein could be substituted with that of immunoglobulin μ_m, influenza HA, or IBV E1 (97). Chimeric glycoproteins formed by substitution of the cytoplasmic domain of immunoglobulin μ_m or influenza HA were transported to the Golgi complex (as determined by endo-H resistance) at about the same rate as the molecule from which the cytoplasmic domains were derived. However, other G/HA chimeras traveled at a rate similar to that of the wild-type G protein (74). Also, substitution of the long cytoplasmic domain (124 amino acids) from the E1 glycoprotein of IBV for the same region of the G protein resulted in a chimera that was transported to the cell surface at about the same rate as the wild-type G protein (97).

The effects of modification of the uncharged, hydrophobic amino acids within the transmembrane domain of the G protein have been examined using site-directed mutagenesis (1). Replacement of an isoleucine residue within the hydrophobic domain with a polar, uncharged glutamine had no effect on transport of G protein to the cell surface. However, replacement of the same isoleucine with a charged arginine residue resulted in a mutant protein that was incorporated into intracellular membranes but was inefficiently transported to the cell surface (1). These results demonstrate that a transmembrane domain whose hydrophobic character is distrubed by the presence of a charged amino acid can still anchor a protein in membranes, but the protein will be aberrant in its transport properties.

The role of oligomerization in the intracellular transport of the VSV G protein has also been examined. Using the ts045 mutant of VSV, Kreis & Lodish (63) demonstrated that formation of oligomers (probably trimers) was necessary for intracellular transport from the RER to the Golgi complex.

Retroviruses

The retroviruses can be divided into types B, C, and D based primarily on their morphology as well as on their genomic composition. We focus on the type C viruses because most of the work concerning assembly and maturation of retroviruses has been performed with these viruses. Murine leukemia viruses (MuLV) encode one glycoprotein, which is synthesized as a high-mannose precursor (endo-H sensitive) molecule of 80–85 kd (gPr80^{env}– gPr85^{env}) depending on the virus strain. The protein is synthesized as a precursor with high-mannose oligosaccharides in the RER and is then transported to the Golgi complex, where the oligosaccharides are processed into complex chains. Concomitantly with transport to the Golgi complex, the *env* precursor is cleaved by a host-cell protease into a glycosylated gp70 subunit and a transmembrane Pr15E molecule, which is responsible for anchoring the gp70/p15(E) glycoprotein into the viral envelope using a hydrophobic domain near the C-terminus (94, 126). The Pr15E also undergoes another proteolytic processing event in which it is cleaved into p15(E) and a p2(E) fragment, also known as the R peptide (45). Unlike cleavage of the *env* precursor into gp70 and Pr15E, cleavage of the Pr15E molecule is mediated by a virally encoded protease. Proteolytic processing of Pr15E is not required for the transport of the envelope glycoprotein, since expression of the unprocessed envelope glycoprotein using a vaccinia virus recombinant resulted in its cell surface localization (128).

The internal structural or *gag* proteins of MuLV are synthesized by proteolytic cleavage of a polyprotein precursor molecule designated $Pr65^{gag}$ into four *gag* proteins designated p15, p12, p30, and p10. In addition, a glycosylated *gag* precursor ($gPr80^{gag}$) is translated from an initiation point upstream from the normal nonglycosylated $Pr65^{gag}$, which results in the addition of a hydrophobic amino terminus (29, 92). This N-terminus presumably acts as a signal sequence for the insertion of the *gag* protein into the membrane of the rough endoplasmic reticulum. Evidence has shown that the anchoring of the $gPr80^{gag}$ protein in the membrane via its hydrophobic amino terminus resembles the anchoring of the influenza neuraminidase (92). The $gPr80^{gag}$ is transported to the cell surface but is not incorporated into mature virions. To investigate the role of the glycosylated *gag* in the assembly and maturation of Moloney murine leukemia virus (M-MuLV), mutants were created with deletions in the region between the 5' long terminal repeat and the initiation codon of $Pr65^{gag}$ (118). These mutants failed to synthesize a glycosylated *gag* molecule but replicated normally in infected cells, demonstrating that this *gag* molecule is not required for assembly and release of M-MuLV.

A *ts* mutant of M-MuLV that was defective in the processing of the *env* precursor but produced normal amounts of the *gag* and *pol* precursors produced particles devoid of the envelope glycoprotein (114). Thus, the envelope protein is apparently not necessary for the release of particles from the cell. Mutants expressing the *gag* precursor but not the *pol* or *env* gene continued to generate viral particles (120), which indicates that only $Pr65^{gag}$ is required for virus particle formation at the cell membrane.

The addition of myristic acid to the *gag* precursor is thought to be required for the association of $Pr65^{gag}$ with the plasma membrane (99, 100, 117). Substitution of the N-terminal glycine residue of the M-MuLV $Pr65^{gag}$ with an alanine residue resulted in the synthesis of unmyristylated *gag* polyprotein molecules (99). Most of the $Pr65^{gag}$ synthesized by this mutant remained associated with the cytosol fraction of the cell, and no discernible viral cores were observed by electron microscopy (99). Similarly, in a study with the type D Mason-Pfizer monkey virus (M-PMV), substitution of the N-terminal glycine of the *gag* polyprotein with valine resulted in unmyristylated *gag* polyproteins that were synthesized at normal levels but remained uncleaved

(100). Core structures did accumulate in the cytoplasm with this M-PMV mutant, although no extracellular virus was released from infected cells (100). Although the results with M-MuLV and M-PMV differ with respect to formation of intracellular particles, results with both viruses demonstrate the importance of myristylation in targeting the *gag* proteins to the plasma membrane for assembly and release (99, 100).

Two recent studies have focused on the role of cleavage of the envelope precursor in cell surface transport and virus infectivity (38, 88). The amino acid sequence at the envelope precursor cleavage site, Lys/Arg-X-Lys/Arg-Arg, is highly conserved among retroviruses. A mutant M-MuLV envelope glycoprotein gene in which the carboxyl terminal arginine of the cleavage site was changed to a lysine residue was created using site-directed mutagenesis (38). As a result of this substitution the envelope precursor was cleaved at a frequency 10-fold lower than in the wild-type protein, was expressed on the surface of infected cells in lower amounts, and produced virions with 10-fold lower infectivity than the wild-type virus (38). These results contrast with results obtained when mutations were created in the cleavage site of the gp85/gp37 glycoprotein of Rous sarcoma virus (RSV). Precursor cleavage was completely inhibited when the precursor cleavage site (Arg-Arg-Lys-Arg) was completely removed and reduced 90% when the lysine residue was changed to glutamic acid (88). However, no alteration was found in intracellular transport or surface expression with these RSV glycoprotein mutants. With certain strains of MuLV (35, 60) the uncleaved precursor can also be detected on surfaces of infected cells.

The role of the cytoplasmic domain of retroviral glycoproteins in viral maturation and assembly has also been investigated. An RSV glycoprotein from which the cytoplasmic domain had been deleted was transported and stably incorporated into the plasma membrane (87). When introduced into cells via an infectious proviral DNA, the cytoplasmic-tail deletion mutant generated infectious virus at levels comparable to that of wild RSV. These results demonstrate that the cytoplasmic domain of this retrovirus glycoprotein is not essential for the assembly and release of infectious virus. Other recent results also indicate that the cytoplasmic domain is nonessential for retrovirus glycoprotein transport (61, 125). The gp52 glycoprotein encoded by Friend spleen focus-forming virus (F-SFFV) lacks a cytoplasmic domain and is defective in transport to the cell surface. A chimeric molecule containing the ectodomain of gp52 and transmembrane and cytoplasmic domains of transport-competent gp70/p15E of F-MuLV was also found to be defective in transport to the cell surface. This indicates that the lack of a cytoplasmic domain is probably not responsible for the transport defect of gp52 (125). Conversely, a chimeric protein containing the ectodomain of F-MuLV and the transmembrane domain of SFFV gp52 was efficiently transported to the surface of cells, although it lacks a cytoplasmic domain (61). However, this chimeric molecule was found to be unstable in its membrane association and was released into the culture medium. These results indicate that the lack of a cytoplasmic tail does not affect transport to the cell surface but may affect the stability of anchorage within membranes.

VIRUS ASSEMBLY AT INNER NUCLEAR MEMBRANES: HERPESVIRUSES

The herpesviruses are assembled by budding at the inner nuclear membrane (22). Most studies on the maturation of herpesviruses have been carried out with the herpes simplex viruses (i.e. HSV-1 and HSV-2); we focus on the assembly of these closely related viruses. Both HSV-1 and HSV-2 contain multiple glycoproteins incorporated into the viral envelope (123). At least seven glycoproteins have been described for HSV-1 (gB1, gC1, gD1, gE1, gG1, gH1, and gI1) and five for HSV-2 (gB2, gC2, gD2, gE2, and gG2). gB appears to be unusual in that besides having an amino-terminal signal sequence, it has three regions of hydrophobicity toward the carboxyl terminus that could act as multiple membrane-spanning domains (86). It was therefore proposed that this glycoprotein could span the membrane multiple times, although no structural studies have been presented to support this observation.

The precise functions of each of the glycoproteins of HSV are not well understood. gB appears to be necessary for the fusion of the viral and cell membranes during the initial stages of infection, since *ts* mutants mapping in this gene absorb at the nonpermissive temperature but do not penetrate the cell (116). Antiserum directed against gD1 and gD2 will neutralize infectivity of HSV-1 and HSV-2, respectively, and protects against disease induced by HSV-1 or HSV-2 when administered passively to mice (23). gE exhibits immunoglobulin G Fc-binding activity, and antiserum directed to gE also neutralizes infectivity (82). Unlike other enveloped viruses, it appears that HSV has some glycoproteins that are unnecessary for replication and assembly at cell membranes (27, 48). Mutants of HSV-1 and HSV-2 that do not express gC continue to replicate normally in vitro. Mutants of pseudorabies virus that do not express the equivalent of gC were also found to replicate normally in cell culture (136). However, gC may be important in the pathogenicity of HSV-1, since nearly all clinical isolates express gC, and a gC-deficient mutant of HSV-1 (KOS) was less pathogenic for mice than wild-type virus (51).

Treatment of infected cells with tunicamycin inhibited the glycosylation of herpes simplex virus glycoproteins and prevented the production of infectious virus (85, 95). Electron microscopic studies on TM-treated HSV-infected cells demonstrated that nucleocapsids were abundant in the nucleus and the cytoplasm, but only a small number of enveloped virions were observed. However, the virions that reached the cell surface did not have HSV glycoproteins or glycoprotein precursors incorporated into the viral envelope (85). gC also contains O-linked oligosaccharides (54, 137). In the presence of TM, gC was efficiently labeled with galactose or glucosamine, whereas gB was not. This suggests that only gC has O-linked oligosaccharides (137). Monensin treatment inhibited the addition of these O-linked oligosaccharides to the HSV glycoproteins (54), which suggests that the attachment of O-linked sugars to HSV-1 glycoproteins probably occurs as a posttranslational modification in the Golgi complex. Apparently O-linked oligosaccharides are not important for infectivity, since virions that accumulate within cytoplasmic vacuoles in the presence of monensin, which contain immature glycoproteins, are still infectious (53). When infected cells were incubated at 34°C, the high-mannose core oligosaccharides appeared to be added to the nonglycosylated precursor of gC in a posttranslational fashion (19).

The HSV glycoproteins that are destined to be incorporated into virions must be transported to the inner nuclear membrane. In fractionation studies the nuclear fraction of virus-infected cells was found to contain predominantly high-mannose precursor forms of glycoproteins gB, gC, and gD, although detectable amounts of the fully processed gC and gD were also found (18). However, glycosylation does not appear to be a prerequisite for nuclear transport, since incubation of infected cells at a reduced temperature (34°C) resulted in the accumulation of nonglycosylated precursors to gB and gC in the nuclear fraction (19). Transport to the inner nuclear membrane may occur at structures where the inner and outer nuclear membranes are contiguous, such as the nuclear pore structures. Apparently the HSV glycoproteins contain the necessary sorting signals for transport to the inner nuclear membrane. The expression of gB and gD in the absence of other HSV proteins resulted in localization of these glycoproteins at the inner nuclear membrane as well as at the cell surface (2, 52). Both the gp110 of Epstein-Barr virus and gB of HSV-1 contain stretches of basic amino acids within the predicted inner nuclear membrane domains (Arg-Arg-Arg-Arg for gp110; Arg-Lys-Arg-Arg for gB). These sequences are similar to those of other proteins known to be targeted to the nucleus, such as the large T-antigens of SV40 and polyoma virus (57). It has been suggested that these regions may be important for the interaction with the nuclear matrix (42). It has also been suggested that viral glycoproteins associate with viral nucleocapsid proteins or other viral structural proteins (i.e. tegument proteins) and that this association causes the accumulation of the glycoproteins at the inner nuclear membrane by retarding their transport to the plasma membrane (52). Patches of the inner nuclear membrane appear to become modified during HSV infection and are seen as electron-dense areas in the electron microscope. Presumably these patches represent aggregated viral glycoproteins and viral tegument proteins, which are incorporated into mature virions (103). Envelopment primarily occurs with nucleocapsids that contain DNA, since enveloped empty capsids are rarely found (103). Enveloped virus particles bud into the perinuclear space, which is contiguous with the endoplasmic reticulum. Following the modification of glycoproteins within the Golgi complex, the mature virions are transported via transport vesicles to the cell surface. These transport vesicles presumably fuse with the cell plasma membrane, which results in the release of mature virions into the extracellular space by an exocytotic mechanism.

VIRUS ASSEMBLY IN POLARIZED EPITHELIAL CELLS

A number of epithelial cell lines in culture retain structural features of differentiated cells of epithelial tissues, including junctional complexes that separate the cell surface into two distinct domains, the apical and basolateral membrane domains. Assembly of enveloped viruses that mature at the plasma membrane is restricted to specific membrane domains in such cells. Influenza virus and paramyxoviruses are released at the apical surfaces of polarized Madin-Darby canine kidney (MDCK) cells, whereas VSV and retroviruses mature only at the basolateral surfaces (102, 108). Based on this finding, enveloped viruses have been extensively used to study the molecular mechanisms involved in the transport of membrane proteins to specific membrane domains of epithelial cells. In studies using SV40 vectors, influenza virus HA or NA glycoproteins expressed in the absence of other influenza viral proteins were transported to the apical surface of polarized monkey kidney cells (55, 108). Similar studies using vaccinia virus vectors or SV40 vectors have shown that the G protein of VSV is transported to the basolateral domain in the absence of other VSV proteins (98, 128). The glycoproteins of ecotropic or dualtropic F-MuLV, when expressed in MDCK cells using vaccinia virus vectors, were also transported to the basolateral surface (126, 128). Since glycosylation of viral glycoproteins is not required for polarized virus budding or surface expression of glycoproteins (109), the results from these studies indicate that each of these molecules possess certain protein domains that serve as sorting signals; these signals are recognized by the cellular machinery that targets these glycoproteins to either the apical or basolateral surface. In doubly infected cells the envelope proteins of influenza virus and VSV are colocalized at least through the Golgi complex (39, 101). Sorting probably occurs as the proteins exit the Golgi complex into distinct sets of transport vesicles. Thus, the sorting signals may interact with cellular proteins involved in formation of transport vesicles at the transcisternae of the Golgi complex, and the resulting vesicles may deliver proteins to the apical or basolateral surface. Examination of deduced amino acid sequences of the glycoproteins reveals no regions of homology in proteins destined for the same membrane domain; in fact, proteins with different transmembrane topology and three-dimensional structures such as influenza HA and NA are transported to the same membrane domain. Thus, epithelial cells appear to possess mechanisms to transport a variety of distinct proteins to the same ultimate plasma membrane destination. Such proteins may have common features in their three-dimensional structures, but if so they are as yet unrecognized.

Recently, several groups have utilized recombinant DNA approaches in which deletion mutants or chimeras between apical and basolateral proteins were constructed to identify the protein domains that contain sorting signals that are responsible for transport to the apical or basolateral surface. For several glycoproteins the cytoplasmic domains were deleted, including the N-terminal 10 amino acids of the influenza NA glycoprotein (55), most of the C-terminal cytoplasmic domain of the p62 glycoprotein of Semliki Forest virus (104), and the entire cytoplasmic domain of murine retrovirus gp70/p15E (61). None of these modifications affected the polarized expression of the resulting glycoprotein, which indicates that the cytoplasmic domains of these proteins do not contain essential information for polarized transport.

In initial studies of chimeric glycoproteins, sequences derived from influenza HA and the VSV G protein were fused within the coding region for the external domain; these proteins were glycosylated but blocked in transport from the RER (72). It was suggested that the failure of these chimeric proteins to be expressed on the surface may have been the result of disruption of the three-dimensional structure of the ectodomain. Based on these results, more recent efforts have focused on maintaining the structural integrity of the ectodomain by construction of chimeras with junctions precisely at one side or the other of the hydrophobic transmembrane domain. Replacement of the transmembrane and cytoplasmic domains of influenza HA with the corresponding region from the VSV G protein resulted in molecules that were fully transport-competent and were expressed on the apical surfaces of epithelial cells (73, 110). These results indicated that the external domain of HA may contain a signal to direct this glycoprotein to the apical membrane. The cleaved N-terminal signal sequence of the HA glycoprotein apparently does not have a role in sorting, since replacement of the HA signal with the corresponding region from the G protein resulted in a molecule that was also transported to the apical surface (74). A construct in which the ectodomain of the G protein was joined to the transmembrane and cytoplasmic domains of HA resulted in a chimeric glycoprotein that was transported to the basolateral surface (74). In another study, a chimera was generated in which the cytoplasmic domain of G was replaced with the corresponding region of HA (98). Using an SV40 vector system, continuous MDCK cell lines were produced that expressed this glycoprotein on both the apical and basolateral surfaces. Although these two studies may appear to have generated conflicting data, the G/HA chimeric proteins differed in their transmembrane portions and in the kinetics of transport from the RER to the Golgi complex. A chimeric DNA clone has also been constructed in which the sequences encoding the transmembrane and cytoplasmic domains of the p15E protein of Friend mink cell focus-forming virus gp70/p15E were replaced with the corresponding regions from influenza HA (E. B. Stephens & R. W. Compans, submitted). Infection of MDCK cells with a vaccinia virus recombinant expressing the chimeric protein resulted in the expression of this glycoprotein exclusively on the apical surface. Thus in this case the ectodomain did not determine the protein's destination. The available information does not present a unified picture; in some chimeric proteins the ectodomain appears to determine the transport pathway, whereas in other constructs the membrane anchor and/or cytoplasmic domain appears to be of primary importance. The explanation for these differences remains to be determined. One possibility is that some glycoproteins contain multiple sorting signals in different domains, one of which has a dominant effect in each construct. Alternatively, some molecules may lack any specific sorting signals, and their site of surface expression may be determined by a default pathway.

Several studies of polarized expression have been carried out with secreted forms of glycoproteins lacking hydrophobic transmembrane domains. Stephens et al (128) removed both the cytoplasmic and transmembrane domains from the basolaterally expressed dualtropic F-MuLV gp70/Pr15E. The resulting glycoprotein was secreted from both the apical and basolateral surfaces of MDCK cells grown on permeable supports. These results indicate either that the membrane anchor of the gp70/Pr15E contains a sorting signal or that the mechanisms involved in sorting of membrane glycoproteins do not recognize secreted proteins. Truncated VSV G proteins that lack both the transmembrane and cytoplasmic domains have also been expressed in MDCK cells grown on permeable supports and were secreted from both the apical and basolateral surfaces (43). In the same study, an HA construct lacking the transmembrane and cytoplasmic domains was also found to be secreted from both the apical and basolateral domains (43). However, another truncated molecule of influenza HA lacking the transmembrane and cytoplasmic domains (110) was reported to be released from the apical surface but not the basolateral surface at early times postinfection. The nondirectional release of several foreign secretory proteins from polarized epithelial cells contrasts with the polarized secretion of endogenous proteins by these cells (44, 62). These observations indicate that, at least for secretory proteins, the default pathway is probably nondirectional (71).

Viruses that are assembled at intracellular membranes of polarized epithelial cells are released by vesicular transport. It has been reported that porcine coronavirus was released from both apical and basolateral surfaces of porcine intestinal epithelial cells (24). Recently, a study using Rift Valley fever virus has provided evidence that a bunyavirus is preferentially released from a specific plasma membrane domain (4). In addition to maturation at the smooth membranes of the Golgi complex and the endoplasmic reticulum, some budding was observed at the basolateral surface of rat hepatocytes. Evidence has also been obtained that the herpesviruses are preferentially released at restricted membrane domains of polarized epithelial cells. Using monoclonal antibodies to the different envelope viral glycoproteins, Srinivas et al (124) showed that four different herpesvirus glycoproteins were expressed on the basolateral surfaces of polarized epithelial cells. Electron microscopic studies on Madin-Darby bovine kidney (MDBK) cells infected with HSV-1 or HSV-2 demonstrated that most virus particles were also released at the basolateral surface. Presently, it is unknown whether the information for sorting these glycoproteins to the basolateral membrane resides within the envelope proteins or whether other viral proteins are required for transport to this surface. However, if the transport of viral glycoproteins occurs in the same vesicles that shuttle the mature virus to the plasma membrane, perhaps a sorting signal is recognized within the lumen of these vesicular structures, since the mature virus is totally contained within these vesicles.

CONCLUDING REMARKS

Although much progress has been made in determining the structural organization of enveloped viruses and the biosynthetic events in their replication, many questions about the assembly process remain to be answered. The precise factors that determine the site of virus assembly are not well understood. The structural features of viral proteins that are essential for incorporation into virions also remain to be determined. Almost no information is available on the possible role of virus-coded nonstructural membrane proteins in virus assembly. The mechanism that results in the exclusion of host-cell membrane proteins from viral envelopes is also not understood.

Two alternative models have been proposed for the sorting of membrane-bound as well as secretory proteins: selective retention versus selective export. The available information provides more evidence of specific signals involved in protein retention; for example, the KDEL sequence is involved in retention of soluble proteins in the RER, and specific hydrophobic domains are involved in retention of viral glycoproteins at intracellular membranes.

However, there is also evidence for positive signals involved in protein transport, such as the mannose-6-phosphate present on cellular proteins destined for lysosomes. The selective retention model predicts that membrane proteins that lack specific retention signals are carried to the cell surface by default. However, at least in the case of membrane proteins destined for distinct plasma membrane domains of polarized epithelial cells, specific signals must exist for the sorting of the proteins destined for apical versus basolateral membranes. Whether both pathways require specific signals, or whether one of the two pathways involves transport by default, remains unanswered. In the case of secretion of proteins from polarized epithelial cells, evidence has been obtained that the default pathway is nondirectional. However, it is not known whether the mechanisms for directional transport of secreted proteins are similar to the sorting events for membrane-bound proteins.

A selection mechanism must exist for the incorporation of proteins into the viral envelope, because host-cell membrane proteins are effectively excluded from the virion even in instances where viral proteins constitute only a minor portion of the proteins of the membranes where assembly takes place. On the other hand, envelope glycoproteins of one virus may be readily incorporated into the envelopes of unrelated viruses, forming phenotypically mixed virions or pseudotypes (14, 142). The glycoproteins of viruses known to undergo phenotypic mixing exhibit considerable variation in their primary amino acid sequences, which indicates a lack of any linear sequence that specifies a recognition signal for virus assembly. Taken together, these observations suggest that there may be a general difference between cellular and viral membrane glycoproteins that determines their ability to be assembled into virions. The nature of this difference remains to be determined. One possibility may be a difference in mobility. Cellular membrane glycoproteins may be prevented from incorporation into virions because of interactions with one another or with other sets of cellular proteins underlying the membrane. Another possibility might be related to the oligomeric structure of the viral proteins, which may have an important role in the lateral interactions needed to form a domain on the plasma membrane from which host-cell proteins are excluded.

ACKNOWLEDGMENTS

Research by the authors was supported by grants AI 12680, AI 23611, AI 26453, and AI 25784 from the National Institute of Allergy and Infectious Diseases and by CA 18611 and CA 47100 from the National Cancer Institute. We thank Betty Jeffrey for assistance in preparing the manuscript and Dr. R. V. Srinivas for preparing Figure 1.

Literature Cited

1. Adams, G., Rose, J. K. 1985. Incorporation of a charged amino acid into the membrane-spanning domain blocks cell surface transport but not membrane anchoring of a viral glycoprotein. *Mol. Cell. Biol.* 5:1442–48
2. Ali, M. A., Butcher, M., Ghosh, H. P. 1987. Expression and nuclear envelope localization of biologically active fusion glycoprotein gB of herpes simplex virus in mammalian cells using cloned DNA. *Proc. Natl. Acad. Sci. USA* 84:5675–79
3. Almeida, J. D. 1979. Morphology and antigenicity of rotavirus. *INSERM* 90:379–92
4. Anderson, G. W., Smith, J. F. 1987. Immunoelectron microscopy of Rift Valley fever viral morphogenesis in primary rat hepatocytes. *Virology* 161:91–100
5. Armstrong, J., Smeekens, S., Rottier, P. J. M., Spann, W., van der Zeijst, B. A. M. 1983. Cloning and sequencing the nucleocapsid and E1 genes of coronavirus MHV-A59. *Adv. Exp. Med. Biol.* 173:155–62
6. Bergmann, J. E., Tokuyasu, K. T., Singer, S. J. 1981. Passage of an integral membrane protein, the vesicular stomatitis virus glycoprotein, through the Golgi apparatus en route to the plasma membrane. *Proc. Natl. Acad. Sci. USA* 78:1746–50
7. Bishop, D. H. L. 1986. Ambisense RNA genomes of arenaviruses and phleboviruses. *Adv. Virus Res.* 31:1–51
8. Blok, J., Air, G. M., Laver, W. G., Ward, C. W., Liley, G. G., et al. 1982. Studies on the size, chemical composition, and partial sequence of the neuraminidase (NA) from type A influenza viruses show that the N-terminal region of the NA is not processed and serves to anchor NA in the viral membrane. *Virology* 119:109–21
9. Both, G. W., Mattick, J. S., Bellamy, A. R. 1983. Serotype-specific glycoprotein of simian 11 rotavirus: coding assignment and gene sequence. *Proc. Natl. Acad. Sci. USA* 80:3091–95
10. Both, G., Seigman, L. J., Bellamy, R. A., Atkinson, P. H. 1983. Coding assignment and nucleotide sequence of simian rotavirus SA11 gene segment 10: Location of glycosylation sites suggests that the signal peptide is not cleaved. *J. Virol.* 48:335–39
11. Cash, P., Hendershot, L., Bishop, D. H. L. 1980. The effects of glycosylation inhibitors on the maturation and intracellular polypeptide synthesis induced by snowshoe hare bunyavirus. *Virology* 103:235–40
12. Chan, W.-K., Penaranda, M. E., Crawford, S. E., Estes, M. E. 1986. Two glycoproteins are produced from the rotavirus neutralization gene. *Virology* 151:243–52
13. Chasey, D. 1980. Investigation of immunoperoxidase-labelled rotavirus in tissue culture by light and electron microscopy. *J. Gen. Virol.* 50:195–200
14. Choppin, P. W., Compans, R. W. 1970. Phenotypic mixing of envelope proteins of the parainfluenza virus SV-5 and vesicular stomatitis virus. *J. Virol.* 5:609–16
15. Collett, M. S., Purichio, A. F., Keegan, K., Frazier, S., Hays, W., et al. 1985. Complete nucleotide sequence of the M RNA segment of Rift Valley fever virus. *Virology* 144:228–45
16. Compans, R. W., Choppin, P. W. 1975. Reproduction of myxoviruses. *Compr. Virol.* 4:174–254
17. Compans, R. W., Klenk, H.-D. 1979. Viral membranes. *Compr. Virol.* 13:293–407
18. Compton, T., Courtney, R. J. 1984. Virus-specific glycoproteins associated with the nuclear fraction of herpes simplex virus type 1–infected cells. *J. Virol.* 49:594–97
19. Compton, T., Courtney, R. J. 1984. Evidence for post-translational glycosylation of a nonglycosylated precursor protein of herpes simplex virus type 1. *J. Virol.* 52:630–37
20. Copeland, C. S., Doms, R. W., Boulau, E. M., Webster, R. G., Helenius, A. 1986. Assembly of influenza hemagglutinin trimers and its role in intracellular transport. *J. Cell Biol.* 103:1179–91
21. Daniels, R. S., Downie, J. C., Hay, A. J., Knossow, M., Skehel, J. J., et al. 1985. Fusion mutants of influenza virus hemagglutinin. *Cell* 40:431–39
22. Darlington, R. W., Moss, L. H. III. 1968. Herpesvirus envelopment. *J. Virol.* 2:48–55
23. Dix, R. D., Pereira, L., Baringer, J. R. 1981. Use of monoclonal antibody directed against herpes simplex virus glycoproteins to protect mice against acute virus-induced neurological disease. *Infect. Immun.* 34:192–99
24. Doughri, A. M., Storz, J. 1977. Light and ultrastructural pathologic changes in intestinal coronavirus infection of newborn calves. *Zentralbl. Veterinaermed. Reihe B* 24:367–87

25. Doyle, C., Roth, M. G., Sambrook, J., Gething, M.-J. 1985. Mutations in the cytoplasmic domain of the influenza virus hemagglutinin affect different stages of intracellular transport. *J. Cell Biol.* 100:704–14

26. Doyle, C., Sambrook, J., Gething, M.-J. 1986. Analysis of progressive deletions of the transmembrane and cytoplasmic domains of influenza hemagglutinin. *J. Cell Biol.* 103:1193–204

27. Draper, K. G., Costa, R. H., Lee, G. T.-Y., Spear, P. G., Wagner, E. K. 1984. Molecular basis of the glycoprotein-C negative phenotypes of herpes simplex type 1 macroplaque strain. *J. Virol.* 51:578–85

28. Dubois-Dalcq, M., Holmes, K., Rentier, B. 1984. *Assembly of Enveloped RNA Viruses.* New York: Springer-Verlag

29. Edwards, S. A., Fan, H. 1979. *gag*-related polyproteins of Moloney murine leukemia virus: evidence for independent synthesis of glycosylated and unglycosylated forms. *J. Virol.* 30:551–63

30. Ericson, B. L., Graham, D. Y., Mason, B. B., Estes, M. K. 1983. Two types of glycoprotein precursors are produced by simian rotavirus SA11. *Virology* 127:320–32

31. Eshita, Y., Bishop, D. H. L. 1984. The complete sequence of the M RNA of snowshoe hare bunyavirus reveals the presence of internal hydrophobic domains in the viral glycoprotein. *Virology* 137:227–40

32. Espejo, R. T., Lopez, S., Arias, C. 1981. Structural polypeptides of simian rotavirus SA11 and the effect of trypsin. *J. Virol.* 37:156–60

33. Estes, M. K., Graham, D. Y., Mason, B. B. 1981. Proteolytic enhancement of rotavirus infectivity: molecular mechanisms. *J. Virol.* 39:879–88

34. Estes, M. K., Palmer, E. L., Obijeski, J. F. 1983. Rotaviruses: a review. *Curr. Top. Microbiol. Immunol.* 105:123–87

35. Famulari, N., English, K. J. 1981. *env* gene products of AKR dual tropic viruses: examination of peptide maps and cell surface expression. *J. Virol.* 40:971–76

36. Fields, S., Winter, G., Brownlee, G. 1981. Structure of the neuraminidase gene in human influenza virus A/PR/8/34. *Nature* 290:213–17

37. Flamand, A. 1970. Etude génétique du virus de la stomatite vésicularie: classment de mutants thermosensibles spontanés engroupe de complémentation. *J. Gen. Virol.* 8:187–95

38. Freed, E. O., Risser, R. 1987. The role of envelope glycoprotein processing in murine leukemia virus infection. *J. Virol.* 61:2852–56

39. Fuller, S. D., Bravo, R., Simons, K. 1985. An enzymatic assay reveals that proteins destined for the apical or basolateral domains of an epithelial cell line share the same late Golgi compartments. *EMBO J.* 4:297–307

40. Gallione, C. J., Rose, J. K. 1985. A single amino acid substitution in a hydrophobic domain causes temperature sensitive cell-surface transport of a mutant viral glycoprotein. *J. Virol.* 54:374–82

41. Gething, M.-J., McCammon, K., Sambrook, J. 1986. Expression of wild type and mutant forms of influenza hemagglutinin: the role of folding in intracellular transport. *Cell* 46:939–50

42. Gong, M., Ooka, T., Matsuo, T., Kieff, E. 1987. Epstein-Barr virus glycoprotein homologous to herpes simplex virus gB. *J. Virol.* 61:499–508

43. Gonzalez, A., Rizzolo, L., Rindler, M., Adesnik, M., Sabatini, D. D., Gottlieb, T. 1987. Nonpolarized secretion of truncated forms of the influenza hemagglutinin and the vesicular stomatitis virus G protein from MDCK cells. *Proc. Natl. Acad. Sci. USA* 84:3738–42

44. Gottlieb, T. A., Beaudry, G., Rizzulo, L., Colman, A., Rindler, M., et al. 1986. Secretion of endogenous and exogenous proteins from polarized MDCK cell monolayers. *Proc. Natl. Acad. Sci. USA* 83:2100–4

45. Green, N., Shinnick, T. M., Witte, O. N., Ponticelli, A., Sutcliffe, J. G., et al. 1981. Sequence-specific antibodies show that maturation of Moloney leukemia virus envelope polyprotein involves removal of a COOH-terminal peptide. *Proc. Natl. Acad. Sci. USA* 78:6023–27

46. Greenberg, H. B., Valdesuso, J., Van Wyke, K., Midthun, K., Walsh, M., et al. 1983. Production and characterization of monoclonal antibodies directed at two surface proteins of rhesus rotavirus. *J. Virol.* 47:267–75

47. Hiti, A. L., Nayak, D. P. 1982. Complete nucleotide sequence of the neuraminidase gene of human influenza virus A/WSN/33. *J. Virol.* 41:730–34

48. Holland, T. C., Homa, F. L., Marlin, S. D., Levine, M., Glorioso, J. 1984. Herpes simplex virus type 1 glycoprotein C–negative mutants exhibit multiple

phenotypes, including secretion of truncated glycoproteins. *J. Virol.* 52:566–74

49. Holmes, K. V., Doller, E. W., Sturman, L. S. 1981. Tunicamycin resistant glycosylation of a coronavirus glycoprotein: demonstration of a novel type of viral glycoprotein. *Virology* 115:334–44

50. Ihara, T., Smith, J., Dalrymple, J. M., Bishop, D. H. L. 1985. Complete sequence of the glycoproteins and M RNA of Punta Toro phlebovirus compared to those of Rift Valley Fever. *Virology* 144:246–59

51. Johnson, D. C., McDermott, M. R., Chrisp, C., Glorioso, J. C. 1986. Pathogenicity in mice of herpes simplex virus type 2 mutants unable to express glycoprotein C. *J. Virol.* 58:36–42

52. Johnson, D. C., Smilley, J. R. 1985. Intracellular transport of herpes simplex virus gD occurs more rapidly in uninfected cells than in infected cells. *J. Virol.* 54:682–89

53. Johnson, D. C., Spear, P. G. 1982. Monensin inhibits the processing of herpes simplex virus glycoproteins, their transport to the cell surface, and the egress of virions from infected cells. *J. Virol.* 43:1102–12

54. Johnson, D. C., Spear, P. G. 1983. O-Linked oligosaccharides are acquired by herpes simplex virus glycoproteins in the Golgi apparatus. *Cell* 32:987–97

55. Jones, L. V., Compans, R. W., Davis, A. R., Bos, T. J., Nayak, D. P. 1985. Surface expression of influenza virus neuraminidase, an amino-terminally anchored viral membrane glycoprotein, in polarized epithelial cells. *Mol. Cell. Biol.* 5:2181–89

56. Kabcenell, A. K., Atkinson, P. H. 1985. Processing of the rough endoplasmic reticulum membrane glycoproteins of rotavirus SA11. *J. Cell Biol.* 101:1270–80

57. Kalderon, D., Richardson, W. D., Markham, A. F., Smith, A. E. 1984. Sequence requirements for nuclear localization of the simian virus large T-antigen. *Nature* 311:33–38

58. Kalica, A. R., Flores, J., Greenberg, H. B. 1983. Identification of the rotaviral gene that codes for the hemagglutination and protease-enhanced plaque formation. *Virology* 125:194–205

59. Kalica, A. R., Greenberg, H. B., Wyatt, R. G., Flores, J., Sereno, M. M., et al. 1981. Genes of human (strain Wa) and bovine (strain UK) rotaviruses that code for neutralization and subgroup antigens. *Virology* 112:385–90

60. Kemp, M. C., Famulari, N. G., Compans, R. W. 1981. Glycopeptides of murine leukemia viruses. III. Glycosylation of *env* precursor glycoproteins. *J. Virol.* 39:463–70

61. Kilpatrick, D. K., Srinivas, R. V., Stephens, E. B., Compans, R. W. 1987. Effects of deletion of the cytoplasmic domain upon surface expression and membrane stability of a viral envelope glycoprotein. *J. Biol. Chem.* 262:16116–21

62. Kondor-Koch, C., Bravo, R., Fuller, S. D., Cutler, D., Garoff, H. 1986. Exocytic pathways exist to both the apical and basolateral cell surface of the polarized epithelial cell MDCK. *Cell* 43:297–306

63. Kreis, T., Lodish, H. F. 1986. Oligomerization is essential for transport of vesicular stomatitis viral glycoprotein to the cell surface. *Cell* 46:929–37

64. Kuismamen, E., Hedman, K., Saraste, J., Pettersson, R. F. 1982. Uukuniemi virus maturation: accumulation of virus particles and viral antigens in the Golgi complex. *Mol. Cell. Biol.* 2:1444–58

65. Lamb, R. A., Zebedee, S. L., Richardson, C. D. 1985. Influenza virus M_2 protein is an integral membrane protein expressed on the infected cell surface. *Cell* 40:627–33

66. Lenard, J. 1978. Virus envelope and plasma membranes. *Ann. Rev. Biophys. Bioeng.* 7:139–65

67. Lenard, J., Compans, R. W. 1974. The membrane structure of lipid-containing viruses. *Biochem. Biophys. Acta* 344:51–94

68. Machamer, C., Rose, J. K. 1987. A specific transmembrane domain of a coronavirus E1 glycoprotein is required for its retention in the Golgi region. *J. Cell Biol.* 105:1205–14

69. Madoff, D. H., Lenard, J. 1982. A membrane glycoprotein that accumulates intracellularly: cellular processing of the large glycoprotein of La Crosse virus. *Cell* 28:821–29

70. Mason, B. B., Graham, D. Y., Estes, M. K. 1983. Biochemical mapping of the simian rotavirus SA11 genome. *J. Virol.* 46:413–23

71. Matlin, K. S. 1986. The sorting of proteins to the plasma membrane in epithelial cells: a review. *J. Cell Biol.* 103:2565–68

72. McQueen, N. L., Nayak, D. P., Jones, L. V., Compans, R. W. 1984. Chimeric influenza virus hemagglutinin containing either the NH_2 terminus or the COOH terminus of G protein of vesicular stomatitis virus is defective in trans-

port to the cell surface. *Proc. Natl. Acad. Sci. USA* 81:395–99

73. McQueen, N. L., Nayak, D. P., Stephens, E. B., Compans, R. W. 1986. Polarized expression of a chimeric protein in which the transmembrane and cytoplasmic domains of the influenza virus hemagglutinin have been replaced by those of vesicular stomatitis virus G protein. *Proc. Natl. Acad. Sci. USA* 83:9318–22

74. McQueen, N. L., Nayak, D. P., Stephens, E. B., Compans, R. W. 1987. The ectodomains of the influenza hemagglutinin and vesicular stomatitis virus G protein contain signals for polarized expression. *J. Biol. Chem.* 262:16233–40

75. Metsikko, K., Simons, K. 1986. The budding mechanism of spikeless vesicular stomatitis virus particles. *EMBO J.* 5:1913–20

76. Munro, S., Pelham, H. R. B. 1987. A C-terminal signal prevents secretion of luminal ER proteins. *Cell* 48:899–907

77. Murphy, F. A., Harrison, A. K., Whitfield, S. G. 1973. Bunyaviridae: morphologic and morphogenetic similarities of Bunyamwera serologic supergroup viruses and several other arthropod-borne viruses. *Intervirology* 1:297–316

78. Nayak, D. P., ed. 1981. *ICN-UCLA Symp. Mol. Cell. Biol.* Vol. 21. New York: Academic. 672 pp.

79. Neimann, H., Klenk, H.-D. 1981. Coronavirus glycoprotein E1, a new type of viral glycoprotein. *J. Mol. Biol.* 153:993–1010

80. Ono, K., Dubois-Dalcq, M. E., Schubert, M., Lazzarini, R. A. 1987. A mutated membrane protein of vesicular stomatitis virus has an abnormal distribution within the infected cell and causes defective budding. *J. Virol.* 61:1332–41

81. Palese, P., Kingsbury, D. W., eds. 1983. Genetics of influenza viruses. New York: Springer-Verlag. 360 pp.

82. Para, M. F., Baucke, R. B., Spear, P. G. 1980. Immunoglobulin G (Fc)-binding receptors on the virions of herpes simplex virus type 1 and transfer of these receptors to the cell surface by infection. *J. Virol.* 34:512–20

83. Paterson, R. G., Lamb, R. A. 1987. Ability of the hydrophobic fusion-related external domain of a paramyxovirus F protein to act as a membrane anchor. *Cell* 48:441–52

84. Pattnaik, A. K., Brown, D., Nayak, D. P. 1986. Formation of influenza virus particles lacking hemagglutinin on the viral envelope. *J. Virol.* 60:994–1001

85. Peake, M. L., Nystrom, P., Pizer, L. I. 1982. Herpesvirus glycoprotein synthesis and insertion into plasma membranes. *J. Virol.* 42:678–90

86. Pellett, P. E., Kousoulas, K. G., Pereira, L., Roizman, B. 1985. Anatomy of the herpes simplex virus type 1 strain F glycoprotein B gene: primary sequence and predicted protein structure of wild type and of monoclonal antibody resistant mutants. *J. Virol.* 53:243–53

87. Perez, L., Davis, G., Hunter, E. 1987. Mutants of the Rous sarcoma virus envelope glycoprotein that lack the transmembrane anchor and cytoplasmic domains: analysis of intracellular transport and assembly into virions. *J. Virol.* 61:2981–88

88. Perez, L., Hunter, E. 1987. Mutations within the proteolytic cleavage site of Rous sarcoma virus glycoprotein that block processing to gp85 and gp37. *J. Virol.* 61:1609–14

89. Petrie, B. L., Estes, M. K., Graham, D. Y. 1983. Effects of tunicamycin on rotavirus morphogenesis and infectivity. *J. Virol.* 46:270–74

90. Petrie, B. L., Graham, D. Y., Hannsen, H., Estes, M. K. 1982. Localization of rotavirus antigens in infected cells by ultrastructural immunocytochemistry. *J. Gen. Virol.* 63:457–67

91. Petrie, B. L., Greenberg, H. B., Graham, D. Y., Estes, M. K. 1984. Ultrastructural localization of rotavirus antigens using colloidal gold. *Virus Res.* 1:133–52

92. Pillemer, E. A., Kooistra, D. A., Witte, O. N., Weisman, I. L. 1986. Monoclonal antibody to the amino-terminal L sequence of murine leukemia virus glycosylated *gag* polyproteins demonstrates their unusual orientation in the cell membrane. *J. Virol.* 57:413–21

93. Deleted in proof

94. Pinter, A., Honnen, W. J. 1983. Topography of murine leukemia virus envelope proteins: characterization of transmembrane components. *J. Virol.* 46:1056–60

95. Pizer, L. I., Cohen, G. H., Eisenberg, R. J. 1980. Effect of tunicamycin on herpes simplex virus glycoproteins and infectious virus production. *J. Virol.* 34:142–53

96. Poruchynsky, M. S., Tyndall, C., Both, G. W., Sato, F., Bellamy, A. R., et al. 1985. Deletions into an NH_2-terminal hydrophobic domain result in secretion of rotavirus VP7, a resident endoplasmic

reticulum membrane protein. *J. Cell Biol.* 101:2199–209

97. Puddington, L., Machamer, C. E., Rose, J. K. 1986. Cytoplasmic domains of cellular and viral integral membrane proteins substitute for the cytoplasmic domain of the vesicular stomatitis virus glycoprotein in transport of the plasma membrane. *J. Cell Biol.* 102:2147–57

98. Puddington, L., Woodgett, C., Rose, J. K. 1987. Replacement of the cytoplasmic domain alters sorting of a viral glycoprotein in polarized cells. *Proc. Natl. Acad. Sci. USA* 84:2756–60

99. Rein, A., McClure, M. R., Rice, N. R., Luftig, R. B., Schultz, A. M. 1986. Myristalization site in Pr65^{gag} is essential for virus particle formation by Moloney murine leukemia virus. *Proc. Natl. Acad. Sci. USA* 83:7246–50

100. Rhee, S. S., Hunter, E. 1987. Myristylation is required for intracellular transport but not for assembly of D-type retrovirus capsids. *J. Virol.* 61:1045–53

101. Rindler, M. J., Ivanov, I. E., Plesken, H., Rodriguez-Boulan, E. J., Sabatini, D. D. 1984. Viral glycoproteins destined for apical or basolateral membrane domains traverse the same Golgi apparatus during their intracellular transport in Madin-Darby canine kidney cells. *J. Cell Biol.* 98:1304–19

102. Rodriguez-Boulan, E., Sabatini, D. D. 1978. Asymmetric budding of viruses in epithelial monolayers: a model system for study of epithelial polarity. *Proc. Natl. Acad. Sci. USA* 75:5071–75

103. Roizman, B., Batterson, W. 1986. Herpesviruses and their replication. In *Fundamental Virology*, eds. B. N. Fields, D. M. Knipe, pp. 607–37. New York: Plenum. 768 pp.

104. Roman, L. M., Garoff, H. 1986. Alteration of the cytoplasmic domain of the membrane spanning glycoprotein p62 of Semliki Forest virus does not affect its polar distribution in established lines of Madin-Darby canine kidney cells. *J. Cell Biol.* 103:2607–18

105. Rose, J. K., Bergmann, J. E. 1982. Expression from cloned cDNA of cell-surface and secreted forms of the glycoprotein of vesicular stomatitis virus. *Cell* 30:753–62

106. Rose, J. K., Bergmann, J. E. 1983. Altered cytoplasmic domains affect intracellular transport of the vesicular stomatitis virus glycoprotein. *Cell* 34:513–14

107. Rose, J. K., Gallione, C. J. 1981. Nucleotide sequences of the mRNA's encoding the vesicular stomatitis G and

M proteins as determined from cDNA clones containing the complete coding regions. *J. Virol.* 39:519–28

108. Roth, M. G., Compans, R. W., Guisti, L., Davis, A. R., Nayak, D. P., et al. 1983. Influenza virus hemagglutinin expression is polarized in cells infected with recombinant SV40 vectors carrying cloned hemagglutinin DNA. *Cell* 33:435–43

109. Roth, M. G., Fitzpatrick, J. P., Compans, R. W. 1979. Polarity of influenza and vesicular stomatitis virus maturation in MDCK cells: lack of a requirement for glycosylation of viral glycoproteins. *Proc. Natl. Acad. Sci. USA* 76:6430–34

110. Roth, M. G., Gundersen, D., Patil, N., Rodriguez-Boulan, E. 1987. The large external domain is sufficient for the correct sorting of secreted or chimeric influenza virus hemagglutinins in polarized monkey kidney cells. *J. Cell. Biol.* 104:769–82

111. Rottier, P. J. M., Armstrong, J., Meyer, D. L. 1985. Signal recognition particle-dependent insertion of coronavirus E1, an intracellular membrane glycoprotein. *J. Biol. Chem.* 260:4648–52

112. Rottier, P. J. M., Rose, J. K. 1987. Coronavirus E1 glycoprotein expressed from cloned cDNA localizes in the Golgi region. *J. Virol.* 61:2042–45

113. Rottier, P. J. M., Welling, G. W., Welling-Wester, S., Niesters, H. G. M., Lenstra, J. A., et al. 1986. Predicted membrane topology of the coronavirus protein E1. *Biochemistry* 25:1335–39

114. Ruta, M., Murray, M. J., Webb, M. C., Kabat, D. 1979. A murine leukemia virus mutant with a temperature sensitive defect in membrane glycoprotein synthesis. *Cell* 16:77–88

115. Sabara, M., Babiuk, L. A., Gilchrist, J., Misra, V. 1982. Effect of tunicamycin on rotavirus assembly and infectivity. *J. Virol.* 43:1082–90

116. Sarmiento, M., Haffey, M., Spear, P. G. 1979. Membrane protein specified by herpes simplex viruses III. Role of glycoprotein VP7(B$_2$) in virion infectivity. *J. Virol.* 29:1149–58

117. Schultz, A. M., Orozslan, S. 1983. In vivo modification of retroviral *gag* gene–encoded polyproteins by myristic acid. *J. Virol.* 46:355–61

118. Schwartzberg, P., Collicelli, J., Goff, S. P. 1983. Deletion mutants of Moloney murine leukemia virus which lack glycosylated *gag* protein are replication competent. *J. Virol.* 46:538–46

119. Shaw, M. W., Choppin, P. W. 1984. Studies on the synthesis of the influenza B virus NB glycoprotein. *Virology* 139:178–84

120. Shields, A., Witte, O. N., Rothenberg, E., Baltimore, D. 1978. High frequency of aberrant expression of Moloney murine leukemia virus in clonal infections. *Cell* 14:601–9

121. Simons, K., Garoff, H. 1980. The budding mechanisms of enveloped animal viruses. *J. Gen. Virol.* 50:1–21

122. Sivasubramanian, N., Nayak, D. P. 1987. Mutational analysis of the signal-anchor domain of influenza virus neuraminidase. *Proc. Natl. Acad. Sci. USA* 84:1–5

123. Spear, P. 1986. Glycoproteins specified by herpes simplex virus. In *The Herpesviruses,* ed. B. Roizman, 3:315–56. New York: Plenum. 416 pp.

124. Srinivas, R. V., Balachandran, N., Alonso-Caplen, F. V., Compans, R. W. 1986. Expression of herpes simplex glycoproteins in polarized epithelial cells. *J. Virol.* 58:689–93

125. Srinivas, R. V., Kilpatrick, D. R., Compans, R. W. 1987. Intracellular transport and leukemogenicity of spleen focus-forming virus envelope glycoproteins with altered transmembrane domains. *J. Virol.* 61:4007–11

126. Stephens, E. B., Compans, R. W. 1986. Non-polarized expression of a secreted murine leukemia virus glycoprotein. *Cell* 47:1053–59

127. Deleted in proof

128. Stephens, E. B., Compans, R. W., Earl, P., Moss, B. 1986. Surface expression of viral glycoproteins in polarized epithelial cells using vaccinia virus vectors. *EMBO J.* 5:237–45

129. Sturman, L. S., Holmes, K. V. 1981. Proteolytic cleavage of peplomeric glycoprotein E2 of MHV yields two 90 K subunits and activates cell fusion. *Adv. Exp. Med. Biol.* 173:25–35

130. Sturman, L. S., Holmes, K. V. 1983. The molecular biology of coronaviruses. *Adv. Virus Res.* 28:35–112

131. Tooze, J., Tooze, S. A. 1985. Infection of AtT20 murine pituitary tumour cells by mouse hepatitis virus strain A59: Virus budding is restricted to the Golgi region. *Eur. J. Cell Biol.* 37:203–12

132. Tooze, J., Tooze, S., Warren, G. 1984. Replication of coronavirus MHV-A59 in Sac⁻ cells: determination of the first site of budding of progeny virions. *Eur. J. Cell Biol.* 33:281–93

133. Ulmanen, I., Seppala, P., Pettersson, R. P. 1981. In vitro translation of Uukuniemi virus–specific RNAs: identification of a nonstructural protein and a precursor to the membrane glycoproteins. *J. Virol.* 37:72–79

134. Varghese, J. N., Laver, W. G., Colman, P. M. 1983. Structure of the influenza virus glycoprotein neuraminidase at 2.9 Å resolution. *Nature* 303:35–40

135. Wagner, R. R., ed. 1987. *The Rhabdoviruses.* New York: Plenum. 562 pp.

136. Wathen, M. W., Wathen, L. M. K. 1986. Characterization and mapping of a non-essential pseudorabies virus glycoprotein. *J. Virol.* 58:173–78

137. Wenske, E. A., Courtney, R. J. 1983. Glycosylation of herpes simplex virus type 1 gC in the presence of tunicamycin. *J. Virol.* 46:297–301

138. White, J., Matlin, K., Helenius, A. 1981. Cell fusion by Semliki Forest, influenza, and vesicular stomatitis viruses. *J. Cell Biol.* 89:674–79

139. Wiley, D. C., Skehel, J. J. 1987. The structure and function of the hemagglutinin membrane glycoprotein of influenza virus. *Ann. Rev. Biochem.* 56:365–94

140. Williams, M. A., Lamb, R. A. 1986. Determination of the orientation of an integral membrane protein and sites of glycosylation by oligonucleotide-directed mutagenesis: Influenza B virus NB glycoprotein lacks a cleavable signal sequence and has an extracellular NH₂-terminal region. *Mol. Cell. Biol.* 6:4317–28

141. Wilson, I. A., Skehel, J. J., Wiley, D. C. 1981. The hemagglutinin membrane glycoprotein of influenza virus: structure at 3 Å resolution. *Nature* 289:366–73

142. Zavada, J. 1982. The pseudotypic paradox. *J. Gen. Virol.* 63:15–24

143. Zebedee, S. L., Richardson, C. D., Lamb, R. A. 1985. Characterization of the influenza virus M₂ integral membrane protein and expression at infected cell surface from cloned cDNA. *J. Virol.* 56:502–11

Ann. Rev. Microbiol. 1988. 42:517–45

NOSOCOMIAL FUNGAL INFECTIONS: A Classification for Hospital-Acquired Fungal Infections and Mycoses Arising from Endogenous Flora or Reactivation[1]

Thomas J. Walsh and Philip A. Pizzo

Section of Infectious Diseases, Pediatric Branch, Clinical Oncology Program, Division of Cancer Treatment, National Cancer Institute, Bethesda, Maryland 20892

CONTENTS

[1]The US Government has the right to retain a nonexclusive, royalty-free license in and to any copyright covering this paper.

INTRODUCTION

Nosocomial fungal infections have emerged during the past two decades as a frequent cause of morbidity and mortality in hospitalized patients. A study (38) from the Centers for Disease Control (CDC) found that from 1970 to 1976 there was a 158% increase in the incidence of nosocomial aspergillosis owing to an increasing number of patients receiving immunosuppressive therapy. A conservative projection of the 1979 nationwide cost of hospitalizations for nosocomial candidiasis and aspergillosis was $5.3 million. The CDC's National Nosocomial Infection Survey (NNIS) (18–21) found that the rate of nosocomial candidemia increased progressively by 114% from 1980 to 1984. These estimates of nosocomial candidiasis and aspergillosis are likely to be quite conservative compared to the true magnitude of nosocomial fungal infection, since invasive candidiasis and aspergillosis are often diagnosed only at postmortem examinations, which are now conducted for only about 20% of deaths nationwide. Nosocomial mycoses have developed especially in association with or as a consequence of the extraordinary progress in the management of seriously ill patients during the past three decades. However, despite this increase, there has been comparatively little progress in elucidating the pathogenesis of nosocomial fungal infections or in their prevention, diagnosis, and treatment. As delineated by Bullock & Deepe (15), there is a critical need for more efforts to be directed toward medical mycology, especially in clinical research for diagnosis, prevention, and treatment and in laboratory investigation of host defense mechanisms and virulence factors of the medically important fungi.

DEFINITIONS AND CLASSIFICATION

Numerous systemic, opportunistic, and dermatophytic fungal pathogens have been reported to cause infection in hospitalized patients. The current literature is ambiguous as to which fungal infections are truly nosocomial. Moreover, the designation of a particular mycosis as a nosocomial infection often provides little insight into the mechanism of acquisition of the organism or the pathogenesis of the infection. For example, the epidemiology, environmental microbiology, pathogenesis, and implications for infection control differ markedly for a patient with asymptomatic gastrointestinal *Candida* colonization who is admitted for newly diagnosed leukemia and subsequently develops disseminated candidiasis during treatment and another patient with newly diagnosed leukemia in whom pulmonary aspergillosis is acquired from a contaminated ventilation system.

There is a need for classification and definition of terms in order to better understand nosocomial fungal infections. The following classification and definitions of nosocomial fungal infections are applied throughout this paper (Table 1). A nosocomial fungal infection is any superficial, locally invasive, or disseminated mycosis acquired from the hospital environment (Type I, hospital-acquired) or developing during the course of a patient's hospitalization from a fungus previously colonizing or latently infecting a patient (Type II, hospital-associated). Type II fungal infections are further classified as infections developing from endogenous colonizing fungi (Subtype IIA) or infections developing from reactivation of latently infecting fungi (Subtype IIB).

ASPERGILLOSIS

Mycology

The genus *Aspergillus* is characterized in culture by hyaline septate hyphae from which conidiophores with vesicles develop. Phialides arising from the vesicles give rise to conidia (phialoconidia), which measure 2.5–3.5 μm in diameter in most pathogenic strains. While numerous species of *Aspergillus* exist in nature, the most commonly infecting species in most institutions is *Aspergillus fumigatus,* followed by *Aspergillus flavus*. *A. flavus* may predominate in some institutions. Occasionally *Aspergillus niger* and *Aspergillus terreus* also cause infection.

Environmental Microbiology

Aspergillus species are ubiquitous saprophytes in nature. Air is the principal route of transmission of *Aspergillus* within the hospital environment, and the respiratory tract is the most common portal of entry. The small diameter of

Table 1 Classification of principal nosocomial fungal infections

Hospital acquired (Type I)	Hospital associated (Type II)	
	Endogenous colonization (Type IIA)	Latent infection (Type IIB)
Aspergillus spp.	*Candida* spp.[a]	*Histoplasma capsulatum*
Mucoraceae:	*Trichosporon* spp.	*Cryptococcus neoformans*
Mucor spp.	*Malassezia furfur*	*Coccidioides immitis*
Rhizopus spp.	*Trichophyton rubrum*	
Absidia spp.	*Saccharomyces cere-*	
Rhizomucor spp.	*visiae*	
Pseudallescheria boydii		
Fusarium spp.		
Dematiaceous hyphomycetes		
Candida spp.[b]		
Coccidioides immitis[c]		
Histoplasma capsulatum[d]		
Cryptococcus neoformans[d]		
Blastomyces dermatitidis[d]		
Dermatophytes[d]		

[a]Principal mechanism of nosocomial infection.
[b]*Candida* spp. have been transmitted from the hospital environment to patients in several well-characterized outbreaks.
[c]Laboratory acquisition and transmission from wound dressings.
[d]Laboratory acquisition.

Aspergillus conidia permits them to reach the pulmonary alveolar spaces, where they may germinate to form hyphae. Generation of conidia usually occurs outside the hospital (112, 118). Conidia can enter the hospital when there is an inadequate filtration system for outside air (94–96). Indeed, Noble & Clayton (77) observed that *A. fumigatus* was easily recovered from the inanimate environment of an unventilated hospital and that autumn and winter were the peak seasons for recovery of this fungus.

Conidia can also be generated within the hospital environment (2, 119, 134). When prospectively monitored corridor air samples at a bone marrow transplant station at the University of Minnesota disclosed a marked increase in airborne thermotolerant *Penicillium* conidia, the source was traced to rotting cabinet wood enclosing a sink with leaking pipes in a medication room (119). Although *Aspergillus* spp. were not recovered from this environmental source, further studies showed that *A. fumigatus* and *A. flavus* could grow on this woody substrate, which was thus a potential source of nosocomial generation of conidia (119). The authors cautioned against permitting wet organic substrates in a hospital's environment because of the risk of propagation of conidiogenous fungi.

Another environmental study (134) reviewed a pseudoepidemic that arose from stored respiratory culture isolates of *Aspergillus* spp. from bone marrow transplant recipients. The cross-contamination of new blood and other cultures that were processed in the same room initially raised concerns about a true outbreak of aspergillosis in a cancer population.

Epidemiology

We classify most cases of nosocomial aspergillosis as hospital-acquired or Type I nosocomial fungal infections. Two principal mechanisms of transmission of *Aspergillus* spp. to hospitalized patients are known: (*a*) airborne transmission through the respiratory tract or into an operative site and (*b*) contact transmission, such as transmission through direct inoculation from occlusive materials. The former is the most common mechanism of nosocomial transmission of most filamentous fungi, including *Aspergillus* spp.

AIRBORNE TRANSMISSION   *Aspergillus* spp. may be cultured from various sources within the hospital environment: (*a*) unfiltered air, (*b*) ventilation systems contaminated at intake and/or exhaust ducts, (*c*) dust dislodged during hospital renovation and construction, and (*d*) horizontal surfaces, food, and ornamental plants (116).

Nonfiltered, nonventilated air   The importance of nonfiltered, nonventilated air was demonstrated indirectly. Rose & Hirsch (99) observed that when a hospital moved from an older building without filtered air to a new building with prefiltered, nonrecirculating air, the rate of new cases of nosocomially acquired pulmonary aspergillosis declined from 11 cases per 5 yr to none. Rosen & Sternberg (100) noted that the frequency of aspergillosis and zygomycosis declined from 145 cases in a 3-yr period in an older hospital with no central ventilation to 30 cases over the next 3 yr in a new hospital with a centrally filtered ventilation system.

Filtered, ventilated air   Ventilation systems may also provide a nidus for growth and transmission of *Aspergillus* conidia. Ventilation systems may be colonized with *Aspergillus* spp. along intake ducts, exhaust ducts, or filters. Involvement of the intake arm of a ventilation system is usually associated with a source of *Aspergillus* conidia on or near the exterior portion of an intake duct. When exhaust ducts are involved, a faulty exhaust fan or air hammer is usually implicated.

Air intake ducts have been identified as the source of several outbreaks of nosocomial aspergillosis. An outbreak of prosthetic endocarditis caused by an *Aspergillus* sp. was associated with contamination of an air intake system (39). Investigators observed that a luxuriant patch of moss on the roof and collections of pigeon feces on the window ledge adjacent to the operating room air intake duct harbored *Aspergillus* species. After removal of the moss and pigeon excreta, cleaning of the air ducts, and replacement of the intake filters by a more efficient particle extraction grade, no new cases of aspergillosis appeared. A cluster of four cases of aspergillosis in renal transplant recipients was associated with pigeon excreta at the air intake ports of the transplantation unit's ventilation system. The external duct inlets were modified to prevent access by birds. An outbreak of prosthetic valve endocarditis due to *A. fumigatus* in seven patients was attributed to contamination of the air intake ducts, including the intake filters of the operating room (85). The first three cases prompted an environmental investigation, which revealed dust and fibers within the ducts, which harbored profuse growth of *A. fumigatus*. Although the ventilation ducts were cleaned and the operating rooms were painted, four more cases of *A. fumigatus* prosthetic valve endocarditis developed. The sources of *Aspergillus* conidia in these subsequent cases were contaminated air intake filters, which had not been changed after the initial investigation. Limited efficiency of air intake filters may also have contributed to the nosocomial aspergillosis observed in allogeneic bone marrow transplant recipients at the Roswell Park Memorial Institute in Buffalo, New York (94, 103).

The air supply ducts to four operating rooms at The Johns Hopkins Hospital in Baltimore were identified recently as the source of the *A. flavus* that caused primary cutaneous aspergillosis associated with insertion of Hickman intravenous catheters in nine granulocytopenic patients with leukemia or aplastic anemia within a 5-mo period (4). Following renovation of the air supply system for the four operating rooms and vacuuming of the air supply ducts to all other operating rooms, no viable fungal particles were found and no additional cases of mycotic postoperative wound infection or colonization were identified.

Exhaust ducts are also sources of nosocomial aspergillosis. Backflow of air from a common exhaust duct into patients' rooms on a renal transplant unit was implicated as the source of nosocomial pulmonary aspergillosis in three renal transplant recipients (57). A bird screen had been removed from the exhaust duct opening, and bird excreta had accumulated and become the substrate for an *Aspergillus* sp. When the exhaust fan malfunctioned, the air flow reversed and evidently carried *Aspergillus* conidia into the patients' rooms. Another outbreak of aspergillosis, related to exhaust ducts and air backflow, developed 10 days after an exhaust fan for four rooms on a pediatric oncology floor was shut down for maintenance (62). The air-conditioning system became contaminated with *Aspergillus* conidia, and five leukemic children in these four rooms acquired inavasive pulmonary, sinopulmonary, or sino-orbital aspergillosis over the next 2 mo. After the contaminated equipment and rooms were cleaned, only two cases of aspergillosis occurred over a 12-mo period, and these occurred in other rooms.

The foregoing outbreaks of nosocomial aspergillosis reinforce the importance of appropriate construction, maintenance, and monitoring of air filtration and ventilation systems. Bone marrow transplant units should be ventilated with HEPA filters. Ventilation systems of bone marrow transplant units and operating rooms should be routinely monitored for such thermotolerant fungi as the *Aspergillus* spp. A cluster of nosocomial aspergillosis cases should prompt a close inspection of ventilation systems as a potential source of infectious conidia.

Construction-associated transmission   Hospital construction and renovation are yet another source of airborne transmission of *Aspergillus* conidia and outbreaks of nosocomial aspergillosis. An outbreak was described when the Baltimore Cancer Research Program moved into a new hospital (2). A clustering of eight new cases of aspergillosis prompted an environmental investigation, which implicated *Aspergillus* conidia–laden cellulose-based fireproofing material coating steel girders and dust on pipes and ceiling panels. Further studies indicated that the *Aspergillus* sp. had probably colonized the fireproofing material after it had been sprayed on the girders. In another outbreak, Arnow et al (8) found that dust that was jarred loose by hospital renovations on one floor and filtered through holes in ceiling acoustical tiles was the probable source of the *A. fumigatus* that caused pulmonary aspergillosis in renal transplant recipients located on the floor below. Arnow et al (8) recommended removing patients from renovation sites, establishing impermeable plastic barriers between patient floors and renovation sites, and vacuuming and damp dusting horizontal surfaces and false ceiling tiles for infection control. Another study (55) attributed nosocomial aspergillosis and zygomycosis in two premature infants to inadequate barriers having permitted a high density of conidial transmission from a construction site to the newborn special care unit. Opal et al (81) reported an outbreak of invasive aspergillosis that developed during the course of hospital renovation in a large military center. The incidence of nosocomial aspergillosis increased from 4 cases in a 30-mo preconstruction period to 11 cases in 24 mo during construction without control measures ($p < 0.05$). Following the installation of airtight plastic and dry-wall barriers around the construction sites, negative-pressure ventilation in the work area, and environmental decontamination with copper8-quinolinolate and HEPA filters in rooms for immunocompromised patients, the incidence of invasive aspergillosis decreased to one case in 18 mo. On the other hand, construction is not invariably associated with increases in nosocomial aspergillosis. Considerable renovation at the National Institutes of Health Clinical Center has taken place during the last 6 yr with no measurable increase in the incidence of nosocomial aspergillosis in immunocompromised patients.

Construction outside the hospital, especially construction that involves excavation of large volumes of soil, has been strongly associated with concurrent nosocomial aspergillosis. Road construction outside the Milwaukee County Medical Center and contaminated air conditioners were correlated by retrospective autopsy review with a cluster of ten cases of invasive aspergillosis in immunocompromised patients within a 27-mo period (60).

Streifel et al (118) demonstrated that during demolition of a seven-story building within the University of Minnesota's hospital complex the outdoor ambient concentration of *A. fumigatus* and *A. niger* increased 1000-fold, whereas indoor concentrations of airborne *A. fumigatus* and *A. niger* conidia increased only 10-fold. Windows and doors had been sealed, and air-handling systems were turned off or adjusted for complete recirculation of internal air to minimize airborne transmission of conidia. Both excavation of massive volumes of soil and demolition of an apartment building on the windward side of the Roswell Park Memorial Institute were identified as possible sources of airborne *Aspergillus* conidia during a concurrent outbreak of nosocomial aspergillosis in allogeneic bone marrow transplant recipients (94).

Several variables may contribute simultaneously to an increase in *Aspergillus* conidia within a hospital's environment. Sarubbi et al (105), for example, found that increased respiratory colonization by *A. flavus* at a North Carolina hospital was associated with construction adjacent to the hospital and with a defective mechanical ventilation and filtration system. Environmental and host variables may be simultaneously involved such that construction or excavation activity can only be linked to some but not all clusters of nosocomial aspergillosis (51, 136).

CONTACT TRANSMISSION   *Aspergillus* conidia may occur on surfaces of the inanimate hospital environment such as dressings and other materials applied to the skin. Although this mechanism of transmission is less common than the airborne transmission of conidia, the clinical consequences of established cutaneous aspergillosis are potentially lethal in granulocytopenic patients. Prystowsky et al (92) described locally invasive cutaneous aspergillosis that developed under adhesive tape or paper-covered extremity boards in three of four granulocytopenic children. The lesions of nosocomial cutaneous aspergillosis were locally invasive and destructive of skin, subcutaneous tissue, and bone. Cutaneous aspergillosis was the apparent portal of entry for fatal disseminated aspergillosis.

Nosocomial burn wound infection may occur through either contact or airborne transmission. After the introduction of topical mafenide acetate (Sulfamylon cream), a 10-fold increase in burn wound infections due to *Aspergillus* and the Mucoraceae was observed at the Brooke Army Medical

Center at Fort Sam Houston, Texas (75). Deep biopsies of suspicious lesions are necessary to distinguish between colonization and infection.

Nosocomial postoperative osteomyelitis may also develop as the result of airborne or contact transmission of *Aspergillus* conidia. The two most common clinical forms of postoperative osteomyelitis due to *Aspergillus* spp. are postthoracotomy sternal osteomyelitis (9, 120, 137) and vertebral osteomyelitis following initial surgery for a noninfectious vertebral process (108, 120). Mycotic pseudoaneurysm of an aortic bypass graft with contiguous vertebral osteomyelitis is an unusual but recently noted form of *Aspergillus* osteomyelitis (14, 41). Nosocomial postoperative lumbar disc space infection with *Aspergillus* spp. (65) may develop under conditions similar to those of postoperative *Aspergillus* vertebral osteomyelitis. Early diagnosis, systemic amphotericin B, and aggressive wound and bone debridement afford the best therapeutic outcome for this infection.

Inadvertant percutaneous intraperitoneal administration of *A. fumigatus* was reported in a 22-year old woman with *Aspergillus* peritonitis who was undergoing peritoneal dialysis for renal failure due to eclampsia (101). *A. fumigatus* was isolated from the labor room, and numerous filamentous fungi were found in dialysis fluid bags, several of which were found to be leaking (117).

Clinical Manifestations

The paranasal sinuses are frequently infected in immunocompromised patients, especially those with profound and protracted granulocytopenia. A recent review by Viollier et al (125) reported that the major factors apparently predisposing to nosocomial *Aspergillus* sinusitis were profound and protracted (mean of 42 days) granulocytopenia ($<$500 granulocytes μl^{-1}) and prolonged antibiotic therapy (mean duration of 22 days). While treatment with amphotericin B was initially successful, the fungal sinusitis frequently recurred when the primary neoplastic disease relapsed or more chemotherapy was required.

Pulmonary involvement by *Aspergillus* spp. may be manifested as allergic bronchopulmonary aspergillosis, aspergilloma (usually involving a preformed tuberculous or sarcoid cavity), or invasive pulmonary aspergillosis. Nosocomial pulmonary aspergillosis is essentially restricted to the last of these manifestations.

The primary host defenses against invasive pulmonary aspergillosis are phagocytic cells, including pulmonary alveolar macrophages and granulocytes. The *Aspergillus* spp. are opportunistic pathogens to which patients with quantitative or qualitative defects in granulocytes are especially susceptible (26, 113, 140). Granulocytopenia is the most common factor predisposing to nosocomial aspergillosis in most institutions. The predisposition of patients with granulocytopenia was recently reinforced in an analysis of risk factors in cancer patients with aspergillosis (40a). Protracted granulocytopenia, especially beyond 21 days, was found to be the single most important independent condition predisposing to pulmonary aspergillosis. A study conducted at the University of Minnesota (84) found nosocomial pulmonary aspergillosis to be the single most important infection causing death in allogeneic bone marrow transplant recipients.

Pulmonary aspergillosis with spiking temperatures may be seen in an immunosuppressed patient receiving broad-spectrum antibacterial antibiotics (72). Pulmonary infiltrates may be absent early in the course of aspergillosis in granulocytopenic patients. Nonproductive cough and tachypnea often develop. The propensity of the *Aspergillus* spp. to invade blood vessels often results in clinical manifestations of pulmonary infarction, including fever, pleuritic pain, pleural rubs, and segmental pulmonary infiltrates.

Invasive pulmonary aspergillosis is a necrotizing pneumonia resulting from pulmonary infarction and from direct tissue invasion by *Aspergillus* spp. Pulmonary cavitation may ensue, even during the course of administration of amphotericin B. A particularly catastrophic consequence of pulmonary aspergillosis is massive hemoptysis. This frequently lethal complication may occur either during granulocytopenia (82) or as the granulocytopenia is resolving (3, 82).

In cases of aspergillosis, hematogenous dissemination of fungal fragments to various tissues occurs in immunocompromised patients. Involvement of the central nervous system (CNS) may present as abrupt onset of a focal neurologic deficit or as focal seizure activity (127). Computerized tomography usually reveals a ring-enhancing lesion or a pattern consistent with infarction. CNS aspergillosis is often misdiagnosed as a progression of the underlying disease. *Aspergillus* spp. may infect the gastrointestinal tract, with resultant necrosis and infarction leading to bleeding and perforation. *Aspergillus* spp. may also disseminate to the heart, resulting in endocarditis, myocarditis, or pericarditis.

The diagnosis of invasive aspergillosis is based upon clinical manifestations and confirmation of the presence of an *Aspergillus* sp. A positive nasal surveillance culture or positive sputum culture for *Aspergillus* spp. in the febrile granulocytopenic patient is strongly indicative of invasive aspergillosis; bronchoscopy or open-lung biopsy may be necessary to establish the diagnosis (141). Immunodiagnostic assays are promising but require additional evaluation (11, 27, 121).

Prevention and Treatment

Measures to prevent nosocomial aspergillosis are based upon elimination of *Aspergillus* conidia from the environment of high-risk patients. The cost and inconvenience of laminar air-flow protected environments are considerable (7). However, such facilities are highly effective in filtering out *Aspergillus* conidia and preventing pulmonary aspergillosis (12, 97). Preliminary results of a study conducted by Meunier et al (70) suggest that the use of an intranasal aerosol of amphotericin B may be effective in reducing the incidence of nosocomial aspergillosis. Systemic antifungal prophylaxis against *Aspergillus* spp. is under investigation (115).

Once a diagnosis of invasive aspergillosis is established, appropriate treatment, in our opinion, consists of intravenous (i.v.) amphotericin B at 0.6–1.2 mg kg^{-1} per day for the duration of granulocytopenia, for a total dose of 1.5–4.0 of amphotericin B. Survival of the patient with pulmonary aspergillosis is critically dependent upon recovery from granulocytopenia or reversal of other immunologic deficits. If the patient becomes granulocytopenic again, aspergillosis frequently recurs in the previously diseased sites, and amphotericin B therapy must be reinstituted. A recent study (15a) found that high doses of amphotericin B (1.0–1.5 mg kg^{-1} per day) or high doses of amphotericin B plus flucytosine (dosed to peak serum level of 30–60 μg ml^{-1}) were associated with a high survival rate in 13 of 15 patients (87%) with invasive aspergillosis and acute leukemia.

In persistently granulocytopenic patients who develop fewer and new pulmonary infiltrates while receiving broad-spectrum antibiotics, the probability is 50% that the new pulmonary infiltrate has a fungal etiology (24). The early initiation of amphotericin B treatments in these patients, even if a microbiologic or histopathologic diagnosis cannot be carried out, improves the likelihood of success of antifungal therapy.

ZYGOMYCOSIS (MUCORMYCOSIS)

Mycology

The term "zygomycosis" (mucormycosis) is defined as fungal infection due to members of the order Mucorales, which includes the families Cunninghamellaceae, Mortierellaceae, Mucoraceae, Saksenaeaceae, and Syncephalastraceae. The first four families are unusual causes of nosocomial zygomycosis (13, 56, 79). Members of the Mucoraceae are more frequently identified as causes of nosocomial infection (71). This family consists of the genera *Mucor, Rhizomucor, Rhizopus,* and *Absidia.* In culture, the species in these genera develop stolons, rhizoids (in *Rhizopus, Absidia,* and *Rhizomucor*), and sporangia atop the aerial sporangiophores. Sporangia contain myriads of sporangiospores, which serve as the means of disseminating the Mucoraceae species into the environment.

Environmental Microbiology

The Mucoraceae are saprophytic hyphomycetes found ubiquitously in soil and dust and on bread and fruit. The distribution of their sporangiospores parallels that of *Aspergillus* conidia in the hospital environment. Most similarly, sporangiospores also enter the hospital from outside sources.

The Mucoraceae are not part of the normal human flora. Nosocomial mucormycosis, like nosocomial aspergillosis, is a hospital-acquired (Type I) nosocomial fungal infection. Sporangiospores are carried from the hospital environment to susceptible patients by airborne or contact transmission (36, 37, 40, 55, 107). The source of an outbreak of nosocomial mucormycosis of surgical wounds, including sternotomy wounds, became apparent when a brand of elasticized adhesive-tape dressing was found to be contaminated by *Rhizopus rhizopodiformis* (37, 40).

Clinical Manifestations

The most common infections due to the Mucoraceae are rhinocerebral zygomycosis and pulmonary mucormycosis, which involve the upper and lower respiratory tracts, respectively (59). Rhinocerebral zygomycosis classically develops in uncontrolled keto-acidotic diabetics, but it may also occur in granulocytopenic patients and organ transplant recipients. This infection is highly aggressive, resulting in rapid destruction of the maxillary and ethmoid sinuses, orbital cavity, orbit, cranial nerves, blood vessels, and eventually the brain.

Pulmonary zygomycosis usually develops in granulocytopenic patients and organ transplant recipients (59, 71). The clinical course is similar to that of pulmonary aspergillosis, with persistent fever, progressive pulmonary infiltrates, and features consistent with pulmonary infarction in susceptible hosts receiving antibacterial antibiotic therapy. Dissemination to the brain presents as focal neurologic deficits or seizures. Intestinal zygomycosis may be complicated by infarction or perforation. Surgical wounds and burn wounds (76) may be infected by zygomycetes through airborne or contact transmission.

Uremia is a well-known, albeit uncommon, condition that increases the risk of zygomycosis. Administration of deferoxamine was recently recognized to predispose patients with chronic renal failure who are undergoing hemodialysis to rhinocerebral and disseminated zygomycosis (42, 138).

Treatment

Early and aggressive surgical debridement, i.v. amphotericin B, and correction of underlying metabolic abnormalities or immunologic deficits are essential for cure of rhinocerebral zygomycosis. Hyperbaric oxygen has been suggested as an adjunct to treatment of rhinocerebral zygomycosis. Optimal treatment of deep zygomycosis of surgical and burn wounds requires debridement of all infected tissue and administration of amphotericin B. Treatment of pulmonary and disseminated zygomycosis requires early initiation of i.v. amphotericin B therapy and reversal of immunologic defects.

OTHER HOSPITAL-ACQUIRED NOSOCOMIAL MYCOSES

Pseudallescheria boydii

Pseudallescheria boydii is a homothallic ascomycete whose conidial counterpart is *Scedosporium (Monosporium) apiospermum*. *P. boydii* is a soil saprophyte and a common cause of mycetoma. However, various medical centers have increasingly reported *P. boydii* as a cause of lethal disseminated infection in immunocompromised patients (summarized in 96a, 109). The host range, clinical manifestations, and histopathology of disseminated pseudallescheriasis are similar to those of disseminated aspergillosis. However, *P. boydii* is usually resistant to amphotericin B and is more susceptible to imidazoles. Disseminated pseudallescheriasis is one of the few remaining indications for i.v. miconazole (61).

Fusarium *Species*

Fusarium spp. are soil saprophytes that usually produce both micro- and macrophialoconidia. *Fusarium* spp. have been recognized with increased frequency as causes of disseminated infection in immunocompromised patients (6, 22, 49, 52). Disseminated *Fusarium* infection is often associated with prominent erythematous maculopapular to nodular lesions, from which the organism may be cultured. Skin biopsy generally reveals branching septate hyphae invading dermal blood vessels. While amphotericin B remains the preferred antifungal agent, some clinical isolates have been found resistant to it.

Paecilomyces *Species*

Paecilomyces variotii is a common soil saprophyte with phialoconidia. This species is known to cause nosocomial infections of prostheses, including cerebrospinal fluid shunts, orthopedic appliances, and prosthetic valves (50). More recently *Paecilomyces lilacinus* was reported as the cause of cellulitis in an immunocompromised patient (47). *P. variotii* is often susceptible to amphotericin B.

Dematiaceous Hyphomycetes

Deep-tissue infection by septate dematiaceous hyphae, which is termed phaeohyphomycosis, is being reported increasingly. Causative organisms include species of the genera *Exserohilum, Bipolaris, Curvularia, Xylohypha,* and *Alternaria* (1, 6). These organisms occur as saprophytes in nature and may be found in the hospital environment. The extent to which they contribute to nosocomial fungal infections remains to be established. For example, phaeomycotic cysts, which are localized infections of skin by dematiaceous hyphomycetes, often develop at the site of i.v. cannula insertion and may be underdiagnosed. Geographic variability in distribution may also contribute to differences in the incidence of nosocomial phaeohyphomycosis.

CANDIDIASIS

Mycology

Candida spp. are characterized by blastoconidia, pseudohyphae, and hyphae. The development of germ tubes is indicative of *Candida albicans*. Germ-tube negative *Candida* spp. are identified by their biochemical characteristics, which include the assimilation and fermentation of carbohydrates.

Identification of specific strains of *Candida* provides important information regarding the epidemiology of nosocomial candidiasis. Burnie et al (16) demonstrated the first outbreak of systemic candidiasis due to cross-infection with a particular strain of *C. albicans*. Yeast isolates were characterized by a biochemical phenotypic marking system. The same strain was identified as the cause of systemic candidiasis in 13 patients in an intensive care unit. Colonization by the same strain was demonstrated in hospital personnel. Two nurses had acquired the strain on their hands after caring for systemically infected patients. No environmental source was identified.

Other techniques that offer potential for further elucidation of nosocomial candidiasis and strain differentiation include analysis of endonuclease restriction patterns of *Candida* DNA (68, 106), immunotyping (58), and strain differentiation using the yeast killer factor system (90).

Environmental Microbiology

In contrast to the *Aspergillus* spp., the Mucoraceae, and the other hyphomycetes previously described, *C. albicans* is a part of the normal gastrointestinal and cutaneous flora of most humans (80). During hospitalization, patients exposed to broad-spectrum antibiotics, which suppress gastrointestinal bacterial flora, may become extensively colonized by *C. albicans*. It may then become invasive and disseminate from the gastrointestinal tract in surgical patients and patients receiving cytotoxic chemotherapy, which often disrupts the normal integrity of the mucosal epithelia. Cutaneous colonization by *Candida* spp. may serve as the source of infection associated with percutaneous intravascular catheters. Dermal and subcutaneous invasion by *C. albicans* in the burn patient may lead to nosocomial disseminated candidiasis (76, 114). Thus, invasive candidiasis is usually a Type II nosocomial fungal infection. This contrasts with the airborne and contact transmissions of conidia generally applicable to Type I nosocomial fungal infections due to *Aspergillus* spp., the Mucoraceae, *P. boydii*, *Fusarium* spp., *Paecilomyces* spp., and dematiaceous hyphomycetes. However, there have also been well-documented outbreaks of nosocomial candidiasis in which transmission from the hospital environment (i.e. Type I nosocomial fungal infection) was clearly substantiated.

C. albicans is the *Candida* species most frequently isolated from hospitals. *Candida tropicalis* has been described in some medical centers as a prevalent and virulent pathogen in granulocytopenic patients (46, 131, 139). *Candida krusei* has also been associated with granulocytopenic patients (69). *Candida parapsilosis* is strongly associated with infections of vascular catheters and total parenteral nutrition (TPN) catheters. *Candida guillermondi* and *Candida lusitaniae* as well as other *Candida* species have been described as resistant to polyenes (28–30, 83, 104); none of these species has caused a nosocomial outbreak. Nosocomial candidiasis may involve virtually any organ system. Edwards et al (33) reviewed many of the clinical syndromes of disseminated candidiasis and their relation to immunologic host defense.

Candidemia

The rate of fungemia due to *Candida* spp. at hospitals surveyed in the NNIS project increased from 0.7 per 10,000 discharges in 1980 to 1.5 per 10,000 discharges in 1984 (18–21). Horn et al (46) studied 200 episodes of fungemia at the Memorial Sloan-Kettering Cancer Center in New York between 1978 and 1982, mostly due to *Candida* spp., and found that the incidence had increased 31% over that of the period from 1974 through 1977. Candidemia incidence increased 73 and 95% in patients with a lymphoma or a solid tumor, respectively. Fungemia was associated with 76% mortality. Among those patients autopsied, 72% had disseminated fungal infection. Maksymiuk et al (64) studied 188 episodes of systemic candidiasis over 4 yr in a cancer hospital; they found that patients with acute leukemia had a higher frequency of candidiasis (11/100 patients) than those with lymphoma (1/100 patients). Blood cultures were positive in only 52% of the cases of systemic candidiasis. The survival rate was inversely related to the number of positive blood cultures. That many patients with disseminated or locally invasive visceral candidiasis have negative blood cultures is a very important finding that markedly influences therapeutic decisions. The true magnititude of systemic candidiasis in an institution may be underestimated by evaluation only of candidemias.

Vascular catheters, especially central venous catheters and chronic indwelling silastic catheters, contribute substantially to the problem of nosocomial candidiasis. The pathogenesis of nosocomial infections of vascular catheters has been reviewed by Maki (63). *Candida* sp. may acquire access to the intravascular portion of the catheter via contaminated solutions (Type I nosocomial fungal infection), from colonization of the percutaneous catheter wound (Type II), or via blood from another infected site. Klein & Watanakunakorn (54) found that 19 of 56 patients with candidemia with two or more positive blood cultures had a contaminated TPN catheter. Fungemia resolved slowly in 33 patients with removal of TPN or other vascular catheters and no antifungal therapy. Twenty of these 33 had only one positive blood culture. However, four of the 33 patients progressed to *Candida* endophthalmitis, which was diagnosed 8–66 days after the last positive blood culture. Sixteen patients who received no amphotericin B died of systemic candidiasis. Turner et al (124) also found a strong association between the use of vascular catheters and development of candidemia at two university hospitals over 5 yr in a pediatric population. Rose (98) studied 69 patients with systemic candidiasis over 13 yr at a Veterans Administation Hospital. Fifty-five of these cases were associated with infected venous catheters. Following removal of the catheter 26 became afebrile; four of these patients later developed endophthalmitis and required amphotericin B. Five other patients remained febrile and also required amphotericin B; the remaining 20 patients died. Central venous catheters have been the source of cardiac and disseminated candidiasis in children undergoing cardiac or gastrointestinal surgery (130). *Candida* mural endocarditis and myocarditis were both associated with *Candida* infections of central venous catheters. Peripheral venous catheters are also sources of candidemia. Seven patients who developed *Candida* peripheral suppurative thrombophlebitis at a Veterans Administration Medical Center were all male surgical patients and had received multiple (three or more) antibiotics for ≥ 2 wk. These patients were older (median of 64 years) and were hospitalized longer (mean of 27 days) than patients with non-*Candida* thrombophlebitis (126). This study (126) and that of Torres-Rojas et al (123) found that candidemia was often sustained after simple removal of the vascular catheter, possibly because of extensive invasion of the venous wall by *Candida* spp.

Several studies have convincingly demonstrated that candidemia may be detected earlier and in a higher percentage of infected patients by using lysis-centrifugation techniques than by using conventional broth techniques (45, 53).

As previously discussed, blood cultures are frequently negative in many cases of disseminated candidiasis. Moreover, blood cultures often become positive only after the infection has become well established in the deep viscera. Accordingly, the development of immunodiagnostic systems, such as tests for *Candida* cell-wall or cytoplasmic antigens, and chromatographic assays for detection of fungal metabolites are especially important for improving the early diagnosis of disseminated candidiasis (11, 27). Currently a latex agglutination kit is the only commercially available rapid detection system for *Candida* spp. This system, as well as other immunodiagnostic and chromatographic methods under investigation, must complement a meticulous clinical evaluation of the patient with suspected disseminated candidiasis.

Rotrosen et al (102) recently reviewed the predilection of *C. albicans* to infect vascular catheters. *C. albicans* adhered more avidly to polyvinyl chloride than to Teflon catheter material. *C. albicans* also has strong propensity to attach to fibrin-platelet matrices and vascular endothelial cells. Mannoproteins of the *Candida* cell wall are likely to be important determinants of adherence to intravascular catheters.

Optimal treatment of *Candida* suppurative peripheral thrombophlebitis consists of resection of the septic thrombosed vein and administration of i.v. amphotericin B. Treatment of candidemia associated with all types of central venous catheters, in our opinion, should consist of removal of the catheter and administration of i.v. amphotericin B; the total dose depends upon the probability of other sites being actively infected.

Candida *Infection of Prosthetic Cardiac Valves*

In a comprehensive review of fungal infection of prosthetic valves, McLeod & Remington (67) found that *Candida* spp. and *Aspergillus* spp. are the first and second most common fungal causes of prosthetic valve endocarditis (PVE). *Candida* PVE is usually associated with a history of prolonged administration of parenteral antibiotics.

Candidemia may develop as an early or late PVE. Clinical manifestations include fever, new or changing murmurs, cutaneous lesions, chorioretinitis, myocarditis, and embolic events involving major (cerebral, femoral, and brachial) arteries and their branches. Vegetations consisting of platelet-fibrin thrombi, blastospores, and pseudohyphae are usually large (approximately 2 cm in diameter or larger) and friable. Such vegetations may obstruct the prosthetic valve orifice leading to rapid onset of pulmonary edema, or they may embolize to major arteries, causing stroke or ischemia to all or part of an extremity. Histopathologic examination of material resected at embolectomy may yield the first evidence of *Candida* PVE. Most cases have positive blood cultures. However, serial blood cultures and cardiac ultrasound may be necessary early in the course of infection to establish a diagnosis.

Treatment of PVE includes removal of the prosthetic valve, i.v. amphotericin B (minimum total dose ≥ 1 g), and followup for a minimum of 2 yr.

Hematogenous Candida *Endophthalmitis*

Hematogenous *Candida* endophthalmitis (HCE) is a common complication of candidemia associated with vascular catheters. Montgomerie & Edwards (73) found that five of 25 patients receiving TPN acquired ocular lesions clinically consistent with HCE. Lesions of *Candida* endophthalmitis are white and cottonball-like in appearance and extend from the chorioretina into the vitreous humor. The number of such ocular lesions observed funduscopically correlates directly with the extent of disseminated visceral infection (32).

Recognition of these lesions mandates treatment with amphotericin B, since HCE may cause irreversible loss of vision or complete blindness in the involved eye.

Hepatosplenic Candidiasis

Hepatosplenic candidiasis is a complication of disseminated candidiasis that has only been well reported within this decade. It is apparently a Type 2 nosocomial fungal infection. To clarify this increasingly recognized complication of nosocomial candidiasis, Thaler et al (122) analyzed eight cases from the National Cancer Institute and 65 cases described in the literature in a comprehensive review. All but five of the 73 patients had a history of granulocytopenia, due either to a primary hematologic disease or to chemotherapy. Among 68 of the granulocytopenic patients, 59 had acute leukemia. However, the clinical manifestations of hepatosplenic candidiasis generally become apparent only after recovery from granulocytopenia and often after remission of the neoplastic disease. Indeed, 21 (31%) of 68 evaluable patients had leukocytosis. Thirty-four (50%) patients had a history of receiving some antifungal therapy (for systemic candidiasis in six cases and for local *Candida* infection or fever with granulocytopenia in 28 cases). Fifty-eight (85%) of these 68 patients presented with fever and 39 (57%) with abdominal pain. Elevated serum alkaline phosphatase was the most consistent biochemical abnormality. Serum transaminases were usually normal. Computerized tomography (CT) scans appeared to be more sensitive than ultrasonography in detecting hepatic lesions, although the two techniques may be complementary in certain cases. Magnetic resonance imaging may offer another means by which to identify these lesions. The hepatosplenic lesions range in diameter from 1 mm (detectable only by peritoneoscopy or laparotomy) to several centimeters (detectable by ultrasonography or CT scanning). Diagnosis is best established by biopsy and culture of the lesions. At the National Cancer Institute the optimal therapy for hepatosplenic candidiasis appeared to be aggressive treatment with total doses of at least 2 g of amphotericin B and early concomitant administration of flucytosine. Total doses of as much as 9 g of amphotericin B have been necessary.

Splenic candidiasis may be the predominant clinically overt component of hepatosplenic candidiasis (44, 48). Left subphrenic abscess due to extension of fungi beyond the splenic capsule may be the presenting manifestation. A recent review of splenic candidiasis (48) found that 12 of 16 patients who underwent laparotomy and splenectomy for splenic candidiasis received postoperative i.v. amphotericin B, and all but one patient survived. The one death in this group was a patient with disseminated candidiasis whose amphotericin B was discontinued prematurely.

Candida *Peritonitis*

Candida peritonitis may develop as a complication of gastrointestinal surgery, spontaneous rupture of a viscus, or peritoneal dialysis (10) and thus may be a Type I or Type II nosocomial fungal infection. Fungal peritonitis in patients receiving peritoneal dialysis has emerged as an increasingly common complication and was the subject of a recent comprehensive review (34). *Candida* spp. were the most common pathogens causing fungal peritonitis. Etiologic agents of fungal peritonitis complicating peritoneal dialysis in this study of 88 cases were *C. albicans* in 37, *C. tropicalis* in 12, *C. parapsilosis* in 7, other *Candida* spp. in 11, *Fusarium* spp. in 6, and *Rhodotorula rubra* in 3. *R. rubra* is noteworthy in that it is a common waterborne saprophytic yeast. The most apparent risk factors for fungal peritonitis in 55 evaluable cases were antibiotic use within the past month in 69% and previous bacterial peritonitis within the past month in 51%. Only 7% of patients had received an immunosuppressive agent. The most common clinical manifestations were clouding of the peritoneal dialysate (86%), abdominal pain (73%), and physical signs of peritonitis, usually abdominal tenderness or rebound pain (65%). Recovery was most successful with early initiation of systemic antifungal therapy and removal of the infected catheter.

Candida *Epiglottitis*

Candida epiglottitis has been described recently in granulocytopenic patients (23, 129). Patients present initially with refractory odynophagia in the hypopharyngeal and epiglottic region. Advanced cases may manifest laryngeal stridor. Intravenous amphotericin B is warranted in the persistently granulocytopenic patient. Diagnosis is established by indirect laryngoscopy and by swab, stains, and cultures of the epiglottis.

Outbreaks of Nosocomial Candidiasis

As previously discussed, most cases of nosocomial candidiasis are Type IIA infections, the result of a progression from colonization to infection with the patient's endogenous *Candida* flora. However, there have been clearly defined outbreaks of nosocomial candidiasis that were truly hospital acquired (Type I).

Intravascular pressure-monitoring devices contaminated with *C. parapsilosis* were the source of fungemia in two pediatric intensive care units (110, 135). Repeated exposure of the pressure transducers to TPN solution was thought to be the source of the *C. parapsilosis*. Infrequent sterilization of the devices allowed for persistence of the fungi. The nosocomial source of *C. parapsilosis* fungemia in two other outbreaks was the inadvertent contamination of TPN solutions during preparation by the hospital pharmacy services

(89, 111). Daisy et al (25) reported an inadvertent infusion of i.v. fluids with a mycelium "fungus ball" that was visibly suspended in the solution. An undetected crack in the glass i.v. bottle was the principle route of access of fungi into the i.v. solution.

 C. parapsilosis has also been linked to nosocomial endophthalmitis due to manufacturer contamination of an intraoperative ophthalmic irrigating solution (66).

Prevention and Treatment

Therapeutic management of disseminated candidiasis is most challenging in persistently granulocytopenic patients, who are highly susceptible. Current antifungal measures for prevention of nosocomial candidiasis are limited (133). Established disseminated candidiasis is difficult to diagnose and is frequently lethal, despite antifungal therapy. Studies conducted at the National Cancer Institute (88) and later confirmed by others (45a) demonstrated that in persistently febrile granulocytopenic patients, initiation of empirical amphotericin B therapy on or after the seventh day of treatment with other antibiotics reduces the patients' risk of developing a superimposed fungal infection. Empirical amphotericin B was also successfully incorporated into a modifiable empirical antibiotic program (87). This approach to antifungal therapy provides early treatment for high risk patients with early clinically occult invasive candidiasis.

TRICHOSPORONOSIS

Trichosporon beigelii is a urease-positive basidiomycetous yeast that causes white piedra and is the most common cause of disseminated or locally invasive tissue infection known as trichosporonosis (6, 132). *T. beigelii* has been isolated in several studies as an uncommon but consistently identified component of the human alimentary tract flora. Disseminated trichosporonosis thus progresses as a Type IIA nosocomial infection from endogenous colonization to active infection. The lungs, kidneys, skin, and eyes are the main targets of this pathogen, which may completely destroy renal glomeruli. Serum plasma and cerebrospinal fluid from patients with disseminated trichosporonosis may react with the cryptococcal latex agglutination test reagent because of antigens shared between *T. beigelii* and *Cryptococcus neoformans* (67a). Successful treatment is dependent upon early initiation of treatment with amphotericin B and recovery from granulocytopenia.

MALASSEZIA FURFUR CATHETER-ASSOCIATED SEPSIS

Malassezia furfur (Pityrosporum obiculare) is part of the normal skin flora in humans whether or not tinea versicolor is present. It causes a usually superficial cutaneous mycosis, and it may infect central venous catheters, especially those used for TPN, as another Type IIA nosocomial fungal infection. Since *M. furfur* is lipophilic, the organism usually infects those patients receiving a parenteral fat emulsion. The syndrome in neonates consists of progressive pulmonary infiltrates, hypoxia, and thrombocytopenia (91, 93). Histopathology reveals *M. furfur* along the subendothelial regions of pulmonary arteries laden with lipid deposits from the TPN solution. Treatment of this potentially lethal syndrome includes discontinuing the parenteral lipid emulsion and removal of the central venous catheter. Systemic antifungal therapy is seldom effective unless these two measures are completed. Moreover, nonimmunocompromised patients seldom require systemic antifungal therapy once the lipid emulsion is discontinued and the central catheter is removed.

NOSOCOMIAL DERMATOPHYTOSIS

T. rubrum is regarded as the most ubiquitous cutaneous fungal pathogen causing superficial dermatophyte infections. It is also the fungal infection most commonly acquired by the technical staff of many hospital laboratories.

Trichophyton rubrum may infect the immunocompromised host via deep local invasion or potentially fatal dissemination (reviewed in 78). Griseofulvin or ketoconazole are effective against *T. rubrum* and are considered the drugs of choice even in the immunocompromised host.

Nosocomial dermatophytosis due to *Microsporum canis* was contracted by seven nurses working in an infested newborn nursery. Each nurse contracted the *M. canis* infection on the left forearm, which corresponded to the location where each rested an infected infant's head during bottle feeding (74). All nurses received griseofulvin for 2 wk and were instructed to wear long sleeves. The infected baby received tolnaftate for the scalp lesion.

SACCHAROMYCES CEREVISIAE

Saccharomyces cerevisiae is an asporogenous yeast utilized to prepare bread, beer (hence "brewer's yeast"), and wine. Viable yeast is also often consumed as a health food. Consequently *S. cerevisiae,* which is usually considered nonpathogenic, may colonize the gastrointestinal tract and skin. Eng and associates (35) reported five cases and reviewed the literature of nosocomial *S. cerevisiae* infections that developed when colonization progressed to infection in susceptible hosts (Type IIA nosocomial fungal infection). Infections included postoperative periurethral fistula formation, polymicrobial empyema, renal parenchymal infection, and cystitis. The last infection was effectively treated with ketoconazole after an unsuccessful course of flucytosine, to which the *S. cerevisiae* isolate from urine was resistant.

REACTIVATION OF LATENT INFECTION DUE TO *HISTOPLASMA CAPSULATUM, COCCIDIOIDES IMMITIS,* AND *CRYPTOCOCCUS NEOFORMANS*

Reactivation of a latent primary pathogenic fungal infection in the hospital-ized immunocompromised patient would be considered a Type IIB nosoco-mial fungal infection. Awareness of the risk of reactivation of latent fungal infection should be heightened by considering the patient's travel history, occupational history, chest radiograph, and any other features that would predict the presence of a latent fungal infection. These clinical findings are especially relevant to coccidioidomycosis (5) and histoplasmosis (128). Di-agnosis of a reactivated latent fungal infection may be obfuscated in im-munocompromised patients by the erroneous attribution of the patient's clini-cal deterioration to the progression of an underlying disease or a resistant bacterial infection.

LABORATORY-ACQUIRED FUNGAL INFECTIONS

These infections are clearly Type I fungal infections. However, hospital personnel rather than patients are the hosts at risk. Laboratory-acquired dermatophyte infections have been previously discussed. Some laboratory-acquired fungal infections are potentially lethal to hospital microbiology staff. *Coccidioides immitis* and *Histoplasma capsulatum* were the first and second most frequently acquired systemic fungi in the medical microbiology labora-tory (43, 86). Both infections are acquired in the laboratory either by inhala-tion or, less commonly, by direct inoculation of *C. immitis* arthroconidia or *H. capsulatum* conidia after these dimorphic fungi have been converted in the laboratory to the mold or mycelial form.

All 93 coccidioidal laboratory-acquired infections recorded by Pike (86) were true clinical infections and not simply conversion of skin tests. If skin test conversions were included, the rate of asymptomatic infection would be considerably higher. Primary cutaneous and pulmonary infections are the most frequent forms of laboratory-acquired infection.

Nosocomial transmission of *C. immitis* to six hospital staff members resulted from conversion of spherules (tissue form) to arthroconidia (in-fectious mycelial phase) in the plaster cast material over a draining coccidioi-dal lesion of a patient's left leg (31). The six staff members caring for this patient had symptomatic pneumonitis. Frequent changing of dressings and avoidance of closed casts over coccidioidal wounds may avert the problem of nosocomial transmission of *C. immitis* from draining wounds.

FUTURE DIRECTIONS

The trend of increasing numbers of nosocomial fungal infections will continue as medical facilities, predisposing interventions, and high-risk patient populations continue to expand. Comprehensive research efforts must be directed toward (*a*) environmental and nosocomial infection control, including the development of molecular biological probes for identification of strains of yeasts, especially of the *Candida* spp., (*b*) augmentation of host defenses with immunomodulatory agents such as colony-stimulating factors for granulocytes and macrophages, (*c*) development of more effective and less toxic antifungal agents, especially those that could be administered to high-risk patients for prevention of nosocomial fungal infections, (*d*) development of rapid, sensitive, and specific immunodiagnostic and biochemical assays for early diagnosis of invasive fungal infections, and (*e*) more incisive understanding of the pathogenesis and biology of the common and newly emerging nosocomial fungal infections.

Literature Cited

1. Adam, R. D., Paquin, M. L., Petersen, E. A., Sauballe, M. A., Rinaldi, M. G. 1986. Phaeohyphomycosis caused by the fungal genera *Bipolaris* and *Exserohilum*. A report of 9 cases and review of the literature. *Medicine Baltimore* 65:203–17
2. Aisner, J., Schimpff, S. C., Bennett, J. E., Young, V. M., Wiernik, P. H. 1976. *Aspergillus* infections in cancer patients: association with fireproofing materials in a new hospital. *J. Am. Med. Assoc.* 235:411–12
3. Albelda, S. M., Talbot, G. H., Gerson, S. L., Miller, W. T., Cassileth, P. A. 1985. Pulmonary cavitation and massive hemoptysis in invasive pulmonary aspergillosis. *Am. Rev. Respir. Dis.* 131:115–20
4. Allo, M. D., Miller, J., Towsend, T., Tan, C. 1987. Primary cutaneous aspergillosis associated with Hickman intravenous catheters. *N. Engl. J. Med.* 317:1105–8
5. Ampel, N. M., Ryan, K. J., Carry, P. J., Wieden, M. A., Schifman, R. B. 1986. Fungemia due to *Coccidioides immitis*. An analysis of 16 episodes in 15 patients and a review of the literature. *Medicine Baltimore* 65:312–21
6. Anaissie, E. A., Hoy, J., Jones, P., Rolston, K., Bodey, G. P. 1985. New fungal pathogens in immunocompromised patients. *Abstr. 25th Intersci. Conf. Antimicrob. Agents Chemother, Minneapolis, Minn*, p. 253. Washington, DC: Am. Soc. Microbiol.
7. Armstrong, D. A. 1984. Protected environments are discomforting and expensive and do not offer meaningful protection. *Am. J. Med.* 76:685–89
8. Arnow, P. M., Anderson, R. L., Mainous, P. D., Smith, E. J. 1978. Pulmonary aspergillosis during hospital renovation. *Am. Rev. Respir. Dis.* 118:49–53
9. Attah, C. A., Cerruti, M. M. 1979. *Aspergillus* osteomyelitis of sternum after cardiac surgery. *NY State J. Med.* 79:1420–21
10. Bayer, A. S., Blumenkrantz, M. J., Montgomerie, J. Z., Galpin, J. E., Coburn, J. W., Guzer, L. B. 1976. Candida peritonitis. Report of 22 cases and review of the English literature. *Am. J. Med.* 61:832–40
11. Bennett, J. E. 1987. Rapid diagnosis of candidiasis and aspergillosis *Rev. Infect. Dis.* 9:398–402
12. Bodey, G. P. 1984. Current status of prophylaxis of infection with protected environments. *Am. J. Med.* 76:678–84
13. Boyce, J. M., Lawson, L. A., Lockwood, W. R., Hughes, J. L. 1981. *Cunninghamella bertholletiae* wound infec-

tion of probable nosocomial origin. *South. Med. J.* 74:1132–35

14. Brandt, S. J., Thompson, R. L., Wenzel, R. P. 1985. Mycotic pseudoaneurysm of anaortic bypass graft and contiguous vertebral osteomyelitis due to *Aspergillus* osteomyelitis. *Am. J. Med.* 79:259–62

15. Bullock, W. E., Deepe, G. S. 1983. Medical mycology in crisis. *J. Lab. Clin. Med.* 102:685–93

15a. Burch, P. A., Karp, J. E., Merz, W. G., Kuhlman, J. E., Fishman, E. K. 1987. Favorable outcome of invasive aspergillosis in patients with acute leukemia. *J. Clin. Oncol.* 5:1985–93

16. Burnie, J. P., Odds, F. C., Lee, W., Webster, C., Williams, J. D. 1985. Outbreak of systemic *Candida albicans* in intensive care unit caused by cross infection. *Br. Med. J.* 290:746–48

17. Burton, J. R., Zachary, J. B., Bessin, R., Rathbun, H. K., Greenough, W. B., et al. 1972. Aspergillosis in four renal transplant recipients. Diagnosis and effective treatment with amphotericin B. *Ann. Intern. Med.* 77:383–88

18. Centers for Disease Control. 1977. *National Nosocomial Infections Study Report: Annual Summary 1975.* Atlanta, Ga: CDC

19. Centers for Disease Control. 1982. *National Nosocomial Infections Study Report: Annual Summary 1979.* Atlanta, Ga: CDC

20. Centers for Disease Control. 1983. Nosocomial infection surveillance 1980–82. *CDC Surv. Summ.* 32(4):1–16

21. Centers for Disease Control. 1984. Nosocomial infection surveillance 1983. *CDC Surv. Summ.* 33(2):9–22

22. Chaulk, C. P., Smith, P. W., Feagler, J. R., Verdirame, J., Commers, J. R. 1986. Fungemia due to *Fusarium solani* in an immunocompromised child. *Pediatr. Infect. Dis.* 5:363–66

23. Cole, S., Zawin, M., Lundberg, B., Hoffman, J., Bailey, L., et al. 1987. *Candida* epiglottitis in an adult with acute non-lymphocytic leukemia. *Am. J. Med.* 82:662–63

24. Commers, J., Robichaud, K. J., Pizzo, P. A. 1984. New pulmonary infiltrates in granulocytopenic patients being treated with antibiotics. *Pediatr. Infect. Dis.* 3:423–28

25. Daisy, J. A., Abrutyn, E. A., MacGregor, R. R. 1979. Inadvertent administration of intravenous fluids contaminated with fungus. *Ann. Intern. Med.* 91:563–65

26. De Gregorio, M. W., Lee, W. M., Linker, C. A., Jacobs, R. A., Ries, C. A. 1982. Fungal infections in patients with acute leukemia. *Am. J. Med.* 73:543–48

27. de Repentigny, L., Reiss, E. 1984. Current trends in immunodiagnosis of candidiasis and aspergillosis. *Rev. Infect. Dis.* 6:301–12

28. Dick, J. D., Merz, W. G., Saral, R. 1980. Incidence of polyene-resistant yeasts recovered from clinical specimens. *Antimicrob. Agents Chemother.* 18:158–63

29. Dick, J. D., Rosengard, B. R., Merz, W. G., Stuart, R. K., Hutchins, G. M., et al. 1985. Fatal disseminated candidiasis due to amphotericin B resistant *Candida guillermondi.* *Ann. Intern. Med.* 102:67–68

30. Drutz, D. J., Lehrer, R. I. 1978. Development of amphotericin B–resistant *Candida tropicalis* in a patient with defective leukocyte function. *Am. J. Med.* 276:77–92

31. Eckmann, B. H., Schaefer, G. L., Huppert, M. 1964. Bedside interhuman transmission of coccidioidomycosis via growth on fomites. An epidemic involving six persons. *Am. Rev. Respir. Dis.* 89:175–85

32. Edwards, J. E. Jr., Foos, R. Y., Montgomerie, J. Z., Guze, L. B. 1974. Ocular manifestations of *Candida* septicemia: review of seventy-six cases of hematogenous *Candida* endophthalmitis. *Medicine Baltimore* 53:47–75

33. Edwards, J. E. Jr., Lehrer, R. I., Stiehm, E. R., Fischer, T. J., Young, L. S. 1978. Severe candidal infections. Clinical perspective, immune defense mechanisms, and current concepts of therapy. *Ann. Intern. Med.* 89:91–106

34. Eisenberg, E. S., Leviton, I., Soeiro, R. 1986. Fungal peritonitis in patients receiving peritoneal dialysis: experience with 11 patients and review of the literature. *Rev. Infect. Dis.* 8:309–21

35. Eng, R. H. K., Drehmel, R., Smith, S. M., Goldstein, E. J. C. 1984. *Saccharomyces cerevisiae* infections in man. *Sabouraudia J. Med. Vet. Mycol.* 22:403–7

36. England, A. C., Weinstein, M., Ellner, J. J., Ajello, L. 1981. Two cases of rhinocerebral zygomycosis (mucormycosis) with common epidemiologic and environmental features. *Am. Rev. Respir. Dis.* 124:497–98

37. Everett, E. D., Pearson, S., Rogers, W. 1979. *Rhizopus* surgical wound infection associated with elasticized adhesive tape dressings. *Arch. Surg. Chicago* 114:738–39

38. Fraser, D. W., Ward, J. I., Ajello, L.,

Plikaytis, B. D. 1979. Aspergillosis and other systemic mycoses: the growing problem. *J. Am. Med. Assoc.* 242:1631–35

39. Gage, A. A., Dean, D. C., Schimert, G., Minsely, N. 1970. *Aspergillus* infection after cardiac surgery. *Arch. Surg. Chicago* 101:384–87

40. Gartenberg, G., Bottone, E. J., Keusch, G. T., Weitzman, I. 1978. Hospital-acquired mucormycosis *(Rhizopus rhizopodiformis)* of skin and subcutaneous tissue: epidemiology, mycology, and treatment. *N. Engl. J. Med.* 299:1115–18

40a. Gerson, S. L., Talbot, G. H., Hurwitz, S., Strom, B. L., Lusk, E. J. 1984. Prolonged granulocytopenia: the major risk factor for invasive pulmonary aspergillosis in patients with acute leukemia. *Ann. Intern. Med.* 100:345–51

41. Glotzbach, R. E. 1982. *Aspergillus terreus* infection of pseudoaneurysm of aortofemoral vascular graft with contiguous vertebral osteomyelitis. *Am. J. Clin. Pathol.* 77:224–27

42. Goodill, J. J., Abuelo, J. G. 1987. Mucormycosis—a new risk of deferoxamine therapy in dialysis patients with aluminum or iron overload? *N. Engl. J. Med.* 317:54

43. Hamel, E., Kruse, R. H. 1967. Laboratory-acquired mycoses. *Fort Detrick Misc. Publ. No. 28. Ad-665376,* Fort Detrick, Md.

44. Helton, W. S., Carrico, C. J., Zaveruha, P. A., Schaller, R. 1986. Diagnosis and treatment of splenic fungal abscesses in the immunosuppressed patient. *Arch. Surg. Chicago* 121:580–86

45. Henry, N. K., McLimans, C. A., Wright, A. J., Thompson, R. L., Wilson, W. R., et al. 1983. Microbiologic and clinical evaluation of the isolator lysis-centrifugation blood culture tube. *J. Clin. Microbiol.* 17:864–69

45a. Holleran, W. M., Wilbur, J. R., De Gregorio, M. W. 1985. Empiric amphotericin B therapy in patients with acute leukemia. *Rev. Infect. Dis.* 7:619–24

46. Horn, R., Wong, B., Kiehn, T. E., Armstrong, D. 1985. Fungemia in a cancer hospital: changing frequency, earlier onset, and results of therapy. *Rev. Infect. Dis.* 7:646–55

47. Jade, K. B., Lyons, M. F., Gnann, J. W. Jr. 1986. *Paecilomyces lilacinus* cellulitis in an immunocompromised patient. *Arch. Dermatol.* 122:1169–70

48. Johnson, J. D., Raff, M. J. 1984. Fungal splenic abscess. *Arch. Intern. Med.* 144:1987–93

49. June, C. H., Beatty, P. G., Schulman, H. M., Rinaldi, M. G. 1986. Disseminated *Fusarium moniliforme* infection after allogeneic marrow transplantation. *South. Med. J.* 79:513–15

50. Kalish, S. B., Goldschmidt, R., Li, C., Knop, R., Cook, F. V., et al. 1982. Infective endocarditis caused by *Paecilomyces varioti. Am. J. Clin. Pathol.* 78:249–52

51. Kallenback, J. J., Dusheiko, J., Block, C. S., Bethlehem, B., Koornhoff, H. J., et al. 1977. *Aspergillus* pneumonia–cluster of four cases in an intensive care unit. *S. Afr. Med. J.* 52:919–23

52. Kiehn, T. E., Nelson, P. E., Bernard, E. M., Edwards, F. F., Koziner, B., et al. 1985. Catheter-associated fungemia caused by *Fusarium chlamydosporum* in a patient with lymphocytic lymphoma. *J. Clin. Microbiol.* 21:501–4

53. Kiehn, T. E., Wong, B., Edwards, F. F., Armstrong, D. 1983. Comparative recovery of bacteria and yeasts from lysis-centrifugation and a conventional blood culture system. *J. Clin. Microbiol.* 18:300–4

54. Klein, J. J., Watanakunakorn, C. 1979. Hospital-acquired fungemia: its natural course and clinical significance. *Am. J. Med.* 67:51–58

55. Krasinski, K., Holzman, R. S., Hanna, B., Greco, M. A., Graff, M., et al. 1985. Nosocomial fungal infection during hospital renovation. *Infect. Control* 6:278–82

56. Kwon-Chung, K. J., Young, R. C., Orlando, M. V. 1975. Pulmonary mucormycosis caused by *Cunninghamella elegans* in a patient with chronic myelogenous leukemia. *Am. J. Clin. Pathol.* 64:544

57. Kyriakides, G. K., Zimmerman, H. H., Hall, W. H., Arora, V. K., Lifton, J., et al. 1976. Immunologic monitoring and aspergillosis in renal transplant patients. *Am. J. Surg.* 131:246–52

58. Lee, W., Burnie, J., Matthews, R. 1986. Fingerprinting *Candida albicans. J. Immunol. Methods* 93:177–82

59. Lehrer, R. I., Howard, D. H., Sypherd, P. S., Edwards, J. E., Segal, G. P., et al. 1980. Mucormycosis. *Ann. Intern. Med.* 93:93–108

60. Lentino, J. R., Rosenkranz, M. A., Michaels, J. A., Kurup, V. P., Rose, H. D., et al. 1982. A retrospective review of airborne disease secondary to road construction and contaminated air conditioners. *Am. J. Epidemiol.* 116:430–37

61. Lutwick, L. I., Rytel, M. W., Yanëz, J. P., Galgiani, J. N., Stevens, D. A.

1979. Deep infections from *Petriellidium boydii* treated with miconazole. *J. Am. Med. Assoc.* 241:272–73

62. Mahoney, D. H. Jr., Stenber, C. P., Starling, K. A., Barrett, F. F., Goldberg, J., et al. 1979. An outbreak of aspergillosis in children with acute leukemia. *J. Pediatr.* 95:70–72

63. Maki, D. G. 1982. Infections associated with intravascular lines. *Curr. Clin. Top. Infect. Dis.* 3:309–63

64. Maksymiuk, A. W., Thongprasert, S., Hopfer, R., Luna, M., Fainstein, V., et al. 1984. Systemic candidiasis in cancer patients. *Am J. Med.* 77(4D):20–27

65. Mawk, J. R., Erickson, D. L., Chou, S. N., Seljeskog, E. L. 1983. *Aspergillus* infections of the lumbar disc spaces. *J. Neurosurg.* 58:270–74

66. McCray, E., Rampell, N., Solomon, S. L., Bond, W. W., Martone, W. J., O'Day, D. 1986. Outbreak of *Candida parapsilosis* endophthalmitis after cataract extraction and intraocular lens implantation. *J. Clin. Microbiol.* 24:625–28

67. McLeod, R., Remington, J. S. 1977. Postoperative fungal endocarditis. Infections of prosthetic heart valves and vascular grafts: prevention, diagnosis, and treatment. In *Infections of Prosthetic Heart Valves and Vascular Grafts*, ed. R. J. Duma, pp. 163–236. Baltimore, Md: Univ. Park Press

67a. McManus, E. J., Jones, J. M. 1985. Detection of a *Trichosporon beigelii* antigen cross-reactive with *Cryptococcus neoformans* capsular polysaccharide in serum from a patient with disseminated *Trichosporon* infection. *J. Clin. Microbiol.* 21:681–84

68. McManus, E. J., Olivio, P. D., Jones, J. M. 1986. Mitochondrial DNA restriction patterns of *Candida albicans* isolates from critical care patients. *Abstr. 1986 Intersci. Conf. Antimicrob. Agents Chemother. New Orleans, La*, p. 241. Washington, DC: Am. Soc. Microbiol.

69. Merz, W. G., Karp, J. E., Schron, D., Saral, R. 1986. Increased incidence of fungemia caused by *Candida krusei*. *J. Clin. Microbiol.* 24:581–84

70. Meunier, F., Leleux, J., Gerain, J., Ninove, D., Snoeck, R., et al. 1987. Prophylaxis of aspergillosis in neutropenic cancer patients with nasal spray of amphotericin B: a prospective randomized study. *Abstr. 1987 Intersci. Conf. Antimicrob. Agents Chemother., New York*, p. 331. Washington, DC: Am. Soc. Microbiol.

71. Meyer, R. D., Rosen, P., Armstrong, D. 1972. Phycomycosis complicating leukemia and lymphoma. *Ann. Intern. Med.* 77:871

72. Meyer, R. D., Young, L. S., Armstrong, D. 1973. Aspergillosis complicating neoplastic disease. *Am. J. Med.* 54:6–15

73. Montgomerie, J. Z., Edwards, J. E. Jr. 1987. Association of infection due to *Candida albicans* with intravenous hyperalimentation. *J. Infect. Dis.* 137:197–201

74. Mossovitch, M., Mossovitch, B., Alkun, M. 1986. Noscomial dermatophytosis caused by *Microsporum canis* in a newborn department. *Infect. Control* 7:593–95

75. Nash, G., Foley, F. D., Goodwin, M. N., Bruck, H. M., Greenwald, K. A., Pruitt, B. A. 1971. Fungal burn wound infection. *J. Am. Med. Assoc.* 215:1664–66

76. Nash, G., Foley, F. D., Pruitt, B. A. 1970. *Candida* burn wound invasion. A cause of systemic candidiasis. *Arch. Pathol.* 90:75–78

77. Noble, W. C., Clayton, Y. M. 1963. Fungi in the air of general hospital wards. *J. Gen. Microbiol.* 32:397–402

78. Novick, M. L., Tapia, L., Bottone, E. J. 1987. Invasive *Trichophyton rubrum* infection in an immunocompromised host. Case report and review of the literature. *Am. J. Med.* 82:321–25

79. Oberle, A. D., Penn, R. L. 1983. Nosocomial *Saksenaea vasiformis* infection. *Am. J. Clin. Pathol.* 80:885–88

80. Odds, F. C. 1984. Ecology and epidemiology of *Candida* species. *Zentralbl. Bakteriol. Parasitenkd. Infektionskr. Hyg. Abt. 1 Reihe A* 257:207–12

81. Opal, S. M., Asp, A. A., Cannady, P. B., Morse, P. L., Burton, L. J., et al. 1986. Efficacy of infection control measures during a nosocomial outbreak of disseminated aspergillosis associated with hospital construction. *J. Infect. Dis.* 153:634–37

82. Panos, R., Barr, L., Walsh, T. J., Silverman, H. J. 1988. Factors associated with fatal hemoptysis in cancer patients. *Chest* In press

83. Pappagianis, D., Collins, M. S., Hector, R., Remington, J. 1979. Development of resistance to amphotericin B in *Candida lusitaniae* infection in a human. *Antimicrob. Agents Chemother.* 16:123–26

84. Petersen, P. K., McGlave, P., Ramsay, N. K. C., Rhame, F., Cohen, E., et al. 1983. A prospective study of infectious diseases following bone marrow transplantation: emergence of *Aspergil-*

lus and *Cytomegalovirus* as the major causes of mortality. *Infect. Control* 4:81–89

85. Petheram, I. S., Seal, R. M. E. 1976. *Aspergillus* prosthetic valve endocarditis. *Thorax* 31:380–90

86. Pike, R. M. 1976. Laboratory-associated infections. Summary and analysis of 3921 cases. *Health Lab. Sci.* 13:105–14

87. Pizzo, P. A., Hathorn, J., Hiemenz, J., Browne, M., Commers, J., et al. 1986. A randomized trial comparing ceftazidime with combination antibiotic therapy in cancer patients with fever and neutropenia. *N. Engl. J. Med.* 315:552–58

88. Pizzo, P. A., Robichaud, K. J., Gill, F. A., Witebsky, F. G. 1982. Empiric antibiotic and antifungal therapy for cancer patients with prolonged fever and granulocytopenia. *Am. J. Med.* 72:101–11

89. Plouffe, J. F., Brown, D. G., Silva, J. Jr., Eck, T., Stricof, R. L., et al. 1977. Nosocomial outbreak of *Candida parapsilosis* fungemia related to intravenous infusions. *Arch. Intern. Med.* 137:1686

90. Polonelli, L., Castagnola, M., Rosetti, D. V., Morace, G. 1985. Use of killer toxins for computer-aided differentiation of *Candida albicans* strains. *Mycopathologia* 91:175–79

91. Powell, D. A., Aungst, J., Snedden, S., Hansen, N., Brady, M. 1984. Broviac catheter–related *Malassezia furfur* sepsis in five infants receiving intravenous fat emulsions. *J. Pediatr.* 105:987–90

92. Prystowsky, S. D., Vogelstein, B., Ettinger, D. S., Merz, W. G., Kaiser, H., et al. 1976. Invasive aspergillosis. *N. Engl. J. Med.* 295:655–58

93. Redline, R. W., Redline, S. S., Boxenbaum, B., Dahms, B. B. 1985. Systemic *Malassezia furfur* infection in patients receiving intralipid therapy. *Hum. Pathol.* 16:815–22

94. Rhame, F. S. 1985. Lessons from the Roswell Park bone marrow transplant aspergillosis outbreak. *Infect. Control* 6:345–46

95. Rhame, F. S. 1986. Endemic nosocomial filamentous fungal disease: a proposed structure for conceptualizing and studying the environmental hazard. *Infect. Control* 7:124–25 (Suppl.)

96. Rhame, F. S., Streifel, A. J., Kersey, J. H. Jr., McGlave, P. B. 1984. Extrinsic risk factors for pneumonia in the patient at high risk of infection. *Am. J. Med.* 76(5A):42–52

96a. Rippon, J. W. 1988. Pseudallescheriasis. *Medical Mycology*, pp. 651–80. Philadelphia: Saunders. 3rd ed.

97. Rogers, T. R. 1986. Infections in hematologic malignancy. *Infect. Control* 7:140–42 (Suppl.)

98. Rose, H. D. 1978. Venous catheter-associated candidemia. *Am. J. Med. Sci.* 275:265–70

99. Rose, H. D., Hirsch, S. R. 1979. Filtering hospital air decreases *Aspergillus* counts. *Am. Rev. Respir. Dis.* 119:511–13

100. Rosen, P. P., Sternberg, S. S. 1976. Decreased frequency of aspergillosis and mucormycosis. *N. Engl. J. Med.* 295:1319–20

101. Ross, D. A., Anderson, D. C., MacNaughton, M. C., Stewart, W. K. 1968. Fulminating disseminated aspergillosis complicating peritoneal dialysis in eclampsia. *Arch. Intern. Med.* 121:183–88

102. Rotrosen, D., Calderone, R. A., Edwards, J. E. Jr. 1986. Adherence of *Candida* species to host tissues and plastic surfaces. *Rev. Infect. Dis.* 8:73–85

103. Rotstein, C., Cummings, K. M., Tidings, J., Killion, K., Powell, E., et al. 1985. An outbreak of invasive aspergillosis among allogeneic bone marrow transplants: a case control study. *Infect. Control* 6:347–55

104. Safe, L. M., Safe, S. H., Subden, R. E., Morris, D. C. 1977. Sterol content and polyene antibiotic resistance in isolates of *Candida krusei, Candida parakrusei,* and *Candida tropicalis. Can. J. Microbiol.* 23:398–401

105. Sarubbi, F. H. Jr., Kopf, H. B., Wilson, M. B., McGinnis, M. R., Rutala, W. A. 1982. Increased recovery of *Aspergillus flavus* from respiratory specimens during hospital construction. *Am. Rev. Respir. Dis.* 125:33–38

106. Scherer, S., Stevens, D. A. 1987. Application of DNA typing methods to epidemiology and taxonomy of *Candida* species. *J. Clin. Microbiol.* 25:675–79

107. Sheldon, D. L., Johnson, W. C. 1979. Cutaneous mucormycosis. Two documented cases of suspected nosocomial cause. *J. Am. Med. Assoc.* 241:1032–34

108. Simpson, M. B., Merz, W. G., Kurlinski, S. P., Solomon, M. H. 1977. Opportunistic mycotic osteomyelitis. Bone infections due to *Aspergillus* and *Candida* species. *Medicine Baltimore* 56:475–81

109. Smith, A. G., Crain, S. M., Dejongh, C., Thomas, G. M., Vigorito, R. D. 1985. Systemic pseudallescheriasis in a patient with acute myelocytic leukemia. *Mycopathologia* 90:85–89

110. Solomon, S. L., Alexander, H., Eley, J.

W., Anderson, R. L., Goodpasture, H. C., et al. 1986. Nosocomial fungemia in neonates associated with intravascular pressure-monitoring devices. *Pediatr. Infect. Dis.* 5:680–85

111. Solomon, S. L., Khabbaz, R. F., Parker, R. H., Anderson, R. L., Geraghty, M. A., et al. 1984. An outbreak of *Candida parapsilosis* bloodstream infections in patients receiving parenteral nutrition. *J. Infect. Dis.* 149:98–107

112. Solomon, W. R., Burge, H. P., Boise, J. R. 1978. Airborne *Aspergillus fumigatus* levels outside and within a large clinical center. *J. Allergy Clin. Immunol.* 62:56–60

113. Spearing, R. L., Pamphilon, D. H., Prentice, A. G. 1986. Pulmonary aspergillosis in immunosuppressed patients with hematologic malignancies. *Q. J. Med.* 59:611–25

114. Spebar, M. J., Lindberg, R. B. 1979. Fungal infection of the burn wound. *Am. J. Surg.* 138:879–82

115. Speller, D. C. E. 1986. Other approaches to the prevention of aspergillosis. *Infect. Control* 7:125–26

116. Staib, F. 1984. Ecological and epidemiological aspects of *Aspergilli* pathogenic for man and animal in Berlin (West). *Zentralbl. Bakteriol. Parasitenkd. Infektionskr. Hyg. Abt. 1 Reihe A* 257:240–45

117. Stewart, W. K., Anderson, D. C., Wilson, M. I. L. 1967. Hazard of peritoneal dialysis: contaminated fluid. *Br. Med. J.* 1:606–7

118. Streifel, A. J., Lauer, J. L., Vesley, D., Juni, B., Rhame, F. S. 1983. *Aspergillus fumigatus* and other thermotolerant fungi generated by hospital building demolition. *Appl. Environ. Microbiol.* 46:375–78

119. Streifel, A. J., Stevens, P. P., Rhame, F. S. 1987. In-hospital source of airborne *Penicillium* spores. *J. Clin. Microbiol.* 25:1–4

120. Tack, K. L., Rhane, F. S., Brown, B., Thompson, R. C. 1982. *Aspergillus* osteomyelitis: report of four cases and review of the literature. *Am. J. Med.* 73:295–300

121. Talbot, G. H., Weiner, M. H., Gerson, S. L., Provencher, M., Hurwitz, S. 1987. Serodiagnosis of invasive aspergillosis in patients with hematologic malignancy: validation of the *Aspergillus fumigatus* antigen radioimmunoassay. *J. Infect. Dis.* 155:12–27

122. Thaler, M., Pastakia, B., Shawker, T. H., O'Leary, T. O., Pizzo, P. A. 1988. Hepatic candidiasis in immunocompromised patients: a new or evolving syndrome. *Ann. Intern. Med.* 108:88–100

123. Torres-Rojas, J. R., Stratton, C. W., Sanders, C. V., Horsman, T. A., Hawley, H. B., et al. 1982. *Candida* suppurative peripheral thrombophlebitis. *Ann. Intern. Med.* 96:431–35

124. Turner, R. B., Donowitz, L. G., Hendley, J. O. 1985. Consequences of candidemia for pediatric patients. *Am. J. Dis. Child.* 139:178–80

125. Viollier, A. F., Peterson, D. E., de Jongh, C. A., Newman, K. A., Gray, W. C., et al. 1986. *Aspergillus* sinusitis in cancer patients. *Cancer Philadelphia* 58:366–71

126. Walsh, T. J., Bustamente, C. I., Vlahov, D., Standiford, H. C. 1986. Candidal suppurative peripheral thrombophlebitis: recognition, prevention, and management. *Infect. Control* 7:16–22

127. Walsh, T. J., Caplan, L. R., Hier, D. B. 1985. *Aspergillus* infections of the central nervous system: a clinicopathologic analysis. *Ann. Neurol.* 18:574–82

128. Walsh, T. J., Catchatourian, R., Cohen, H. 1983. Disseminated histoplasmosis complicating bone marrow transplantation. *Am. J. Clin. Pathol.* 79:509–11

129. Walsh, T. J., Gray, W. 1987. *Candida* epiglottitis in immunocompromised patients. *Chest* 9:482–85

130. Walsh, T. J., Hutchins, G. M. 1980. Postoperative *Candida* infections of the heart in children. *J. Pediatr. Surg.* 15:325–31

131. Walsh, T. J., Merz, W. G. 1986. Pathologic features in the human alimentary tract associated with invasiveness of *Candida tropicalis*. *Am. J. Clin. Pathol.* 85:498–502

132. Walsh, T. J., Newman, K. R., Moody, M., Wharton, R. C., Wade, J. C. 1986. Trichosporonosis in patients with neoplastic disease. *Medicine Baltimore* 65:268–79

133. Walsh, T. J., Schimpff, S. C. 1983. Prevention of infections among patients with cancer. *Eur. J. Cancer Clin. Oncol.* 19:1333–44

134. Weems, J. J. Jr., Andermont, A., Davis, B. J., Tancrede, C. H., Guiget, M., et al. 1987. Pseudoepidemic of aspergillosis after development of pulmonary infiltrates in a group of bone marrow transplant patients. *J. Clin. Microbiol.* 25:1459–62

135. Weems, J. J., Chamberland, M. E., Ward, J., Willy, M., Padhye, A. A., Solomon, S. L. 1987. *Candida parapsilosis* fungemia associated with parenteral nutrition and contaminated blood pressure transducers. *J. Clin. Microbiol.* 25:1029–32

136. Weiland, D., Ferguson, R. M., Peterson, P. K., Snover, D. C., Simmons, R. L., et al. 1983. Aspergillosis in 25 renal transplant patients. *Ann. Surg.* 198:622–24

137. Wellens, F., Potuliege, C., Deuvart, F. E., Primo, G. 1982. *Aspergillus* osteochondritis after median sternotomy. Combined operative treatment and drug therapy with amphotericin B. *Thorac. Cardiovasc. Surg.* 30:322–24

138. Windus, D. W., Stokes, T. J., Julian, B. A., Fenues, A. Z. 1987. Fatal *Rhizopus* infections in hemodialysis patients receiving deferoxamine. *Ann. Intern. Med.* 107:678–80

139. Wingard, J. R., Merz, W. G., Saral, R. 1979. *Candida tropicalis:* a major pathogen in immunocompromised patients. *Ann. Intern. Med.* 91:539–43

140. Young, R. C., Bennett, J. E., Vogel, C. L., Carbone, P. P., De Vita, V. T. 1970. Aspergillosis. The spectrum of the disease in 98 patients. *Medicine Baltimore* 49:147–73

141. Yu, V. L., Muder, R. R., Poorsatter, A. 1986. Significance of isolation of *Aspergillus* from the respiratory tract in diagnosis of invasive pulmonary apsergillosis. Results from a three-year prospective study. *Am. J. Med.* 81:249–54

Ann. Rev. Microbiol. 1988. 42:547–74

GENETICS OF *STREPTOMYCES FRADIAE* AND TYLOSIN BIOSYNTHESIS[1]

Richard H. Baltz and Eugene T. Seno

Eli Lilly and Company, Indianapolis, Indiana 46285

CONTENTS

[1]Abbreviations: AUD, amplifiable unit of DNA; *car,* carbomycin biosynthetic or resistance gene; CM, chloramphenicol; EMS, ethyl methanesulfonate; HA, hydroxylamine; HM, hygromycin; KM, kanamycin; MC, mitomycin C; *mcr,* mitomycin C resistance gene; MMS, methyl methanesulfonate; MNNG, *N*-methyl-*N'*-nitro-*N*-nitrosoguanidine; NQO, 4-nitroquinoline-1-oxide; PEG, polyethylene glycol; R$^+$, restriction proficient; R$^-$, restriction defective; RS, repeat sequence; Spc, spectinomycin; Spo, sporulation; *srm,* spiramycin biosynthetic or resistance gene; *tlr,* tylosin resistance gene; Tyl, tylosin; Tyl^r, tylosin resistant, Tyl^s, tylosin sensitive; Tyl$^-$, defective in tylosin production; *tylA, tylB,* etc, tylosin biosynthetic genes; UV, ultraviolet light.

0066-4227/88/1001-0547$02.00

INTRODUCTION

We first started working on the genetics of tylosin biosynthesis in *Streptomyces fradiae* nearly 15 years ago. Since we were primarily interested in developing procedures to improve the production of tylosin in *S. fradiae*, we set out (*a*) to understand the fundamental mechanisms of mutation in *S. fradiae* and to optimize mutagenesis procedures to maximize the probability of isolating mutants with improved tylosin production or with other specific phenotypes; (*b*) to develop efficient recombination systems to reassort mutations that affect antibiotic yields in additive or synergistic ways; (*c*) to define the pathway for tylosin biosynthesis and to determine the rate-limiting steps; (*d*) to define the structural organization and regulation of tylosin biosynthetic genes; and (*e*) to develop gene cloning methodologies to improve the levels or catalytic properties of enzymes involved in rate-limiting steps by gene amplification, promoter replacement, or site-directed mutagenesis.

During the course of these studies, it became apparent that the mutagenesis and gene cloning technologies being developed might also be used to produce novel or hybrid antibiotics. It was also observed that *S. fradiae* produced potent restriction endonucleases that cut incoming plasmid or phage DNA. Thus, if heterologous cloning of antibiotic biosynthetic genes was to be successful, methods to bypass restriction would need to be developed. In this chapter we review the development of methodologies for gene mutation, recombination, and cloning, and we describe the current understanding of the genetics and biochemistry of *S. fradiae* and tylosin biosynthesis. We also speculate on how this knowledge and the genetic methods developed can be applied to improve tylosin yields and to produce novel macrolide antibiotics.

MUTAGENIC MECHANISMS

General Approach

The approach to exploring the fundamental mechanisms of mutation in *S. fradiae* and to applying this knowledge to improve tylosin production was to carry out comparative studies drawing upon the wealth of information available from studies with *E. coli*. Much insight on the fundamental mutagenic mechanisms was gained by studying mutants of *E. coli* defective in the repair of damage induced by mutagenic agents. Several pathways of DNA repair and mutagenesis have been characterized, including error-free and error-prone repair and an adaptive response to low doses of alkylating agents (32, 88, 89). Studies were undertaken, therefore, to determine whether similar DNA repair pathways were active in *S. fradiae*.

Isolation of Mutants

Mutants of *S. fradiae* M1 defective in the repair of damage caused by mitomycin C (MC) were isolated and characterized, along with their parent, for their relative sensitivities to potential lethal damage and to mutagenesis by several other agents including ultraviolet light (UV), *N*-methyl-*N'*-nitro-*N*-nitrosoguanide (MNNG), methyl methanesulfonate (MMS), 4-nitroquinoline-1-oxide (NQO), hydroxylamine (HA), and ethyl methanesulfonate (EMS). The responses of several mutants were compared to the responses observed with well-characterized mutants of *E. coli* defective in excision repair or error-prone repair. One of the mutants, *S. fradiae* JS2, containing the *mcr-2* mutation, was determined to be blocked in error-free DNA repair, whereas *S. fradiae* JS6, containing the *mcr-6* mutation, was blocked in error-prone (or mutagenic) DNA repair.

Error-Free DNA Repair

S. fradiae JS2 *(mcr-2)* was more sensitive than strain M1 to potentially lethal damage induced by NQO, UV, and MC, but showed normal levels of resistance to EMS, MMS, MNNG, and HA (8, 15). JS2 also expressed normal levels of mutation induction by HA, MMS, MNNG, and EMS, but showed an enhanced level of mutagenesis at low doses of UV light (8, 18). All but one of these responses were identical to those expressed by *uvrA* mutants of *E. coli* (8, 47, 54). The only exception was that JS2 did not show increased mutation induction at low doses of NQO. These results suggested that JS2 was defective in a major excision repair system for bulky lesions, comparable to the *uvrABC* system in *E. coli*.

Error-Prone DNA Repair

S. fradiae JS6 *(mcr-6)* was partially defective in the repair of DNA damage induced by MC, UV, NQO, HA, MNNG, and MMS (8, 9, 84). It also showed reduced levels of mutagenesis induced by UV, NQO, HA, MNNG, EMS, and MMS (8, 84). A spontaneous revertant of JS6 expressed wild-type levels of induced mutagenesis and resistance to all the mutagenic agents, which indicated that the multiple defects in repair and mutagenesis were caused by a single mutation (84). JS6, therefore, appeared to be defective in a function analogous to the *recA* function in *E. coli* (88, 89) and the *recE* function in *Bacillus subtilis* (93). The *recA* gene of *E. coli* was cloned into JS6 and was shown to complement defects in DNA repair and mutagenesis (60).

These studies indicated that most induced mutagenesis in *S. fradiae* is regulated by a protein identical in function and mechanism to the *recA* protein of *E. coli*. *S. fradiae*, therefore, must express genes similar to the *umuC* and *umuD* genes of *E. coli* and encode a repressor analogous to the *lexA* protein of *E. coli,* which is cleaveable by the *E. coli recA* protein (60, 88, 89).

Adaptive Response

In *E. coli,* treatment of cells with a low dose of MNNG induces an adaptive response that results in the repair of O^6-methylguanine residues (56). The enzyme involved in adaptive repair is a methyltransferase that specifically removes methyl groups from O^6-methlguanine or ethyl groups from O^6-ethylguanine. Adapted cells are less susceptible to MNNG mutagenesis at low doses.

Streptomyces fradiae expresses an adaptive response after treatment of cells with 0.5 or 1.0 μg of MNNG ml^{-1} for 4 hr (19). Adapted cells were not mutable by up to 20 μg of MNNG ml^{-1}, whereas unadapted cells were. No adaptive response was observed when cells were treated with 5 μg of MNNG ml^{-1} for 4 hr, but a high frequency of mutants was induced during the treatment period. Thus it appears that adaptive repair in *S. fradiae* is relevant only at very low doses of MNNG.

Comparative Mutagenesis

Since *S. fradiae* expresses at least three DNA repair systems similar to those expressed in *E. coli, S. fradiae* was expected to show mutagenic responses to chemical mutagens and UV light similar to those observed in *E. coli.* This similarity was observed with one prominent exception. Whereas UV light is a very efficient mutagen for *E. coli* (91), it is only marginally mutagenic for *S. fradiae* (7, 8, 18, 84).

The efficiency of mutagenesis per lethal hit in *S. fradiae* ranged over 1000-fold (7, 8) in the order MNNG > NQO > EMS = MMS > HA >> UV. MNNG-induced mutation frequencies were further increased about 10-fold by mutagenizing cells in the presence of chloramphenicol (8, 19). Thus MNNG has been the mutagen of choice for induction of many kinds of mutants, including auxotrophic mutants (4), mutants blocked in antibiotic production (14), mutants defective in restriction (63; P. Matsushima & R. H. Baltz, manuscript in preparation), and mutants that overproduce tylosin (79).

RECOMBINATION

Background Information

When we initiated genetic studies on *S. fradiae* around 1975, very little was known about gene transfer and recombination in *Streptomyces.* No generalized transducing phages had been identified for any species, no transformation systems had been reported, and conjugation had been described in only a limited number of species, most notably in *Streptomyces coelicolor* (42). We first attempted to isolate generalized transducing phages and also to demonstrate conjugal transfer of chromosomal genes (as defined by auxotrophic mutations induced by MNNG). Since our preliminary attempts to demonstrate transduction or conjugation were not successful (A. K. Radue & R. H. Baltz, unpublished), we turned our attention to a potentially more general approach to recombination, that of protoplast fusion and cell regeneration. This was a considerable departure from the conventional wisdom learned from the *E. coli* systems, but it was encouraged by advancements made in genetic recombination by protoplast fusion in several fungal genera (1, 35) and in *Bacillus* species (39, 77). We reasoned that if we could demonstrate genetic recombination by protoplast fusion in *S. fradiae,* this methodology might be generally applicable to many other species of *Streptomyces* and might also be used to generate interspecies recombinants. We did not abandon the approaches of conjugation, transformation, and transduction, but all of these techniques required further advances in technology. In the following sections we describe the development of all four genetic methodologies in *S. fradiae,* in approximately chronological order.

Protoplast Fusion

It had been demonstrated by 1974 that certain *Streptomyces* species could be converted to protoplasts and that the protoplasts could regenerate viable cells on hypertonic media optimized for cell regeneration (70). These observations suggested (*a*) that protoplasts of *Streptomyces* might be fused to produce genetic recombinants and (*b*) that *Streptomyces* protoplasts might be transformed or transfected by plasmid or phage DNA by techniques well developed for transfection in *E. coli* (2, 3, 10, 22).

In the initial studies, we found that protoplasts could be readily prepared from *S. fradiae* cells grown in tryptic soy broth containing 0.4% glycine (4). The addition of glycine to growth media had been shown previously to make streptomycete cells much more susceptible to protoplast formation by lysozyme and lytic-enzyme treatment (70). Protoplasts prepared from cells grown at 29°C to a transition phase between the exponential and stationary growth phases regenerated viable cells very efficiently on modified R2 agar at 29°C (12). Much poorer results were obtained if cells were grown at higher temperatures, if cells were grown to the mid-exponential or late stationary phase, or if protoplasts were regenerated at elevated temperatures (12). For instance, cell growth and protoplast regeneration at 37°C resulted in a 10^4-fold reduction in regeneration efficiency compared to that at 29°C. These observations had important implications for later studies on transformation and transduction (see below).

In the early studies on protoplast regeneration, it was also observed that if large numbers of protoplasts of *S. fradiae* were plated on modified R2 agar, the colonies that developed earliest inhibited the regeneration of nearby protoplasts (4). This autoinhibition of cell regeneration was eliminated by dehydrating the surface of the modified R2 agar plates (12). The dehydration procedure was also important for the development of an efficient protoplast transformation system for *S. fradiae*.

Protoplasts prepared from cells grown under optimal conditions for cell regeneration were readily induced to fuse by suspension in polyethylene glycol (PEG) 1000 or 6000. Maximum yields of genetic recombinants (usually 10% or more of total progeny) were obtained when 40–60% PEG 1000 was used as the fusing agent (4, 12). Protoplast fusion was used to construct a rudimentary circular genetic map of *S. fradiae* (5). However, attempts to map certain tylosin biosynthetic genes gave uncertain results, since most genetic recombinants were no longer able to produce tylosin or the products of mutants (see below) blocked in different steps in tylosin biosynthesis (5). This curious observation suggested that tylosin genes might be located on an element that is readily lost during protoplast formation, regeneration, or genetic recombination.

Protoplast fusion has also been used to rescue genetic markers from heat-inactivated protoplasts (13, 63). This technique has been used to construct complex genotypes under conditions where the donor strain cannot be readily counterselected (63).

Protoplast Transformation

Transformation of streptomycete protoplasts was first demonstrated by Bibb et al (25) using the pock-forming plasmid, SCP2*. Subsequently, several antibiotic resistance genes were cloned, and the first streptomycete plasmid cloning vectors became available between 1980 and 1982 (50, 76, 86). We initiated transformation studies in several strains of *S. fradiae,* including a sporulation-defective mutant, M1 (8, 13, 58; see Table 1 for the derivation of M1). M1 was thought to be less restricting than most sporulating strains, since it was a better host for propagating many different bacteriophages. For example, many streptomycete phages formed larger plaques and plated at higher efficiencies on M1 than on other strains (K. Cox & R. H. Baltz, unpublished). Also, from protoplast fusion studies, M1 was shown to form more viable protoplasts per cell mass, and the resulting protoplasts regenerated more rapidly than protoplasts prepared from most other *S. fradiae* strains (P. Matsushima & R. H. Baltz, unpublished).

pFJ105, a 5-kb plasmid carrying the thiostrepton resistance gene *(tsr)* (76) from *Streptomyces azureus,* transformed M1 at a frequency of about one transformant per microgram of DNA (13, 58) when the procedure of Bibb et al was followed (25) and the conditions for cell regeneration were optimal as determined in the protoplast fusion studies. DNA prepared from a transfor-

Table 1 Derivation of *Streptomyces fradiae* strains

Strain	Relevant phenotype or property[a]	Derivation[b]	Reference
T59235	Tyl^r Tyl^+ Spo^+ R^+; highly restricting	Wild type	79
JS82	Tyl^s Tyl^- AUD^{500+}	Regenerated protoplast of T59235	15, 20, 37
JS85	Tyl^s Tyl^- Spo^+ AUD^{500+} Spc^r R^+; highly restricting	Spontaneous Spc^r mutant of JS82	15, 20, 37, 63
PM32	Tyl^s Tyl^- Spo^- AUD^+ Spc^r; less restricting than JS85	MNNG mutagenesis (5 rounds) from JS85	63
PM41	Tyl^s Tyl^- Spo^+ AUD^+ Spc^r; less restricting than JS85	Protoplast fusion between JS85 and PM32	63
PM60	Tyl^r $TylF^-$ Spo^+ AUD^+ Spc^r; less restricting than GS15	PM41 × GS15	63
PM61	Tyl^r $TylB^-$ Spo^+ AUD^+ Spc^r; less restricting than GS50	PM42 × GS50	63
PM63	Tyl^r $TylB^-$ Spo^- AUD^+ Spc^r; less restricting than PM61	MNNG mutagenesis of PM61	63
PM65	Tyl^r $TylB^-$ Spo^+ AUD^+ Spc^r	Protoplast fusion between PM61 and PM63	63
PM71	Tyl^r $TylF^-$ Spo^- AUD^+ Spc^r; less restricting than PM60	MNNG mutagenesis of PM60	63
PM73	Tyl^r $TylB^-$ Spo^+ AUD^+ Spc^r Ery^r; less restricting than PM61	Spontaneous mutagenesis of PM65	63
PM76	Tyl^s Tyl^- Spo^+ Spc^r R^-; less restricting than PM41	MNNG mutagensis of PM41	c
C4	Tyl^r Tyl^+ Nar^- AUD^+; high tylosin production	Sequential mutation of T59235	79, 85
JS51	Tyl^r Tyl^+ Nar^- AUD^+ Rif^r	Spontaneous Rif^r mutant of C4	85
JS87	Tyl^s Tyl^- Nar^- AUD^-	Regenerated protoplast of JS51	20, 29
GS93	Tyl^s Tyl^- Nar^- AUD^-	Regenerated protoplast of C4	29
GS14	Tyl^r $TylA^-$ Nar^- AUD^+	MNNG mutagensis of C4	14, 85
GS50	Tyl^r $TylB^-$ Nar^- AUD^+	MNNG mutagensis of C4	14, 85
GS52	Tyl^r $TylC^-$ Nar^- AUD^+	MNNG mutagensis of C4	14, 85
GS48	Tyl^r $TylD^-$ Nar^- AUD^+	MNNG mutagensis of C4	14, 85
GS16	Tyl^r $TylE^-$ Nar^-	MNNG mutagensis of C4	14
GS15	Tyl^r $TylF^-$ Nar^- AUD^+	MNNG mutagensis of C4	14, 85
GS28	Tyl^r $TylF^-$ Nar^-	MNNG mutagensis of C4	14
GS76	Tyl^r $TylH^-$ $TylD^-$ Nar^-	MNNG mutagensis of GS48	14
GS77	Tyl^r $TylI^-$ $TylD^-$ Nar^- AUD^+	MNNG mutagensis of GS48	14, 85
GS22	Tyl^r $TylG^-$ Nar^- AUD^+	MNNG mutagensis of C4	14, 85
GS18	Tyl^r $TylG^-$ Nar^-	MNNG mutagensis of C4	14
GS40	Tyl^s Tyl^- Nar^-	MNNG mutagensis of C4	14
GS41	Tyl^r Tyl^- Nar^- AUD^+	MNNG mutagensis of C4	14, 85
GS88	Tyl^r $TylJ^-$	MNNG mutagensis of C4	29
GS85	Tyl^r $TylK^-$	MNNG mutagensis of C4	29
GS33	Tyl^r $TylL^-$	MNNG mutagensis of C4	29
GS62	Tyl^r $TylM^-$	MNNG mutagensis of C4	29
JS90	Tyl^r Tyl^+ Nar^+ Spc^r AUD^+	JS51 × JS85	85
PM78	Tyl^r $TylD^-$ R^-	GS48 × PM76	c

Table 1   (*Continued*)

Strain	Relevant phenotype or property[a]	Derivation[b]	Reference
PM79	Tyl^r TylE$^-$ R$^-$	GS16 × PM76	c
PM77	Tyl^r TylF$^-$ R$^-$	GS15 × PM76	c
PM82	Tyl^r TylJ$^-$ R$^-$	GS88 × PM76	c
PM80	Tyl^r TylD$^-$ TylH$^-$ R$^-$	GS76 × PM76	c
M1	Tyl^r Tyl$^+$ Spo$^-$	Spontaneous Spo$^-$ mutant of C4	8, 58
JS2	UV^s MC^s; defective in error-free repair	MNNG mutagenesis of M1	8, 18
JS6	UV^s MC^s; defective in error-prone repair	MNNG mutagenesis of M1	8, 84

[a]Tyl^r, resistant to tylosin at 400 μg ml^{-1}; Tyl$^+$, produces tylosin; Spo$^+$, produces aerial mycelia and spores; AUD$^+$, contains low copy number of amplifiable DNA; AUD^{500+}, contains about 500 copies of amplifiable DNA reiterated in tandem; Spc^r, resistant to spectinomycin at 50 μg ml^{-1}; Ery^r, resistant to erythromycin at 500 μg ml^{-1}; TylA, TylB, etc, blocked in specific steps in tylosin biosynthesis (see Table 4 and Figure 3); R$^+$, contains about five restriction systems; R$^-$, contains mutations blocking about five restriction systems.

[b]Conjugal crosses are designated A × B. Protoplast fusion crosses are stated as such. MNNG mutagenesis was carried out as described (7, 84).

[c]P. Matsushima & R. H. Baltz, manuscript in preparation.

mant of M1 transformed M1 at a frequency of about 1000 transformants per microgram of DNA, which suggests that *S. fradiae* produces functional restriction and modification enzymes (see below). Subsequent optimization studies with plasmid pFJ105 and pIJ702 (50) modified for *S. fradiae* restriction resulted in the development of a highly efficient transformation procedure for *S. fradiae* (58). Under optimum conditions, *S. fradiae* M1 has been transformed at efficiencies ranging from 10^6 to 10^7 transformants per microgram of modified DNA. Unmodified DNA transforms at frequencies from 10^3- to 10^5-fold lower, depending on the plasmid. In many cases, plasmids containing inserts of up to 35 kb could be transformed into M1 at detectable frequencies, but could not be transformed into other *S. fradiae* strains. The plasmids subsequently isolated from M1 could transform other more restricting strains of *S. fradiae*. This two-step cloning procedure, while time consuming, has been applied successfully to clone many tylosin biosynthetic genes (see below).

Conjugation and the Tylosin Element

We describe the conjugation experiments here since they introduce the interesting and useful properties of a tylosin-defective strain, JS85. As mentioned above, early attempts to demonstrate conjugal transfer of chromosomal genes were unsuccessful. However, the experiments on protoplast fusion–induced recombination suggested that tylosin genes might not exist as normal chromosomal genes and might be readily cured by protoplast manipulations. We reasoned that if the tylosin genes were located on a plasmid or similar element, they might be self-transmissible into strains devoid of the element.

Therefore, attempts were made to obtain *S. fradiae* strains cured of the putative tylosin element. Protoplasts from two *S. fradiae* strains were plated on regeneration medium, and colonies from regenerated protoplasts were screened for mutants that had lost tylosin production (Tyl$^-$) and tylosin resistance (Tyl^s). Two prototrophic variants with these traits were obtained and further characterized (20). Both mutants exhibited other phenotypes not observed in the parental strains. One Tyl$^-$ Tyl^s strain, JS87 (see Table 1), was more sensitive to chloramphenicol (CM) than wild-type strains. This strain also displayed an unusual genetic instability associated with high-frequency segregation of auxotrophic mutants requiring histidine or methionine (but not both), and with further segregation of variants highly resistant to MC. JS87 was subsequently found to have a deletion including many tylosin biosynthetic genes and the four repeat sequences normally present in *S. fradiae* strains (see below). JS87 was a poor recipient for conjugal transfer of tylosin genes (85).

The second Tyl$^-$ Tyl^s strain, JS82 (see Table 1), was more sensitive than the wild type to CM, MC, hygromycin B (HM), and kanamycin (KM) (20). It contained a large deletion encompassing many tylosin biosynthetic genes (29; see below). It also contained a 500-fold amplification of a 10-kb segment of DNA bounded by two of the four repeat sequences deleted in JS87 (37, 38, 40). A spontaneous spectinomycin-resistant mutant of JS82 (called JS85) was shown to be a highly efficient recipient for the conjugal transfer of many tylosin biosynthetic genes independently of chromosomal gene recombination (15, 16, 85). The recombinants also expressed wild-type levels of resistance to CM, MC, HM, and KM and contained single copies of the amplifiable unit of DNA. These studies indicated that the tylosin biosynthetic genes were genetically linked and present on a self-transmissable, plasmid-like element. However, numerous attempts to isolate the element have been unsuccessful (37, 85).

Restriction-Deficient Cloning Hosts

S. fradiae produces restriction endonucleases that inhibit plasmid transformation (58) and infection with certain broad–host range bacteriophages (28, 63). Baltz and coworkers (6, 11) proposed that a set of nonrestricting strains containing specific *tyl* gene mutations could be developed for complementation studies and for heterologous gene cloning to produce hybrid macrolide antibiotic–producing strains. Since it was not known a priori how many restriction systems would be encountered, Matsushima and coworkers (63; P. Matsushima & R. H. Baltz, manuscript in preparation) developed a strategy in which a nonrestricting derivative of JS85 would be selected, and this strain would be a recipient for the conjugative transfer of blocks of tylosin biosynthetic genes containing different mutant alleles. JS85 was mutagenized by MNNG, converted to protoplasts, and transformed with several different plasmid vectors. Transformants were pooled and cured of plasmids, and the mutagenesis, transformation, and selection procedure was repeated several times (63). After six rounds of mutagenesis and selection, two highly transformable and essentially nonrestricting mutants, PM73 and PM76, were isolated (63; P. Matsushima & R. H. Baltz, manuscript in preparation; Table 1). From the restriction enzyme cutting patterns of JS85 and PM73, it appeared that four modification systems, for isoschizomers of ScaI, PstI, EcoRI, and XhoI modification, were inactivated after mutagenesis and selection. The phage plating efficiencies and plasmid transformation efficiencies of the lineage of strains from JS85 to PM73 suggested that at least five restriction systems were inactivated in PM73 and PM76. PM76 was shown to be an efficient recipient for conjugal transfer of tylosin genes, and phage plating efficiency and transformation studies suggested that none of the genes encoding restriction endonucleases were transferred with tyl genes during conjugation. All of the exconjugants were transformable with unmodified DNA at efficiencies about 10^4-fold higher than that of JS85, and all efficiently plated several phages that normally do not form plaques on S. fradiae (63; P. Matsushima & R. H. Baltz, manuscript in preparation; Table 2). These relatively nonrestricting recombinant strains containing specific tyl mutations are currently being evaluated as potential hosts for cloning and expression of antibiotic biosynthetic genes.

Transduction of Plasmid DNA

A versatile new technique for rapid transfer of plasmid DNA between different Streptomyces species is bacteriophage FP43–mediated transduction (65). A segment of FP43 DNA, designated hft for high frequency transduction, was cloned into plasmid pIJ702. This segment causes the resulting plasmid, pRHB101, to be efficiently packaged into FP43 phage heads as linear concatemers; the transducing particles attach and inject the DNA, and circular monomers are formed, presumably by host recombination systems (65). It seemed that this transduction system might be very useful in S. fradiae for at least two reasons. First, transduction could offer a means for rapid transfer of cloned genes into a variety of strains, bypassing protoplast transformation. Secondly, it seemed that transduction might offer an alternative route for bypassing restriction. For instance, it was conceivable that at high multiplicity of infection, competition by the excess phage DNA might protect the plasmid from restriction endonucleases; or concatenated linear plasmid DNA might be less susceptible to inactivation by restriction endonucleases than the monomeric plasmid. Also, if restriction endonuclease gene expression is regulated during cell growth and differentiation, then growth conditions might be found that could minimize the expression of restriction.

The efficiency of transduction of different S. fradiae strains varies over 1000-fold with cell growth phase and growth temperature (M. A. McHenney

& R. H. Baltz, manuscript in preparation). Maximum transduction frequencies in M1 were obtained using cells grown at 39°C to mid-exponential phase (Figure 1, *A*). Interestingly, this condition could not be used for protoplast transformation, since protoplasts from cells in exponential growth phase do not regenerate well (4), and since cells grown at elevated temperatures yield protoplasts that regenerate very poorly (12). The most important observation was that under optimum conditions for transduction, the highly restricting strains JS85 and GS15 were no more than threefold less transducible than the nonrestricting derivative strains PM76 and PM77 (Figure 1, *B* and *C*; Table 2). *S. fradiae* M1 actually gave higher frequencies of transductants than the nonrestricting strain PM76 (Figure 1, *A* and *B*). It is not known how transduction of exponential phase cells grown at 39°C bypasses restriction; a combination of events may be responsible (M. A. McHenney & R. H. Baltz, manuscript in preparation). It is also not known how linear concatemers of the plasmid DNA injected from transducing particles reform circles during the process of transduction. It is interesting that *S. fradiae* JS6, a strain defective in a function analogous to the *recA* function in *E. coli* (see section on mutagenesis), was transduced by pRHB101 nearly as efficiently as its parent, M1 (Figure 1, *A*). This result suggests either that JS6 is defective only

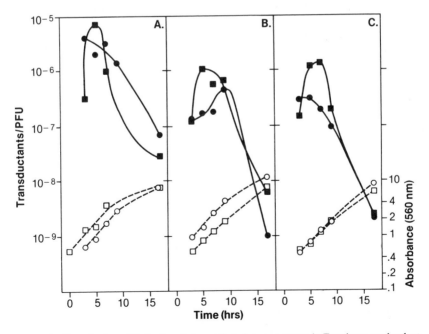

Figure 1   Transduction of *S. fradiae* strains. All strains were grown in Trypticase soy broth at 39°C. Cells were taken at various times for analysis of bacteriophage FP43–mediated transduction of plasmid pRHB101. (*A*) M1, *squares;* JS6, *circles.* (*B*) PM76, *squares;* JS85, *circles.* (*C*) PM77, *squares;* GS15, *circles.*

in *recA* protease and not in recombinase, or that plasmid circle formation during transduction occurs independently of host *recA* recombinase function. Further work is needed to determine if JS6 is proficient or defective in other modes of recombination and to determine if plasmid circle formation occurs by a *recA*-dependent mechanism.

Since FP43 transduction works in about 80% of the *Streptomyces* species tested (65), it should be an effective gene transfer method, especially for many other restricting strains for which transformation systems have not been developed.

TYLOSIN BIOSYNTHESIS

Biochemical and Physiological Studies

Tylosin is a macrolide antibiotic composed of a 16-membered branched lactone, tylonolide, and three deoxy-sugar residues: the amino sugar mycaminose, attached at the C-5 hydroxyl position; the neutral sugar mycarose, attached to the C-4 hydroxyl of mycaminose; and the neutral sugar mycinose, linked to the lactone at the C-23 hydroxyl position (Figure 2). Early studies on

Table 2 Expression of restriction during phage infection, transformation, and transduction in different *S. fradiae* strains[a]

Strain[b]	Phenotype[c]	Bacteriophage EOP		Transformation frequency[d]		Transduction frequency[e]
		FP43	FP55	pIJ702	pHJL281	pRHB101
JS85	R+ Tyl−	$<2 \times 10^{-9}$	$<2 \times 10^{-8}$	$<4 \times 10^{1}$		5×10^{-7}
GS50	R+ TylB−	$<6 \times 10^{-9}$	$<2 \times 10^{-8}$			
GS15	R+ TylF−	$<4 \times 10^{-9}$	$<2 \times 10^{-8}$		<2	3×10^{-7}
GS16	R+ TylE−	$<6 \times 10^{-9}$	$<2 \times 10^{-8}$		6	
GS48	R+ TylD−	$<6 \times 10^{-9}$	$<2 \times 10^{-8}$		<2	
GS76	R+ TylH− TylD−	$<6 \times 10^{-9}$	$<2 \times 10^{-8}$		2	
GS88	R+ TylJ−	4×10^{-6}	$<2 \times 10^{-8}$		<2	
PM76	R− Tyl−	7×10^{-1}	3×10^{-1}			1×10^{-6}
PM73	R− TylB−	4×10^{-2}	2×10^{-3}	9×10^{4}		
PM77	R− TylF−	3×10^{-1}	8×10^{-2}		1×10^{4}	1×10^{-6}
PM79	R− TylE−	1×10^{-1}	7×10^{-3}		5×10^{4}	
PM78	R− TylD−	5×10^{-2}	1×10^{-3}		2×10^{4}	
PM80	R− TylH− TylD−	1×10^{-1}	4×10^{-2}		4×10^{3}	
PM82	R− TylJ−	4×10^{-2}	5×10^{-2}		2×10^{4}	

[a]Data from P. Matsushima & R. H. Baltz, manuscript in preparation; M. A. McHenney & R. H. Baltz, manuscript in preparation.
[b]*S. fradiae* strain derivations are shown in Table 1.
[c]R+, expresses ~5 restriction systems; R−, defective in ~5 restriction systems. TylA, TylB, etc, contains mutations in *tylA*, *tylB*, etc (see Table 4 and Figure 3).
[d]Transformants per microgram of plasmid pIJ702 (50) or pHJL281 (29) DNA.
[e]Transduction of plasmid pRHB101 per PFU of bacteriophage FP43; maximum frequencies from Figure 1

the biosynthesis of the lactone core of the molecule were influenced by the similarity of tylactone (Figures 2 and 3) to a cyclic branched fatty acid (81). This notion of similarity was reinforced by the observation that cerulenin, an inhibitor of the condensing enzyme in fatty acid biosynthesis (71), inhibited the biosynthesis of tylactone and other macrolide lactones (72). Experiments evaluating the incorporation of ^{13}C-labeled putative precursors into tylactone suggested that tylactone was derived by condensation of five propionate units, two acetate units, and one butyrate unit (73). By analogy with fatty acid biosynthesis, tylactone biosynthesis would be initiated by a propionyl-CoA primer (C-15, C-16 and C-17) (Figure 2), with successive condensations of two molecules of methylmalonyl-CoA (C-13, C-14, C-23; C-11, C-12, C-22), one acetyl-CoA (C-9, C-10), a methylmalonyl-CoA (C-7, C-8, C-21), an ethylmalonyl-CoA (C-5, C-6, C-19, C-20), a methylmalonyl-CoA (C-3, C-4, C-18), and a final acetyl-CoA molecule (C-1, C-2). The intermediates in tylactone biosynthesis appear to be derived primarily from amino acid catabolism; valine, isoleucine, and methionine provide the three-carbon units, and leucine and phenylalanine provide the two- and four-carbon units (31). The catabolism of fatty acids from oils in the fermentation medium probably also contributes to the two- and four-carbon unit pools.

Fermentation of *S. fradiae,* under certain conditions, causes accumulation of other macrolide compounds in addition to tylosin. The major nontylosin products are macrocin (Figure 3) and relomycin (Table 3). (Table 3 details the

Tylactone

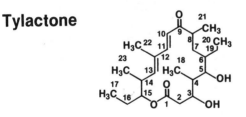

Tylosin

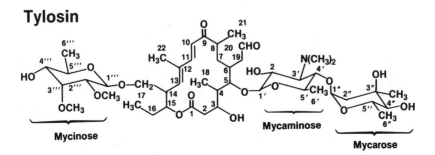

Figure 2 The structures of tylosin and tylactone.

Table 3 Structures of tylosin intermediates and shunt products[a]

Compound	Oxidation level		Sugars present			O-Methyl groups present	
	C-20	C-23[b]	Mycaminose	Mycarose	6-Deoxy-D-allose	2'''	3'''
Tylactone	CH_3	CH_3	–	–	–	–	–
O-Mycaminosyl tylactone	CH_3	CH_3	+	–	–	–	–
20-Deoxy-20-dihydro-O-mycaminosyltylonolide	CH_3	CH_2OH	+	–	–	–	–
20-Dihydro-23-deoxy-O-mycaminosyltylonolide	CH_2OH	CH_3	+	–	–	–	–
20-Dihydro-O-mycaminosyltylonolide	CH_2OH	CH_2OH	+	–	–	–	–
23-Deoxy-O-mycaminosyltylonolide	CHO	CH_3	+	–	–	–	–
O-Mycaminosyltylonolide	CHO	CH_2OH	+	–	–	–	–
20-Dihydro-23-deoxydemycinosyltylosin	CH_2OH	CH_3	+	+	–	–	–
23-Deoxydemycinosyltylosin	CHO	CH_3	+	+	–	–	–
20-Deoxy-20-dihydrodemycinosyltylosin	CH_3	CH_2OH	+	+	–	–	–
20-Dihydrodemycinosyltylosin	CH_2OH	CH_2OH	+	+	–	–	–
Demycinosyltylosin	CHO	CH_2OH	+	+	–	–	–
20-Dihydrodemethyllactenocin	CH_2OH	CH_2OR	+	–	+	–	–
Demethyllactenocin	CHO	CH_2OR	+	–	+	–	–
20-Dihydrodemethylmacrocin	CH_2OH	CH_2OR	+	+	+	–	–
Demethylmacrocin	CHO	CH_2OR	+	+	+	–	–
20-Dihydrolactenocin	CH_2OH	CH_2OR	+	–	+	+	–
Lactenocin	CHO	CH_2OR	+	–	+	+	–
20-Dihydromacrocin	CH_2OH	CH_2OR	+	+	+	+	–
Macrocin	CHO	CH_2OR	+	+	+	+	–
20-Dihydrodesmycosin	CH_2OH	CH_2OR	+	–	+	+	+
Desmycosin	CHO	CH_2OR	+	–	+	+	+
Relomycin	CH_2OH	CH_2OR	+	+	+	+	+
Tylosin	CHO	CH_2OR	+	+	+	+	+

[a]See Figure 2 for locations of positions of oxidation, sugar attachment, and O-methylation.
[b]R, 6-Deoxy-D-allose or the O-methylated derivatives demethylmycinose or mycinose.

relevant structural features of all of the tylosin intermediates and analogs discussed in this paper.) Studies with radioactively labeled tylosin intermediates suggested that macrocin is a precursor to tylosin, while relomycin is a product of tylosin reduction (82). These experiments also suggested that desmycosin is converted to tylosin and that lactenocin (Table 3) is converted to tylosin via macrocin and desmycosin.

The enzyme macrocin *O*-methyltransferase, which catalyzes the conversion

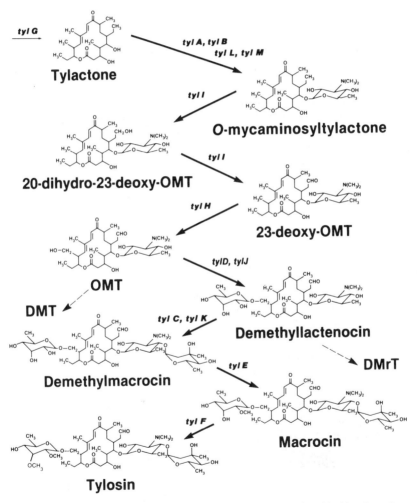

Figure 3 The biosynthetic pathway from tylactone to tylosin. Mutations blocking the pathway at the different steps are indicated. *OMT:* *O*-mycaminosyltylonolide; *DMT:* demycinosyltylosin; *DMrT:* demycarosyltylosin (desmycosin).

of macrocin to tylosin, was measurable in cell-free extracts of *S. fradiae* (82). The specific activity of this enzyme was low early in the fermentation and peaked 2–3 days after inoculation (78, 79), when tylosin biosynthesis was occurring at its maximum rate. The enzyme was inhibited by its product, tylosin, and by other macrolide intermediates with a sugar residue at C-23 (78). Successive selection for increased tylosin productivity in a lineage of *S. fradiae* strains was associated with an increase in the specific activity of macrocin *O*-methyltransferase (79). This enzyme has been purified; it has a molecular mass of 65 kd and consists of two identical subunits (N. J. Bauer, A. J. Kreuzman, J. E. Dotzlaf & W.-K. Yeh, manuscript submitted). Studies with the purified enzyme generally confirmed the macrolide inhibition patterns observed in cell-free extracts.

A second tylosin-specific *O*-methyltransferase activity was identified in cell-free extracts (78) after the mutant GS16 (see below), which lacked this activity and accumulated demethylmacrocin, was isolated (14). This enzyme, demethylmacrocin *O*-methyltransferase, catalyzed the conversion of demethylmacrocin to macrocin by methylation of the 2-hydroxyl group of the 6-deoxy-D-allose residue in the substrate (Figures 2 and 3). Demethylmacrocin *O*-methyltransferase was recently purified and was shown to have a molecular mass of 118–126 kd (A. J. Kreuzman, J. R. Turner & W.-K. Yeh, manuscript in preparation). The subunit size was estimated at 42 kd, which suggests that the native enzyme is a trimer. The purified enzyme was subject to inhibition by its product, macrocin, and by a number of other macrolide intermediates in tylosin biosynthesis (A. J. Kreuzman, J. R. Turner & W.-K. Yeh, manuscript in preparation).

Some limited studies on the biosynthesis of tylosin sugars in cell-free extracts have been reported. A cell-free extract of the tylosin-producing strain *Streptomyces rimosus* catalyzed the conversion of TDP-D-glucose to TDP-mycarose in the presence of *S*-adenosyl-L-methionine (74). The reaction proceeded through the intermediate TDP-4-keto-6-deoxy-D-glucose. The enzyme TDP-D-glucose oxidoreductase, which was responsible for the conversion of TDP-D-glucose to this intermediate, was purified, and the time course of its activity was shown to parallel that of the TDP-mycarose–synthesizing activity and tylosin biosynthesis (57). It is likely that the biosynthesis of mycarose in *S. fradiae* proceeds in a similar manner from TDP-D-glucose through 4-keto-6-deoxyglucose.

Analysis of Blocked Mutants

Genetic analysis of tylosin biosynthesis required the isolation of *tyl* mutants blocked at many different steps in the pathway. Nitrosoguanidine was known to be a potent mutagen of *S. fradiae* (7, 8, 14; see section on comparative mutagenesis, above) and was therefore the mutagen of choice for generating *tyl* mutants. Since many tylosin intermediates were biologically active, a simple screen for loss of antibiotic activity would have been inadequate for isolating mutations at steps throughout the pathway. However, much information was available on the separation of tylosin intermediates by thin layer chromatography. This technique, coupled with the use of acidic vanillin to visualize macrolide intermediates, was employed to screen fermentation broths of *S. fradiae* mutants to detect strains blocked in tylosin biosynthesis.

About 6000 fermentation broths were screened in the initial experiments, and 72 *tyl* mutants were identified (14); novel compounds produced by several mutants were characterized (48, 52, 53). Several other *tyl* mutants, which have been identified in the course of routine strain improvement procedures, have since been added to the collection. About 80% of the mutants, designated *tylG*, were unable to produce the first detectable intermediate in the pathway, tylactone (Table 4). The remaining mutants (Table 4) produced tylactone, desmycosin, demycinosyltylosin, 23-deoxydemycinosyltylosin, demethylmacrocin, macrocin, and 20-deoxy-20-dihydrodemycinosyltylosin as major products (see Table 3 for relevant structural details). In several cases the phenotypic classes could be further subdivided by either of two techniques: cosynthesis of tylosin with other *tyl* mutants or complementation with cloned *tyl* DNA sequences (see below). The results of cosynthesis tests indicated that the *tylA* and *tylL* mutants were different from the *tylB* and *tylM* mutants; the former two were defective in the production or attachment of all three tylosin sugars, while the latter two were deficient in only mycaminose synthesis or attachment. Complementation by cloned *tyl* DNA indicated that *tylA* and *tylL* were different loci, as were *tylB* and *tylM* (29, 36). Complementation with cloned *tyl* genes also revealed that two loci, *tylC* and *tylK*, were represented in mutants accumulating desmycosin; these mutants were presumed to be defective in mycarose biosynthesis or attachment. Two other loci, *tylD* and *tylJ*, were defective in mutants producing demycinosyltylosin; the defects were probably in the synthesis or attachment of the mycinose precursor, 6-deoxy-D-allose. The *tylE* and *tylF* mutations prevented methylation of the 2- and 3-hydroxyl positions, respectively, in the 6-deoxy-D-allose residue. The strains containing the *tylE* and *tylF* mutations lacked demethylmacrocin *O*-methyltransferase activity and macrocin *O*-methyltransferase activity, respectively (78). The *tylH* and *tylI* mutations blocked oxidation of C-23 and C-20, respectively; therefore these genes were presumed to encode, or control the expression of, oxidizing enzymes.

The accumulation of tylactone by the *tylB* and *tylM* mutants implied that the attachment of mycaminose was the obligate first step in the conversion of tylactone to tylosin. The production of macrocin by the *tylF* mutant and the conversion of macrocin to tylosin by methylation in cell-free extracts of *S. fradiae* indicated that the methylation of macrocin was the final step in tylosin

Table 4 Summary of classes of S. *fradiae* mutants blocked in tylosin biosynthesis

Mutant class	Biochemical block	Compound produced
tylA	Mycaminose, mycarose, and 6-deoxy-D-allose	Tylactone
tylL	Mycaminose, mycarose, and 6-deoxy-D-allose	Tylactone
tylB	Mycaminose[a]	Tylactone
tylM	Mycaminose[a]	Tylactone
tylC	Mycarose[a]	Desmycosin
tylK	Mycarose[a]	Desmycosin
tylD	6-Deoxy-D-allose[a]	Demycinosyltylosin
tylJ	6-Deoxy-D-allose[a]	Demycinosyltylosin
tylE	2'''-O-methylation	Demethylmacrocin
tylF	3'''-O-methylation	Macrocin
tylG	Tylactone formation	None
tylH	C-23 oxidation	23-Deoxydemycinosyltylosin
tylI	C-20 oxidation(s)	Not known[b]

[a]May be blocked in sugar biosynthesis or in glycosyltransferase reactions.
[b]No *tylI* single mutant is available. The *tylI tylD* double mutant produces 20-deoxy-20-dihydro-O-mycaminosyltylonolide and 20-deoxy-20-dihydrodemycinosyl tylosin.

biosynthesis. Similarly, the accumulation of demethylmacrocin by the *tylE* mutant (14) and the conversion of demethylmacrocin to macrocin by cell-free extracts of the *tylF* mutant (78) indicated that the methylation of demethylmacrocin was the penultimate step in the pathway. A pulse-chase experiment using demethylmacrocin as substrate with wild-type, cell-free extracts revealed the conversion of demethylmacrocin to tylosin through the intermediate macrocin (17).

The remaining steps in the bioconversion of tylactone to tylosin were deduced by experiments based on the observation that most *tylG* mutants could make tylosin if provided with tylactone or other macrolide biosynthetic intermediates of tylosin. A large number of putative tylosin intermediates (Table 3), obtained directly from mutant fermentation cultures or by chemical modification of mutant macrolide compounds, were added to fermentations of the *tylG* mutant GS22, and the relative efficiencies of conversion of the compounds to tylosin were assessed (17). The results of these experiments revealed a preferred pathway for the bioconversion of tylactone to tylosin in S. *fradiae* (17; Figure 3).

Cloning and Analysis of Tylosin Resistance Genes

It was clear that a better understanding of the genetics and regulation of tylosin biosynthesis could be attained by cloning the genes involved in tylosin biosynthesis. A number of approaches to cloning the *tyl* genes were undertaken. The results of the conjugal mating experiments had suggested that genes controlling tylosin resistance were closely linked to tylosin biosynthetic genes (85). This, in turn, suggested that a cloned tylosin resistance gene might be used as a probe to locate neighboring biosynthetic genes in a library of *S. fradiae* DNA. This approach was especially appealing because the availability of the cloned resistance gene(s) would also facilitate studies of both the mechanism and control of expression of tylosin resistance.

Two routes were taken in attempting to clone the *S. fradiae* tylosin resistance gene(s). One involved shotgun cloning *S. fradiae* DNA fragments into *Streptomyces lividans* in the plasmid vector pIJ702 (26; (V. A. Birmingham, K. L. Cox & E. T. Seno, manuscript in preparation). The other employed the same vector, but used the tylosin-sensitive *S. fradiae* mutant JS87 (20; Table 1) as the cloning host (K. L. Cox & E. T. Seno, manuscript in preparation). In both cases, the transformants were screened for resistance to 500 μg of tylosin ml^{-1}. Tylosin-resistant transformants of *S. lividans* contained either of two cloned, nonoverlapping sequences of *S. fradiae* DNA. One of the sequences was designated *tlrA* (26) and the other *tlrB* (V. A. Birmingham, K. L. Cox & E. T. Seno, manuscript in preparation). The cloning experiment in JS87 yielded a third tylosin resistance sequence, *tlrC* (K. L. Cox & E. T. Seno, manuscript in preparation).

The *tlrA* gene was used as a probe to identify large fragments of DNA containing this gene in a library of *S. fradiae* DNA (35) prepared in a bacteriophage vector (26). Nearly 30 kb of DNA containing *tlrA* were tested for the presence of *tyl* biosynthetic genes by complementation of *tyl* mutants with plasmids containing the subcloned DNA. No *tyl* genes were found, which suggested, but did not prove, that *tlrA* is not clustered with *tyl* biosynthetic genes. The other two genes, *tlrB* and *tlrC*, were located close to *tyl* genes, as described in the following section.

The presence in *S. fradiae* of three genes involved in tylosin resistance was unexpected. The *tlrA* gene conferred high levels of tylosin and erythromycin resistance in *S. lividans*, and it conferred resistance to tylosin, a number of other macrolide antibiotics, and lincomycin in *Streptomyces griseofuscus* (26; V. A. Birmingham, K. L. Cox & E. T. Seno, manuscript in preparation). This phenotype was typical of that shown by MLS (macrolide, lincosamide, streptogramin B) resistance genes found in other gram-positive bacteria (90). These genes code for enzymes that methylate a specific residue in 23S ribosomal RNA (rRNA), thereby inhibiting the binding of the MLS antibiotics. Cell extracts of *S. griseofuscus* carrying the cloned *tlrA* gene contained an rRNA-methylating activity that was active on *S. griseofuscus* rRNA but inactive on rRNA from the *S. griseofuscus* carrying the cloned *tlrA* gene (V. Birmingham & E. T. Seno, unpublished data). A gene with a restriction cleavage pattern indistinguishable from that of *tlrA*, *ermSF*, was independently cloned from *S. fradiae* (49) by virtue of its homology with a

MLS resistance gene, *ermE* (24, 86a, 87), cloned from *Saccharopolyspora erythrea*. The amino acid sequence of the *ermSF* product, as deduced from the DNA sequence of the gene, strongly resembled that of other confirmed MLS resistance–conferring methylases. Thus, it seems very likely that *tlrA* encodes an MLS rRNA–methylating activity.

The *tlrB* gene conferred lower levels of tylosin resistance in *S. lividans* than *tlrA*, and it did not enhance erythromycin resistance. The *tlrB* gene conferred no resistance phenotype in *S. griseofuscus* (V. A. Birmingham, K. L. Cox & E. T. Seno, manuscript in preparation). Thus, the *tlrB* phenotype seems to be more tylosin specific than *tlrA*, and its expression may be more host specific.

The *tlrC* gene conferred tylosin resistance in JS87, but it conferred no resistance in *S. lividans*, and it was not tested in other hosts (V. A. Birmingham, K. L. Cox & E. T. Seno, manuscript in preparation). The *tlrC* gene was deleted in JS87, and both *tlrC* and *tlrB* were deleted in another similar tylosin-sensitive mutant, GS93. Both mutants contained *tlrA*. The mutants were sensitive to tylosin, erythromycin, and lincomycin. However, exposure to low tylosin concentrations induced strong erythromycin resistance and some lincomycin resistance, but no tylosin resistance. Introduction of either the cloned *tlrB* or *tlrC* gene restored tylosin-inducible tylosin resistance, but introduction of the cloned *tlrA* gene into JS87 had no effect on tylosin resistance.

Thus, the *tlrB* and *tlrC* genes are likely to be very important for tylosin resistance in *S. fradiae* based on their apparent specificity for tylosin, their activity in tylosin-sensitive mutants, and their location near tylosin biosynthetic genes (see below). The role of *tlrA* in *S. fradiae* is unclear. The observation that *tlrA* conferred high-level tylosin resistance in two other hosts but not in the tylosin-sensitive *S. fradiae* mutants suggests that the *tlrA* gene is not expressed in the mutants.

Cloning of Tylosin Biosynthetic Genes and Genomic Organization

It was implied in the previous section that the approach of cloning tylosin resistance genes as probes for nearby biosynthetic genes could have been successful in the cases of *tlrB* and *tlrC*. By the time these two genes were identified, however, a more direct approach to cloning *tyl* genes became available. The macrocin *O*-methyltransferase enzyme had been purified (N. J. Bauer, A. J. Kreuzman, J. E. Dotzlaf & W.-K. Yeh, manuscript submitted), and the sequence of 35 amino acids at its amino terminus had been determined (29, 36). A 44-base oligonucleotide probe was constructed on the basis of the amino acid sequence. Streptomycete DNA has very high (~73%) guanine plus cytosine (G+C) content. Since this proportion of G+C is near the upper limit of what could be tolerated without a substantial change in the amino acid composition of proteins (8), the third positions of degenerate codons exhibit a very high frequency ($\geq$90%) of G+C (23, 80). This codon bias was exploited in the construction of the macrocin *O*-methyltransferase probe; degenerate codons containing guanine or cytosine in the third position were used to the exclusion of those containing adenine or thymine (36). The probe was hybridized against both bacteriophage and cosmid libraries of *S. fradiae* DNA. DNA inserts homologous to the probe, totaling about 60 kb (the pMOMT3 and pMOMT4 regions in Figure 4), were subcloned into plasmid vectors and introduced into the *tyl* mutants to determine the presence of genes complementing *tyl* mutations. In these initial experiments genes complementing 9 of the 13 *tyl* mutations were identified (29, 36; Figure 4). These genes, which generally were involved in later steps in the biosynthetic pathway (refer to Figure 3), were located in a ~40-kb segment of DNA flanked on the left by the resistance gene *tlrB* and on the right by the amplifiable unit of DNA (AUD), which is amplified 500-fold (37) in the JS85 mutant. Subsequent attempts to locate genes complementing the remaining four mutant classes, *tylG, tylA, tylB,* and *tylI,* extended the cloned DNA ~60–70 kb to the right and located the *tlrC* gene and two other repeated sequences (RS) (R. Beckmann & E. T. Seno, unpublished data). Previous work had indicated that a pair of RS flanked the AUD and that another pair was present elsewhere in the genome (38). The locations of the remaining *tyl* genes remain to be established.

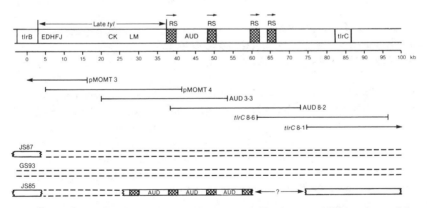

Figure 4 The *tyl* region of the *S. fradiae* genome. The figure shows a ~100-kb segment of the *S. fradiae* genome containing the nine cloned *tyl* biosynthetic genes (*tylE, tylD, tylH, tylF, tylJ, tylC, tylK, tylL,* and *tylM*), two tylosin resistance genes (*tlrB* and *tlrC*), an amplifiable DNA sequence (AUD), and four direct repeat sequences (RS). pMOMT3, pMOMT4, AUD 3-3, AUD 8-2, *tlrC* 8-6, and *tlrC* 8-1 are cosmid inserts isolated with different probes. JS87, GS93, and JS85 are pleiotropic Tyl⁻ Tylˢ mutants isolated after protoplast formation and regeneration. The *broken lines* indicate deleted DNA sequences. A 500-fold tandem amplification of AUD/RS in JS85 is also depicted. *?* indicates uncertainty about the DNA structure at the right end of the amplification.

Some functional clustering is apparent among the nine *tyl* genes identified (Figure 4). The *tylE, tylD, tylH, tylF,* and *tylJ* genes are tightly grouped on a 5–6–kb fragment. The products of all five genes are involved with reactions occurring at the C-23 position of tylactone or modifications to the 6-deoxy-D-allose residue attached at the C-23 hydroxyl position (see Figures 2 and 3). Thus *tylH* is required for oxidation of C-23, *tylD* and *tylJ* are involved with 6-deoxy-D-allose biosynthesis or attachment, and *tylE* and *tylF* encode enzymes that methylate the 2-hydroxyl and 3-hydroxyl positions of the attached 6-deoxy-D-allose residue. Similarly, *tylC* and *tylK,* which are also closely linked to each other, are required for mycarose biosynthesis or attachment. Both the *tylL* and *tylM* genes are required for further modification of tylactone involving biosynthesis or attachment of mycaminose alone *(tylM)* or of all three *tyl* sugars *(tylL)*.

The linkage of *tyl* genes to tylosin resistance genes and amplifiable DNA sequences is consistent with genetic data obtained from conjugal crosses between JS85 and some *tyl* mutants (85; see section on conjugation). The results of the crosses indicated high-frequency transfer of the *tylA, tylB, tylC, tylF, tylD, tylI,* and *tylG* mutations with genes controlling tylosin resistance and DNA amplification, in the apparent absence of chromosomal recombination. Since the *tylG, tylA, tylB,* and *tylI* genes were among those cotransferred, they must be located on the same transmissible element as the other *tyl* genes. This suggests that cloning DNA sequences adjacent to those already isolated (i.e. chromosome walking) might be a reasonable approach for locating the remaining *tyl* genes.

The use of cloned *tyl* and *tlr* DNA sequences as probes has indicated that JS87 and GS93 (Table 1) contain large deletions encompassing most, if not all, of the DNA shown in Figure 4 (40; K. Cox, R. Beckmann, V. Birmingham, E. Seno, S. Fishman & C. Hershberger, unpublished data). One deletion endpoint in JS87 is located just right of *tlrB,* and neither endpoint has been located in GS93. JS85 contains a large amplification of the RS/AUD/RS sequence (37) and is deleted for *tylF* and other sequences between the amplification and *tlrB* (S. Fishman, C. Hershberger, R. Stanzak, V. Birmingham & E. Seno, unpublished data). The nature of the DNA to the right of the amplification in JS85 is not clear. The physical basis for the instability in this region of the *S. fradiae* genome remains to be determined.

SUMMARY AND PERSPECTIVES

We chose *S. fradiae* as a system in which to study the genetics of an antibiotic-producing streptomycete not because of the ease of manipulating it, but because it produced an economically important antibiotic, tylosin. Since we wanted to apply the knowledge gained from *S. fradiae* to other antibiotic-producing actinomycetes, we attempted, whenever possible, to develop genetic methods that could be widely applicable to other streptomycetes and possibly to other actinomycetes. Thus, when we initiated studies on mutation induction in *S. fradiae,* we chose to explore fundamental mechanisms rather than simply to catalog comparative rates of mutagenesis by different chemical mutagens. We found that the general mutagenic mechanisms in *S. fradiae* were surprisingly similar to those observed in the gram-negative eubacterium *E. coli.* The finding that most induced mutagenesis in the gram-positive actinomycete *S. fradiae* occurs by a *recA*-regulated error-prone DNA repair system emphasizes the evolutionary conservation of this important system, and suggests that the mutagenic procedures and principles developed for *S. fradiae* should be readily applicable to other actinomycetes.

The recombination procedures developed for *S. fradiae* include protoplast fusion, protoplast transformation, plasmid transduction, and, to a limited extent, conjugation. The protoplast fusion techniques developed in our laboratories (4, 12) and at the John Innes Institute (43–45) have proven to be widely applicable to other *Streptomyces* species and have been used successfully for a number of different applications (46, 51, 59, 67–69, 92). Protoplast transformation was difficult to accomplish in *S. fradiae,* but the development of an efficient transformation system helped define important parameters, which aided in the subsequent development of methods to transform species of *Amycolatopsis* (64) and *Micromonospora* (61). The transformation studies in *S. fradiae* also highlighted the functional role of restriction endonucleases in inhibiting transformation of unmodified plasmid DNA, and they established mutation and selection procedures to isolate mutants defective in restriction. The success obtained with *S. fradiae,* which has at least five restriction systems, suggests that the procedures developed should be applicable for eliminating restriction in other streptomycetes containing fewer restriction and modification systems. The studies of plasmid transduction by bacteriophage FP43 have demonstrated a widely applicable method to bypass restriction, which may eliminate the need to develop transformation procedures for all streptomycete strains of interest. The studies in *S. fradiae* clearly demonstrated that even highly restricting strains are vulnerable to transduction if cells are grown to the appropriate physiological phase at the appropriate temperature. The conjugation studies established that many tylosin genes were genetically linked on a self-transmissible element. These studies suggested that macrolide antibiotic genes would be linked to resistance genes; this theme has been used for the successful cloning of genes for the biosynthesis of tylosin (29, 36), erythromycin (83), and carbomycin (33; J. K. Epp & B. E. Schoner, manuscript in preparation).

The analysis of tylosin biosynthesis in *S. fradiae* has contributed substantially to the overall understanding of antibiotic biosynthesis and regulation in streptomycetes. The tylosin biosynthetic pathway is the best characterized antibiotic biosynthetic pathway, based upon the number and variety of genes identified by mutation and the number of genes cloned and identified. Further analysis of cloned *tyl* DNA sequences should enhance our knowledge of antibiotic gene organization and regulation, particularly with respect to DNA sequences controlling transcription and translation and regulatory proteins or other molecules acting on these sequences.

The identification of three tylosin resistance genes in *S. fradiae* and the analysis of tylosin-sensitive mutants have already suggested that the monolithic model of MLS resistance as defined in *Staphylococcus* and other gram-positive bacteria (90) will not apply to the macrolide producers. It now seems likely that inducible mechanisms conferring resistance to one macrolide antibiotic will not necessarily cause resistance to another in the macrolide producers. Modifications specific for the antibiotic product appear to be operative. The role of the multiple tylosin resistance genes in the development of tylosin resistance in *S. fradiae* is not clear, but the presence of multiple genes for resistance to the self-made antibiotic is not unique to *S. fradiae* (reviewed in 80). Among other macrolide producers, there are data indicating the presence of at least two carbomycin resistance genes *(car)* in the carbomycin producer *Streptomyces thermotolerans* (J. K. Epp & B. E. Schoner, manuscript in preparation) and three spiramycin resistance genes *(srm)* in the spiramycin producer *Streptomyces ambofaciens* (75). How multiple resistance genes contribute to the survival of the antibiotic producers is an interesting topic for future research.

The knowledge gained and tools developed through the research on *S. fradiae* and tylosin biosynthesis will probably have practical applications in the production of hybrid macrolide antibiotics and in the improvement of the efficiency of tylosin production. The *tyl* mutations in nonrestricting backgrounds and the transformation and transduction methods developed for *S. fradiae* may be employed directly in the cloning and identification of other macrolide biosynthetic genes. Some of these genes, when maintained in *S. fradiae* or in specific *tyl* mutants, might cause the production of hybrid antibiotics with improved biological properties.

The work on tylosin biosynthesis has identified biosynthetic steps that are rate limiting for tylosin production. These bottlenecks in tylosin biosynthesis can probably be relieved by supplying multiple copies of the cloned gene(s) that code for the rate-limiting enzyme(s) or by exploiting the regulatory mechanisms that control the expression of such genes. As the antibiotic productivity of the organism is enhanced, it is possible that the normal resistance mechanism(s) might be overcome, which would result in inhibition of antibiotic production. Knowledge of the mode of action of the products of the cloned tylosin resistance genes, and an understanding of the regulation of their expression, might help us enhance the tylosin resistance in *S. fradiae* to permit higher levels of tylosin production.

ACKNOWLEDGMENTS

We thank C. Hershberger, R. N. Rao, B. Schoner, W.-K. Yeh, and B. Weisblum for communicating results prior to publication, and B. Fogleman for typing the manuscript. We also thank the management of Eli Lilly and Company for supporting this work.

Literature Cited

1. Anne, J., Peberdy, J. F. 1976. Induced fusion of fungal protoplasts following treatment with polyethylene glycol. *J. Gen. Microbiol.* 92:413–17
2. Baltz, R. H. 1971. Infectious DNA of bacteriophage T4. *J. Mol. Biol.* 62:425–37
3. Baltz, R. H. 1976. Biological properties of an improved transformation assay for native and denatured T4 DNA. *Virology* 70:52–64
4. Baltz, R. H. 1978. Genetic recombination in *Streptomyces fradiae* by protoplast fusion and cell regeneration. *J. Gen. Microbiol.* 107:93–102
5. Baltz, R. H. 1980. Genetic recombination by protoplast fusion in *Streptomyces. Dev. Ind. Microbiol.* 21:43–54
6. Baltz, R. H. 1982. Genetics and biochemistry of tylosin production: a model for genetic engineering in antibiotic-producing *Streptomyces.* In *Genetic Engineering of Microorganisms for Chemicals,* ed. A. Hollaender, pp. 431–44. New York: Plenum
7. Baltz, R. H. 1986. Mutagenesis in *Streptomyces.* In *Manual of Industrial Microbiology and Biotechnology,* ed. A. L. Demain, N. A. Solomon, pp. 184–90. Washington, DC: Am. Soc. Microbiol.
8. Baltz, R. H. 1986. Mutation in *Streptomyces.* In *The Bacteria,* Vol. 9, *Antibiotic-Producing* Streptomyces, ed. S. W. Queener, L. E. Day, pp. 61–94. Orlando, Fla: Academic
9. Baltz, R. H. 1987. Mechanisms of mutation and DNA repair in *Streptomyces fradiae.* In *Genetics of Industrial Microorganisms,* ed. M. Alačević, D. Hranueli, Z. Toman, pp. 85–94. Zagreb, Yugoslavia: Pliva
10. Baltz, R. H., Drake, J. W. 1972. Bacteriophage T4 transformation: an assay for mutations induced *in vitro. Virology* 49:462–74
11. Baltz, R. H., Fayerman, J. T., Ingolia,

T. D., Rao, R. N. 1986. Production of novel antibiotics by gene cloning and protein engineering. In *Protein Engineering: Applications in Science, Medicine, and Industry,* ed. M. Inouye, R. Sarma, pp. 365–81. New York: Academic
12. Baltz, R. H., Matsushima, P. 1981. Protoplast fusion in *Streptomyces:* Conditions for efficient genetic recombination and cell regeneration. *J. Gen. Microbiol.* 127:137–46
13. Baltz, R. H., Matsushima, P. 1983. Advances in protoplast fusion and transformation in *Streptomyces. Experientia Suppl.* 46:143–48
14. Baltz, R. H., Seno, E. T. 1981. Properties of *Streptomyces fradiae* mutants blocked in biosynthesis of the macrolide antibiotic tylosin. *Antimicrob. Agents Chemother.* 20:214–25
15. Baltz, R. H., Seno, E. T., Stonesifer, J., Matsushima, P., Wild, G. M. 1981. Genetics and biochemistry of tylosin production by *Streptomyces fradiae.* In *Microbiology–1981,* ed. D. Schlessinger, pp. 371–75. Washington, DC: Am. Soc. Microbiol.
16. Baltz, R. H., Seno, E. T., Stonesifer, J., Matsushima, P., Wild, G. M. 1982. Genetics and biochemistry of tylosin production. In *Trends in Antibiotics Research—Genetics, Biosyntheses, Actions and New Substances,* ed. H. Umezawa, A. L. Demain, T. Hata, C. R. Hutchinson, pp. 65–72. Tokyo: Jpn. Antibiot. Res. Assoc.
17. Baltz, R. H., Seno, E. T., Stonesifer, J., Wild, G. M. 1983. Biosynthesis of the macrolide antibiotic tylosin. A preferred pathway from tylactone to tylosin. *J. Antibiot.* 36:131–41
18. Baltz, R. H., Stonesifer, J. 1985. Mutagenic and error-free DNA repair in *Streptomyces. Mol. Gen. Genet.* 200:351–55
19. Baltz, R. H., Stonesifer, J. 1985. Adap-

tive response and enhancement of *N*-methyl-*N'*-nitro-*N*-nitrosoguanidine mutagenesis by chloramphenicol in *Streptomyces fradiae*. *J. Bacteriol*. 164: 944–46

20. Baltz, R. H., Stonesifer, J. 1985. Phenotypic changes associated with loss of expression of tylosin biosynthesis and resistance genes in *Streptomyces fradiae*. *J. Antibiot*. 38:1226–36

21. Deleted in proof

22. Benzinger, R. 1978. Transfection of *Enterobacteriaceae* and its applications. *Microbiol. Rev*. 42:194–236

23. Bibb, M. J., Findlay, P. R., Johnson, M. W. 1984. The relationship between base composition and codon usage in bacterial genes and its use for the simple and reliable identification of protein coding sequences. *Gene* 30:157–66

24. Bibb, M. J., Janssen, G. R., Ward, J. M. 1985. Cloning and analysis of the promoter region of the erythromycin resistance gene *(ermE)* of *Streptomyces erythraeus*. *Gene* 38:215–26

25. Bibb, M. J., Ward, J. M., Hopwood, D. A. 1978. Transformation of plasmid DNA into *Streptomyces* at high frequency. *Nature* 274:398–400

26. Birmingham, V. A., Cox, K. L., Larson, J. L., Fishman, S. E., Hershberger, C. L., Seno, E. T. 1986. Cloning and expression of a tylosin resistance gene from a tylosin-producing strain of *Streptomyces fradiae*. *Mol. Gen. Genet*. 204:532–39

27. Deleted in proof

28. Cox, K. L., Baltz, R. H. 1984. Restriction of bacteriophage plaque formation in *Streptomyces* spp. *J. Bacteriol*. 159:499–504

29. Cox, K. L., Fishman, S. E., Larson, J. L., Stanzak, R., Reynolds, P. A., et al. 1986. The use of recombinant DNA techniques to study tylosin biosynthesis and resistance in *Streptomyces fradiae*. *J. Nat. Prod*. 49:971–80

30. Deleted in proof

31. Dotzlaf, J. E., Metzger, L. S., Foglesong, M. A. 1984. Incorporation of amino acid–derived carbon into tylactone by *Streptomyces fradiae* GS14. *Antimicrob. Agents Chemother*. 25:216–20

32. Drake, J. W., Baltz, R. H. 1976. The biochemistry of mutagenesis. *Ann. Rev. Biochem*. 45:11–37

33. Epp, J. K., Burgett, S. G., Schoner, B. E. 1987. Cloning and nucleotide sequence of a carbomycin-resistance gene from *Streptomyces thermotolerans*. *Gene* 53:73–83

34. Deleted in proof

35. Ferenczy, L., Kevei, F., Szegedi, M.

1975. High-frequency fusion of fungal protoplasts. *Experientia* 31:1028–30

36. Fishman, S. E., Cox, K., Larson, J. L., Reynolds, P. A., Seno, E. T. et al. 1987. Cloning genes for the biosynthesis of a macrolide antibiotic. *Proc. Natl. Acad. Sci. USA* 84:8248–52

37. Fishman, S. E., Hershberger, C. L. 1983. Amplified DNA in *Streptomyces fradiae*. *J. Bacteriol*. 155:459–66

38. Fishman, S. E., Rosteck, P. R. Jr., Hershberger, C. L. 1985. A 2.2 kilobase repeated DNA segment is associated with DNA amplification in *Streptomyces fradiae*. *J. Bacteriol*. 161:199–206

39. Fodor, K., Alföldi, L. 1976. Fusion of protoplasts of *Bacillus megaterium*. *Proc. Natl. Acad. Sci. USA* 73:2147–50

40. Hershberger, C. L., Fishman, S. E. 1985. Amplified DNA: structure and significance. In *Microbiology–1985*, ed. D. Schlessinger, pp. 427–30. Washington, DC: Am. Soc. Microbiol.

41. Deleted in proof

42. Hopwood, D. A. 1967. Genetic analysis and genome structure in *Streptomyces coelicolor*. *Bacteriol. Rev*. 31:373–403

43. Hopwood, D. A., Wright, H. M. 1978. Bacterial protoplast fusion: recombination in fused protoplasts of *Streptomyces coelicolor*. *Mol. Gen. Genet*. 162:307–17

44. Hopwood, D. A., Wright, H. M. 1979. Factors affecting frequency in protoplast fusions of *Streptomyces coelicolor*. *J. Gen. Microbiol*. 111:137–43

45. Hopwood, D. A., Wright, H. M., Bibb, M. J., Cohen, S. N. 1977. Genetic recombination through protoplast fusion in *Streptomyces*. *Nature* 268:171–74

46. Ideka, H., Inoue, M., Tanaka, H., Ōmura, S. 1984. Interspecies protoplast fusion among macrolide-producing streptomycetes. *J. Antibiot*. 37:1224–30

47. Ishii, Y., Kondo, S. 1975. Comparative analysis of deletion and base-change mutabilities of *Escherichia coli* B strains differing in DNA repair capacity (wild type, *uvrA⁻*, *polA⁻*, *recA⁻*) by various mutagens. *Mutat. Res*. 27:27–44

48. Jones, N. D., Chaney, M. O., Kirst, H. A., Wild, G. M., Baltz, R. H., et al. 1982. Novel fermentation products from *Streptomyces fradiae*; X-ray crystal structure of 5-*O*-mycarosyltylactone and proof of the absolute configuration of tylosin. *J. Antibiot*. 35:420–25

49. Kamimiya, S., Weisblum, B. 1988. Translational attenuation control of *ermSF*, and inducible ribosomal RNA *N*-methyl transferase from *Streptomyces fradiae*. *J. Bacteriol*. 170:1800–11

50. Katz, E., Thompson, C. J., Hopwood,

D. A. 1982. Cloning and expression of the tyrosinase gene from *Streptomyces antibioticus* in *Streptomyces lividans*. *J. Gen. Microbiol.* 129:2703–14

51. Keller, U., Pöschmann, S., Krengel, U., Kleinkauf, H., Kraepelin, G. 1983. Studies of protoplast fusion in *Streptomyces chrysomallus*. *J. Gen. Microbiol.* 129:1725–31

52. Kirst, H. A., Wild, G. M., Baltz, R. H., Hamill, R. L., Ott, J. L., et al. 1982. Structure-activity studies among 16-membered macrolide antibiotics related to tylosin. *J. Antibiot.* 35:1675–82

53. Kirst, H. A., Wild, G. M., Baltz, R. H., Hamill, R. L., Paschal, J. W., et al. 1983. Elucidation of structure of novel macrolide antibiotics produced by mutant strains of *Streptomyces fradiae*. *J. Antibiot.* 36:376–82

54. Kondo, S., Ichikawa, H., Iwo, K., Kato, T. 1970. Base-change mutagenesis and prophage induction in strains of *Escherichia coli* with different DNA repair capacities. *Genetics* 66:187–217

55. Deleted in proof

56. Lindahl, T. 1982. DNA repair enzymes. *Ann. Rev. Biochem.* 51:61–87

57. Matern, H., Brillinger, G. U., Pape, H. 1973. Metabolic products of microorganisms. 114. Thymidine diphospho-D-glucose oxidoreductase from *Streptomyces rimosus*. *Arch. Microbiol.* 88:37–48

58. Matsushima, P., Baltz, R. H. 1985. Efficient plasmid transformation of *Streptomyces ambofaciens* and *Streptomyces fradiae* protoplasts. *J. Bacteriol.* 163:180–85

59. Matsushima, P., Baltz, R. H. 1986. Protoplast fusion. See Ref. 7, pp. 170–83

60. Matsushima, P., Baltz, R. H. 1987. The *recA* gene of *Escherichia coli* complements defects in DNA repair and mutagenesis in *Streptomyces fradiae* JS6 (*mcr-6*). *J. Bacteriol.* 169:4834–36

61. Matsushima, P., Baltz, R. H. 1988. Genetic transformation of *Micromonospora rosaria* by the *Streptomyces* plasmid pIJ702. *J. Antibiot.* 41:583–85

62. Deleted in proof

63. Matsushima, P., Cox, K. L., Baltz, R. H. 1987. Highly transformable mutants of *Streptomyces fradiae* defective in several restriction systems. *Mol. Gen. Genet.* 206:393–400

64. Matsushima, P., McHenney, M. A., Baltz, R. H. 1987. Efficient transformation of *Amycolatopsis orientalis (Nocardia orientalis)* protoplasts by *Streptomyces* plasmids. *J. Bacteriol.* 169:2298–300

65. McHenney, M. A., Baltz, R. H. 1988. Transduction of plasmid DNA in *Strep-*

tomyces and related genera by bacteriophage FP43. *J. Bacteriol.* 170:2276–82

66. Deleted in proof

67. Nakano, M. M., Ishihara, H., Ogawara, H. 1982. Fusion of protoplasts of *Streptomyces lavendulae*. *J. Antibiot.* 35: 359–63

68. Ochi, K., Hitchcock, M. J. M., Katz, E. 1979. High frequency fusion of *Streptomyces parvulus* or *Streptomyces antibioticus* protoplasts induced by polyethylene glycol. *J. Bacteriol.* 139:984–92

69. Ogata, S., Yamada, S., Hayashida, S. 1985. Genetic recombination in *Streptomyces azureus* by protoplast fusion and high production of thiostrepton by the recombinants. *J. Gen. Appl. Microbiol.* 31:187–91

70. Okanishi, M., Suzuki, K., Umezawa, H. 1974. Formation and reversion of streptomycete protoplasts: cultural conditions and morphological study. *J. Gen. Microbiol.* 80:389–400

71. Ōmura, S. 1976. The antibiotic cerulenin, a novel tool for biochemistry as an inhibitor of fatty acid synthesis. *Bacteriol. Rev.* 40:681–97

72. Ōmura, S., Kitao, C., Miyazawa, J., Imai, H., Takeshima, H. 1978. Bioconversion and biosynthesis of 16-membered macrolide antibiotic, tylosin, using enzyme inhibitor: cerulenin. *J. Antibiot.* 31:254–56

73. Ōmura, S., Nakagawa, A., Takeshima, H., Miyazawa, J., Kitao, C. 1975. A ^{13}C nuclear magnetic resonance study of the biosynthesis of the 16-membered macrolide antibiotic tylosin. *Tetrahedron Lett.* 50:4503–6

74. Pape, H., Brillinger, G. U. 1973. Metabolic products of microorganisms. 113. Biosynthesis of thymidine di-phospho mycarose in a cell-free system from *Streptomyces rimosus*. *Arch. Microbiol.* 88:25–35

75. Richardson, M. A., Kuhstoss, S., Solenberg, P., Schaus, N. A., Rao, R. N. 1987. A new shuttle cosmid vector, pKC505, for streptomycetes: its use in the cloning of three different spiramycin-resistance genes from a *Streptomyces ambofaciens* library. *Gene* 61:231–41

76. Richardson, M. A., Mabe, J. A., Beerman, N. E., Nakatsukasa, W. A., Fayerman, J. T. 1982. Development of cloning vehicles from the *Streptomyces* plasmid pFJ103. *Gene* 20:451–57

77. Schaeffer, P., Cami, B., Hotchkiss, R. D. 1976. Fusion of bacterial protoplasts. *Proc. Natl. Acad. Sci. USA* 73:2151–55

78. Seno, E. T., Baltz, R. H. 1981. Properties of S-adenosyl-L-methionine:macrocin O-methyltransferase in extracts of *Streptomyces fradiae* strains which produce normal or elevated levels of tylosin and in mutants blocked in specific O-methylations. *Antimicrob. Agents Chemother.* 20:370–77

79. Seno, E. T., Baltz, R. H. 1982. S-adenosyl-L-methionine:macrocin O-methyltransferase activities in a series of *Streptomyces fradiae* mutants that produce different levels of the macrolide antibiotic tylosin. *Antimicrob. Agents Chemother.* 21:758–63

80. Seno, E. T., Baltz, R. H. 1988. Structural organization and regulation of antibiotic biosynthesis and resistance genes in actinomycetes. In *Regulation of Secondary Metabolism in Actinomycetes*, ed. S. Shapiro. Boca Raton, Fla. CRC. In press

81. Seno, E. T., Hutchinson, C. R. 1986. The biosynthesis of tylosin and erythromycin: model systems for studies of the genetics and biochemistry of antibiotic formation. See Ref. 8, pp. 231–79

82. Seno, E. T., Pieper, R. L., Huber, F. M. 1977. Terminal stages in the biosynthesis of tylosin. *Antimicrob. Agents Chemother.* 11:455–61

83. Stanzak, R., Matsushima, P., Baltz, R. H., Rao, R. N. 1986. Cloning and expression in *Streptomyces lividans* of clustered erythromycin biosynthesis genes from *Streptomyces erythreus*. *Bio-Technology* 4:229–32

84. Stonesifer, J., Baltz, R. H. 1985. Mutagenic DNA repair in *Streptomyces*. *Proc. Natl. Acad. Sci. USA* 82:1180–83

85. Stonesifer, J., Matsushima P., Baltz, R. H. 1986. High frequency conjugal transfer of tylosin genes and amplifiable DNA in *Streptomyces fradiae*. *Mol. Gen. Genet.* 202:348–55

86. Thompson, C. J., Ward, J. M., Hopwood, D. A. 1980. DNA cloning in *Streptomyces:* resistance genes from antibiotic producing species. *Nature* 286:525–27

86a. Thompson, C. J., Ward, J. M., Hopwood, D. A. 1982. Cloning of antibiotic resistance and nutritional genes in streptomycetes. *J. Bacteriol.* 151:668–77

87. Uchiyama, H., Weisblum, B. 1985. Conservation of amino acid sequence in RNA N-methyltransferases that confer resistance to the macrolide-lincosamide-streptogramin-B antibiotics: the N-methyltransferase of *Streptomyces erythraeus*. *Gene* 38:103–10

88. Walker, G. C. 1984. Mutagenesis and inducible responses to deoxyribonucleic acid damage in *Escherichia coli*. *Microbiol. Rev.* 48:60–93

89. Walker, G. C. 1985. Inducible DNA repair systems. *Ann. Rev. Biochem.* 54:425–57

90. Weisblum, B. 1984. Inducible erythromycin resistance in bacteria. *Br. Med. Bull.* 40:47–53

91. Witkin, E. M. 1976. Ultraviolet mutagenesis and inducible DNA repair in *Escherichia coli*. *Bacteriol. Rev.* 40:869–907

92. Yamashita, F., Hotta, K., Kurasawa, S., Okami, Y., Umezawa, H. 1985. New antibiotic-producing streptomycetes, selected by antibiotic resistance marker. 1. New antibiotic production generated by protoplast fusion treatment between *Streptomyces griseus* and *S. tenjimariensis*. *J. Antibiot.* 38:59–63

93. Yasbin, R. E., Miehl-Lester, R., Love, P. E. 1987. Inducible error prone repair in *Bacillus subtilis*. See Ref. 9, pp. 73–83

Ann. Rev. Microbiol. 1988. 42:575–606

CELL BIOLOGY OF *AGROBACTERIUM* INFECTION AND TRANSFORMATION OF PLANTS

Andrew N. Binns

Department of Biology, University of Pennsylvania, Philadelphia, Pennsylvania 19104

Michael F. Thomashow

Department of Crop and Soil Sciences and Department of Microbiology and Public Health, Michigan State University, East Lansing, Michigan 48824

CONTENTS

INTRODUCTION

Virulent strains of *Agrobacterium tumefaciens* and *Agrobacterium rhizogenes* elicit crown gall tumors and "hairy roots," respectively, in infected plant tissues. These pathogenic responses result from the expression of genetic information, the Ti or Ri plasmid T-DNAs, transferred from the bacteria into host cells (32, 56, 79, 129, 190). The resultant transformed plant cells produce novel sugar and amino acid conjugates, termed opines, which can be

575

used by the inciting bacteria in two ways: as a source of carbon and nitrogen and as an inducer of Ti or Ri plasmid transfer between bacteria. Different *Agrobacterium* strains carry different types of Ti plasmids, which can be categorized by the type of opine produced by the transformed plant cells and metabolized by the bacteria. For example, pTiA6 encodes octopine production and utilization, whereas pTiC58 encodes nopaline production and utilization.

In addition to opine synthesis, the T-DNA genes encode the biosynthesis of the plant hormones auxin and cytokinin (1, 9, 24, 153, 171, 172, 179). The expression of these genes in the transformed cells results in the unregulated accumulation of the hormones and is responsible for the non–self-limiting growths that we recognize as tumors. Thus, the DNA inserted into plant cells by *Agrobacterium* species results in the synthesis of opines for the bacteria and causes multiplication of plant cells carrying out such production. This genetic transfer between members of different kingdoms has profound evolutionary implications as well as obvious practical uses in both basic and applied research.

Numerous recent reviews have focused on the utilization of *Agrobacterium*-based vectors for plant genetic engineering and on the expression and functions of the transferred genetic information of wild-type *Agrobacterium* (100, 152, 186). However, the infection process itself is a complex series of events whose temporal sequence is defined by cellular activities of the interacting partners (for earlier reviews see 10, 21, 56, 79, 111, 129, 167). These activities include bacterial colonization and attachment to cells at or near wound sites; the plant wound response; the mobilization of bacterial DNA into plant cells; and the integration of the T-DNA into plant DNA and its expression (Figure 1). Our objective is to focus on the cellular responses of both the bacterium and plant that lead to transformation. We address the following questions: How is the interaction initiated? What signals or other cellular events are required for the onset of a virulent interaction? What cellular activities are requisite for transfer of genetic information? How does the T-DNA achieve opine and plant hormone production? How can perturbations in the normal sequence of infection affect host range? The answers to these questions are incomplete. Yet it is certain that such basic information about the control of genetic transfer between organisms will be crucial in understanding not only this pathogenic interaction, but plant/microbe interactions in general.

INITIATION OF INTERACTION

Colonization and Attachment

Agrobacterium sp. are commonly found in both cultivated and nonagricultural soils, are routinely isolated from the roots of plants, and have even been found

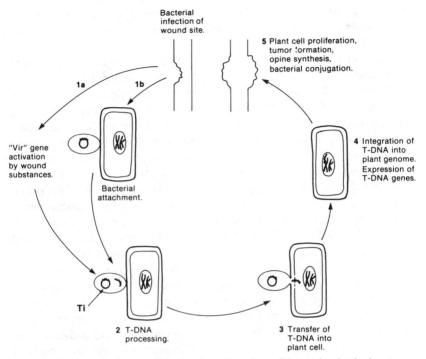

Figure 1   A schematic diagram of the *Agrobacterium*–plant cell interaction that leads to tumor development. The *wavy line* of the Ti plasmid represents the T-DNA and its possible intermediates (see text for details).

in the vascular systems of plants (15, 25, 95, 105, 130, 154). While virulent strains have been recovered from all of these environments, the vast majority of *Agrobacterium* isolates from soils and roots appear to be nonpathogenic (130, 154). Even in soils where the incidence of crown gall is high, the nonpathogenic isolates outnumber the pathogenic isolates by 10–100 fold. Whether these strains are completely avirulent, are incapable of transfoming plants until they acquire a Ti plasmid from a virulent strain (94), or simply have a very restricted host range remains to be determined. The role of these *Agrobacterium* sp. in the soil community also remains to be established.

One of the earliest stages in the interaction between *Agrobacterium* and plants is the attachment of the bacteria to the plant cell surface. Such binding can be observed by either light microscopy (e.g. 45, 118) or scanning electron microscopy (e.g. 120, 145) and can be quantitated by a variety of direct binding assays (127, 144). The initial attachment of single bacterial cells, often in a polar fashion, is followed by massive aggregation of the bacteria at the plant cell surface. These aggregates result from the formation of cellulose fibrils by the bacteria (116). Cellulose-deficient mutants still attach to plant cells, which indicates that the fibrils are not required for attachment per se. Rather, their role appears to be in firmly anchoring the bacteria to the plant cell surface.

Lippincott & Lippincott (109) first proposed that site-specific attachment of *A. tumefaciens* to host cells is an essential stage in crown gall tumorigenesis. This hypothesis was based on the finding that tumor induction on pinto bean leaves by *A. tumefaciens* B6 was inhibited by coinoculation with certain avirulent strains of *A. tumefaciens* or with strain B6 that had been inactivated by heat or ultraviolet light. Other bacteria, including *Rhizobium meliloti, Pseudomonas savastanoi, Pseudomonas aeruginosa, Escherichia coli,* and *Corynebacterium fascians,* had little effect on tumor induction by *A. tumefaciens.* Similar experiments by other investigators using tomato (95), potato (58), and Jerusalem artichoke (168) led to the same general conclusion.

While the site-specific attachment hypothesis is not yet proven, it is supported by the finding that binding is saturable. Both suspension-cultured tomato cells (127) and carrot cells (63) proved to have about 200 attachment sites per cell. Specificity of binding was indicated by the observations that less than three *E. coli* cells bound per tomato cell and that prior incubation of the tomato cells with saturating amounts of *A. tumefaciens* strongly inhibited (60–75%) subsequent binding by *A. tumefaciens* (127). The fact that *Agrobacterium* mutants that are defective in attachment are either avirulent or severely attenuated in virulence is also consistent with the hypothesis of site-specific attachment (27, 45, 118, 173). At a minimum, these results argue that attachment, site-specific or otherwise, is important for virulence.

A number of articles have appeared that review the earlier literature on attachment in detail (112, 117, 142). Therefore, we focus here on the results of recent efforts to elucidate the nature of the plant cell and bacterial receptors.

PLANT CELL RECEPTOR   The saturable *Agrobacterium* binding site appears to be in the cell wall. This was initially suggested by the finding that tumor induction was inhibited when cell wall preparations were added along with the bacteria in virulence tests (110, 147). More recently, Gurlitz et al (63) presented evidence indicating that *Agrobacterium* binds to plasmolyzed carrot cells, even in areas in which the plasmalemma does not make contact with the cell wall. However, Firoozabody & Galbraith (52) found that in the presence of 2,6-dichlorobenzonitrile (DB), tobacco protoplasts do not regenerate their cell wall, as measured by calcolflour staining and protoplast shape, but are still transformable by *Agrobacterium.* These results suggest that attachment of bacteria to an intact cell wall may not be a prerequisite for transformation. However, the possibility that certain wall components involved in attachment are deposited in the presence of DB has not been excluded.

The nature of the putative cell wall receptor is not yet known. Tumor induction on pinto bean leaves (147) and potato slices (143) was inhibited by both polygalacturonic acid (PGA) and citrus pectin, which suggested that the plant receptor might be composed of a galacturonan. Direct adsorption assays, however, indicated that PGA and pectin actually stimulated bacterial binding to potato tissues (144). This result suggested that PGA and pectin were not inhibiting tumor induction by inhibiting binding, which in turn cast doubt on the hypothesis that a galacturonan was the receptor. New evidence suggests that a pectin or pectin-associated receptor does indeed exist. Neff and colleagues (127, 128) have prepared a pectin-enriched soluble cell wall fraction (CWF) from suspension-cultured tomato cells and have shown that it inhibits both *A. tumefaciens* binding (tomato, tobacco, and sugar cane cells) and tumor induction (potato tuber slices) in a dose-dependent manner. Furthermore, the concentrations of CWF required to inhibit binding and tumorigenesis were similar.

Gurlitz et al (63) have proposed that the plant receptor may be a protein. This is an attractive hypothesis, although the data in support of it are indirect and not yet conclusive. Carrot cells treated with Triton X-100 and a variety of proteases bound fewer bacteria than control cells, and the recovery of binding sites that normally occurs upon subsequent culture of protease-treated cells did not occur when protein synthesis was inhibited. Because the CWF isolated by Neff & Binns (127) contained protein and its binding-inhibitory activity was partially abolished by protease treatment, it is possible that a glycoprotein is the active component in these preparations. It is hoped that additional fractionation and biochemical characterization of the CWF will define the active moiety in the preparation and reveal the plant receptor.

BACTERIAL RECEPTOR   *A. tumefaciens* attachment presumably involves the interaction of one or more of its surface molecules with the plant cell wall. One molecule that has been implicated is lipopolysaccharide (LPS), although the data are equivocal (compare References 121 and 189 with 144). A very useful approach has been to isolate and characterize bacterial mutants that cannot attach to plant cells in the usual fashion (27, 45, 46, 118, 173). These studies implicate molecules other than LPS that may have important roles in this physical interaction.

To date, three genetic loci have been defined as having roles in *A. tumefaciens* attachment to plant cells: *chvA* (46), *chvB* (46), and *pscA (exoC)* (27, 173). *chvA* and *chvB* are 1.5 and 5.0 kb, respectively, and each consists of one transcriptional unit. They are both encoded by the chromosome and are physically linked (they are separated by only about 4 kb of DNA in *A. tumefaciens* strain C58). Experiments to date indicate that both loci are constitutively expressed (C. Douglas & C. Thienes, unpublished results). The *pscA* locus (173), also referred to as *exoC* (27), is approximately 3.0 kb, and like *chvA* and *chvB,* it is chromosomally encoded and appears to be composed of a single transcriptional unit, which is constitutively expressed (J. Karlinsey & M. Thomashow, unpublished results). Evidence for a fourth region of the chromosome that encodes functions important in attachment and virulence has been presented by Matthysse (118).

chvA *and* chvB   The first *chvA* and *chvB* mutants were isolated by Garfinkel & Nester (54) in a random screen for Tn*5*-induced mutations that resulted in avirulence. The mutants were subsequently shown by Douglas et al (45) to be defective in plant cell attachment. Direct binding assays indicated that the mutants were markedly reduced in their abilities to attach to both tobacco tissue culture cells and freshly isolated *Zinnia* leaf mesophyll cells. It was also shown that the mutants could not inhibit tumor induction by the parent strain when both were coinoculated onto Jerusalam artichoke slices; avirulent strains that could bind did inhibit tumorigenesis. These results led to the conclusion that the reduced virulence phenotype of the *chvA* and *chvB* mutants is due to their attachment-defective phenotype (45).

While these mutants were initially described as avirulent, they are, in fact, not completely avirulent, but rather are attenuated in virulence. Using the sensitive potato disc assay for tumorigenesis, *chvA* and *chvB* mutants produce up to 50% and 7% the number of tumors, respectively, as wild-type strain A348 inoculated at equivalent concentrations (N. T. Neff & A. N. Binns, unpublished observations). Thus it seems that the defects of *chvA* and *chvB* may prevent *Agrobacterium* species from efficiently forming a productive physical interaction with plant cells, rather than preventing such an interaction altogether.

The molecular basis for the attachment-defective phenotype of the *chvA* and *chvB* mutants is not known but presumably involves alterations in surface chemistry. Indeed, both *chvA* and *chvB* mutants are unable to transfer the *Agrobacterium* plasmid pAgK84 conjugatively, which suggests surface alterations (A. Kerr, unpublished results). More specifically, *chvB* mutants are nonmotile, are resistant to specific bacteriophage, lack flagella, and do not produce β-1,2-D-glucan, a cyclic glucose polymer that is found both in the periplasmic space and extracellularly (46, 146). The *chvA* mutants are less well characterized, but it appears that they, too, are defective in β-1,2-D-glucan production. There is evidence that this polysaccharide is synthesized but not secreted (J. Hendelson, personal communication; G. A. Cangelosi & E. W. Nester, personal communication).

Which, if any, of these phenotypes is responsible for defective attachment remains to be determined. The lack of flagella, which is responsible for the nonmotility and bacteriophage resistance, does not appear to be the cause of the impaired attachment, since other flagella-negative mutants were shown to be virulent and unaffected in binding (16). The fact that *chvA* and *chvB* mutants are defective in the production of β-1,2-D-glucan suggests that this polysaccharide may have a direct role in attachment. An alternative possibility is that the defect in β-1,2-D-glucan synthesis is indirectly responsible for the inability to bind. Miller et al (126) have shown that the synthesis of β-1,2-D-glucan is osmotically regulated in both *Agrobacterium* and *Rhizobium*. Perhaps a defect in the ability to adapt to prevailing osmotic conditions leads to membrane alterations which in turn affect membrane associated processes such as the production of flagella and attachment.

The gene product encoded by *chvA* has not been characterized. In contrast, it has been established that *chvB* encodes at least one protein involved in the synthesis of the β-1,2-D-glucan. Zorreguieta & Ugalde (207) have shown that β-1,2-D-glucan production proceeds through the formation of a protein-bound intermediate. The protein was shown to have a molecular mass of approximately 235 kd and to be located in the cytoplasmic membrane. Analysis of *chvB* mutants indicated that they did not synthesize the 235-kd polypeptide. It has recently been shown that *chvB* encodes this polypeptide (206a).

The functions encoded by the *chvB* loci are important not only in *Agrobacterium* attachment and virulence, but also in *Rhizobium* nodulation. Two genetic loci in *R. meliloti*, *ndvA* and *ndvB*, have been shown to be structurally and functionally related to *chvA* and *chvB*, respectively (48). DNA hybridization data suggest that similar loci exist not only in other *Rhizobium* species but in *Azospirillum* as well (182). *R. meliloti* strains with mutations at either *ndvA* or *ndvB* cannot invade root hairs and induce the formation of defective nodules that lack infection threads, are devoid of bacteroids, and are incapable of fixing nitrogen (48).

pscA (exoC)   *pscA* mutants are either avirulent or severely attenuated in virulence, depending on the host plant tested, and they are defective in attachment to plant cells (27, 173). The gene product encoded by *pscA* has not been characterized, but it is clear that it has an important role in the production of surface polysaccharides; *pscA* mutants are defective in the production of cellulose fibrils, do not make the major acidic extracellular polysaccharide succinylglycan, and do not appear to make β-1,2-D-glucan (27, 173). In addition, the mutants are nonmotile. Whether the *pscA* locus has a direct role in polysaccharide production, e.g. by encoding a gene product required for a synthetic or secretory step common to surface polysaccharides, or whether it has an indirect role, perhaps encoding a protein that imparts a general membrane structure required for polysaccharide synthesis, remains to be determined.

The precise defect in the *pscA* mutants that accounts for their inability to attach to plant cells also remains to be established. As mentioned above, neither the cellulose-negative nor the nonmotile phenotype appears to be responsible for avirulence (16, 116). In addition, several mutants deficient in succinylglycan have been found to be fully virulent (27). Thus this polysaccharide does not appear to have a critical role in tumorigenesis. The finding that *pscA* mutants, like *chvA* and *chvB* mutants, appear to lack β-1,2-D-glucan is consistent with the notion that this polysaccharide has a role in virulence. However, the pleiotropic nature of the *pscA* mutations rules out firm conclusions on this point. Additional classes of mutations affecting β-1,2-D-glucan synthesis are needed to establish conclusively a link between this polysaccharide and attachment and virulence.

Other attachment and virulence mutants   Matthysse (118) has isolated five Tn5-induced attachment mutants of strain C58 by screening directly for mutants that did not bind to carrot suspension cells. All of the strains proved to be avirulent. Mapping data indicated that in each mutant Tn5 was inserted into a 12-kb *Eco*RI fragment of chromosomal origin. Biochemical analysis did not reveal any changes in β-1,2-D-glucan or cellulose production, motility, flagella, or hydrophobicity. In addition, the LPS did not appear to be altered. Differences were found, however, in the polypeptides released upon preparation of spheroplasts of the wild-type and mutant strains: Three mutants lacked a 34-kd polypeptide, one mutant lacked both a 34- and a 38-kd polypeptide, and the fifth mutant lacked a 33-kd polypeptide. Whether these polypeptides have a role in binding remains to be determined. From the site of the Tn5 inserts (the 12-kb *Eco*RI fragment) and the biochemical characteristics of the mutants, it appears that they are not *chvA*, *chvB*, or *pscA* mutants.

Wounding and Competence for Transformation

Early experiments with *Agrobacterium* indicated that plant wounding is required for tumor development. It was shown, for example, that the delivery of bacteria to plant tissues in the absence of wounding rarely resulted in tumor formation and that tumors only developed if tissues were inoculated with bacteria within a few days of the wounding event (see 21, 111 for reviews). One of the reasons for this wound requirement is now clearly established: Expression of a specific set of bacterial genes that are required for tumor induction is induced by compounds produced in wounded plant tissues. The wounding process, however, is a complex series of metabolic and cellular activities (92). Whether other wound-associated events are also required for tumor induction is not certain, but a number of observations suggest that wound-induced cell division and/or DNA replication may have a role.

BACTERIAL ACTIVATION   Once it was established that the Ti plasmid was required for *Agrobacterium* transformation, many studies were conducted to define the regions of the plasmid that were involved in tumor induction. In addition to the T-DNA, a second set of genes essential for transformation was identified (70, 85, 99, 101). These genes, referred to as *vir* genes, do not enter the plant cell, but must be expressed in the bacterium for T-DNA transfer to occur (see below for further detail). The critical finding was that most of the *vir* genes are not expressed under normal bacterial growth conditions, but are induced by plant cell exudates (6, 160, 163). Further analysis has shown that the molecules responsible for *vir* induction, acetosyringone and related phenolic compounds, are present at very low levels in uninjured plants, but are significantly more abundant in wounded tissues (14, 161). Thus as part of the normal response to wounding, the plant makes molecules that signal the *A. tumefaciens* to initiate processes required for plant cell transformation. Recent experiments also indicate that *A. tumefaciens* is chemotactic toward many of the phenolics that induce the *vir* genes (7, 134), which presumably results in a recruitment of agrobacteria to the wounded plant tissues.

CELL DIVISION AND COMPETENCE   Is the induction of *vir* genes the sole role of wounding in the transformation process? Why have agrobacteria developed the strategy of keying into plant molecules that are made during wounding? Are specific physiological events that occur during the wound response perhaps critical to the transformation process?

Braun (17) and Braun & Mandle (20) presented evidence indicating that a "window of competence," a time during which the cells are suseptible to *Agrobacterium* transformation, follows the wounding event. In experiments with *Kalenchoe daigremontiana* the window was between 24 and 120 hr postwounding (17). Histological analysis of the wound tissues indicated that a burst of cell division was initiated and ceased during this time unless virulent agrobacteria were present (21, 108). The correlation of cell division with maximal transformation suggested that cell division and/or DNA synthesis might be important to the process (19).

Recent results are consistent with this notion. *A. tumefaciens* transformed tobacco suspension cultured cells at high frequency (6). While cells from an early stage in the culture period were capable of inducing the bacterial *vir* genes, maximum transformation frequencies were correlated with the rapid growth phase of the culture. Similarly, several studies have indicated that maximum transformation of protoplasts occurs when regenerating protoplasts have entered the DNA synthesis phase of the cell cycle and are beginning to proliferate (52, 115, 197). Studies using a soybean petiole transformation assay (103) were also consistent with the hypothesis that transformation frequency is the greatest during the period of wound-induced cell division.

In summary, a distinct correlation exists between the timing of wound-induced cell divisions and the competence of such cells to be transformed by *Agrobacterium*. It seems likely that processes related to DNA synthesis and cell division would be required for incorporating the T-DNA into the host genome (see below).

T-DNA TRANSFER, INTEGRATION, AND EXPRESSION

Three important sets of Ti plasmid sequences are required for tumor induction by *A. tumefaciens*. Two of these, the *vir* genes and the 25-bp border sequences that flank the T-DNA, are required for appropriate processing of the T-DNA in the bacterium and its transfer into the plant cell (Figure 2). Once the T-DNA genes are integrated into the nuclear genome of the host, the expression of the oncogenes results in tumorigenesis. Elucidating the functions of the *vir* genes, border sequences, and oncogenes has been the subject of intense experimental activity over the past few years, and remarkable progress has been made.

T-DNA Transfer

vir GENE IDENTIFICATION AND ORGANIZATION   The *vir* region is approximately 35 kb and encodes six transcriptional loci: *virA, virB, virC, virD, virE,* and *virG* (149, 162). Four of these loci, *virA, virB, virD,* and *virG,* are absolutely required for agrobacteria to incite tumors, while mutations at *virC* and *virE* affect tumor formation on some plants but not others (71, 72, 202). Mutations at each of the six loci can be complemented in *trans* (70, 71, 78, 85, 99, 113, 162). In addition, the *vir* genes can function in *trans* with respect to the T-DNA; if the T-DNA plus borders is carried on a plasmid separate from the *vir* genes, productive T-DNA transfer will occur (42, 74). This has allowed the development of "binary vectors," which contain only the border repeats and T-DNA and are considerably smaller and easier to manipulate than intact Ti plasmids (see 100 for review).

The direction of transcription of each *vir* locus has been established, primarily through the use of *vir* gene fusions to *lacZ, cat,* and *lux* (which encode β-galactosidase, chloramphenicol acetyl transferase, and luciferase, respectively) (36, 149, 160, 162). In addition, the DNA sequence for each locus has been determined (37, 72, 86, 107, 124, 185, 195, 196, 203). The data for pTiA6 indicate that *virA* and *virG* are monocistronic, while all of the rest encode multiple gene products: *virB* encodes 11 open reading frames (ORFs), *virC* encodes two ORFs, *virD* encodes five ORFs, and *virE* encodes two ORFs.

The *vir* loci are highly conserved among Ti plasmids. Indeed, the physical organization of the *vir* loci in nopaline and octopine Ti plasmids is essentially identical (50, 149, 162). Moreover, the activities of the *vir* genes from a

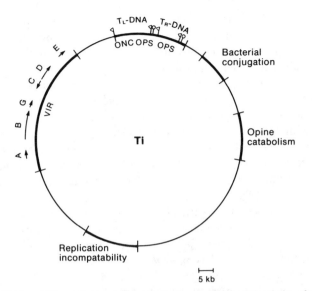

Figure 2   Ti plasmid (octopine type) showing the location, size, and polarity of the *vir* loci transcripts (162), the location and general functions of the T-DNA and border elements (138, 174), and the approximate locations encoding plasmid replication, bacterial conjugation, and opine catabolism (50). *Small triangles* are border repeats; *small circles* represent overdrive sequences. *onc,* oncogenes; *ops,* opine synthesis genes.

nopaline plasmid can complement mutations in an octopine plasmid, and vice versa (75, 78). It has also been shown that the *vir* region of the Ri plasmid of *A. rhizogenes* is homologous to that of the Ti plasmid and that it can complement *vir* mutations in the Ti plasmid (75, 78, 83). These findings indicate that the general mechanism of T-DNA transfer is highly conserved. However, not all *vir* regions are identical. For example, the limited host range Ti plasmid pTiAg162 (202) does not have *virC* sequences, and its *virA* gene product, although related to the VirA of octopine and nopaline Ti plasmids, does not recognize the same plant inducer molecules (see discussion of host range below).

vir GENE EXPRESSION   As discussed earlier, *vir* gene expression is induced by plant signal molecules that are synthesized in wounded plant tissues. The first inducer to be characterized was acetosyringone, a naturally occurring metabolite of the wound-inducible phenylpropanoid pathway (92, 161). Other phenolic compounds that can also serve as *vir* inducers include sinapinic acid, *p*-hydroxybenzoic acid, and vanillin (14).

The basic regulatory mechanism that controls *vir* expression is now established. Stachel & Zambryski (166), in an elegant series of experiments, showed that *vir* induction by plant inducers is mediated through the action of the *virA* and *virG* proteins. Mutations at *virG* in an octopine plasmid totally eliminated the induction of *virB, virC, virD,* and *virE* by acetosyringone, whereas mutations at *virA* resulted in a severe attenuation of their induction. Similar results have been obtained in an investigation of the regulation of *vir* gene expression in a nopaline type plasmid (149). Regulation of *virA* and *virG* is complex. While both of these genes are expressed constitutively, the level of expression increases substantially in response to the phenolic *vir* inducers, demonstrating that these genes are self-regulated (149, 166; S. C. Winans & E. W. Nester, personal communication). From these experiments it has been proposed that VirA might serve to sense or transport the signal molecules, thereby activating VirG, which in turn is a positive regulator of the *vir* loci (149, 166, 167, 196).

Support for a positive *virA/virG* regulatory mechanism has come from comparative DNA sequence analyses. The product encoded by *virG* shares extensive amino acid sequence homology with a family of positive regulatory proteins including the products of *ompR (E. coli), spoOA (Bacillus subtillus),* and *ntrC (Klebsiella pneumoniae)* (107, 124, 196). Interestingly, all of these regulatory proteins share the feature of acting in response to environmental stimuli to induce the expression of genes that are needed to adjust to or to take advantage of a particular environmental situation. Most, if not all, of these positive regulators require the action of a second protein in order to function: OmpR requires the *envZ* product, SpoOA requires the product of *spoOB,* and NtrC requires the *ntrB* product (47, 65, 176). These second proteins, which also comprise a family, are believed to receive the environmental cues and then activate the positive regulatory protein. Comparative sequence analysis has shown that the *virA* gene product is in fact homologous with this second family of proteins (107). Further, the VirA protein has been localized to the inner membrane of *Agrobacterium* species (107). Thus it has been proposed that the protein acts as an environmental sensor of plant inducer molecules and, in an as yet unknown way, transmits an activating signal to VirG.

A second mechanism for controlling *vir* gene expression may also exist. Close et al (35, 36) have isolated spontaneous *Agrobacterium* mutants, termed Ros, that constitutively express *virC* and *virD*. The levels of expression of these loci in the mutants are approximately 10 and 100 times greater, respectively, than in the wild-type strain under noninduced conditions (i.e. without plant inducers). Interestingly, the expression of the other *vir* loci is not affected by the *ros* mutation, nor is the positive regulation of *virC* and *virD* by the *virA/virG* system; acetosyringone induces increased levels of expression of both loci in the *ros* mutants. Genetic analysis has indicated that the *ros* locus is chromosomally encoded and has suggested that *ros* may encode a negative regulator of *virC* and *virD* expression. The *ros* allele has pleiotropic effects; *ros* mutants have a rough surface, do not form colonies at

24°C, and do not produce the major extracellular polysaccharide, succinylglycan. The effects of the *ros* allele, however, do not interfere with T-DNA transfer; *ros* mutants still incite tumors on plants at about the same frequency as wild-type strains (35).

T-DNA PROCESSING   Substantial evidence demonstrates that the T-DNA is excised from the Ti plasmid following *vir* region induction. This processing involves the action of *vir* gene products and the *cis*-acting border sequence elements, imperfect 25-bp direct repeats that flank all T-DNA regions from either Ri or Ti plasmids (8, 157, 158, 183, 199). Soon after *vir* induction, processing of the T-DNA begins, and a variety of intermediates are produced.

Several laboratories have documented the formation of linear single-stranded DNA molecules, termed T-strands, which correspond to the bottom strand of the T-DNA (3, 86, 164, 165, 181a). They are formed by nicking of the bottom strand of the T-DNA at both the right and left borders, followed by release of the bottom strand. Nicks were found to occur between the fourth and fifth bases in both the right and left borders of the pTiA6 TL-DNA (see below) (3) and between the third and fourth bases in the right border of pTiC58 (184). The border nicking is brought about by the concerted action of the 16- and 47-kd polypeptides encoded by *virD* (86, 165, 204). On average there are 0.1–1 T-strands produced per bacterium (86, 164, 165).

The mechanism responsible for release of the T-strand from the Ti plasmid is not yet known. Release could result from the action of a DNA helicase or perhaps from degradation of the top strand of the T-DNA. However, Stachel et al (164) and Albright et al (3) have proposed a more interesting hypothesis, that T-strand formation results from DNA replication and strand displacement. DNA synthesis would be primed by the 3' OH group at the nick in the bottom strand of the right border and would proceed from right to left (5'→3') using the top strand as template. The bottom strand would be displaced during replication and would finally be released when replication reached the left border.

The DNA replication and strand displacement model offers an explanation of the observed polarity of border action in Ti plasmids. Reversing the orientation of the right border essentially abolishes tumor induction (139, 183), as would be expected if the model were correct; the T-strand formed would be composed of the sequences to the right of the T-DNA. Only if DNA synthesis proceeded completely around the Ti plasmid would the T-DNA oncogenes, the genes responsible for tumor induction (see below), be included in the T-strand. The model also explains why only a right border is needed on the Ti plasmid for plant transformation and why border polarity is not important for small binary vectors (53, 81, 87, 151); DNA synthesis could proceed around such molecules at high efficiency.

While it is true that only a border repeat to the right of the T-DNA is required for T-DNA transfer to plants (139, 155, 183), it has been shown that sequences to the right of the natural right border greatly stimulate its function. Peralta et al (138) found that a 15-bp element with a highly conserved 6-bp core, termed *overdrive*, lies just to the right of the right border from three different Ti plasmids. The sequence stimulates T-DNA transfer in an orientation-independent manner (138). Recent evidence suggests that T-strand formation and right-border nicking are stimulated by the presence of *overdrive* (181a; N. Torro & E. W. Nester, personal communication). Thus, *overdrive* may increase the efficiency of T-DNA transfer (53, 69, 82) by assisting *virD* gene products to gain access to right-border sequences.

T-DNA MOVEMENT   The polarity of border utilization and the formation of a single-stranded T-DNA intermediate are intriguing, as they suggest that T-DNA transfer to plants may occur by a mechanism analogous to bacterial conjugation (164, 167, 183). In the case of the *E. coli* fertility factor F, for example, a single-strand nick occurs at *oriT* owing to the action of *traY* along with either *traI* or *traZ* (191). This strand is then transferred from donor to recipient as a linear single-strand intermediate. DNA replication in the donor produces a replacement strand, and DNA replication in the recipient converts the single-strand intermediate into duplex DNA. In T-DNA transfer the T-DNA borders are analogous to *oriT;* the *virD* 16- and 47-kd polypeptides to *traYI* and *traYZ;* and the T-strand to the single-strand intermediate. DNA replication in the donor is another obvious analogy.

In bacterial conjugation the transferred single strand is believed to be coated with a single-strand DNA-binding protein. Indeed, it has recently been shown that the *virE* 65-kd polypeptide is a single-strand DNA-binding protein (33a, 34a, 57a). Interestingly, coinfection of *virE* mutants carrying oncogenic T-DNA with agrobacteria carrying a wild-type *vir* region but no T-DNA resulted in a restoration of oncogenicity (133). This effect was not observed, however, if the complementing strain contained a mutation at *virE* or any other *vir* locus. Otten et al (133) suggested that the *virE* product must be modified by other components of the *vir* region; the product is then secreted, and it has its activity outside the bacterium. Alternatively, it is possible that the complementing strain establishes an effective mating complex with the plant cell and can deliver the *virE* product into the cell, thereby facilitating T-strand movement, protection, or integration.

If T-DNA transfer does proceed by a conjugative mechanism, then one might expect that some of the *vir* gene products would be analogous to pilin, pilin assembly proteins, and other mating complex proteins encoded by F *tra* genes. It remains to be proven whether the *vir* region encodes such gene products. However, Engström et al (51) have determined that three of the *virB* genes encode proteins of about 33, 25, and 80 kd and that these polypeptides are located in the bacterial cell envelope. While there is no obvious homology between *virB* and available bacterial conjugative transfer protein sequences, the complexity of the *virB* operon, the membrane localization of *virB*-encoded proteins, and the analogy between bacterial conjugation and T-DNA transfer all suggest that the *virB* products are involved in forming the putative T-DNA transfer complex.

A conjugative mechanism is also suggested by the recent exciting findings of Buchanan-Wollaston et al (22). Certain nonconjugative bacterial plasmids can utilize the *tra* fuctions of conjugative plasmids for transfer to other bacteria (191). Such plasmids encode their own *oriT* and mobilization *(mob)* proteins. The *mob* proteins nick the plasmid at *oriT* and initiate mobilization, but require the *tra* gene functions of the conjugative plasmid for transfer to the recipient cell. Buchanan-Wollaston et al (22) found that the *mob* genes and *oriT* of the broad–host range plasmid RSF1010 can mediate the transfer of plasmid DNA from agrobacteria to plants. Apparently, the *mob* genes nick the plasmid at *oriT* to initiate mobilization and use the Ti plasmid *vir* region functions to transfer the RSF1010 plasmid to plant cells. These data support the hypothesis that the T-DNA transfer occurs by a conjugative mechanism. They also raise profound questions regarding the extent of bacteria-plant gene flow in nature. If a wide variety of conjugative broad–host range plasmids can interact with the Ti plasmid *vir* region to mobilize DNA into plants, it would follow that plants are potentially exposed to a vast bacterial gene pool.

While a bacterial conjugation–like mechanism involving T-strand formation is an attractive hypothesis, other possible intermediates besides T-strands have in fact been observed after *vir* induction and must be considered. Circular T-DNA molecules have been detected at a frequency of about 10^{-6}–10^{-5} in *A. tumefaciens* (102, 114) and have been observed in *E. coli* carrying the *virC* and *virD* loci (4). Their formation requires *virD* activity (4, 201) and is apparently due to recombination at the border sequences. Further, most investigators have detected double-strand breaks at the T-DNA borders in addition to single-strand breaks; these double-stranded breaks are mediated by the two 5' *virD* polypeptides (86, 164, 165, 181). However, there is a disparity in the relative frequencies of these events observed among laboratories: The single strand/double strand ratios run from 1000/1 to 1/1. Most certainly these structures are providing clues as to the mechanism of T-DNA transfer. Whether they have a direct role or are incidental to the process remains to be determined.

T-DNA INTEGRATION   The process of T-DNA integration is perhaps the least understood aspect of *Agrobacterium*-mediated transformation. T-DNA intermediates have not been identified in plant cells, and very little is known about either the specific sites or mechanism of T-DNA integration. The data gathered to date, however, allow for a few generalities to be drawn. The primary site of T-DNA insertion is the nuclear genome of plant cells (31, 175, 192). [There has been a single report of T-DNA integration into chloroplast DNA (39).] Genetic mapping studies and molecular analysis indicate that inserts can occur throughout the genome and that several independent insertion events can take place in an individual plant cell (5, 34, 41, 44, 123, 159). Individual inserts can occur in either repeated or single-copy DNA sequences (57, 76, 158, 198, 199, 205). Whether there are specific nucleotide sequence requirements for integration remains to be determined. It has been noted, however, that the plant sequences at the junctions of T-DNA inserts are often AT rich. Gheysen et al (57) suggested that this characteristic may result in local destabilization of duplex plant DNA and may thereby enhance opportunites for an integration event.

While the mechanism of T-DNA integration is not yet known, results to date suggest that it does not result from a transposition-like event. T-DNA inserts are not precise DNA segments, as transposon inserts generally are. The right T-DNA/plant junction, which shows the most consistency, usually occurs within 10–40 bp of 'he right border, while the left junction can be spread over 30–2000 bp (76, 91, 156, 158, 199, 205). T-DNA structures missing large segments of either the right or left region of the T-DNA are not uncommon (67, 131, 135, 178). Moreover, inverted or direct tandem repeats of T-DNA inserts have been observed (76, 89, 91, 159, 178, 205, 206); such structures are not characteristic of transposon inserts.

Gheysen et al (57) have proposed an alternative model to transposition. They determined the DNA sequences of both the right and left T-DNA border/plant DNA junctions as well as the plant target sequences before insertion, and they found that a 158-bp direct repeat of the plant target sequences occurred upon T-DNA integration. They also found that a number of small insertions and deletions occurred at the ends of right and left copies of the 158-bp repeat. These data, along with previous analyses of T-DNA insertions, led the investigators to suggest that T-DNA integration is a multi-step process of recombination accompanied by local replicative and repair reactions. They proposed the following general model (57): (*a*) The T-strand is transferred to the plant cell as a protein-DNA complex. (*b*) The right border and accompanying protein interact with a nick in the plant DNA. (*c*) Local torsional strain on the plant DNA produces a second nick on the opposite strand at varying distances from the first nick. (*d*) The T-strand is ligated to the plant DNA, and the homologous strand replicates. (*e*) Replication and repair of the staggered nick in the plant target DNA results in both the production of a repeated sequence and additional sequence rearrangements at the ends of the inserted T-DNA element. It is not clear whether this model can account for the inverted and direct repeats of T-DNA that are often observed in transformed plant cells.

Since there is no evidence that any bacterial genes are required specifically for T-DNA integration, it appears that the proposed replication and repair activities are plant encoded. This notion is consistent with the original suggestion of Braun (19) that *Agrobacterium* transformation is dependent on plant processes associated with wound cell division and/or DNA replication.

T-DNA Expression

Once the T-DNA is integrated into plant DNA its expression is required for crown gall tumor development. The genes encoded by the T-DNA include the T-DNA oncogenes, which are responsible for the characteristic uncontrolled proliferation of crown gall tissues, and the opine biosynthetic genes, which catalyze the synthesis of unusual amino acid and sugar derivatives.

ONCOGENES   To facilitate discussion we focus on the closely related octopine Ti plasmids, because with some exceptions (see below), the oncogenes of the various Ti plasmids are highly conserved in both structure and function (50, 55, 90, 106, 132, 193, 194). The octopine plasmids pTiA6, pTiAch5, and pTi15955 each have two separate T-DNA regions, the TL-DNA and the TR-DNA (174). Genetic (55, 106, 132), transcription (193, 194), and DNA sequence analyses (8) indicate that the TL-DNA and TR-DNA regions encode eight and five genes, respectively. Three of the TL-DNA genes, *tms1* (also referred to as *gene1, iaaM*), *tms2 (gene2, iaaH)*, and *tmr (gene4, ipt)*, code for the biosynthesis of an auxin and a cytokinin, two classes of potent plant hormones (for recent reviews see 38). The biochemical activities of these oncogenes have been characterized. The *tmr* gene encodes an isopentenyl transferase, which catalyzes the synthesis of the cytokinin iso-pentenyladenosine 5'-monophosphate from dimethylallyl-pyrophosphate and 5'-AMP (1, 9, 24) . The *tms1* and *tms2* genes encode a two-step pathway for the production of indole-3-acetic acid, the primary auxin in plants. The *tms1* gene product converts tryptophan to indole-3-acetamide (172, 179), and the *tms2* gene product converts the indole-3-acetamide to indole-3-acetic acid (84, 153, 171). Interestingly, the same pathway operates in *Pseudomonas syringae* pv. *savastanoi,* which causes galls on oleander and olive. In this case the conversion of tryptophan to indole-3-acetic acid occurs in the bacterium and is catalyzed by enzymes that share significant sequence homology with the gene products of *tms1* and *tms2* (200).

Two additional TL-DNA genes, *tml (gene6b)* and *gene5,* have an influence on the phenotypes of tumors but are not required for tumorigenesis. Mutants at *tml* give rise to abnormally large tumors on some host plants (148), but at present the mode of action of the gene product is unknown. Mutations at *gene5* have little effect on tumor induction by themselves, but they appear to accentuate, through unknown mechanism, the effects of mutations at either *tms1* or *tms2* (106).

The Ri plasmid of *A. rhizogenes* also carries oncogenes. These are required for the formation of hairy roots by transformed tissues. One of the three classes of Ri plasmids carries two T-DNA segments, TR and TL. The TR-DNA contains genes homologous to the auxin biosynthetic loci of the Ti plasmids (28, 83). All three classes of Ri plasmids identified to date have a conserved region of T-DNA, referred to as TL-DNA, that contains oncogenes required for hairy root formation and opine biosynthetic genes (43, 104). Recent experiments have indicated that the oncogenes of the TL-DNA confer auxin responsiveness to the transformed cells; cells carrying these genes will respond to auxin by forming roots (29). The molecular basis of this activity is not known.

OPINE GENES   The TL-DNA encodes one gene known to be involved in opine synthesis, *ocs (gene3)*. It codes for octopine synthase, an enzyme that catalyzes a reductive condensation between pyruvate and arginine to yield octopine (93, 169). A second gene, *ons (gene6a),* appears to code for a protein that transports octopine across plant membranes (125). Three additional genes that have roles in opine synthesis, *gene0'*, *gene1'*, and *gene2'*, are located in the TR-DNA and encode a pathway for mannopine and agropine synthesis (49).

In contrast to the conserved oncogene complement of Ti plasmids, the opine biosynthetic genes are diverse. Whereas the octopine T-DNA genes encode the synthesis of octopine and agropine, the nopaline T-DNA encodes the production of nopaline, a condensation product between arginine and α-ketoglutarate, and agrocinopine, a phosphorylated sugar derivitive (93, 169). The T-DNA harbored by *A. tumefaciens* 542, A2, and AT1 does not code for the production of octopine or nopaline, but does encode the synthesis of leucinopine and succinamopine (30, 33). In addition, a number of transcribed regions from various Ti plasmids presumably encode functional gene products whose roles are unknown. Perhaps these genes encode additional opine synthetic genes. The T-DNA of Ri plasmids also contains opine biosynthetic genes, including those for mannopine, agropine, and cucumopine (140).

RESPONSE TO T-DNA EXPRESSION   The response of the plant cells and the inciting bacteria to the expression of the T-DNA defines the final cellular interactions under discussion. Analyses of tumors on the plant or in culture have indicated that the response of the cells to the T-DNA expression can be complex. Expression of the hormone biosynthetic loci usually causes elevated levels of auxin and cytokinin (2, 136, 180, 187). Cells that are normally quiescent then exhibit uncontrolled proliferation, which results in a tumor. Such transformed tissue is also capable of growing in culture in a hormone-independent, autonomous fashion (18).

The ultimate phenotype resulting from the transfer of T-DNA is defined not only by the expression of that DNA, but by the plant cells' response to the hormones produced. If, for example, the cell type in question requires higher levels of hormone for growth and division than can be provided as a result of T-DNA expression, then a tumor will not develop. In contrast, some crown gall tumors contain very low levels of auxin and/or cytokinin, yet exhibit an autonomous phenotype (137, 187). Finally, the metabolism of these hormones may vary from host to host (e.g. 122) and could have an important role in defining the ultimate host response.

Agrobacteria respond to the expression of the T-DNA by plant cells in two important ways: The opines induce both opine degradation and conjugative transfer of Ti plasmids (see 96, 98, 170 for reviews). There is a correlation between the type of opine encoded for by the T-DNA of a Ti plasmid and the type of opine that is catabolized by that plasmid and that induces its conjugative transfer (97, 141). When multiple opines are encoded by a T-DNA (e.g. agrocinopine and nopaline), often only one of the opines will induce transfer (96).

It has been suggested that the genetic engineering of plants by agrobacteria creates a niche providing a nutritional resource, the opines, that only the agrobacteria can utilize (169). It is becoming apparent, however, that other bacteria can compete for these resources. A variety of agrobacteria carry plasmids that are nononcogenic, but carry opine catabolism loci and opine-inducible conjugal transfer loci (96). More surprising is the evidence that pseudomonads found in crown gall tumors and in the soil can utilize opines as nitrogen and carbon sources (e.g. 150).

CELLULAR BASIS OF HOST RANGE

The induction of tumors by agrobacteria is clearly a complex process. It is therefore remarkable that the host range of *Agrobacterium* species is so broad (40); a wide range of dicotyledonous plants can serve as hosts, as well as a number of gymnosperms and a few monocots. The implication is that the transformation process revolves around physiological and molecular parameters that are highly conserved among plants. However, host-range differences are observed among individual *Agrobacterium* isolates. In addition, certain plants, including most monocots, are refractory to tumor induction. What accounts for these limitations in host range?

An *Agrobacterium* isolate could fail to incite a tumor on a particular plant for one of two basic reasons. First, the T-DNA might not be transferred to a given host because of poor attachment, lack of *vir* region induction, defective *vir* genes, or inefficient T-DNA transfer or integration. Secondly, the T-DNA oncogenes might not induce tumorigenesis because of poor expression of the oncogenes or because of host responses to oncogene expression. The alteration in phytohormone levels brought about by oncogene expression might not induce uncontrolled proliferation of certain plant cells. There is evidence that a number of these factors do indeed affect host range.

Quantitative comparison of attachment by various strains of *A. tumefaciens* to a range of different hosts has been the subject of several investigations (66, 119, 147). In general, there is a rough correlation between the capacitiy of a strain to attach to plant cells and its ability to incite tumors. From these studies and from the fact that attachment-defective *Agrobacterium* mutants (see above) are severely attenuated in virulence, it appears that ineffective attachment results in inefficient T-DNA transfer and thus poor tumor induction.

The induction of *vir* genes can also affect *Agrobacterium* host range. Yanofsky et al (202) found that the *virA* gene product encoded by the limited–host range (LHR) Ti plasmid pTiAg162 could not complement *virA* mutations of the wide–host range (WHR) Ti plasmid pTiA6 on some host plants. This suggested that the *virA* gene product of pTiAg162 might not recognize the same plant-inducing molecules as the *virA* gene product of pTiA6. Indeed, Leroux et al (107) found that although these two *vir* genes are structurally related, the *virA* gene product of pTiAg162 did not induce *vir* gene expression in response to acetosyringone or cocultivation with tobacco cells.

The induction of *vir* genes is also responsible for the supervirulent phenotype of *A. tumefaciens* strains carrying pTiBo542. Such strains induce tumors larger than those induced by strains carrying the more typical WHR Ti plasmids such as pTiA6 (64, 77, 88). In addition, these supervirulent strains induce tumors on some hosts that are usually recalcitrant to tumor induction. The genetic basis for this increased virulence is that strains carrying pTiBo542 express *virG* at higher levels than strains carrying pTiA6 (88). This elevated expression of *virG* in turn results in higher levels of expression of the other *vir* loci. By transferring either the *vir* region or the *virG* locus in particular from pTiBo542 into strains carrying pTiA6, it was possible to extend the host range of these strains and increase their tumorigenic potential (77, 88).

As described above, mutations at *virC* or *virE* affect tumor induction on some hosts but not others. Results of a number of studies suggest that these host-range differences result from decreased efficiency of T-DNA transfer or integration (69, 82). Interestingly, decreases in transfer efficiency appear to have positive effects on tumor induction with some hosts. *A. tumefaciens* strains carrying the WHR plasmid pTiA6 formed small necrotic tumors on grapevines, but when the *virC* locus was inactivated, the strains induced normal tumors on this host (202). This result is particularly satisfying, since the LHR plasmid pTiAg162, a plasmid carried by an *Agrobacterium* strain that was isolated from grapevines and that can incite tumors in this host, does not have a functional *virC* locus (202).

A number of studies indicate that the T-DNA oncogene complement encoded by a Ti plasmid can also have an effect on host range. The LHR strains that form tumors on grapevines are avirulent on common *Agrobacterium* hosts including tomato. This characteristic is not due to lack of T-DNA transfer to tomato, since addition of a *tmr* oncogene from a WHR plasmid to the T-DNA region of the LHR Ti plasmid resulted in tumor induction on this host (23, 73, 202). Genetic analysis has revealed that the LHR Ti plasmids pTiAg162 and pTiAg63 do not have active *tmr* genes. Thus, neoplastic growth of tomato cells requires both the auxin and cytokinin genes. In contrast, either neoplastic growth of grape cells does not require cytokinin, or the transfer of the auxin genes is suffcient to induce cytokinin accumulation. Similarly, the *tmr* gene alone is capable of inducing both cytokinin and auxin autonomy in *Nicotiana glutinosa* (11, 12); *N. glutinosa* cells carrying only the *tmr* gene accumulate auxin to levels near those of cells transformed by strains carrying typical WHR Ti plasmids (11). These results clearly demonstrate that transformed cells ultimately exhibit a combination of host- and T-DNA-specified activities.

A final question then is why are most monocots refractory to tumor induction by *Agrobacterium?* Is the block at the level of T-DNA transfer or integration, or does it have to do with T-DNA expression or host responses to T-DNA expression? This question is of practical importance. While Ti plasmid–based vectors are now available for introducing genes of choice into a variety of dicotyledonous plants, they have not yet proven useful for transformation of cereals and other agronomically important monocots. Understanding why tumor induction does not generally occur in monocots should assist in the discovery of means of *Agrobacterium*-mediated transformation of these plants.

A. tumefaciens can stably transform at least one monocot, *Asparagus officinalis*, a member of the family Liliaceae. Asparagus tissues exposed to wild-type *A. tumefaciens* can grow on hormone-free medium and produce opines (68). Furthermore, DNA hybridization data have indicated the presence of T-DNA in the transformed cells (26). T-DNA–based cloning vectors could be used to transform asparagus, and plants containing the foreign DNA were regenerated (26). Thus there is not an absolute block in *Agrobacterium*-mediated monocot transfomation.

Evidence for T-DNA transfer and expression in other monocots has also been reported. *Agrobacterium* inoculation of *Chlorophytum capense* (Liliaceae), *Narcissus* cv. Paperwhite (Amaryllidaceae), *Gladiolus* sp. (Iridaceae), and *Zea mays* (Graminaceae) results in the production of opines around the inoculation site (59, 60, 80). Transfer of T-DNA to maize has also been demonstrated using the sensitive technique of agroinfection. In this procedure, partially or completely duplicated genomes of viruses are incorporated into the Ti plasmid T-DNA or binary vector. Upon transfer of the T-DNA to the plants, single copies of the viral genomes can escape from the T-DNA, and in compatible host-virus combinations the viruses can replicate and spread systemically (53, 61, 69). To test for T-DNA transfer to maize, Grimsley et al (62) inserted the maize streak virus genome into the T-DNA and inoculated maize seedlings. Almost all of the inoculated plants developed viral symptoms. Control experiments indicated that symptom development was dependent on the presence of T-DNA border sequences and a functional *virA* locus.

From these experiments it would appear that T-DNA transfer can occur in a number of monocots. What then limits tumor development? One possibility is that T-DNA transfer is inefficient. The fact that *Agrobacterium* attachment to monocot cells is generally much less efficient than attachment to dicot cells is consistent with this notion (66, 147). A potentially more profound possibility is that poor tumor induction might be related to differences in physiological and growth characteristics of monocot and dicot cells. For example, Usami et al (177) have reported that a variety of monocot seedlings do not secrete *vir*-inducing molecules that are secreted by dicot seedlings.

Perhaps more fundamentally, the wound response of many monocots differs from that of dicots. In many monocots the cells around the wound site differentiate into lignified or sclerified cells without apparent cell division (13, 92). In addition, while it is generally easy to establish growth of various dicot tissues in culture, many monocot tissues (especially mature somatic tissues) are difficult to culture. For example, studies on corn leaves of different developmental stages have shown a narrow window of competence for proliferation in culture (188). In this regard, it is interesting that tissues of the one monocot that has proven susceptible to *Agrobacterium* transformation, asparagus, can be grown relatively easily in culture (26, 68). Consideration of these facts suggests that T-DNA might be delivered to monocot cells, but that the recipient cells either cannot integrate the T-DNA owing to the lack of wound divisions or are incapable of proliferating. In either case the transfer of T-DNA would not result in an expansion of the transformed population of cells.

CONCLUDING REMARKS

Over the past five years remarkable progress has been made in developing an understanding of the *Agrobacterium* infection and transformation process. The literature revewed here provides strong evidence that these bacteria attach and respond to competent plant cells, transferring DNA into the plant cells, possibly by a mechanism similar to the conjugative system in bacteria. While this provides an attractive working model, a variety of questions related to the transformation process remain unanswered. For example, what is the cellular and biochemical nature of a productive physical interaction between virulent agrobacteria and the plant cell? How do the *vir* proteins facilitate the putative conjugative event? What happens to the T-DNA intermediates (and perhaps accompanying *vir* proteins) inside the plant cell? How is T-DNA stably integrated, and in what frequency in relation to the delivery of the T-DNA into the plant cell? The answers to such questions will not only help explain *Agrobacterium*-mediated transformation, but may provide clues for improving direct DNA transfer methodology as well.

ACKNOWLEDGMENTS

We would like to thank Timothy Close, Peter Rogowski, John Ward, and Patricia Zambryski for sending us preprints of various manuscripts. In addition we would like to thank Stan Gelvin, Jo Hendelsman, Harry Klee, Eugene Nester, Nicola Neff, Walt Ream, John Ward, and Patricia Zambryski for helpful discussions, and Robert Donald, Gary Kuleck, Jean Labriola, and Rita Teutonico for reading early versions of the manuscript.

Literature Cited

1. Akiyoshi, D. E., Klee, H., Amasino, R. M., Nester, E. W., Gordon, M. P. 1984. T-DNA of *Agrobacterium tumefaciens* encodes an enzyme of cytokinin biosynthesis. *Proc. Natl. Acad. Sci. USA* 81:5994–98

2. Akiyoshi, D. E., Morris, R. O., Hinz, R., Mischke, B. S., Kosuge, T., et al. 1983. Cytokinin-auxin balance in crown gall tumors is regulated by specific loci in the T-DNA. *Proc. Natl. Acad. Sci. USA* 80:407–11

3. Albright, L. M., Yanofsky, M. F., Leroux, B., Ma, D., Nester, E. W. 1987. Processing of the T-DNA of *Agrobacterium tumefaciens* generates border nicks and linear single-stranded T-DNA. *J. Bacteriol.* 169:1046–55

4. Alt-Morebe, J., Rak, B., Schröder, J. 1986. A 3.6 kbp segment from the *vir* region of Ti plasmids contains genes responsible for border sequence-directed production of T-region circles in *E. coli*. *EMBO J.* 5:1129–35

5. Ambros, P. F., Matzke, A. J. M., Matzke, M. A. 1986. Localization of *Agrobacterium rhizogenes* T-DNA in plant chromosomes by *in situ* hybridization. *EMBO J.* 5:2073–77

6. An, G. 1985. High efficiency of transformation of cultured tobacco cells. *Plant Physiol.* 79:568–70

7. Ashby, A. M., Watson, M. D., Shaw, C. H. 1987. A Ti-plasmid determined function is responsible for chemotaxis of *Agrobacterium tumefaciens* towards the plant wound product acetosyringone. *FEMS Microbiol. Lett.* 41:189–92

8. Barker, R. F., Idler, K. B., Thompson, D. V., Kemp, J. D. 1983. Nucleotide sequence of the T-DNA region from the *Agrobacterium tumefaciens* octopine Ti plasmid pTil5955. *Plant Mol. Biol.* 2:335–50

9. Barry, G. F., Rogers, S. G., Fraley, R. T., Brand, L. 1984. Identification of a cloned cytokinin biosynthetic gene. *Proc. Natl. Acad. Sci. USA* 81:4776–80

10. Binns, A. N. 1984. The biology and molecular biology of plant cells infected by *Agrobacterium tumefaciens*. *Oxford Surv. Plant Mol. Cell Biol.* 1:133–60

11. Binns, A. N., Labriola, J., Black, R. C. 1987. Initiation of auxin autonomy in *Nicotiana glutinosa* cells by the cytokinin biosynthesis gene from *Agrobacterium tumefaciens*. *Planta* 171:539–48

12. Binns, A. N., Sciaky, D., Wood, H. N.

1982. Variation in hormone autonomy and regenerative potential of cells transformed by strain A66 of *Agrobacterium tumefaciens*. *Cell* 31:605–12

13. Bloch, R. 1941. Wound healing in higher plants. *Bot. Rev.* 7:110–46

14. Bolton, G. W., Nester, E. W., Gordon, M. P. 1986. Plant phenolic compounds induce expression of the *Agrobacterium tumefaciens* loci needed for virulence. *Science* 232:983–85

15. Bouzar, H., Moore, L. W. 1987. Isolation of different *Agrobacterium* biovars from a natural oak savanna and tall grass prairie. *Appl. Environ. Microbiol.* 53: 717–21

16. Bradley, D. E., Douglas, C. J., Peschon, J. 1984. Flagella specific bacteriophages of *Agrobacterium tumefaciens:* demonstration of virulence of nonmotile mutants. *Can. J. Microbiol.* 30: 676–81

17. Braun, A. C. 1952. Conditioning of the host cell as a factor in the transformation process in crown gall. *Growth* 16:65–74

18. Braun, A. C. 1958. A physiological basis for the autonomous growth of the crown gall tumor cell. *Proc. Natl. Acad. Sci. USA* 44:344–49

19. Braun, A. C. 1975. The cell cycle and tumorigenesis in plants. *Results Probl. Cell Differ.* 7:177–96

20. Braun, A. C., Mandle, R. J. 1948 Studies on the inactivation of the tumor inducing principle in crown gall. *Growth* 12:255–69

21. Braun, A. C., Stonier, T. 1958. Morphology and physiology of plant tumors. *Protoplasmatalogia* 10(5a):1–93

22. Buchanan-Wollaston, V., Passiatore, J. E., Cannon, F. 1987. The *mob* and *oriT* mobilization functions of a bacterial plasmid promote its transfer to plant cells. *Nature* 328:172–75

23. Buchholz, W. G., Thomashow, M. W. 1984. Host range encoded by the *Agrobacterium tumefaciens* tumor inducing plasmid pTiAg63 can be expanded by modification if its T-DNA oncogene complement. *J. Bacteriol.* 160:327–32

24. Buchmann, I., Marner, F. J., Schröder, G., Waffenschmidt, S., Schröder, J. 1985. Tumor genes in plants: T-DNA encoded cytokinin biosynthesis. *EMBO J.* 4:853–59

25. Burr, T. J., Katz, B. H. 1983. Isolation of *Agrobacterium tumefaciens* biovar 3 from grapevine galls and sap and from vineyard soil. *Phytopathology* 73:163–65

26. Bytebier, B., Deboeck, F., DeGreve, H., Van Montagu, M., Hernalsteens, J.

P. 1987. T-DNA organization in tumor cultures and in transgenic plants of the monocotyledon *Asparagus officinalis*. *Proc. Natl. Acad. Sci. USA* 84:5345–49

27. Cangelosi, G. A., Hung, L., Puvanesarajah, V., Stacey, G., Ozga, D. A., et al. 1987. Common loci for *Agrobacterium tumefaciens* and *Rhizobium meliloti* exopolysaccharide synthesis and their roles in plant interactions. *J. Bacteriol.* 169: 2086–91

28. Cardarelli, M., Spanò, L., DePaolis, A., Mauro, M. L., Vitali, G., et al. 1985. Identification of the genetic locus responsible for non-polar root induction by *Agrobacterium rhizogenes* 1855. *Plant Mol. Biol.* 5:385–91

29. Cardarelli, M., Spanò, L., Mariotti, D., Mauro, M. L., Van Sluys, M., Costantino, P. 1987. The role of auxin in hairy root induction. *Mol. Gen. Genet.* 208: 457–63

30. Chang, C.-C., Chen, C.-M., Adams, D. R., Trost, B. M. 1983. Leucinopine, a characteristic compound of some crowngall tumors. *Proc. Natl. Acad. Sci. USA* 80:3573–76

31. Chilton, M. D., Saiki, R. K., Yadav, N., Gordon, M. P., Quetier, F. 1980. T-DNA from *Agrobacterium* Ti plasmid is in the nuclear DNA fraction of crown gall tumor cells. *Proc. Natl. Acad. Sci. USA* 77:4060–64

32. Chilton, M. D., Tepfer, D. A., Petit, A., Casse-Delbart, F., Tempé, J. 1982. *Agrobacterium rhizogenes* inserts T-DNA into the genomes of host plant root cells. *Nature* 295:432–34

33. Chilton, W. S., Tempé, J., Matzke, M., Chilton, M. D. 1984. Succinamopine: a new crown gall opine. *J. Bacteriol.* 157:357–62

33a. Christie, P. J., Ward, J. E., Winans, S. C., Nester, E. W. 1988. The *Agrobacterium tumefaciens virE2* gene product is a single-stranded DNA-binding protein that associates with T-DNA. *J. Bacteriol.* 170:2659–67

34. Chyi, Y. S., Jorgensen, R. A., Goldstein, D., Tanksley, S. D., Loaiza-Figueroa, F. 1986. Locations and stability of *Agrobacterium* mediated T-DNA insertions in the *Lycopersicon* genome. *Mol. Gen. Genet.* 204:64–69

34a. Citovsky, V., De Vos, G., Zambryski, P. 1988. Single stranded DNA binding protein encoded by the *virE* locus of *Agrobacterium tumefaciens*. *Science* 240:501–4

35. Close, T. J., Rogowsky, P. M., Kado, C. I., Winans, S., Yanofsky, M., et al. 1987. Dual control of *Agrobacterium*

tumefaciens Ti plasmid virulence genes. *J. Bacteriol.* 169:5113–18

36. Close, T. J., Tait, R. C., Kado, C. I. 1985. Regulation of Ti plasmid virulence genes by a chromosomal locus of *Agrobacterium tumefaciens. J. Bacteriol.* 164:774–81

37. Close, T. J., Tait, R. C., Rempel, H. C., Hirooka, T., Kim, L., et al. 1987. Molecular characterization of the *virC* genes of the Ti plasmid. *J. Bacteriol.* 169:2336–44

38. Davies, P. J., ed. 1987. *Plant Hormones and Their Role in Plant Growth and Development.* Dordrecht, the Netherlands: Nijhoff

39. DeBlock, M., Schell, J., Van Montagu, M. 1985. Chloroplast transformation by *Agrobacterium tumefaciens. EMBO J.* 4:1367–72

40. DeCleene, M., DeLey, J. 1976. The host range of crown gall. *Bot. Rev.* 42:389–466

41. deFramond, A. J., Back, E. W., Chilton, W. S., Kayes, L., Chilton, M. D. 1986. Two unlinked T-DNAs can transform the same tobacco cell and segregate in the F_1 generation. *Mol. Gen. Genet.* 202:125–31

42. deFramond, A. J., Barton, K. A., Chilton, M. D. 1983. Mini-Ti: a new vector strategy for plant genetic engineering. *Bio-Technology* 1:262–69

43. DePaolis, A., Mauro, M. L., Pomponi, M., Cardarelli, M., Spano, L., et al. 1985. Localization of agropine synthesizing functions in the TR region of the root-inducing plasmid of *Agrobacterium rhizogenes* 1855. *Plasmid* 13:1–7

44. Depicker, A., Herman, L., Jacobs, A., Schell, J., Van Montagu, M. 1985. Frequencies of simultaneous transformation with different T-DNA's and their relevance to the *Agrobacterium*/plant cell interaction. *Mol. Gen. Genet.* 201: 477–84

45. Douglas, C., Halperin, W., Nester, E. W. 1982. *Agrobacterium tumefaciens* mutants affected in attachment to plant cells. *J. Bacteriol.* 152:1265–75

46. Douglas, C. J., Staneloni, R. J., Rubin, R. A., Nester, E. W. 1985. Identification and genetic analysis of an *Agrobacterium tumefaciens* chromosomal virulence region. *J. Bacteriol.* 161:850–60

47. Drummond, M., Whitty, P., Wootton, J. 1986. Sequence and domain relationships of *ntrC* and *nifA* from *Klebsiella pneumoniae:* homologies to other regulatory proteins. *EMBO J.* 5:441–47

48. Dylan, T., Ielpi, L., Stanfield, S., Kashyap, L., Douglas, C., et al. 1986. *Rhizobium meliloti* genes required for

nodule development are related to chromosomal virulence genes in *Agrobacterium tumefaciens. Proc. Natl. Acad. Sci. USA* 83:4403–7

49. Ellis, J. G., Ryder, M. H., Tate, M. E. 1984. *Agrobacterium tumefaciens* TR-DNA encodes a pathway for agropine biosynthesis. *Mol. Gen. Genet.* 195: 466–73

50. Engler, G., Depicker, A., Maenhaut, R., Villarroel, R., Van Montagu, M., et al. 1981. Physical mapping of DNA base sequence homologies between an octopine and a nopaline Ti plasmid of *Agrobacterium tumefaciens. J. Mol. Biol.* 152:183–208

51. Engström, P., Zambryski, P., Van Montagu, M., Stachel, S. 1987. Characterization of *Agrobacterium tumefaciens* virulence proteins induced by the plant factor acetosyringone. *J. Mol. Biol.* 197:635–46

52. Firoozabody, E., Galbraith, D. W. 1984. Presence of a plant cell wall is not required for transformation of *Nicotiana* by *Agrobacterium tumefaciens. Plant Cell Tissue Organ Cult.* 3:175–84

53. Gardner, R. C., Knauf, V. C. 1986. Transfer of *Agrobacterium* DNA to plants requires a T-DNA border but not the *virE* locus. *Science* 231:725–27

54. Garfinkel, D. J., Nester, E. W. 1980. *Agrobacterium tumefaciens* mutants affected in crown gall tumorigenesis and octopine catabolism. *J. Bacteriol.* 144: 732–43

55. Garfinkel, D. J., Simpson, R. B., Ream, L. W., White, F. F., Gordon, M. P., et al. 1981. Genetic analysis of crown gall: a fine structure map of the T-DNA by site-directed mutagenesis. *Cell* 27:143–53

56. Gheysen, G., Dhaese, P., Van Montagu, M., Schell, J. 1985. DNA flux across genetic barriers: the crown gall phenomenon. In *Advances in Plant Gene Research,* ed. B. Hohn, E. S. Dennis, 2:11–47. Vienna: Springer-Verlag

57. Gheysen, G., Van Montagu, M., Zambryski, P. 1987. Integration of *Agrobacterium tumefaciens* T-DNA involves rearrangements of target plant DNA sequences. *Proc. Natl. Acad. Sci USA* 84:6169–73

57a. Gietl, C., Koukolikova-Nicola, Z., Hohn, B. 1987. Mobilization of T-DNA from *Agrobacterium* to plant cells involves a protein that binds single-stranded DNA. *Proc. Natl. Acad. Sci. USA* 84:9006–10

58. Glogowski, W., Galsky, A. G. 1978. *Agrobacterium tumefaciens* site attachment as a necessary prerequisite for

crown gall tumor formation on potato discs. *Plant Physiol.* 61:1031–33

59. Graves, A. C., Goldman, S. L. 1986. The transformation of *Zea mays* seedlings with *Agrobacterium tumefaciens:* detection of T-DNA specified enzyme activities. *Plant Mol. Biol.* 7:43–50

60. Graves, A. C. F., Goldman, S. L. 1987. *Agrobacterium tumefaciens* mediated formation of the monocot genus *Gladiolus:* detection of the expression of T-DNA encoded genes. *J. Bacteriol.* 169:1745–46

61. Grimsley, N., Hohn, B., Hohn, T., Walden, R. 1986. Agroinfection, a novel route for plant viral infection using Ti plasmid. *Proc. Natl. Acad. Sci. USA* 83:3282–86

62. Grimsley, N., Hohn, T., Davies, J. W., Hohn, B. 1987. *Agrobacterium* mediated delivery of infectious maize streak virus into maize plants. *Nature* 325:177–79

63. Gurlitz, R. H. G., Lamb, P. W., Matthysse, A. G. 1987. Involvement of carrot cell surface proteins in attachment of *Agrobacterium tumefaciens. Plant Physiol.* 83:564–68

64. Guyon, P., Chilton, M. D., Petit, A., Tempé, J. 1980. Agropine in "null" type crown gall tumors: evidence for the generality of the opine concept. *Proc. Natl. Acad. Sci. USA* 80:4803–7

65. Hall, M. N., Silhavy, T. J. 1981. Genetic analysis of the major outer membrane proteins of *Escherichia coli. Ann. Rev. Genet.* 15:91–142

66. Hawes, M. C., Pueppke, S. G. 1985. Relationship between binding of *Agrobacterium tumefaciens* to isolated root cap cells and susceptibility to crown gall. *Curr. Top. Plant Biochem. Physiol.* 4:244–51

67. Hepburn, A. G., White, J. 1985. The effect of right terminal repeat deletion on the oncogenicity of the T-region of pTiT37. *Plant Mol. Biol.* 5:3–12

68. Hernalsteens, J. P., Thia-Toong, L., Schell, J., Van Montagu, M. 1984. An *Agrobacterium* transformed cell culture from the monocot *Asparagus officinalis. EMBO J.* 13:3039–41

69. Hille, J., Dekker, M., Luttighuis, H. O., van Kammen, A., Zabel, P. 1986. Detection of T-DNA transfer to plant cells by *A. tumefaciens* virulence mutants using agroinfection. *Mol. Gen. Genet.* 205:411–16

70. Hille, J., Klasen, I., Schilperoort, R. A. 1982. Construction and application of R prime plasmids, carrying different segments of the octopine Ti plasmid from *Agrobacterium tumefaciens,* for complementation of *vir* genes. *Plasmid* 7:107–18

71. Hille, J., van Kan, J., Schilperoort, R. A. 1984. *trans*-Acting virulence functions of the octopine Ti plasmid from *Agrobacterium tumefaciens. J. Bacteriol.* 158:754–56

72. Hirooka, T., Rogowsky, P. M., Kado, C. I. 1987. Characterization of the *virE* locus of *Agrobacterium tumefaciens* plasmid pTiC58. *J. Bacteriol.* 169:1529–36

73. Hoekema, A., dePater, B. S., Fellinger, A. J., Hooykaas, P. J. J., Schilperoort, R. A. 1984. The limited host range of an *Agrobacterium tumefaciens* strain extended by a cytokinin gene from a wide host range T-region. *EMBO J.* 3:3043–47

74. Hoekema, A., Hirsch, P. R., Hooykaas, P. J. J., Schilperoort, R. A. 1983. A binary vector strategy based on separation of *vir* and T-region of the *Agrobacterium tumefaciens* Ti plasmid. *Nature* 303:179–80

75. Hoekema, A., Hooykaas, P. J. J., Schilperoort, R. A. 1984. Transfer of octopine T-DNA segment to plant cells mediated by different types of *Agrobacterium* tumor or root-inducing plasmids: generality of virulence systems. *J. Bacteriol.* 158:383–85

76. Holsters, M., Villarroel, R., Gielen, J., Seurinck, J., DeGreve, H., et al. 1983. An analysis of the boundaries of octopine TL-DNA in tumors induced by *Agrobacterium tumefaciens. Mol. Gen. Genet.* 190:35–41

77. Hood, E. E., Helmer, G. L., Fraley, R. T., Chilton, M. D. 1986. The hypervirulence of *Agrobacterium tumefaciens* A281 is encoded in a region of pTiBo542 outside of T-DNA. *J. Bacteriol.* 168:1291–301

78. Hooykaas, P. J. J., Hoftker, M., den Dulk-Ras, H., Schilperoort, R. A. 1984. A comparison of virulence determinants in an octopine Ti plasmid, a nopaline Ti plasmid, and an Ri plasmid by complementation analysis of *Agrobacterium tumefaciens. Plasmid* 11:195–205

79. Hooykaas, P. J. J., Schilperoort, R. A. 1984. The molecular genetics of crown gall tumorigenesis. *Adv. Genet.* 22:210–83

80. Hooykaas-VanSlogteren, G. M. S., Hooykaas, P. J. J., Schilperoort, R. A. 1984. Expression of Ti plasmid genes in monocotyledonous plants infected with *Agrobacterium tumefaciens. Nature* 311:763–64

81. Horsch, R. B., Klee, H. J. 1986. Rapid

assay of foreign gene expression in leaf discs transformed by *Agrobacterium tumefaciens:* role of T-DNA borders in the transfer process. *Proc. Natl. Acad. Sci. USA* 83:4428–32

82. Horsch, R. B., Klee, H. J., Stachel, S., Winans, S. C., Nester, E. W., et al. 1986. Analysis of *Agrobacterium tumefaciens* virulence mutants in leaf discs. *Proc. Natl. Acad. Sci. USA* 83:2571–75

83. Huffman, G. A., White, F. F., Gordon, M. P., Nester, E. W. 1984. Hairy root inducing plasmid: physical map and homology to tumor-inducing plasmids. *J. Bacteriol.* 157:269–76

84. Inzé, A., Follin, A., Van Lijsebettens, C., Simoens, C., Genetello, C., et al. 1984. Genetic analysis of individual genes of *Agrobacterium tumefaciens:* further evidence that two genes are involved in indole-3-acetic acid synthesis. *Mol. Gen. Genet.* 194:265–74

85. Iyer, V. N., Klee, H. J., Nester, E. W. 1982. Units of genetic expression in the virulence region of a plant tumor-inducing plasmid of *Agrobacterium tumefaciens. Mol. Gen. Genet.* 188:418–24

86. Jayaswal, R. K., Veluthambi, K., Gelvin, S. B., Slightom, J. L. 1987. Double-stranded cleavage of T-DNA and generation of single stranded T-DNA molecules in *Escherichia coli* by a *virD* encoded border-specific endonuclease from *Agrobacterium tumefaciens. J. Bacteriol.* 169:5035–45

87. Jen, G. C., Chilton, M.-D. 1986. The right border region of pTiT37 T-DNA is intrinsically more active than the left border in promoting T-DNA transformation. *Proc. Natl. Acad. Sci. USA* 83:3895–99

88. Jin, S., Komari, T., Gordon, M. P., Nester, E. W. 1987. Genes responsible for the supervirulence phenotype of *Agrobacterium tumefaciens* A281. *J. Bacteriol.* 169:4417–25

89. Jones, J. D. G., Gilbert, D. E., Grady, K. L., Jorgensen, R. A. 1987. T-DNA structure and gene expression in petunia plants transformed by *Agrobacterium tumefaciens* C58 derivatives. *Mol. Gen. Genet.* 207:478–85

90. Joos, H., Inzé, D., Caplan, A., Sormann, M., Van Montagu, M., et al. 1983. Genetic analysis of T-DNA transcripts in nopaline crown galls. *Cell* 32:1057–67

91. Jorgensen, R., Snyder, C., Jones, J. D. G. 1987. T-DNA is organized predominantly in inverted repeats in plants transformed with *Agrobacterium tume-*

faciens C58 derivatives. *Mol. Gen. Genet.* 207:471–77

92. Kahl, G. 1982. Molecular biology of wound healing: the conditioning phenomenon. In *Molecular Biology of Plant Tumors,* ed. G. Kahl, J. Schell, pp. 211–67. New York: Academic. 615 pp.

93. Kemp, J. D. 1982. Enzymes in octopine and nopaline metabolism. See Ref. 92, pp. 461–74

94. Kerr, A. 1969. Transfer of virulence between isolates of *Agrobacterium. Nature* 223:1175–76

95. Kerr, A. 1969. Crown gall of stone fruit. I. Isolation of *Agrobacterium tumefaciens* and related species. *Aust. J. Biol. Sci.* 22:111–16

96. Kerr, A., Ellis, J. G. 1982. Conjugation and transfer of Ti plasmids in *Agrobacterium tumefaciens.* See Ref. 92, pp. 321–44

97. Kerr, A., Mannigault, P., Tempé, J. 1977. Transfer of virulence *in vivo* and *in vitro* in *Agrobacterium. Nature* 265:560–61

98. Klapwijk, P. M., Schilperoort, R. A. 1982. Genetic determination of octopine degradation. See Ref. 92, pp. 475–95

99. Klee, H. J., Gordon, M. P., Nester, E. W. 1982. Complementation analysis of *Agrobacterium tumefaciens* Ti plasmid mutations affecting oncogenicity. *J. Bacteriol.* 150:327–31

100. Klee, H., Horsch, R., Rogers, S. 1987. *Agrobacterium* mediated plant transformation and its further applications to plant biology. *Ann. Rev. Plant Physiol.* 38:467–86

101. Klee, H. J., White, F. F., Iyer, V. N., Gordon, M. P., Nester, E. W. 1983. Mutational analysis of the virulence region of an *Agrobacterium tumefaciens* Ti plasmid. *J. Bacteriol.* 153:878–83

102. Koukolikova-Nicola, Z., Shillito, R., Hohn, B., Wang, K., Van Montagu, M., et al. 1985. Involvement of circular intermediates in the transfer of T-DNA from *Agrobacterium tumefaciens* to plant cells. *Nature* 313:191–96

103. Kudirka, D. T., Colburn, S. M., Hinchee, M. A., Wright, M. S. 1986. Interactions of *Agrobacterium tumefaciens* with soybean (*Glycine max* (L.) Merr.) leaf explants in tissue cultures. *Can. J. Genet. Cytol.* 28:808–17

104. Lahners, K., Byrne, M. C., Chilton, M.-D. 1984. T-DNA fragments of hairy root plasmid pRi8196 are distantly related to octopine and nopaline Ti plasmid T-DNA. *Plasmid* 11:130–40

105. Lechoczky, J. 1971. Further evidence concerning the systemic spreading of *Agrobacterium tumefaciens* in the vascu-

lar system of grape vine. *Vitis* 10:215–21

106. Leemans, J., Deblaere, R., Willmitzer, L., DeGreve, H., Hernalsteens, J. P., et al. 1982. Genetic identification of functions of T_L-DNA transcripts in octopine crown galls. *EMBO J.* 1:147–52

107. Leroux, B., Yanofsky, M. F., Winans, S. C., Ward, J. E., Ziegler, S. F., et al. 1987. Characterization of the *virA* locus of *Agrobacterium tumefaciens:* a transcriptional regulator and host range determinant. *EMBO J.* 6:849–56

108. Lipetz, J. 1966. Crown gall tumorigenesis II. Relations between wound healing and the tumorigenic response. *Cancer Res.* 26:1597–605

109. Lippincott, B. B., Lippincott, J. A. 1969. Bacterial attachment to a specific wound site as an essential stage in tumor initiation by *Agrobacterium tumefaciens. J. Bacteriol.* 97:620–28

110. Lippincott, B. B., Whatley, M. W., Lippincott, J. A. 1977. Tumor induction by *Agrobacterium* involves attachment of the bacterium to a site on the host plant cell wall. *Plant Physiol.* 59:388–90

111. Lippincott, J. A., Lippincott, B. B. 1975. The genus *Agrobacterium* and plant tumorigenesis. *Ann. Rev. Microbiol.* 29:377–405

112. Lippincott, J. A., Lippincott, B. B., Scott, J. J. 1984. Adherence and host recognition in *Agrobacterium* infection. In *Current Perspectives in Microbial Ecology,* ed. M. J. Klug, C. N. Reddy, pp. 230–36. Washington, DC: Am. Soc. Microbiol.

113. Lundquist, R. C., Close, T. J., Kado, C. I. 1984. Genetic complementation of *Agrobacterium tumefaciens* Ti plasmid mutants in the virulence region. *Mol. Gen. Genet.* 193:1–7

114. Machida, Y., Usami, S., Yamamoto, A., Takebe, I. 1986. Plant inducible recombination between the 25 base pair border sequences of T-DNA in *Agrobacterium tumefaciens. Mol. Gen. Genet.* 204:374–82

115. Márton, L., Wullems, G. J., Molendijk, L., Schilperoort, R. A. 1979. *In vitro* transformation of cultured cells from *Nicotiana tabacum* by *Agrobacterium tumefaciens. Nature* 277:129–31

116. Matthysse, A. G. 1983. Role of bacterial cellulose fibrils in *Agrobacterium tumefaciens* infection. *J. Bacteriol.* 154:906–15

117. Matthysse, A. G. 1984. Interaction of *Agrobacterium tumefaciens* with plant cell surface. In *Plant Gene Research. Genes Involved in Microbe-Plant In-*

teractions, ed. D. P. S. Verma, T. Hohn, pp. 33–54. New York: Springer Verlag

118. Matthysse, A. 1987. Characterization of non-attaching mutants of *Agrobacterium tumefaciens. J. Bacteriol.* 169:313–23

119. Matthysse, A. G., Gurlitz, R. H. G. 1982. Plant cell range for attachment of *Agrobacterium tumefaciens* to tissue culture cells. *Physiol. Plant Pathol.* 21:381–87

120. Matthysse, A. G., Homes, K. V., Gurlitz, R. H. 1981. Elaboration of cellulose fibrils by *Agrobacterium tumefaciens* during attachment to carrot cells. *J. Bacteriol.* 145:583–95

121. Matthysse, A. G., Wyman, P. M., Holmes, K. V. 1978. Plasmid-dependent attachment of *Agrobacterium tumefaciens* to plant tissue culture cells. *Infect. Immun.* 22:516–22

122. McGaw, B. A. 1987. Cytokinin biosynthesis and metabolism. See Ref. 38, pp. 94–112

123. McKnight, T. D., Lillis, M. T., Simpson, R. B. 1987. Segregation of genes transferred to one plant cell from two separate *Agrobacterium* strains. *Plant Mol. Biol.* 8:439–45

124. Melchers, L. S., Thompson, D. V., Idler, K. B., Schilperoort, R. A., Hooykaas, P. J. J. 1986. Nucleotide sequence of the virulence gene *virG* of the *Agrobacterium tumefaciens* octopine Ti plasmid: significant homology between *virG* and the regulatory genes *ompR, phoB* and *dye* of *E. coli. Nucleic Acids Res.* 14:9933–42

125. Messens, E., Lenaerts, A., Van Montagu, M., Hedges, R. W. 1985. Genetic basis for opine secretion from crown gall tumor cells. *Mol. Gen. Genet.* 199:344–48

126. Miller, V. J., Kennedy, E. P., Reinhold, V. N., 1986. Osmotic adaptation of gram-negative bacteria: possible role for periplasmic oligosaccharides. *Science* 231:48–51

127. Neff, N. T., Binns, A. N. 1985. *Agrobacterium tumefaciens* interaction with suspension-cultured tomato cells. *Plant Physiol.* 77:35–42

128. Neff, N. T., Binns, A. N., Brandt, C. 1987. Inhibitory effects of a pectin-enriched tomato cell wall fraction on *Agrobacterium tumefaciens* binding and tumor formation. *Plant Physiol.* 83:525–28

129. Nester, E. W., Gordon, M. P., Amasino, R. M., Yanofsky, M. F. 1984. Crown gall: a molecular and physiological analysis. *Ann. Rev. Plant Physiol.* 35:387–413

130. New, P. B., Kerr, A. 1971. A selective medium for *Agrobacterium radiobacter* biotype 2. *J. Appl. Bacteriol.* 34:233–36
131. Ooms, G., Hooykaas, P. J. J., Molendijk, L., Wullems, G. J., Gordon, M. P., et al. 1982. T-DNA organization in homogeneous and heterogeneous octopine type crown gall tissues of *Nicotiana tabacum*. *Cell* 30:589–97
132. Ooms, G., Hooykaas, P. J. J., Noleman, G., Schilperoort, R. A. 1981. Crown gall tumors of abnormal morphology induced by *Agrobacterium tumefaciens* carrying mutated octopine Ti plasmids: analysis of T-DNA functions. *Gene* 14:33–50
133. Otten, L., De Greve, H., Hain, R., Hooykaas, P., Schell, J. 1984. Restoration of virulence of *vir* region mutants of *Agrobacterium tumefaciens* strain B6S3 by coinfection with normal and mutant *Agrobacterium* strains. *Mol. Gen. Genet.* 195:159–63
134. Parke, D., Ornston, L. N., Nester, E. W. 1987. Chemotaxis to plant phenolic inducers of virulence genes is constitutively expressed in the absence of the Ti plasmid in *Agrobacterium tumefaciens*. *J. Bacteriol.* 169:5336–38
135. Peerbolte, R., Leenhouts, K., Hooykaas-Van Slogteren, G. M. S., Hoge, J. H. C., Wullems, G. J., et al. 1986. Clones from a shooty tobacco crown gall tumor I: deletions, rearrangements and amplifications resulting in irregular T-DNA structures and organizations. *Plant Mol. Biol.* 7:265–84
136. Pengelly, W. L., Meins, F. Jr. 1982. The relationship of indole-3-acetic acid content and growth of crown gall tumor tissue of tobacco in culture. *Differentiation* 21:27–31
137. Pengelly, W. L., Vijayarghavan, S. J., Sciaky, D. 1986. Neoplastic progression in crown gall in tobacco without elevated auxin levels. *Planta* 169:454–61
138. Peralta, E. G., Helmiss, R., Ream, W. 1986. *Overdrive,* a T-DNA transmission enhancer on the *A. tumefaciens* tumour-inducing plasmid. *EMBO J.* 5:1137–42
139. Peralta, E. G., Ream, L. W. 1985. T-DNA border sequences required for crown gall tumorigenesis. *Proc. Natl. Acad. Sci. USA* 82:5112–16
140. Petit, A., David, C., Dahl, G., Ellis, J. G., Guyon, P., et al. 1983. Further extension of the opine concept: plasmids in *Agrobacterium rhizogenes* cooperate for opine degradation. *Mol. Gen. Genet.* 190:204–14
141. Petit, A., Delhaye, S., Tempé, J., Morel, G. 1970. Recherches surs les guanidines des tissus de crown gall. Mise en évidence d'une relation biochemique spécifique entre les souches d'*Agrobacterium tumefaciens* et les tumeurs qu'elles induisent. *Physiol. Veg.* 8:205–13
142. Pueppke, S. G. 1984. Adsorption of bacteria to plant surfaces. In *Plant-Microbe Interactions: Molecular Genetic Perspectives,* ed. T. Kosuge, E. W. Nester, 1:215–61. New York: MacMillan
143. Pueppke, S. G., Benny, U. K. 1981. Induction of tumors on *Solanum tuberosum* by *Agrobacterium tumefaciens:* quantitative analysis, inhibition by carbohydrates, and virulence of selected strains. *Physiol. Plant Pathol.* 18:169–79
144. Pueppke, S. G., Benny, U. K. 1984. Adsorption of tumorigenic *Agrobacterium tumefaciens* cells to susceptible potato tuber tissues. *Can. J. Microbiol.* 30:1030–37
145. Pueppke, S. G., Hawes, M. C. 1985. Understanding the binding of bacteria to plant surfaces. *Trends Biotechnol.* 3:310–13
146. Puvanisarajah, V., Schell, F. M., Stacey, G., Douglas, C. J., Nester, E. W. 1985. Role for 2-linked-β-D-glucan in the virulence of *Agrobacterium tumefaciens*. *J. Bacteriol.* 164:102–6
147. Rao, S. S., Lippincott, B. B. Lippincott, J. A. 1982. *Agrobacterium* adherence involves the pectic portion of the host cell wall and is sensitive to the degree of pectin methylation. *Physiol. Plant.* 56:374–80
148. Ream, L. W., Gordon, M. P., Nester, E. W. 1983. Multiple mutations in the T-regions of the *Agrobacterium tumefaciens* tumor inducing plasmid. *Proc. Natl. Acad. Sci. USA* 80:1660–64
149. Rogowsky, P. M., Close, T. J., Chimera, J. A., Shaw, J. J., Kado, C. I. 1987. Regulation of *vir* genes of *Agrobacterium tumefaciens* plasmid pTiC58. *J. Bacteriol.* 169:5101–12
150. Rossignol, G., Dion, P. 1985. Octopine, nopaline and octopinic acid utilization in *Pseudomonas*. *Can. J. Microbiol.* 31:68–74
151. Rubin, R. A. 1986. Genetic studies on the role of octopine T-DNA border regions in crown gall tumor formation. *Mol. Gen. Genet.* 202:312–20
152. Schell, J. 1987. Transgenic plants as tools to study the molecular organization of plant genes. *Science* 237:1176–83
153. Schröder, G., Waffenschmidt, S., Weiler, E. W., Schröder, J. 1984. The T-region of Ti plasmids codes for an enzyme synthesizing indole-3-acetic acid. *Eur. J. Biochem.* 138:387–91

154. Schroth, M. N., Weinhold, A. R., McCain, A. H., Hildebrand, D. C., Ross, N. 1971. Biology and control of *Agrobacterium tumefaciens*. *Hilgardia* 40:537–52

155. Shaw, C. H., Watson, M. D., Carter, G. H., Shaw, C. H. 1984. The right hand copy of the nopaline Ti-plasmid 25 bp repeat is required for tumor formation. *Nucleic Acids Res.* 12:6031–41

156. Simpson, R. B., O'Hara, P. J., Kwok, W., Montoya, A. L., Lichtenstein, C., et al. 1982. DNA from the A6S/2 crown gall tumor contains scrambled Ti plasmid sequences near its junctions with plant DNA. *Cell* 29:1005–14

157. Slightom, J. L., Durand-Tardif, M., Jouanin, L., Tepfer, D. 1986. Nucleotide sequence analysis of TL-DNA of *Agrobacterium rhizogenes* agropine type plasmid: identification of open-reading frames. *J. Biol. Chem.* 261:108–21

158. Slightom, J. L., Jouanin, L., Leach, F., Drong, R. F., Tepfer, D. 1985. Isolation and identification of TL-DNA/plant junctions in *Convolvulus arvensis* transformed by *Agrobacterium rhizogenes*. *EMBO J.* 4:3069–77

159. Spielmann, A., Simpson, R. B., 1986. T-DNA structure in transgenic tobacco plants with multiple independent integration sites. *Mol. Gen. Genet.* 205:34–41

160. Stachel, S., An, G., Flores, C., Nester, E. W., 1985. A Tn3 *lacZ* transposon for the random generation of β-galactosidase gene fusions: application to the analysis of gene expression of *Agrobacterium*. *EMBO J.* 4:891–98

161. Stachel, S. E., Messens, E., Van Montagu, M., Zambryski, P. 1985. Identification of the signal molecules produced by wounded plant cells that activate T-DNA transfer in *Agrobacterium tumefaciens*. *Nature* 318:624–29

162. Stachel, S. E., Nester, E. W. 1986. The genetic and transcriptional organization of the *vir* region of the A6 Ti plasmid of *Agrobacterium tumefaciens*. *EMBO J.* 5:1445–54

163. Stachel, S. E., Nester, E. W., Zambryski, P. C. 1986. A plant cell facter induces *Agrobacterium tumefaciens vir* gene expression. *Proc. Natl. Acad. Sci. USA* 83:379–83

164. Stachel, S. E., Timmerman, B., Zambryski, P. 1986. Generation of single-stranded T-DNA molecules during the initial stages of T-DNA transfer from *Agrobacterium tumefaciens* to plant cells. *Nature* 322:706–12

165. Stachel, S. E., Timmerman, B., Zambryski, P. 1987. Activation of *Agrobac-*terium tumefaciens *vir* gene expression generates multiple single stranded T-strand molecules from the pTiA6 T-region: requirement for the 5' *virD* gene products. *EMBO J.* 6:857–63

166. Stachel, S. E., Zambryski, P. C. 1986. *virA* and *virG* control the plant induced activation of the T-DNA transfer process of *A. tumefaciens*. *Cell* 46:325–33

167. Stachel, S. E., Zambryski, P. 1986. *Agrobacterium tumefaciens* and the susceptible plant cell: A novel adaptation of extracellular recognition and DNA conjugation. *Cell* 47:155–57

168. Tanimoto, E., Douglas, C., Halperin, W. 1979. Factors affecting crown gall tumorigenesis in tuber slices of Jerusalem artichoke (*Helianthus tuberosus* L.). *Plant Physiol.* 63:989–94

169. Tempé, J., Goldman, A. 1982. Occurrence and biosynthesis of opines. See Ref. 92, pp. 427–49

170. Tempé, J., Petit, A. 1982. Opine utilization by *Agrobacterium*. See Ref. 92, pp. 451–59

171. Thomashow, L. S., Reeves, S., Thomashow, M. F. 1984. Crown gall oncogenesis: evidence that a T-DNA gene from the *Agrobacterium* Ti plasmid pTiA6 encodes an enzyme that catalyzes synthesis of indoleacetic acid. *Proc. Natl. Acad. Sci. USA* 81:5071–75

172. Thomashow, M. F., Hugly, S., Buchholz, W. G., Thomashow, L. S. 1986. Molecular basis for the auxin-independent phenotype of crown gall tumor tissues. *Science* 231:616–18

173. Thomashow, M. F., Karlinsey, J. E., Marks, J. R., Hurlbert, R. E. 1987. Identification of a new virulence locus in *Agrobacterium tumefaciens* that affects polysaccharide composition and plant cell attachment. *J. Bacteriol.* 169:3209–16

174. Thomashow, M. F., Nutter, R., Montoya, A. L., Gordon, M. P., Nester, E. W. 1980. Integration and organization of Ti-plasmid sequences in crown gall tumors. *Cell* 19:729–39

175. Thomashow, M. F., Nutter, R., Postle, K, Chilton, M.-D., Blattner, F. R., et al. 1980. Recombination between higher plant DNA and the Ti plasmid of *Agrobacterium tumefaciens*. *Proc. Natl. Acad. Sci. USA* 77:6448–52

176. Trach, K. A., Chapman, J. W., Piggot, P. J., Hoch, J. A. 1985. Deduced product of the stage O sporulation gene *spoOF* shares homology with the SpoOA, OmpR, and SfrA proteins. *Proc. Natl. Acad. Sci. USA* 82:7260–64

177. Usami, S., Morkawa, S., Takebe, I.,

Machida, Y. 1987. Absence in monocotyledonous plants of the *diffusible* plant factors inducing T-DNA circularization and *vir* gene expression in *Agrobacterium. Mol. Gen. Genet.* 209:221–26

178. Van Lijsebettens, M., Inzé, D., Schell, J., Van Montagu, M. 1986. Transformed cell clones as a tool to study T-DNA integration mediated by *Agrobacterium tumefaciens. J. Mol. Biol.* 188:129–45

179. Van Onckelen, H., Prinsen, E., Inzé, D., Rudelsheim, P., Van Lijsebettens, M., et al. 1986. *Agrobacterium* T-DNA gene codes for tryptophan 2-monooxygenase activity in tobacco crown gall cells. *FEBS Lett.* 198:357–60

180. Van Onckelen, H., Rudelsheim, P., Hermans, R., Horemans, S., Messens, E., et al. 1984. Kinetics of endogenous cytokinin, IAA and ABA levels in relation to the growth and morphology of tobacco crown gall tissue. *Plant Cell Physiol.* 25:1017–25

181. Veluthambi, K., Jayaswal, R. K., Gelvin, S. B. 1987. Virulence genes A, G, and D mediate the double stranded border cleavage of T-DNA from *Agrobacterium* Ti plasmid. *Proc. Natl. Acad. Sci. USA* 84:1881–85

181a. Veluthambi, K., Ream, W., Gelvin, S. B. 1988. Virulence genes, borders, and overdrive generate single-stranded T-DNA molecules from the A6 Ti plasmid of *Agrobacterium tumefaciens. J. Bacteriol.* 170:1523–32

182. Waelkens, F., Maris, M., Verreth, C., Vanderleyden, J., Van Gool, A. 1987. *Azospirillum* DNA shows homology with *Agrobacterium* chromosomal virulence genes. *FEMS Microbiol. Lett.* 43:241–46

183. Wang, K., Herrara-Estrella, L., Van Montagu, M., Zambryski, P. 1984. Right 25 bp terminus sequence of the nopaline T-DNA is essential for and determines direction of DNA transfer from *Agrobacterium* to the plant genome. *Cell* 38:455–62

184. Wang, K., Stachel, S. E., Timmerman, B., Van Montagu, M., Zambryski, P. 1987. Site specific nick in the T-DNA border sequence as a result of *Agrobacterium vir* gene expression. *Science* 235:587–91

185. Ward, J. E., Akiyoshi, D. E., Regier, D., Datta, A., Gordon, M. P., et al. 1988. Characterization of the *virB* operon from an *Agrobacterium tumefaciens* Ti plasmid. *J. Biol. Chem.* 263:5804–14

186. Weiler, E. W., Schröder, J. 1987. Hormone genes and the crown gall disease. *TIBS* 12:271–75

187. Weiler, E. W., Spanier, K. 1981. Phytohormones in the formation of crown gall tumors. *Planta* 153:326–37

188. Wenzler, H., Meins, F. Jr. 1986. Mapping regions of the maize leaf capable of proliferation in culture. *Protoplasma* 131:103–5

189. Whatley, M. H., Bodwin, J. S., Lippincott, B. B., Lippincott, J. A. 1976. Role for *Agrobacterium* cell envelope lipopolysaccharide in infection site attachment. *Infect. Immun.* 13:1080–83

190. White, F. F., Ghidossi, G., Gordon, M. P., Nester, E. W. 1982. Tumor induction by *Agrobacterium rhizogenes* involves the transfer of plasmid DNA to the plant genome. *Proc. Natl. Acad. Sci. USA* 79:3193–97

191. Willetts, N., Wilkins, B. 1984. Processing of plasmid DNA during bacterial conjugation. *Microbiol. Rev.* 48:24–41

192. Willmitzer, L., DeBeukeleer, M., Lemmers, M., Van Montagu, M., Schell, J. 1980. The Ti-plasmid derived T-DNA is present in the nucleus and absent from plastids of plant crown gall cells. *Nature* 287:359–61

193. Willmitzer, L., Dhaese, P., Schreider, P. H., Schmalenbach, W., Van Montagu, M., et al. 1983. Size, location, and polarity of T-DNA encoded transcripts in nopaline crown gall tumors: evidence for common transcripts present in both octopine and nopaline tumors. *Cell* 32:1045–56

194. Willmitzer, L., Simons, G., Schell, J. 1982. The TL-DNA of octopine crown gall tumors codes for seven well defined polyadenylated transcripts. *EMBO J.* 1:139–46

195. Winans, S. C., Allenza, P., Stachel, S. E., McBride, K. E., Nester, E. W. 1987. Characterization of the *virE* operon of the *Agrobacterium* Ti plasmid pTiA6. *Nucleic Acids Res.* 15:825–37

196. Winans, S. C., Ebert, P. R., Stachel, S. E., Gordon, M. P., Nester, E. W. 1986. A gene essential for *Agrobacterium* virulence is homologous to a family of positive regulatory loci. *Proc. Natl. Acad. Sci. USA* 83:8278–82

197. Wullems, G. J., Molendijk, L., Ooms, G., Schilperoort, R. A. 1981. Differential expression of crown gall tumor markers in transformants obtained after *in vitro Agrobacterium tumefaciens*–induced transformation of cell wall regenerating protoplasts derived from *Nicotiana tabacum. Proc. Natl. Acad. Sci. USA* 78:4344–48

198. Yadav, N. S., Postle, K., Saiki, R. K.,

Thomashow, M. F., Chilton, M.-D. 1980. DNA of crown gall teratoma is covalently joined to host plant DNA. *Nature* 287:458–61

199. Yadav, N. S., Vanderleyden, J., Bennett, D. R., Barnes, W. M., Chilton, M.-D. 1982. Short direct repeats flank the T-DNA on a nopaline Ti plasmid. *Proc. Natl. Acad. Sci. USA* 79:6322–26

200. Yamada, T., Palm, C. J., Brooks, B., Kosuge, T. 1985. Nucleotide sequences of the *Pseudomonas savastanoi* indoleacetic acid genes show homology with *Agrobacterium tumefaciens* T-DNA. *Proc. Natl. Acad. Sci. USA* 82:6522–26

201. Yamamoto, A., Iwahashi, M., Yanofsky, M. F., Nester, E. W., Takebe, I., et al. 1987. The promoter proximal region in the *virD* locus of *Agrobacterium tumefaciens* is necessary for the plant-inducible circularization of T-DNA. *Mol. Gen. Genet.* 206:174–77

202. Yanofsky, M., Lowe, B., Montoya, A., Rubin, R., Krul, W., et al. 1985. Molecular and genetic analysis of factors controlling host range in *Agrobacterium tumefaciens*. *Mol. Gen. Genet.* 201: 237–46

203. Yanofsky, M. F., Nester, E. W. 1986.

Molecular characterization of a host-range determining locus from *Agrobacterium tumefaciens*. *J. Bacteriol.* 168: 244–50

204. Yanofsky, M. F., Porter, S. G., Young, C., Albright, L. M., Gordon, M. P., et al. 1986. The *virD* operon of *Agrobacterium tumerfaciens* encodes a site-specific endonuclease. *Cell* 47:471–77

205. Zambryski, P., Depicker, A., Kruger, K., Goodman, H. 1982. Tumor induction by *Agrobacterium tumefaciens:* analysis of the boundaries of T-DNA. *J. Mol. Appl. Genet.* 1:361–70

206. Zambryski, P., Holsters, M., Kruger, K., Depicker, A., Schell, J., et al. 1980. Tumor DNA structure in plant cells transformed by *A. tumefaciens. Science* 209:1385–91

206a. Zorreguieta, A., Geremia, R. A., Cavaignac, S., Cangelosi, G. A., Nester, E. W., et al. 1988. Identification of the product of an *Agrobacterium tumefaciens* chromosomal virulence gene. *Mol. Plant-Microbe Interaction* In press

207. Zorreguieta, A., Ugalde, R. A. 1986. Formation in *Rhizobium* and *Agrobacterium* spp. of a 235 kilodalton protein intermediate in β-D(1-2) glucan synthesis. *J. Bacteriol.* 167:947–51

Ann. Rev. Microbiol. 1988. 42:607–25
Copyright © 1988 by Annual Reviews Inc. All rights reserved

SIMIAN IMMUNODEFICIENCY VIRUSES

Ronald C. Desrosiers

New England Regional Primate Research Center, Harvard Medical School, Southborough, Massachusetts 01772

CONTENTS

607

SUBCLASSIFICATION OF RETROVIRUSES

Retroviruses can be subclassified according to simple morphologic criteria (Figure 1). Such morphologic classification is consistent with the biology of the viruses in each group as well as with molecular and genetic characteristics. Thus, morphology remains a convenient means for classifying retroviruses.

Type C Oncoviruses

Type C oncoviruses, which probably have been the most studied of the retroviruses, include the familiar murine, feline, and avian leukemia and sarcoma viruses. These viruses bud as a crescent shape at the cellular membrane without preformed subviral particles visible in the cytoplasm or at the membrane prior to budding. Type C oncoviruses can have either an endogenous or an exogenous mode of transmission and are noted for their capacity to induce a variety of cancers in animals. Mechanisms of cell growth transformation that have been identified to date include oncogene capture (acutely transforming viruses) and insertional mutagenesis (chronic tumor viruses). The human T-cell leukemia viruses type I and type II (HTLV-I and HTLV-II), the simian counterpart STLV-I, and bovine leukemia virus (BLV) apparently form a distinct subgroup of the type C oncoviruses having an

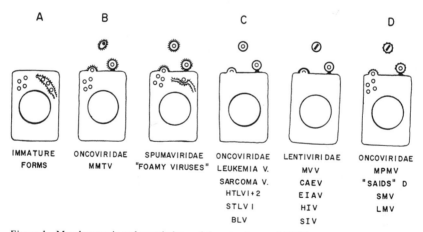

Figure 1 Morphogenesis and morphology of the retroviruses. *MMTV*, mouse mammary tumor virus; *HTLV-1*, human T-lymphotropic virus type 1; *STLV-1*, simian T-lymphotropic virus type 1; *BLV*, bovine leukemia virus; *MVV*, Maedi-Visna virus of sheep; *CAEV*, caprine arthritis encephalitis virus; *EIAV*, equine infectious anemia virus; *HIV*, human immunodeficiency virus; *SIV*, simian immunodeficiency virus; *MPMV*, Mason-Pfizer monkey virus; *SMV*, squirrel monkey type D retrovirus; *LMV*, langur monkey type D retrovirus. Modified from a figure of Norval W. King.

additional gene for transactivation of transcription *(tat)*. It is generally believed that this *tat* gene has an important role in cell growth transformation and in the slowly evolving leukemias and lymphomas observed in vivo.

Type B Oncoviruses

Mouse mammary tumor virus (MMTV) is the only well characterized member of the B type retroviruses. With MMTV, preformed type A subviral particles are found in the cytoplasm and at the membrane prior to budding of the virus. Mature, extracellular particles usually have an eccentric nucleoid.

Type D Oncoviruses

The type D viruses also have preformed type A subviral particles in the cytoplasm and at the membrane prior to budding of the virus. Mature particles, however, have a cylindrical or rod-shaped nucleoid. Mason-Pfizer monkey virus (MPMV), originally isolated in 1969 by Chopra & Mason (5) from a rhesus macaque, is the prototype of the D type retroviruses.

Lentiviruses

The lentiviruses have been little studied in the past because of difficulties in growing virus in vitro and because of the need for cumbersome experimental animals. This group includes the classic ungulate lentiviruses (Maedi-Visna virus of sheep, caprine arthritis encephalitis virus, and equine infectious anemia virus) and viruses that cause acquired immunodeficiency syndrome (AIDS) in humans (human immunodeficiency virus, HIV) and monkeys (simian immunodeficiency virus, SIV). More recently, lentiviruses have been isolated from cats (45) and from cattle (23). All lentiviruses have characteristic morphogenesis and morphology that distinguish them from the other retrovirus groups. The lentiviruses bud in the manner of type C viruses, with perhaps a slightly thicker crescent at the membrane. Mature, extracellular particles typically have a cylindrical or rod-shaped nucleoid.

In addition to morphologic similarity, the lentiviruses share a distinctive genome organization and biological properties. Not only do they have the standard retroviral *gag, pol,* and *env* genes, but they also have additional unusual genes not found with other retroviruses. These genes include *3'orf* or *f, sor* or *q, tat, art,* and *r*. Although it is not yet clear whether all lentiviruses have each of these genes, the lentiviruses that have been examined to date do contain at least some of them. Furthermore, lentiviruses are not oncogenic, but characteristically produce long-term, persistent infections in the host which eventually lead to chronic, debilitating diseases. All lentiviruses appear to replicate in and persist in cells of the monocyte/macrophage lineage, although other cell types (such as $CD4^+$ lymphocytes for HIV and SIV) can also be permissive.

Both type D retroviruses (8, 13, 40, 41, 49) and lentiviruses (2, 9, 11) have been associated with acquired immunodeficiency syndromes in macaque monkeys. Diseases in macaque monkeys caused by these two distinct viruses have been referred to as simian AIDS or SAIDS. Macaque type D retrovirus isolates have often been referred to as SAIDS retrovirus types 1 and 2 (SRV-1, SRV-2). The simian lentivirus was originally called STLV-III but is now generally referred to as SIV. SIV is clearly similar to HIV in genome organization and genetic relatedness (see below), while the type D retroviruses are not. When examining the scientific literature in this field, readers need to discern carefully which virus, SIV or type D, is being described.

ANIMAL MODELS FOR AIDS

Animal models currently available for studying AIDS can be classified into four groups based on increasing direct relevance to the disease in humans: (*a*) infections with nonlentivirus retroviruses that can induce chronic disease, including immunodeficiency; (*b*) infections with ungulate, feline, and possibly other lentiviruses; (*c*) HIV-related lentivirus infections of nonhuman primates (SIV); and (*d*) HIV infection of chimpanzees and gibbons.

Retroviruses That Are Not Lentiviruses

Included in this category are feline leukemia virus (FeLV) and the macaque type D retroviruses. In addition to subclinicial infection and tumors that generally take a long time to develop, FeLV can be responsible for a wide spectrum of chronic nonneoplastic conditions in cats including immunosuppression, wasting, severe diarrhea, and anemia. Similarly, type D retroviruses have been associated with immunodeficiency and chronic wasting syndromes, opportunistic infections, necrotizing gingivitis, and retroperitoneal fibromatosis in United States Regional Primate Research Centers (8, 13, 40, 41, 49). It is hoped that continued investigation of these and similar systems will provide insight into basic and molecular mechanisms of retrovirus-induced chronic disease and that these will be relevant to our understanding of AIDS. It is also hoped that related studies will identify useful strategies for the prevention or treatment of these diseases.

Ungulate, Feline, and Other Lentiviruses

The lentiviruses and their respective natural hosts provide important, relevant animal models for AIDS because of the demonstrated genetic relatedness of these viruses to HIV and their basic similarities in biological properties. Ungulate lentiviruses share a sequence homology and a similar genomic organization with HIV that other retroviruses do not have (23, 48). Furthermore, the long-term persistent infection and chronic disease produced by ungulate lentiviruses provide an opportunity to understand the basic mechanisms underlying persistent infection and pathogenesis. The similarities between these ungulate lentiviruses and HIV suggest that therapeutic approaches and vaccination strategies successful for control of the former should be similarly applicable to AIDS (15).

Infection of cats with the newly described feline lentivirus shows particular promise as an animal model for AIDS. This virus has relevance as a lentivirus, it has a small animal as host, it is lymphotropic, and it is associated with a disease in cats that is similar to AIDS. It was first isolated by Pedersen and colleagues in Davis, California (45). Specific pathogen-free cats inoculated with this virus have developed transient fever, leukopenia, and persistent lymphadenopathy. As of 9 mo, however, none of the experimentally infected cats have developed the AIDS-like syndrome seen with natural infection by this agent (N. Pedersen, personal communication). It is hoped that detailed characterization of the virus and further development of the feline lentivirus model will proceed rapidly.

Since lentiviruses have now been found in humans, nonhuman primates, four different ungulates, and cats, it would not be surprising to find wild mice naturally harboring a lentivirus. Research should be encouraged to find such lentiviruses in feral mice. The ability to study lentivirus-induced chronic disease in laboratory mice would enormously facilitate both practical and basic aspects of AIDS research.

HIV-Related Lentiviruses of Old World Primates (SIV)

SIV, the focus of this review, is more closely related to HIV than all other nonhuman viruses identified to date. Most importantly, SIV is able to induce in macaques long-term persistent infection and chronic disease remarkably similar to AIDS in humans (see below). Macaques breed well in captivity and are readily available for experimental studies.

HIV Infection of Chimpanzees and Gibbons

The ideal animal model for AIDS would be one in which HIV infects and induces an AIDS-like disease in a common laboratory animal. Since only humans, chimpanzees, and gibbons have been found to be susceptible to HIV infection (1, 20–22; P. Markham, personal communication), an ideal animal model for AIDS has not yet been found. Although chimpanzees and possibly gibbons will be needed for research related to AIDS vaccine development, they are available to researchers in only very limited numbers. Furthermore, although chimpanzees and gibbons have become persistently infected following HIV inoculation, none have died from the infection and none have developed disease suggestive of AIDS. The ability of HIV to infect other species, particularly common New World primates, warrants further investigation (42).

ISOLATION AND PROPERTIES OF SIV

Macaques

Macaques are Asian Old World primates. SIV was first isolated from captive rhesus macaques *(Macaca mulatta)* at the New England Regional Primate Research Center (NERPRC) (9). The original isolates were called simian T-lymphotropic virus type III (STLV-III), but these viruses are now generally referred to as SIV based on recommendations of the International Committee on the Taxonomy of Viruses. SIV has now also been isolated from macaques at the Washington Regional Primate Research Center (WRPRC) (2), California Regional Primate Research Center (CRPRC) (L. J. Lowenstine & M. Gardner, personal communication), and Delta Regional Primate Research Center (DRPRC) (43). At NERPRC four independent SIV isolates have been obtained from four different rhesus macaques, and one was obtained from a cynomolgus macaque *(Macaca fascicularis)*. Of the five SIV-infected macaques from which virus was recovered, four had clearly become infected while at NERPRC and were never inoculated with tissue or virus-containing material (11). These four were not in contact with any animals other than macaques. At WRPRC the single isolate of SIV was from a pig-tailed macaque *(Macaca nemestrina)*. At CRPRC SIV was isolated from stored tissue samples from stump-tailed macaques *(Macaca arctoides)*. The macaques at DRPRC had received tissue material from infected mangabeys, so these isolates are almost certainly of mangabey origin.

Macaques from which SIV was isolated very often died with clinical signs and necropsy findings reminiscent of AIDS. Opportunistic infections, diarrhea, wasting, and lymphoid depletion have been observed associated with SIV infection. Curiously, lymphomas and lymphoproliferative diseases (LPD) have often been associated with macaque SIV infections at all four of the regional primate centers described above. At NERPRC three cases of lymphoma and LPD were found associated with eleven known SIV infections (11); these lymphomas have not been associated with infection by STLV-I or an Epstein-Barr virus (EBV)–related virus (16). At DRPRC the lymphomas were associated with infection by an EBV-related virus of macaques (43, 46). In humans, AIDS-associated lymphomas can be either EBV negative or EBV positive (38).

SIV isolates from macaques (SIVmac) have growth properties and a CD4$^+$ cell tropism similar to those of HIV (2, 9, 34). SIV can be isolated and grown using continuous human CD4$^+$ tumor cell lines such as HUT-78, H9, and MT4, and it is often cytopathic for such cells. Normal human and macaque peripheral blood T lymphocytes growing in the presence of interleukin-2 (IL-2) have also been used successfully for isolation and growth of SIVmac.

Replication of SIVmac has not been detected in monolayer fibroblast or epithelial cells such as Vero, HeLa, rhesus fibroblast, or owl monkey kidney cells. SIVmac can be isolated from the adherent blood cell population of infected macaques and grown in adherent monocyte/macrophage cultures; attempts to grow SIVmac in the continuous U937 monocyte/macrophage cell line have so far been unsuccessful (M. D. Daniel & R. C. Desrosiers, unpublished).

Although SIVmac can replicate efficiently in T-lymphocyte cells of monkey origin, including those from macaques, baboons, and gibbons (2, 34), human cells have generally been used for technical reasons. First, a variety of continuously growing CD4$^+$ tumor cell lines of human origin are available. Such lines are extremely useful for continuous production of virus of high titer for laboratory use. The only similar lines of nonhuman primate origin are infected with STLV-I. Secondly, normal peripheral blood lymphocytes (PBL) of human origin are much cleaner than those of macaque origin and grow longer in culture in the pesence of IL-2. Macaques harbor other agents such as foamy viruses (a retrovirus) and parasites in their blood. Normal PBL growing in IL-2 are often the most sensitive indicator cells for SIV recovery from infected animals.

When PBL were fractionated into T4-enriched and T8-enriched cell populations, preferential replication occurred in the former population (9). Furthermore, OKT4A and OKT4B monoclonal antibodies were able to block infection by SIVmac in PBL cultures (34).

Antigenic relatedness was originally demonstrated between HIV-1 and SIVmac isolates from NERPRC using radioimmune precipitation–SDS polyacrylamide gel electrophoresis (32). Antigenic relatedness between HIV and these and other SIVmac isolates has now also been demonstrated using Western blot hybridization, competitive radioimmunoassay, and radioimmune precipitation–SDS gel analysis (2, 16, 43). Serologic cross-reactivity of the group-specific antigen *(gag)* protein is easily demonstrated, while cross-reactivity of the *env* protein is weak.

Green Monkeys

Green monkeys *(Cercopithecus aethiops)* are African Old World primates. Subsequent to the initial isolation of SIV from macaques, SIV-reactive antibodies were demonstrated in 30–50% of green monkeys from Africa (31). Soon thereafter Kanki et al (29) reported the isolation of an SIV called STLV-III $_{AGM}$ from African green monkeys. However, these initial STLV-III$_{AGM}$ isolates were apparently derived from SIVmac strain 251–infected cell cultures.

The following can be cited as evidence that isolates called STLV-III$_{AGM}$ by others were actually derived from SIVmac-251–infected cell cultures. (*a*)

Five original isolates of STLV-III$_{AGM}$ (29) and two isolates of HTLV-IV described by Kanki et al (30) did not vary as one would expect when restriction endonuclease maps were compared (36). Furthermore, the restriction maps of the HTLV-IV isolates were identical to those of STLV-III$_{AGM}$, and 99% conservation was observed in the regions that were sequenced (27, 28, 36). Such conservation would be highly unusual for authentic independent isolates. Not only do independent isolates of HIV or SIV vary one from another, even independent isolates of HIV-1 from the same individual can vary (26). Successive SIVmac isolates following in vivo transmission can also differ in their restriction endoculease maps (35). (b) Although six isolates of SIVmac at NERPRC were readily distinguished, one of them, SIVmac-251, had an identical restriction endonuclease map with STLV-III$_{AGM}$ and HTLV-IV and had greater than 99% sequence identity within *env* when compared to published sequences for STLV-III$_{AGM}$ and HTLV-IV (35). (c) SIVmac-251 was one of two isolates originally provided to Kanki et al in October and November, 1984 (14). (d) SIVmac-251 was the prototype used by Kanki et al in their early studies (32). (e) The nucleotide sequence of HIV-2 isolated by Clavel et al (6, 7) has only about 75% nucleotide conservation when compared to SIVmac (24). Restriction-site conservation between HIV-2ROD and SIVmac-142 is less than 20%. Isolates of HIV-2 obtained by Clavel et al (7) from humans in West Africa do exhibit intratypic variation. Furthermore, a virus isolated by Hahn et al (25) from the same group of prostitutes that was the source of Kanki et al's isolates (30) resembled LAV-2 (HIV-2) of the French researchers, not HTLV-IV. (f) More recent isolates of SIV from green monkeys (35, 44; M. D. Daniel, Y. Li, Y. M. Naidu, P. J. Durda, D. K. Schmidt, et al, submitted for publication) are quite distinct from SIVmac.

Since authentic SIV isolates from green monkeys have only recently been obtained, important data on biological and molecular properties and pathogenic potentials will only now emerge. These SIVagm isolates do have a lentivirus morphology, growth properties, and CD4$^+$ cell tropism similar to those of SIVmac. CD4$^+$ cell lines that support the replication of SIVagm, however, differ from those that support SIVmac replication. These SIVagm isolates are related to but distinct from SIVmac and HIV-2.

Mangabeys

Mangabeys are African Old World primates. SIV isolates have been obtained from captive mangabeys at DRPRC (43) and at Yerkes Regional Primate Research Center (YRPRC) (19) and from zoo mangabeys by CRPRC researchers (39). These SIV mangabey isolates have a lentivirus morphology, growth properties, and CD4$^+$ cell tropism similar to those described above, and they demonstrate antigenic cross-reactivity with HIV.

Other Nonhuman Primates

SIV has been isolated from mandrills (genus *Mandrillus* or *Papio*) from Gabon (M. Hayami, personal communication).

SEQUENCE, GENOME ORGANIZATION, AND PROTEIN PRODUCTS

The complete nucleotide sequence of SIVmac isolate 142 from NERPRC has been published (3). The SIVmac genome is similar in its basic organization to the HIV-1 and HIV-2 genomes and is generally typical of a lentivirus (Figure 2). In addition to the standard *gag, pol*, and *env* genes, SIV macaque has open reading frames for *sor* (q), 3'orf (f), tat, art, and r. Two differences, however, are worth noting. SIV has a premature translation termination signal in *env*, resulting in a truncated form of the transmembrane protein. This premature stop codon was previously observed in SIV (28) and in HIV-2 (24). It is not an artifact of cloning, since it is present in infectious molecular clones. The second difference is an additional open reading frame between

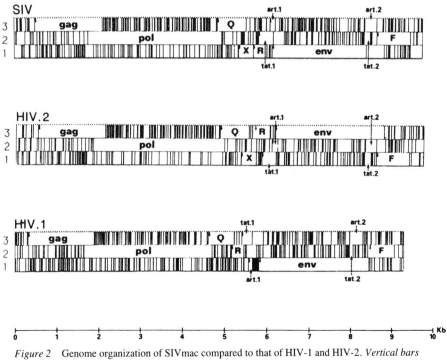

Figure 2 Genome organization of SIVmac compared to that of HIV-1 and HIV-2. *Vertical bars* represent translation termination signals. Reprinted from Reference 3.

Table 1 Comparison of overall nucleotide sequence identity of SIVmac, HIV-1, and HIV-2[a]

	HIV-1	HIV-2
SIVmac	40%	75%
HIV-1		40%

[a]Derived from Reference 3.

sor and *r* called *x*, which is absent in HIV-1 and differently positioned relative to *r* in HIV-2.

Sequences have also been published from two laboratories for the STLV-IIIagm isolates discussed above (18, 28, 36). These sequences appear to be of SIVmac isolate 251 (35; see above). They are > 95% homologous with the SIVmac 142 sequence over most of the genome, although sequence homologies in some regions of *env* are less than 95%.

Table 1 summarizes the overall nucleotide conservation in HIV-1, HIV-2, and SIVmac. Both SIVmac and HIV-2 share only about 40% overall nucleotide identity with HIV-1. SIVmac and HIV-2 exhibit about 75% overall nucleotide conservation. Thus HIV-2 is indeed more closely related to the monkey virus SIVmac than it is to the other human lentivirus, HIV-1. Amino acid conservation is highest in *gag* and *pol* and lower in *env* (Table 2).

The in-frame stop codon within *env* results in a transmembrane protein of 30–32 kd rather than the 41-kd protein that is observed with HIV-1 and that would be observed with SIV in the absence of the premature stop codon. The sequences beyond the stop codon in the same reading frame are still preserved as an open reading frame, which suggests that the carboxyl terminal portion of the transmembrane protein sequences might be differentially utilized by a number of possible mechanisms. The three nucleotides of the stop codon exactly precede presumed splice acceptor sites (in different reading frames) for *tat* and *art*. In one infectious SIVmac clone isolated at NERPRC, a single base change converting the stop codon to a glutamine codon has been compensated by a another single base change three nucleotides to the left

Table 2 Comparison of amino acid identities in different regions of the genome

| vc | | | *env*[b] | | | | | | | |
SIVmac vs	*gag*	*pol*	EGP	TMP	*f*	*q*	*r*	*x*	*tat*	*art*
HIV-1	57.3	59	34	49.1	45.7	37.8	57	—	48.1	33.3
HIV-2	87	83.6	75.4	74.1	59.8	73.4	70.3	85.7	59.2	61.0

[a]Percentage amino acid identities (3).
[b]EGP, external glycoprotein; TMP, transmembrane protein.

converting a glutamine to a stop codon (35). Removal of the stop codon in an infectious molecular clone by site-specific mutagenesis resulted in inability of the virus to replicate significantly (V. Hirsch & J. I. Mullins, personal communication).

The *tat* sequences in SIVmac appear to be functional since SIVmac can transactivate its own long terminal repeat (LTR). Transactivation by SIVmac is best with its own LTR, quite good with that of HIV-2, and poor with that of HIV-1 (17; M. Emerman, personal communication; G. Viglianti & J. Mullins, personal communication).

Restriction endonuclease maps have been derived for five SIV isolates from rhesus macaques *(M. mulatta)* and one SIV isolate from a cynomolgus macaque *(M. fascicularis)* at the NERPRC (35). All six isolates were quite similar but easily distinguished. The single isolate from a cynomolgus macaque (SIVmac-186) was the most different. Restriction-site conservation ranged from 86% with 18 of 21 sites conserved between SIVmac isolates 239 and 309 to only 56% conservation with only 14 of 25 sites conserved between isolates 251 and 186.

Information is currently lacking on the genetic similarity of SIVmac isolates from other regional primate research centers.

Preliminary nucleotide sequence data of one isolate of SIVagm indicate that this SIVagm is approximately equidistant to HIV-1, HIV-2, and SIVmac; the closest relatedness was observed in *pol,* but even in this region SIVagm had only about 60% amino acid identity with the other viruses (M. Hayami, personal communication).

Information on the genome organization of SIVmangabey and its genetic relatedness to other SIV isolates has not yet been obtained.

Three different infectious molecular clones of SIVmac have been isolated at NERPRC (Y. M. Naidu, H. W. Kestler, Y. Li, C. V. Butler, D. P. Silva, et al, manuscript in preparation), one from WRPRC (R. Benveniste, personal communication), and one from CRPRC (P. Luciw, personal communication). Infectious molecular clones described as HTLV-4 are probably clones of SIVmac isolate 251 (36). These infectious molecular clones, which are now being intensively characterized, will be enormously useful for future studies of the molecular basis of persistence, pathogenicity, tropism, and species specificity. They will also be useful for vaccine studies.

PREVALENCE AND DIVERSITY

Macaques

Natural infection of macaques in captivity with SIV appears to be rare. In a 1986 survey at the NERPRC only three macaques of 848 (0.35%) had antibodies to SIV (11). From analysis of stored sera and previous studies, only 11 natural (i.e. not transmitted via inoculation of virus or tissue materials) macaque infections could be documented at NERPRC. From these, five SIVmac isolates were obtained. Infection of macaques at WRPRC appears similarly rare (M. Thouless, B. Morton, & R. Benveniste, personal communication). While infection at CRPRC is also rare or nonexistent, an outbreak of lymphoma and AIDS in a group of *M. arctoides* about 15 years ago has been traced to SIV infection (L. J. Lowenstine & M. Gardner, personal communication).

The rarity of infection of macaques in captivity with SIV raises questions as to whether macaques are indeed the natural hosts of the viruses being isolated or whether they have acquired these viruses from some other species in the process of importation or while in captivity. No antibodies to SIV were detected in sera from wild caught macaques in two surveys (39, 44). However, in neither case was SIVmac used as antigen; Lowenstine et al (39) used HIV as antigen and Ohta et al (44) used SIVagm as antigen in indirect immunofluorescence tests. These weakly cross-reactive antigens may not have been sensitive enough to detect antibodies if they were present, especially if the percentage of antibody-positive animals was low. Furthermore, it is possible that infection of macaques in their natural environment is confined to selected species or geographic habitats; once in captivity, SIV-infected macaques could transmit virus to other macaques to which they are not normally exposed.

The origin of SIV in captive macaque monkey colonies is thus not known at this time. Although it is possible that animals of some other genus could have been a source for captive macaque infection, we do not know if this is actually so. If macaques at New England, Washington, and California primate centers have acquired SIV from some other species, the potential source species has not been identified. The SIVagm and SIVmangabey isolates that have been examined to date are so different from the SIVmac isolates that such viruses would not be a likely source for these captive macaque infections. Much more work is needed to detail the genetic makeup of SIV isolates from a variety of feral as well as captive Old World primates.

Green Monkeys

Antibodies to SIV in African green monkeys were first reported by Kanki et al (31); 42% of green monkeys in this survey were SIV antibody positive. These findings were confirmed by Ohta et al (44), who found 26% SIV antibody positives, and by Daniel et al (11), who found approximately 30% SIV antibody positives. Most of these green monkeys originated from Kenya and Ethiopia. Thus a high percentage of green monkeys in their natural habitats in Kenya and Ethiopia have been naturally infected with their own SIV.

Green monkeys inhabit virtually all of sub-Saharan Africa. At present there is no information regarding SIV prevalence in green monkeys from different geographic regions of Africa. Furthermore, since authentic SIVagm isolates currently being analyzed have all been derived from green monkeys from Kenya and Ethiopia (44; M. Hayami, personal communication; M. D. Daniel, Y. Li, Y. M. Naidu, P. J. Durda, D. K. Schmidt, et al, submitted for publication), future work will need to examine genetic variability in SIVagm derived from different geographic regions.

In the seventeenth and eighteenth centuries, sailors involved in the slave trade often brought African green monkeys with them to the New World; since that time, free-roaming troups of green monkeys have become established on three islands in the Caribbean: Barbados, Saint Kitts, and Nevis (12). Daniel et al (11) found no antibodies to SIV in the 160 Caribbean green monkeys from Barbados that were examined. It is tempting to conclude from these studies that SIV was introduced into the African green monkey population at some time since the seventeenth century. However, such a conclusion is not justified for at least two reasons. First, it is not known how complex the genetic stock was that gave rise to the free-roaming green monkeys present in the Caribbean area today. It is possible that all Caribbean green monkeys are descendents of only a few animals, and these could all have been SIV negative. Second, it is not known from what region of Africa the green monkeys that populated these Caribbean islands came. SIV prevalence might vary significantly among green monkeys in different regions of Africa.

Mangabeys

Attempts to demonstrate SIV antibodies in mangabeys in Africa have so far been unsuccessful (39, 44; P. Fultz, personal communication). In contrast, 70% or more of adult sooty mangabeys in captivity are antibody positive. As with SIV infection in macaques, it is not known whether mangabeys are actually the natural host of the SIV being isolated or, if they are not, what the source species is.

Other Nonhuman Primates

Serological surveys have indicated that additional nonhuman primate species in the wild and in captivity are infected with SIV (39, 44). SIV-positive sera have been identified in *Cercopithecus* relatives of the green monkey as well as in feral mandrills. Guenons, talapoins, and colobus monkeys in American zoos have also been found to be seropositive. One difficulty in performing such broad surveys is the limitation imposed by the specificity of the antigen being used; antibodies to SIV of some species may be difficult to detect using SIVmac, SIVagm, or HIV as antigen.

ANIMAL STUDIES

Macaques have been infected experimentally by inoculation of SIVmac at NERPRC (4, 10, 33, 37, 47) and WRPRC (R. Benveniste, personal communication) and by inoculation of SIVmangabey at DRPRC (L. Martin, G. Baskin & M. Murphey-Corb, personal communication) and YRPRC (P. Fultz and H. McClure, personal communcation). Results from NERPRC are summarized in the most detail here, but the results of others have been similar.

1. All 27 rhesus macaques inoculated with undiluted or diluted SIVmac at NERPRC became persistently infected. SIV was recovered from peripheral blood mononuclear cells on numerous occasions months after SIV inoculation. SIV grown in human PBL and in the HUT-78 cell line has successfully infected macaques and induced AIDS (10).
2. Of 27 macaques infected experimentally, 20 (74%) have died to date. The median time to death was 266 days; the shortest was 62 days; the longest was 1061 days (10).
3. Macaques that died displayed clinical signs and necropsy findings remarkably similar to those of AIDS in humans (37). Macaques have exhibited diarrhea and wasting; some animals have lost up to 60% of their body weight. Immune abnormalities, including decreases in peripheral T4 lymphocyte numbers and mitogen-proliferative responses, have been noted. Opportunistic infections have been common; agents that have been identified include *Pneumocystis carinii,* cytomegalovirus, *Cryptosporidium, Candida,* adenovirus, and *Mycobacterium avium intracellulare.* Approximately 50% of SIV-inoculated macaques have died with a characteristic granulomatous encephalitis very similar to that often seen in human AIDS.
4. The ability of macaques to survive SIVmac infection has been directly correlated with the strength of the antibody response (10, 33).
5. There has been no correlation between the dose of virus inoculum and either the strength of the antibody response or the clinical outcome (10).

Although lymphomas and lymphoproliferative disease have often been noted in macaques naturally infected with SIV (see above), these have not been seen in any experimentally infected macaques at NERPRC. Lymphomas have been observed, however, in macaques inoculated with SIV at DRPRC (M. Murphey-Corb, personal communication) and CRPRC (L. J. Lowenstine, personal communication). Although brain lesions have been observed in about 50% of experimentally infected macaques at NERPRC, they were curiously not seen in any of the 11 naturally infected macaques that died there (N. W. King, personal communication). DRPRC researchers appear to have one strain of SIV that is more apt to induce brain lesions than other strains they have used. AT WRPRC, disease and death were induced in both rhesus and pig-tailed macaques, but no infection was observed in inoculated baboons (R. Benveniste, personal communication).

Because of difficulties in getting seronegative mangabeys, researchers at DRPRC and YRPRC have tried to establish experimental infections with SIVmangabey in macaques. Macaques became infected with SIVmangabey following inoculation, and many died with lesions consistent with lentivirus-induced disease as described above. At YRPRC, SIV recovered from a macaque dying from SIVmangabey-induced disease is now reproducibly killing macaques in 8–12 days with severe diarrhea and giant cell disease.

PROSPECTS

The similarities between SIV and HIV are many (Table 3). Both HIV and SIV are tropic for cells bearing the CD4 antigen, and both viruses can be cytopathic for these cells. Both viruses also have the characteristic morphogenesis and morphology typical of the lentivirus subfamily of retroviruses. While the major modes of transmission of HIV are known to be intimate sexual contact and the exchange of blood products, the primary routes of SIV transmission are not known. HIV-1 is highly unusual when compared to nonlentivirus retroviruses in that it contains five extra genes in addition to the standard *gag, pol,* and *env;* SIVmac also contains these five genes, and it may contain one more if the *x* open reading frame is shown to code for a protein. Both HIV and SIV have the ability to transactivate their own LTRs. The question of rapid emergence of antigenic variants has not yet been adequately addressed. HIV causes AIDS in humans and SIV causes AIDS in monkeys.

Study of the HIV-related lentiviruses of primates is most important for the following three areas of AIDS research.

Table 3 Properties of human and simian lentiviruses

Property	HIV-1	SIV
Cell tropism	CD4$^+$	CD4$^+$
Cytopathic	Yes	Yes
Morphology	Lentivirus	Lentivirus
Mode of transmission	Sex, blood	?
Additional genes	Yes (5)	Yes (6)
Transactivation	Yes	Yes
Rapid emergence of antigentic variants	?	?
AIDS in monkeys	No	Yes
AIDS in humans	Yes	?

Origins

Quantitation of genetic relatedness among human and nonhuman primate lentiviruses should lead to better understanding of the history and evolution of this group of viruses and may provide insight into the origins of viruses that cause AIDS in humans. We need to know which species harbor SIV viruses in the wild and we need to determine the precise genetic makeup of these viruses. In the case of green monkeys, whose habitat is all of sub-Saharan Africa, we need to know whether there is geographic or subspecies variation in the genetic makeup of the SIV that they harbor.

Pathogenesis

SIV infection of macaques is ideally suited for basic research issues requiring in vivo infection. This system can thus be used to better understand the pathogenesis of disease and the mechanisms of tropism, persistence, and pathogenicity. Understanding of these issues at the basic and molecular levels will, in the long run, fuel vaccine and therapy efforts. Knowledge of the mechanisms by which these viruses are able to persist without control by the host immune system is most critical for understanding the chronic disease.

Treatment and Prevention

In vivo SIV infection of macaques is also ideally suited for (a) investigation of therapies and combination therapies (e.g. antiviral agents and biological response modifiers) for AIDS treatment and (b) evaluation of the best approaches for vaccine development. If development of a vaccine for AIDS were to be easy, all we would probably need would be the HIV-chimpanzee system. However, it is not going to be easy. The macaque SIV system will allow investigators to compare a variety of different vaccine approaches with numerous variables in each, which is probably impossible with chimpanzees or humans. We can then take the best of what we learn in macaques to chimpanzee or human trials.

Acknowledgments

I thank Norval W. King for permission to publish Figure 1. The following investigators kindly provided information prior to publication: Masanori Hayami; James Mullins; Michael Murphey-Corb and Lou Martin; Margaret Thouless, Bill Morton, and Raoul Benveniste; Patricia Fultz and Harold McClure; Linda Lowenstine and Murray Gardner. The manuscript was prepared for publication by Nancy Adams and Joanne Newton. I thank N. W. King and B. J. Blake for critical reading of the manuscript.

Literature Cited

1. Alter, H. J., Eichberg, J. W., Masur, H., Saxinger, W. C., Gallo, R., et al. 1984. Transmission of HTLV-III infection from human plasma to chimpanzees: an animal model for AIDS. *Science* 226:549–52
2. Benveniste, R. E., Arthur, L. O., Tsai, C.-C., Sowder, R., Copeland, T. D., et al. 1986. Isolation of a lentivirus from a macaque with lymphoma: comparison with HTLV-III/LAV and other lentiviruses. *J. Virol.* 60:483–90
3. Chakrabarti, L., Guyader, M., Alizon, M., Daniel, M. D., Desrosiers, R. C., et al. 1987. Sequence of simian immunodeficiency virus from macaque and its relationship to other human and simian retroviruses. *Nature* 328:543–47
4. Chalifoux, L. V., Ringler, D. J., King, N. W., Sehgal, P. K., Desrosiers, R. C., et al. 1987. Lymphadenopathy in macaques experimentally infected with the simian immunodeficiency virus (SIV). *Am. J. Pathol.* 128:104–10
5. Chopra, H. C., Mason, M. M. 1970. A new virus in a spontaneous mammary tumor of a rhesus monkey. *Cancer Res.* 30:2081–86
6. Clavel, F., Guetard, D., Brun-Vezinet, F., Chamaret, S., Rey, M.-A., et al. 1986. Isolation of a new human retrovirus from West African patients with AIDS. *Science* 233:343–46
7. Clavel, F., Guyader, M., Guetard, D., Salle, M., Montagnier, L., et al. 1986. Molecular cloning and polymorphism of the human immune deficiency virus type 2. *Nature* 324:691–95
8. Daniel, M. D., King, N. W., Letvin, N. L., Hunt, R. D., Sehgal, P. K., et al. 1984. A new type D retrovirus isolated from macaques with an immunodeficiency syndrome. *Science* 223:602–5
9. Daniel, M. D., Letvin, N. L., King, N. W., Kannagi, M., Sehgal, P. K., et al. 1985. Isolation of T-cell tropic HTLV-III-like retrovirus from macaques. *Science* 228:1201–4
10. Daniel, M. D., Letvin, N. L., Sehgal, P. K., Hunsmann, G., Schmidt, D. K., et al. 1987. Long-term persistent infection of macaque monkeys with the simian immunodeficiency virus. *J. Gen. Virol.* 68:3183–89
11. Daniel, M. D., Letvin, N. L., Sehgal, P. K., Schmidt, D. K., Silva, D. P., et al. 1988. Prevalence of antibodies to three retroviruses in a captive colony of macaque monkeys. *Int. J. Cancer* 41:601–8
12. Denham, W. W. 1981. History of green monkeys in the West Indies. Part I. Migration from Africa. Part II. Population dynamics of Barbadian monkeys. *J. Barbados Mus. Hist. Soc.* 36:211–28, 353–71
13. Desrosiers, R. C., Daniel, M. D., Butler, C. V., Schmidt, D. K., Letvin, N. L., et al. 1985. Retrovirus D/New England and its relation to Mason-Pfizer monkey virus. *J. Virol.* 54:552–60
14. Desrosiers, R. C., Daniel, M. D., Letvin, N. L., King, N. W., Hunt, R. D. 1987. Origins of HTLV-4. *Nature* 327:107
15. Desrosiers, R. C., Letvin, N. L. 1987. Animal models for acquired immunodeficiency syndrome. *Rev. Infect. Dis.* 9:438–46
16. Desrosiers, R. C., Letvin, N. L., King, N. W., Hunt, R. D., Blake, B. J., et al. 1987. Three retroviruses infecting macaques at the New England Regional Primate Research Center. *UCLA Symp. Mol. Cell. Biol.* (NS) 43:451–66
17. Emerman, M., Guyader, M., Montagnier, L., Baltimore, D., Muesing, M. A. 1987. The specificity of the human immunodeficiency virus type 2 transactivator is different from that of human immunodeficiency virus type 1. *Embo J.* 6:3755–60
18. Franchini, G., Gurgo, C., Guo, H.-G., Gallo, R. C., Collalti, E., et al. 1987. Sequence of simian immunodeficiency virus and its relationship to the human immunodeficiency viruses. *Nature* 328:539–43
19. Fultz, P. N., McClure, H. M., Anderson, D. C., Swenson, R. B., Anand, R., et al. 1986. Isolation of a T-lymphotropic retrovirus from naturally infected sooty mangabey monkeys (*Cercocebus atys*). *Proc. Natl. Acad. Sci. USA* 83:5286–90
20. Fultz, P. N., McClure, H. M., Swenson, R. B., McGrath, C. R., Brodie, A., et al. 1986. Persistent infection of chimpanzees with human T-lymphotropic virus type III/lymphadenopathy-associated virus: a potential model for acquired immunodeficiency syndrome. *J. Virol.* 58:116–24
21. Gajdusek, D. C., Amyx, H. L., Gibbs, C. J. Jr., Asher, D. M., Rodgers-Johnson, P., et al. 1985. Infection of chimpanzees by human T-lymphotropic retroviruses in brain and other tissues from AIDS patients (letter). *Lancet* 1:55–56
22. Gajdusek, D. C., Amyx, H. L., Gibbs,

C. J. Jr., Asher, D. M., Yanagihara, R. T., et al. 1984. Transmission experiments with human T-lymphotropic retroviruses and human AIDS tissue (letter). *Lancet* 1:1415–16

23. Gonda, M. A., Braun, M. J., Carter, S. G., Kost, T. A., Bess, J. W. Jr., et al. 1987. Characterization and molecular cloning of a bovine lentivirus related to human immunodeficiency virus. *Nature* 330:388–91

24. Guyader, M., Emerman, M., Sonigo, P., Clavel, F., Montagnier, L., et al. 1987. Genome organization and transactivation of the human immunodeficiency virus type 2. *Nature* 326:662–69

25. Hahn, B. H., Kong, L. I., Lee, S.-W., Kumar, P., Taylor, M. E., et al. 1987. Relation of HTLV-4 to simian and human immunodeficiency-associated viruses. *Nature* 330:184–86

26. Hahn, B. H., Shaw, G. M., Taylor, M. E., Redfield, R. R., Markham, P. D., et al. 1986. Genetic variation in HTLV-III/LAV over time in patients with AIDS or at risk for AIDS. *Science* 232:1548–53

27. Hirsch, V., Riedel, N., Kornfeld, H., Kanki, P. J., Essex, M., et al. 1986. Cross-reactivity to human T-lymphotropic virus type III/lymphadenopathy-associated virus and molecular cloning of simian T-cell lymphotropic virus type III from African green monkeys. *Proc. Natl. Acad. Sci. USA* 83:9754–58

28. Hirsch, V., Riedel, N., Mullins, J. I. 1987. The genome organization of STLV-3 is similar to that of the AIDS virus except for a truncated transmembrane protein. *Cell* 49:307–19

29. Kanki, P. J., Alroy, J., Essex, M. 1985. Isolation of T-lymphotropic retrovirus related to HTLV-III/LAV from wild-caught African green monkeys. *Science* 230:951–54

30. Kanki, P. J., Barin, F., M'Boup, S., Allan, J. S., Romet-Lemonne, J. L., et al. 1986. New human T-lymphotropic retrovirus related to simian T-lymphotropic virus type III (STLV-III$_{AGM}$). *Science* 232:238–43

31. Kanki, P. J., Kurth, R., Becker, W., Dreesman, G., McLane, M. F., et al. 1985. Antibodies to simian T-lymphotropic retrovirus type III in African green monkeys and recognition of STLV-III viral proteins by AIDS and related sera. *Lancet* 1:1330–32

32. Kanki, P. J., McLane, M. F., King, N. W. Jr., Letvin, N. L., Hunt, R. D., et al. 1985. Serologic identification and characterization of a macaque T-lymphotropic retrovirus closely related to HTLV-III. *Science* 228:1199–201

33. Kannagi, M., Kiyotaki, M., Desrosiers, R. C., Reimann, K. A., King, N. W., et al. 1986. Humoral immune responses to T cell tropic retrovirus simian T lymphotropic virus type III in monkeys with experimentally induced acquired immune deficiency–like syndrome. *J. Clin. Invest.* 78:1229–36

34. Kannagi, M., Yetz, J. M., Letvin, N. L. 1985. *In vitro* growth characteristics of simian T-lymphotropic virus type III. *Proc. Natl. Acad. Sci. USA* 82:7053–57

35. Kestle., H. W. III, Li, Y., Naidu, Y. M., Butler, C. V., Ochs, M. F., et al. 1988. Genetic variability among simian immunodeficiency virus isolates. *Nature* 331:619–22

36. Kornfeld, H., Riedel, N., Viglianti, G. A., Hirsch, V., Mullins, J. I. 1987. Cloning of HTLV-4 and its relation to simian and human immunodeficiency viruses. *Nature* 326:610–13

37. Letvin, N. L., Daniel, M. D., Sehgal, P. K., Desrosiers, R. C., Hunt, R. D., et al. 1985. Induction of AIDS-like disease in macaque monkeys with T-cell tropic retrovirus STLV-III. *Science* 230:71–73

38. Levine, A. M., Gill, P. S., Meyer, P. R., Burkes, R. L., Ross, R. et al. 1985. Retrovirus and malignant lymphoma in homosexual men. *J. Am. Med. Assoc.* 254:1921–25

39. Lowenstine, L. J., Pedersen, N. C., Higgins, J., Pallis, K. C., Uyeda, A., et al. 1986. Seroepidemiologic survey of captive Old-World primates for antibodies to human and simian retroviruses, and isolation of a lentivirus from sooty mangabeys *(Cercocebus atys)*. *Int. J. Cancer* 38:563–74

40. Marx, P. A., Bryant, M. L., Osborn, K. G., Maul, D. H., Lerche, N. W., et al. 1985. Isolation of a new serotype of simian acquired immune deficiency syndrome type D retrovirus from Celebes black macaques *(Macaca nigra)* with immune deficiency and retroperitoneal fibromatosis. *J. Virol.* 56:571–78

41. Marx, P. A., Maul, D. H., Osborn, K. G., Lerche, N. W., Moody, P., et al. 1984. Simian AIDS: isolation of a type D retrovirus and transmission of the disease. *Science* 223:1083–86

42. McClure, M. O., Sattentau, Q. J., Beverley, P. C. L., Hearn, J. P., Fitzgerald, A. K., et al. 1987. HIV infection of primate lymphocytes and conservation of the CD4 receptor. *Nature* 330:487–89

43. Murphey-Corb, M., Martin, L. N., Rangan, S. R. S., Baskin, G. B., Gormus, B. J., et al. 1986. Isolation of an HTLV-III–related retrovirus from macaques with simian AIDS and its possible origin in asymptomatic mangabeys. *Nature* 321:435–37

44. Ohta, M., Ohta, K., Mori, F., Nishitani, H., Saida, T. 1986. Sera from patients with multiple sclerosis react with human T cell lymphotropic virus-I gag proteins but not env proteins—Western blotting analysis. *J. Immunol.* 137:3440–43

45. Pedersen, N. C., Ho, E. W., Brown, M. L., Yamamoto, J. K. 1987. Isolation of a T-lymphotropic virus from domestic cats with an immunodeficiency-like syndrome. *Science* 235:790–93

46. Rangan, S. R. S., Martin, L. N., Bozelka, B. E., Wang, N., Gormus, B. J.

1986. Epstein-Barr virus–related herpesvirus from a rhesus monkey *(Macaca mulatta)* with malignant lymphoma. *Int. J. Cancer* 38:425–32

47. Ringler, D. J., Hancock, W. W., King, N. W., Letvin, N. L., Daniel, M. D., et al. 1987. Immunophenotypic characterization of the cutaneous exanthem of SIV-infected rhesus monkeys. *Am. J. Pathol.* 126:199–207

48. Sonigo, P., Alizon, M., Staskus, K., Klatzmann, D., Cole, S., et al. 1985. Nucleotide sequence of the visna lentivirus: relationship to the AIDS virus. *Cell* 42:369–82

49. Stromberg, K., Benveniste, R. E., Arthur, L. O., Rabin, H., Giddens, W. E. Jr., et al. 1984. Characterization of exogenous type D retrovirus from a fibroma of a macaque with simian AIDS and fibromatosis. *Science* 224:289–92

Ann. Rev. Microbiol. 1988. 42:627–56
Copyright © 1987 by Annual Reviews Inc. All rights reserved

INTERGENERIC COAGGREGATION AMONG HUMAN ORAL BACTERIA AND ECOLOGY OF DENTAL PLAQUE[1,2]

Paul E. Kolenbrander

Laboratory of Microbial Ecology, National Institute of Dental Research, National Institutes of Health, Bethesda, Maryland 20892

CONTENTS

[1]This review is dedicated to my friend and colleague, Dr. Charles Lewis Wittenberger, who died unexpectedly on June 27, 1987. He encouraged and supported my study of coaggregation from the time of its inception.

[2]The US Government has the right to retain a nonexclusive, royalty-free license in and to any copyright covering this paper.

INTRODUCTION

Coaggregation is defined as the recognition between surface molecules on two different bacterial cell types so that a mixed-cell aggregate is formed. It is likely that most if not all human oral bacteria participate in coaggregation, and that these interactions have a major role in bacterial adherence and colonization on host surfaces. The interactions are highly specific in that only certain cell types are partners. Viability is not required of either partner; cell walls of the two cell types clump together, and isolated adhesins agglutinate partner cells. This review is limited to covering the extent of coaggregation among oral bacteria and its potential role in the accretion of dental plaque. It does not cover adherence of oral bacteria to solid surfaces or to epithelial cells, except where this is related to tissue tropism and oral microbial ecology. A number of reviews have covered adherence and oral ecology (10, 46, 47, 53), and two reviews have focused on microbial lectins among oral bacteria (15, 16).

Coaggregation appears to be mediated by specific recognition between complementary lectin-carbohydrate molecules on the participating partners. The recognition is immediate; coaggregates are usually visible within seconds of cell mixing. Coaggregation is a bacteria-bacteria interaction; it is not bacterial adherence to a eukaryotic cell or to an inanimate surface, nor is it caused by soluble molecules or suspended substances.

Coaggregation in the oral ecosystem appears to be unique, since interactions among bacteria from other ecosystems involve nutritional (55, 119) or predatory (132) relationships. Coaggregates also differ from the massive cellular aggregates of metabolic consortia observed in anaerobic biodegradative ecosystems (126, 159) because the latter do not require cell-to-cell contact, but rather involve metabolic communication within an extracellular biopolymer matrix. In fact, except in conjugation systems, reports of bacteria-bacteria interactions outside of the oral microflora are few, and the adherence properties of the participating cell types have not been described (8, 45).

BACKGROUND AND BASIC PROPERTIES

Coaggregation was first reported in 1970 by Gibbons & Nygaard (50), who called it interbacterial aggregation. They paired all combinations of 23 strains and found that 5 of 253 pairs formed large coaggregates. Some other pairs weakly coaggregated, but most paired suspensions remained evenly turbid. Historically, the term coaggregation was coined to describe a clumping phenomenon that occurred when sucrose-grown cells of *Streptococcus mutans* or *Streptococcus sanguis* were mixed with *Actinomyces viscosus* (9). The term was used to distinguish that intergeneric kind of clumping from the dextran-mediated intraspecies aggregation of *A. viscosus*. A few other reports of interbacterial aggregation appeared (38, 65), but it was not until 1978, when McIntire et al (107) advanced the idea that coaggregation was mediated by lectin-carbohydrate interactions, that extensive investigation of coaggregation began.

Coaggregation Assay Procedures

Coaggregation is readily observed with the naked eye (21, 65). Screening for the ability of large numbers of strains to coaggregate can take place in small test tubes (Figure 1) or microtiter wells (66, 79). In the most convenient assay, equal volumes (0.1 ml) of a dense suspension (about 10^9 cells ml^{-1}) of each cell type (Figure 1, tubes 1 and 2) are mixed in a 10×75 mm test tube. The extent of coaggregation is scored on a scale of zero (unchanged evenly turbid suspension) to four plus (maximal clumping with large settling coaggregates). An example of a score of four plus is shown in Figure 1. Immediately after mixing (tube 3) coaggregates are visible, which settle within seconds to the bottom of the tube (tube 4). After adding lactose (tube 5) or EDTA the suspension returns to an evenly turbid consistency, becoming indistinguishable from the suspensions of the individual partner strains (tubes

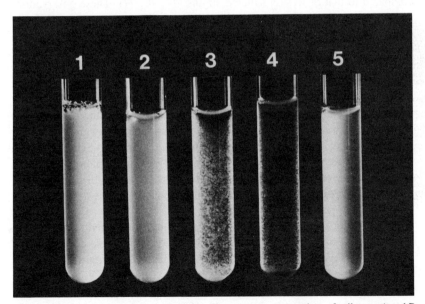

Figure 1   The visual assay for coaggregation. Homogeneous suspensions of cell types A and B are shown before mixing (tubes 1 and 2, respectively) and immediately after mixing of equal volumes (tube 3). Within seconds coaggregates settle to the bottom of the tube, leaving a clear supernatant (tube 4). Addition of inhibitor (sugar or chelating agent) completely reverses the interaction (tube 5).

1 and 2). The tubes in this figure clearly depict the ease of scoring strong coaggregations. Although only semiquantitative, this procedure yields an accurate representation of the results obtained quantitatively by turbidometric methods (38, 50, 64, 107) or radioactivity measurements (69, 82; P. Kolenbrander, unpublished observations). The visual assay is also useful for initial screening of inhibitory compounds such as sugars (21, 68) or monoclonal antibodies that block coaggregation (152).

Factors Affecting the Ability of Cells to Coaggregate

Coaggregation is visibly strongest when about equal numbers of the partners are present, presumably because of the development of a large network of interacting cells, each with two or more recognition sites on its cell surface. Coaggregates are visible but much smaller when the cell ratio is 10:1 (9, 75). At higher ratios, coaggregates are in the form of rosettes that are detectable only with a microscope or a radioactivity-based assay (70).

Cell age and culture medium do not seem to be factors in most coaggregating pairs. Neither physiological aging (9, 21, 38, 50, 65) nor calendar aging (culture transfers, freezing and thawing of stock cultures, and shipping of cultures to other laboratories) (21) affects the ability of cells to coaggregate compared to that of freshly isolated strains (74, 76, 77). These observations suggest a genetic stability of encoding regions for coaggregation-mediating surface components. In addition, the observation that the streptococci and actinomyces suspensions used in the first survey in 1978–1979 (21) still coaggregate after having been stored in 0.02% sodium azide buffer at 4°C for 10 years indicates the functional stability of these mediators (P. Kolenbrander & R. Andersen, unpublished observations). Cells routinely grown in a tryptone-based medium containing yeast extract, Tween 80, and glucose (99) have been used as a standard (21) for comparison with cells grown in other media including modified Schaedler broth (13), commercial Schaedler broth (151), trypticase soy broth with yeast extract and neopeptone (139), and Todd-Hewitt broth (P. Kolenbrander & R. Andersen, unpublished observations). In most cases cells grown in any of these media have exhibited similar coaggregation properties.

Either one or both partners of all coaggregating pairs tested were found to be inactivated by heating (85°C for 30 min) (65) or protease digestion, and many pairings are prevented in the presence of lactose or certain other simple sugars. The inhibitory effect of lactose on the *A. viscosus* T14V–*S. sanguis* 34 pair, coupled with the inactivation of the actinomyces cells but not the streptococci by heat, led McIntire et al (107) to propose that this coaggregation was mediated by a lectin on *A. viscosus* that recognized a carbohydrate on *S. sanguis*. Since that original report, sugar-inhibitable coaggregations among at least seven genera of oral bacteria have been noted (21, 61, 64, 67, 71, 72, 76, 85).

Coaggregation occurs in buffer or in saliva (75). Although it may be somewhat reduced by certain salivary molecules (78), it is equally sensitive to the same inhibitors that are effective with buffer-suspended cells (14). Coaggregation has no distinct pH optimum over a range of values from about 5.0 to 9.0 (9, 21, 50, 65), but at pH 4.5 addition of Ca^{2+} markedly enhances coaggregation. Chelating agents such as EDTA and EGTA inhibit many coaggregations.

Coaggregation Among the Primary Colonizers

Among the first organisms to repopulate a freshly cleaned tooth are *S. sanguis* and related bacteria as well as *A. viscosus* and related bacteria (37, 52, 131, 133). Three surveys of the coaggregation properties of these bacteria were performed. First, various laboratory stock cultures including some from the original reporting laboratory (50) were tested (21). All of the nine human *A. viscosus* and nine *Actinomyces naeslundii* strains coaggregated with some of the *S. sanguis* and *Streptococcus mitis* strains. None of the four animal *A. viscosus* strains or four strains of *Actinomyces israelii* or strains of *Actinomyces parabifidus* and *Actinomyces odontolyticus* showed coaggregation patterns similar to those of the human *A. viscosus* and *A. naeslundii* strains. Of 25 strains of *S. sanguis* and *S. mitis*, 17 coaggregated with actinomyces.

The second survey dealt with 473 fresh isolates from dental plaque of 16 persons (76, 77). The survey was conducted as a double blind in which the taxonomic identification was done independently of the test for coaggregation properties. The isolates were separated into two morphological groups, and all rod-shaped isolates were tested for their ability to coaggregate with a set of coaggregation-positive streptococci from the previous survey. Similarly, the spherical isolates were tested with coaggregation-positive actinomyces from the previous survey. Among the rod-shaped bacteria (presumed to be actinomyces-related isolates), strains belonging to the genera *Actinomyces, Arachnia, Bifidobacterium,* and *Bacterionema* were found, but only members of *Actinomyces* coaggregated with streptococci. *A. viscosus* and *A. naeslundii* isolates were identified in 15 of the 16 samples. In 14 of these, one or more of the isolates coaggregated with the *S. sanguis–S. mitis* test groups delineated in the previous survey. In fact, all 24 *A. viscosus* and 37 of 43 *A. naeslundii* isolates coaggregated. Of these 61 isolates, 87% exhibited lactose-inhibitable coaggregation with the reference streptococcal strains.

Of the 110 spherical isolates, 30 were identified as either *S. mutans, Streptococcus anginosus constellatus,* or *Veillonella parvula,* and none coaggregated with actinomyces. The remaining 80 isolates were identified as *S. sanguis, S. mitis, Streptococcus MG-intermedius,* or *Streptococcus morbillorum,* and 61% coaggregated with actinomyces reference strains. In addition, nearly 90% (43 of 49 isolates) exhibited lactose-inhibitable coaggregation. The high percentage of both actinomyces and streptococci that exhibited lactose-sensitive coaggregation suggests that these coaggregations are prevalent in vivo and are mediated by a network of lectin-carbohydrate interactions similar to those noted with stock cultures of actinomyces and streptococci.

The third survey focused on the coaggregation properties of actinomyces and streptococci from the same site on the tooth surface (74). All were recent isolates from subgingival samples. All 34 actinomyces isolates (*A. viscosus* and *A. naeslundii*) coaggregated with reference streptococci, but only 8 of 14 (57%) recent *S. sanguis* isolates coaggregated with the reference actinomyces. However, four of the six nonreactive streptococci coaggregated with actinomyces from the same site in a new, highly specific coaggregation pattern for *S. sanguis*. The actinomyces-streptococcus pairs from seven of the ten tooth sites sampled from eight patients coaggregated. The other three sites either contained streptococci that failed to coaggregate with any actinomyces or contained the highly specific streptococci and no corresponding actinomyces partner.

Collectively, the results of all three surveys indicate that nearly all strains of *A. viscosus, A. naeslundii, S. sanguis, S. mitis,* and *S. morbillorum* coaggregate. Most fresh isolates exhibited the same coaggregation patterns as stock cultures maintained in various laboratory collections for up to 17 years, which reflects the genetic stability of these mediators of coaggregation. Thus, these coaggregations are not random, but have highly specific patterns, which are shown diagrammatically in Figure 2. From more than 100 strains tested of each cell type, only six streptococcal coaggregation groups (1–6 in the figure) and six actinomyces groups (A–F) emerged. The interactions of each group are represented by a set of complementary symbols. The symbols with a line attached represent heat-inactivated, protease-sensitive surface components and their complementary symbols represent components that are insensitive to these treatments. For example, the highly specific streptococcal group 6, which was discovered in the survey of same-site coaggregating pairs, coaggregates only with members of actinomyces coaggregation group D. The streptococci are heat inactivated; the actinomyces are not. Identical symbols on different coaggregation groups, such as those on streptococcal coaggregation groups 1, 3, 4, 5, and 6 that complement the obelisk-shaped structure on group D actinomyces, do not indicate identical surface components on the different cells; identical symbols are used to keep the diagram as simple as possible.

Additional information suggests that most of these identical symbols represent functionally closely related molecules. For example, coaggregation-defective (Cog⁻) mutants of group 6 streptococcus that were selected for failing to coaggregate with group D actinomyces did not coaggregate with any other actinomyces (P. Kolenbrander & R. Andersen, unpublished results).

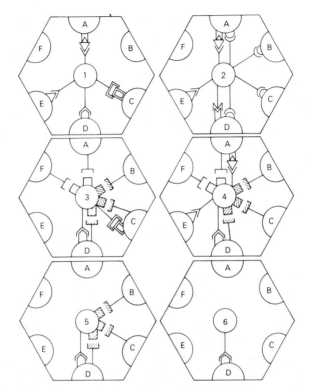

Figure 2 Diagrammatic representation of the coaggregations between members of the six streptococcal coaggregation groups *(numbered circles)* and the six actinomyces coaggregation groups *(lettered semicircles).* See text for details.

Furthermore, Cog⁻ mutants of group D actinomyces that were selected for failing to coaggregate with group 6 streptococci were also unable to coaggregate with group 1 streptococci. While their coaggregation with group 2 streptococci was unchanged, their coaggregation with streptococcal groups 3, 4, and 5 became lactose inhibitable (2). This unmasking of a lactose-inhibitable coaggregation (rectangular symbols) presumably occurs through the loss of the ability to synthesize the structure represented by the obelisk-shaped symbol. Prior to isolation of these Cog⁻ mutants, the lactose inhibition could be observed only if the group 3, 4, or 5 streptococci were heat inactivated or protease treated, which would have destroyed the structure represented by the obelisk complement on the streptococcus. Since the coaggregations between these three streptococcal groups and group D actinomyces are bimodal (prior heat treatment or protease digestion of both cell types is required to prevent coaggregation completely), inactivation of the mode that could not be inhibited by lactose (obelisk symbols) left the other, lactose-inhibitable mode (rectangular symbols) available for study. Inactivation of only the actinomyces yielded coaggregations that were visibly identical to those observed with control unheated cells because the lactose-inhibitable coaggregation was masked by the complementary structures that could not be inhibited by lactose. While bimodal coaggregations can only be prevented by inactivating both cell types before mixing, unimodal coaggregating pairs such as group 6 streptococci–group D actinomyces or group F actinomyces–group 3 streptococci were totally abolished by heating the first member of each pair at 85°C for 30 min.

The results from the surveys of the coaggregation properties of oral streptococci and actinomyces form the foundation on which the surveys of other oral bacteria are based. These are the primary colonizers of a freshly cleaned tooth surface, and other bacteria can adhere to them. Over 100 strains of *S. sanguis* and *S. morbillorum* have been screened, and 64% were coaggregation positive with either *A. viscosus* or *A. naeslundii*. This percentage increased to 86% when an actinomyces from the same site was included in the screening process (74). In total from the three surveys, 65 of the 76 coaggregation-positive isolates (86% of total tested) exhibited lactose-inhibitable coaggregations.

For the actinomyces the numbers and percentages are even higher. All 40 of the *A. viscosus* strains tested in the three surveys belonged to coaggregation group A, which means they exhibited lactose-inhibitable coaggregations. Furthermore, 95% of all actinomyces isolates were coaggregation positive, and of these, 93% exhibited lactose-inhibitable coaggregation with streptococci. On the basis of these surveys, it appears that most of the *S. sanguis* and *S. morbillorum* isolates, most of the *A. naeslundii* isolates, and most if not all of the *A. viscosus* isolates from human dental plaque samples express coaggregation mediators similar or identical to those that characterize the six streptococcal and six actinomyces coaggregation groups.

The high specificity among streptococci and their partners was seen in an extensive survey of 70 potential partners for 22 strains of mutans streptococci (31). Only two strains of *A. naeslundii* and one strain of *A. odontolyticus* exhibited strong coaggregations, and only with *S. cricetus*. The paucity of partners was also observed in two earlier surveys (21, 50).

Genera of Oral Bacteria that Coaggregate

Using representative strains of the six actinomyces and six streptococcal coaggregation groups as a reference set of potential partners, a large number of strains within a genus could be surveyed for their ability to coaggregate. Currently, we use representative strains of *Actinobacillus, Actinomyces, Bacteroides, Capnocytophaga, Fusobacterium, Rothia, Streptococcus,* and *Veil-*

lonella as potential partners for each new group of oral bacteria to be studied. In addition, members of *Bacterionema* (114), *Propionibacterium* (76), *Peptostreptococcus* (P. Kolenbrander & R. Andersen, unpublished results), *Eubacterium* (100), *Haemophilus* (85), and *Selenomonas* (71) are known to coaggregate.

Surveys of the coaggregation properties of more than 800 strains representing most of these genera have added many fresh ideas to the working model of the role of coaggregation in the ecology of the oral cavity (61, 67, 72, 73, 76, 77, 85). Some *Haemophilus* strains, although they are among the early colonizers of the tooth surface, coaggregate with *S. sanguis* and *S. mitis* but not with *A. viscosus* (85). Certain strains of *Bacteroides loescheii* exhibit similar coaggregation patterns, and as with the haemophili, their coaggregations are lactose-inhibitable (72). Both *B. loescheii* and *B. intermedius* have many partners. Thus Kolenbrander et al (72) proposed that two bacterial cell types that do not coaggregate with each other may recognize a common partner, which becomes a coaggregation bridge. Bridges that involve three or more cell types may be integral units in the network of cell-to-cell interactions of accreting bacterial plaque.

In sharp contrast to bridging, competition occurs when two noncoaggregating cell types recognize a common partner by the same mechanisms of adherence. This concept was first advanced on the basis of studies with Cog⁻ mutants of actinomyces (68), and was later extended during a survey of *Bacteroides* species (72). It has obvious implications in any ecosystem.

Capnocytophaga gingivalis coaggregates with two strains of *A. israelii*, *Capnocytophaga sputigena* coaggregates with these two actinomyces strains plus 11 more, and *Capnocytophaga ochracea* coaggregates with the 13 actinomyces strains and three *S. sanguis* strains (67). Through the use of Cog⁻ mutants and careful analyses of sugar inhibitors of these coaggregations, Weiss et al (151) proposed that the lectinlike surface component on *C. ochracea* possessed three adhesin sites. Further investigation with anti-capnocytophaga antisera suggested a phylogenetic resemblance of the surface adhesins among the three *Capnocytophaga* species (E. Weiss, P. Kolenbrander & J. London, unpublished results).

A survey of more than 200 oral veillonellae from subgingival plaque and the tongue dorsum revealed that they coaggregate with partners originating from the subgingiva and tongue habitats, respectively (61), which indicates a clear role for coaggregation in oral surface adherence. Of 41 strains of the motile *Selenomonas* surveyed, only a few coaggregated with a few fusobacteria. This is not surprising, since (*a*) selenomonads are motile and may not be dependent on adherence mechanisms and (*b*) no group of oral bacteria tested to date coaggregates with more numerous or varied oral bacteria than the 28 strains of *Fusobacterium nucleatum* examined in an earlier survey (71). The ability of fusobacteria to adhere to a wide variety of bacteria may be reflected in their numerical dominance in both states of health and periodontal disease (110; W. E. C. Moore, personal communication).

Use of Coaggregation-Defective Mutants

Naturally occurring Cog⁻ mutants are readily isolated by mixing a suspension of parent cells with an appropriate coaggregation partner (66). The coaggregates settle to the bottom, and the supernatant is treated again with more partner cells. Using a parent strain that is resistant to two antibiotics and an antibiotic-sensitive partner strain makes selection easy on antibiotic-containing solid media. Numerous Cog⁻ mutants have been isolated by this method and have been used to study the mechanisms of coaggregation of *A. viscosus* (19), *A. naeslundii* (68), *A. israelii* (E. Weiss, P. Kolenbrander & J. London, unpublished results), *B. loescheii* (150), *C. gingivalis* (63), *C. ochracea* (151), and *S. sanguis* (66, 68, 108; E. Weiss, P. Kolenbrander & J. London, unpublished results).

A second method of obtaining Cog⁻ mutants relies on the multiple functions of surface components. Accordingly, bacteriophage-resistant mutants of *A. viscosus* MG-1 have been isolated and were shown to exhibit altered coaggregation properties with certain streptococci (139).

ECOLOGY OF DENTAL PLAQUE

Composition of Dental Plaque

The primary constituents of dental plaque are bacteria in a matrix composed of extracellular bacterial polymers and salivary products. Immediately after a tooth is cleaned, the enamel surface is coated with an acquired pellicle, which is an amorphous covering of 0.1 to several microns thick that consists mainly of salivary glycoproteins. Supragingival plaque is constantly bathed with saliva, but subgingival plaque is in contact with gingival crevicular fluid, which is of serous origin and which contains plasma protein as well as a wide variety of antibodies to many of the potential periodontal pathogens (136). Saliva and crevicular fluid have pH of about 7.4 and 8.0 (4), respectively, but the pH of dental plaque is about 5.5–6.5. The bacterial species present in dental plaque are heterogeneous and progressively change as the clinical condition goes from normal health through gingivitis to advanced stages of periodontitis. Several model systems have been developed to investigate the bacterial composition and to identify potential periodontal pathogenic bacteria. This review does not cover the formation of smooth-surface plaque and the generally accepted etiologic agent of dental caries, *Streptococcus mutans*. The focus here is on changes in bacterial populations in several model systems, and where possible, these changes are related to adherence properties of the bacteria.

Morphological Observations

Early ultrastructural studies showed by electron microscopy that dense bacterial layers are associated with the acquired pellicle at the enamel surface and that the bacterial density diminishes distally. The bacteria were packed closely together, but some special cellular arrangements such as a central cell surrounded by cells of a different type were observed in thin sections (3, 30). Examination by scanning electron microscopy revealed corn-cob formations, where a central rod-shaped cell was surrounded by spherical cells (62). One such arrangement was dissected by micromanipulation into a rod-shaped *Bacterionema matruchotii* cell and spherical streptococci (115). Corn-cob formations were shown to be common in dental plaque, along with certain other cell-to-cell arrangements (e.g. bristle brush) between unlike cells (91–93). Similar cellular arrangements have been made in the laboratory; corn cobs have been formed with *B. matruchotii–S. sanguis* (82) and *F. nucleatum–S. sanguis* pairs (83; P. Kolenbrander & R. Andersen, unpublished observations), and rosettes have been formed with *S. sanguis* spheres surrounded by *B. loescheii* (70).

Succession of Bacterial Population as Plaque Matures

CLEANED TEETH MODEL (0–24 HOURS)   Within minutes after teeth are professionally cleaned, the enamel surface is repopulated with bacteria (131). Among the predominant primary colonizers are *S. sanguis, A. viscosus,* and *A. naeslundii,* followed by *Haemophilus* spp. (87), *Veillonella* spp. (121), and *Peptostreptococcus* spp. (131). Bacteria must attach or they will be washed away by normal salivary flow. Consequently, when all the sites on the tooth are filled, the incoming bacteria must recognize and adhere to already attached bacteria.

EXPERIMENTAL GINGIVITIS MODEL (1–30 DAYS)   After having their teeth professionally cleaned, human subjects are asked to refrain from normal oral hygiene methods for several days, during which time the tooth surface is sampled. Early studies indicated that the predominant flora initially consists of gram-positive bacteria such as streptococci and actinomyces and then changes to mainly gram-negative bacteria (94, 137). These changes in microbial flora are accompanied by gingival inflammation (e.g. gingival bleeding). More chronic and severe forms of periodontal disease follow, characterized by loss of the supporting bone and ligament surrounding the teeth. More recent, more refined analyses have all pointed to the same principle of overall changes in bacterial populations without any one bacterium clearly identified as the causative agent (95, 112, 113, 124, 129, 133, 142). The flora of gingivitis is intermediate in composition between that of oral health and that of more severe forms of periodontal disease (111). Moreover, the predominant species of severe periodontitis are present in low numbers during gingivitis, but are usually absent during oral health.

POTENTIAL PERIODONTAL PATHOGENS   There is no strain of oral bacteria that by itself is known to cause human periodontal disease. Some bacteria are often present in high numbers with a specific clinical condition, e.g. *Actinobacillus actinomycetemcomitans* in localized juvenile periodontitis (98, 130), but in many instances the bacteria are either not detected or present in very low numbers (110, 134, 155). Some bacteria appear to proliferate more frequently than others under conditions of active periodontal lesions (35, 96, 110, 127; W. E. C. Moore & L. V. H. Moore, personal communication). The bacteria most commonly associated with periodontal disease are *A. actinomycetemcomitans, Bacteroides intermedius, Bacteroides gingivalis, C. gingivalis, Eikenella corrodens, Eubacterium nodatum, Eubacterium brachy, F. nucleatum, Wolinella recta,* and spirochetes. A notion that is consistent with available data is that episodes of periodontal destruction may be due to successive assaults by different species (155). When the host develops resistance to one species, another proliferates and causes additional damage.

GNOTOBIOTIC ANIMAL MODEL   Gnotobiotic rats are fed pure cultures of bacteria, and the teeth are analyzed for the presence of adherent bacteria at various intervals. Adherence of *Veillonella alcalescens* to *S. mutans*–coated rat teeth matched its coaggregation with *S. mutans* (102). Those veillonellae that coaggregated were found in large numbers on the teeth, whereas those that did not coaggregate also did not adhere. Even those that coaggregated could not colonize in the absence of streptococci. Using the same gnotobiotic rat model, the lactose-inhibitable coaggregation between *S. sanguis* and *A. viscosus* was tested (140). A distinction was made between adherent cells (those present 2 hr after oral administration of bacteria to the animal) and established cells (those on the teeth 2 wk later). First either the parent *A. viscosus* or a Cog⁻ mutant that failed to coaggregate with *S. sanguis* was administered and was allowed to establish itself for 2 wk. Both strains became equally established. Next, the streptococci were inoculated into the rats, and only some of the rats were given lactose in their drinking water. Increased numbers of adherent *S. sanguis* were found only with the parent *A. viscosus* in the absence of lactose, which indicates that lactose has a measurable effect on in vivo adherence for lactose-inhibitable coaggregations. However, this sugar inhibitory effect was transitory, since after 2 wk no difference in the numbers of streptococci was found in any of the four groups of actinomyces-pretreated animals. *S. sanguis* became established more effectively in the absence of actinomyces than in their presence, which may reflect the absence of competitor bacteria in the monoinfected animals.

The two examples given here illustrate the usefulness of an animal system where even transitory inhibitions can easily be monitored *(Actinomyces–Streptococcus)* and bacteria that cannot become established by themselves are quickly washed out *(Streptococcus–Veillonella)*.

FLUORESCENT ANTIBODY MODEL   To test for the presence of coaggregating pairs in dental plaque, samples are mixed with fluorescent dye–labeled antibodies against surface components on the two cell types. This method was used to study lactose-inhibitable coaggregations between streptococci (fluorescein-conjugated antibody) and actinomyces (rhodamine-conjugated antibody). The results convincingly demonstrated that both fluorescent dyes were in close proximity in dental plaque samples (18). It is inferred from these results that coaggregating pairs exist in plaque and furthermore that interbacterial adherence is likely to be essential for bacterial colonization of dental plaque.

HYDROXYAPATITE MODEL   This model follows the same principle as the cleaned teeth model: Bacteria adhere or are washed away. Spherical hydroxyapatite has been used as the compound that most closely imitates tooth enamel, and commercially available forms are of consistent quality. Generally, it is coated with clarified saliva (centrifuged at $12,000 \times g$ for 15 min to sediment cells and debris). It is then mixed with a radioactively labeled bacterial suspension, and the number of adherent bacteria is determined (24). The effects of various agents and conditions on bacterial adherence are used to correlate the model to teeth. Combinations of coaggregating pairs have been added simultaneously to saliva-coated hydroxyapatite (SHA) (81, 86). The contribution of the pair or of each cell type alone to the total adherence to SHA is difficult to assess, because both members of the pair attach to SHA independently. However, SHA is a particularly useful model system because many of the bacteria that adhere well to hydroxyapatite are also primary colonizers of the tooth surface.

Recent investigations have included studies of attachment of a second cell type to SHA precoated with a coaggregation partner cell type (14, 125). The binding of *B. gingivalis* was always greater to SHA that was pretreated with a suspension of *A. viscosus,* its usual partner (5, 128), than to SHA alone (125). The primary colonization by *A. viscosus* and later colonization by *B. gingivalis,* which is considered a potential periodontal pathogen, supports the contention that this model system reflects in vivo occurrences.

Ciardi et al (14) studied the attachment of *Propionibacterium acnes* to *S. sanguis*–coated SHA. The two cell types were known to coaggregate by a lactose-inhibitable mechanism (76). About 400-fold more *S. sanguis* bound to

SHA than *P. acnes,* but with *S. sanguis*–coated SHA the number of *P. acnes* bound was nearly unity with the prebound number of *S. sanguis.* In addition, the *P. acnes* were prevented from binding to *S. sanguis*–coated SHA by lactose and *N*-acetylgalactosamine, which had no effect on the binding of *S. sanguis* to SHA. Thus, this model can be used to study lectin-mediated cell-to-cell interactions with bacteria already attached to the hydroxyapatite surface.

MULTIGENERIC AGGREGATES: SIMULATED DENTAL PLAQUE   On the basis of the coaggregation properties of certain bacterial pairs, predictions can be made about the interactions among cells of three or more types. Kolenbrander & Andersen (69) developed a model that simulated dental plaque by using the kinds of cells found in dental plaque samples. When only one cell type in the multigeneric mixture was radioactively labeled, it was possible to monitor the distribution of that cell type in either the aggregates (pellet) or the unbound state (supernatant) after a low-speed centrifugation ($500 \times g$ for 1 min). Cell types were chosen to test specific principles of cell-to-cell interactions in a millieu of potentially interfering interactions. The results clearly indicated that noncoaggregating cell types are not trapped by multigeneric aggregates being formed in the same cell suspension. Thus cells act independently. Furthermore, the cell surface components of any given cell type act independently. For example, coaggregations that are inhibitable and noninhibitable by lactose are independent. *S. sanguis* J22, which exhibited lactose-inhibitable coaggregation with *A. naeslundii* PK947 and noninhibitable coaggregation with *A. israelii* ATCC 10048 and several other cell types also included in the mutigeneric aggregate, did not become unbound when lactose was added to the mixture. In contrast, *A. naeslundii* PK947, which was bound to the aggregate only by lactose-sensitive adherence, was completely released from the aggregate by this treatment.

Bacterial plaque is thought to accumulate mass by a combination of the sequential attachment of different cell types and the growth of microcolonies of attached bacteria. The hypothesis of sequential addition or accretion of different cell types to preexisting coaggregates was tested by sequentially adding cell types that would only coaggregate with the previously added cell type. Although the aggregate became progressively more complex in composition, accretion of a new cell type was dependent upon the presence of the previously added cell type. This observation illustrates the concept of coaggregation bridging (72) among oral bacteria that may be functionally active in the progression of periodontal disease. Multigeneric aggregates were found to be networks of coaggregating pairs acting independently. However, while surface-to-surface interaction may be independent, metabolic interactions among cells in human dental plaque may be interdependent and may be vital to successful colonization by the bacterial cell.

Specificity of Bacterial Attachment to Certain Oral Surfaces

ORAL STREPTOCOCCI   In any lotic environment bacterial attachment is a prerequisite to colonization of available surfaces. In several elegant studies, the relative adherence of various streptococci to different oral surfaces was correlated with their proportional distribution found naturally on those surfaces (51, 88, 89, 141). *Streptococcus salivarius* constitutes a major proportion of the oral streptococci in saliva and on the cheek and tongue but only a small percentage of the bacteria in dental plaque (80). *S. sanguis* is found in high numbers in dental plaque, but considerably fewer were observed on the cheek and especially few on the tongue (11, 141). Thus, the primary ecological determinant for successful colonization of a surface appears to be the selective adherence by bacteria to that surface (51). This concept was confirmed by orally administering streptomycin-resistant streptococci to human volunteers and showing that the streptococci in the inoculum distributed onto the oral surfaces in the proportions that were naturally found on these surfaces.

ORAL VEILLONELLAE   The distribution of streptomycin-resistant veillonellae on oral surfaces in human volunteers was monitored by the same techniques described above for oral streptococci (88). They attach poorly to cleaned teeth but markedly better than *S. sanguis* or even *S. salivarius* to the tongue dorsum. Veillonellae constitute 5 and 8% of the cultivable bacteria in saliva and on the tongue dorsum, respectively, but average less than 1% of supragingival flora. *V. parvula* comprises about 5 and 4% of the bacteria in supragingival and subgingival plaque in normal healthy sites (113; W. E. C. Moore, personal communication).

Of 29 subgingival veillonella isolates screened, 24 coaggregated with the reference strains of the 12 streptococcal and actinomyces coaggregation groups, which are common subgingival *S. sanguis*, *S. morbillorum*, *A. viscosus*, and *A. naeslundii* isolates (61). The other five veillonella isolates coaggregated only with *S. salivarius*, a common tongue inhabitant. Although 58 of 59 fresh veillonella isolates from the tongue coaggregated with *S. salivarius*, none of these recognized the same battery of reference strains that were partners of the veillonellae of subgingival origin. Furthermore, none were *V. parvula* isolates; all were either *Veillonella atypica* or *Veillonella dispar*. The relation of the coaggregation properties of taxonomically diverse veillonella isolates with the site from which they are isolated provides strong evidence that coaggregation has a critical role in the bacterial ecology of the oral cavity.

SURFACE COMPONENTS THAT MEDIATE COAGGREGATION

Fimbriae

Fuzzy coats, fuzzy components, fibrillar structures, delicate fibrils, and threadlike structures were early names given to the fibrous-appearing surface layer of oral bacteria observed by electron microscopy (3, 38, 54, 84, 89, 116, 117, 138). This layer was generally considered to be important in adherence of the bacteria to other cells or surfaces. Today the structures are called fuzzy coats, fibrils, or fimbriae. The multiple names reflect how little is known about the outer surface of oral bacteria. Handley et al (56) have attempted to distinguish fibrils from fimbriae on the basis of electron microscopic observations. According to the definition of Ottow (118), fimbriae are nonflagellar bacterial appendages other than those clearly involved in the transfer of nucleic acids, which are called pili. Even so, this distinction is very difficult purely on the basis of morphology (118). The structures on *A. viscosus* are usually flexuous (22) and are over 200 nm in length (R. Ellen, personal communication). Those on oral streptococci can be up to nearly 200 nm long, with an estimated width of only 2 nm (149). A clearer distinction will be possible when some of these structures are purified and a definite correlation can be made between their composition, structure, and function. Because of these uncertainties, the single term fimbriae is used in this review to describe cell surface fibrillar structures of suspected protein (glycoprotein) composition. Many of these appear to be involved in adherence functions.

ACTINOMYCES FIMBRIAE   The first purified fimbriae were from *A. viscosus* (22, 23, 101, 154). The purified structures from *A. viscosus* WVU627 were resolved as a single band on SDS-polyacrylamide gels (SDS-PAGE) of approximately 64 kd (101). Another *A viscosus* strain, T14V, was found to have two kinds of fimbriae (17, 22, 154), which were resistant to dissociation by standard SDS-PAGE procedures. The subunits of these fimbriae have been identified as proteins of approximately 60 kd by cloning and expression of the respective *A. viscosus* T14V genes in *E. coli* (34, 156). The presence of both fimbrial antigens on strain WVU627 was suggested by the reaction of this organism with monospecific antisera against *A. viscosus* T14V type 1 and type 2 fimbriae (20). Further analysis of surface-structure preparations of the two strains by SDS-PAGE and immunoblotting with monospecific antisera revealed identical multiple bands in each fimbrial sample, including a band migrating at approximately 64 kd (J. Cisar, personal communication). Thus, the fimbriae of *A. viscosus* strains T14V and WVU627 appear to be structurally similar.

In a set of thorough studies, the two types of fimbriae were distinguished by (*a*) crossed immunoelectrophoresis with monoclonal antibodies against type 2 fimbriae (Ag2) (17); (*b*) the observation that lactose-inhibitable coaggregation with streptococci was blocked by Fab fragments of anti-Ag2 but not by Fab fragments of anti-Ag1 (120); (*c*) immunoelectronmicroscopy showing that Cog⁻ mutants (66) bore Ag1 but not Ag2 (19); (*d*) polymorphonuclear leukocyte phagocytosis of Ag2- but not Ag1-bearing actinomyces (123); (*e*) adherence of Ag2- but not Ag1-bearing cells to human epithelial cell lines (6); (*f*) the observation that anti-Ag1 or its Fab fragment, but not anti-Ag2, blocked adsorption of Ag1- but not Ag2-bearing actinomyces to SHA (25, 26); (*g*) adherence of Ag1 but not Ag2 to purified proline-rich protein from saliva (49); and (*h*) cloning of the structural gene *(fimA)* for type 2 fimbriae (34) and of the type 1 fimbrial subunit of *A. viscosus* T14V (156). Clearly, Ag1 and Ag2 are different fimbriae that mediate different adherence functions. Ag1 mediates adherence to specific salivary molecules on hydroxyapatite, and Ag2 bears the lactose-sensitive lectin activity observed in certain coaggregations with streptococci and neuraminidase-treated erythrocytes (27, 39). Using monospecific antisera to Ag1 and to Ag2 and a series of adherence-defective mutants lacking either Ag1 or Ag2 or both antigens (kindly provided by J. Cisar), morphological differences in the length of type 1 and type 2 fimbriae were noted (R. Ellen, personal communication). The average maximum labeling distance from the cell was 207 nm for Ag1 and 324 nm for Ag2 of *A. viscosus* T14VJ1 (R. Ellen, personal communication).

Type 2 fimbriae with the lactose-sensitive lectin activity are present on both *A. viscosus* and *A. naeslundii,* and both species attach to human epithelial cells (6). Binding of *A. naeslundii* to epithelial cells was inhibited most effectively by D-galactose (Gal), *N*-acetyl-D-galactosamine (GalNAc), lactose, and methyl-β-D-galactoside (6). These results agree well with the evidence for binding of these bacteria to Gal-β-(1→3)GalNAc-bearing glycosphingolipids (7). They also support the evidence that Gal-β-(1→3)GalNAc is the most effective inhibitor both of lactose-inhibitable coaggregation with streptococci (105, 106) and of binding of *A. viscosus* T14V to glycoprotein-coated latex beads (44, 58, 59).

STREPTOCOCCUS FIMBRIAE   Weerkamp & McBride (145) showed that protease digestion of *S. salivarius* prevented its adherence to human erythrocytes and buccal epithelial cells but not its coaggregation with several oral bacteria including *Veillonella alcalescens* V1. Two classes of mutants of *S. salivarius* HB were isolated (145, 146). Class I, represented by strain HB-7, failed to adhere to buccal epithelium but retained its coaggregation properties. Class II, represented by strain HB-V5, no longer coaggregated with *V. alcalescens* V1 but did adhere to buccal epithelial cells. Mutanolysin digestion of the parent *S. salivarius* HB and subsequent chromatographic techniques separated the cell walls into three antigens, AgB, AgC, and AgD, of 320, 220–280, and two bands of 121 and 129 kd, respectively (144). The three antigens contained relatively low amounts of basic amino acids and high levels of nonpolar amino acids. Purified AgB retained biological activity in that it agglutinated *Veillonella* strains, and it was called veillonella-binding protein. Purified AgC did not retain biological activity, but comparison of the properties of the two classes of mutants with the parent indicated that it was a host attachment factor of fibrillar morphology (148). Immunoelectronmicroscopy showed that AgC was located mainly on the fimbriae in the parent strain (147), but accumulated intracellularly in mutants without wall-associated AgC (143). Antigen D has no known function, but it was found on the fimbriae of the parent strain (143). Interestingly, AgB, which mediates coaggregation with veillonella, is found at the distal portion of the fimbriae (143) in an arrangement similar to that reported for coaggregation-mediating adhesins of *Bacteroides loescheii* (153).

Fimbrae mediate the adherence of streptococci to SHA (41, 48); antibodies against fimbriae block adherence (40) and were used to isolate an adhesin (36). A structural gene for adhesion fimbriae has been cloned in *Escherichia coli* (42). The fimbriated parent strain *S. sanguis* FW213 did not coaggregate with actinomyces, and only one of 17 nonadherent, nonfimbriated mutants exhibited lactose-inhibitable coaggregation (43). While fimbriae do not seem to be important to coaggregation with *S. sanguis* FW213, many fimbriated fresh isolates of *S. sanguis* do coaggregate with *A. viscosus*, *A. naeslundii*, and *Fusobacterium nucleatum* (57). Six of the 36 *S. sanguis* isolates carried tufts of fimbriae, and none of these strains coaggregated with the actinomyces. Thus the role of *S. sanguis* fimbriae in mediating coaggregation is uncertain.

BACTEROIDES FIMBRIAE   Selecting monoclonal antibodies (MAbs) that specifically blocked coaggregation rather than screening MAbs that agglutinated cells was a cornerstone for obtaining the evidence that adhesins are fimbriae-associated proteins but not actual fimbrial subunits (152). This technique was a major advance in the study of mechanisms of coaggregation because it directly measured interference with the activity of the desired molecule, rather than measuring agglutination, which could be caused by antigen-antibody recognitions between surface molecules that are totally unrelated to coaggregation.

The study of fimbriae-associated adhesins was initiated with *B. loescheii* PK1295, a coaggregation bridge between *S. sanguis* 34 and *A. israelii* PK14 (72). One side of the bridge was lactose inhibitable and the other was not, which suggested that two distinct adhesins may mediate these two different coaggregations. Two types of Cog$^-$ mutants were isolated (150); type 1 mutants failed to coaggregate with *S. sanguis* 34 but retained coaggregation with *A. israelii* PK14, and type 2 mutants were unable to coaggregate with either partner. Rabbit antisera raised against whole cells of *B. loescheii* blocked coaggregation with both partners at dilutions of antiserum that did not agglutinate bacteroides. Adsorbing the antiserum with whole cells of type 1 mutant yielded anitserum that partially blocked (approximately 75%) coaggregation between bacteroides and streptococci, whereas antiserum adsorbed with type 2 mutant cells blocked coaggregation completely with both gram-positive partners. Both types of mutants possessed fimbriae. Fimbriae from the parent and mutants exhibited the same coaggregation properties as the respective whole cells. Parent fimbriae formed aggregates with both *S. sanguis* and *A. israelii;* only the streptococcal partnership was lactose inhibitable. Fimbriae from the type 1 mutant aggregated only *A. israelii*, whereas type 2 mutant fimbriae had no aggregating activity. Both a 75-kd protein and a 45-kd protein were associated with fimbriae from the parent, but the type 1 mutant possessed only the 45-kd protein, and neither polypeptide was found in fimbrial preparations from the type 2 mutant. It was proposed that 75- and 45-kd fimbriae-associated proteins on the surface of *B. loescheii* mediate coaggregation with *S. sanguis* and *A. israelii*, respectively (150).

Fimbria-enriched preparations were used to raise adhesin-specific MAbs (152). Ten MAbs were isolated. They represented three distinct functional classes: Four MAbs specifically blocked bacteroides-streptococcus coaggregation, two MAbs specifically blocked bacteroides-actinomyces interactions, and four MAbs agglutinated bacteroides but did not inhibit either coaggregating pair. None of the six MAbs that inhibited adherence were capable of agglutinating bacteroides. Fab fragments from the two classes of the MAbs that inhibited adherence blocked the same coaggregating pairs that were blocked by the whole immunoglobulin. Between 0.5 and 2.0 μg protein ml^{-1} of either MAbs or Fabs blocked coaggregation with the respective partners by 50%. Results from immunoblotting and immunoprecipitation procedures revealed that the *S. sanguis* adhesin-specific MAbs reacted with a fimbria-associated 75-kd polypeptide and the *A. israelii* adhesin-specific MAbs recognized a fimbria-associated 45-kd polypeptide. These results confirmed and extended the earlier proposal (150) that the *B. loescheii–S. sanguis* and *B. loescheii–A. israelii* coaggregations are mediated by different adhesins. The observation that both Fabs and MAbs block coaggregation at the same very low concentration of protein suggests that the antibody reacts at or near the active site of the adhesin molecule.

MAbs that agglutinated whole bacteroides cells could not detect adhesin molecules. When agglutination was used as the screening assay for MAbs, several MAbs were observed that were highly reactive as determined by reactions in an ELISA assay system with whole bacteroides cells (152). In fact, they were much more reactive than the coaggregation-blocking MAbs. The reaction in the ELISA assay is often used as the basis for selecting the most reactive MAbs for further study, but had it been used in this study the adhesins would not have been identified. It is therefore important to use a functional assay that measures the activity of the desired adhesin rather than agglutination, which is a general surface phenomenon that may or may not specifically involve the desired adhesin.

Weiss et al (153) enumerated and localized the adhesins that mediate the two kinds of coaggregation using the MAbs that specifically block these coaggregations. Maxima of about 400 adhesin molecules specific for *S. sanguis* and 310 *A. israelii*–specific adhesins per cell were calculated from binding studies with radiolabeled MAbs. Immunoelectronmicroscopy with colloidal gold–conjugated goat anti–mouse IgG revealed that the adhesins were usually on the distal portion of the fimbrial structure and were not part of the fimbrial subunit. Type 1 mutants did not bind gold particles following treatment with anti–*S. sanguis* adhesin MAb, but they did bind gold particles after treatment with anti–*A. israelii* adhesin MAb. Type 2 mutants did not bind gold particles when pretreated with either MAb. The association of adhesins with fimbriae and the localization of adhesins at or near the fimbrial tip of *B. loescheii* are in agreement with the results from other gram-negative bacteria (1, 90, 109).

The fimbriae from *B. gingivalis* have been isolated but have no known function (157, 158). They are not antigenically related to those on *B. loescheii* (F. Yoshimura, personal communicaticn).

Outer Membrane

Some gram-negative oral bacteria, including *Capnocytophaga* spp., do not have fimbriae (60). Members of all three *Capnocytophaga* species coaggregate with various oral bacteria (67); coaggregation appears to be mediated by membranous excrescences on their surface (73, 97). A protein of about 155 kd was isolated from *C. gingivalis* DR2001 and found to have a lectinlike activity in coaggregations with *A. israelii* PK16 (63). Antiserum raised against the parent cells but adsorbed with Cog$^-$ mutant cells blocked coaggregation. It was also used in immunoblot analyses to identify the 155-kd polypeptide, which appears to be the monomeric form of the lectin. Monoclonal antibodies against the lectinlike protein have recently been prepared, and 50% inhibition of coaggregation was achieved with the MAb or its Fab fragment at about 1 μg ml^{-1} (135).

Fusobacterium nucleatum, a gram-negative bacterium, forms corn cob arrangements with many streptococci (32; P. Kolenbrander & R. Andersen, unpublished observations). Two types of receptors on the fusobacteria were proposed. The type I receptor is a loosely bound surface protein that binds lipoteichoic acid on the streptococcal surface (32). The type II receptor does not bind lipoteichoic acid or deacylated lipoteichoic acid; it appears to be associated with a 41-kd protein, is firmly anchored in the outer membrane, and is exposed on the cell surface (33). The type II receptor may be involved in a specific protein-protein interaction, which may be another major form of cell-to-cell recognition among oral bacteria.

Carbohydrate

The most complete analysis of any coaggregation-mediating polysaccharide was recently published (104). The major constituent is a unique structure: α-D-Gal$p$NAc-(1→3)-β-L-Rha$p$-(1→4)-β-D-Glc$p$-(1→6)-β-D-Gal$f$-(1→6)-β-D-Gal$p$NAc-(1→3)-D-Gal. The polysaccharide is the receptor on *S. sanguis* 34 that is recognized by the lactose-sensitive lectin on type 2 fimbriae of *A. viscosus* T14V. Receptor-containing suspensions prepared from the cell walls of *S. sanguis* 34 were 19-fold more potent than lactose as coaggregation inhibitors (103). The receptor had a molecular mass of about 30 kd and contained about 6% phosphate. It appears to comprise one major and two very minor oligosaccharides joined by phosphoric diester bridges.

A coaggregation-inhibiting polysaccharide from *S. sanguis* H1 that participates in a lactose-inhibitable coaggregation with *C. ochracea* ATCC 33596 has recently been characterized (12). It was prepared by methods similar to those used to prepare the receptor from *S. sanguis* 34. Its molecular mass is about 45 kd. It consists of rhamnose, galactose, glucose, and phosphate.

Besides coaggregating with *C. ochracea*, *S. sanguis* H1 also coaggregates with *A. viscosus* T14V. In this latter pair, however, the carbohydrate (which is heat and protease resistant) is expressed on the actinomyces, and the coaggregation is not lactose inhibitable (21). A carbohydrate-containing fraction was prepared from *A. viscosus;* it contained some cell wall and protein components but also a polysaccharide consisting of *N*-acetylgalactosamine, rhamnose, and 6-deoxytalose (108). The fraction specifically inhibited coaggregation with *S. sanguis* H1 but not with other streptococcal partners of *A. viscosus* T14V. However, although it only partially inhibited (35%) streptococcal coaggregation with *A. viscosus* T14V at 2 mg ml^{-1}, it nearly completely (85%) inhibited coaggregation with *A. naeslundii* WVU45 at the same concentration. The latter strain is a member of actinomyces coaggregation group B, whereas T14V is a member of group A (see Figure 2; *S. sanguis* H1 is the representative strain for streptococcal coaggregation group 2; the aggregation factor is represented by a semicircle on actinomyces coaggregation groups A, B, C, and D). These data strongly support an earlier suggestion (66) that T14V expresses at least two distinct receptors for different *S. sanguis*

H1 adhesins, whereas *A. naeslundii* WVU45 displays only one of those receptors.

PRINCIPLES OF COAGGREGATION

From the investigations of a wide variety of pairs (thousands have been tested) and more complex aggregates, the main features of coaggregates have been identified:

1. Coaggregation is a result of recognition between accessible surface components on bacterial partners. Cell walls and isolated adhesin or receptor molecules mimic whole cells.
2. Adherence between partners is tenacious.
3. Coaggregation occurs between specific partners. Cell viability is not required.
4. Coaggregation specificity is a stable property. Older stock cultures maintained in various laboratories exhibit coaggregation properties identical to those of freshly isolated strains of the same species.
5. Different kinds of surface components mediate coaggregations. Carbohydrates, fimbriae, and outer-membrane proteins have been identified.
6. Simple sugars such as lactose often inhibit coaggregations.
7. A partner that is recognized by two or more noncoaggregating cell types can be involved in (*a*) competition when the cell types interact with the same component on the common partner, or (*b*) bridging when the cell types interact with different components on the common partner.
8. Multigeneric aggregates are composed of a network of coaggregating pairs acting independently. No unexpected surface interaction is detected when a coaggregating pair becomes part of a multigeneric aggregate.

From an ecological viewpoint, one of the most promising concepts that has evolved from these studies of multigeneric aggregates is that the complementary molecules on each coaggregating pair can describe the surface interactions of a complex group of bacteria such as that constituting dental plaque.

CONCLUSIONS

Coaggregation is widespread among human oral bacteria, but it is highly specific. It is proposed that the specificities reside in the complementary adhesin-receptor surface components on partner cell types. Cell types with similar specificities compete for their common coaggregation partner. Dissimilarity of specificities encourages bridging among partners and accretion, yielding a network of interacting bacteria. Although each econiche in the oral cavity is likely to be a community of interacting metabolisms, coaggregations exhibit independence, which may confer on the consortium a dynamic mobility within the population. It is also clear from the studies on *Veillonella* in subgingival plaque and on the tongue dorsum that coaggregation may be a critical factor in adherence to very different tissues that contain equally distinct inhabitants. Coaggregation is a great advantage in a lotic environment. It is not certain whether the oral cavity is unique in exhibiting widespread interbacterial adherence among its members or whether other ecosystems are also composed of bacteria with similar interactions that have yet to be discovered.

Sugar-inhibitable coaggregations are the most prevalent kinds of interactions observed in extensive surveys of the genera that account for nearly all of the predominant human oral isolates. In addition, most coaggregations are prevented by protease digestion of one cell type while the partner is protease insensitive. These observations suggest that complementary lectin and carbohydrate molecules are primarily responsible for mediating coaggregations. Indeed, adhesins with lectin activity have been isolated from *B. loescheii* and several other oral bacteria.

ACKNOWLEDGMENTS

I thank R. N. Andersen for helping with drawing and developing the final form of the figures. J. London, J. Cisar, and C. Hughes provided helpful discussions about the content of the review. I am especially thankful to the above individuals and to E. Weiss, F. Cassels, F. Yoshimura, R. Ellen, W. E. C. Moore, and L. V. H. Moore for making available unpublished information, which gives a current status to this review. I thank Sarah Smith and Elizabeth Walter for their care and expertise in typing this manuscript.

Literature Cited

1. Abraham, S. N., Goguen, J. D., Sun, D., Klemm, P., Beachey, E. H. 1987. Identification of two ancillary subunits of *Escherichia coli* Type 1 fimbriae by using antibodies against synthetic oligopeptides of *fim* gene products. *J. Bacteriol.* 169:5530–36

2. Andersen, R. N., Kolenbrander, P. E. 1986. Masked lactose-reversible coaggregations of oral streptococci with *Actinomyces naeslundii* detected with coaggregation-defective mutants. *Abstr. Annu. Meet. Am. Soc. Microbiol.* 1986:95

3. Barkin, M. E. 1970. Ultrastructural cytochemical localization of polysaccharides in dental plaque. *J. Dent. Res.* 49:979–85

4. Bickel, M., Munoz, J. L., Giovannini, P. 1985. Acid-base properties of human gingival crevicular fluid. *J. Dent. Res.* 64:1218–20

5. Boyd, J., McBride, B. C. 1984. Fractionation of hemagglutinating and bacterial binding adhesins of *Bacteroides gingivalis. Infect. Immun.* 45:403–9

6. Brennan, M. J., Cisar, J. O., Vatter, A. E., Sandberg, A. L. 1984. Lectin-dependent attachment of *Actinomyces naeslundii* to receptors on epithelial cells. *Infect. Immun.* 46:459–64

7. Brennan, M. J., Joralmon, R. A., Cisar, J. O., Sandberg, A. L. 1987. Binding of *Actinomyces naeslundii* to glycosphingolipids. *Infect. Immun.* 55:487–89

8. Breznak, J. A., Pankratz, H. S. 1977. In situ morphology of the gut microbiota of wood-eating termites [*Reticulitermes flavipes* (Kollar) and *Coptotermes for-*

mosanus Shiraki]. *Appl. Environ. Microbiol.* 33:406–26

9. Bourgeau, G., McBride, B. C. 1976. Dextran-mediated interbacterial aggregation between dextran-synthesizing streptococci and *Actinomyces viscosus*. *Infect. Immun.* 13:1228–34

10. Bowden, G. H. W., Ellwood, D. C., Hamilton, I. R. 1979. Microbial ecology of the oral cavity. *Adv. Microb. Ecol.* 3:135–217

11. Carlsson, J. 1965. Zooglea-forming streptococci, resembling *Streptococcus sanguis,* isolated from dental plaque in man. *Odontol. Revy* 16:348–58

12. Cassels, F., Allen, J., London, J. 1988. A coaggregation-inhibiting polysaccharide from *Streptococcus sanguis* H1: purification and characterization. *Abstr. Annu. Meet. Am. Soc. Microbiol.* 1988:89

13. Celesk, R. A., London, J. 1980. Attachment of oral *Cytophaga* species to hydroxyapatite-containing surfaces. *Infect. Immun.* 29:768–77

14. Ciardi, J. E., McCray, G. F. A., Kolenbrander, P. E., Lau, A. 1987. Cell-to-cell interaction of *Streptococcus sanguis* and *Propionibacterium acnes* on saliva-coated hydroxyapatite. *Infect. Immun.* 55:1441–46

15. Cisar, J. O. 1982. Coaggregation reactions between oral bacteria: studies of specific cell-to-cell adherence mediated by microbial lectins. In *Host-Parasite Interactions in Periodontal Diseases,* ed. R. J. Genco, S. E. Mergenhagen, pp. 121–31. Washington, DC: Am. Soc. Microbiol.

16. Cisar, J. O. 1986. Fimbrial lectins of the oral actinomyces. In *Microbial Lectins and Agglutinins: Properties and Biological Activity,* ed. D. Mirelman, pp. 183–96. New York: Wiley

17. Cisar, J. O., Barsumian, E. L., Curl, S. H., Vatter, A. E., Sandberg, A. L., Siraganian, R. P. 1981. Detection and localization of a lectin on *Actinomyces viscosus* T14V by monoclonal antibodies. *J. Immunol.* 127:1318–22

18. Cisar, J. O., Brennan, M. J., Sandberg, A. L. 1985. Lectin-specific interaction of *Actinomyces* fimbriae with oral streptococci. In *Molecular Basis of Oral Microbial Adhesion,* ed. S. E. Mergenhagen, B. Rosan, pp. 159–63. Washington, DC: Am. Soc. Microbiol.

19. Cisar, J. O., Curl, S. H., Kolenbrander, P. E., Vatter, A. E. 1983. Specific absence of type 2 fimbriae on a coaggregation defective mutant of *Actinomyces viscosus* T14V. *Infect. Immun.* 40:759–65

20. Cisar, J. O., David, V. A., Curl, S. H., Vatter, A. E. 1984. Exclusive presence of lactose-sensitive fimbriae on a typical strain (WVU45) of *Actinomyces naeslundii.* *Infect. Immun.* 46:453–58

21. Cisar, J. O., Kolenbrander, P. E., McIntire, F. C. 1979. Specificity of coaggregation reactions between human oral streptococci and strains of *Actinomyces viscosus* or *Actinomyces naeslundii.* *Infect. Immun.* 24:742–52

22. Cisar, J. O., A. E. Vatter, 1979. Surface fibrils (fimbriae) of *Actinomyces viscosus* T14V. *Infect. Immun.* 24:523–31

23. Cisar, J. O., Vatter, A. E., McIntire, F. C. 1978. Identification of the virulence-associated antigen on the surface fibrils of *Actinomyces viscosus* T14. *Infect. Immun.* 19:312–19

24. Clark, W. B., Bammann, L. L., Gibbons, R. J. 1978. Comparative estimates of bacterial affinities and adsorption sites on hydroxyapatite surfaces. *Infect. Immun.* 19:846–53

25. Clark, W. B., Wheeler, T. T., Cisar, J. O. 1984. Specific inhibition of adsorption of *Actinomyces viscosus* T14V to saliva-treated hydroxyapatite by antibody against type 1 fimbriae. *Infect. Immun.* 43:497–501

26. Clark, W. B., Wheeler, T. T., Lane, M. D., Cisar, J. O. 1986. Actinomyces adsorption mediated by Type-1 fimbriae. *J. Dent. Res.* 65:1166–68

27. Costello, A. H., Cisar, J. O., Kolenbrander, P. E., Gabriel, O. 1979. Neuraminidase-dependent hemagglutination of human erythrocytes by human strains of *Actinomyces viscosus* and *Actinomyces naeslundii.* *Infect. Immun.* 26:563–72

28. Deleted in proof

29. Deleted in proof

30. Critchley, P., Saxton, C. A. 1970. The metabolism of gingival plaque. *Int. Dent. J.* 20:408–25

31. Crowley, P. J., Fischlschweiger, W., Coleman, S. E., Bleiweis, A. S. 1987. Intergeneric bacterial coaggregations involving mutans streptococci and oral actinomyces. *Infect. Immun.* 55:2695–700

32. DiRienzo, J. M., Porter-Kaufman, J., Haller, J., Rosan, B. 1985. Corncob formation: a morphological model for molecular studies of bacterial interactions. See Ref. 18, pp. 172–76

33. DiRienzo, J. M., Rosan, B. 1984. Isolation of a major cell envelope protein from *Fusobacterium nucleatum. Infect. Immun.* 44:386–93

34. Donkersloot, J. A., Cisar, J. O., Wax, M. E., Harr, R. J., Chassy, B. M. 1985.

Expression of *Actinomyces viscosus* antigens in *Escherichia coli:* cloning of a structural gene *(fimA)* for type 2 fimbriae. *J. Bacteriol.* 162:1075–78

35. Dzink, J. L., Tanner, A. C. R., Haffajee, A. D., Socransky, S. S. 1985. Gram negative species associated with active destructive periodontal lesions. *J. Clin. Periodontol.* 12:648–59

36. Elder, B. L., Fives-Taylor, P. 1986. Characterization of monoclonal antibodies specific for adhesion: isolation of an adhesin of *Streptococcus sanguis* FW213. *Infect. Immun.* 54:421–27

37. Ellen, R. P. 1982. Oral colonization by gram-positive bacteria significant to periodontal disease. See Ref. 15, pp. 98–111.

38. Ellen, R. P., Balcerzak-Raczkowski, I. B. 1977. Interbacterial aggregation of *Actinomyces naeslundii* and dental plaque streptococci. *J. Periodontal Res.* 12:11–20

39. Ellen, R. P., Fillery, E. D., Chan, K. H., Grove, D. A. 1980. Sialidase-enhanced lectin-like mechanism for *Actinomyces viscosus* and *Actinomyces naeslundii* hemagglutination. *Infect. Immun.* 27:335–43

40. Fachon-Kalweit, S., Elder, B. L., Fives-Taylor, P. 1985. Antibodies that bind to fimbriae block adhesion of *Streptococcus sanguis* to saliva-coated hydroxyapatite. *Infect. Immun.* 48:617–24

41. Fives-Taylor, P. 1982. Isolation and characterization of a *Streptococcus sanguis* FW213 mutant nonadherent to saliva-coated hydroxyapatite beads. In *Microbiology,* ed. D. Schlessinger, pp. 206–9. Washington, DC: Am. Soc. Microbiol.

42. Fives-Taylor, P. M., Macrina, F. L., Pritchard, T. J., Peene, S. S. 1987. Expression of *Streptococcus sanguis* antigens in *Escherichia coli:* cloning of a structural gene for adhesion fimbriae. *Infect. Immun.* 55:123–28

43. Fives-Taylor, P. M., Thompson, D. W. 1985. Surface properties of *Streptococcus sanguis* FW213 mutants nonadherent to saliva-coated hydroxyapatite. *Infect. Immun.* 47:752–59

44. Gabriel, O., Heeb, M. J., Hinrichs, M. B. 1985. Interaction of the surface adhesins of the oral *Actinomyces* spp. with mammalian cells. See Ref. 18, pp. 45–52.

45. Geesey, G. G., Richardson, W. T., Yeomans, H. G., Irvin, R. T., Costerton, J. W. 1977. Microscopic examination of natural sessile bacterial populations from an alpine stream. *Can. J. Microbiol.* 23:1733–36

46. Gibbons, R. J. 1980. Adhesion of bacteria to the surfaces of the mouth. In *Microbiol Adhesion to Surfaces,* ed. R. C. W. Berkeley, J. M. Lynch, J. Melling, P. R. Rutter, B. Vincent, pp. 351–88. Chichester, England: Horwood

47. Gibbons, R. J. 1984. Adherent interactions which may affect microbial ecology in the mouth. *J. Dent. Res.* 63:378–85

48. Gibbons, R. J., Etherden, I., Skobe, Z. 1983. Association of fimbriae with the hydrophobicity of *Streptococcus sanguis* FC-1 and adherence to salivary pellicles. *Infect. Immun.* 41:414–17

49. Gibbons, R. J., Hay, D. I., Schluckebier, S. K. 1986. Proline-rich proteins are pellicle receptors for type 1 fimbriae of *A. viscosus. J. Dent. Res.* 65 (Special Issue):179 (Abstr.)

50. Gibbons, R. J., Nygaard, M. 1970. Interbacterial aggregation of plaque bacteria. *Arch. Oral Biol.* 15:1397–400

51. Gibbons, R. J., van Houte, J. 1971. Selective bacterial adherence to oral epithelial surfaces and its role as an ecological determinant. *Infect. Immun.* 3:567–73

52. Gibbons, R. J., van Houte, J. 1973. On the formation of dental plaques. *J. Periodontol.* 44:347–60

53. Gibbons, R. J., van Houte, J. 1980. Bacterial adherence and the formation of dental plaques. In *Bacterial Adherence: Receptors and Recognition,* ed. E. H. Beachey, pp. 63–104. London: Chapman & Hall

54. Girard, A. E., Jacius, B. H. 1974. Ultrastructure of *Actinomyces viscosus* and *Actinomyces naeslundii. Arch. Oral Biol.* 19:71–79

55. Güde, H. 1982. Interactions between floc-forming and nonfloc-forming bacterial populations from activated sludge. *Curr. Microbiol.* 7:347–50

56. Handley, P. S., Carter, P. L., Fielding, J. 1984. *Streptococcus salivarius* strains carry either fibrils or fimbriae on the cell surface. *J. Bacteriol.* 157:64–72

57. Handley, P. S., Carter, P. L., Wyatt, J. E., Hesketh, L. M. 1985. Surface structures (peritrichous fibrils and tufts of fibrils) found on *Streptococcus sanguis* strains may be related to their ability to coaggregate with other oral genera. *Infect. Immun.* 47:217–27

58. Heeb, M. J., Costello, A. H., Gabriel, O. 1982. Characterization of a galactose-specific lectin from *Actinomyces viscosus* by a model aggregation system. *Infect. Immun.* 38:993–1002

59. Heeb, M. J., Marini, A. M., Gabriel, O. 1985. Factors affecting binding of galac-

to ligands to *Actinomyces viscosus* lectin. *Infect. Immun.* 47:61–67

60. Holt, S. C., Leadbetter, E. R., Socransky, S. S. 1979. *Capnocytophaga:* new genus of gram-negative gliding bacteria. II. Morphology and ultrastructure. *Arch. Microbiol.* 122:17–27

61. Hughes, C. V., Kolenbrander, P. E., Andersen, R. N., Moore, L. V. H. 1988. Coaggregation properties of human oral *Veillonella:* relationship to colonization site and oral ecology. *Appl. Environ. Microbiol.* 54:In press

62. Jones, S. J. 1972. A special relationship between spherical and filamentous microorganisms in mature human dental plaque. *Arch. Oral Biol.* 17:613–16

63. Kagermeier, A., London, J. 1986. Identification and preliminary characterization of a lectinlike protein from *Capnocytophaga gingivalis* (emended). *Infect. Immun.* 51:490–94

64. Kagermeier, A. S., London, J., Kolenbrander, P. E. 1984. Evidence for the participation of *N*-acetylated amino sugars in the coaggregation between *Cytophaga* species strain DR2001 and *Actinomyces israelii* PK16. *Infect. Immun.* 44:299–305

65. Kelstrup, J., Funder-Nielsen, T. D. 1974. Aggregation of oral streptococci with *Fusobacterium* and *Actinomyces*. *J. Biol. Buccale* 2:347–62

66. Kolenbrander, P. E. 1982. Isolation and characterization of coaggregation-defective mutants of *Actinomyces viscosus, Actinomyces naeslundii,* and *Streptococcus sanguis*. *Infect. Immun.* 37:1200–8

67. Kolenbrander, P. E., Andersen, R. N. 1984. Cell to cell interactions of *Capnocytophaga* and *Bacteroides* species with other oral bacteria and their potential role in development of plaque. *J. Periodontal Res.* 19:564–69

68. Kolenbrander, P. E., and Andersen, R. N. 1985. Use of coaggregation-defective mutants to study the relationship of cell-to-cell interactions and oral microbial ecology. See Ref. 18, pp. 164–71

69. Kolenbrander, P. E., Andersen, R. N. 1986. Multigeneric aggregations among oral bacteria: a network of independent cell-to-cell interactions. *J. Bacteriol.* 168:851–59

70. Kolenbrander, P. E., Andersen, R. N. 1988. Intergeneric rosettes: sequestered surface recognition among human periodontal bacteria. *Appl. Environ. Microbiol.* 54:1046–50

71. Kolenbrander, P. E., Andersen, R. N., Holdeman, L. V. 1985. Coaggregation of oral *Fusobacterium, Selenomonas,*

and *Veillonella* with other oral gram-negative and gram-positive bacteria. *Abstr. Annu. Meet. Am. Soc. Microbiol.* 1985:26

72. Kolenbrander, P. E., Andersen, R. N., Holdeman, L. V. 1985. Coaggregation of oral *Bacteroides* species with other bacteria: central role in coaggregation bridges and competitions. *Infect. Immun.* 48:741–46

73. Kolenbrander, P. E., Celesk, R. A. 1983. Coaggregation of human oral *Cytophaga* species and *Actinomyces israelii*. *Infect. Immun.* 40:1178–85

74. Kolenbrander, P. E., Inouye, Y., Holdeman, L. V. 1983. New *Actinomyces* and *Streptococcus* coaggregation groups among human oral isolates from the same site. *Infect. Immun.* 41:501–6

75. Kolenbrander, P. E., Phucus, C. S. 1984. Effect of saliva on coaggregation of oral *Actinomyces* and *Streptococcus* species. *Infect. Immun.* 44:228–33

76. Kolenbrander, P. E., Williams, B. L. 1981. Lactose-reversible coaggregation between oral actinomycetes and *Streptococcus sanguis*. *Infect. Immun.* 33:95–102

77. Kolenbrander, P. E., Williams, B. L. 1983. Prevalence of viridans streptococci exhibiting lactose-inhibitable coaggregation with oral actinomycetes. *Infect. Immun.* 41:449–52

78. Komiyama, K., Gibbons, R. J. 1984. Inhibition of lactose-reversible adherence between *Actinomyces viscosus* and oral streptococci by salivary components. *Caries Res.* 18:193–200

79. Komiyama, K., Habbick, B. F., Gibbons, R. J. 1987. Interbacterial adhesion between *Pseudomonas aeruginosa* and indigenous oral bacteria isolated from patients with cystic fibrosis. *Can. J. Microbiol.* 33:27–32

80. Krasse, B. 1953. The proportional distribution of different types of streptococci in saliva and plaque material. *Odont. Revy* 4:304–12

81. Kuramitsu, H. K., Paul, A. 1980. Role of bacterial interactions in the colonization of oral surfaces by *Actinomyces viscosus*. *Infect. Immun.* 29:83–90

82. Lancy, P. Jr., Appelbaum, B., Holt, S. C., Rosan, B. 1980. Quantitative *in vitro* assay for "corncob" formation. *Infect. Immun.* 29:663–70

83. Lancy, P. Jr., Dirienzo, J. M., Appelbaum, B., Rosan, B., Holt, S. C. 1983. Corncob formation between *Fusobacterium nucleatum* and *Streptococcus sanguis*. *Infect. Immun.* 40:303–9

84. Lie, T. 1978. Ultrastructural study of

early dental plaque formation. *J. Peri-odontal Res.* 13:391–409

85. Liljemark, W. F., Bloomquist, C. G., Fenner, L. J. 1985. Characteristics of the adherence of oral *Haemophilus* species to an experimental salivary pellicle and to other oral bacteria. See Ref. 18, pp. 94–102

86. Liljemark, W. F., Bloomquist, C. G., Germaine, G. R. 1981. Effect of bacterial aggregations on the adherence of oral streptococci to hydroxyapatite. *Infect. Immun.* 31:935–41

87. Liljemark, W. F., Fenner, L. J., Bloomquist, C. G. 1986. In vivo colonization of salivary pellicle by *Haemophilus, Actinomyces* and *Streptococcus* species. *Caries Res.* 20:481–97

88. Liljemark, W. F., Gibbons, R. J. 1971. Ability of *Veillonella* and *Neisseria* species to attach to oral surfaces and their proportions present indigenously. *Infect. Immun.* 4:264–68

89. Liljemark, W. F., Gibbons, R. J. 1972. Proportional distribution and relative adherence of *Streptococcus miteor (mitis)* on various surfaces in the human oral cavity. *Infect. Immun.* 6:852–59

90. Lindberg, F., Lund, B., Johansson, L., Normark, S. 1987. Localization of the receptor-binding protein adhesin at the tip of the bacterial pilus. *Nature* 328:84–87

91. Listgarten, M. A. 1976. Structure of the microbial flora associated with periodontal health and disease in man. *J. Periodontol.* 47:1–18

92. Listgarten, M. A., Mayo, H., Amsterdam, M. 1973. Ultrastructure of the attachment device between coccal and filamentous microorganisms in "corn cob" formation of dental plaque. *Arch. Oral Biol.* 18:651–56

93. Listgarten, M. A., Mayo, H. E., Tremblay, R. 1975. Development of dental plaque on epoxy resin crowns in man. *J. Periodontol.* 46:10–25

94. Löe, H., Theilade, E., Jensen, S. B. 1965. Experimental gingivitis in man. *J. Periodontol.* 36:177–87

95. Loesche, W. J., Syed, S. A. 1978. Bacteriology of human experimental gingivitis: effect of plaque and gingivitis score. *Infect. Immun.* 21:830–39

96. Loesche, W. J., Syed, S. A., Schmidt, E., Morrison, E. C. 1985. Bacterial profiles of subgingival plaques in periodontitis. *J. Periodontol.* 56:447–56

97. London, J., Celesk, R., Kolenbrander, P. 1982. Physiological and ecological properties of the oral gram-negative gliding bacteria capable of attaching to hydroxyapatite. See Ref. 15, pp. 76–85

98. Mandell, R. L., Socransky, S. S. 1981. A selective medium for *Actinobacillus actinomycetemcomitans* and the incidence of the organism in juvenile periodontitis. *J. Periodontol.* 52:593–98

99. Maryanski, J. H., Wittenberger, C. L. 1975. Mannitol transport in *Streptococcus mutans*. *J. Bacteriol.* 124:1475–81

100. Mashimo, P. A., Murayama, Y., Reynolds, H., Mouton, C., Ellison, S. A., Genco, R. J. 1981. *Eubacterium saburreum* and *Veillonella parvula:* a symbiotic association of oral strains. *J. Periodontol.* 52:374–79

101. Masuda, N., Ellen, R. P., Grove, D. A. 1981. Purification and characterization of surface fibrils from taxonomically typical *Actinomyces viscosus* WVU627. *J. Bacteriol.* 147:1095–104

102. McBride, B. C., van der Hoeven, J. S. 1981. Role of interbacterial adherence in colonization of the oral cavities of gnotobiotic rats infected with *Streptococcus mutans* and *Veillonella alcalescens*. *Infect. Immun.* 33:467–72

103. McIntire, F. C. 1985. Specific surface components and microbial coaggregation. See Ref. 18, pp. 153–58

104. McIntire, F. C., Bush, C. A., Wu, S.-S., Li, S.-C., Li, Y.-T., et al. 1987. Structure of a new hexasaccharide from the coaggregation polysaccharide of *Streptococcus sanguis* 34. *Carbohydr. Res.* 166:133–43

105. McIntire, F. C., Crosby, L. K., Barlow, J. J., Matta, K. L. 1983. Structural preferences of β-galactoside-reactive lectins on *Actinomyces viscosus* T14V and *Actinomyces naeslundii* WVU45. *Infect. Immun.* 41:848–50

106. McIntire, F. C., Crosby, L. K., Vatter, A. E. 1982. Inhibitors of coaggregation between *Actinomyces viscosus* T14V and *Streptococcus sanguis* 34: β-galactosides, related sugars, and anionic amphipathic compounds. *Infect. Immun.* 36:371–78

107. McIntire, F. C., Vatter, A. E., Baros, J., Arnold, J. 1978. Mechanism of coaggregation between *Actinomyces viscosus* T14 and *Streptococcus sanguis* 34. *Infect. Immun.* 21:978–88

108. Mizuno, J., Cisar, J. O., Vatter, A. E., Fennessey, P. V., McIntire, F. C. 1983. A factor from *Actinomyces viscosus* T14V that specifically aggregates *Streptococcus sanguis* H1. *Infect. Immun.* 40:1204–13

109. Moch, T., Hoschutzky, H., Hacker, J., Kroncke, K.-D., Jann, K. 1987., Isolation and characterization of the α-sialyl-β-2,3-galactosyl specific adhesion from

fimbriated *Escherichia coli. Proc. Natl. Acad. Sci. USA* 84:3462–66

110. Moore, W. E. C., Holdeman, L. V., Cato, E. P., Smibert, R. M., Burmeister, J. A., et al. 1985. Comparative bacteriology of juvenile periodontitis. *Infect. Immun.* 48:507–19

111. Moore, W. E. C., Holdeman, L. V., Cato, E. P., Smibert, R. M., Burmeister, J. A., et al. 1987. Bacteriology of human gingivitis. *J. Dent. Res.* 66:989–95

112. Moore, W. E. C., Holdeman, L. V., Smibert, R. M., Cato, E. P., Burmeister, J. A., et al. 1984. Bacteriology of experimental gingivitis in children. *Infect. Immun.* 46:1–6

113. Moore, W. E. C., Holdeman, L. V., Smibert, R. M., Good, I. J., Burmeister, J. A., et al. 1982. Bacteriology of experimental gingivitis in young adult humans. *Infect. Immun.* 38:651–67

114. Mouton, C., Reynolds, H. S., Gasiecki, E. A., Genco, R. J. 1979. In vitro adhesion of tufted oral streptococci to *Bacterionema matruchotii. Curr. Microbiol.* 3:181–6

115. Mouton, C., Reynolds, H., Genco, R. J. 1977. Combined micromanipulation, culture and immunofluorescent techniques for the isolation of the coccal organisms comprising the "corn-cob" configuration of human dental plaque. *J. Biol. Buccale* 5:321–32

116. Nalbandian, J., Freedman, M. L., Tanzer, J. M., Lovelace, S. M. 1974. Ultrastructure of mutants of *Streptococcus mutans* with reference to agglutination, adhesion, and extracellular polysaccharide. *Infect. Immun.* 10:1170–79

117. Newman, H. N. 1972. Structure of approximal human dental plaque as observed by scanning electron microscopy. *Arch. Oral. Biol.* 17:1445–53

118. Ottow, J. C. G. 1975. Ecology, physiology, and genetics of fimbriae and pili. *Ann. Rev. Microbiol.* 29:79–108

119. Paerl, H. W., Gallucci, K. K. 1985. Role of chemotaxis in establishing a specific nitrogen-fixing cyanobacterial-bacterial association. *Science* 227:647–49

120. Revis, G. J., Vatter, A. E., Crowle, A. J., Cisar, J. O. 1982. Antibodies against the Ag2 fimbriae of *Actinomyces viscosus* T14V inhibit lactose-sensitive bacterial adherence. *Infect. Immun.* 36:1217–22

121. Ritz, H. L. 1967. Microbial population shifts in developing human dental plaque. *Arch. Oral Biol.* 12:1561–68

122. Deleted in proof

123. Sandberg, A. L., Mudrick, L. L., Cisar, J. O., Brennan, M. J., Mergenhagen, S. E., Vatter, A. E. 1986. Type 2 fimbrial lectin mediated phagocytosis of oral *Actinomyces* spp. by polymorphonuclear leukocytes. *Infect. Immun.* 54:472–76

124. Savitt, E. D., Socransky, S. S. 1984. Distribution of certain subgingival microbial species in selected periodontal conditions. *J. Periodont. Res.* 19:111–23

125. Schwarz, S., Ellen, R. P., Grove, D. A. 1987. *Bacteroides gingivalis–Actinomyces viscosus* cohesive interactions as measured by a quantitative binding assay. *Infect. Immun.* 55:2391–97

126. Shelton, D. R., Tiedje, J. M. 1984. Isolation and partial characterization of bacteria in an anaerobic consortium that mineralizes 3-chlorobenzoic acid. *Appl. Environ. Microbiol.* 48:840–48

127. Slots, J., Emrich, L. J., Genco, R. J., Rosling, B. G. 1985. Relationship between some subgingival bacteria and periodontal pocket depth and gain or loss of periodontal attachment after treatment of adult periodontitis. *J. Clin. Periodontol.* 12:540–52

128. Slots, J., Gibbons, R. J. 1978. Attachment of *Bacteroides melaninogenicus* subsp. *asaccharolyticus* to oral surfaces and its possible role in colonization of the mouth and of periodontal pockets. *Infect. Immun.* 19:254–64

129. Slots, J., Möenbo, D., Lanebaek, J., Frandsen, A. 1978. Microbiota of gingivitis in man. *Scand. J. Dent. Res.* 86:174–81

130. Slots, J., Reynolds, H., Genco, R. J. 1980. *Actinobacillus actinomycetemcomitans* in human periodontal disease: a cross-sectional microbiological investigation. *Infect. Immun.* 29:1013–20

131. Socransky, S. S., Manganiello, A. D., Propas, D., Oram, V., van Houte, J. 1977. Bacteriological studies of developing supragingival dental plaque. *J. Periodont. Res.* 12:90–106

132. Stolp, H., Starr, M. P. 1963. *Bdellovibrio bacteriovorus* gen. et sp. n., a predatory ectoparasitic, and bacteriolytic microorganism. *Antonie van Leeuwenhoek J. Microbiol. Seriol.* 29:217–48

133. Syed, S. A., Loesche, W. J. 1978. Bacteriology of human experimental gingivitis: effect of plaque age. *Infect. Immun.* 21:821–29

134. Tanner, A. C. R., Haffer, C., Bratthall, G. T., Visconti, R. A., Socranksy, S. S. 1979. A study of bacteria associated with advancing periodontitis in man. *J. Clin. Periodontol.* 6:278–307

135. Tempro, P. J., London, J., Siraganian, R. P., Hand, A. R., Cassels, F. 1988. Characterization of monoclonal antibodies against *Capnocytophaga gingivalis* outer membrane antigens. *J. Dent. Res.* 67:126

136. Tew, J. G., Marshall, D. R., Burmeister, J. A., Ranney, R. R. 1985. Relationship between gingival crevicular fluid and serum antibody titers in young adults with generalized and localized periodontitis. *Infect. Immun.* 49:487–93

137. Theildade, E., Wright, W. H., Jensen, S. B., Löe, H. 1966. Experimental gingivitis in man. II. A longitudinal clinical and bacteriological investigation. *J. Periodont. Res.* 1:1–13

138. Tinanoff, N., Gross, A., Brady, J. M. 1976. Development of plaque on enamel: parallel investigations. *J. Periodontal Res.* 11:197–209

139. Tylenda, C. A., Enriquez, E., Kolenbrander, P. E., Delisle, A. L. 1985. Simultaneous loss of bacteriophage receptor and coaggregation mediator activities in *Actinomyces viscosus* MG-1. *Infect. Immun.* 48:228–33

140. van der Hoeven, J. S., de Jong, M. H., Kolenbrander, P. E. 1985. In vivo studies of microbial adherence in dental plaque. See Ref. 18, pp. 220–27

141. van Houte, J., Gibbons, R. J., Banghart, S. B. 1970. Adherence as a determinant of the presence of *Streptococcus salivarius* and *Streptococcus sanguis* on the human tooth surface. *Arch. Oral Biol.* 15:1025–34

142. van Palenstein Helderman, W. H. 1975. Total viable count and differential count of *Vibrio (Campylobacter) sputorum, Fusobacterium nucleatum, Selenomonas sputigena, Bacteroides ochraceus* and *Veillonella* in the inflamed and noninflamed human gingival crevice. *J. Periodont. Res.* 10:294–305

143. Weerkamp, A. H., Handley, P. S., Baars, A., Slot, J. W. 1986. Negative staining and immunoelectron microscopy of adhesion-deficient mutants of *Streptococcus salivarius* reveal that the adhesive protein antigens are separate classes of cell surface fibril. *J. Bacteriol.* 165:746–55

144. Weerkamp, A. H., Jacobs, T. 1982. Cell wall–associated protein antigens of *Streptococcus salivarius:* purification, properties, and function in adherence. *Infect. Immun.* 38:233–42

145. Weerkamp, A. H., McBride, B. C. 1980. Characterization of the adherence properties of *Streptococcus salivarius*. *Infect. Immun.* 29:459–68

146. Weerkamp, A. H., McBride, B. C. 1980. Adherence of *Streptococcus salivarius* HB and HB-7 to oral surfaces and saliva-coated hydroxyapatite. *Infect. Immun.* 30:150–58

147. Weerkamp, A. H., van der Mei, H. C., Engelen, D. P. E., de Windt, C. E. A. 1984. Adhesin receptors (adhesins) of oral streptococci. In *Bacterial Adhesion and Preventive Dentistry*, ed. J. M. tenCate, S. A. Leach, J. Arends, pp. 85–97. Oxford, England: IRL

148. Weerkamp, A. H., van der Mei, H. C., Liem, R. S. 1984. Adhesive cell wall–associated glycoprotein of *Streptococcus salivarius* (K$^+$) is a cell surface fibril. *FEMS Microbiol. Lett.* 23:163–66

149. Weerkamp, A. H., van der Mei, H. C., Liem, R. S. 1986. Structural properties of fibrillar proteins isolated from the cell surface and cytoplasm of *Streptococcus salivarius* (K$^+$) cells and nonadhesive mutants. *J. Bacteriol.* 165:756–62

150. Weiss, E. I., Kolenbrander, P. E., London, J., Hand, A. R., Andersen, R. N. 1987. Fimbria-associated proteins of *Bacteroides loescheii* PK1295 mediate intergeneric coaggregations. *J. Bacteriol.* 169:4215–22

151. Weiss, E. I., London, J., Kolenbrander, P. E., Kagermeier, A. S., Andersen, R. N. 1987. Characterization of lectin-like surface components on *Capnocytophaga ochracea* ATCC33596 that mediate coaggregation with gram-positive oral bacteria. *Infect. Immun.* 55:1198–202

152. Weiss, E. I., London, J., Kolenbrander, P. E., Andersen, R. N., Fischler, C., Siraganian, R. P. 1988. Characterization of monoclonal antibodies to fimbria-associated adhesins of *Bacteroides loescheii* PK1295. *Infect. Immun.* 56:219–24

153. Weiss, E. I., London, J., Kolenbrander, P. E., Hand, A. R., Siraganian, R. 1988. Localization and enumeration of fimbria-associated adhesins of *Bacteroides loescheii*. *J. Bacteriol.* 170:1123–28

154. Wheeler, T. T., Clark, W. B. 1980. Fibril-mediated adherence of *Actinomyces viscosus* to saliva-treated hydroxyapatite. *Infect. Immun.* 28:577–84

155. Williams, B. L., Ebersole, J. L., Spektor, M. D., Page, R. C. 1985. Assessment of serum antibody patterns and analysis of subgingival microflora of members of a family with a high prevalence of early-onset periodontitis. *Infect. Immun.* 49:742–50

156. Yeung, M. K., Chassy, B. M., Cisar, J. O. 1987. Cloning and expression of a

type 1 fimbrial subunit of *Actinomyces viscosus* T14V. *J. Bacteriol.* 169:1678–83

157. Yoshimura, F., Takahashi, K., Nodasaka, Y., Suzuki, T. 1984. Purification and characterization of a novel type of fimbriae from the oral anaerobe *Bacteroides gingivalis*. *J. Bacteriol.* 160:949–57

158. Yoshimura, F., Takasawa, T., Yone-yama, M., Yamaguchi, T., Shiokawa, H., Suzuki, T. 1985. Fimbriae from the oral anaerobe *Bacteroides gingivalis:* physical, chemical, and immunological properties. *J. Bacteriol.* 163:730–34

159. Zeikus, J. G. 1983. Metabolic communication between biodegradative populations in nature. *Symp. Soc. Gen. Microbiol.* 34:423–62

Ann. Rev. Microbiol. 1988. 42:657–83
Copyright © 1988 by Annual Reviews Inc. All rights reserved

EVOLUTION OF RNA VIRUSES

James H. Strauss and Ellen G. Strauss

Division of Biology, California Institute of Technology, Pasadena, California 91125

CONTENTS

INTRODUCTION

The evolution of RNA viruses has received a great deal of attention in the past several years. The subject is based on information very different from that considered in other evolutionary discussions. As obligate intracellular parasites, viruses leave no fossil record; indeed the oldest historical accounts describing symptoms believed to be caused by viruses are only a millenium or two old (51, 73). Recently, however, the complete nucleotide sequences of the genomic RNAs of a number of viruses belonging to different virus families have been obtained, and these make possible detailed comparisons of the families on a molecular level. These comparisons have led to new insights

657

into RNA virus evolution, including the beginning of taxonomy based on the relatedness of their genomes, a better understanding of the radiation of viruses to different hosts, and renewed speculation on the origin of viruses. In addition, partial or complete sequences have been obtained for many different strains of certain viruses or for different members of a given family. The comparative sequence data have led to a greater understanding of the rate of divergence of RNA viruses in nature and in cell culture. All of these studies, however, have been of viruses currently extant; the oldest virus isolates date back only to the turn of this century.

Until recently viruses were grouped and classified by such parameters as the structure of the virion; host range (in particular whether plant or animal); the transmitting vector, if relevant; disease syndrome; and for closely related viruses, antigenic cross-reaction. Later, viral taxonomy was refined using protein structure and the differential stability of viruses in the presence of various chemical and physical agents. By such methods most known viruses have been grouped into families (71). However, recent comparisons on the sequence level have led to super groupings containing more than one family, which are thought to reflect ancestral relationships among seemingly divergent viruses. These superfamily relationships were first hinted at by similarities in genome organization and structure (reviewed in 101) such as the order, number, and type of genes along the genome, the nature of modifications at the 3' and 5' ends of the RNA and the RNA polarity, whether posttranslational processing occurs, whether subgenomic RNAs are produced, and whether readthrough of stop codons occurs. With complete nucleotide sequences available, it has become clear that there are long stretches of amino acid sequence similarity in the replicase proteins of certain groups of viruses of both plants and animals (1, 22, 31, 46), which suggests that these superfamilies have descended from a common ancestor. Furthermore, X-ray diffraction studies at high resolution have revealed that many icosahedral viruses, hitherto considered totally unrelated, have capsid proteins whose folding in three-dimensional space is remarkably similar despite the absence of sequence similarities in the proteins (reviewed in 93). Thus, on the basis of sequence similarities and higher order similarities, we speculate that all of the positive-stranded RNA viruses, i.e. those that contain the message sense RNA as their genome, have evolved from a common ancestor; mutation, recombination, and selection have produced the present members of this group, which infect many different hosts (plants, insects, and higher animals) and have a variety of divergent morphologies. Furthermore, all of the currently known negative-stranded viruses (which package the nonmessage sense RNA as their genome) appear to form a fairly homogeneous group, which suggests that they too descended from a common ancestor, although nucleotide sequence homology and amino acid sequence homology are limited in these families. It remains to be seen whether the positive- and negative-stranded viruses (and, for that matter, the double-stranded RNA viruses) also shared a common ancestor at some more distant time in the past; however, the flaviviruses have characteristics in common with both the positive- and negative-stranded viruses and may represent a bridge between ancestral positive- and negative-stranded RNA viruses.

Longitudinal studies of virus strains isolated in nature have shown that the genome of RNA viruses is remarkably plastic (reviewed in 50, 99). In a number of different studies, mutations have been shown to accumulate in virus genomes at the rate of 0.1%–1% per year. In view of this extraordinarily high rate of divergent evolution, it is remarkable that amino acid sequence homologies between RNA viruses are detectable at all.

A number of review articles have recently appeared covering different aspects of RNA virus evolution in detail, including genome organization of RNA viruses (101), amino acid sequence similarities in viral replicase proteins (35, 38, 49), and the rapid divergence of RNA genomes (50, 98, 99).

SUPERFAMILIES OF RNA VIRUSES: EVOLUTIONARY RELATIONSHIPS

The RNA-containing viruses are a diverse group. Different families vary in morphology, genome organization, and host range. Recently, however, on the basis of deduced amino acid sequence homology, primarily in the regions of the replicase proteins, and other considerations (discussed in 101), a number of RNA viruses have been grouped into several superfamilies. These superfamilies are listed in Table 1 together with certain characteristics and a representative member for each component family.

Superfamily I: Alphavirus-like Families

Replicase proteins from several groups of plant viruses, including the tobamoviruses, the bromoviruses, the ilarviruses, the tobraviruses, and carnation mottle virus (CarMV), have been shown to share amino acid sequence homology with each other and with the replicase proteins of the alphaviruses (1, 7, 22, 41, 45, 46). Several morphologies are represented: tobacco mosaic virus (TMV) is a helical rod, alfalfa mosaic virus (AlMV) is bacilliform, brome mosaic virus (BMV) is icosahedral, and Sindbis virus (SIN) is enveloped. Furthermore, CarMV has a single piece of RNA of only 4 kb as its genome, TMV has a 7-kb genome, and SIN has a genome of 11.7 kb. In constrast, the genome of tobacco rattle virus (TRV) is bipartite, whereas those of BMV, AlMV, and cucumber mosaic virus (CMV) (76) are tripartite. Three regions of amino acid sequence homology, each extending over several hundred amino acids, are present in whole or in part in the replicase proteins

Table 1 Relationships of RNA virus superfamilies[a]

Taxon	Virus[b]	Genome structure				Translation strategies			Virion structure[e]	Other characters	
		Polarity	5' End	3' End	mRNAs[c]/origin of caps	Polyprotein processing	RT[d]	Overlapping ORFs		Size[f]/Number of segments	Primary host phyla/vectors
Superfamily 1: Sindbis-like											
Alphavirus (*Togaviridae*)	SIN	+	CAP	polyA	1g + 1sg/viral	+	+	–	I core; E	11.7/1	mammals, birds/mosquitoes
Tobamovirus	TMV	+	CAP	$tRNA_{His}$	1g + 2sg/viral	–	+	–	rod	6.4/1	plants
Bromovirus	BMV	+	CAP	$tRNA_{Tyr}$	3g + 1sg/viral	–	–	–	S; I	8.2/3	plants
Cucumovirus	CMV	+	CAP	$tRNA_{Tyr}$	3g + 1sg/viral	–	–	–	S; I	8/3	plants
Ilarvirus	AIMV	+	CAP	X_{OH}	3g + 1sg/viral	–	–	–	S; bacilliform	8.3/3	plants
Tobravirus	TRV	+	CAP	C_{OH}	2g + 2sg/viral	–	+	–	rod	8.5/2	plants
Unclassified	CarMV	+	?	X_{OH}	1g + 1sg/viral	–	+	–	I	4.3/1	plants
Superfamily 2: Picornavirus-like											
Picornaviridae	POLIO	+	Vpg	polyA	1g/none	+	–	–	I	7.4/1	mammals, insect/none
Caliciviridae	VESV	+	Vpg	polyA	1g + 1sg(?)/none	+	–	–	I	7.8/1	mammals/none
Comovirus	CPMV	+	Vpg	polyA	2g/none	+	–	–	S; I	9.4/2	plants/none
Nepovirus	TBRV	+	Vpg	polyA	2g/none	+	–	–	S; I	12.6/2	plants/nematodes
Potyvirus	TEV	+	Vpg	polyA	1g/none	+	–	–	rod	9.5/1	plants/none
Superfamily 3: Negative-stranded											
Paramyxoviridae (3 genera)	RSV	–	self-complementary		8–10/viral	–	–	+	H core; E	~15.4/1	mammals, birds/none
Rhabdoviridae (2 genera)	VSV	–	self-complementary		5/viral	–	–	+	H core; E; bulletshaped	11.2/1	vertebrates, arthropods, plants/mosquitoes, aphids

Orthomyxoviridae (3 genera)	FLU	–	self-complementary	10/host	–	+(splicing)	H core; E	13.6/8	mammals, birds/none
Bunyaviridae (5 genera)	LAC	–	self-complementary	3–4/host	+	Ambisense/+	H core; E	14–17/3	vertebrates, arthropods/mosquitoes, ticks
Arenaviridae	LCM	–	self-complementary	3/?	+	Ambisense	E	10–12/2	rodents/none
Superfamily 4: Double-stranded									
Reoviridae (6 genera)	REO	DS	CAP X_{OH}	10/viral	–	+	I; double protein	~23/10–12	vertebrates, plants, insects/mosquitoes, ticks, leafhoppers
Fungal viruses		DS		?/?	+		S	?/1 to 3	fungi/none
Birnaviruses	IBDV	DS	?	2/viral	+		I; double protein	~6/2	insects, mollusks, fish, birds
Unassigned Viruses									
Flaviviridae	YF	+	CAP structure	none	+	–	I; E	11/1	vertebrates/mosquitoes, ticks
Coronaviridae	IBV	+	CAP polyA	6 nested sg/viral	–	+	H core; E	28–30/1	vertebrates/none
Nodaviridae	BBV	+	CAP polyA	1sg/viral	+	+	I	4.5/2	insects

[a]Data are from References 38, 39, 53, 58, 59, 71, 75, and 101.
[b]SIN = Sindbis; TMV = tobacco mosaic; BMV = brome mosaic; CMV = cucumber mosaic; AlMV = alfalfa mosaic; TRV = tobacco rattle; CarMV = carnation mottle; VESV = vesicular exanthema of swine; CPMV = cowpea mosaic; TBRV = tomato black ring; TEV = tobacco etch; RSV = respiratory syncytial; VSV = vesicular stomatitis; REO = reovirus; IBDV = infectious bursal disease; FLU = influenza; LAC = La Crosse; LCM = lymphocytic choriomeningitis; YF = yellow fever; IBV = avian infectious bronchitis; BBV = black beetle.
[c]g = genomic RNA; sg = subgenomic RNA.
[d]RT = readthrough of termination codons during translation.
[e]H = helical; I = isometric; E = enveloped; S = separate encapsidation.
[f]Entire genome size in kilobases.

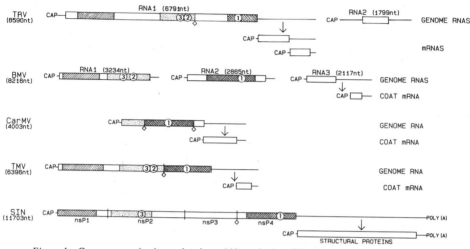

Figure 1   Genome organization and amino acid homologies of Sindbis virus and four families of plant viruses. Virus families and abbreviations are given in Table 1. Untranslated regions are shown as *lines,* translated regions as *open boxes.* Readthrough *(open diamonds)* of termination codons produces downstream products. Within the translated regions three areas of homology are indicated with different types of shading. The figure contains data from References 1, 41, and 45. Circled numbers *1, 2,* and *3* indicate the location of shared sequence motifs: the GDD motif (45, 60), the Hodgman motif (48), and the GKS/T motif (49), respectively.

of each of these families (Figure 1). Domains 1 and 2 are found in the RNA 1 segment of AlMV and BMV, and comparable domains are translated from RNA 1 of TRV and from the genomic RNAs of TMV and SIN upstream of a termination codon (amber in TMV, opal in TRV and SIN). These domains remain associated in one polypeptide in the plant viruses, but are separated by posttranslational cleavage as nsP1 and nsP2 in SIN. CarMV contains a truncated version of the nsP2 domain, and the nsP1 domain is absent. Domain 3 is present on the RNA 2 segments of AlMV and BMV and is present downstream of the amber termination codon in TMV and CarMV and downstream of the opal codon in TRV and SIN (in polypeptide nsP4 in SIN). Domain 3 is thus either downstream of a termination codon (and therefore producible only by readthrough) or on an independent segment, separated from domains 1 and 2, which indicates that domain 3 may be under different regulatory control from domains 1 and 2. [In some alphaviruses, however, the opal codon is replaced by a sense codon (100, 103) and the precise function of the termination codon is not clear.] In addition, all of the viruses in this group produce their capsid proteins from a subgenomic message located near the 3' end of the genome (or of a genome segment). These similarities in sequence, genome organization, and translation strategy have led to the suggestion that all of these families are descended from a common ancestor.

Each family in turn consists of a number of individual viruses or strains that are closely related and that have arisen by divergent evolution from a common ancestor or, at least in a few cases, by recombination between members of the family (discussed below). For example, the Alphavirus genus of the family *Togaviridae* comprises about 25 known animal viruses. These viruses have been extensively studied by sequence analysis (reviewed in 102), and it is clear that they diverged from one another fairly recently (see section on rate of divergence). All of their proteins, both structural and nonstructural, exhibit clear sequence homologies with one another. The extent of the amino acid sequence conservation varies between 25 and 90%, depending upon the protein domains being compared and the viruses under consideration. The most highly conserved domains are domains 1, 2, and 3 of the replicase proteins that are shared with the plant viruses (Figure 1). The many different strains and isolates of each alphavirus are even more closely related.

The two other genera of *Togaviridae* are Pestivirus, containing bovine diarrhea virus and hog cholera virus, and Rubivirus, whose sole member is rubella (the virus for German measles). Rubella virus (RUB) has now been sequenced in part, and its genome organization appears to be identical to that of alphaviruses (18, 32, 33, 77, 80, 106); RUB clearly belongs to the *Togaviridae*. Even so, no amino acid sequence homology has been demonstrated between the structural proteins of RUB and any of the alphaviruses, even though these related viruses are morphologically similar if not identical. However, limited amino acid sequence similarity between RUB and the alphaviruses can be seen in the C terminus of nsP4, the only nonstructural protein sequence currently available for rubella (32). Sequence homology at the nucleotide level can be seen in the junction region between the nonstructural and structural proteins (Figure 2). This sequence is believed to form a binding site or promoter element for transcription of the subgenomic mRNA (reviewed in 102). The conservation of this nucleotide sequence despite the extensive divergence at the level of both structural and nonstructural proteins is noteworthy and has implications for the mechanisms involved in transcription of the subgenomic mRNAs.

The various members of this superfamily have distinct morphologies. Since viruses belonging to the same family, such as SIN and RUB, show no sequence similarity in their structural proteins, it is unlikely that sequence similarity in the virion proteins will be found among viruses belonging to different families, and none has been found within this superfamily. Three-dimensional similarities in capsid proteins do exist among isometric viruses (see below), and such higher order similarities almost certainly exist between alphaviruses and RUB as well. Some of the structural proteins, however, appear to be completely unrelated; the coat protein of TMV (37, 78) has no characteristics in common with alphavirus structural proteins or with capsid proteins of the icosahedral plant viruses. Similarly, the envelope of togavi-

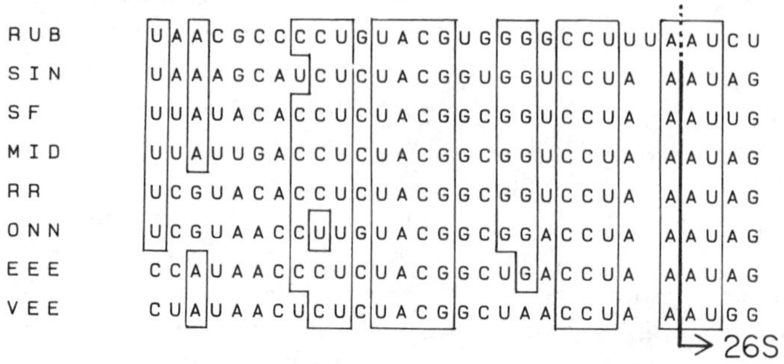

Figure 2 Nucleotide sequence immediately upstream of the start of the subgenomic 26S RNA for rubella virus (RUB) compared to those of seven alphaviruses: Sindbis virus (SIN), Semliki Forest virus (SF), Middelburg virus (MID), Ross River virus (RR), O'Nyong-nyong virus (ONN), Eastern equine encephalitis virus (EEE), and Venezuelan equine encephalitis virus (VEE). Data for RUB are from Reference 32, and those for alphaviruses are from References 17, 57, 100, and 102. Nucleotides in RUB that are shared with at least four other viruses are *boxed*. The start of the alphavirus subgenomic 26S RNA is indicated; the start of the RUB subgenomic RNA is unknown, but it is postulated from homology and indicated by the *dotted* extension of the line.

ruses has no parallel in the other families. These comparisons indicate that the various families of this superfamily may have diverged in part by recombination, acquiring wholly new structural proteins and new morphologies (see also section on recombination, below).

Superfamily II: Picornavirus-like Families

Amino acid sequence homology in several domains between the proteins of the animal picornaviruses and those of the plant comoviruses (3, 31) forms the basis for superfamily II (reviewed in 38, 49). Also included in this family are the nepoviruses, whose structure and organization are virtually identical to those of the comoviruses, and for which amino acid sequence similarities have been demonstrated in the polymerase protein (74); the potyviruses, whose genome organization is similar to that of the picornaviruses and whose polymerase shares sequence similarities with other members of the group (2, 25), although the virion structure is quite distinct; and the caliciviruses, whose assignment is based on similarities in genome organization and replication strategy only (101). The genome organizations and regions of sequence homology between the picornaviruses and the comoviruses are illustrated in Figure 3. The animal picornaviruses contain a single molecule of RNA as their genome, while the plant comoviruses have segmented genomes, with two pieces of RNA separately encapsidated. Separate encapsidation of independent genome segments does not appear to confer a selective disadvantage on plant viruses and is found commonly among them (Table 1).

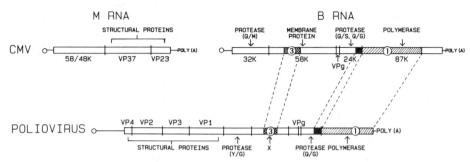

Figure 3 Genome organization and comparison of a comovirus (CMV) and a picornavirus (poliovirus). Conventions are the same as those in Figure 1. The circled *1* and *3* are the same sequence motifs highlighted in Figure 1. Three regions of extensive sequence similarity have been shown with three types of shading. Data are from Reference 31.

The family *Picornaviridae* contains four genera of animal RNA viruses with single-stranded genomes of approximately 7.5 kb (reviewed in 64, 81), which lack 5' terminal caps but are polyadenylated at the 3' end. In place of a cap structure, a small protein (VPg) encoded in the viral genome and probably involved in initiation of RNA replication is covalently linked to the 5' terminus of the RNA. The genome has a single long open reading frame and is translated as one large polyprotein, which is posttranslationally processed by proteolytic cleavage using two or three virus-encoded proteases.

The plant comoviruses are similar in their genome organization, except the genome is divided into two segments. Each segment has a VPg covalently attached to the 5' end, and each segment has a poly(A) tract on the 3' end; both features are unusual in plant viruses. These features are also shared with the plant nepoviruses. In the comoviruses and nepoviruses each genome segment is translated as a polyprotein and processed posttranslationally by cleavage, presumably by one or more virus-encoded proteases (40).

Three regions of amino acid homology have been established between the comoviruses and the picornaviruses (Figure 3). These three regions include one domain within the replicase protein, a domain within one of the viral proteases, and a domain in a protein of unknown function. It is likely that the picornaviruses and comoviruses shared a common ancestor at some time in the past, and that upon radiation to plants the picornavirus genome was divided into two segments, one encoding the replicase proteins and one encoding primarily the structural proteins of the virion.

The animal caliciviruses also contain a VPg and a poly(A) tract, and on this basis are also probably closely related to picornaviruses and comoviruses.

Superfamily III: Negative-Stranded Viruses

The negative-stranded viruses comprise a more homogeneous group than the positive-stranded viruses. Five families of negative-stranded viruses are currently recognized, and they have numerous structural properties in common (101). Firstly, the nucleocapsid in each case is helical, consisting of the RNA, one predominant nucleocapsid protein, and minor amounts of replicase proteins. Secondly, all of the negative-stranded viruses are enveloped and contain in their envelopes glycoproteins encoded in the respective viral genomes. A more detailed analysis of their genome organization as well as limited sequence similarity suggests that all negative-stranded viruses descended from a common ancestor.

RHABDOVIRUSES AND PARAMYXOVIRUSES   The rhabdoviruses and paramyxoviruses have nonsegmented negative-strand genomes of approximately 11 and 15 kb, respectively (19, 83, 90). There are two genera of animal *Rhabdoviridae* (Vesiculovirus and Lyssavirus) plus a large number of unclassified rhabdoviruses (including viruses of plants and insects), and there are three genera of *Paramyxoviridae* (Paramyxovirus, Morbillivirus, and Pneumovirus). Transcription begins at the 3' end of the genome, sequentially producing a small nontranslated leader RNA followed by a series of capped and polyadenylated mRNAs that are almost all monocistronic. These mRNAs encode a nucleocapsid protein, a matrix protein, a fusion protein (in paramyxoviruses), a major glycoprotein, one or more nonstructural proteins, and a large replicase protein. The location and order of these genes for *Paramyxoviridae* and *Rhabdoviridae* are illustrated in Figure 4a. Paramyxoviruses are larger than rhabdoviruses and encode more proteins, but the proteins are aligned in the same order along the genome such that each gene in the rhabdoviruses has a direct counterpart in the paramyxoviruses. It is also notable that in both these groups overlapping open reading frames are used (albeit sparingly) to increase the information content of the genomic RNA (36, 47). Viewed from this level, the similar genome organization suggests that these two families are evolutionarily related. Sequencing studies on representatives of the two groups of viruses reveal little amino acid similarity between corresponding proteins, although some similarity is found in the L proteins (109) and there are nucleotide sequence homologies within the intergenic regions (42) and in the leader sequences (66).

Once again, each family within the group contains many members. The rhabdoviruses are a particularly diverse group (86), with members that infect plants (56), insects, and vertebrates (96). The two recognized genera of animal *Rhabdoviridae*, Vesiculovirus [type virus vesicular stomatitis virus (VSV)] and Lyssavirus (whose best known member is rabies virus) exhibit only weak amino acid sequence similarity, although the homology is clear

a

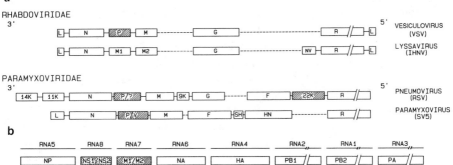

Figure 4 Genome organization of negative-strand viruses. (*a*) Comparison of *Rhabdoviridae* and *Paramyxoviridae*, drawn roughly to scale. For VSV the proteins are N, the nucleocapsid protein; P, the phosphoprotein (formerly called NS); M, the matrix protein; G, the glycoprotein; and R, the replicase (formerly L, the large protein). The untranslated leaders are shown as L (90). For IHNV, infectious hemapoietic necrosis virus, M1 is a phosphoprotein corresponding to P, M2 corresponds to M, and there is also a small nonvirion protein NV (65). For *Paramyxoviridae*, N is the nucleocapsid protein, P is the phosphoprotein, M is the matrix protein, F is the fusion protein, and R is the replicase. For RSV several additional genes are known (19). Simian virus 5 (SV5) contains HN, the hemagglutinin-neuraminidase glycoprotein, and a small hydrophobic nonstructural protein (SH) located between F and HN (83). *Hatched* genes are those for which two proteins appear to be encoded in the same gene (either two products encoded in different frames or two products from independent initiation at different places in the same frame) (36, 47). (*b*) The genome segments of influenza virus aligned to correspond to the gene order in *a*. NP, the nucleocapsid protein; NS_1 and NS_2, nonstructural proteins; M1, matrix protein; M2, a nonstructural protein; NA, neuraminidase; HA, hemagglutinin; PB1, PB2, and PA, components of the replicase. For RNAs 7 and 8, two products are encoded in each segment in different frames, translated from differentially spliced mRNAs (67).

(89). It is of interest that an insect rhabdovirus, sigma virus, appears to be closer to VSV (which infects higher vertebrates, although it is vectored by mosquitoes) than VSV is to rabies virus (104).

SEGMENTED NEGATIVE-STRAND VIRUSES   The remaining three families of negative-strand viruses, the *Orthomyxoviridae,* the *Bunyaviridae,* and the *Arenaviridae,* all have segmented genomes and share a number of features. The family *Orthomyxoviridae* contains three genera, Influenzavirus A, B, and C. All influenza viruses contain eight segments of negative-stranded RNA (67). Six segments are transcribed into monocistronic mRNAs, which are translated into three replicase proteins, P1, P2, and P3; two envelope proteins, HA and NA; and a nucleocapsid protein, NP. The two smallest segments, 7 and 8, each give rise to two mRNAs owing to differential splicing of the message causing the use of different reading frames. Segment 7 encodes the matrix protein M and, upon splicing, a nonstructural protein known as M2. Segment 8 encodes nonstructural proteins NS1 and NS2. None of the mRNAs are exact complements of the corresponding genome segments; they contain a 5' terminal cap and an oligonucleotide 12–14 nucleotides long derived from a host mRNA that serves as a primer for transcription, and they end with a poly(A) tract not present in the genomic RNA. Most of the proteins encoded by the influenza viruses have counterparts in the paramyxovirus and rhabdovirus genomes; it is possible to align the independent segments to correspond to the gene order found for the *Paramyxoviridae* and *Rhabdoviridae* (see Figure 4*b*). This suggests that the influenza viruses may have evolved from a protoparamyxovirus that became segmented during evolution to permit reassortment and greater diversity (see section on recombination, below). It is also possible that independent genome segments from a protoinfluenza virus were assembled into a paramyxovirus-like genome to allow concomitant regulation of all the genes. However, little amino acid sequence similarity now remains between these diverse families, although the influenza M protein is homologous to the M protein of the rhabdoviruses (89) and the HN protein of the paramyxovirus Sendai virus has domains homologous to domains within the influenza HA and NA glycoproteins (10).

The *Bunyaviridae* (Bunyavirus, Hantavirus, Nairovirus, Phlebovirus, and Uukuvirus) contain three genome segments (9). The largest RNA encodes the replicase, the middle segment encodes two external virion glycoproteins, and the small segment encodes the nucleocapsid protein (54, 55). Bunyaviruses also use 5' terminal caps and attendant short oligonucleotides derived from host mRNAs as primers for their polyadenylated messenger RNAs (8, 30, 84). This suggests a relationship to the influenza viruses, although transcription of the mRNAs occurs in the cytoplasm in bunyaviruses (91), rather than in the nucleus as in influenza virus.

The arenaviruses contain two genome segments, an L segment and an S segment (20). The L segment encodes the replicase and corresponds to the L segment of the bunyaviruses. The S segment encodes two glycoproteins and the capsid protein by an ambisense strategy (5) and corresponds to the M and S segments of bunyaviruses linked head to head. In arenaviruses one message is transcribed from the virion S RNA (nominally negative stranded) and is translated into the two glycoproteins. The nucleocapsid protein is translated from a second mRNA transcribed from RNA complementary to S RNA (vcSRNA); this mRNA is of genomic polarity, although the genomic S RNA itself is not translated. The two regions transcribed from S RNA do not overlap. A similar ambisense strategy is used for S RNA of the Phlebovirus genus, but not in the other four Bunyaviridae genera (54, 55). Thus the arenaviruses and the bunyaviruses may have had a common ancestor in which the S segment of the arenaviruses was split to form the two segments in the current bunyaviruses. Conversely, two segments of a protobunyavirus may have fused to form the single ambisense S RNA of the arenaviruses.

An additional similarity shared by all of the negative-strand viruses is a short stretch of self-complementary nucleotide sequence at the ends of the genome (or at the ends of the genome segments in the segmented viruses) (reviewed in 101). Thus the genome RNA (or genome segments) can cyclize to form circles of greater or lesser stability. Of greater import, however, is the probable role of these terminal sequences in replication. Firstly, all segments of a segmented virus contain highly conserved self-complementary terminal sequences. Secondly, the same short oligonucleotide sequence is found both at the 3' terminus of the genome RNA and at the 3' terminus of the virion complementary RNA. Thirdly, such conserved sequence elements may serve as initiation sites or replicase binding sites for RNA replication (101). Thus in negative-strand viruses initiation of replication is similar for both positive and negative strands, whereas in positive-stranded viruses (except the flaviviruses, which are discussed below) the terminal sequences of the positive and negative strands are very different.

Double-Stranded RNA Viruses

The double-stranded RNA viruses comprise, for the most part, a remarkably homogeneous group of viruses, which are currently classified as a single family, the *Reoviridae* (58). Within the *Reoviridae* there are six genera: Orthoreovirus, animal viruses that infect higher vertebrates, including humans; Rotavirus, vertebrate viruses that include important human pathogens; Orbivirus, animal viruses that infect vertebrates and insects; Phytoreovirus, plant viruses, some of which are vectored by insects; Fijivirus, plant viruses vectored by insects; and Cypovirus, insect viruses. There are also numerous ungrouped double-stranded RNA viruses. These viruses all have a genome of 10–12 segments of double-stranded RNA encased in a double-layered protein capsid with cubic symmetry. Subviral core particles containing RNA polymerase activity are the sites of mRNA transcription. Single-stranded mRNAs transcribed from each segment are translated to produce the virus products and then assembled with the coat proteins into provirions. The final stage of viral maturation consists of the transcription of a negative-stranded complement of these single-stranded messages to form the double-stranded RNA genome segments. All of these double-stranded viruses undergo efficient reassortment of segments during replication to generate recombinant viruses, at least in the laboratory. The rate of sequence divergence of these viruses, where examined, has been found similar to that of the single-stranded RNA viruses.

There is also a second class of double-stranded virus-like agents, the double-stranded particles of fungi (68, 105). These agents differ in several ways from the reoviruses. For example, they never form complete infectious particles that are found outside their host cells, but they are instead passed from cytoplasm to cytoplasm during mating or by vertical transmission during spore formation. They also differ in having a genome that consists of one, two, or three pieces of RNA. However, the replication strategy of these particles appears to be otherwise similar in many ways to that of the reoviruses (79, 85), and the fungal viruses are probably evolutionarily related to the other double-stranded RNA viruses.

A third family of double-stranded RNA viruses comprises the birnaviruses, which cause disease in insects, mollusks, and birds. These agents have two segments of double-stranded RNA. One encodes the viral polymerase. The other encodes three structural proteins, which are translated as a polyprotein and proteolytically processed posttranslationally. These agents are morphologically similar to reoviruses, but little is known about their RNA replication (6, 53, 71).

Unassigned Groups: Missing Links?

There are a number of families of plant viruses that we have not attempted to classify into any of the superfamilies discussed above (reviewed in 38, 39). In addition, there are two well-known and highly successful families of animal viruses that do not fit easily into any of the above classifications, the coronaviruses and the flaviviruses.

The family *Flaviviridae* is monogeneric and contains about 70 animal viruses, many of which are important human pathogens (reviewed in 88). Their genome is single stranded and of plus polarity, and it shares many features with the genomes of several other groups of RNA viruses. The RNA is capped on the 5' end, but like the genomes of negative-strand viruses and reoviruses it lacks a 3' terminal poly(A) tail. Instead, the RNA forms a remarkably stable hydrogen-bonded stem-and-loop structure at the 3' terminus, which is reminiscent of plant-virus 3' termini. The flavivirus genome is organized as a single long open reading frame translated as one polyprotein, which is processed by posttranslational cleavage. The structural proteins are found at the 5' end and the nonstructural replicase components at the 3' end; this organization is reminiscent of the picornaviruses. On the other hand, the morphology of the flaviviruses is very similar to that of alphaviruses. Finally, the extreme 5' and 3' termini of the genome RNA contain short self-complementary nucleotide sequences reminiscent of those found in negative-strand viruses. Thus, the *Flaviviridae* seem to combine features characteristic of many other virus groups and may be descended from a protovirus that has since radiated to form many other contemporary virus families. With the exception of the brief sequence motifs discussed in the following section, no sequence homologies have been demonstrated between the flaviviruses and any other group of RNA viruses.

The *Coronaviridae* is a family of animal viruses that are in many ways distinct (97). They contain a very large genome, up to 30 kb in size, of single-stranded RNA of positive polarity (12). The nucleocapsid is a quasihelix and is enveloped in a lipoprotein envelope. The genome RNA is transcribed into a nested set of capped and polyadenylated subgenomic mRNAs that are 3' coterminal. Only the 5' domain of any message is translated into protein. A short sequence at the 5' end of the genomic RNA is used as a primer to initiate transcription of all subgenomic messenger RNAs (70). The complete genome of avian infectious bronchitis virus has now been sequenced (12), and the polymerase or replicase gene occupies 20 kb. Throughout the entire coronavirus genome no detectable sequence homologies have been found to any other group of RNA viruses.

One final group of viruses with mixed characteristics is the *Nodaviridae*. These are positive-stranded insect viruses with two genome segments (as in several plant viruses) encapsidated together in an isometric particle. The two RNAs are capped and polyadenylated, and the capsid protein is translated from a subgenomic message (23). The identification of an insect virus family with characteristics of both plant and animal viruses has led to the suggestion that the prototype positive-strand virus originated in insects.

CONSERVED MOTIFS AND SHORT SEQUENCE SIMILARITIES

Outside the examples cited above, extensive regions of amino acid sequence homology are not found between members of diverse virus families; however, a number of relatively short stretches of homology or sequence motifs have been identified, particularly in the replicase proteins. These short regions are composed of 10–15 contiguous amino acids. It is unclear whether these regions represent examples of convergent evolution for a particular enzymatic function or to create a particular three-dimensional domain, whether they are the remnants of genes that were all originally obtained from virus hosts, or whether they indicate that all of these divergent viruses are indeed related to one another. One motif, which was first identified by Kamer & Argos (60), consists of the sequence Gly-Asp-Asp surrounded by hydrophobic amino acids. This motif is found in the Sindbis-like viruses (at the position of the circled number 1 in Figure 1), in the picornavirus/comovirus group (at the circled 1 in Figure 3), and in the flaviviruses in nonstructural protein NS5 (45, 60, 88). The identification of this motif led to the suggestion that the picornavirus-like families and the alphavirus-like families (as well as the flaviviruses) were all related and that perhaps all RNA viruses had a common ancestor. However, this motif has not been found in the replicase gene of the coronaviruses or among the negative-stranded viruses, although a related motif consisting of Asp-Asp has been found in some of these viruses (95, 109). A second motif, approximately 14–15 amino acids long, has been identified by Hodgman (48) and is centered about the sequence Tyr/Phe-Gly-Asp-Thr-Asp/Glu. This motif is found at the circled 2 in Figure 1. In addition, a very similar motif has been found in the DNA polymerases of adenovirus 2 and three members of the herpes virus family, but the function of this domain is unknown. A third motif of the same general variety, consisting of Gly-Lys-Ser/Thr, has been recognized by Hodgman & Zimmern (49). It is found in the second conserved domain of Figure 1 at the circled 3 and also at the circled 3 in Figure 3. This motif is not only shared by the picorna-like viruses and the alpha-like viruses, but is also found in ATP-binding proteins such as adenylate kinase, *Escherichia coli* ATPase β, and rabbit myosin. A related motif is seen in the GTP-binding sites of elongation factors Tu and G as well.

SIMILARITIES IN THREE-DIMENSIONAL STRUCTURE OF ICOSAHEDRAL CAPSID PROTEINS

The three-dimensional structures of several isometric plant viruses and two members of the picornavirus family have been determined by X-ray crystallography to atomic resolution (reviewed in 93). The capsid proteins of the picornaviruses and of plant viruses such as tomato bushy stunt and southern bean mosaic viruses, although sharing no detectable amino acid sequence similarity with each other, have a remarkably similar three-dimensional structure, which has been described as an eight-stranded anti-parallel β-barrel. The capsid protein of black beetle virus (an insect Nodavirus) also has similar folding (52). Because of these structural similarities, the proteins appear to have descended from a common ancestral protein under selection pressures such that the only conservation is at the level of protein structure. This suggests that it is possible for the linear amino acid sequence to diverge until no trace of homology is left while the structure is little changed (see, for example, 92). Similarly in the picornaviruses, three of the four monomeric units, VP1, VP2, and VP3, all share the common eight-fold β-barrel configuration, even though the linear amino acid sequences are markedly divergent; this suggests that gene duplication events gave rise to three copies of the structural protein, which subsequently diverged independently (93).

These similarities in structure lend support to the hypothesis that all of these positive-stranded viruses have descended from a common ancestral protovirus. It has been suggested that the capsid protein of the alphaviruses may also be folded in the same way (34) and that all cubic viruses may be packaged by proteins of a common origin (see 4). Some authors have proposed that the attachment proteins of all viruses arose from one protocapsid protein (93).

RECOMBINATION AS A FORCE IN RNA VIRUS EVOLUTION

Recombination is certainly an important force in RNA virus evolution. As in other organisms, it allows more rapid testing of new combinations of genes than does simple mutation, even though the mutation frequency is very high in RNA viruses (see section on rate of divergence). Three levels of recombination can be distinguished: recombination between two virus strains or virus species producing a virus that has an assortment of features from the two parents but that is still a member of the same genus; modular evolution, in which the replicase genes, the capsid genes, or other genes evolve quasi-independently and are reassorted into new viruses to form new genera or families; and the acquisition of new properties from the host cell, although this level is still speculative.

Recombination Among Members of a Genus

The most common form of recombination among the RNA viruses is the independent reassortment of different viral genome segments during a mixed infection to produce progeny with characteristics from both parents. Such reassortment is readily demonstrable in the laboratory for most segmented viruses. The best studied case of reassortment in nature is influenza virus; it has been shown that new epidemic strains arise when the genes allowing the virus to replicate efficiently in humans (primarily the replicase genes) are combined with genes from another host that encode new surface antigens. The recombinants (reassortants) can replicate in humans, but the human population has no immunological resistance to them (24, 107). Successful reassorted viruses occur rarely (on average every 10 years or so) but can cause great pandemics. It can even be argued that the segmented genome arose in part because of the selective advantage conferred by the ability to recombine readily. In particular, recombination allows viruses to acquire new structural proteins to evade immune pressures; it also allows plant viruses to swap genes that allow efficient transmission by vectors or spread from cell to cell.

Recombination in nonsegmented viruses or within a segment in segmented viruses, in which the progeny genome consists of covalently linked polynucleotides that were derived from more than one parent, is more difficult to demonstrate for RNA viruses. Recombination has occurred in cell culture for picornaviruses (21) and coronaviruses (69), and was recently achieved by forced selection for brome mosaic virus (15). In poliovirus, recombination has been shown to occur by a copy-choice mechanism during RNA replication (63); this is presumably the mechanism in the other viruses as well. With poliovirus, recombination has also occurred in human vaccinees (61) who received simultaneous high doses of three different polio strains, all of which had been impaired to some extent. The importance of such recombination in nature as a general mechanism for generating new successful strains has not been proven, although one clear-cut case in the alphaviruses has recently been described. Recombination between alphaviruses has not been demonstrated in the laboratory, which indicates that the frequency of such events is quite low, but we have recently found that western equine encephalitis virus (WEE) arose by recombination between eastern equine encephalitis virus (EEE) and a Sindbis-like virus (43). WEE had always been somewhat of a mystery because it cross-reacts antigenically with Sindbis virus, but like EEE it is a New World virus that causes encephalitis in humans and animals. Nucleotide sequence data indicate that the 5' region of the genome, presumably including all of the nonstructural domain, the junction region, and the nucleocapsid protein, is derived from the EEE parent, while the glycoproteins E1 and E2 (which form the major antigenic determinants) are derived from the Sindbis-like parent. A sequence comparison dot plot that illustrates this point is shown in Figure 5. WEE is today widely disseminated throughout western North America from Canada to Mexico and is also present in much of South America. Thus this recombination event has given rise to a highly successful new virus.

Modular Evolution

The concept of modular evolution, in which the various genes of a virus evolve quasi-independently and are reassorted into new viruses, was first proposed for bacteriophages, but parts of this model are applicable to RNA virus evolution as well (reviewed recently in 49). For example, TMV has been grouped into the Sindbis-like superfamily because of homologies in the replicases. However, the coat protein of TMV has no similarity in either

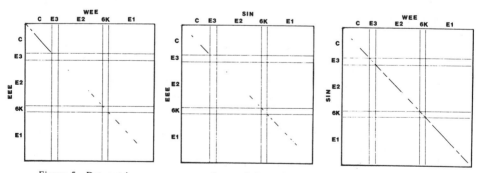

Figure 5 Dot matrix sequence comparisons of the amino acid sequences of the structural proteins of SIN, EEE, and WEE. The string length for the comparison was 14, and 10 matches were required to put a dot on the diagonal. Axes are labeled with the names of the structural proteins C, E2, and E1. E3 and 6K are signal peptides not found in mature virus.

sequence or three-dimensional structure with the coat proteins of any other member of the superfamily. On the other hand, TMV has a protein of 30 kd that is believed to facilitate virus spread from cell to cell; this protein is homologous to a 29-kd protein in TRV, to a protein encoded on RNA 3 of BMV and AlMV, and to a domain in RNA 1 of two members of the picornavirus superfamily, cowpea mosaic virus (CPMV) and tomato black ring virus (TBRV). The sharing of particular functional domains is easier to explain by recombination than by divergent (or convergent) evolution.

Viral Proteases

Many biologists would argue that viruses arose from cellular elements that acquired the ability to be packaged and to be transmitted from cell to cell. In this model all virus functions could ultimately be traced back to a cellular origin. The question remains whether during evolution viruses have continued to acquire new functions from their hosts. We and others have speculated that viral proteases, which process viral precursor polyproteins, might have been acquired from the host cell. In the alphaviruses, one of the proteases appears to be homologous to serine proteases, although convergent evolution cannot be excluded as a possible source of the sequence similarities (44). Homologies have been found between enzymes encoded in DNA viruses and cellular proteins (14, 72), suggesting that DNA viruses have acquired new functions from their hosts.

RATE OF DIVERGENCE OF RNA VIRUSES

The genomes of RNA viruses are quite plastic and diverge very rapidly. (This is not to imply that DNA viruses are more stable, but this review is limited to RNA viruses.) The rate of mutation has been estimated as on the order of 10^{-4} per nucleotide per round of replication, although this may vary from virus to virus (98, 99). The rapidity has been attributed, at least in part, to the lack of proofreading activity in RNA replicases. It has been argued that this high rate of mutation sets an upper limit to the size of RNA virus genomes (87). Because of the very high rate of mutation and the large number of rounds of replication of an RNA virus per year, the potential rate of divergence in the nucleotide sequence is very rapid. However, the actual rate of divergence, i.e. the fixing in the genome of mutated residues, is limited by selection, since most mutations are deleterious. Viruses containing deleterious mutations would either fail to replicate or would be placed at selective disadvantage with respect to the parental population. Attempts have been made to measure the actual rate of divergence in a number of systems and to compare the divergence of viruses during passage in cell culture with the divergence of the same virus during natural passage. The latter studies are of more interest, although the measured divergence rates in culture and in nature do not differ very much (reviewed in 50, 99). Figure 6 illustrates one attempt to quantitate the rate of RNA sequence divergence in a virus in nature. In this study of influenza virus isolates the rate of divergence was 1.0% per year in the third codon position (where many of the changes are silent) and 0.50% per year in both the first and second codon positions. Studies of other RNA viruses have found divergence rates of 0.03–2.0% per year, depending upon the virus and the gene studied (98). Although the methods used may overestimate the basal rate of divergence, the results are in striking contrast to the rate of divergence in animals of 1% per million years (13). This enormously high rate of RNA sequence divergence has important implications for

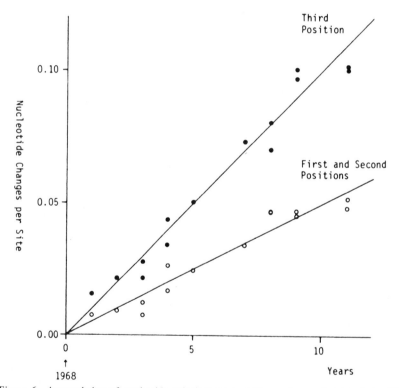

Figure 6 Accumulation of nucleotide substitutions in influenza hemagglutinin of the H3 subtype. The number of changes accumulating in first and second codon positions *(open circles)* and third codon position *(filled circles)* are plotted as a function of time. Regression lines with a coefficient of 0.0050 per site per year and 0.0104 per site per year are superimposed for the data of the first and second positions and the third position, respectively. The ancestral strain was isolated in 1968. Reproduced from Saitou & Nei (94), with permission of the authors and University of Chicago Press.

thinking about the evolutionary origin of the current families of RNA-containing viruses. It is clear that two present-day virus families that demonstrate amino acid sequence homology and that infect plants and vertebrates, respectively, could not have coevolved with their hosts. On the other hand, viruses that today contain no detectable sequence similarity could conceivably have shared a common ancestor within a few thousand years; lack of sequence similarity does not necessarily imply that two viruses are unrelated. Viruses whose genome organizations are very similar may have diverged very recently even if they lack sequence similarity. Similarities in organization therefore may be more useful as indicators of evolutionary relationships than sequence homologies.

The very high rate of mutation also has important implications for RNA viruses as disease pathogens, in that new viruses arise continuously. Furthermore, the structural proteins, which are subject to immune pressure, diverge more rapidly than the nonstructural proteins. The rapid drift of many viruses means that vaccine strains must continually be modified to keep abreast of the current virus. This has been particularly important for influenza virus (11, 16, 94) and foot and mouth disease virus (26, 62). Other viruses appear to diverge somewhat more slowly. The parental strain of yellow fever virus used to generate the 17D vaccine strain was isolated more than 50 years ago, but the virus has not yet changed sufficiently in the field to necessitate a new vaccine. Perhaps viruses that replicate alternately in a vertebrate host and an invertebrate vector have a lower rate of sequence divergence than viruses with only a human host such as influenza or poliovirus. It is not clear if the rate is lower because replication in both vertebrates and arthropods places two sets of different and not completely compatible constraints on the virus, or because fewer replication cycles per unit time occur in the cold-blooded vector in the absence of immune pressure, or both.

Because of the high rate of mutation, any population of RNA viruses, even a population from a single clone, consists of individuals that differ from one another at many nucleotide positions (the quasispecies concept) (27). Under stable conditions, selection maintains an average sequence, but in the presence of changing environmental circumstances many variants preexisting in the population may acquire selective advantage.

The concept of an RNA virus diverging a million times as fast as its DNA-containing host does not meet with uniform enthusiasm in the scientific community. In a counter example, Gibbs (35) has described a strain of turnip yellow mosaic virus found in the mountains of Australia that may only have changed by 1% at the nucleotide sequence level during the past 10,000 years. This would be an extraordinarily slow rate of RNA sequence divergence in comparison with the rates of other RNA viruses that have been examined, even after allowance for the effect of colder temperatures on the rate of replication and the small number of passages per year. A much lower rate of mutation, on the order of 10^{-7}, even in the apparent absence of selection, has also been reported for particular mutations in poliovirus (82) and Sindbis virus (28). This suggests that the rate of change at least at certain nucleotide positions can be much lower than the average mutation rate observed for the population as a whole.

CONCLUDING REMARKS

The origin of RNA viruses is a topic that has generated a great deal of discussion and controversy. Many believe that viruses arose from cellular elements that acquired the ability to self-replicate and self-assemble. Once a primitive protovirus existed, it could diversify relatively rapidly by acquiring point mutations sequentially. It could also acquire wholly new capabilities from rare events of recombination with other cellular elements, which would thus expand the pool of genes available. The question of when RNA viruses first arose is interesting. A recent paper has speculated that the 3' terminal structures present on many viral RNAs, especially the tRNA-like structures in plant viruses, represent remnants from the RNA world (108) that existed before DNA and proteins arose (29). The fact that current RNA viruses infect such diverse groups as bacteria, plants, and animals also suggests that these viruses have been around for a long time. However, given the high rate of divergence and recombination between RNA viruses, it seems unlikely that we will ever be able to trace the evolutionary history of the extant viruses back more than a few thousand years (or at best a few million years if we have overestimated the current rate of divergence). Although the recent flood of nucleotide sequence information has greatly explanded our understanding of virus evolution, other approaches may be needed to unravel the more tenuous associations between viruses that have diverged so completely that sequence similarities no longer exist. The determination of three-dimensional structures has been of great value. Examination of the three-dimensional structures of viral replicases and comparison of these structures with those of cellular RNA and DNA polymerases may go a long way toward illuminating the evolutionary events that have occurred.

Given the extraordinary replication efficiency of RNA viruses, it seems inherently unlikely that the various families of RNA viruses arose independently. It is most likely that they arose from an ancestral protovirus through evolutionary divergence, including recombinational events. Moreover, we know that the current rate of sequence divergence among RNA viruses would permit all of the currently extant RNA viruses to have diverged from a common ancestor within a relatively short period, measured in thousands (or at most millions) of years. Furthermore, the continued divergence of RNA viruses poses a clear threat to our physical well being. Within the last 200 years most of the more serious RNA viral diseases have been controlled by the now classical method of vaccination. Alternative therapies such as antiviral chemotherapy have been of remarkably limited use. However, we now recognize that RNA viruses will continue to evolve rapidly as they have over the millennia. As the recent epidemic of AIDS makes clear, new pathogens can and will arise.

ACKNOWLEDGMENTS

We are grateful to many colleagues for stimulating discussions and for sharing preprints with us prior to publication. We particularly want to thank R. Goldbach, D. Zimmern, A. Gibbs, and T. Frey. Work of the authors is supported by Grants AI10793 and AI20612 from the National Institutes of Health and Grant DMB8617372 from the National Science Foundation.

Literature Cited

1. Ahlquist, P., Strauss, E. G., Rice, C. M., Strauss, J. H., Haseloff, J., Zimmern, D. 1985. Sindbis virus proteins nsP1 and nsP2 contain homology to nonstructural proteins from several RNA plant viruses. *J. Virol.* 35:536–42

2. Allison, R., Johnston, R. E., Dougherty, W. G. 1986. The nucleotide sequence of the coding region of tobacco etch virus genomic RNA: evidence for the synthesis of a single polyprotein. *Virology* 154:9–20

3. Argos, P., Kamer, G., Nicklin, M. J. H., Wimmer, E. 1984. Similarity in gene organization and homology between proteins of animal picornaviruses and a plant comovirus suggest common ancestry of these virus families. *Nucleic Acids Res.* 12:7251–67

4. Argos, P., Tsukihara, T., Rossmann, M. G. 1980. A structural comparison of concanavalin A and tomato bushy stunt virus protein. *J. Mol. Evol.* 15:169–79

5. Auperin, D. D., Romanowski, V., Galinski, M., Bishop, D. H. L. 1984. Sequencing studies of Pichinde arenavirus S RNA indicate a novel coding strategy, an ambisense viral S RNA. *J. Virol.* 52:897–904

6. Becht, H. 1980. Infectious bursal disease virus. *Curr. Top. Microbiol. Immunol.* 90:107–21

7. Bergh, S. T., Koziel, M. G., Huang, S.-C., Thomas, R. A., Gilley, D. P., Siegel, A. 1985. The nucleotide sequence of tobacco rattle virus RNA-2 (CAM strain). *Nucleic Acids Res.* 13:8507–18

8. Bishop, D. H. L., Gay, M. E., Matsuoko, Y. 1983. Nonviral heterogeneous sequences are present at the 5' ends of one species of snowshoe hare bunyavirus S complementary RNA. *Nucleic Acids Res.* 11:6409–18

9. Bishop, D. H. L., Shope, R. E. 1979. Bunyaviridae. In *Comprehensive Virology*, Vol. 14, *Newly Characterized Vertebrate Viruses*, ed. H. Fraenkel-Conrat, R. R. Wagner, pp. 1–156. New York: Plenum

10. Blumberg, B., Giorgi, C., Roux, L., Raju, R., Dowling, P., et al. 1985. Sequence determination of the Sendai virus HN gene and its comparison to the influenza virus glycoproteins. *Cell* 41:269–78

11. Both, G. W., Sleigh, M. J., Cox, N. J., Kendal, A. P. 1983. Antigenic drift in influenza virus H3 hemagglutinin from 1968 to 1980: multiple evolutionary pathways and sequential amino acid changes at key antigenic sites. *J. Virol.* 48:52–60

12. Boursnell, M. E. G., Brown, T. D. K., Foulds, I. J., Green, P. F., Tomley, F. M., Binns, M. M. 1987. Completion of the sequence of the genome of the coronavirus avian infectious bronchitis virus. *J. Gen. Virol.* 68:57–77

13. Britten, R. J. 1986. Rates of DNA sequence evolution differ between taxonomic groups. *Science* 231:1393–98

14. Broyles, S. S., Moss, B. 1986. Homology between RNA polymerases of poxviruses, prokaryotes, and eukaryotes: nucleotide sequence and transcriptional

analysis of vaccinia virus genes encoding 147-kDa and 22-kDa subunits. *Proc. Natl. Acad. Sci. USA* 83:3141–45

15. Bujarski, J. J., Kaesberg, P. 1986. Genetic recombination between RNA components of a multipartite plant virus. *Nature* 321:528–31

16. Buonagurio, D. A., Nakada, S., Parvin, J. D., Krystal, M., Palese, P., Fitch, W. M. 1986. Evolution of human influenza A viruses over 50 years: rapid, uniform rate of change in NS gene. *Science* 232:980–82

17. Chang, G.-J. J., Trent, D. W. 1987. Nucleotide sequence of the genome region encoding the 26S mRNA of Eastern equine encephalomyelitis virus and the deduced amino acid sequence of the viral structural proteins. *J. Gen. Virol.* 68:2129–42

18. Clarke, D. M., Loo, T. W., Hui, I., Chong, P., Gillam, S. 1987. Nucleotide sequence and *in vitro* expression of rubella virus 24S subgenomic messenger RNA encoding the structural proteins E1, E2, and C. *Nucleic Acids Res.* 15:3041–57

19. Collins, P. L., Huang, Y. T., Wertz, G. W. 1984. Identification of a tenth mRNA of respiratory syncytial virus and assignment of polypeptides to the 10 viral genes. *J. Virol.* 49:572–78

20. Compans, R. W., Bishop, D. H. L. 1985. Biochemistry of arenaviruses. *Curr. Top. Microbiol. Immunol.* 114:153–75

21. Cooper, P. D. 1977, Genetics of picornaviruses. See Ref. 9, 9:133–207

22. Cornelissen, B. J. C., Bol, J. F. 1984. Homology between proteins encoded by tobacco mosaic virus and two tricornaviruses. *Plant Mol. Biol.* 3:379–84

23. Dasmahapatra, B., Dasgupta, R., Ghosh, A., Kaesberg, P. 1985. Structure of the black beetle virus genome and its functional implications. *J. Mol. Biol.* 182:183–89

24. Desselberger, U., Nakajima, K., Alfino, P., Pedersen, F. S., Haseltine, W. A., et al. 1978. Biochemical evidence that "new" influenza virus strains in nature may arise by recombination (reassortment). *Proc. Natl. Acad. Sci. USA* 75:3341–45

25. Domier, L. L., Shaw, J. G., Rhoads, R. E. 1987. Potyviral proteins share amino acid sequence homology with picorna-, como-, and caulimoviral proteins. *Virology* 158:20–27

26. Domingo, E., Davila, M., Ortin, J. 1980. Nucleotide sequence heterogeneity of the RNA from a natural population of foot-and-mouth-disease virus. *Gene* 11:333–46

27. Domingo, E., Martinez-Salas, E., Sobrino, F., de la Torre, J.-C., Portela, A., et al. 1985. The quasispecies (extremely heterogeneous) nature of viral RNA genome populations: biological relevance—a review. *Gene* 40:1–8

28. Durbin, R. K., Stollar, V. 1986. Sequence analysis of the E2 gene of a hyperglycosylated, host restricted mutant of Sindbis virus and estimation of mutation rate from frequency of revertants. *Virology* 154:135–43

29. Eigen, M., Gardiner, W. 1984. Evolutionary molecular engineering based on RNA replication. *Pure Appl. Chem.* 56:967–78

30. Eshita, Y., Ericson, B., Romanowski, V., Bishop, D. H. L. 1985. Analyses of the mRNA transcription processes of snowshoe hare bunyavirus S and M RNA species. *J. Virol.* 55:681–89

31. Franssen, H., Leunissen, J., Goldbach, R., Lomonossoff, G., Zimmern, D. 1984. Homologous sequences in nonstructural proteins from cowpea mosaic virus and picornaviruses. *EMBO J.* 3:855–61

32. Frey, T. K., Marr, L. D. 1988. Sequence of the region coding for virion proteins C and E2 and the carboxyterminus of the nonstructural proteins of rubella virus. *Gene* 62:85–99

33. Frey, T. K., Marr, L. D., Hemphill, M. L., Dominguez, G. 1986. Molecular cloning and sequencing of the region of the rubella virus genome coding for glycoprotein E1. *Virology* 154:228–32

34. Fuller, S. D., Argos, P. 1987. Is Sindbis a simple picornavirus with an envelope? *EMBO J.* 6:1099–105

35. Gibbs, A. 1987. Molecular evolution of viruses: "trees," "clocks," and "modules." In *Proc. 7th John Innes Symp.* Cambridge: Co. Biol./J. Cell Sci. In press

36. Giorgi, C., Blumberg, B. M., Kolakovsky, D. 1983. Sendai virus contains overlapping genes expressed from a single mRNA. *Cell* 35:829–36

37. Goelet, P., Lomonossoff, G. P., Butler, P. J. G., Akam, M. E., Gait, M. J., Karn, J. 1982. Nucleotide sequence of tobacco mosaic virus RNA. *Proc. Natl. Acad. Sci. USA* 79:5818–22

38. Goldbach, R. 1986. Molecular evolution of plant RNA viruses. *Ann. Rev. Phytopathol.* 24:289–310

39. Goldbach, R. 1987. Genome similarities between plant and animal RNA viruses. *Microbiol. Sci.* 4:197–202

40. Goldbach, R., Van Kammen, A. 1985. Structure, replication and expression of the bipartite genome of cowpea mosaic virus. In *Molecular Plant Virology,* ed.

J. Davis, 2:83–120. Boca Raton, Fla: CRC

41. Guilley, H., Carrington, J. C., Balazs, E., Jonard, G., Richards, K., Morris, T. J. 1985. Nucleotide sequence and genome organization of carnation mottle virus RNA. *Nucleic Acids Res.* 13: 6663–77

42. Gupta, K. C., Kingsbury, D. W. 1984. Complete sequences of the intergenic and mRNA start signals in the Sendai virus genome: homologies with the genome of vesicular stomatitis virus. *Nucleic Acids Res.* 12:3829–41

43. Hahn, C. S., Lustig, S., Strauss, E. G., Strauss, J. H. 1988. Western equine encephalitis virus is a recombinant virus. *Proc. Natl. Acad. Sci. USA* In press

44. Hahn, C. S., Strauss, E. G., Strauss, J. H. 1985. Sequence analysis of three Sindbis virus mutants temperature-sensitive in the capsid autoprotease. *Proc. Natl. Acad. Sci. USA* 82:4648–52

45. Hamilton, W. D. O., Bocarro, M., Robinson, D. J., Baulcombe, D. C. 1987. The complete nucleotide sequence of tobacco rattle virus RNA-1. *J. Gen. Virol.* 68:2563–75

46. Haseloff, J., Goelet, P., Zimmern, D., Ahlquist, P., Dasgupta, R., Kaesberg, P. 1984. Striking similarities in amino acid sequence among nonstructural proteins encoded by RNA viruses that have dissimilar genomic organization. *Proc. Natl. Acad. Sci. USA* 81:4358–62

47. Herman, R. C. 1986. Internal initiation of translation on the vesicular stomatitis virus phosphoprotein mRNA yields a second protein. *J. Virol.* 58:797–804

48. Hodgman, T. C. 1986. An amino acid sequence motif linking viral DNA polymerases and plant virus proteins involved in RNA replication. *Nucleic Acids Res.* 14:6769

49. Hodgman, T. C., Zimmern, D. 1987. Evolution of RNA Viruses. In *RNA Genetics*, ed. J. J. Holland, E. Domingo, P. Ahlquist. Boca Raton, Fla: CRC. In press

50. Holland, J., Spindler, K., Horodyski, F., Grabau, E., Nichol, S., VandePol, S. 1982. Rapid evolution of RNA genomes. *Science* 215:1577–85

51. Hopkins, D. R. 1983. *Princes and Peasants: Smallpox in History.* Chicago: Chicago Univ. Press

52. Hosur, M. V., Staffacher, J. E., Usha, R., Schmidt, T., Harrington, M., et al. 1986. The structure of cowpea mosaic virus and black beetle virus determined by single crystal X-ray diffraction methods. *Abstr. EMBO Workshop Mol. Plant Virol., Wageningen, the Netherlands,* p. 22

53. Hudson, P. J., McKern, N. M., Power, B. E., Azad, A. A. 1986. Genomic structure of the large RNA segment of infectious bursal disease virus. *Nucleic Acids Res.* 14:5001–12

54. Ihara, T., Akashi, H., Bishop, D. H. L. 1984. Novel coding strategy (ambisense genomic RNA) revealed by sequence analyses of Punta Toro phlebovirus S RNA. *Virology* 136:293–306

55. Ihara, T., Smith, J., Dalrymple, J. M., Bishop, D. H. L. 1985. Complete sequences of the glycoproteins and M RNA of Punta Toro phlebovirus compared to those of Rift Valley fever virus. *Virology* 144:246–59

56. Jackson, A. O., Francki, R. I. B., Zuidema, D. 1987. Biology, structure and replication of plant rhabdoviruses. In *The Rhabdoviruses,* ed. R. R. Wagner, pp. 427–508. New York: Plenum

57. Johnson, B. J. B., Kinney, R. M., Kost, C. L., Trent, D. W. 1986. Molecular determinants of alphavirus neurovirulence: nucleotide and deduced protein sequence changes during attenuation of Venezuelan equine encephalitis virus. *J. Gen. Virol.* 67:1951–60

58. Joklik, W. K. 1985. Recent progress in reovirus research. *Ann. Rev. Genet.* 19: 537–75

59. Joshi, S., Haenni, A.-L. 1984. Plant RNA viruses: strategies of expression and regulation of viral genes. *FEBS Lett.* 177:163–77

60. Kamer, G., Argos, P. 1984. Primary structural comparison of RNA-dependent polymerases from plant, animal, and bacterial viruses. *Nucleic Acids Res.* 12:7269–82

61. Kew, O. M., Nottay, B. K. 1984. Evolution of the oral polio vaccine strains in humans occurs by both mutation and intramolecular recombination. In *Modern Approaches to Vaccines: Molecular and Chemical Basis of Virus Virulence,* ed. R. M. Chanock, pp. 357–62. New York: Cold Spring Harbor Lab.

62. King, A. M. Q., Underwood, B. O., McCahon, D., Newman, J. W. I., Brown, F. 1981. Biochemical identification of viruses causing the 1981 outbreaks of foot-and-mouth disease in the UK. *Nature* 293:479–80

63. Kirkegaard, K., Baltimore, D. 1986. The mechanism of RNA recombination in poliovirus. *Cell* 47:433–43

64. Kuhn, R. J., Wimmer, E. 1987. The replication of picornaviruses In *The Molecular Biology of the Positive Strand Viruses,* ed. D. J. Rowlands, M. A. Mayo, B. W. J. Mahy, pp. 17–51. London: Academic

65. Kurath, G., Ahern, K. G., Pearson, G.

D., Leong, J. C. 1985. Molecular cloning of the six mRNA species of infectious hematopoietic necrosis virus, a fish rhabdovirus, and gene order determination by R-loop mapping. *J. Virol.* 53:469–76

66. Kurilla, M. G., Stone, H. O., Keene, J. D. 1985. RNA sequence and transcriptional properties of the 3' end of the Newcastle disease virus genome. *Virology* 145:203–12

67. Lamb, R. A. 1983. The influenza virus RNA segments and their encoded proteins. In *Genetics of Influenza Viruses*, ed. P. Palese, D. W. Kingsbury, pp. 21–69. Vienna: Springer-Verlag

68. Lemke, P. A., ed. 1979. *Viruses and Plasmids in Fungi*. New York: Dekker

69. Makino, S., Keck, J. G., Stohlman, S. T., Lai, M. M. C. 1986. High-frequency RNA recombination of murine coronaviruses. *J. Virol.* 57:729–37

70. Makino, S., Stohlman, S. A., Lai, M. M. C. 1986. Leader sequences of murine coronavirus mRNAs can be freely reassorted: evidence for the role of free leader RNA in transcription. *Proc. Natl. Acad. Sci. USA* 83:4204–8

71. Matthews, R. E. F. 1982. Classification and nomenclature of viruses. *Intervirology* 17:1–199

72. McGeoch, D. J., Davison, A. J. 1986. Alphaherpesviruses possess a gene homologous to the protein kinase gene family of eukaryotes and retroviruses. *Nucleic Acids Res.* 14:1765–77

73. Melnick, J. L. 1983. Portraits of viruses: the picornaviruses. *Intervirology* 20:61–100

74. Meyer, M., Hemmer, O., Mayo, M. A., Fritsch, C. 1986. The nucleotide sequence of tomato black ring virus RNA-2. *J. Gen. Virol.* 67:1257–71

75. Morgan, E. M., Rakestraw, K. M. 1986. Sequence of the Sendai virus *L* gene: open reading frames upstream of the main coding region suggest that the gene may be polycistronic. *Virology* 154:31–40

76. Murthy, M. R. N. 1983. Comparison of the nucleotide sequences of cucumber mosaic virus and brome mosaic virus. *J. Mol. Biol.* 168:469–75

77. Nakhasi, H. L., Meyer, B. C., Liu, T.-Y. 1986. Rubella virus cDNA: sequence and expression of E1 envelope protein. *J. Biol. Chem.* 261:16616–21

78. Namba, K., Stubbs, G. 1986. Structure of tobacco mosaic virus at 3.6Å resolution: implications for assembly. *Science* 231:1401–6

79. Nemeroff, M. E., Bruenn, J. A. 1986. Conservative replication and transcription of *Saccharomyces cerevisiae* viral double-stranded RNA in vitro. *J. Virol.* 57:754–58

80. Oker-Blom, C. 1984. The gene order for rubella virus structural proteins is NH_2-C-E2-E1-COOH. *J. Virol.* 51:354–58

81. Palmenberg, A. C. 1987. Genome organization, translation and processing in picornaviruses. See Ref. 64, pp. 1–15

82. Parvin, J. D., Moscona, A., Pan, W. T., Leider, J. M., Palese, P. 1986. Measurement of the mutation rates of animal viruses: influenza A virus and poliovirus type 1. *J. Virol.* 59:377–83

83. Paterson, R. G., Harris, T. J. R., Lamb, R. A. 1984. Analysis and gene assignment of mRNAs of a paramyxovirus, simian virus 5. *Virology* 138:310–23

84. Patterson, J. L., Kolakovsky, D. 1984. Characterization of La Crosse virus small-genome transcripts. *J. Virol.* 49:680–85

85. Podila, G. K., Flurkey, W. H., Bozarth, R. F. 1987. Identification and comparison of viral genes coding for capsid proteins of *Ustilago maydis* virus. *J. Gen. Virol.* 68:2741–50

86. Pringle, C. R. 1987. Rhabdovirus genetics. See Ref. 56, 167–244

87. Reanney, D. C. 1982. The evolution of RNA viruses. *Ann. Rev. Microbiol.* 36:47–73

88. Rice, C. M., Strauss, E. G., Strauss, J. H. 1986. Structure of the flavivirus genome. In *The Togaviridae and Flaviviridae*, ed. S. Schlesinger, M. J. Schlesinger, pp. 279–326. New York: Plenum

89. Rose, J. K., Doolittle, R. F., Anilionis, A., Curtis, P. J., Wunner, W. H. 1982. Homology between the glycoproteins of vesicular stomatitis virus and rabies virus. *J. Virol.* 43:361–64

90. Rose, J., Schubert, M. 1987. Rhabdovirus genomes and their products. See Ref. 56, pp. 129–66

91. Rossier, C., Patterson, J., Kolakovsky, D. 1986. La Crosse virus small genome mRNA is made in the cytoplasm. *J. Virol.* 58:647–50

92. Rossmann, M. G., Moras, D., Olsen, K. W. 1974. Chemical and biological evolution of a nucleotide-binding protein. *Nature* 250:194–99

93. Rossmann, M. G., Rueckert, R. R. 1987. What does the molecular structure of viruses tell us about viral functions? *Microbiol. Sci.* 4:206–14

94. Saitou, N., Nei, M. 1986. Polymorphism and evolution of influenza A virus genes. *Mol. Biol. Evol.* 3:57–74

95. Shioda, T., Iwasaki, K., Shibuta, H.

1986. Determination of the complete nucleotide sequence of the Sendai virus genome RNA and the predicted amino acid sequences of the F, HN, and L proteins. *Nucleic Acids Res.* 14:1545–63

96. Shope, R. E., Tesh, R. B. 1987. The ecology of rhabdoviruses that infect vertebrates. See Ref. 56, pp. 509–34

97. Siddell, S., Wege, H., ter Meulen, V. 1982. The structure and replication of coronaviruses. *Curr. Top. Microbiol. Immunol.* 99:131–63

98. Smith, D. B., Inglis, S. C. 1987. The mutation rate and variability of eukaryotic viruses: an analytical review. *J. Gen. Virol.* 68:2729–40

99. Steinhauer, D. A., Holland, J. J. 1987. Rapid evolution of RNA viruses. *Ann. Rev. Microbiol.* 41:409–33

100. Strauss, E. G., Levinson, R., Rice, C. M., Dalrymple, J., Strauss, J. H. 1988. Nonstructural proteins nsP3 and nsP4 of Ross River and O'Nyong-nyong viruses: sequence and comparison with those of other alphaviruses. *Virology* 164:265–74

101. Strauss, E. G., Strauss, J. H. 1983. Replication strategies of the single-stranded RNA viruses of eukaryotes. *Curr. Top. Microbiol. Immunol.* 105:1–98

102. Strauss, E. G., Strauss, J. H. 1986. Structure and replication of the alphavirus genome. See Ref. 88, pp. 35–90

103. Takkinen, K. 1986. Complete nucleotide sequence of the nonstructural protein genes of Semliki Forest virus. *Nucleic Acids Res.* 14:5667–82

104. Teninges, D., Bras-Herreng, F. 1987. Rhabdovirus sigma, the hereditary CO_2 sensitivity agent of *Drosophila:* nucleotide sequence of a cDNA clone encoding the glycoprotein. *J. Gen. Virol.* 68:2625–38

105. Tipper, D. J., Bostian, K. A. 1984. Double-stranded ribonucleic acid killer systems in yeasts. *Microbiol. Rev.* 48:125–56

106. Vidgren, G., Takkinen, K., Kalkkinen, N., Kaariainen, L., Pettersson, R. F. 1987. Nucleotide sequence of the genes coding for the membrane glycoproteins E1 and E2 of rubella virus. *J. Gen. Virol.* 68:2347–57

107. Webster, R. G., Laver, W. G., Air, G. M., Schild, G. C. 1982. Molecular mechanisms of variation in influenza viruses. *Nature* 296:115–21

108. Weiner, A. M., Maizels, N. 1987. tRNA-like structures tag the 3' ends of genomic RNA molecules for replication: implications for the origin of protein synthesis. *Proc. Natl. Acad. Sci. USA* 84:7383–87

109. Yusoff, K., Millar, N. S., Chambers, P., Emmerson, P. T. 1987. Nucleotide sequence analysis of the L gene of Newcastle disease virus: homologies with Sendai and vesicular stomatitis virus. *Nucleic Acids Res.* 15:3961–76

Ann. Rev. Microbiol. 1988. 42:685–716

IMMUNITY IN FILARIASIS:
Perspectives for Vaccine Development

Mario Philipp, Theodore B. Davis, Neil Storey, and Clotilde K. S. Carlow

Molecular Parasitology Group, New England Biolabs Inc., 32 Tozer Road, Beverly, Massachusetts 01915

CONTENTS

INTRODUCTION

Our aim is to review the immunobiology of filariae that infect humans, in both natural and experimental infections, and the available strategies for the

0066-4227/88/1001-0685$02.00

identification of host-protective antigens. With this information as a background, we analyze the prospects for vaccine development.

BIOLOGY OF THE PARASITE AND THE DISEASE

Of the filarial genera that infect humans, the most pathogenic and prevalent are *Wuchereria, Brugia,* and *Onchocerca* (71). The first two are termed lymphatic filariae because the lymphatic system is the main site of pathology and of adult parasite localization. *Wuchereria bancrofti* is endemic throughout the tropics and is the *Wuchereria* species most frequently found in humans (102). It is remarkably host specific, in that natural infections only occur in humans. *Brugia* parasites, in contrast, are more cosmopolitan and include zoonotic species (21). *Brugia malayi* and *Brugia timori* are confirmed human parasites endemic in areas of South and Southeast Asia. They are responsible for almost a third of the 90 million worldwide cases of lymphatic filariasis (128). *Onchocerca volvulus,* a tissue-dwelling filarid that causes blindness and skin disease in Africa and South and Central America, infects an estimated 17.6 million people (129). It is the only species within the genus that naturally infects humans, and it otherwise infects only certain nonhuman primates.

The filariases are characteristically debilitating diseases that cause little direct mortality but provoke a spectrum of clinical symptoms (102). Depending on the infecting species, disease morbidity may range from transient fevers or mild itching to extensive and virtually immobilizing edema of the limbs or skin atrophy and blindness.

Of the five developmental stages in the two-host filarial life cycle, it is the third stage, or infective larva, that is transmitted from a hematophagous arthropod vector during feeding. Following maturation and sexual reproduction, female adult worms liberate first-stage larvae, commonly called microfilariae, which are infective to the vector and continue the life cycle.

The microfilariae, and possibly the immune response to them, are responsible for skin and eye pathology in onchocerciasis (74), whereas in lymphatic filariasis this stage is linked only with the rarer allergic syndrome of tropical pulmonary eosinophilia (TPE). Lymphatic pathology is associated with the presence of, and the immune response to, developing larvae and adult worms in afferent lymphatic ducts (77). In contrast, *Onchocerca* adult worms cause little harm. They are confined to benign granulomatous nodules, onchocercomata, found either deep in the connective tissues or subcutaneously (74).

The pursuit of antifilarial vaccines is complicated by three main features of the parasites' immunobiology: the possible link between immunity and pathology, the strict primate host specificity of *W. bancrofti* and *O. volvulus,* and the diversity of clinical manifestations of infection. The immunological basis of this diversity is the subject of the next section.

IMMUNITY IN FILARIASIS   687

IMMUNITY IN HUMANS

As mentioned above, in areas where human filariases are endemic the resident population exhibits a spectrum of clinical and parasitological manifestations of infection. For the purpose of delineating strategies of immunoprophylaxis, the immunological analysis of three categories within this spectrum is essential: "endemic normals" (77), i.e. people with no symptoms of disease or infection; asymptomatic microfilaria carriers; and people with filarial pathologies.

Endemic Normals

This group is the least studied, perhaps because it is still so difficult to distinguish accurately between its true members and individuals with occult filariasis, who are infected but do not carry a detectable load of microfilariae (86). The available information, however, from studies of carefully chosen populations, is consistent with a state of protective immunity. Both cellular and humoral responses to filarial antigens were higher in endemic normals from areas with lymphatic filariasis than in asymptomatic microfilaremics. Blastogenic responses of peripheral blood mononuclear cells (PBMC) to parasite antigen were especially augmented in children, but not as high in older groups (82). In contrast, children and adults had similarly enhanced levels of antifilarial specific IgG, IgM, and IgE antibodies compared to infected subjects (83).

Endemic normal populations have also been studied in areas where onchocerciasis is endemic (125a). PBMC of subjects with no determinable medical history of onchocerciasis produced, when stimulated with *Onchocerca* antigens, significantly greater levels of interleukin-2 (IL-2) than did cells from normal individuals unexposed to infection or cells from infected subjects. In contrast, anti-*Onchocerca* antibody responses were slightly higher in infected individuals than in endemic normals. This finding does not preclude a protective role for antibody, since the predominance of a particular antibody specificity may have been diluted amid the polyspecific response measured (125a). Clarification of this issue will have to await more complete analyses of the humoral response, especially with defined (purified or recombinant) antigens. The described immunological correlates, albeit from a small study population, clearly suggest that an absence of filarial infection among people exposed to filariae or the presence of an infection level below detection thresholds may result from immune resistance to infection.

Asymptomatic Microfilaria Carriers

Asymptomatic carriers of microfilariae form perhaps the most studied group. However, the precise immunological mechanism leading to the lack of cellular immune response among its members is a matter of renewed controversy.

The original observation, made by Ottesen and coworkers (82) and further substantiated by others (75, 94), was that the proliferative response of PBMC stimulated by filarial antigens in vitro was much lower in asymptomatic carriers of *W. bancrofti* microfilariae than in endemic normals or individuals with clinical and pathological manifestations of bancroftian filariasis. The blastogenic response to antigens of bacterial origin such as tuberculin purified protein derivative (PPD) or to mitogens was comparable in all groups (82, 94), indicating that the lack of immune response was relatively specific for filarial antigens.

In patients with the generalized form of onchocerciasis, characterized by high densities of skin microfilariae but little or no skin pathology, a decreased cellular immune reactivity to both parasite-related and unrelated antigens has been observed (28, 29). However, evidence of the latter is somewhat contradictory. For example, studies performed on Liberian patients demonstrated a reduced intradermal delayed hypersensitivity response to PPD (29) and a diminished PBMC blastogenic response to the bacterial antigen streptokinase-streptodornase (28). In contrast, in a similar study in Guatemala, normal PBMC blastogenesis and IL-2 production were found in response to PPD, while IL-2 responses to *Onchocerca* antigens were significantly diminished in the infected individuals (125a).

There have been as yet no reports of a mechanism explaining the parasite antigen–independent immune unresponsiveness; studies of the filarial antigen–specific phenomenon observed in lymphatic filariasis have yielded disparate results. Depressed PMBC blastogenic responses to *Brugia* microfilarial antigen in asymptomatic microfilaremics with brugian filariasis could be restored to normal levels by removal of plastic adherent cells (98) or by reducing the number of cluster of differentiation 8–positive (CD-8$^+$) T cells (suppressor/cytotoxic) from about 50% to 25% (104). In asymptomatic microfilaremics infected with *W. bancrofti* similar procedures did not augment the blastogenic responsiveness of PBMC or the diminished levels of IL-2 or gamma interferon (IFN-γ) secreted (75). Thus the lack of antigen-specific cellular immune responsiveness could be caused either by active suppression or by a state of anergy or tolerance.

Humoral responses are also lower in asymptomatic microfilaremics than in patients with chronic pathology in lymphatic filariasis (76, 83, 95). For instance, titers of antibody against soluble *B. malayi* antigens measured by enzyme-linked immunosorbent assay were about eightfold lower than in people with chronic lymphatic pathology and as much as 400-fold lower than in people with TPE (76). Because similar differences were found in titers of antibody directed to the microfilarial surface, it was speculated that antibody had been removed from the peripheral circulation by passive absorption onto the worms' surface (95). This was probably not so, for in contrast to PBMC of amicrofilaremics, cells from microfilarial carriers infected with *W. bancrofti* were unable to secrete IgM or IgG antibodies specific to *B. malayi* antigen when stimulated in vitro with either pokeweed mitogen (PWM) or antigen. Interestingly, neither removal of monocytes nor removal of CD-8-positive cells restored the parasite antigen–driven antibody production, which was, however, critically dependent on the simultaneous presence of T cells, macrophages, and of course B cells (76). Thus a parasite-specific anergy may also explain the diminished humoral responses of asymptomatic microfilaremics.

Filarial Immunopathologies

The third and final epidemiological category to consider is that of people with immunopathologies. The notion that symptoms such as papular pruritic dermatitis and ocular punctate keratitis in onchocerciasis or chronic lymphatic obstruction in bancroftian or brugian filariasis are immunologically determined stems largely from their association with cellular infiltration and inflammatory reactions around dead or dying *Onchocerca* microfilariae in the skin or eyes (65) or adult worms of *Wuchereria* or *Brugia* in the lymphatics (77). In support of this notion there is additional circumstantial evidence. For example, unlike cells from asymptomatic microfilaremics, lymphocytes from people with chronic lymphatic pathology proliferated or secreted IL-2 or IFN-γ in vitro in response to parasite antigen just as well as endemic normals (75, 94). Interestingly, male adult worm antigens stimulated a higher PBMC response than did microfilarial antigens, which is in agreement with the absence of a pathogenic response to microfilariae in diseased lymphatic vessels (94). The low levels of skin microfilariae and the intense skin pathology often found in patients with the so-called localized form of onchocerciasis are also considered to be causally linked to as-yet-unidentified immune responses (19).

Several attempts have been made to find more precise immunological correlates of pathology in the antibody responses of patients with lymphatic filariasis. So far, no strict correlation has been found between disease state and antifilarial antibody specificities, apart from more frequent IgG1 and IgG3 antibodies reactive with antigens of >68 kd and low levels of IgG4 reactive with antigens of any molecular mass in patients with chronic lymphatic pathology, compared to those of subjects with asymptomatic microfilaremia (40).

More meaningful information may be associated with the quantitative isotype pattern of antibody responses, in particular that of IgG4. On average, the level of filaria-specific IgG4 is 2–20 times higher in patients with asymptomatic microfilaremia than in patients with chronic pathology (40, 81). In patients with TPE, however, specific IgG4 reaches levels as high as 3000 μg ml^{-1} (77, 81). The biological significance of such elevated IgG4 levels has not been elucidated (104). There are, however, some hints of a possible role in preventing allergic responses in lymphatic filariasis. Antigen-specific IgE antibody titers in lymphatic filariasis range between 0.03 and 0.07 μg ml^{-1} in the majority of cases in the absence of allergic symptoms, but are as high as 8–9 μg ml^{-1} in patients with the TPE allergic syndrome (41, 77). First, a naturally occurring filaria-specific IgG antibody was shown to block in vitro antigen-triggered release of histamine by basophils from infected patients (79). Secondly, there is a striking similarity between the filarial antigen recognition pattern of IgE and IgG serum antibodies in filariasis patients of all epidemiological types (42), which is due to IgG4 antibodies (43). Thirdly, the ratio of filarial-specific IgG4 to IgE correlates inversely with allergic symptoms in the different epidemiological categories; its average value is as high as 2500 in asymptomatic microfilaremics, but only 350 in patients with TPE (79, 81). Thus, varying concentrations of IgG4 blocking antibody could modulate the potentially pathogenic effect of high concentrations of parasite-specific IgE in vivo.

Levels of blocking antibodies higher than the above have been observed in some cases of onchocerciasis. Although these levels correlate with the concentration of specific IgE antibodies in individual patients, the role of this isotype in the pathogenesis of onchocerciasis is not clear (78). Recently, antiretinal autoantibodies were found in the serum and ocular fluids of onchocerciasis patients with a frequency 2.5 times higher than in normal controls. This observation indicates that humoral responses may cause pathology. However, the immunoglobulin isotype of these autoantibodies was not determined (16).

Immunity in Humans and Immunoprophylaxis

When analyzing the feasibility of an antiparasite vaccine, the question of whether there is naturally acquired protective immunity to infection often arises. Although the evidence for sterile immunity in filariasis is not conclusive, it is likely that some levels of naturally induced protective immunity exist. Even if all individuals considered endemic normals were in fact extreme cases of occult filariasis, they would seem capable at least of having a powerful antimicrofilarial immunity. While the absence of naturally acquired resistance against a parasitic infection does not imply that vaccination is impossible, its existence indicates that the adaptational stratagems protecting the parasite can be circumvented. The question, of course, is how. The simplest answer may lie in analyzing the pattern of antigen recognition by antibody and cells from endemic normals and contrasting it with that of the other two groups. This approach is currently being used in the pursuit of putative protective antigens.

It is also important to discern which is the predominant mechanism leading to immune unresponsiveness in filariasis. If the predominant mechanism is active antigen-dependent immune suppression, as found by Piessens and coworkers (94, 96) in patients with malayan filariasis, treatment with the antifilarial drug diethylcarbamazine may restore the patient's capacity to respond to antigen (97) and thus to a prophylactic vaccine. If immune unresponsiveness is brought about instead by a state of parasite-specific anergy, as suggested by the results of Nutman and coworkers (75, 76) with patients infected with bancroftian filariasis, the above approach may not be feasible. What remains to be determined, however, is the degree of stage specificity of the unresponsiveness, which so far has only been assayed with antigen extracts of male or female adult worms or of microfilariae alone, but not with larval antigens. It is possible that the antigen-specific immune unresponsiveness may exclude larval antigenic specificities that could be host protective.

Comparisons of the antigens recognized by antibodies from people with chronic pathologies, such as elephantiasis, and those of the other epidemiological categories have not yielded a simple distinction defining potentially pathogenic antigens to be avoided in vaccine development. While these molecules may still be found after further technical refinements such as two-dimensional Western blot analysis of appropriate antisera and antigen extraction procedures more exhaustive than those used so far, other avenues should also be explored. Differences in effector mechanisms may explain the contrast between the harmless consequences of immune responses in endemic normals and the pathology that these responses sometimes appear to induce. Unfortunately, studies of immunity in humans have not yet significantly advanced our understanding of the mechanisms that kill filariae or of those that elicit or prevent pathology, apart from suggesting the role of IgG4 in the modulation of allergic responses and revealing the positive correlation between microfilaremia and the absence of antisurface antibodies. In the next section we discuss how animal models may help to elucidate the mechanisms of protective immunity, immune unresponsiveness, and immunopathology and contribute to the identification of protective antigens.

IMMUNITY IN EXPERIMENTAL ANIMALS

Problems and Compromises in the Study of Experimental Immunity

The perfect animal model of a human parasitic infection would faithfully mimic the infection's parasitological, immunological, and pathological aspects. The parasite should be available in the laboratory in large quantities, and the animal's immune system must be amenable to detailed experimental analysis. For the human helminthiases, the model that perhaps best approaches this definition is *Schistosoma mansoni* in mice, in which susceptibility, course of infection, and the ensuing pathology are similar to those in humans (108, 126). The murine immune system is the most thoroughly investigated of all experimental animal immune systems. In addition, the laboratory-adapted life cycle of *S. mansoni* yields large numbers of parasites. This fortunate combination has undoubtedly facilitated the significant progress made toward the identification of an antigenically defined vaccine against schistosomiasis (45).

The situation vis à vis animal models and parasite availability is quite different in filariasis. *O. volvulus* and *W. bancrofti* will only reach reproductive maturity in primates (the chimpanzee and the silvered leaf monkey, respectively) (27, 85). Therefore, more practical animal models are required. On the other hand, alternative means for obtaining large numbers of microfilariae for larval production also need to be developed.

Scarce availability of parasites is the first obstacle to overcome in addressing experimental immunoprophylaxis against bancroftian filariasis or onchocerciasis. Presently, infective larvae are reared by feeding or injection of fresh or cryopreserved microfilariae isolated mainly from human sources into the appropriate vector. In a year, with great logistic difficulties, up to 10^6 skin microfilariae of *O. volvulus* may be collected and cryopreserved for future use (M. Karam, personal communication). Recovery rates of cryopreserved microfilariae, assessed by motility, are within 50–75% (32). Recovered microfilariae produce infective larvae at an efficiency rate of 15–20% in subarctic species of *Simulium* (E. W. Cupp, personal communication). Alternatively, the more direct procedure of feeding laboratory-reared Liberian *Simulium yahense* and *Simulium sanctipauli* flies on infected volunteers has yielded $5-7 \times 10^3$ larvae in 9 days (E. W. Cupp, M. J. Bernardo, A. E. Kizewski, M. Trpis & H. R. Taylor, personal communication). This is an adequate number of parasites for limited experiments in immunoprophylaxis, but unfortunately the approach seems difficult to implement on a continual basis and is not available to most research laboratories.

Production of *W. bancrofti* larvae entails similar problems. Although recovery rates of infective larvae from the filaria-susceptible (fm) strain of *Aedes aegypti* (63) are commonly about 25% (E. W. Cupp, personal communication) and can be as high as 47% (61), microfilarial densities reported in monkeys are extremely low (less than 16 microfilariae ml^{-1} blood) (20). Furthermore, there are no large laboratory colonies of infected monkeys nor, to our knowledge, well-funded systematic efforts to collect microfilariae for cryopreservation. Nevertheless the procedures available for production of *O. volvulus* and *W. bancrofti* larvae will certainly yield enough material for biochemical analyses of putatively relevant antigens and production of polyspecific antisera and monoclonal antibodies. With the current yields of bacteriophage lambda vector packaging extracts, these methods may even supply enough larval mRNA for larval cDNA libraries. However, they seem ill suited to provide the thousands of parasites that will be needed for actual vaccination experiments, which will inevitably incur many trial-and-error cycles.

The second problem in experimental immunoprophylaxis of onchocerciasis and bancroftian filariasis is the absence of animal models. The failure of attempts to obtain full development of *W. bancrofti* in congenitally athymic (nude) mice and in normal and immune-suppressed gerbils[1] implies that, unlike murine infections with *Brugia* larvae (see below), *W. bancrofti* infections are not established in these rodents despite the probable absence of lethal immune effector mechanisms (reviewed in 93). *Onchocerca* larvae also failed to develop in both normal CBA mice and immune-deficient CBA/N or nude mice (117) and also in gerbils (110). However, this result may be the consequence of a recovery artifact, since in CBA mice *Onchocerca lienalis* third-stage larvae molted to the fourth stage 2–5 days after infection and survived for at least 48 days, with some development, when the worms were enclosed in diffusion chambers implanted subcutaneously (A. E. Bianco, personal communication). While selective experimental immunoprophylaxis trials could be done using implanted chambers in rodents, the technique is too labor intensive to be extensively employed in surveys of antigen preparations and/or immunization protocols. Cows infected with cattle *Onchocerca* species are a natural infection model, but are even less practical.

The lack of convenient laboratory hosts and the scarcity of parasites seem to preclude any immediate extensive experimental immunity studies or testing of candidate vaccines against infection with larval stages of *W. bancrofti* or *Onchocerca* species. How then can we obtain the needed information? While much about immune effector mechanisms may be learned vicariously from rodent filariae such as *Acanthocheilonema viteae* (formerly *Dipetalonema viteae*) (reviewed in 93), it is from models involving human filarial genera that we may gain insight into the mechanisms of resistance to these parasites and the antigens by which they are induced. Undoubtedly the parasite of choice for the study of lymphatic filariasis is the zoophilic type of *B. malayi*. Infective larvae are obtainable in relatively large quantities. The parasite develops partially or fully in inbred rodents, depending on species, and the cat is a natural host (53). Male gerbils are fully susceptible to *B. malayi* infection (3), and in combination with the fm strain of *A. aegypti* as many as 4×10^4 larvae can be harvested weekly (J. McCall, personal communication). As regards onchocerciasis, more attention should be paid to the study of anti-microfilarial immunity, not only because it is a feasible option, but also because of the importance of microfilariae in the pathology of the disease. In

---

[1]*Meriones unguiculatus*, MON/Tum strain when inbred, throughout this text.

one year about 10^8 microfilariae of *O. lienalis* (cattle) or *Onchocerca cervicalis* (horse) can be cryopreserved for future use (E. R. James, personal communication). In addition, murine models have been developed to investigate mechanisms controlling skin microfilarial burdens and the antigens stimulating these responses. Thus, while the search for an inbred rodent host for *Onchocerca* larvae continues and the methods to produce these worms in the laboratory improve, it seems worthwhile to devote some effort to investigating immunity to *Onchocerca* microfilariae of animals. Finally, the often-voiced concern that filarial immunoprophylaxis may accentuate pathologic responses may now be addressed objectively, through rodent models of immunopathology.

Immunity to Brugia Larvae

Although scarcity of parasites is less of an issue with *B. malayi* than with the other filariae infecting humans, modeling of lymphatic filariasis is complicated by the diversity with which the infection is expressed in the natural host. No single inbred species of laboratory animal mimics such a spectrum of susceptibilities and antipodal immunological correlates (99), for which not only environmental (5, 88) but also genetic causes have been implicated (80). Therefore, we submit that *B. malayi* infections should be investigated in two separate inbred rodent species: gerbils and mice. Cats, where available as laboratory animals, will remain the preferred model in advanced trials of a candidate vaccine. However, their usefulness is limited for analysis of the cellular basis of the immune response to *Brugia* parasites, for there are, to our knowledge, no rigorously inbred strains of cats. Gerbils appear to be most suitable for the investigation of mechanisms of immune unresponsiveness, but they can also be used to investigate protective immunity (18, 130). Mice, whose relative insusceptibility to infection (17) is largely based on T-cell dependent antilarval immune responses (109, 111, 120, 121, 123), may mimic the process by which endemic normals appear to resist infection (120).

BRUGIA LARVAE IN MICE: A MODEL FOR NATURALLY ACQUIRED PROTECTIVE IMMUNITY?

Subcutaneous infections   Although it has long been known that parasites of the genus *Brugia* rarely reach maturity in mice (17), we have only recently begun to understand the immunological basis of this relative insusceptibility. The observation that larvae of the animal parasite *Brugia pahangi* could fully develop to sexual maturity in T cell–depleted (109) or congenitally athymic (nude) mice (111, 123) was the first indication that T cell–dependent phenomena were important in determining murine resistance to infection with *Brugia* larvae. Since then, the following additional information has been gathered: Susceptibility to infection was reduced by 70–100% in nude C3H/ HeN mice by immune reconstitution with intravenous injections of spleen cells or neonatal thymocytes from normal syngeneic mice or by implantation of neonatal thymus grafts several weeks before a subcutaneous challenge with *B. pahangi* larvae (121). Vickery & Nayar (118) transferred more defined cell populations from normal syngeneic animals into nude mice. Transfer of Thy 1.2^+–enriched spleen cell populations from heterozygous C3H/HeN mice, primed in vivo by exposure to larvae, restored resistance to infection. Neither T nor B cell–enriched primed spleen cells treated with anti–Thy-1.2^+ antibody and complement reproduced this effect (118). Further involvement of T cells in the murine response to larval infection was shown by delayed type hypersensitivity (DTH) reactions in heterozygous infected mice challenged in the foot pad with soluble antigens extracted from adult worms. Infection also augmented peripheral blood eosinophilia in normal but not nude mice, especially after a secondary exposure to the parasite (120).

The serum antibody response to soluble antigens of *B. pahangi* adult worms was analyzed by ELISA in immune competent heterozygous mice after both primary and secondary subcutaneous infections. Both IgM and IgG antibody responses were detected after a primary infection, and IgG responses were greatly elevated after challenge. IgE antiparasite antibody was also detected, both in the peripheral circulation and by passive cutaneous anaphylaxis. Such antibody responses notwithstanding, transfer of serum from immune animals that had been given a primary infection did not improve the ability of nude mice to resist infection (121). However, transfer of sera from hyperimmunized animals was not attempted. Normal mice of the C3H strain thus seem capable of eliciting strong protective immunity to subcutaneous larval infections with *Brugia* larvae. The induction of this immunity is mediated by Thy 1^+ cells, but not by B cells alone. The role of antibody-dependent effector mechanisms cannot presently be ruled out, and thus far no other effector mechanisms have been identified in this model.

Intraperitoneal infections   The model of intraperitoneal infections in mice has also been investigated, using third-stage larvae of *B. pahangi* and *B. malayi*. This route of inoculation permits an easier and more accurate quantification of parasite burdens than subcutaneous infections, since the worms do not seem to emigrate from the peritoneal cavity in normal mice (15).

Some aspects of immunity to intraperitoneal infections with *B. malayi* larvae have been studied in the BALB/c mouse strain, particularly the potential of this model as an aid in the identification of protective antigens. It was initially observed that strong resistance to challenge by intraperitoneal infections could be obtained by subcutaneous vaccination with gamma-irradiated third-stage larvae (35). Subsequently, normal third- and fourth-stage larvae, microfilariae, and adult worms were also found to be effective immunogens. The resistance induced was systemic, and sensitizing doses as small as two larvae were effective stimulators of resistance even when administered just 2 wk before the challenge infection (15). Therefore, these mice can respond quickly to very small amounts of sensitizing antigens, which is an important quality of a model to test putative immunogens. In addition, the more abundant material obtainable from adult worms could be used to identify and purify functional antigens (15).

There has been little investigation of the effector mechanisms that eliminate intraperitoneal infections with either *B. pahangi* or *B. malayi*. In the only study, intraperitoneal infections of CBA/Ca mice with third-stage larvae of *B. pahangi* lasted longer and developed further than infections with *B. malayi* and resulted in the production of substantial numbers of microfilariae (64, 84). Only peritoneal cells from infected animals were able to kill the microfilariae in vitro, even though cell adherence was quantitatively comparable to that of cells from uninfected donors. The investigators then inferred that activated macrophages had a role in the killing of both microfilariae and adult worms (84). Macrophages were activated during the infection, as evidenced by their increased ability to phagocytose latex beads compared to that of normal peritoneal macrophages. However, it is difficult to conclude that macrophages were the only effector cells, since the peritoneal cells were unfractionated; moreover, eosinophils were also observed to adhere to microfilariae. In addition, increased phagocytic and bactericidal activities have also been observed during intraperitoneal infections with *B. pahangi* in the more susceptible gerbil (47). This observation suggests that activated macrophages may be just bystanders, not killer cells. Alternatively, macrophages or granulocytes could be responsible for the initial attrition of parasites that occurs even in gerbils, in which worm recoveries rarely exceed 15% after subcutaneous infections (3) or 50% after intraperitoneal infections (68a). Immune responses involving other cells or mechanisms may then effect killing of the remaining parasites.

We have begun to investigate the involvement of macrophages in the termination of an intraperitoneal infection with *B. malayi* larvae by comparing the course of infection in mice of the BALB/c and P strains (N. Storey & M. Philipp, unpublished data). P mice have a T-cell dysfunction at the level of the production of macrophage-activating lymphokines (46) and a decreased macrophage responsiveness to lymphokine-induced activation in vitro (6). As a consequence, these mice have a diminished tumoricidal and microbicidal capacity. We found that the courses of both primary and secondary intraperitoneal infections were virtually parallel in both mouse strains (Figure 1). While this preliminary result argues against the involvement of activated

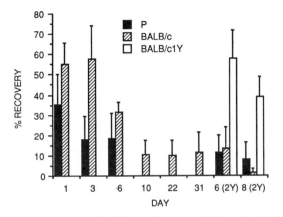

Figure 1   Resistance to experimental infection by *Brugia malayi* larvae in BALB/c and P mice. Bars show mean percentage recoveries (±SD) of living *B. malayi* larvae from the peritoneal cavities of mice during primary (days 1–31) and secondary (2Y, days 6 and 8) infections. Animals received 50 parasites on each occasion.

macrophages as effector cells in this host-parasite system, it remains to be determined whether the P macrophages were activated at all, since the activation defect in these mice is not complete (6) and might have been overcome by the sensitizing dose of 50 parasites.

BRUGIA LARVAE IN GERBILS: A MODEL FOR ASYMPTOMATIC IN-FECTIONS?   Several studies examining immune unresponsiveness have recently been conducted in inbred gerbils infected with *Brugia* larvae.

Both antigen-specific and nonspecific immunoregulatory mechanisms have been observed during infection with *B. pahangi* (100). In animals chronically infected with this parastie (>5 mo), both spleen and peripheral lymph node cells were unable to mount a proliferative response to the B-cell mitogens lipopolysaccharide (from *Escherichia coli*) and pokeweed mitogen in vitro, unlike cells from uninfected animals. This nonspecific immune unresponsiveness was anatomically compartmentalized, however, since spleen but not lymph node cells failed to respond to the T-cell mitogens concanavalin A (ConA) and phytohemagglutinin (56). In general, normal and infected animals mounted a comparable DTH response to dinitrofluorobenzene and produced similar numbers of direct hemolytic plaque-forming spleen cells secreting antibody to sheep red blood cells. However, during the first 4 wk after the appearance of circulating microfilariae, infected animals had fivefold fewer plaque-forming cells than controls. Therefore, there was only a transient inhibition of nonspecific humoral or T cell–mediated cellular immunity (57). Onset of *Brugia* antigen–specific spleen cell immune unresponsiveness also coincided with the appearance of circulating microfilariae. This unresponsiveness was not transient, but persisted throughout the remaining 24-wk observation period. In contrast, peripheral lymph node cells did respond to parasite antigen at all times (57), but were rendered unresponsive if cocultured with spleen cells from infected animals. This finding emphasizes the anatomical localization of the immunoregulatory effects induced by the infection (58).

Different types of cells appeared to induce the parasite antigen–specific and nonspecific unresponsiveness. The ability of spleen cells to respond to T-cell mitogens was restored by removing nylon wool–adherent, glass wool–adherent, or plastic-adherent cells (56, 100). In contrast, parasite-specific responsiveness of spleen cells to parasite antigen was restored by removing cells bearing receptors for histamine, peanut agglutinin, soybean agglutinin, or jird immunoglobulin or cells adherent to nylon wool, but not plastic–adherent cells (58). The evidence is consistent with an antigen-specific unresponsiveness mediated by T cells and B cells, but not plastic-adherent cells, while responsiveness to mitogens was probably suppressed by a plastic-adherent splenic macrophage. Interestingly, the nonadherent suppressor cell appeared at the onset of microfilaremia, whereas the plastic-adherent one was found only several weeks later. Unfortunately, in the absence of defined cell-surface phenotypic markers for gerbil lymphocytes, a more precise definition of the suppressor cells involved was not possible.

In addition to antigen-specific immune suppression, parasite-specific anergy has also been described in asymptomatic microfilaremics (75). This observation is consistent with epidemiological data suggesting that this state is favored in long-term residents of endemic areas because they may have acquired tolerance to infection by pre- or perinatal exposure to parasite antigen (5, 88). A phenomenon that may be interpreted in these terms has also been observed experimentally. Female gerbils born to microfilaremic mothers were more susceptible to infection than the progeny of uninfected mothers (103). Unlike similar experiments described in the *A. viteae*/rat model (33, 93), these experiments have not to our knowledge included any immunological studies to substantiate a parasite-specific immunologic anergy in these gerbils.

PATHOLOGY IN GERBILS AND IN MICE   Subcutaneous infections with *Brugia* larvae in gerbils and in nude mice induce lymphatic pathology. In gerbils vessel dilation, lymphatic and perilymphatic cellular infiltrates, intravascular granulomas, and irregular fibrosis of some valves and portions of the lymphatic walls are well established by the fourth month of infection (122). Since by this time there is no measurable resistance to reinfection (52), resistance is not a sine qua non condition for the development of pathology in these animals. However, immunization with adult worm extracts prior to infection has been reported to accentuate signs of pathology (51). In nude mice lymphatic fibrosis, intralymphatic thrombi, lymphangiectasis, and cellular infiltration, including eosinophils around dead worms, have been observed. These responses developed more slowly and were in general less pronounced than in normal mice (124). Interestingly, nude mice chronically infected with *B. malayi,* but not those infected with *B. pahangi* or *Brugia patei,* developed signs of elephantiasis. This syndrome did not correlate with microfilaremia, and it was reversed by removal of the adult worms from the lymphatic vessels (119).

Immunity to Onchocerca *Microfilariae in Mice*

Murine models for investigating immunity to microfilariae of *B. malayi* have already shown their worth in a number of studies (reviewed in 93). In early work it was observed that the course of subcutaneous infection with microfilariae of *O. lienalis* in CBA/HT6T6 mice could be reliably quantified by measuring the parasite burden in the ears of the mice only. The peak level and longevity of microfilariae from a secondary infection were significantly diminished compared to those of primary infections (13, 115). Subsequently it was shown that immunity is apparently T-cell dependent, as demonstrated by the extended duration of a primary infection in surgically athymic mice and the absence in these animals of the anamnestic resistance to challenge infections exhibited by euthymic mice (11, 116). Cells participate in this antimicrofilarial immunity, since significant levels of immunity could be transferred to syngeneic recipients in a dose-dependent fashion with as few as 4×10^6 spleen cells from immune animals (10). The role of antibody in microfilarial clearance is less defined, however, since the success of serum-transfer experiments was crucially dependent on the time at which the sera were collected from the donor mice after secondary infection. Neither primary-infection sera from mice nor sera from rabbits or calves hyperimmunized with homologous or heterologous *Onchocerca* microfilariae were effective in transferring immunity in this system (10).

OCULAR PATHOLOGY IN MICE   The mouse is currently under investigation as a possible model for ocular pathology in onchocerciasis. Microfilariae of *Onchocerca cervicalis* invade ocular tissues after subcutaneous injections (44) and can be directed to the retina by intravenous inoculations (E. R. James, personal communication). Preliminary results suggest that ocular pathology is host-strain related, the C57Bl strain being the most prone to anterior segment changes of five mouse strains evaluated so far (E. R. James, personal communication). Although such evidence is lacking in humans and mice, studies in guinea pigs suggest that IgE antibodies mediate eye pathology following injection of microfilariae into the eye (23).

Immunity in Experimental Animals and Immunoprophylaxis

Our understanding of the immune response of mice and of endemic normals to *Brugia* infection is still very limited. The predictive value of the mouse model can only broadly be assessed with the present data, in that murine resistance to infection and some of the immunological correlates of resistance in humans are both T-cell dependent. Therefore, until more data are available to sustain it, the analogy between mice and humans that we and others (120) have postulated is only phenomenological. However, the available information does illustrate the potential of the murine model. For instance, preliminary results with the P mouse strain have suggested that an absence of activated peritoneal macrophages does not affect the course of infection. If these results are confirmed, then it will be less tempting to pursue the characterization of antigens such as internal or secreted worm components, which would likely be protective if nonspecific immune effector mechanisms were elicited. The simplified parasite quantification made possible by intraperitoneal infections is undoubtedly useful whenever many antigens and/or immunization protocols are to be tested. Furthermore, the intraperitoneal milieu can be reproduced in vitro more faithfully than that of other body cavities, and this in vitro model simplifies the investigation of antilarval immune effector mechanisms. However, intraperitoneal effector mechanisms may differ from those operating in subcutaneous infections and thus from those in humans.

Further studies on the strain dependence of murine susceptibility to infections with *Brugia* larvae (39) should be undertaken, including studies on mouse strains possessing defined immune defects less all-encompassing than the nude mouse defect. A good example of the usefulness of this approach is the investigation that led to the discovery of an effector mechanism and a protective antigen (paramyosin) in schistosomiasis (45).

Brugia infections in gerbils and in humans appear to be similar in other ways. Especially striking are the similarities in the source of the immune suppressive antigens and the cell types involved in immune unresponsiveness. As reviewed in the previous section, parasite antigen–specific immune unresponsiveness of PBMC in humans is clearly associated with chronic microfilaremia in an infection that is otherwise asymptomatic. In gerbils, the appearance in the spleen of *Brugia* antigen–specific cells with suppressor activity is claimed to coincide with the appearance of circulating microfilariae (58). Microfilarial and female adult worm antigen extracts have been used in humans (96) and in gerbils (59), respectively, to induce suppressor cells from infected donors in vitro. More precisely, a high–molecular weight protein fraction that inhibited ConA-induced proliferation of human PBMC has been identified in material released both in vivo and in vitro by microfilariae (125). Therefore, both in gerbils and in humans, immune suppression is most likely induced by antigens of microfilarial origin and perhaps also by adult worm components.

Plastic-adherent antigen-specific suppressor cells, and also T cells of the CD-8 suppressor/cytotoxic phenotype, have been identified among PBMC of asymptomatic microfilaremic patients infected with *B. malayi*. The analogy with gerbil spleen cells is of course not strict, since the cells were from different immunological compartments and the gerbil T-suppressor cell phenotype is undefined. Furthermore, suppression mediated by the human plastic-adherent cell was parasite-antigen specific, whereas that mediated by the gerbil cell was not.

Immunosuppression is very likely to be the cause of the reported failures to vaccinate gerbils by single or multiple infections with *B. pahangi* and of the increased susceptibility to reinfection found in chronically infected animals (22, 52). However, gerbils can be protected against reinfection with *B. pahangi* (18) or *B. malayi* (130) if they are sensitized with larvae that are developmentally abbreviated by irradiation so that they cannot attain reproductive maturity. Thus it seems that in the absence of microfilariae, resistance rather than immune unresponsiveness is induced.

The contribution of immune sensitization to the lymphatic pathology induced by *Brugia* infections in gerbils and in mice is more quantitative than qualitative. This is relevant for the prospects of vaccine development, since pathogenesis seems to be associated more directly with the presence of healthy adult worms in the vessels than with the host's immune competence in responding to infection. Immune intervention resulting in significant reductions in adult worm burdens could conceivably diminish rather than accentuate pathology.

Several features of the *Onchocerca* microfilariae/mouse model substantiate its potential. As with the *B. malayi*/mouse model, it should be possible to extend the list of selectively immunodeficient mice already used (11, 115), to elucidate the effector mechanisms responsible for the attrition of skin microfilariae, and to refine further the cell-transfer experiments so that mechanisms of induction of protective immune responses may be understood. Protection in this model can be induced with as few as 20 microfilariae (13). This level of sensitivity is essential in vaccination trials with purified antigens from parasite or recombinant sources, which are always initially in short supply. Antigens from diverse sources, such as whole eggs and embryos from homologous and heterologous *Onchocerca* species and their detergent extracts, were protective against infections with *O. lienalis* microfilariae (10, 116). Finally, surface antigens of *O. cervicalis* microfilariae were shown to cross-react with sera from onchocerciasis patients (P. Busto & M. Philipp, unpublished observation). Therefore, there should be no shortage of parasite material from which to purify putatively relevant antigens to test in this model and from which to generate both antibody and DNA probes to screen genomic and cDNA expression libraries of *O. volvulus*.

STRATEGIES FOR IDENTIFICATION OF PROTECTIVE ANTIGENS

Our limited understanding of the immunobiology of human filariae still precludes a fully rational approach to the selection of protective antigens. A pragmatic approach is being taken instead. We summarize the main strategies here.

Surface Antigens

By analogy with other microbial infections, where surface antigens are the targets of lethal immune effector mechanisms, antigens expressed on the cuticular surface of filariae are considered prima facie vaccine candidates. Surface antigens of parasitic helminths (e.g. larval stages of *S. mansoni* and the intestinal nematode *Trichinella spiralis*) have recently been shown to elicit partially protective immune responses in animal models (9, 30, 34). Rather than providing a full list of the several filarial surface antigens already identified, we discuss a few observations that we believe are topical to the subject of immunoprophylaxis.

Two interesting properties have been described in a mouse IgG1 monoclonal antibody (MF1) that mediates attachment of human buffy coat cells or gerbil spleen cells to the surface of *B. malayi* microfilariae in vitro. First, a significant reduction in the density of circulating microfilariae was transiently induced in gerbils by intraperitoneal injections of MF1 (8). Secondly, treatment of microfilariae with the antibody prior to their ingestion by *A. aegypti* reduced the number of worms penetrating the insects' midgut by as much as 96% compared to controls. This inhibition was also induced by $F(ab')_2$ fragments of the antibody in the absence of contaminating gerbil cells (J. A. Fuhrman, S. S. Urioste, A. Spielman & W. F. Piessens, personal communication). MF1 binds to two antigens of 70 and 75 kd expressed only by microfilariae (8). An IgM monoclonal antibody binding to a surface antigen of *B. malayi* microfilariae (110 kd) was also able to clear microfilaremia in 70% of infected mice (1).

Surface antigens are likely to be protective not only if immune effector mechanisms are antibody dependent, but also in cell-mediated responses elicited by specific antigens yet targeted nonspecifically. In each case, however, the dynamic properties of the antigens have to be different. Molecules with a long residence time on the parasite surface should be good targets for antibody-dependent mechanisms, whereas rapidly released antigens are better suited to elicit cell-mediated nonspecific immunity (107). Larval stages of ascarid, strongyloid, and trichuroid nematodes have been shown to shed surface-bound antibodies or antigens in vitro, and surface antigens of adult worms and microfilariae of *B. malayi* have also been identified among the released components (25, 48, 92). Recently, we found that skin microfilariae of *O. cervicalis* (M. K. Edwards, E. R. James, C. K. S. Carlow & M. Philipp, unpublished observation) and third-stage larvae of *Dirofilaria immitis* and *B. malayi* (14; T. B. Davis & M. Philipp, unpublished observation) also shed surface-bound polyclonal and monoclonal antibodies, respectively, in vitro. Release of an anti–*B. malayi* monoclonal IgM antibody (NEB-D_1E_5) in vitro at 37°C occurred at a very fast rate, with a half-life of only 2.5 min (Figure 2). We found no evidence to suggest that bound antibody was partially digested by proteolytic enzymes and thus removed from the worms. Therefore, antigen shedding is probably the cause of the observed phenomenon. Once the original antibody was entirely shed, more antibody could be bound, which suggests that the target antigen is replaced on the larval surface or that several antigenic layers preexist. Complete loss of the surface epitope occurred after 4–6 days in vitro and after 2–3 days in vivo (14). This dynamic behavior may have contributed to our failure to protect mice against challenge

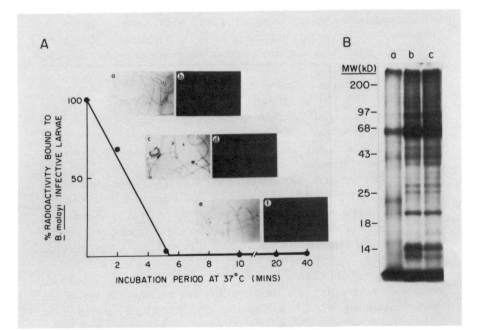

Figure 2 Rapid changes in the surface of *Brugia malayi* third-stage larvae. *A.* Time course of antigen/antibody release measured by worm bound radioactivity *(graph)* or by fluorescence *(photos)*; light and UV after 0- *(a, b)*, 5- *(c, d)*, or 10-min *(e, f)* incubation at 37°C. *B.* Autoradiograh of radiolabeled worm extracts; larvae were surface-labeled with ^{125}I after 0 (lane *a*), 1 (lane *b*), or 2 (lane *c*) days' residence in gerbils, extracted with SDS and β-mercaptoethanol, and subjected to SDS-PAGE.

infection with *B. malayi* larvae by intraperitoneal injections of monoclonal antibody NEB-D₁E₅ (C. K. S. Carlow & M. Philipp, unpublished data).

In apparent concert with the loss of their outermost layer, shortly after infection vector-derived third-stage larvae of *B. malayi* and *D. immitis* exhibit a qualitative change in the repertoire of antigens accessible to surface iodination. For instance, a prominent 35-kd surface antigen expressed by *D. immitis* larvae (91) was no longer detectable on the worms following 24-hr residence inside micropore chambers implanted into mice (T. B. Davis & M. Philipp, unpublished data). Similar dramatic changes were observed on *B. malayi* larvae after one or two days in the peritoneal cavity of gerbils (Figure 2). Since some of the newly expressed molecules were immunoprecipitable with sera from infected hosts, it is unlikely that they all originated from the host.

Exported Antigens

Filarial nematodes release a variety of antigens, which are not all surface derived. The recent discovery that some excretory/secretory (E/S) products have enzymatic activity has prompted several investigators to attempt the biochemical and functional characterization of these molecules. Three types of enzymes have been described: acetylcholinesterases, superoxide dismutases, and proteases.

Acetylcholinesterases of various molecular weights have been found in the E/S products of adult worms (100 and 200 kd) and microfilariae (200, 40, and 30 kd) of *B. malayi* (101). This diversity suggests the enzyme either is secreted in multiple molecular forms (101) or may functionally withstand partial degradation by proteases also secreted by *B. malayi* microfilariae (89; J. H. McKerrow, personal communication).

Microfilariae and adult worms of *Onchocerca* and *Dirofilaria* contain large amounts of catalase, glutathione peroxidase, and superoxide dismutase (7). The last enzyme has also been found in the material secreted by microfilariae of *O. cervicalis* in vitro, where it protected the worms against free radicals generated by the xanthine-xanthine oxidase system (E. R. James, personal communication).

Elastases have been identified in the secretions of skin-penetrating helminth larvae such as cercariae of *S. mansoni* (69) and third-stage larvae of *Ancylostoma caninum* (37, 70). Collagenase activity has been reported in the secretions of *O. volvulus* microfilariae (89). Extracts and secretions of microfilariae of *O. cervicalis* both contained a 60-kd metalloprotease and a 30-kd serine protease. The former could be homologous to the collagenase of *O. volvulus* (J. A. Sakanari, E. R. James, A. E. Bianco, S. D. Resnick & J. H. McKerrow, personal communication). Two proteases of 22 and 76 kd were found in extracts of *D. immitis* microfilariae, but the E/S products were not examined (113). Proteolytic activity has also been observed, but not characterized, in third-stage larvae of several filarial species including *B. pahangi* and *B. malayi* (70). Finally, it is possible that "molting enzymes" such as leucine aminopeptidase, secreted during parasite development, could have protective value (127). Nelson et al (73) have therefore tried to clone these molecules, taking advantage of the antigenic cross-reactivity between antibodies to *W. bancrofti* and molting enzymes of the fish nematode *Phocanema decipiens*.

Animal Models as Aids in the Identification of Protective Antigens

Animal models have been used directly to assess the immunogenicity of complex antigen preparations. In addition, cells and antibodies from immune animals are currently being employed as indicators of putatively protective antigens. The work of Kazura et al (49, 50) offers an example of the former approach. Kazura & Davies (50) initially established that immunization of mice with a saline extract of microfilariae of *B. malayi* could significantly reduce the levels of a parasitemia established by an intravenous injection of microfilariae. The same preparation was later used to vaccinate gerbils against a subcutaneous larval challenge. Both the resulting microfilaremia and female, but not male, adult worm burdens were significantly reduced in sensitized animals. Analysis of the antibody specificities in serum of immunized gerbils yielded bands of 150, 76, 42, and 25 kd on Western blots of microfilarial extracts (49). However, no immunization experiments have yet been reported using purified components of these relative molecular masses. Paramyosin of *B. malayi*, which was recently shown to induce resistance to microfilariae in mice, has a molecular mass of 97 kd (72). Therefore either it was not a major component of the extracts described above, or the immune gerbils did not respond to it.

Fletcher et al (26) analyzed antigens of microfilariae, adult worms, and infective larvae of *B. pahangi*, which were differentially recognized by antibodies from microfilaremic cats and from cats that had spontaneously become amicrofilaremic. Several components that reacted only with sera from the putatively immune animals were identified by Western blotting, but there have been as yet no reports of vaccination experiments with purified antigens of the same identity.

We are using spleen cells from BALB/c mice to identify antigens that stimulate immunity to *Brugia* infections in these animals (N. Storey & M. Philipp, unpublished data), much as described with human PBMC (55). Aqueous extracts of third-stage larvae and adult worms stimulated between 70 and 80% protection against a larval intraperitoneal infection. High performance liquid chromatography of these extracts on ion-exchange and gel-filtration columns yielded several fractions. Their ability to stimulate blastogenic responses in spleen cells from normal and immune mice was assayed in vitro by a standard thymidine incorporation method. Both mitogenic and antigenic fractions were identified and are being analyzed using the same in vitro assay after further resolution by two-dimensional gel electroblotting. We have also used BALB/c mice to assess in vivo the immunogenicity of pools of recombinant antigens from a *B. malayi* genomic DNA library made in the expression vector λgt11 (2). So far no protective pool of antigens has been identified (P. Arasu, C. K. S. Carlow, F. Perler & M. Philipp, unpublished data).

Screening with Cells and Antisera From Putatively Resistant Individuals

Sera and cells from individuals with apparent signs of resistance either to infection or just to microfilariae are being used to screen parasite antigen extracts, proteins translated in vitro, and recombinant antigens. Serum samples from individuals with the localized and generalized forms of onchocerciasis who had mean microfilaridermias of 1.4 and 95.2 microfilariae mg^{-1} skin respectively were analyzed by Western blotting using detergent extracts of *O. volvulus* female adult worms. Despite this extreme contrast between average parasite burdens, no distinct pattern of antigen recognition by IgM or IgG antibodies could be ascribed to either group (62). To reveal a difference it may be necessary to investigate individual antibody isotypes, rather than whole IgG antibody responses. For instance, only sera from patients with the localized form of onchocerciasis had IgG3 antibodies to a 20-kd antigen of *Onchocerca* adult worms (87).

Material translated in vitro from mRNA of adult *B. pahangi* adult worms was immunoprecipitated with pools of sera from people residing in areas where *B. malayi* is endemic. A 75-kd peptide was selectively precipitated by pooled serum samples from amicrofilaremic individuals suffering from elephantiasis or lymphadenopathies and from endemic normals, but not by sera from microfilaremic patients with similar pathology or from asymptomatic microfilaremics (105). No similar differential reactivity was found with surface antigens of adult worms or microfilariae or with collagenous nonsurface cuticular components or phosphorylcholine-bearing glycoconjugates (67). Several recombinant antigens from a *B. pahangi* adult worm cDNA expression library were also recognized by the above serum pool, but not by sera that lacked reactivity to the 75-kd molecule (105). More detailed analysis of individual sera with gel-purified recombinant antigens revealed that the overall reactivity of the pool of eight sera was determined in some cases by the high titer of antibodies of only two contributors (106). Sera from endemic normal residents of areas where onchocerciasis is prevalent are also being employed to screen both genomic and adult worm cDNA expression libraries of *O. volvulus* (F. Perler, personal communication). Reacting recombinants are then further investigated for their ability to stimulate PBMC from these patients in vitro.

Advantages and Shortcomings of the Different Approaches, and Some Technical Problems

The demonstration that antisurface monoclonal antibodies can reduce *B. malayi* microfilaremia implies that the microfilarial sheath may not possess the dynamic properties of the cuticle. Furthermore, it substantiates the possibility that the negative correlation observed in humans between anti-sheath antibodies and microfilarial density is of a causal nature. Although immunoprophylaxis of transmission is not practical in brugian filariasis because of its zoonotic nature, the observations made with the MF1 antibody have important implications. Immunization protocols with surface antigens of *O. volvulus* or *W. bancrofti* microfilariae that result in high antibody titers could reduce worm densities and at the same time block vector infection. This would be enormously effective in reducing transmission of these parasites, for which there is no known animal reservoir.

In retrospect, the dynamic properties of surface antigens of vector-derived third-stage larvae may explain an interesting observation made some years ago by Tanner & Weiss (114). Serum from highly resistant gerbils immunized with irradiated larvae of *A. viteae* contained antibodies that were reactive with the surface of vector-derived larvae, third-stage larvae recovered from gerbils after 5 days of infection, and fourth-stage larvae. However, only the last two types of larvae were killed in vitro by immune serum and complement. Vector-derived larvae were resistant to this effector mechanism. Thus it is conceivable that vector-derived larvae are coated by an antigenic layer that is rapidly shed and eventually lost after infection, and that postparasitic third- and fourth-stage larvae have more stable surface antigens. Indeed, Marshall & Howells (68) have shown a rapid loss of radioactivity from surface-labeled *B. pahangi* infective larvae in vivo, whereas fourth-stage larvae and adult worms retained the label for many days. Larval stages of filariae, and especially third-stage larvae, which possess a nonfunctional gut, probably absorb most of their nutrients transcuticularly (38). Viewed as a nutritional organ, the cuticle should undergo rapid changes upon infection in response to the constraints of a drastically different new environment. The observations described in Figure 2 may be a reflection, at the molecular level, of this adaptation to a vertebrate host. Therefore, surface antigens of postparasitic third-stage and fourth-stage larvae not only are more stable targets for antibody-dependent effector mechanisms, but are probably more important functionally in the vertebrate host than the surface antigens of vector-derived third-stage larvae. Fourth-stage larvae of *D. immitis*, for instance, express the same surface antigens for at least 3 wk after the third molt (T. B. Davis & M.

Philipp, unpublished observation). We should keep in mind, however, that the rapidly released outermost layer of the epicuticle of the invading larva could be a most effective inducer of T cell–dependent, nonspecific effector mechanisms.

Although much has been speculated about the possible role of secreted acetylcholinesterase in the host-parasite relationship of intestinal nematodes (60, 90), there has been no definitive experiment assessing the consequences of enhanced immune responses directed against it. However, since acetylcholine is now known to activate a variety of cells of the immune system (reviewed in 101), the contention that acetylcholinesterase may down-regulate crucial immune networks is well worth exploring. Similarly, there is no direct evidence that secreted superoxide dismutases neutralize peroxides and free radicals of phagocytes or granulocytes, but this is an attractive postulate deserving further study. Finally, immunological neutralization of secreted proteases could simply result in parasite death or interfere with penetration of the lymphatics by *Brugia* larvae or intradermal localization by *Onchocerca* microfilariae, and may thus give rise to the interesting alternative of an antipathology vaccine (J. McKerrow, personal communication).

Two of the observations made during the vaccination of gerbils with microfilarial extracts could be of general applicability: First, an antigen preparation initially characterized as protective in an insusceptible host retained this property when later tested in gerbils. Secondly, stimulation of antimicrofilarial immunity may also target gravid female worms. No comparative study of the serum antibody specificities of gerbils vaccinated with irradiated *B. malayi* larvae (130) and those of chronically infected animals has been reported. We await this comparison with interest in light of the preliminary communication that immunity could be passively transferred with serum from vaccinated animals (131).

No strategy for the selection of putative protective antigens has proven so far to be more successful than any other, but some of the technical problems to be faced in the pursuit of these antigens are already apparent. The most serious problem we have encountered stems, not surprisingly, from paucity of parasite material. We estimated that about 1 pg per larva of a 35-kd surface antigen was produced by third-stage larvae of *D. immitis* (T. B. Davis & M. Philipp, unpublished observation). To purify several micrograms of this protein to generate probes for the screening of DNA libraries, i.e. monospecific polyclonal antisera or amino acid sequencing data for synthesis of oligonucleotide DNA probes, several million larvae would be necessary. This is a prohibitive number for even the most prolific laboratory-run filarial life cycle. The two simplest solutions for this problem are to generate a battery of monoclonal antibodies to the antigen of interest, screening hybrid cell supernatants by immunoprecipitation with radiolabeled larval antigen, or to identify the larval antigen also among the components expressed by the much larger adult worms. Two of the larval surface antigens that we are investigating were also expressed by adult worms. A 16-kd molecule of *B. malayi*, which was initially identified on the surface of adult worms (112), was also expressed on the surface of fourth-stage larvae and on the surface of third-stage larvae extracted from the peritoneal cavity of gerbils 3 days after infection (N. Storey & M. Philipp, unpublished data) (Figure 3*A*). A 25-kd surface antigen of fourth-stage larvae of *D. immitis*, shown on an auto-radiograph of a two-dimensional gel (Figure 3*B, b, arrow*), overlapped with a Coomassie blue–stained component of an adult worm extract with which it was coelectrophoresed (Figure 3*B, a, arrow*). Spots containing the adult worm 25-kd antigen were excised from the gel and used to immunize rabbits by intrapopliteal node injections. Screening of a λgt11 adult worm cDNA library with the antiserum obtained in this way yielded several positive clones, which are currently being analyzed (A. Grandea, T. B. Davis, L. McReynolds & M. Philipp, unpublished data).

An alternative way to investigate interesting but scarce antigens is to use internal-image anti-idiotypic antibodies. Carlow et al (12) recently described a monoclonal antibody that binds to a species- and stage-specific surface antigen of *B. malayi* third-stage larvae. The antibody was unsuitable for immunoprecipitation or for antigen purification by affinity chromatography. It also failed to bind to whole-worm extracts attached to nitrocellulose or related solid phases, and it is therefore unlikely to be good for screening DNA

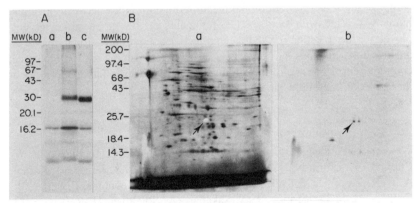

Figure 3   Coexpression of filarial surface antigens by several developmental stages. *A*. Auto-radiograph of radiolabeled *Brugia malayi* extracts; day 3 postparasitic third-stage larvae (lane *a*), day 10 fourth-stage larvae (lane *b*), and adult male worms (lane *c*) were surface-labeled with ^{125}I and extracted with 1.5% octyl glucoside, and the extracts were analyzed by SDS-PAGE. *B*. Coomassie protein stain (*a*) and autoradiograph (*b*) of *Dirofilaria immitis* adult somatic antigen and ^{125}I surface-labeled fourth-stage larval extract coelectrophoresed in two dimensions. *Arrow* marks overlapping 25-kd antigen shared by fourth stage larvae and adult worms.

expression libraries. We have produced an anti-idiotypic antibody in rabbits, which when used as an immunogen in mice stimulated antibodies to the larval surface with the same specificity as the parent monoclonal antibody (C. K. S. Carlow & M. Philipp, unpublished observation). This approach is thus useful not only for investigating the protective capacity of carbohydrate epitopes (31), of which there are many among surface and secreted filarial antigens (66), but also in the analysis of immunological properties of otherwise not easily purified epitopes of any composition.

CONCLUDING REMARKS

As in other parasitic infections, the prospects for vaccine development in filariasis are inextricably linked to progress in understanding immunity to infection, both in humans and in experimental animals. This link has been dramatically illustrated for malaria by the results of the recent sporozoite vaccine trials (4, 24, 36, 54). The parallelism uncovered between the cellular basis of the immune response to *Brugia* parasites in gerbils and in susceptible humans, in connection with the fact that this animal can be vaccinated by exposure to developmentally abbreviated larvae, should stimulate further research with this model. Equally encouraging is the finding that murine resistance to larval and microfilarial infections and several immunological correlates of resistance in humans are both T-cell dependent. The availability of a panoply of reagents to dissect the immune response of mice should stimulate the search for additional meaningful similarities between mice and humans infected with filariae. The rationales of the varied strategies employed for the selection of candidate protective antigens are strongly based on analogy with methods successfully used in other systems and on biological common sense. These efforts should continue in the hope that the diversity of the approach may hold the key to future success.

ACKNOWLEDGMENTS

We thank our colleagues Drs. P. Arasu, M. J. Bernardo, A. E. Bianco, E. W. Cupp, M. K. Edwards, J. A. Fuhrman, A. Grandea, E. R. James, M. Karam, A. E. Kizewski, J. McCall, J. H. McKerrow, L. McReynolds, F. Perler, W. F. Piessens, S. D. Resnick, J. A. Sakanari, A. Spielman, H. R. Taylor, M. Trpis, and S. S. Urioste for generously allowing us to cite some of their data before publication.

Drs. E. A. Ottesen, R. M. E. Parkhouse, and E. J. Winchell made many useful comments on this manuscript, for which we are grateful.

The support of Dr. Donald G. Comb is gratefully acknowledged.

Literature Cited

1. Aggarwal, A., Cuna, W., Haque, A., Dissous, C., Capron, A. 1985. Resistance against *Brugia malayi* microfilariae induced by a monoclonal antibody which promotes killing by macrophages and recognizes surface antigen(s). *Immunology* 54:655–63
2. Arasu, P., Philipp, M., Perler, F. 1987. *Brugia malayi:* recombinant antigens expressed by genomic DNA clones. *Exp. Parasitol.* 64:281–91
3. Ash, L. R. 1973. Chronic *Brugia pahangi* and *Brugia malayi* infections in *Meriones unguiculatus. J. Parasitol.* 59: 442–47
4. Ballou, W. R., Hoffman, S. L., Sherwood, J. A., Hollingdale, M. R., Neva, F. A., et al. 1987. Safety and efficacy of a recombinant DNA *Plasmodium falciparum* sporozoite vaccine. *Lancet* 1: 1277–81
5. Beaver, P. C. 1970. Filariasis without microfilaremia. *Am. J. Trop. Med. Hyg.* 19:181–89
6. Boraschi, D., Meltzer, M. S. 1980. Defective tumoricidal capacity of macrophages from P/J mice: characterization of the macrophage cytotoxic defect after *in vivo* and *in vitro* activation stimuli. *J. Immunol.* 125:777–82
7. Callahan, H. L., James, E. R., Crouch, R. K. 1988. *Anti-oxidant enzymes in two filarial nematodes.* Presented at UCLA Symp. Oxy-Radicals Mol. Biol. Pathol., Park City, Utah
8. Canlas, M., Wadee, A., Lamontagne, L., Piessens, W. P. 1984. A monoclonal antibody to surface antigens on microfilariae of *Brugia malayi* reduces microfilaremia in infected jirds. *Am. J. Trop. Med. Hyg.* 33:420–24
9. Capron, A., Dessaint, J. P., Capron, M., Ouma, J. H., Butterworth, A. E. 1987. Immunity to schistosomes: progress toward vaccine. *Science* 238: 1065–72
10. Carlow, C. K. S., Bianco, E. A. 1987. Transfer of immunity to the microfilariae of *Onchocerca lienalis* in mice. *Trop. Med. Parasitol.* 39:283–86
11. Carlow, C. K. S., Dobinson, A. R., Bianco, A. E. 1988. Parasite-specific immune responses to *Onchocerca lienalis* microfilariae in normal and immunodeficient mice. *Parasite Immunol.* In press
12. Carlow, C. K. S., Franke, E. D., Lowrie, R. C., Partono, F., Philipp, M. 1987. Monoclonal antibody to a unique surface epitope of the human filaria *Bru-gia malayi* identifies infective larvae in mosquito vectors. *Proc. Natl. Acad. Sci. USA* 84:6914–18
13. Carlow, C. K. S., Muller, R., Bianco, E. A. 1986. Further studies on the resistance to *Onchocerca* microfilariae in CBA mice. *Trop. Med. Parasitol.* 37: 276–81
14. Carlow, C. K. S., Perrone, J., Spielman, A., Philipp, M. 1987. A developmentally regulated surface epitope expressed by the infective larva of *Brugia malayi* which is rapidly lost after infection. *UCLA Symp. Mol. Cell. Biol.* 60:301–10
15. Carlow, C. K. S., Philipp, M. 1987. Protective immunity to *Brugia malayi* larvae in BALB/c mice: potential of this model for the investigation of protective antigens. *Am. J. Trop. Med. Hyg.* 37: 597–604
16. Chan, C. C., Nussenblatt, R. B., Kim, M. K., Palestine, A. G., Awadzi, K., et al. 1987. Immunopathology of ocular onchocerciasis. 2. Anti-retinal autoantibodies in serum and ocular fluids. *Ophthalmology* 94:439–43
17. Chong, L. K., Wong, M. M. 1967. Experimental infection of laboratory mice with *Brugia pahangi. Med. J. Malaya* 21:382
18. Chusattayanond, W., Denham, D. A. 1986. Attempted vaccination of jirds *(Meriones unguiculatus)* against *Brugia pahangi* with radiation attenuated infective larvae. *J. Helminthol.* 60:149–55
19. Connor, D. H., George, G. H., Gibson, D. W. 1985. Pathologic changes of human onchocerciasis: implications for future research. *Rev. Infect. Dis.* 7:809–19
20. Cross, J. H., Partono, F., Hsu, M. Y. K., Ash, L. R., Oemijati, S. 1979. Experimental transmission of *Wuchereria bancrofti* to monkeys. *Am. J. Trop. Med. Hyg.* 28:56–66
21. Denham, D. A., McGreevy, P. B. 1977. Brugian filariasis: epidemiological and parasitological studies. *Adv. Parasitol.* 15:243–309
22. Denham, D. A., Suswillo, R. R., Chusattayanond, W. 1984. Parasitological observations on *Meriones unguiculatus* singly or multiply infected with *Brugia pahangi. Parasitology* 88:295–301
23. Donelly, J. J., Rockey, J. H., Taylor, H. R., Soulsby, E. J. 1985. Onchocerciasis: experimental models of ocular disease. *Rev. Infect. Dis.* 7:820–25
24. Egan, J. E., Weber, J. L., Ballou, W. R., Hollingdale, M. R., Majarian, W.

R., et al. 1987. Efficacy of murine malaria sporozoite vaccines: implications for human vaccine development. *Science* 236:453–56

25. Egwang, T. G., Kazura, J. W. 1987. Immunochemical characterization and biosynthesis of major antigens of lodobead surface-labeled *Brugia malayi* microfilariae. *Mol. Biochem. Parasitol.* 22:159–68

26. Fletcher, C., Birch, D. W., Samad, R., Denham, D. A. 1986. *Brugia pahangi* infections in cats: antibody responses which correlate with the change from microfilaraemic to the amicrofilaraemic state. *Parasite Immunol.* 8:345–57

27. Greene, B. M. 1987. Primate model for onchocerciasis research. *Ciba Found. Symp.* 127:236–43

28. Greene, B. M., Fanning, M. M., Ellner, J. J. 1983. Non-specific suppression of antigen-induced lymphocyte blastogenesis in *Onchocerca volvulus* infection in man. *Clin. Exp. Immunol.* 52:259–65

29. Greene, B. M., Gbakima, A. A., Albiez, E. J., Taylor, H. R. 1985. Humoral and cellular immune responses to *Onchocerca volvulus* infections in humans. *Rev. Infect. Dis.* 7:789–95

30. Grencis, R. K., Crawford, C., Pritchard, D. I., Behnke, J. M., Wakelin, D. 1986. Immunization of mice with surface antigens from the muscle larvae of *Trichinella spiralis*. *Parasite Immunol.* 8:587–96

31. Grzych, J. M., Capron, M., Lambert, P. H., Dissous, C., Torres, S., et al. 1985. An anti-idiotype vaccine against experimental schistosomiasis. *Nature* 316:74–76

32. Ham, P. J., James, E. R., Bianco, A. E. 1979. *Onchocerca* spp.: Cryopreservation of microfilariae and subsequent development in the insect host. *Exp. Parasitol.* 47:384–91

33. Haque, A., Capron, A. 1982. Transplacental transfer of rodent microfilariae induces antigen-specific tolerance in rats. *Nature* 299:361–63

34. Harn, D. A. 1987. Immunization with schistosome membrane antigens. *Acta Trop.* 44(Suppl. 12):46–49

35. Hayashi, Y., Noda, K., Shirasaka, A., Nogami, S., Nakamura, M. 1984. Vaccination of BALB/c mice against *Brugia malayi* and *Brugia pahangi* with larvae attenuated by gamma irradiation. *Jpn. J. Exp. Med.* 54:177–81

36. Herrington, D. A., Clyde, D. F., Losonsky, G., Cortesia, M., Murphy, J. R., et al. 1987. Safety and immunogenicity in man of a synthetic peptide malaria vaccine against *Plasmodium*

falciparum sporozoites. *Nature* 328:257–59

37. Hotez, P. J., Cerami, A. 1983. Secretion of a proteolytic anticoagulant by *Ancylostoma* hookworms. *J. Exp. Med.* 157:1594–603

38. Howells, R. E. 1987. Dynamics of the filarial surface. *Ciba Found. Symp.* 127:94–106

39. Howells, R. E., Devaney, E., Smith, G., Hedges, T. 1983. The susceptibility of BALB/c and other inbred mouse strains to *Brugia pahangi*. *Acta Trop.* 40:341–50

40. Hussain, R., Grögl, M., Ottesen, E. A. 1987. IgG antibody subclasses in human filariasis. Differential subclass recognition of parasite antigens correlates with different clinical manifestations of infection. *J. Immunol.* 139:2794–98

41. Hussain, R., Hamilton, R. G., Kumaraswami, V., Adkinson, N. F. Jr., Ottesen, E. A. 1981. IgE responses in human filariasis. 1. Quantitation of filaria-specific IgE. *J. Immunol.* 127:1623–29

42. Hussain, R., Ottesen, E. A. 1985. IgE responses in human filariasis. III. Specificities of IgE and IgG antibodies compared by immunoblot analysis. *J. Immunol.* 135:1415–20

43. Hussain, R., Ottesen, E. A. 1986. IgE responses in human filariasis. IV. Parallel antigen recognition by IgE and IgG4 subclass antibodies. *J. Immunol.* 136:1859–63

44. James, E. R., Smith, B., Donelly, J. 1986. Invasion of the mouse eye by *Onchocerca* microfilariae. *Trop. Med. Parasitol.* 37:359–60

45. James, S. L. 1987. *Schistosoma* spp. Progress towards a defined vaccine. *Exp. Parasitol.* 63:247–52

46. James, S. L., Deblois, L. A., Al-Zamel, F., Glaven, J., Langhorne, J. 1986. Defective vaccine-induced resistance to *Schistosoma mansoni* in P strain mice. III. Specificity of the associated defect in cell-mediated immunity. *J. Immunol.* 137:3959–67

47. Jeffers, G. W., Klei, T. R., Enright, F. M. 1984. Activation of jird (*Meriones unguiculatus*) macrophages by the filarial parasite *Brugia pahangi*. *Infect. Immun.* 43:43–48

48. Kaushal, N. A., Hussain, R., Nash, T. E., Ottesen, E. A. 1982. Identification and characterization of excretory-secretory products of *Brugia malayi*, adult filarial parasites. *J. Immunol.* 129:338–43

49. Kazura, J. W., Cicirello, H., McCall, J. W. 1986. Induction of protection against

Brugia malayi infection in jirds by microfilarial antigens. *J. Immunol.* 136: 1422–26

50. Kazura, J. W., Davies, R. S. 1982. Soluble *Brugia malayi* microfilarial antigens protect mice against challenge by an antibody-dependent mechanism. *J. Immunol.* 128:1792–96

51. Klei, T. R., Enright, F. M., Blanchard, D. P., Uhl, S. A. 1982. Effects of presensitization on the development of lymphatic lessions in *Brugia pahangi*–infected jirds. *Am. J. Trop. Med. Hyg.* 31:280–91

52. Klei, T. R., McCall, J. W., Malone, J. B. 1980. Evidence for increased susceptibility of *Brugia pahangi*–infected jirds *(Meriones unguiculatus)* to subsequent homologous infections. *J. Helminthol.* 54:161–65

53. Laing, A. B. G., Edeson, J. F. B., Wharton, R. H. 1960. Studies on filariasis in Malaya: the vertebrate hosts of *Brugia malayi* and *B. pahangi*. *Ann. Trop. Med. Parasitol.* 54:92–99

54. Lal, A. A., de la Cruz, V. F., Godd, M. F., Weiss, W. R., Lunde, M., et al. 1987. In vivo testing of subunit vaccines against malaria sporozoites using a rodent system. *Proc. Natl. Acad. Sci. USA* 84:8647–51

55. Lal, R. B., Lynch, T. J., Nutman, T. B. 1987. *Brugia malayi* antigens associated with lymphocyte activation in filariasis. *J. Immunol.* 139:1652–57

56. Lammie, P. J., Katz, S. P. 1983. Immunoregulation in experimental filariasis. I. *In vitro* suppression of mitogen-induced blastogenesis by adherent cells from jirds chronically infected with *Brugia pahangi*. *J. Immunol.* 130:1382–85

57. Lammie, P. J., Katz, S. P. 1983. Immunoregulation in experimental filariasis. II. Responses to parasite and nonparasite antigens in jirds with *Brugia pahangi*. *J. Immunol.* 130:1386–89

58. Lammie, P. J., Katz, S. P. 1984. Immunoregulation in experimental filariasis. III. Demonstration and characterization of antigen-specific suppressor cells in the spleen of *Brugia pahangi*-infected jirds. *Immunology* 52: 211–19

59. Lammie, P. J., Katz, S. P. 1984. Immunoregulation in experimental filariasis. IV. Induction of non-specific suppression following in vitro exposure of spleen cells from infected jirds to *Brugia pahangi* antigen. *Immunology* 52: 221–29

60. Lee, D. L. 1970. The fine structure of the excretory system in adult *Nippostrongylus brasiliensis* (Nematoda) and a suggested function for the excretory glands. *Tissue Cell* 2:225–31

61. Lowrie, R. C. Jr. 1983.Cryopreservation of the microfilariae of *Brugia malayi*, *Dirofilaria corynodes*, and *Wuchereria bancrofti*. *Am. J. Trop. Med. Hyg.* 32:138–45

62. Lucius, R., Büttner, D. W., Kirsten, C., Diesfeld, H. J. 1986. A study on antigen recognition by onchocerciasis patients with different clinical forms of disease. *Parasitology* 92:569–80

63. MacDonald, W. W. 1962. The selection of a strain of *Aedes aegypti* susceptible to infection with semi-periodic *Brugia malayi*. *Ann. Trop. Med. Parasitol.* 56: 368–72

64. Mackenzie, C. D., Oxenham, S. L., Liron, D. A., Grennan, D., Denham, D. A. 1985. The induction of functional mononuclear and multinuclear macrophages in murine Brugian filariasis: morphological and immunological properties. *Trop. Med. Parasitol.* 36:163–70

65. Mackenzie, C. D., Williams, J. F., Guderian, R. H., O'Day, J. 1987. Clinical responses in human onchocerciasis: parasitological and immunological implications. *Ciba Found. Symp.* 127:46–72

66. Maizels, R. M. Bianco, A. E., Flint, J. E., Gregory, W. F., Kennedy, M. W., et al. 1987. Glycoconjugate antigens from parasitic nematodes. *UCLA Symp. Mol. Cell. Biol.* 60:267–79

67. Maizels, R. M., Selkirk, M. E., Sutanto, I., Partono, F. 1987. Antibody responses to human lymphatic filarial parasites. *Ciba Found. Symp.* 127:189–202

68. Marshall, E., Howells, R. E. 1986. Turnover of the surface proteins of adult and third and fourth stage larval *Brugia pahangi*. *Mol. Biochem. Parasitol.* 18:17–24

68a. McCall, J. W., Malone, J. B., Ah, H., Thompson, P. E. 1973. Mongolian jirds *(Meriones unguiculatus)* infected with *Brugia pahangi* by the intraperitoneal route: a rich source of developing larvae, adult filariae and microfilariae. *J. Parasitol.* 59:436

69. McKerrow, J. H., Pino-Heiss, S., Lindquist, R. L., Werb, Z. 1985. Purification and characterization of an elastinolytic proteinase secreted by cercariae of *Schistosoma mansoni*. *J. Biol. Chem.* 260:3703–7

70. McKerrow, J. H., Sakanari, J., Brown, M., Brindley, P., Railey, J., et al. 1988. Proteolytic enzymes as mediators of parasite invasion of skin. In *Models in Dermatology*, ed. H. Maibach. Basel,

Switzerland: Karger. In press
71. Muller, R. 1975. *Worms and Disease. A Manual of Medical Helminthology.* London: Heinemann Med. 161 pp.
72. Nanduri, J., Sher, A., Kazura, J. 1987. A Brugia malayi 97 kD molecule reactive with anti-paramyosin antibodies induces resistance to microfilaraemia in mice. Presented at Ann. Meet. Am. Soc. Trop. Med. Hyg., 13th, Los Angeles
73. Nelson, F. K., Davey, K. G., Rajan, T. V. 1987. *An approach to cloning genes for nematode molting enzymes.* Presented at Ann. Meet. Am. Soc. Trop. Med. Hyg., 13th, Los Angeles
74. Nelson, G. S. 1970. Onchocerciasis. *Adv. Parasitol.* 8:173–224
75. Nutman, T. B., Kumaraswami, V., Ottesen, E. A. 1987. Parasite-specific anergy in human filariasis: insights after analysis of parasite antigen-driven lymphokine production. *J. Clin. Invest.* 79:1516–23
76. Nutman, T. B., Kumaraswami, V., Pao, L., Narayanan, P. R., Ottesen, E. A. 1987. An analysis of in vitro B-cell immune responsiveness in human lymphatic filariasis. *J. Immunol.* 138:3954–59
77. Ottesen, E. A. 1980. Immunopathology of lymphatic filariasis in man. *Springer Sem. Immunopathol.* 2:373–85
78. Ottesen, E. A. 1985. Immediate hypersensitivity responses in the immunopathogenesis of human onchocerciasis. *Rev. Infect. Dis.* 7:796–80
79. Ottesen, E. A., Kumaraswami, V., Paranjape, R., Poindexter, R. W., Tripathy, S. P. 1981. Naturally occurring blocking antibodies modulate immediate hypersensitivity responses in human filariasis. *J. Immunol.* 127:2014–20
80. Ottesen, E. A., Mendell, N. R., MacQueen, J. M., Weller, P. F., Amos, D. B., et al. 1981. Familial predisposition to filarial infection—not linked to HLA-A or -B locus specificities. *Acta Trop.* 38:205–16
81. Ottesen, E. A., Skvaril, F., Tripathy, S. P., Poindexter, R. W., Hussain, R. 1985. Prominence of IgG4 in the IgG antibody response to human filariasis. *J. Immunol.* 134:2707–12
82. Ottesen, E. A., Weller, P. F., Heck, L. 1977. Specific cellular immune unresponsiveness in human filariasis. *Immunology* 33:413–21
83. Ottesen, E. A., Weller, P. F., Lunde, M. N., Hussain, R. 1982. Endemic filariasis on a Pacific island. II. Immunologic aspects: immunoglobulin, complement, and specific antifilarial IgG, IgM, and IgE antibodies. *Ann. J. Trop. Med. Hyg.* 81:953–56

84. Oxenham, S. L., Mackenzie, C. D., Denham, D. A. 1984. Increased activity of macrophages from mice infected with *Brugia pahangi: in vitro* adherence to microfilariae. *Parasite Immunol.* 6:141–56
85. Palmieri, J. R., Connor, D. H., Purnomo, Dennis, D. T., Marwoto, H. 1982. Experimental infection of *Wuchereria bancrofti* in the silvered leaf monkey *Presbytis cristatus* Eschscholtz, 1821. *J. Helminthol.* 56:243–45
86. Paranjape, R. S., Hussain, R., Nutman, T. B., Hamilton, R. G., Ottesen, E. A. 1986. Identification of circulating parasite antigen in patients with bancroftian filariasis. *Clin. Exp. Immunol.* 63:508–16
87. Parkhouse, R. M. E. 1987. *Cloning the genes and producing the critical antigens for protective immunity.* Presented at FIL/SWG Meet. Prot. Immun. Vaccination Onchocerciasis Lymphatic Filariasis, 13th, Woods Hole, Mass.
88. Partono, F. 1982. Elephantiasis and its relation to filarial immunity. *Southeast Asian J. Trop. Med. Public Health* 13:275–79
89. Petralanda, I., Yarzábal, L., Piessens, W. F. 1986. Studies on a filarial antigen with collagenase activity. *Mol. Biochem. Parasitol.* 19:51–59
90. Philipp, M. 1984. Acetylcholinesterase secreted by intestinal nematodes: a reinterpretation of its putative role of "biochemical holdfast." *Trans. R. Soc. Trop. Med. Hyg.* 78:138–39
91. Philipp, M., Davis, T. B. 1986. Biochemical and immunologic characterization of a major surface antigen of *Dirofilaria immitis* infective larvae. *J. Immunol.* 136:2621–27
92. Philipp, M., Rumjaneck, F. D. 1984. Antigenic and dynamic properties of helminth surface structures. *Mol. Biochem. Parasitol.* 10:245–68
93. Philipp, M., Worms, M. J., Maizels, R. M., Ogilvie, B. M. 1984. Rodent models of filariasis. *Contemp. Top. Immunobiol.* 12:275–321
94. Piessens, W. F., McGreevy, P. B., Piessens, P. W., McGreevy, M., Koiman, I., et al. 1980. Immune responses in human infections with *Brugia malayi.* Specific cellular unresponsiveness to filarial antigen. *J. Clin. Invest.* 65:172–79
95. Piessens, W. F., McGreevy, P. B., Ratiwayanto, S., McGreevy, M., Piessens, P. W., et al. 1980. Immune responses in human infections with *Brugia malayi:* correlation of cellular and humoral reactions to microfilarial anti-

gens with clinical status. *Am. J. Trop. Med. Hyg.* 29:563–70

96. Piessens, W. F., Partono, F., Hoffman, S. L., Ratiwayanto, S., Piessens, P. W., et al. 1982. Antigen-specific suppressor T lymphocytes in human lymphatic filariasis. *N. Engl. J. Med.* 307:144–48

97. Piessens, W. F., Ratiwayanto, S., Piessens, P. W., Tuti, S., McGreevey, P. B., et al. 1981. Effect of treatment with diethylcarbamazine on immune responses to filarial antigens in patients infected with *Brugia malayi. Acta Trop.* 38:227–34

98. Piessens, W. F., Ratiwayanto, S., Tuti, S., Palmieri, J. H., Piessens, P. W., et al. 1980. Antigen-specific suppressor cells and suppressor factors in human filariasis with *Brugia malayi. N. Engl. J. Med.* 302:833–37

99. Piessens, W. W., Wadee, A. A., Kurniawan, L. 1987. Regulation of immune responses in lymphatic filariasis. *Ciba Found. Symp.* 127:164–79

100. Portaro, J. K., Britton, S., Ash, L. R. 1976. *Brugia pahangi:* depressed mitogen reactivity in filarial infection in jirds, *Meriones unguiculatus. Exp. Parasitol.* 40:438–46

101. Rathaur, S., Robertson, B. D., Selkirk, M. E., Maizels, R. M. 1987. Secretory acetylcholinesterases from *Brugia malayi* adult and microfilarial parasites. *Mol. Biochem. Parasitol.* 26:257–65

102. Sasa, M. 1976. *Human Filariasis.* Baltimore, Md: Univ. Park

103. Schrater, A. F., Spielman, A., Piessens, W. F. 1983. Predisposition to *Brugia malayi* microfilaraemia in progeny of infected gerbils. *Am. J. Trop. Med. Hyg.* 32:1306–8

104. Schur, P. H. 1987. IgG subclasses—a review. *Ann. Allergy* 58:89–99

105. Selkirk, M. E., Denham, D. A., Partono, F., Sutanto, I., Maizels, R. M. 1986. Molecular characterization of antigens of lymphatic filarial parasites. *Parasitology* 9:S15–S38

106. Selkirk, M. E., Rutherford, P. J., Bianco, E. A., Flint, J. E., Denham, D. A., et al. 1987. Characterization and cloning of lymphatic filarial antigens. *UCLA Symp. Mol. Cell. Biol.* 60:169–81

107. Sher, A. F., Pearce, E. J. 1987. Strategies for induction of protective immunity against schistosomes. *Acta Trop.* 44(Suppl. 12):41–45

108. Smithers, S. R., Simpson, A. J. G., Yi, X., Omer-Ali, P., Kelly, C., et al. 1987. The mouse model of schistosome immunity. *Acta Trop.* 44(Suppl. 12):21–30

109. Suswillo, R. R., Doenhoff, M. J.,

Denham, D. A. 1981. Successful development of *Brugia pahangi* in T-cell deprived CBA mice. *Acta Trop.* 38:305–8

110. Suswillo, R. R., Nelson, G. S., Muller, R., McGreevy, P. B., Duke, B. O. L., et al. 1977. Attempts to infect jirds *(Meriones unguiculatus)* with *Wuchereria bancrofti, Onchocerca volvulus, Loa loa,* and *Mansonella ozzardi. J. Helminthol.* 51:132–34

111. Suswillo, R. R., Owen, D. G., Denham, D. A. 1980. Infections of *Brugia pahangi* in conventional and nude (athymic) mice. *Acta Trop.* 37:327–35

112. Sutanto, I., Maizels, R. M., Denham, D. A. 1985. Surface antigens of a filarial nematode: analysis of adult *Brugia pahangi* surface components and their use in monoclonal antibody production. *Mol. Biochem. Parasitol.* 15:203–14

113. Tamashiro, W. K., Rao, M., Scott, A. L. 1987. Proteolytic cleavage of IgG and other protein substrates by *Dirofilaria immitis* microfilarial extracts. *J. Parasitol.* 73:149–54

114. Tanner, M., Weiss, N. 1981. *Dipetalonema viteae* (Filarioideae): evidence for a serum-dependent cytotoxicity against developing third and fourth stage larvae in vitro. *Acta Trop.* 38:325–28

115. Townson, S., Bianco, A. E. 1982. Experimental infection of mice with the microfilariae of *Onchocerca lienalis. J. Helminthol.* 85:283–93

116. Townson, S., Bianco, A. E., Doenhoff, M. J., Muller, R. 1984. Immunity to *Onchocerca lienalis* microfilariae in mice. I. Resistance induced by the homologous parasite. *Trop. Med. Parasitol.* 35:202–8

117. Townson, S., Bianco, A. E., Owen, D. 1981. Attempts to infect small laboratory animals with infective larvae of *Onchocerca lienalis. J. Helminthol.* 55:247–49

118. Vickery, A. C., Nayar, J. K. 1987. *Brugia pahangi* in nude mice: protective immunity to infective larvae is Thy 1.2+ cell dependent and cyclosporin A resistant. *J. Helminthol.* 61:19–27

119. Vickery, A. C., Nayar, J. K., Albertine, K. H. 1985. Differential pathogenicity of *Brugia malayi, B. patei,* and *B. pahangi* in immunodeficient nude mice. *Acta Trop.* 42:353–63

120. Vickery, A. C., Vincent, A. L. 1984. Immunity to *Brugia pahangi* in athymic nude and normal mice: eosinophilia, antibody and hypersensitivity responses. *Parasite Immunol.* 6:545–59

121. Vickery, A. C., Vincent, A. L., Sode-

man, W. A. 1983. Effect of immune reconstitution on resistance to *Brugia pahangi* in congenitally athymic nude mice. *J. Parasitol.* 69:478–85

122. Vincent, A. L., Ash, L. R., Rodrick, G. E., Sodeman, W. A. 1980. The lymphatic pathology of *Brugia pahangi* in the Mongolian jird. *J. Parasitol.* 66:613–20

123. Vincent, A. L., Sodeman, W. A., Winters, A. 1980. Development of *Brugia pahangi* in nude (athymic) and thymic mice C3H/HeN. *J. Parasitol.* 66:448

124. Vincent, A. L., Vickery, A. C., Lotz, M. J., Desai, U. 1984. The lymphatic pathology of *Brugia pahangi* in nude (athymic) and thymic mice C3H/HeN. *J. Parasitol.* 70:48–56

125. Wadee, A. A., Vickery, A. C., Piessens, W. F. 1987. Characterization of immunosuppressive proteins of *Brugia malayi* microfilariae. *Acta Trop.* 44:343–52

125a. Ward, D. J., Nutman, T. B., Zea-Flores, G., Portocarrero, C., Lujan, A., et al. 1988. Onchocerciasis and immunity in humans: enhanced T-cell responsiveness to parasite antigen in putatively immune individuals. *J. Infect. Dis.* 157:536–43

126. Warren, K. S. 1982. The secret of the immunopathogenesis of schistosomiasis: in vivo models. *Immunol. Rev.* 6:189–213

127. Wong, M. M. 1987. *Successful strategies in veterinary anti-helminth vaccines.* Presented at FIL/SWG Meet. Prot. Immun. Vaccination Onchocerciasis Lymphatic Filariasis, 13th, Woods Hole, Mass.

128. World Health Organization. 1984. *Lymphatic Filariasis. Fourth Report of the WHO Expert Committee on Filariasis. WHO Tech. Rep. Ser.* Vol. 702

129. World Health Organization. 1987. *Third Report of the WHO Expert Committee on Onchocerciasis. WHO Tech. Rep. Ser.* Vol. 752

130. Yates, J. A., Higashi, G. I. 1985. *Brugia malayi:* vaccination of jirds with [60]Cobalt-attenuated infective stage larvae protects against homologous challenge. *Am. J. Trop. Med. Hyg.* 34:1132–37

131. Yates, J. A., Higashi, G. I. 1987. *Experimental induction of immunity to filariae: animal models.* Presented at FIL/SWG Meet. Prot. Immun. Vaccination Onchocerciasis Lymphatic Filariasis, 13th, Woods Hole, Mass.

Ann. Rev. Microbiol. 1988. 42:717–43

PLASMID-MEDIATED HEAVY METAL RESISTANCES

Simon Silver and Tapan K. Misra

Department of Microbiology and Immunology, University of Illinois College of Medicine, Chicago, Illinois 60680

CONTENTS

INTRODUCTION

Many bacterial strains contain genetic determinants of resistances to heavy metals such as Hg^{2+} (and organomercurials), Ag^+, AsO_2^-, AsO_4^{3-}, Bi^{3+}, BO_3^{3-}, Cd^{2+}, Co^{2+}, CrO_4^{2-}, Cu^{2+}, Ni^{2+}, Pb^{3+}, Sb^{2+}, TeO_3^{2-}, Tl^+, Zn^{2+}, and undoubtedly others (91, 100, 106). These resistance determinants are often found on plasmids and transposons (94, 100, 106), which facilitates their analysis by molecular genetic techniques. In the frequent absence of any obvious source of direct selection, these resistances occur with surprisingly

0066-4227/88/1001-0717$02.00

high frequencies. Selective agents in hospitals and other environments are only beginning to be examined. It has been suggested that heavy metal resistances may have been selected in earlier times, and that they are merely carried along today for a free ride with selection for antibiotic resistances. We doubt that there is such a thing as a free ride as far as these determinants are concerned. For example, in Tokyo in the late 1970s both heavy metal resistances and antibiotic resistances were found with high frequencies in *Escherichia coli* isolated from hospital patients, whereas heavy metal resistance plasmids without antibiotic resistance determinants were found in *E. coli* from an industrially polluted river (66). Selection occurs for resistances to both types of agents in the hospital (81), but only for resistance to toxic heavy metals in the river environment. Radford et al (82) found Hg^{2+}-resistant microbes in agricultural soil with no known mercurial input. In such settings resistant microbes may be very rare, but they may come into much greater quantitative prominence after industrial or agricultural pollution.

The topic of plasmid-mediated heavy metal resistance has been frequently reviewed over the last decade (12, 25, 26, 61, 94–96, 99–102, 106, 107). A conference proceedings contained our last, more narrow summary of this topic (102). The major recent progress has consisted of the cloning and DNA sequence analysis of determinants for mercury, arsenic, cadmium, and tellurium resistances and the initial reports of still additional resistances. Since mercury resistance has recently been reviewed (25, 26, 107), we start our consideration with other heavy metal resistance systems and only cover mercury resistance within the constraints of space, emphasizing newer findings and more speculative directions. A review by Summers & Silver published in this series 10 years ago (111) was the first to develop this area.

ARSENIC AND ANTIMONY RESISTANCES

Arsenic and antimony resistances are governed by plasmids that also code for antibiotic and other heavy metal resistances (23, 71, 94, 110). Arsenate, arsenite, and antimony(III) resistances are coded for by an inducible operon-like system in both *Staphylococcus aureus* and *E. coli* (97). Each of the three ions induces all three resistances.

The mechanism of arsenate resistance is reduced accumulation of arsenate by induced resistant cells (97). Arsenate is normally accumulated via the cellular phosphate transport systems (Figure 1). The presence of the resistance plasmid does not alter the kinetic parameters of the cellular phosphate transport systems; even the K_i for arsenate as a competitive inhibitor of phosphate transport (Figure 1) is unchanged. Direct evidence for plasmid-governed energy-dependent efflux of arsenate (64, 98) indicated that the reduced net uptake of arsenic resulted from rapid efflux. Glucose but not succinate could energize arsenate efflux in an *E. coli* strain that could not synthesize ATP from respiratory substrates (64). Thus AsO_4^{3-} efflux appears to involve an ATPase transport system. Arsenite-stimulated ATPase activity has recently been directly measured (17, 84, 86).

An interesting question about the arsenate efflux system concerns its specificity. Arsenate generally functions as a phosphate analog and is accumulated by bacteria via phosphate transport systems (93). The arsenate-resistance efflux system should not excrete phosphate, since phosphate starvation would be no more advantageous to the cells than arsenate inhibition. Energy-dependent efflux systems functioning as resistance mechanisms must be highly specific for the toxic anion or cation to prevent loss of the required nutrient.

The physiological and biochemical studies of the arsenic resistance determinant were followed by molecular genetic studies, cloning, Southern blotting, and minicell polypeptide synthesis (63, 65). The DNA sequence of the *ars* operon of plasmid R773 contains three reading frames, *arsA, arsB,* and *arsC,* in addition to the *arsR* regulatory gene, which is yet to be sequenced. The *arsA* sequence has significant sequence homology at the polypeptide level to adenylate-binding proteins such as nitrogenase and the beta subunit of the oxidative phosphorylation ATPase (17). Thus the *arsA* gene appears to determine the efflux ATPase predicted from transport-inhibition studies (64,

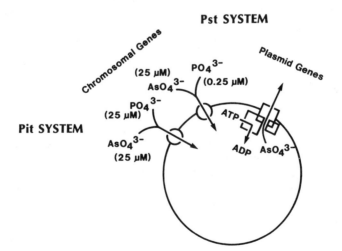

Figure 1   The phosphate (arsenate) transport systems and the arsenate efflux system of *E. coli.* 0.25 and 25 μM, where indicated, are the K_m for PO_4^{3-} and the K_i for AsO_4^{3-} as a competitive inhibitor of PO_4^{3-} transport for the Pit (for P_i transport) and Pst (for phosphate-specific transport) chromosomal phosphate transport systems. The three *boxes* for the plasmid arsenic ATPase represent the ArsA, ArsB, and ArsC polypeptides. Modified from Reference 102.

98). These adenylate-binding regions have also been identified in mammalian membrane transport proteins such as the multidrug resistance (Mdr) glycoprotein associated with tumor lines that have become refractory to chemotherapy (34, 38). This glycoprotein is thought to determine resistance by pumping a variety of structurally unrelated compounds from the cells. Surprisingly, the *arsA* gene seems to have arisen from a gene duplication and fusion, so that adenylate-binding regions appear twice (17). The tumor cell multidrug resistance protein also appears to have arisen by duplication and fusion (38), but the homologous halves of the ArsA protein are much smaller than the homologous halves of the Mdr protein. This pattern of fused tandem dimers for a membrane protein may have arisen more than once in nature to satisfy a structure/function need. Such an idea is quite speculative, supported now only by partial evidence from a variety of rather different membrane transport systems.

The second open reading frame, *arsB*, encodes a very hydrophobic polypeptide, postulated (17) to be the membrane carrier protein responsible for efflux of both arsenate (64, 98) and arsenite (85). In the mammalian Mdr protein the sequences homologous to ArsA and ArsB are fused into a single 1276–amino acid polypeptide (38).

The ArsA ATPase polypeptide has now been purified and identified as an arsenite-stimulated ATPase (86). The isolated ArsA polypeptide binds ATP specifically and appears in solution as a monomeric protein, only loosely associated with the cell membrane. The ArsB protein was identified as an integral inner-membrane protein (M. San Francisco & B. P. Rosen, personal communication), as predicted from its amino acid sequence (17).

The third gene, *arsC*, encodes a smaller polypeptide. The deletions generated for sequencing gave additional information about gene function: Deleting *arsC* from the 3' end of the operon caused loss of arsenate resistance but retention of arsenite and antimony resistances. Further deletion from the right into *arsB* caused additional loss of arsenite and antimony resistances, and deletion from the upstream end entering *arsA* caused simultaneous loss of all three resistances (17). Thus we drew the tentative model (Figure 1). The *arsA* gene encodes a relatively soluble polypeptide that has ATPase activity and is required for both arsenate and arsenite resistances. The *arsB* gene encodes a hydrophobic membrane protein that is required for both arsenate and arsenite resistances and is the membrane carrier for arsenic efflux. ArsC is a small polypeptide that confers substrate specificity upon the ArsB protein. In its absence, only arsenite (and antimony) are transported, but in the presence of ArsC protein, arsenate is transported also. We think of ArsC as analogous to a periplasmic binding protein that provides specificity to an ATP-dependent uptake system. Only here the specificity-conferring binding protein is on the inside of the cell and the transport is outwardly oriented.

The DNA sequence of a *Staphylococcus* arsenic resistance determinant (51) was recently completed (F. Gotz, personal communication). There are again three reading frames, but only the ArsB polypeptide sequence is homologous to that from plasmid R773 in gram-negative bacteria. The plasmid-determined arsenic resistance system has (to date) always had the same biochemical mechanism, reduced uptake (97) due to an ATPase efflux system (64, 98), in both gram-negative and gram-positive bacteria.

TELLURIUM RESISTANCE

Tellurite and tellurate resistances have been detected from plasmid-mediated determinants (108, 114). The resistance mechanism is not known. Early work on tellurite resistance in bacteria showed that although reduction of TeO_3^{2-} to black Te^0 can be seen, it is apparently not the resistance mechanism because it occurs with both sensitive and resistant cells (108). Alkylation of tellurium is apparently also not the mechanism of resistance (108).

Recently, Jobling & Ritchie (49) cloned the tellurium resistance determinant from a large *Alcaligenes* plasmid into *E. coli*, where it was expressed from a 3.55-kb DNA fragment. Insertion mutagenesis created two classes of mutations, one apparently in the regulatory region of this inducible system and the other in structural genes. Maxicell polypeptide analysis (49, 50) indicated that four proteins might be involved in tellurite resistance, but none has been pinned down with regard to function; the first polypeptide (339 amino acids) may be involved in regulation, and at least two may be involved in the actual resistance mechanism (49). The cloned tellurite resistance determinant has now been sequenced (50). Five open reading frames were identified, which correspond to the four polypeptides that were synthesized plus another, apparently a membrane protein (347 amino acids), which has not yet been seen in protein gels (Figure 2). Although the sequence analysis provides much information [e.g. the products of ORF4 and ORF5 are about 65% identical (50)], the complete sequence of this five–open reading frame determinant and available deletions and mutations leave us no closer to an explanation of the basis of tellurite and tellurate resistance. It is a measure of the power (and limitations) of cloning and sequence analysis that sophisticated

Figure 2 Diagram of the tellurium resistance operon and polypeptide pattern from the *Alcaligenes* plasmid (50). *ORF,* open reading frames from DNA sequence, with mass of the polypeptide products in kilodaltons.

analysis can precede even preliminary understanding—and still not necessarily lead to basic understanding.

Bradley & Taylor (10, 113) found a seemingly different tellurite resistance system on a class of plasmids that includes the well-studied broad–host range plasmid RP4. This system is initially "cryptic" in that the cells with the plasmid are tellurite sensitive. However, there is a low-frequency mutation on the plasmid that results in a tellurite resistance determinant on RP4. There is no physical rearrangement associated with the mutation, but tellurite resistance was found on a 4.5-kb transposon, Tn521, that can hop to other plasmids (10, 113). Mutagenesis studies and the finding of other genes in that area of RP4 have limited the tellurite resistance determinant to probably no more than 2 kb (113). The tellurite resistance system of RP4 appears to be constitutive, unlike the inducible resistance with other plasmid groups (49). Further work will be needed to clarify the nature of both chromosome- and plasmid-encoded resistance to tellurite.

CHROMATE RESISTANCE

Bacterial resistance to chromate (CrO_4^{2-}) has been found in several *Pseudomonas* strains (9, 16, 45, 109) and also with a plasmid in *Streptococcus lactis* (23). Horitsu et al (45) showed that a CrO_4^{2-}-sensitive *Pseudomonas ambigua* strain accumulated six times more chromate than a resistant strain.

Ohtake et al (76) recently concluded that the basis for plasmid-mediated chromate resistance in *Pseudomonas* is reduced chromate uptake by the plasmid-bearing strain. With *Pseudomonas fluorescens* containing a chromate resistance plasmid the V_{max} for chromate uptake was reduced (76), the K_m for chromate uptake was unchanged, and the K_i for chromate as a competitive inhibitor of sulfate transport was unchanged. There was no difference in chromate efflux between the sensitive and resistant *P. fluorescens* strains, which suggests that the block may be at the level of uptake rather than efflux.

Ohtake et al (76) demonstrated reduced chromate uptake in resistant *P. fluorescens*, but not in *Pseudomonas aeruginosa*, with plasmid pLMB1 [originally isolated in *P. fluorescens* (9)] and with plasmid pMG6 [originally isolated in *P. aeruginosa* (109)]. Both *P. fluorescens* (76) and *P. aeruginosa* express chromate resistance with the plasmids (H. Ohtake & C. Cervantes, unpublished data). Thus, there is an interplay between plasmid-governed functions and chromosomal gene–determined properties. The sulfur source used for cell growth greatly affects chromate resistance levels (76). The rate of chromate uptake by *P. aeruginosa* also was regulated by the sulfur source; and chromate resistance levels were affected by the sulfur source in parallel to transport rates (H. Ohtake & C. Cervantes, unpublished data).

The chromate resistance determinant from *P. aeruginosa* plasmid pUM505

(16) has been cloned as a 3.8-kb DNA fragment, which is expressed in *P. aeruginosa* but not in *E. coli* (15). The cloned fragment Southern DNA/DNA hybridizes with plasmid pUM505 DNA but not with *Pseudomonas* chromosomal DNA and barely with two other chromate resistance plasmids, pLHB1 (9, 76) and pMG6 (109).

CADMIUM RESISTANCE

There are six or more systems for bacterial cadmium resistance known today. However, little physiological and biochemical work has been accomplished. Only one of these systems *(cadA)* has been cloned (123), and DNA sequencing has just been completed in our laboratory. Therefore, our understanding of bacterial cadmium resistance is preliminary and tentative.

The first two cadmium resistance systems are the *cadA* and *cadB* systems on *S. aureus* plasmids that also confer resistance to several other heavy metals (71). The mechanisms are Cd^{2+} efflux for *cadA* (119) and increased binding for *cadB* (80). A third *S. aureus* cadmium resistance system involves energy-dependent Cd^{2+} efflux, as *cadA* does, but it differs from *cadA* in that it confers Cd^{2+} resistance alone, whereas *cadA* confers both Cd^{2+} and Zn^{2+} resistances (123). A fourth mechanism of cadmium resistance in gram-positive bacteria is found with a chromosomal mutation in *Bacillus subtilis* (112) that resulted in a change in the membrane manganese transport system (which normally transports both Mn^{2+} and Cd^{2+}) (80, 118) so that Cd^{2+} is no longer accumulated (52). The fifth reported mechanism of cadmium resistance in bacteria is the synthesis of a polythiol Cd^{2+}-binding protein analogous to metallothionein of animal cells (40, 90). And the sixth cadmium resistance system is found with a 9.1-kb *Alcaligenes eutrophus* plasmid DNA fragment that simultaneously confers resistances to Cd^{2+}, Zn^{2+}, and Co^{2+} (58, 70). The mechanism involved is still not known.

Perhaps it is time to reevaluate the reports of polythiol-containing cadmium-binding components of bacterial cells grown and adapted to cadmium (43, 44, 62, 77, 78). The problem with these reports has been the absence of follow-up or reproducible work that furthers understanding. Recent reports (77, 78a) indicate that the metallothionein polypeptide from cyanobacteria (78) is rather different in composition and sequence from those of animals (40). Workers in Sadler's laboratory have been unable to confirm their own beautiful and encouraging NMR data (43, 44) demonstrating the existence of such compounds (P. D. Sadler, personal communication).

Recently, Higham et al (44a) reported that the *Pseudomonas* strain that appears to make intercellular metallothionein also produced [when exposed to Au(I) salts only] an extracellular gold-binding yellow polypeptide consisting mostly of cysteine (48%), glutamate or glutamine (19%), and glycine (22%)

residues, which is suggestive of a phytochelatin-like structure. We hope that this work will develop further.

A new opportunity is promised by the discovery of polyglutathione compounds in plant cells (37, 47, 105) exposed to Cd^{2+} and other toxic cations. These compounds have been given the trivial name "phytochelatin" and have the structure (γ-glutamyl-cysteinyl)$_n$-glycine, with $n = 2$–8 (37, 105). Glutathione has this structure with $n = 1$. Phytochelatin is the major intracellular thiol form in Cd^{2+}-exposed plant cells (105). Whether any of the cadmium resistance systems found with bacterial plasmids will involve small peptide intracellular binding factors is yet to be determined.

The basic mechanism of Cd^{2+} uptake and of *cadA* Cd^{2+} resistance in *S. aureus* is shown in Figure 3. Cd^{2+} enters bacterial cells as a toxic alternative substrate for the cellular Mn^{2+} transport system in gram-positive bacteria (80, 118) or for the Zn^{2+} transport system in gram-negative bacteria (54). Both of these systems are normal, nutritionally required cation transport systems (93, 95). Once Cd^{2+} has entered the bacterial cells by this membrane potential–dependent uptake system, it is toxic for the sensitive cells or is rapidly effluxed from resistant cells (Figure 3).

Tynecka et al (119) proposed that this efflux system was a chemiosmotic electroneutral Cd^{2+}-for-$2H^+$ exchange system, based on its insensitivity to agents that disrupt the membrane potential but sensitivity to agents that block or accelerate proton movement. However, with the new DNA sequence analysis of the *cadA* determinant (73), it is apparent that this system includes a cation-translocating ATPase. The *cadA* DNA sequence includes an open

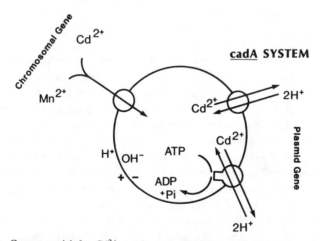

Figure 3   Current model for Cd^{2+} uptake and efflux in gram-positive bacteria. Both the chemiosmotic antiporter (119) and the ATPase alternative (73) for the efflux system are shown.

reading frame that would encode a 727–amino acid polypeptide that is about 25% identical at the sequence level to the K^+-ATPases from the *E. coli kdpB* gene (42) and from *Streptococcus faecalis* (104). There is a weaker but yet highly significant homology between these bacterial cation-translocating ATPases and the sarcoplasmic reticulum Ca^{2+}-ATPase, the Na^+/K^+-exchange ATPases of animal cell membranes, and the fungal cell membrane proton ATPases (discussed in 104, 120). All of these enzymes share a crucial aspartyl residue which is phosphorylated during the transport cycle and additional residues that bind ATP (104, 120). These proteins are primarily cytoplasmic, but have localized hydrophobic regions, which appear to be pairs of transmembrane polypeptides. The *cadA* sequence has an amino-terminal sequence that is homologous to the Hg^{2+}-binding region of the mercuric reductase enzyme and periplasmic *merP* polypeptide (see next section). Two cysteines are conserved, and these may function as soft metal–binding residues.

MERCURY AND ORGANOMERCURIAL RESISTANCES

Genetic Studies

The plasmid R100 mercury resistance system is the system on which the most progress has been made. Cloning and transposon insertion studies established the existence of the *merR* regulatory gene, whose product is a *trans*-acting inducer-repressor (positive as well as negative regulation) (28), followed by a promoter/operator region (*cis*-acting mutations only), followed by two structural genes, *merT* [for a Hg^{2+} transport system that brings extracellular Hg^{2+} into the cell (Figure 4)] and *merA* (for the subunit of mercuric reductase) (30, 31, 67). The *merT* transport system was discovered in "hypersensitive" mutants, which were much more sensitive to Hg^{2+} than the wild-type *E. coli* without a plasmid (29, 30, 67). The mutants only showed the hypersensitive phenotype when induced by subtoxic levels of Hg^{2+}, and such induced cells took up excessive amounts of $^{203}Hg^{2+}$ (67). This seemed a bizarre idea, but it was rationalized by the need for mercuric reductase for a high-energy cofactor NADPH. Since NADPH must be intracellular, any enzyme requiring it must be intracellular. Since Hg^{2+} is toxic and inactivates cell surface enzymes and transport systems, the plasmid system contains a highly selective transport system that traps the Hg^{2+} outside the cell and delivers it to the mercuric reductase enzyme without ever allowing the Hg^{2+} to be free (Figure 4).

Additional mutagenesis studies (69) provided evidence for two additional genes called *merC* and *merD*. [The term *merB* had been used for the gene for the organomercurial lyase (111, 122), which is lacking in the R100 system.] The *merC* gene was defined by mutations between *merT* and *merA*, and *merD*

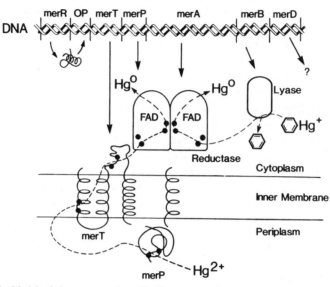

Figure 4 Model of the mercury detoxification system of broad organomercurial resistance plasmid pDU1358. Modified from Reference 102.

was defined by mutations distal to *merA*; both classes of mutations led to reduced resistance to Hg^{2+}.

Early attempts failed to assign any of the $[^{35}S]$methionine-labeled polypeptides to genes other than *merA* (20, 48). Only after the DNA sequencing was complete were NiBhriain & Foster (68) able to identify the MerT, MerP, and MerC polypeptides (and MerA, but not MerR or MerD) on minicell gels.

The increased mercury uptake by Hg^{2+}-hypersensitive mutants (lacking mercuric reductase) provides direct evidence of a genetically determined mercury transport system (30, 67). Deletion of the *merT* and *merP* region eliminates that hypersensitivity (55). The *merT*-*merP* deletion mutants are Hg^{2+}-sensitive, although they contain intracellular mercuric reductase (55).

DNA Sequences From Gram-Negative Sources

Three gram-negative mercuric resistance systems have been sequenced, and the biological understanding that has come from comparison of these sequences has deepened understanding of the *mer* operon. The first surprise from the sequencing of plasmid R100 [which contains transposon Tn*21* (19)] and transposon Tn*501* was that they were so similar (about 85% identical base pairs) (14, 59, 60), although R100 came from a *Shigella* sp. (51% chromosomal G+C content) and Tn*501* originated in a pseudomonad (>65% chromosomal G+C content). The R100 system was sequenced independently by Barrineau et al (3).

The first gene of the *mer* operon is the *merR* regulatory gene, which is transcribed divergently from the structural genes (27). The 144–amino acid MerR protein is helix-turn-helix DNA-binding protein with gene regulatory functions (21, 22, 79). The *merR* gene has been cloned into expression vectors (41, 75). The MerR proteins of R100, pDU1358, and Tn*501* are cross-complementary and about 85% identical at the amino acid level (28, 72).

Next, continuing along the R100, Tn*501*, and pDU1358 *mer* operons, the operator/promoter regions show diad structures and are remarkably similar (59). They are more similar, in fact, than the sequences in any of the genes. This region contains the transcription start site (56) for the 3-kb mRNA for the *mer* operon. The transcript ends immediately after the *merD* gene (41). The first structural gene in all three operons *(merT)* encodes a polypeptide of 116 amino acids (3, 59, 72). The MerT protein is a very hydrophobic membrane protein and appears to pass through the cell membrane three times. From this modeling based on DNA sequence analysis (Figure 4) there is a cysteine pair close to the outside of the membrane toward the amino end of the polypeptide and another cysteine pair perhaps on the inner side of the cell membrane between the second and third transmembrane segments. These four cysteines are hypothesized to be involved in the transport of mercury by the MerT protein (Figure 4).

The next gene in pDU1358, Tn*501*, and R100 encodes the 91–amino acid MerP protein (3, 59, 72) (Figure 4), which appears to be a periplasmic mercury-binding protein. The amino acid sequences start with a good canonical leader signal sequence of two N-proximal lysines followed by a long hydrophobic sequence followed by a pair of alanines as candidates for signal peptidase sites. After the 20–amino acid leader sequence on the *merP* protein, the remaining 71 amino acids are remarkably similar to the first 68 amino acids of the *merA* mercuric reductase sequence (60), with 35% amino acid identities. These *merP* sequences contain a cysteine pair, which we have hypothesized (Figure 4) (12) is involved in Hg^{2+} binding. Thus the Hg^{2+} that is bound initially by the periplasmic MerP Hg^{2+}-scavenging protein is then passed along from cysteine pair to cysteine pair (first one pair in MerP, then two in MerT, then two in MerA). Although these cysteines are kinetically active in the exchange, the Hg^{2+} is never freed and therefore does not inhibit cellular processes (Figure 4).

After *merP* there is an additional gene, *merC*, in the R100 system (60). There is uncertainty about the function of the *merC* gene product, although the protein is made in minicells (48, 68) and there are mutants in the gene (69). Barrineau & Summers (4) suggested that this region might govern a regulatory function, but the amino acid sequence of this polypeptide (3, 60) indicates that it may be a membrane protein. It cannot be essential for Hg^{2+} resistance in the host cells so far tested, since it is absent from Tn501 (60) and pDU1358 (72), which confer identical Hg^{2+} resistance levels.

Still moving gene by gene from the left in Figure 4, the next gene is merA, for the mercuric reductase subunit. The Tn501 and Tn21 polypeptides differ at only 79 of the 561 (Tn501) or 564 (R100) positions (60). The partial pDU1358 merA sequence (72) is quite similar. Comparison of the amino acid sequences from the DNA sequences has been very informative (60). Firstly, the active site region can be readily identified and contains another cysteine pair (Figure 5). The amino acid sequences of the R100, pDU1358, and Tn501 mercuric reductases are identical for the 15 amino acids shown, and there is only one or a pair of isoleucine/valine changes in the comparable amino acid sequence for the merA gene product of gram-positive systems (53; Y. Wang & I. Mahler, personal communication; J. Altenbuchner, personal communication). The active site of the mercuric reductase is very similar to that of glutathione reductase, which is known from a range of organisms (Figure 5); there is even a high-resolution structure for human glutathione reductase from X-ray diffraction crystal analysis (116). The overall amino acid sequences of

Mercuric Reductase

Tn501		T	I	G	G	T	C	V	N	V	G	C	V	P	S	K
Tn21		.	.	.	.	.	.	.	.	.	.	.	.	.	.	.
pDU1358		.	.	.	.	.	.	.	.	.	.	.	.	.	.	.
pI258		.	V	.	.	.	.	.	.	.	I	.	.	.	.	.
Bacillus strain		.	V	.	.	.	.	.	.	.	.	.	.	.	.	.
Streptomyces lividans		.	T	.	.	.	.	.	.	.	.	.	.	.	.	.

Glutatione reductase

Human erythrocytes		K	L	.	.	.	.	.	.	.	.	.	.	.	K	.
Yeast		A	L	.	.	.	.	.	.	.	.	.	.	.	K	
Escherichia coli		Q	L	.	.	.	.	.	.	.	.	.	.	.	K	

Trypanothione reductase

Crithidia fasciculata		A	L	.	.	.	.	.	.	.	.	.	.	.	K	.

Lipoamide dehydrogenase

Pig Heart		.	L	.	.	.	L	.	.	.	I	.	.	.		
Escherichia coli		.	L	.	.	V	.	L	.	.	.	I	.	.	.	.
Bacillus stearothermophilus		L	.	.	V	.	L	.	.	.	I	.	.	.	.	

Figure 5 Comparison of the active site amino acid sequences of six mercuric reductase enzymes (53, 72; Y. Wang & I. Mahler, personal communication; J. Altenbuchner, personal communication) with those of lipoamide dehydrogenase (83), glutathione reductase (83), and trypanothione reductase (92). Dots represent the same amino acids as in the Tn501 sequence.

glutathione reductase and mercuric reductase are sufficiently similar (53, 60) that one can (with a little imagination) fold the mercuric reductase amino acid sequence along the glutathione reductase coordinates. The match is best in the active site region (Figure 5), where 13 amino acids in a row are identical in the three versions of glutathione reductase and in a newly found related enzyme, trypanothione reductase (92), which reduces the glutathione analog trypanothione. Another member of this family of related FAD-containing NAD(P)H-requiring oxidoreductases is lipoamide dehydrogenase, for which sequence information is also available (83). The lipoamide dehydrogenase active site sequences differ from those of mercuric reductase only by replacement of closely related amino acids (Figure 5). The overall mercuric reductase, glutathione reductase, and lipoamide dehydrogenase amino acid sequences are equally related (26–28% amino acid identities) (53, 60).

In addition to the clear evolutionary relationship between the mercuric reductase, glutathione reductase, and lipoamide dehydrogenase active sites (13, 53, 60), another striking evolutionary relationship, between the mercuric reductase N-terminus and the MerP polypeptide sequence (60), became apparent. The 35% amino acid identity between the C-terminal 71 amino acids of the *merP* polypeptide and the N-terminal 68 amino acids of the *merA* polypeptides of R100 and Tn*501* (60) essentially requires a common ancestral gene. We have hypothesized (102) that the ancestral gene for the mercury-binding polypeptide was duplicated; one copy fused with the gene for an ancestral FAD-containing NAD(P)H-dependent oxidoreductase (about 470 amino acids long) to produce the mercuric reductase enzyme (about 550 amino acids long), whereas the other copy of this gene fused with the determinant of a membrane transporting leader signal sequence to produce the current MerP periplasmic protein.

Following the *merA* gene, the Tn*501*, R100, and pDU1358 DNA sequences remain about 85% identical through the last gene, *merD* (14, 36, 69). The gene affects mercury resistance only slightly, and its function is still uncertain. There is a significant sequence homology between the DNA and amino acid sequences of *merD* and *merR* (14), which suggests a regulatory role.

Plasmid pDU1358 has still another gene, *merB*, the determinant of the organomercurial lyase. This gene lies between *merA* and *merD* (Figure 4) and confers resistance to phenylmercury and other organomercurials (36).

DNA Sequences From Gram-Positive Sources

The first *mer* operon sequence from a gram-positive organism was determined from *S. aureus* plasmid pI258 (53). Laddaga et al (53) wanted to compare the most different possible (low versus high G+C; gram-negative versus gram-positive) versions of the *merA* mercuric reductase and the *merB* organomercurial lyase genes. *S. aureus* has low G+C (about 35%), so the *mer* region of pI258 was cloned into *Bacillus subtilis* and *E. coli* cloning vectors; it functioned in *B. subtilis* but not in *E. coli* (53).

Initially, we expected to find repeated base pair sequences at the ends of the 6.4-kb *Bgl*II fragment of plasmid pI258 that contains the *mer* operon (Figure 6). The same 6.4-kb *Bgl*II *mer* fragment occurs on different *S. aureus* mercury resistance plasmids (71) and with chromosomal mercury resistance systems (123), which led to the suggestion that mercury resistance might be determined by a transposon. However, no evidence for transposition of *mer* from pI258 has been seen (R. P. Novick & E. Murphy, personal communication). This situation has been clarified by the identification of putative insertion sequences, called IS*431* (2) or IS*257* (35), that occur on both ends of the *mer* sequence, as shown in Figure 6. The IS*431* Left and Right sequences are 800 and 786 bp respectively in length and are bracketed by 22- or 14-bp inverted repeats. The sequences include open reading frames corresponding to a 234–amino acid polypeptide (Figure 6), which shows sequence homology (40% amino acid identities) to the transposase protein of *Proteus vulgaris* transposon IS*26* (2). A third copy of IS*431* occurs in the DNA of methicillin-resistant clinical isolates (2, 35). Its sequence (including the 14-bp inverted repeat) is essentially identical to that of the IS*431* elements bordering the *mer* operon.

The pI258 *mer* sequence starts at the left with a putative *merR* gene that is transcribed from left to right in Figure 6, i.e. in the same direction as the structural genes and opposite that of R100, pDU1358, and Tn*501* (Figure 4). There is considerable uncertainty about this gene, and therefore it is listed as ORF1. The open reading frame starts to the left of Figure 6, before the *Bgl*II site marking the beginning of the sequence. Removal of this reading frame with restriction endonucleases led to constitutive mercury resistance (R. A. Laddaga & M. Horwitz, unpublished data). This provided the first suggestion that the gram-positive *mer* operon may be negatively controlled, in contrast to the positive control of the *E. coli* mercury resistance systems (27, 28, 69). There is no detectable homology between the *merR* amino acid sequence of

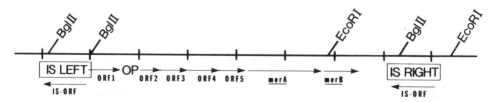

Figure 6   Summary of the DNA sequence information on the mercuric resistance operon of *S. aureus* plasmids (53) and associated insertion sequences (2).

R100 and the regulatory gene sequence for pI258 (53). After a predicted transcriptional start point [with canonical -35 and -10 sequences (53)] and an operator/promoter sequence with promising diad symmetries, there are four large open reading frames (Figure 6), which lack strong homologies to the reading frames of the gram-negative systems.

ORF2 may determine a 135–amino acid soluble protein, but we have no idea of its function. ORF3 has a canonical leader sequence, and therefore the ORF3 protein may be an exported binding protein or a membrane-embedded protein. ORF4 could determine a very hydrophobic protein, although there is no sequence homology to *merT* and the pI258 ORF4 is almost twice the length of the *merT* in plasmids of gram-negative bacteria. Since *merA* deletion mutants of the pI258 mercury resistance system are hypersensitive to Hg^{2+} and phenylmercury (R. A. Laddaga & M. Horwitz, unpublished data), the gram-positive system must contain genes for a mercury transport system upstream from the *merA* gene. The 128–amino acid ORF5 polypeptide has a large central hydrophobic region and may also be involved in membrane transport. The ORF5 amino acid sequence includes a Cys-Cys-*X*-Gly-Pro-*X*-*X*-Leu-Val-Ala-Leu-Gly-*X*-*X*-Gly sequence at positions 47 through 61, which is identical to the sequence at positions 24 to 38 of MerT of plasmid R100. This identity suggests a common transport function for both. Thus, the sequence results have raised questions to be answered by subsequent site-directed mutagenesis.

After these five genes, which are not closely homologous to the R100 system, there is a clearly recognizable mercuric reductase *merA* gene in pI258 (Figure 6) (53). Whereas the Tn*501* and R100 enzymes differ at only 79 of 560 shared positions (60), the pI258 mercuric reductase differs in 66% of its residues from the R100 enzyme (53). When amino acids known from the crystal structure of glutathione reductase (116) to make specific contacts with FAD or NADPH were considered, the identity of the residues between the R100 and the pI258 mercuric reductases increased to more than 90% (53). This demonstrates that essential residues have been conserved during evolutionary drift between the gram-positive and gram-negative versions of this enzyme. It was important to know whether the N-terminal domain of about 80 amino acids that is shared by the enzymes from R100, Tn*501*, and pDU1358 was also present in the pI258 enzyme. The pI258 enzyme does indeed have this N-terminal domain, including the crucial cysteine pair (53), in the sequence Gly-Met-Thr-Cys-*X*-*X*-His-*X*-*X*-*X*-Ala-Leu-Glu. This sequence is also conserved in the *Bacillus* version of the mercuric reductase sequence (see below; Y. Wang & I. Mahler, personal communication).

After the mercuric reductase gene there is a clearly recognizable *merB* gene for the organomercurial lyase in pI258 (Figure 6). The N- and C-terminal 50 amino acids of the pDU1358 (gram-negative) and pI258 (gram-positive) organomercurial lyase enzymes are barely related (Figure 7), whereas the middle 50 amino acids show 62% residue identities, including one stretch of 18/19 identical amino acids and a cysteine pair.

DNA sequence data have just become available for a new mercury resistance system in a soil *Bacillus* sp. (121: Y. Wang & I. Mahler, personal communication). The mercury resistance determinant occurs on the chromosome and not on a plasmid (121). It barely hybridizes in DNA/DNA Southern blotting analysis with the probe from plasmid pI258. This last result became understandable from analysis of the sequence of the *merA* gene for the mercuric reductase subunit, which is only 63% identical to that for pI258 (and only 66% identical in the aligned amino acid sequences). The *merB* gene for the organomercurial lyase is not contiguous with the *merA* gene in this *Bacillus* determinant, but appears to lie some 2 kb distal (121); the separation is analogous to the 11-kb distance of the *merB* gene reported for *E. coli* plasmid R831b (74). The most remarkable finding to date with this new *Bacillus* determinant concerns the length and beginning of the *merA* gene and the corresponding mercuric reductase subunit. The polypeptide sequence appears to be 631 amino acids in length (Y. Wang & I. Mahler, personal communication), compared with 547 amino acids for the mercuric reductase from plasmid pI258. The extra amino acids correspond to an N-terminal extension, so that the first 79 amino acids are 67% identical with the next 79

S. lividans	*******LA***********G*A*S**WL*RPLL**LA*GR	40
S. aureus	**********L******G*A*S**WL*RPLL**LA*G*	38
E. coli	**LA***L***************L**PLL**LA*GR	35
S. lividans	PV*VE*IA**TD******V***L***P*TEYDE*GRI*G*GLT**P	85
S. aureus	PV*VE*IA**T**P*E*V**VL***PS*E*DE*GR**GYGLTL*P	83
E. coli	PV*****A***D*P*E*V**VL****STEYD**G*I*GYGLTL**	80
S. lividans	TPH*FEBDG*QLY*WCALDTLIFPA**GR*AHV*SPCHATG*PVR	130
S. aureus	TPH*FEVDG*QLYAWCALDTL*FPALIGRT*H**SPCH*TG**VR	128
E. coli	T***FE*D***LYAWCALDTLIFPALIGRTA*V*S*C*ATG*PV*	125
S. lividans	LTVEPD*V*SVEPATAVVSIVTPD**AS*R*AFCN*VHFFA*P*A	175
S. aureus	LTVEPD*V*SVEP*TAVVSIVTPDE*ASVR*AFCN*VHFF*SP*A	173
E. coli	LTV*P*****VEPA***VS*V*P*E*A*VR**FC**VHFFAS***	170
S. lividans	****L**HP***VLPV*DA**LGR*L*E*********G*C	215
S. aureus	A*DWL**HP***VLPV*DA*ELGR*L*********T*GSC***	216
E. coli	A*DW***H*****L******E******E*******T**S***	212

Figure 7 Alignment of the three organomercurial lyase sequences from *Streptomyces lividans* (J. Altenbuchner, personal communication), gram-positive plasmid pI258 from *Staphylococcus aureus* (53), and gram-negative plasmid pDU1358 from *Escherichia coli* (36). Amino acids shared by two or three of the sequences are shown.

amino acids and contain the Gly-Met-Thr-Cys-*X*-*X*-Cys-*X*-*X*-His-*X*-*X*-*X*-Ala-Leu-Glu sequence as well. It appears as if the hypothesized mercury-binding domain duplicated and fused in tandem at the DNA level, leading to an enzyme with tandem N-terminal extensions (Figure 8, *right*). If the mercuric reductases from the four previously sequenced systems are related to glutathione reductase (Figure 8, *left*) by having an N-terminal extension of about 80 amino acids (Figure 8, *center*), then the new *Bacillus* enzyme has duplicate fusions of this domain (Figure 8, *right*). The model in Figure 8 is speculative and new enough to call for caution in its consideration, but it seems appropriate here.

During the writing of this review, a sixth DNA sequence for a mercury resistance determinant became available. This sequence, from *Streptomyces lividans* (a high-G+C gram-positive organism, in contrast to low-G+C *S. aureus* and *Bacillus*), contains four open reading frames that are transcribed divergently from recognizable *merA* and *merB* genes (J. Altenbuchner, personal communication). The four thus-far unidentified open reading frames are not homologous at the sequence level to those of *S. aureus* plasmid pI258, although the amino acid sequence of the fourth open reading frame is very hydrophobic, like ORF4 of pI258. The translated *merA* sequence of the *Streptomyces* mercuric reductase (J. Altenbuchner, personal communication) matches those of the *Bacillus* strain (Y. Wang & I. Mahler, personal communication) and pI258 (53) in only about 50% of the amino acid positions. Even more remarkably, the *Streptomyces* mercuric reductase sequence lacks the amino-terminal putative Hg^{2+}-binding domain and is only 474 amino acids long, just like lipoamide dehydrogenase (83). The *Streptomyces merB* sequence starts only 36 bp after *merA* ends (J. Altenbuchner, personal communication), and the *Streptomyces* organomercurial lyase amino acid sequence is closer to the organomercurial lyase from pI258 (55% identical residues) than it is to the lyase from plasmid pDU1358 (38% identical residues) (Figure 7).

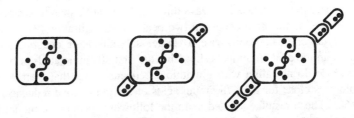

Figure 8   Diagram of the protein structures of glutathione reductase *(left)*, the mercuric reductases from plasmids R100 and pI258 *(center)*, and the chromosomal mercuric reductase determinant from a *Bacillus* species *(right)*. *Pairs of dots* represent N-terminal, C-terminal, and active site cystines.

Enzymology of Mercurial Detoxification

Mechanistic studies on the mercuric reductase enzyme progressed with the efforts of Fox & Walsh (31, 32). The reductase was shown to carry out the two-electron reduction of Hg^{2+} to Hg^0 using NADPH as electron donor. The redox-active cysteine was labeled with iodo[^{14}C]acetamide, and the active site peptide sequence was determined chemically (32). Site-directed mutagenesis was used to convert each of the active site cysteine residues to serines (88). The two Cys→Ser enzyme forms showed different catalytic activities toward Hg(II) complexes and dithiobis(nitrobenzoate) (DTNB), and only the native Cys-Cys enzyme could reduce $Hg(SR)_2$ compounds (88).

The N-terminal 85 amino acids could be removed from the 561 amino acid Tn*501* mercuric reductase without affecting in vitro enzyme activity (32). Surprisingly, the N-proximal cysteines could be changed to alanines by site-specific mutagenesis (M. Moore & C. T. Walsh, personal communication) without affecting the in vivo level of mercury resistance. But the N-terminal sequence is conserved in five mercuric reductase sequences. We hypothesize that this peptide domain functions to bind Hg^{2+} and protect the intracellular environment (Figure 4), but experimental evidence also indicates that it is not necessary.

Schottel (87) reported that the mercuric reductase from plasmid R831 was about 180 kd as estimated from filtration experiments on Sephadex G200 columns and from migration rates on acrylamide gels of varying porosity (a Ferguson plot). Since the denatured subunits of the R831 mercuric reductase were about 60 kd, Schottel (87) concluded that the enzyme was probably trimeric. Silver & Kinscherf (99; S. Silver & T. G. Kinscherf, unpublished data) found similar mobilities indicating a 180-kd mass for mercuric reductase from plasmid R100 and transposon Tn*501*. However, mercuric reductase is closely related to glutathione reductase, whose dimeric structure was established directly from X-ray crystal analysis (116). Since we could not imagine a family of closely related enzymes in which some were dimeric and others trimeric, we set aside the evidence for a number of years and assumed that the mercuric reductase is always dimeric. The native mercuric reductase from a single strain often occurs in two peaks on gel filtration (31, 32, 87), with the larger peak giving rise to larger subunits after denaturation. In addition, amino acid alignments show that the first 80 amino acids of mercuric reductase do not align with glutathione reductase, but that thereafter the homology is close for the 478 amino acids of the glutathione reductase subunit (60). These results have led us to the following speculative model (Figure 8): Glutathione reductase is an approximately spherical protein (Figure 8, *left*) with identical subunits related to one another by a twofold rotation axis. The N-terminal 18 amino acids of glutathione reductase do not have a fixed crystal position (116) but appear to have considerable positional flexibility. For mercuric reductase, the N-terminal 100 amino acids may represent a mercury-binding domain closely homologous to the putative mercury-binding sequence of the *merP* periplasmic protein. If this domain sticks out from the roughly globular flavoprotein structure (Figure 8, *center*), then a parachute effect would account for the greater retardation of the native enzyme on gel filtration and on polyacrylamide gel electrophoresis. Some support for this speculative model has come from the recent results of Wang et al (121; Y. Wang & I. Mahler, personal communication), who sequenced a *merA* gene from a soil *Bacillus* strain. This new *merA* gene determines a 631–amino acid subunit, 85 amino acids longer than the homologous polypeptide from *S. aureus* plasmid pI258 (Y. Wang & I. Mahler, personal communication). The N-terminal sequence of this polypeptide contains a remarkable duplication and fusion (at the DNA level as well as at the protein level), in which the first 79 amino acids are 67% identical to the next 79 amino acids in the sequence (Y. Wang & I. Mahler, personal communication). It appears as if this enzyme consists of tandem mercury-binding domains (Figure 8, *right*) prior to the central enzyme sequence.

Begley et al (5) purified the organomercurial lyase from a plasmid-containing gram-negative microbe. The enzyme is a 22.4-kd monomer, and the directly determined N-terminal amino acid sequence is closely related to that from another plasmid source (36) determined at the DNA level. The lyase catalyzes the cleavage of a wide range of organomercurial Hg-C bonds (6). This lyase has already become the first well-characterized example of an enzyme that cleaves an organometallic bond. The reaction proceeds via a S_E2 pathway (6).

COPPER RESISTANCE

Plasmid-determined copper resistance has been reported on an antibiotic resistance plasmid (46), in *E. coli* isolated from pigs fed copper supplements as growth stimulants (115), and in *Pseudomonas syringae* from plants treated with copper as an antibiotic agent (7, 8, 18). Although the genes determining copper resistance from *P. syringae* have been cloned and transferred into *E. coli* (8), there have been no detailed studies of the mechanism of copper resistance. D. A. Cooksey (personal communication) stated that the copper-resistant bacterial colonies turned bright blue when grown on high copper and that the residual copper in the growth medium was reduced. Thus he proposed that cellular copper sequestration might be the basic mechanism of copper resistance. The repeated polypeptide sequence in two of the polypeptides, deduced from DNA sequencing (57a; see below), contains aspartic acid, histidine, and methionine residues that might be involved in the binding of copper (57a; D. A. Cooksey, personal communication). The copper resistance determinants did not function in *E. coli,* but they functioned when transferred back to *P. syringae.* The copper resistance determinant from *P. syringae* did not hybridize in Southern DNA/DNA analysis with that from *Xanthomonas* sp. It has been localized to a 3.9-kb restriction endonuclease fragment by cloning analysis and by interruption by Tn5 insertions (8). The same copper resistance determinant is found in a wide range of *P. syringae* isolates (18).

Mellano & Cooksey (57a) recently finished determining the 4.5-kb DNA sequence of the copper resistance determinant of a *P. syringae* plasmid. Four open reading frames (ORFs) were identified as probable genes, and deletions of (or frameshift mutations in) two of these led to a loss of copper resistance. A repeated octapeptide sequence [aspartyl-histidyl-seryl-glutaminyl (or lysl)-methionyl-glutaminyl-glycyl-methionine] occurs five times in one predicted polypeptide, and a related octapeptide occurs four times in another. Deletions in the other two ORFs led to a partial loss of resistance. None of the hypothetical polypeptides translated from these ORFs is recognizable by comparison to protein libraries and none has a recognizable function. Thus, as with the tellurite resistance determinant (50; Figure 2), the DNA sequence analysis has preceded biological understanding. Thus far it has not led to any further understanding.

Recently, Eradi et al (24) reported the association of copper tolerance with a large plasmid in *Mycobacterium scrofulaceum.* The resistant bacteria accumulated copper from solution as a black intracellular precipitate of CuS. Thus, although bacterial copper resistance is widespread, it is only beginning to be studied.

SILVER RESISTANCE

Microbial silver toxicity is found in situations of industrial pollution, especially those associated with mining and use of photographic film. In hospitals, silver salts are the preferred antimicrobial agents for topical use on patients with large burns (33). It is thus not surprising that silver-resistant bacteria have been described (1, 11) and found in polluted industrial and mining sites (39, 110) and that silver resistance plasmids (39, 57) have been described. What little is known about microbial resistance to silver salts and bioaccumulation of silver by bacteria has recently been reviewed (117). Plasmid-determined silver resistance is very strong: The ratio of minimal inhibitory Ag^+ concentrations for resistant and sensitive strains can be greater than 100 : 1 (39, 101), and under some conditions resistant cells can grow in concentrations of up to 0.5 M added silver salts (117). The level of resistance depends on silver-complexing components, such as halide ions; without Cl^- there was relatively little difference between cells with and without silver resistance plasmids (39, 101). Both sensitive and resistant cells bind Ag^+ tightly and are killed (not just inhibited) by effects on cell respiration and other cell-surface functions (117). A working hypothesis (101, 117) is that sensitive cells bind Ag^+ so tightly that they extract it from AgCl and other complexed forms, whereas cells with resistance plasmids do not compete successfully with Ag^+-halide precipitates for Ag^+. However, it is unclear how this type of resistance mechanism can explain the accumulation of amounts of silver equivalent to up to 30% of cell mass by resistant cells (117). Apparently silver is reduced to metallic Ag^0 equally by sensitive and by resistant cells, so that reduction of Ag^+ does not appear to be a primary resistance mechanism.

OTHER HEAVY METAL RESISTANCES

Several other plasmid heavy metal resistances have been recognized for years, but less is known about them (71, 100, 101, 103, 110). These include resistances to antimony, bismuth, boron, cobalt, lead, and nickel. Recently, plasmid-mediated thallium resistance has been added to this list (89).

CONCLUDING REMARKS

DNA sequence analyses of several heavy metal resistance systems have recently become available. These have advanced our understanding of mercury, cadmium, and arsenic resistance mechanisms enormously. Understanding of tellurium resistance is less complete because the biochemical mechanism is not known. There remain resistance systems for about ten additional toxic heavy metals, which probably merit physiological and biochemical studies before DNA cloning and sequencing will be useful.

This review has broadly covered plasmid-determined heavy metal resistances. Heavy metal–resistant strains isolated from environmental or clinical sources generally have these resistances on plasmids (66, 82) or on transposons, or at least the genetic determinants are homologous to those found on plasmids (123). Chromosomal mutations to heavy metal resistance can be produced in the laboratory but do not generally occur in nature. In this regard toxic heavy metal resistances are comparable to antibiotic resistances; e.g. streptomycin resistance found in natural isolates is always due to aminoglycoside-inactivating enzymes and not to the modified ribosomal protein that is responsible for laboratory-selected resistance to streptomycin. The chromosomally determined mutations leading to arsenate, cadmium, chromate, and cobalt resistances are due to changes in the membrane transport systems responsible for uptake of the beneficial analog of these toxic materials, i.e. phosphate, manganese, sulfate, and magnesium, respectively. For most resistances except cadmium resistance, and most notably for mercurial and arsenic resistances, single biochemical mechanisms seem to be widespread among all bacterial types tested.

ACKNOWLEDGMENTS

This report summarizes efforts of many people, both in our laboratory (including C. Cervantes, H. Griffin, R. A. Laddaga, Chu Lien, H. Ohtake, and G. Nucifora) and elsewhere (N. L. Brown, T. J. Foster, B. P. Rosen, D. A. Ritchie, and A. O. Summers), who have contributed to the recent molecular genetic analysis of heavy metal resistances. Our studies were supported by NSF grant DMB86-04781 and NIH grant AI24795.

Literature Cited

1. Annear, D. I., Mee, B. J., Bailey, M. 1976. Instability and linkage of silver resistance, lactose fermentation and colony structure in *Enterobacter cloacae* from burn wounds. *J. Clin. Pathol.* 29:441–43

2. Barberis-Maino, L., Berger-Bächi, B., Weber, H., Beck, W. D., Kayser, F. H. 1987. IS431, a staphylococcal insertion sequence-like element related to IS26 from *Proteus vulgaris*. *Gene* 59:107–13

3. Barrineau, P., Gilbert, P., Jackson, W. J., Jones, C. S., Summers, A. O. Wisdom, S. 1984. The DNA sequence of the mercury resistance operon of the IncFII plasmid NR1. *J. Mol. Appl. Genet.* 2:601–19

4. Barrineau, P. O., Summers, A. O. 1983. A second positive regulatory function in the *mer* (mercury resistance) operon. *Gene* 25:209–21

5. Begley, T. P., Walts, A. E., Walsh, C. T. 1986. Bacterial organomercurial lyase: overproduction, isolation and characterization. *Biochemistry* 25:7186–92

6. Begley, T. P., Walts, A. E., Walsh, C. T. 1986. Mechanistic studies of a protonolytic organomercurial cleaving enzyme: bacterial organomercurial lyase. *Biochemistry* 25:7192–200

7. Bender, C. L., Cooksey, D. A. 1986. Indigenous plasmids in *Pseudomonas syringae* pv. *tomato:* conjugative transfer and role in copper resistance. *J. Bacteriol.* 165:534–41

8. Bender, C. L., Cooksey, D. A. 1987. Molecular cloning of copper resistance genes from *Pseudomonas syringae* pv. *tomato*. *J. Bacteriol.* 169:470–74

9. Bopp, L. H., Chakrabarty, A. M., Ehrlich, H. L. 1983. Chromate resistance plasmid in *Pseudomonas fluorescens*. *J. Bacteriol.* 155:1105–9

10. Bradley, D. E., Taylor, D. E. 1987. Transposition from RP4 to other replicons of a tellurite-resistance determinant not normally expressed by IncPα plasmids. *FEMS Microbiol. Lett.* 41:237–40

11. Bridges, K., Kidson, A., Lowbury, E. J. L., Wilkins, M. D. 1979. Gentamicin- and silver-resistant *Pseudomonas* in a burns unit. *Br. Med. J.* 1:446–49

12. Brown, N. L. 1985. Bacterial resistance to mercury—reductio ad absurdum? *Trends Biochem. Sci.* 10:400–3

13. Brown, N. L., Ford, S. J., Pridmore R. D., Fritzinger, D. C. 1983. Nucleotide sequence of a gene from the *Pseudomonas* transposon Tn501 encoding mercuric reductase. *Biochemistry* 22:4089–95

14. Brown, N. L., Misra, T. K., Winnie, J. N., Schmidt, A., Seiff, M., Silver, S. 1986. The nucleotide sequence of the mercuric resistance operons of plasmid R100 and transposon Tn501: further evidence for *mer* genes which enhance the activity of the mercuric ion detoxification system. *Mol. Gen. Genet.* 202:143–51

15. Cervantes, C., Ohtake, H., Silver, S. 1988. Cloning of plasmid-determined chromate resistance from *Pseudomonas aeruginosa*. *Abstr. Ann. Meet. Am. Soc. Microbiol.* 1988:180

16. Cervantes-Vega, C., Chavez, J., Cordova, N. A., de la Mora, P., Velasco, J. A. 1986. Resistance to metals by *Pseudomonas aeruginosa* clinical isolates. *Microbios* 48:159–63

17. Chen, C.-M., Misra, T. K., Silver, S., Rosen, B. P. 1986. The nucleotide sequence of the structural genes for an anion pump. The plasmid encoded arse-

nical resistance operon. *J. Biol. Chem.* 261:15030–38

18. Cooksey, D. A. 1987. Characterization of a copper resistance plasmid conserved in copper-resistant strains of *Pseudomonas syringae* pv. *tomato. Appl. Environ. Microbiol.* 53:454–56

19. de la Cruz, F., Grinsted, J. 1982. Genetic and molecular characterization of Tn*21*, a multiple resistance transposon from R100.1. *J. Bacteriol.* 151:222–28

20. Dempsey, W. B., McIntire, S. A., Willetts, N., Schottel, J., Kinscherf, T. G., et al. 1978. Properties of lambda transducing bacteriophages carrying R100 plasmid DNA: mercury resistance genes. *J. Bacteriol.* 136:1084–93

21. Dodd, I. B., Egan, J. B. 1987. Systematic method for the detection of potential Cro-like DNA-binding regions in proteins. *J. Mol. Biol.* 194:557–64

22. Ebright, R. H., Cossart, P., Gicquel-Sanzey, B., Beckwith, J. 1984. Molecular basis of DNA sequence recognition by the catabolite gene activator protein: detailed inferences from three mutations that alter DNA sequence specificity. *Proc. Natl. Acad. Sci. USA* 81:7274–78

23. Efstathiou, J. D., McKay, L. L. 1977. Inorganic salts resistance associated with a lactose-fermenting plasmid in *Streptococcus lactis. J. Bacteriol.* 130:257–65

24. Eradi, F. X., Failla, M. L., Falkinham, J. O. III. 1987. Plasmid-encoded copper resistance and precipitation by *Mycobacterium scrofulaceum. Appl. Environ. Microbiol.* 53:1951–54

25. Foster, T. J. 1983. Plasmid-determined resistance to antimicrobial drugs and toxic metal ions in bacteria. *Microbiol. Rev.* 47:361–409

26. Foster, T. J. 1987. Genetics and biochemistry of mercury resistance. *CRC Crit. Rev. Microbiol.* 15:117–40

27. Foster, T. J., Brown, N. L. 1985. Identification of the *merR* gene of R100 by using *mer-lac* gene and operon fusions. *J. Bacteriol.* 163:1153–57

28. Foster, T. J., Ginnity, F. 1985. Some mercurial resistance plasmids from different incompatibility groups specify *merR* regulatory functions that both repress and induce the *mer* operon of plasmid R100. *J. Bacteriol.* 162:773–76

29. Foster, T. J., Nakahara, H. 1979. Deletions in the r-determinant *mer* region of plasmid R100-1 selected for loss of mercury hypersensitivity. *J. Bacteriol.* 140:301–5

30. Foster, T. J., Nakahara, H., Weiss, A. A., Silver, S. 1979. Transposon A-

generated mutations in the mercuric resistance genes of plasmid R100-1. *J. Bacteriol.* 140:167–81

31. Fox, B., Walsh, C. T. 1982. Mercuric reductase. Purification and characterization of a transposon-encoded flavoprotein containing an oxidation-reduction active disulfide. *J. Biol. Chem.* 257:2498–503

32. Fox, B. S., Walsh, C. T. 1983. Mercuric reductase: homology to glutathione reductase and lipoamide dehydrogenase. Iodoacetamide alkylation and sequence of the active site peptide. *Biochemistry* 22:4082–88

33. Fox, C. L. Jr., Modak, S. M. 1974. Mechanism of silver sulfadiazine action on burn wound infections. *Antimicrob. Agents Chemother.* 5:582–88

34. Gerlach, J. H., Endicott, J. A., Juranka, P. F., Henderson, G., Sarangi, F., et al. 1986. Homology between P-glycoprotein and a bacterial haemolysin transport protein suggests a model for multidrug resistance. *Nature* 324:485–89

35. Gillespie, M. T., Lyon, B. R., Loo, L. S. L., Matthews, P. R., Stewart, P. R., Skurray, R. A. 1987. Homologous direct repeat sequences associated with mercury, methicillin, tetracycline and trimethoprim resistance determinants in *Staphylococcus aureus. FEMS Microbiol. Lett.* 43:165–71

36. Griffin, H., Foster, T. J., Silver, S., Misra, T. K. 1987. Cloning and DNA sequence analysis of the mercuric and organomercurial resistance determinants of plasmid pDU1358. *Proc. Natl. Acad. Sci. USA* 84:3112–16

37. Grill, E., Winnacker, E. L., Zenk, M. H. 1987. Phytochelatins, a class of heavy-metal-binding peptides from plants, are functionally analogous to metallothioneins. *Proc. Natl. Acad. Sci. USA* 84:439–43

38. Gros, P., Croop, J., Housman, D. 1986. Mammalian multidrug resistance gene: complete cDNA sequence indicates strong homology to bacterial transport proteins. *Cell* 47:371–80

39. Haefeli, C., Franklin, C., Hardy, K. 1984. Plasmid-determining silver resistance in *Pseudomonas stutzeri* isolated from a silver mine. *J. Bacteriol.* 158:389–92

40. Hamer, D. H. 1986. Metallothionein. *Ann. Rev. Biochem.* 55:913–51

41. Heltzel, A., Gambill, D., Jackson, W. J., Totis, P. A., Summers, A. O. 1987. Overexpression and DNA-binding properties of the *mer*-encoded regulatory protein from plasmid NR1 (Tn*21*). *J. Bacteriol.* 169:3379–84

42. Hesse, J. E., Wieczorek, L., Altendorf, K., Reicin, A. S., Dorus, E., Epstein, W. 1984. Sequence homology between two membrane transport ATPases, the Kdp-ATPase of *Escherichia coli* and the Ca^{2+}-ATPase of sarcoplasmic reticulum. *Proc. Natl. Acad. Sci. USA* 81:4746–50

43. Higham, D. P., Sadler, P. J., Scawen, M. D. 1984. Cadmium-resistant *Pseudomonas putida* synthesizes novel cadmium proteins. *Science* 225:1043–46

44. Higham, D. P., Sadler, P. J., Scawen, M. D. 1985. Cadmium-resistance in *Pseudomonas putida:* growth and uptake of cadmium. *J. Gen. Microbiol.* 131:2539–44

44a. Higham, D. P., Sadler, P. J., Scawen, M. D. 1986. Gold-resistant bacteria: excretion of a cysteine-rich protein by *Pseudomonas cepacia* induced by an antiarthritic drug. *J. Inorg. Biochem.* 28:253–61

45. Horitsu, H., Futo, S., Ozawa, K., Kawai, K. 1983. Comparison of characteristics of hexavalent chromium-tolerant bacterium, *Pseudomonas ambigua* G-1, and its hexavalent chromium-sensitive mutant. *Agric. Biol. Chem.* 47:2907–8

46. Ishihara, M., Kamio, Y., Terawaki, Y. 1978. Cupric ion resistance as a new genetic marker of a temperature sensitive R plasmid, Rts1 in *Escherichia coli*. *Biochem. Biophys. Res. Commun.* 82:74–80

47. Jackson, P. J., Unkefer, C. J., Doolen, J. A., Watt, K., Robinson, N. J. 1987. Poly(γ-glutamylcysteinyl)glycine: its role in cadmium resistance in plant cells. *Proc. Natl. Acad. Sci. USA* 84:6619–23

48. Jackson, W. J., Summers, A. O. 1982. Biochemical characterization of the $HgCl_2$-inducible polypeptides encoded by the *mer* operon of plasmid R100. *J. Bacteriol.* 151:962–70

49. Jobling, M. G., Ritchie, D. A. 1987. Genetic and physical analysis of plasmid genes expressing inducible resistance to tellurite in *Escherichia coli*. *Mol. Gen. Genet.* 208:288–93

50. Jobling, M. G., Ritchie, D. A. 1988. The nucleotide determinant encoding resistance to tellurium. *Gene* In press

51. Kreutz, B., Gotz, F. 1984. Construction of *Staphylococcus* plasmid vector pCA43 conferring resistance to chloramphenicol, arsenate, arsenite and antimony. *Gene* 31:301–4

52. Laddaga, R. A., Bessen, R., Silver, S. 1985. Cadmium-resistant mutant of *Bacillus subtilis* 168 with reduced cadmium uptake. *J. Bacteriol.* 162:1106–10

53. Laddaga, R. A., Chu, L., Misra, T. K., Silver, S. 1987. Nucleotide sequence and expression of the mercurial resistance operon from *Staphylococcus aureus* plasmid pI258. *Proc. Natl. Acad. Sci. USA* 84:5106–10

54. Laddaga, R., Silver, S. 1985. Cadmium uptake in *Escherichia coli* K-12. *J. Bacteriol.* 162:1100–5

55. Lund, P. A., Brown, N. L. 1987. Role of the *merT* and *merP* gene products of transposon Tn501 in the induction and expression of resistance to mercuric ions. *Gene* 52:207–14

56. Lund, P. A., Ford, S. J.,. Brown, N. L. 1986. Transcriptional regulation of the mercury-resistance genes of transposon Tn501. *J. Gen. Microbiol.* 132:465–80

57. McHugh, G. L., Moellering, R. C., Hopkins, C. C., Swartz, M. N. 1975. *Salmonella typhimurium* resistant to silver nitrate, chloramphenicol, and ampicillin. *Lancet* 1:235–40

57a. Mellano, M. A., Cooksey, D. A. 1988. Nucleotide sequence and organization of copper resistance genes from *Pseudomonas syringae* pv. *tomato*. *J. Bacteriol.* 170:2879–83

58. Mergeay, M., Nies, D., Schlegel, H. G., Gerits, J., Charles, P., Van Gijsegem, F. 1985. *Alcaligenes eutrophus* CH34 is a facultative chemolithotroph with plasmid-bound resistance to heavy metals. *J. Bacteriol.* 162:328–34

59. Misra, T. K., Brown, N. L., Fritzinger, D. C., Pridmore, R. D., Barnes, W. M., et al. 1984. The mercuric-ion resistance operons of plasmid R100 and transposon Tn501: the beginning of the operon including the regulatory region and the first two structural genes. *Proc. Natl. Acad. Sci. USA* 81:5975–79

60. Misra, T. K., Brown, N. L., Haberstroh, L., Schmidt, A., Goddette, D., Silver, S. 1985. Mercuric reductase structural genes from plasmid R100 and transposon Tn501: functional domains of the enzyme. *Gene* 34:253–62

61. Misra, T. K., Silver, S., Mobley, H. L. T., Rosen, B. P. 1984. Molecular genetics and biochemistry of heavy metal resistance in bacteria. In *Molecular and Cellular Approaches to Understanding Mechanisms of Toxicity*, ed. A. H. Tashjian Jr., pp. 63–81. Boston: Harvard Sch. Public Health

62. Mitra, R. S., Gray, R. H., Chin, B., Bernstein, I. A. 1975. Molecular mechanisms of accommodation in *Escherichia coli* to toxic levels of cadmium. *J. Bacteriol.* 121:1180–88

63. Mobley, H. L. T., Chen, C.-M., Silver, S., Rosen, B. P. 1983. Cloning and ex-

pression of R-factor mediated arsenate resistance in *Escherichia coli. Mol. Gen. Genet.* 191:421–26

64. Mobley, H. L. T., Rosen, B. P. 1982. Energetics of plasmid-mediated arsenate resistance in *Escherichia coli. Proc. Natl. Acad. Sci. USA* 79:6119–22

65. Mobley, H. L. T., Silver, S., Porter, F. D., Rosen, B. P. 1984. Homology among arsenate resistance determinants of R-factors in *Escherichia coli. Antimicrob. Agents Chemother.* 25:157–61

66. Nakahara, H., Kozukue, H. 1982. Volatilization of mercury determined by plasmids in *E. coli* isolated from an aquatic environment. In *Drug Resistance in Bacteria: Genetics, Biochemistry, and Molecular Biology,* ed. S. Mitsuhashi, pp. 337–40. Tokyo: Jpn. Sci. Soc.

67. Nakahara, H., Silver, S., Miki, T., Rownd, R. H. 1979. Hypersensitivity to Hg^{2+} and hyperbinding activity associated with cloned fragments of the mercurial resistance operon of plasmid NR1. *J. Bacteriol.* 140:161–66

68. NíBhriain, N., Foster, T. J. 1986. Polypeptides specified by the mercuric resistance *(mer)* operon of plasmid R100. *Gene* 42:323–30

69. NíBhriain, N., Silver, S., Foster, T. J. 1983. Tn5 insertion mutations in the mercuric ion resistance genes derived from plasmid R100. *J. Bacteriol.* 155:690–703

70. Nies, D., Mergeay, M., Friedrich, B., Schlegel, H. G. 1987. Cloning of plasmid genes encoding resistances to cadmium, zinc and cobalt in *Alcaligenes eutrophus* CH34. *J. Bacteriol.* 169:4865–68

71. Novick, R. P., Murphy, E., Gryczan, T. J., Baron, E., Edelman, I. 1979. Penicillinase plasmids of *Staphylococcus aureus:* restriction-deletion maps. *Plasmid* 2:109–29

72. Nucifora, G., Chu, L., Misra, T. K. 1988. DNA sequence analysis of the organomercurial resistance operon of plasmid pDU1358 from *Serratia marcescens. Abstr. Ann. Meet. Am. Soc. Microbiol.* 1988:180

73. Nucifora, G., Chu, L., Silver, S., Misra, T. K. 1988. Cd^{2+} ATPase: DNA sequence analysis from *Staphylococcus aureus* plasmid pI258. *Abstr. Annu. Meet. Am. Soc. Microbiol.* 1988:179

74. Ogawa, H. I., Tolle, C. L., Summers, A. O. 1984. Physical and genetic map of the organomercury resistance (Omr) and inorganic mercury (Hgr) loci of the IncM plasmid R831b. *Gene* 32:311–20

75. O'Halloran, T., Walsh, C. 1987. Metal-loregulatory DNA-binding protein encoded by the *merR* gene: isolation and characterization. *Science* 235:211–14

76. Ohtake, H., Cervantes, C., Silver, S. 1987. Decreased chromate uptake in *Pseudomonas fluorescens* carrying a chromate-resistance plasmid. *J. Bacteriol.* 169:3853–56

77. Olafson, R. W. 1984. Prokaryotic metallothionein. *Int. J. Pept. Protein Res.* 24:303–8

78. Olafson, R. W., Loya, S., Sim, R. G. 1980. Physiological parameters of prokaryotic metallothionein induction. *Biochem. Biophys. Res. Commun.* 95:1495–503

78a. Olafson, R. W., McCubbin, W. D., Kay, C. M. 1988. Primary- and secondary-structural analysis of a unique prokaryotic metallothionein from a *Synechococcus* sp. cyanobacterium. *Biochem. J.* 251:691–99

79. Pabo, C. O., Sauer, R. T. 1984. Protein-DNA recognition. *Ann. Rev. Biochem.* 53:293–321

80. Perry, R. D., Silver, S. 1982. Cadmium and manganese transport in *Staphylococcus aureus* membrane vesicles. *J. Bacteriol.* 150:973–76

81. Porter, F. D., Ong, C., Silver, S., Nakahara, H. 1982. Selection for mercurial resistance in hospital settings. *Antimicrob. Agents Chemother.* 22:852–58

82. Radford, A. J., Oliver, J., Kelly, W. J., Reanney, D. C. 1981. Translocatable resistance to mercuric and phenylmercuric ions in soil bacteria. *J. Bacteriol.* 147:1110–12

83. Rice, D. W., Schulz, G. E., Guest, J. R. 1984. The structural relationship between glutathione reductase and lipoamide dehydrogenase. *J. Mol. Biol.* 174:483–96

84. Rosen, B. P. 1986. Recent advances in bacterial ion transport. *Ann. Rev. Microbiol.* 40:263–86

85. Rosen, B. P., Borbolla, M. G. 1984. A plasmid-encoded arsenite pump produces arsenite resistance in *Escherichia coli. Biochem. Biophys. Res. Commun.* 124:760–65

86. Rosen, B. P., Weigel, U., Karkaria, C., Gangola, P. 1988. Molecular characterization of an anion pump. The *arsA* gene product is an arsenite(antimonate)-stimulated ATPase. *J. Biol. Chem.* 263:3067–70

87. Schottel, J. L. 1978. The mercuric and organomercurial detoxifying enzymes from a plasmid-bearing strain of *Escherichia coli. J. Biol. Chem.* 253:4341–49

88. Schultz, P. G., Au, K. G., Walsh, C. T. 1985. Directed mutagenesis of the redox-active disulfide in the flavoenzyme mercuric ion reductase. *Biochemistry* 24:6840–48

89. Sensfuss, C., Reh, M., Schlegel, H. G. 1986. No correlation exists between the conjugative transfer of the autotrophic character and that of plasmids in *Norcardia opaca* strains. *J. Gen. Microbiol.* 132:997–1007

90. Sequin, C., Hamer, D. H. 1987. Regulation in vitro of metallothionein gene binding factors. *Science* 235:1383–87

91. Shalita, Z., Murphy, E., Novick, R. P. 1980. Penicillinase plasmids of *Staphylococcus aureus:* structural and evolutionary relationships. *Plasmid* 3:291–311

92. Shames, S. L., Fairlamb, A. H., Cerami, A., Walsh, C. T. 1986. Purification and characterization of trypanothione reductase from *Crithidia fasciculata,* a newly discovered member of the family of disulfide-containing flavoprotein reductases. *Biochemistry* 25:3519–26

93. Silver, S. 1978. Transport of cations and anions. In *Bacterial Transport,* ed. B. P. Rosen, pp. 221–324. New York: Dekker

94. Silver, S. 1981. Mechanisms of plasmid-determined heavy metal resistances. In *Molecular Biology, Pathogenicity and Ecology of Bacterial Plasmids,* ed. S. B. Levy, R. C. Clowes, E. L. Koenig, pp. 179–89. New York: Plenum

95. Silver, S. 1982. Bacterial interactions with mineral cations and anions: good ions and bad. In *Biomineralization and Biological Metal Accumulation,* ed. P. Westbroek, E. W. deJong, pp. 439–57. Dordrecht, the Netherlands: Reidel

96. Silver, S. 1985. Bacterial transformations of and resistances to heavy metals. In *Environmental Inorganic Chemistry,* ed. K. J. Igrolic, A. E. Martell, pp. 513–40. Deerfield Beach, Fla: VCH

97. Silver, S., Budd, K., Leahy, K. M., Shaw, W. V., Hammond, D., et al. 1981. Inducible plasmid-determined resistance to arsenate, arsenite and antimony(III) in *Escherichia coli* and *Staphylococcus aureus. J. Bacteriol.* 146:983–96

98. Silver, S., Keach, D. 1982. Energy-dependent arsenate efflux: the mechanism of plasmid-mediated resistance. *Proc. Natl. Acad. Sci. USA* 79:6114–18

99. Silver, S., Kinscherf, T. G. 1982. Genetic and biochemical basis for microbial transformations and detoxification of mercury and mercurial compounds. In *Biodegradation and Detoxification of Environmental Pollutants,* ed. A. M.

Chakrabarty, pp. 85–103. Boca Raton, Fla: CRC

100. Silver, S., Misra, T. K. 1984. Bacterial transformations of and resistances to heavy metals. In *Genetic Control of Environmental Pollutants,* ed. G. S. Omenn, A. Hollaender, pp. 23–46. New York: Plenum

101. Silver, S., Perry, R. D., Tynecka, Z., Kinscherf, T. J. 1982. Mechanisms of bacterial resistances to the toxic heavy metals antimony, arsenic, cadmium, mercury and silver. See Ref. 66, pp. 347–61

102. Silver, S., Rosen, B. P., Misra, T. K. 1987. DNA sequencing analysis of mercuric and arsenic resistance operons of plasmids from gram negative and gram positive bacteria. In *5th Int. Symp. Genet. Ind. Microorg., Split, Yugoslavia, 1986,* ed. M. Alačević, D. Hranueli, Z. Toman, pp. 357–71. Zagreb, Yugoslavia: Pliva

103. Smith, D. H. 1967. R factors mediate resistances to mercury, nickel, and cobalt. *Science* 156:1114–16

104. Solioz, M., Mathews, S., Furst, P. 1987. Cloning of the K^+-ATPase of *Streptococcus faecalis. J. Biol. Chem.* 262:7358–62

105. Steffens, J. C., Hunt, D. F., Williams, B. G. 1986. Accumulation of non-protein metal-binding polypeptides (γ-glutamyl-cysteinyl)$_n$-glycine in selected cadmium-resistant tomato cells. *J. Biol. Chem.* 261:13879–82

106. Summers, A. O. 1985. Bacterial resistance to toxic elements. *Trends Biotechnol.* 3:122–25

107. Summers, A. O. 1986. Organization, expression, and evolution of genes for mercury resistance. *Ann. Rev. Microbiol.* 40:607–34

108. Summers, A. O., Jacoby, G. A. 1977. Plasmid-determined resistance to tellurium compounds. *J. Bacteriol.* 129:276–81

109. Summers, A. O., Jacoby, G. A. 1978. Plasmid-determined resistance to boron and chromium compounds in *Pseudomonas aeruginosa. Antimicrob. Agents Chemother.* 13:637–40

110. Summers, A. O., Jacoby, G. A., Swartz, M. N., McHugh, G., Sutton, L. 1978. Metal cation and oxyanion resistances in plasmids of gram-negative bacteria. In *Microbiology—1978,* ed. D. Schlessinger, pp. 128–31. Washington, DC: Am. Soc. Microbiol.

111. Summers, A. O., Silver, S. 1978. Microbial transformations of metals. *Ann. Rev. Microbiol.* 32:637–72

112. Surowitz, K. G., Titus, J. A., Pfister,

R. M. 1984. Effects of cadmium accumulation on growth and respiration of a cadmium-sensitive strain of *Bacillus subtilis* and a selected cadmium resistant mutant. *Arch. Microbiol.* 140:107–12

113. Taylor, D. E., Bradley, D. E. 1987. Location on RP4 of a tellurite resistance determinant not normally expressed in IncPα plasmids. *Antimicrob. Agents Chemother.* 31:823–25

114. Taylor, D. E., Summers, A. O. 1979. Association of tellurium resistance and bacteriophage inhibition conferred by R plasmids. *J. Bacteriol.* 137:1430–33

115. Tetaz, T. J., Luke, R. K. J. 1983. Plasmid-controlled resistance to copper in *Escherichia coli. J. Bacteriol.* 154:1263–68

116. Thieme, R., Pai, E. F., Schirmer, R. H., Schulz, G. E. 1981. Three-dimensional structure of glutathione reductase at 2 Å resolution. *J. Mol. Biol.* 152:763–82

117. Trevors, J. T. 1987. Silver resistance and accumulation in bacteria. *Enzyme Microb. Technol.* 9:331–33

118. Tynecka, Z., Gos, Z., Zajac, J. 1981. Reduced cadmium transport determined by a resistance plasmid in *Staphylococcus aureus. J. Bacteriol.* 147:305–12

119. Tynecka, Z., Gos, Z., Zajac, J. 1981. Energy-dependent efflux of cadmium coded by a plasmid resistance determinant in *Staphylococcus aureus. J. Bacteriol.* 147:313–19

120. Walderhaug, M. O., Dosch, D. C., Epstein, W. 1987. Potassium transport in bacteria. In *Ion Transport in Bacteria*, ed. B. P. Rosen, S. Silver, pp. 84–130. San Diego, Calif: Academic

121. Wang, Y., Mahler, I., Levinson, H. S., Halvorson, H. O. 1987. Cloning and expression in *Escherichia coli* of chromosomal mercury resistance genes from a *Bacillus* sp. *J. Bacteriol.* 169:4848–51

122. Weiss, A. A., Murphy, S. D., Silver, S. 1977. Mercury and organomercurial resistances determined by plasmids in *Staphylococcus aureus. J. Bacteriol.* 132:197–208

123. Witte, W., Green, L., Misra, T. K., Silver, S. 1986. Resistance to mercury and to cadmium in chromosomally resistant *Staphylococcus aureus. Antimicrob. Agents Chemother.* 29:663–69

Ann. Rev. Microbiol. 1988. 42:745–763

THE PASTEUR INSTITUTE'S CONTRIBUTIONS TO THE FIELD OF VIROLOGY

Marc Girard

Laboratory of Molecular Virology, Pasteur Institute, 25, rue du Dr. Roux, 75015 Paris, France

CONTENTS

Le Progrès de la science dépend dans une large part de l'anticipation.

René Dubos: *Louis Pasteur, franc tireur de la science*, 1955

The Pasteur Institute was created in 1887 through a public subscription launched in France and abroad for the erection of "an institute . . . for the prevention of rabies." Construction work lasted a year, and the institute was officially inaugurated on November 14, 1888. The 100th anniversary of the institute was celebrated in Paris on October 5–10, 1987.

745

0066-4227/88/1001-0745$02.00

I attempt to relate here contributions of the "Pastoriens" to the development of virology during the past 100 years. This is by no means an exhaustive list of achievements, and I apologize in advance to my colleagues whose work I could not give due justice because of limited space or arbitrary selection of topics.

EARLY DISCOVERY OF VIRUSES

L'homme eut la surprise de constater que sans les microorganismes, ce monde ne serait pas ce qu'il est.

François Jacob: *La logique du Vivant*, 1970

As early as 1876, Louis Pasteur, in collaboration with the French physicist Jules Joubert, used a plaster of paris filter connected to a vacuum pump to demonstrate the unfilterability of the "virus" of anthrax. The use of filters to retain and concentrate microbes became standard practice after 1884, when Charles Chamberland and Emile Roux, experimenting with a broken clay pipe purchased from Chamberland's tobacconist, developed bacteria-proof porcelain filter candles ("bougies de Chamberland").

At the time the Pasteur Institute was established, rabies was the main subject of research (77). There was a lot of frustration in Louis Pasteur's laboratory because the infectious agent of rabies could not be retained by filtration. The conclusion drawn from these experiments was that the causal agent must be a microbe of extremely small size, but neither Pasteur nor his collaborators suspected that it was in any basic way different from the pathogenic bacteria they had isolated previously. It was to take almost half a century before the concept of viruses came of age and the word virus acquired a well-defined meaning of its own (76).

At the turn of the century, four infectious agents were described as filterable viruses: tobacco mosaic virus, discovered a few years earlier by Dimitri Ivanovski in Russia; foot-and-mouth disease virus, identified by Friedrich Loeffler and Paul Frosch in Germany; the agent of bovine pleuropneumonia (not a virus, but a mycoplasma), isolated by Edmond Nocard, Amédée Borrel, and collaborators at the Pasteur Institute in Paris; and the virus of myxomatosis, discovered by Giuseppe Sanarelli, an ex-Pastorien and lifelong friend of Elie Metchnikoff and of the Pasteur Institute staff, in Uruguay. Martinus Beijerinck was the only one to identify tobacco mosaic virus correctly as being different from bacteria (7, 50). His premonitory views did not, however, gain public support. The fact that the agent of pleuropneumonia could be seen under the light microscope (if only barely) and could be grown in vitro (under very special conditions) probably misled the other microbiologists into thinking that the inability to observe the other filterable agents and to grow them in vitro was due to technical difficulties that eventually would be overcome. Thus, in 1903, when the filterability of the rabies virus was finally demonstrated at the Pasteur Institute by Paul Remlinger, Emile Roux still wrote that Beijerinck's theory was "at best, interesting, but unproven" (64).

In 1902, Charles Nicolle[1] showed rinderpest virus to be filterable and Amédée Borrel showed the same for the virus of sheep pox. In the same year, the filterability of the first human virus, yellow fever virus, was demonstrated by Walter Reed and James Carroll in Cuba. A few years later, Karl Landsteiner, in collaboration with Constantin Levaditi (42), then in Metchnikoff's department at the Pasteur Institute, showed poliovirus to be filterable using monkeys as susceptible animals. Landsteiner had discovered the susceptibility of monkeys to poliovirus the year before in Vienna.

Perhaps the first classification of viruses was in 1903, when Amédée Borrel (11) suggested that viruses be grouped on the basis of their pathogenicity. This classification led, however, to bizarre groupings. For example, foot-and-mouth disease virus and rinderpest virus were placed with the poxviruses identified at the time (vaccinia, smallpox, fowlpox, sheep pox) because they were all agents of "infectious epithelioses" (11). It is interesting to note that Borrel tried throughout his scientific career to prove the viral origin of human cancers using poxviruses as a model.

As early as 1913, Constantin Levaditi (46, 47) attempted to grow poliovirus and rabies virus in spinal ganglion cells, but these early attempts met with little success because cell culture techniques were still very primitive. The cells were actually not growing, but only surviving, and persistence of the input virus, rather than replication, was all that could be expected. It was not until 1949, when John Enders and his collaborators (29) opened the way to modern animal virology by demonstrating virus multiplication in cell cultures of nonneuronal origin, that these techniques became standard laboratory practice.

FROM BACTERIOPHAGE TO MOLECULAR BIOLOGY

L'enchantement de la science consiste en ce que, partout et toujours nous pouvons donner la justification de nos principes et la preuve de nos découvertes.

Louis Pasteur, 1882

In 1915, while a captain in the Royal Army Medical Corps, Frederick Twort discovered bacteriophagy (reviewed in 27). The following year, however, Felix d'Herelle (25; reviewed in 27) rediscovered the phenomenon at the

---

[1]In 1918 in Tunis, Charles Nicolle was also the first experimenter to demonstrate the filterability of influenza virus.

Pasteur Institute, was immediately convinced that the agent responsible was a living parasite of the bacterial cell, and coined the word "bacteriophage." He developed methods for the quantitative study of phage and invented the plaque method, which was instrumental in the development of modern phage genetics in the late 1940s.

Throughout the 1920s, a fierce argument persisted between d'Herelle and Jules Bordet, then director of what was to become the Pasteur Institute of Brabant in Brussels. Bordet (10), together with André Gratia, advanced the concept that bacteriophagy was a transmissible autolysis of bacteria due to an enzyme produced by the bacteria themselves. His ideas received considerable support from the discovery he made in 1925, at the same time as Otto Bail did so in Prague, of the existence of lysogenic strains of *Escherichia coli*. The fact that (lysogenic) bacteria could lyse in the absence of exogenous bacteriophage strengthened his belief that the ability to produce phage was a hereditary property of the bacterial strain studied. But Bordet did not guess the true nature of bacteriophage (49, 76).

Eugene and Elisabeth Wollman, at the Pasteur Institute from 1919 onward, came much closer to understanding phage and lysogeny. In 1938 the Wollmans (80) established that bacteriophages were not physically present as such in lysogenic bacteria, but were maintained in a potential, hereditary form. Immediately after infection, phage infectivity disappeared. They reasoned, therefore, that there must be infectious and noninfectious phases in the life cycle of bacteriophage. The Wollmans purchased a micromanipulator to study the behavior of single, isolated lysogenic bacteria. They were not, however, allowed to complete their experiment. In 1943 they were arrested in occupied Paris and deported to Auschwitz, from which they never returned.

The study of lysogeny was resumed by André Lwoff after the war, using single bacteria inoculated into individual microdrops. By then much had been learned of the nature and replication of bacteriophage through the work of the new American school headed by Max Delbruck and Salvador Luria. But lysogeny remained a mystery. In 1949 and 1950, Lwoff (49), in a series of elegant experiments with Louis Siminovitch and Niels Kjeldgaard, showed that lysogenic bacteria *(Bacillus megaterium)* could be induced to produce phage by UV light irradiation. Other inducers were discovered: hydrogen peroxide, X-rays, and certain chemicals. Most of these are known carcinogens. Spontaneous production of phage was also observed, but at a much lower frequency. Lysogenic bacteria did not contain virus particles, but carried and transmitted to their progeny the genetic material of the phage in a noninfectious form, which Lwoff christened "prophage" (49, 51).

The next major step in understanding phage–bacterial host interaction came from the study of a new virus, the lambda phage, naturally carried as a prophage by the K-12 strain of *E. coli*. To study conjugation between lysogenic and nonlysogenic strains of *E. coli,* Elie Wollman and François Jacob (39, 40a, 78, 79) used a blender to interrupt mating at different times. They demonstrated that the bacterial chromosome was circular and that prophage was carried on it as a genetic element. Analogy was drawn between bacteriophage and the fertility factor, F, which can exist either in a free form in the cytoplasm of F^+ bacteria or as an element integrated into the bacterial chromosome of Hfr (high frequency of recombination) strains. In the case of bacteriophage, the genetic material can similarly exist in either of two states: integrated into the host chromosome as a prophage (lysogenic state) or freely replicating in the cell cytoplasm (vegetative state). Jacob & Wollman (40a, 78) coined the word "episome" to designate mobile genetic elements such as F factor or bacteriophage.

François Jacob and Jacques Monod (40) formulated the hypothesis that the prophage in a lysogenic cell was controlled by the phage repressor, the product of one of the prophage genes. Induction resulted from the temporal inactivation of the repressor or from the reduction of its intracellular concentration to levels low enough to permit initiation of the vegetative phase of bacteriophage replication. Induction was thus the consequence of derepression. This system lent itself to the numerous elegant genetic experiments that Jacob (39), together with Elie Wollman, Louis Siminovitch, and others, performed during the 1950s and the early 1960s, which established the bases for the concepts of operons, operators, and replicons.

Another problem solved at the same time was that of control of protein synthesis during bacteriophage replication. Jacob and Monod proposed that ribosomes were nonspecific parts of the protein synthesis machinery of the cell and that information for the synthesis of specific (phage) proteins was carried to the ribosomes by an unstable RNA species, mRNA. Using differential density labeling, Sidney Brenner, François Jacob, and Mat Meselson (13) were able to show that a new, short-lived RNA species with a base composition corresponding to that of the phage DNA was synthesized in phage-infected bacteria and that this new RNA attached to preexisting cellular ribosomes. François Gros and his coworkers obtained similar results using sucrose gradient centrifugation (2a).

Additional evidence for the existence of phage-specific mRNAs was obtained from molecular hybridization experiments, first by Sol Spiegelman in the United States and then by Gros and collaborators at the Pasteur Institute (2a). The concept of messenger RNA, which has now become so familiar, was at the time revolutionary. It had far-reaching and enormous consequences on the development of molecular biology and virology.

DEFINITION OF VIRUSES AND STUDIES ON VIRUS PHYSIOLOGY

Viruses are viruses.
André Lwoff, 1957

The concept of viruses and awareness of the essential differences between viruses and bacteria slowly came of age after Stanley's crystallization of tobacco mosaic virus in 1935 (68a). As described above, the long saga of the bacteriophage contributed much to the evolution of ideas in the field, but progress in biophysical technology (ultracentrifugation, electron microscopy, X-ray diffraction) and biochemistry was crucial to understanding the true nature of viruses.

André Lwoff (50) summarized the characteristics of viruses in 1957: Outside the cell, viruses exist in the form of specialized particles named virions. These contain either DNA or RNA, but not both; they contain no ribosomes; they replicate only inside living cells, from which they derive the energy necessary for their own multiplication; and they are unable to grow or to undergo binary fission. Some viruses integrate their genome into that of the host cell and are thus transferred vertically from cell to cell in the form of a provirus.

Lwoff became actively engaged in the classification and nomenclature of viruses. Together with Thomas Anderson and François Jacob, he coined words such as virion, capsid, capsomere, nucleocapsid, and envelope, which have become part of the virologist's basic vocabulary. In collaboration with R. Horne (52) and Paul Tournier (54), he devised one of the first modern classification systems of viruses. Jacques Maurin, long a Pastorien but now at the University of Rennes, has continued this heritage.

In the 1960s Lwoff showed that the replication of viruses was affected by a variety of effectors, such as pH, temperature, and rate of cell growth (53). He became very interested in the effects of temperature, guanidine, and other inhibitors on the growth of poliovirus. He studied systematically the effect of supraoptimal temperatures and invented the concept of $rt,$ i.e. the temperature at which the yield of a virus is decreased by 90%. He observed that the more virulent a virus strain, the higher its rt value. Lwoff was the first to stress the importance of the role of fever and inflammation in the pathophysiology of viral infections.

It was in Lwoff's laboratory in the attic of the Pasteur Institute in 1966 that I continued the work on poliovirus replication that I had initiated with James Darnell at Albert Einstein College of Medicine in New York and had pursued with David Baltimore at the Salk Institute. This work led to the demonstration that poliovirus replicative intermediate molecules are precursors to single-stranded (genomic) and double-stranded RNA molecules, which is consistent with a semiconservative scheme of viral RNA replication (31).

An important class of effectors that greatly influence virus replication is the interferons. Luc Montagnier, who joined the Pasteur Institute in the early 1970s, and Edward and Jacqueline De Maeyer at the Curie Institute (23) showed that avian and monkey cells transfected with mRNAs from mouse cells that had been treated with an interferon inducer expressed and secreted murine interferons. This phenomenon was observed even in Vero cells, which are naturally unable to produce interferon; this demonstrated that interferons are the products of inducible cellular genes.

The mode of action of interferons has since been studied by Ara Hovanessian, who characterized the enzyme 2-5A synthetase. This enzyme is responsible for the production of $2'-5'$ oligoadenylates in interferon-treated cells (38). These molecules function as cofactors for a RNase that degrades mRNAs, thus arresting protein synthesis and blocking virus replication. A protein kinase is also involved (37).

Current studies of virus physiology at the Pasteur Institute are not limited to the few topics discussed above. Some of the other topics being studied are the human papillomaviruses (20), of which new types are frequently identified in association with human neoplasmas (6, 62); hepatitis A virus and epidemic non A, non B hepatitis virus (65); and the genomic structure and gene expression of bunyaviruses (12), rabies virus, poliovirus, and human cyto-megalovirus (57). Aspects of the work with hepatitis B virus, rabies virus, and poliovirus are described below (see section on research and development of viral vaccines). The pathophysiology of viral diseases of the CNS is studied in two models: rabies virus (73) and Theiler's mouse encephalitis virus (14). In addition, several groups are investigating the process of viral cancerization at the molecular level using hepatitis B virus (21), polyoma virus (35), or papillomaviruses (6, 20, 62) as a model. Finally, but of critical importance, is the recent work on the virus responsible for acquired immunodeficiency syndrome (AIDS).

AIDS VIRUSES

> For the first time in the history of Science, we can keep the certain and sound hope that, concerning epidemics, medicine will be saved from empiricism and placed on real scientific bases.
>
> John Tyndall, letter to Louis Pasteur, 1876 (quoted in 63a)

The first cases of AIDS were reported in 1981. By the end of 1982, there were indications that AIDS was an infectious disease transmitted by blood transfusion or sexual contact, but the causative agent remained unidentified.

Cytomegalovirus, Epstein-Barr virus, hepatitis B virus, and herpesvirus were investigated unsuccessfully. Robert Gallo and his coworkers at the National Institutes of Health in Bethesda focused on the possible role of the human T-lymphotropic virus, HTLV-I (30a). This virus is associated with adult T-cell leukemia, which is endemic in southwestern Japan, the southern United States, Africa, and the Caribbean. At the same time, Luc Montagnier's group at the Pasteur Institute was researching the possible involvement of retro-viruses in human cancers by cultivating T cells from cancer patients in vitro and assaying for reverse transcriptase activity in the culture medium.

At a meeting held at the end of 1982 with Montagnier and Jean-Claude Chermann, Willy Rozenbaum, Françoise Brun, and a group of French physi-cians expressed the view that the best chance to find and isolate the AIDS agent might be at the beginning of the disease, before the patient's T cells were eliminated. They asked Montagnier's group to look for retrovirus in the lymph cells of a patient with persistent lymphadenopathy but little sign of immunodeficiency. After stimulation with phytohemagglutinin, the cells from a lymph-node biopsy from this patient were cultivated in the presence of interleukin-2 (IL-2) and anti-interferon antiserum. Two weeks later, in early January 1983, Françoise Barre-Sinoussi detected the first evidence of reverse transcriptase activity in the cell culture medium. The virus was subsequently observed by electron microscopy; enveloped particles resembling those of a retrovirus could be seen budding from the cell surface or free in the medium. Their morphology was reminiscent of that of equine infectious anemia virus, a retrovirus belonging to the lentivirus subgroup. The virus was christened "LAV-1" for lymphadenopathy-associated virus (4).

An ELISA test to detect antibodies to the virus was developed in Jean-Claude Chermann's laboratory with Françoise Brun and Christine Rouzioux from Hôpital Claude Bernard in Paris. Antibodies to LAV-1 were present in serum from most AIDS patients and from many homosexuals, hemophiliacs, and donors and recipients of blood transfusions. This epidemiological re-search was successful because of an active collaboration between the Pasteur Institute and the Centers for Disease Control in Atlanta, where biological samples from American AIDS cases were collected and stored.

Similar results and new isolates of the virus were reported in 1984 by Robert Gallo, Mika Popovic, and their colleagues at the NIH in Bethesda (30) and by Jay Levy et al in San Francisco (48). All of these viruses were characterized by a common tropism for T4$^+$ cells, fusogenic properties that resulted in giant multinucleated cells in infected cell cultures, and a similar morphology when viewed in the electron microscope.

Unfortunately, each isolate was christened with a different name: LAV-1 in France, HTLV-III in Bethesda, and ARV in San Francisco. A controversy erupted between virologists at the Pasteur Institute and those at the NIH, which concerned, among other things, the name of the virus, the relationship of the different isolates to each other and to other retroviruses, and the authorship of the discovery. This controversy could have ended with the results of the nucleotide sequencing of the virus genome. Simon Wain-Hobson, Pierre Sonigo, Olivier Danos, Stewart Cole, and Marc Alizon at the Pasteur Institute finished sequencing the LAV-1 genome by the end of 1984. The complete nucleotide sequence was reported in early 1985 (75), and the results clearly showed that the genomic organization of LAV-1 was not that of an HTLV. Sonigo, Wain-Hobson, and colleagues demonstrated shortly thereafter that the LAV-1 genome closely resembled that of Visna virus, a lentivirus responsible for progressive demyelinating disease in sheep (68). The nucleotide sequence of HTLV-III, determined soon afterwards, was almost identical to that of LAV-1. The use of the name LAV-1/HTLV-III thus prevailed for some time, until an international committee chaired by Harold Varmus proposed to rechristen the virus HIV (human immunodeficiency virus). The HIVs are now classified in the group of lentiviruses.

In late 1985, Luc Montagnier's group, in collaboration with physicians from Hospital Egas Moniz in Lisbon and virologists from Hôpital Claude Bernard in Paris, discovered a novel AIDS virus in West Africa, which they named LAV-II (HIV-2) (18). An ELISA test for the detection of HIV-2 antibodies was developed. The virus was then cloned and sequenced (18a), and the complete nucleotide sequence was reported by Mireille Guyader and colleagues (33) in early 1987. The sequence of HIV-2 is only about 40% homologous to that of HIV-1, but shows a much closer relationship to SIV (STLV-III), the simian immunodeficiency virus isolated from macaque monkeys by Ronald Desrosiers and colleagues at the New England Primate Center in the United States (45a). The nucleotide sequence of SIV was determined by different groups, including that of Pierre Sonigo and Pierre Tiollais at the Pasteur Institute (15). In addition to the classical retrovirus *gag, pol,* and *env* genes, which encode the core proteins, reverse transcriptase, and the viral envelope glycoproteins, respectively, the genomes of HIV and SIV contain unique genes that encode several proteins involved in the control of gene expression. The *tat* gene, for example, codes for a transactivator protein that amplifies expression of the genome and helps boost the rapid synthesis of viral components (61, 67). In contrast, some data indicate that the *F (3'orf)* gene product might be a repressor of the system. Bruno Guy at Transgene in Strasbourg, working in collaboration with Marie-Paule Kieny, Jean-Pierre Lecocq, Luc Montagnier's group, and Pasteur Vaccins, has found that the *F* gene product is a GTP-binding protein with GTPase activity and some homology to the *ras* oncogene (32). Implications of this finding concerning the pathogenicity of the virus remain to be studied.

Obviously, we are only beginning to understand the molecular biology of the AIDS viruses, and much is still to be learned about the pathophysiology of the disease.

RESEARCH AND DEVELOPMENT OF VIRAL VACCINES

> However violent your attacks, sir, they will remain without success. I confidently await the results that virus attenuation holds in reserve to help mankind in its struggle against the onslaught of disease.
>
> Louis Pasteur, letter to Robert Koch, 1882
> (translation quoted in 63a)

Since the beginning of the century a series of veterinary vaccines have been developed at the Pasteur Institute; these include vaccines against sheep pox, Newcastle disease, avian pox, equine influenza, and infectious bronchitis. Many human vaccines have also been developed, including vaccines against rabies, yellow fever, influenza, poliomyelitis and, more recently, hepatitis B. Below is a brief summary of the history of their development.

Rabies

Louis Pasteur's original method for preparing the rabies vaccine was based on the use of rabid rabbit spinal cords at various degrees of attenuation. This was not practical when several treatments had to be given, since there was no way to store the spinal cords.

At the age of 27, Albert Calmette[2] was sent by Pasteur to Saigon, Vietnam, to establish the first Pasteur Institute overseas. While there, he discovered that attenuated spinal cords could be kept for several days in glycerol in the cold without losing efficacy. This amelioration of Pasteur's method was immediately adopted and was used at the Pasteur Institute in Paris until 1952 (45).

Major progress in the production of rabies vaccine came with the use of rabid suckling mice brains, which was suggested by E. Fuenzalida and R. Palacios in 1955 in Chile. This vaccine, which had the great advantage of being devoid of myelin, was introduced at the Pasteur Institute by André Gamet in 1968 with β-propiolactone as the inactivating agent. Its increased immunogenicity allowed Gamet to reduce successfully the number of injections required for postexposure vaccination to only seven daily injections (as compared to 12–14 in Pasteur's original procedure) followed by three booster injections at days 10, 20, and 90.

---

[2]Calmette later, with Camille Guérin, became the father of bacille Calmette-Guérin (BCG), an attenuated *Mycobacterium boris* strain used for vaccination against tuberculosis.

The next major improvement was the use of animal cell cultures for the production of the virus. In 1954, Vieuchange, Bequignon, and their colleagues had shown that mouse kidney cell cultures could support rabies virus multiplication (8, 69). This observation and other successful attempts to cultivate the virus in cells of nonneuronal origin eventually led to the production of rabies vaccines in primary hamster kidney cell cultures (the method that is still used in the USSR, China, and Eastern Europe) and in primary fetal bovine kidney cell cultures, a method developed by Pascu Atanasiu and colleagues (2) at the Pasteur Institute in the early 1970s. The latter vaccine is still in use in France and several other countries. Only five doses are required for postexposure treatment (on days 0, 3, 7, 14, and 30), with an optional booster injection at day 90.

Although quite efficient, the last vaccine is, however, too expensive for large-scale utilization in developing countries. A new vaccine has been developed recently, using as a substrate for virus growth the Vero cell line, a continuous heteroploid cell line derived from African green monkey *(Cercopithecus aethiops)* kidneys. These cells can be cultivated on microcarrier beads in large bioreactors of several hundred liter capacity, allowing mass production of rabies vaccine. The Vero cell vaccine was developed by Bernard Montagnon and colleagues at Institut Merieux in Lyon, France (58) and by Pierre Reculard at the Pasteur Institute. It is highly immunogenic, which allows a further reduction in the number of injections required for postexposure vaccination (69).

Studies on a new generation of rabies vaccines were recently initiated at the Pasteur Institute using recombinant DNA technology. The successful molecular cloning of the entire rabies virus genome has been achieved (72). A variety of new virus strains that can be lethal for humans, some of which do not show cross-protection with the original rabies virus strain, are also being studied.

Yellow Fever

Inactivated vaccines are efficacious against rabies, but they do not work well against yellow fever. The first successful attempt at immunization against yellow fever was initiated by A. Sellards and J. Laigret (66) in 1932 at the Pasteur Institute in Tunis using a live virus vaccine strain, the French neurotropic strain. This strain was adapted to grow in the brains of mice. The vaccine was initially administered in the form of dessicated mouse brain suspension mixed with egg yolk, as described by Charles Nicolle and J. Laigret (60). However, after the work of M. Peltier and colleagues in 1939 at the Pasteur Institute in Dakar, administration of the vaccine (without egg yolk) by skin scarification became standard practice. The French neurotropic strain, which became known as the Dakar vaccine, was used successfully for more than 20 years in French-speaking African countries. Its use has now been discarded in favor of the more attenuated 17D strain that Max Theiler and H. H. Smith developed at the Rockfeller Institute in 1937.

One of the problems associated with live virus vaccines is that they are heat sensitive. Michel Barme and Christian Bronnert at the Pasteur Institute solved this problem for the 17D vaccine a few years ago by using a mixture of sugars, amino acids, and mineral salts as a stabilizer (3). The resulting vaccine, which is manufactured by Pasteur Vaccins, can be stored at temperatures above 0°C for months, which is of great advantage in developing countries.

The yellow fever virus genome has now been molecularly cloned and sequenced (24). This progress opens the way for the development of genetically engineered vaccines.

Poliomyelitis

The search for a polio vaccine was initiated at the Pasteur Institute in the 1930s, but early attempts to produce poliovirus in cell or tissue cultures failed. After the successful cultivation of virus in monkey kidney cells by Enders and collaborators in 1949 (29), the development of an inactivated poliovirus vaccine resumed with increased vigor under the leadership of Jonas Salk in the United States and Pierre Lepine in Paris. In contrast to Salk, Lepine used two steps to inactivate the virus for his vaccine, formalin treatment followed by β-propiolactone treatment. The latter was used to prevent the persistence of the small amount of residual virus that escaped formalin treatment, which could otherwise eventually produce disease after injection of the vaccine into children (as observed in some of the first clinical trials with Salk's vaccine). Lepine's vaccine was marketed by the Pasteur Institute starting in 1957, and was totally safe from the beginning (44). Today's Pasteur vaccine is a killed vaccine manufactured from virus grown in Vero cells cultivated on microcarrier beads in fermentors of several hundred liter capacity. Owing to its high antigenic content, this vaccine gives close to 100% seroconversion after only two injections and can readily be associated with diphtheria, pertussis, and tetanus vaccines.

In 1954, Lepine, together with Georges Barski, invented what was probably the first multiwell plate for virus titration assays, an araldite plate with 25 12-mm wells for the typing of polioviruses (5).

In recent years, work on poliovirus at the Pasteur Institute by my group in collaboration with that of Florian Horaud has focused on the natural genetic variations of the virus (19) and the identification of antigenic sites important for seroneutralization (9, 19, 36). The type 1 poliovirus genome was cloned and sequenced in collaboration with Eckard Wimmer and colleagues at the State University of New York at Stony Brook (74).

Poliovirus capsid polypeptide VP1 was expressed in *E. coli,* and the major seroneutralization epitope (NIm I site) was identified precisely using appropriate monoclonal antibodies (81). Precise substitution of amino acids at this site has been successfully undertaken and has allowed in vitro construction of chimeric viruses expressing foreign antigenic determinants on their capsid surface. This may lead to the creation of hybrid viruses for the preparation of new multivalent vaccines.

Influenza

A first vaccine against influenza virus was developed at the Pasteur Institute by Rene Dujarric de la Rivière, Genevieve Cateigne, and Claude Hannoun in 1956. The vaccine manufactured today at Pasteur Vaccins is a split purified vaccine developed at the end of the 1970s by Paul Prunet and Philippe Adamowicz at Institut Pasteur Production (now Pasteur Vaccins).

Hepatitis B

The most recent of the viral vaccines to be developed at the Pasteur Institute is against hepatitis B. Infectious, epidemic forms of hepatic diseases have been known since ancient times, but it was only during World War II that enough epidemiological evidence was gathered to establish that there were at least two different kinds of viral hepatitis: hepatitis A, transmitted by the fecal-oral route, and hepatitis B, transmitted by blood or blood products from chronic hepatitis B virus (HBV) carriers. (The existence of non A, non B hepatitis is a much more recent finding.) Saul Krugman's observation in 1972 that a crude inactivated preparation of HBV surface antigen (HBsAg) particles isolated from the plasma of chronic hepatitis B patients could induce protection against the disease made possible the development of hepatitis B vaccines. Philippe Maupas and his collaborators in Tours prepared the first French vaccine using a partially purified preparation of HBsAg particles. Clinical trials were initiated in 1976 and were very successful (55).

The present-day vaccine is manufactured by Pasteur Vaccins and is a redesigned version of Maupas's initial vaccine. The more efficient purification procedure was devised in the early 1980s by Philippe Adamowicz, Paul Prunet, and colleagues (1). This vaccine was shown to be very safe and efficacious. More than six million doses have been marketed as of today, and the vaccine was chosen for the systematic mass vaccination of newborn babies in Taiwan. However, sources of HBsAg-positive plasmas are limited and would be insufficient for a worldwide vaccination campaign. Moreover, the vaccine is expensive to produce because of the cost of the plasmas and the numerous quality-control steps required to monitor and ascertain its safety. For these reasons, much work has been devoted in recent years to the development of new, genetically engineered vaccines.

At the end of the 1970s, Pierre Tiollais and collaborators at the Pasteur

Institute succeeded in cloning and sequencing the HBV genome (70, 71). Four partially overlapping genes were identified on the genome, one of which, the S gene, encodes the HBsAg (17, 71). HBsAg was recently shown to consist of a nested set of three polypeptides, L, M, and S, which share a common S sequence at the C-terminal end but differ on the N-terminal side. The M polypeptide thus contains an additional N-terminal sequence, pre-S2, and the L polypeptide contains a further additional N-terminal sequence, pre-S1. The pre-S1, pre-S2, and S domains each contain at least one seroneutralization epitope (34, 59).

Expression of HBsAg in cultured mammalian cells was initially reported in 1980 (26). To increase the efficiency of expression, Marie-Louise Michel, Pierre Tiollais, Rolf Streeck, and their colleagues at the Pasteur Institute (56) inserted the sequence of the entire S gene downstream from the SV40 promoter in a plasmid containing the *E. coli* dihydrofolate reductase (DHFR) gene. They used the resulting DNA to transform DHFR$^-$ Chinese hamster ovary (CHO) cells and to select DHFR$^+$ transformants. These transformants express the HBsAg gene and continuously synthesize and excrete into the culture medium HBsAg particles containing the S and M polypeptides (the L polypeptide is apparently not exported). Using these CHO cells as the source of HBsAg, a vaccine has now been prepared at Pasteur Vaccins. It was successful in tests for efficacy in chimpanzees and for immunogenicity in humans, and a license application has been filed. It is hoped that the vaccine will be on the market in early 1989.

PERSPECTIVES

La grandeur des actions humaines se mesure à l'inspiration qui les fait naître.

Louis Pasteur

During the time that scientists at the Pasteur Institute were studying new viruses and trying to prepare vaccines to fight them, they were also trying to understand mechanisms of immunization. It is interesting to note that as early as 1886, Emile Duclaux (28), one of Pasteur's early collaborators, wrote that "the explanation of a long-lasting immunity [conferred by vaccination] is more likely to reside in cellular than in humoral theories." It is therefore understandable that Pasteur invited Elie Metchnikoff to work at the Pasteur Institute, for Metchnikoff was a warm partisan of cellular immunity.

Today's research in viral immunology at the Pasteur Institute concentrates on mechanisms of the T-cell immune response, the distinction between self and nonself (41), interleukins, adjuvants and immunomodulators, the use of synthetic peptides as vaccines, the expression of virus neutralization epitopes at the surface of bacteria that might be used as live (oral) vaccines (16), the use of hepatitis B surface antigen particles as vectors for presenting foreign viral antigenic determinants (22), and the production of genetically engineered vaccines, to name a few areas.

But the most formidable challenge is certainly the search for an AIDS vaccine. The AIDS epidemic is spreading, particularly in Africa; thus the development of a vaccine and of effective drugs is urgently needed. At the moment, live vaccinia virus recombinants expressing the envelope or the core antigens, subunit vaccines, and synthetic peptides are being tried. Preliminary data obtained by groups at the Pasteur Institute and Pasteur Vaccins, in collaboration with Transgene in Strasbourg, are encouraging, but strongly suggest that more imaginative approaches may be necessary. Designing an effective vaccine against AIDS may be a long and difficult process.

One could even argue that there is no evidence that a vaccine against AIDS will actually be effective. However, if we were to look back 35 years, we would see that Jonas Salk and Pierre Lepine faced similar uncertainties when they attempted to produce the first vaccine against poliomyelitis (43). Even greater was the doubt that nagged Louis Pasteur when he entered the field of postexposure vaccination against rabies 100 years ago (63). The historical success of these and similar attempts breeds optimism and the belief that a vaccine against AIDS will eventually be found.

Since its foundation in 1887, the Pasteur Institute has been deeply committed to work on public health matters. It now houses nine international reference centers, which collaborate with the World Health Organization on the surveillance of rabies, influenza, viral hepatitis, AIDS, and arboviruses, as well as other pathogens in the field of microbiology. It is at the heart of a network of 27 institutes throughout the world, 19 of which bear the name of Pasteur. Six of these institutes (in Pointe à Pitre, French West Indies; Cayenne, French Guyana; Noumea, New Caledonia; Bangui, Central African Republic; Dakar, Senegal; and Tananarive, Madagascar) are still entirely dependent on the Paris institute. In line with the Pastorian tradition, these institutes pursue public health activities and perform research in coordination with the Pasteur Institute in Paris, most particularly in the fields of viral hemorrhagic fevers, hepatitis B, viral diarrheas, rabies, and AIDS. Their very existence is proof that the Pasteur Institute has remained true to the spirit of its founder and that the Pastorian tradition is indeed a reality.

ACKNOWLEDGMENTS

I would like to thank Raymond Dedonder, Patricia Fultz, Florian Horaud, André Lwoff, Luc Montagnier, Maxime Schwartz, and Pierre Sureau for critical review of the manuscript and encouragement. The excellent secretarial assistance of Michèle Bazelot and Muriel Bombray is gratefully acknowledged.

Literature Cited

1. Adamowicz, P., Chabanier, G., Lucas, G., Prunet, P., Vinas, R. 1984. Elimination of serum proteins and potential virus contaminants during hepatitis B vaccine preparation. *Vaccine* 2:209–14
2. Atanasiu, P., Tsiang, H., Gamet, A. 1974. Nouveau vaccin rabique de cultures cellulaires primaires. *Ann. Inst. Pasteur Microbiol. 125B*:419–32
2a. Attardi, G., Naono, S., Rouvière, J., Jacob, F., Gros, F. 1963. Production of messenger RNA and regulation of protein synthesis. *Cold Spring Harbor Symp. Quant. Biol.* 28:363–74
3. Barme, M., Bronnert, C. 1984. Thermostabilisation du vaccin antiamaril 17D lyophilisé. Essai de substances protectrices. *J. Biol. Stand.* 12:435–42
4. Barre-Sinoussi, F., Cherman, J.-C., Rey, F., Nugeyre, M.-T., Chamaret, S., et al. 1983. Isolation of a T lymphotropic retrovirus from a patient at risk of acquired immunodeficiency syndrome (AIDS). *Science* 220:868–71
5. Barski, G., Lepine, P. 1954. Microméthode et séroneutralisation de la poliomyélite. Emploi de cultures sur plaques moulées de matière plastique. *Ann. Inst. Pasteur Paris* 86:693–701
6. Beaudenon, S., Kremsdorf, D., Croissant, O., Jablonska, S., Wainhobson, S., Orth, G. 1986. A novel type of human papillomavirus associated with genital neoplasias. *Nature* 321:246–49
7. Beijerinck, M. W. 1899. Uber ein contagium vivum fluidum als Ursache der fleckenkrankheit der Tabaksplätter. *Zentralbl. Bakteriol. Parasitenkd. Infektionskr. Hyg. Abt. 2* 5:27–33
8. Bequignon, R., Gruest, J., Viala, C., Vieuchange, J. 1954. Culture du virus rabique *in vitro*. *C. R. Acad. Sci.* 202:702–70
9. Blondel, B., Crainic, R., Fichot, O., Dufraisse, G., Candrea, A., et al. 1986. Mutations conferring resistance to neutralization with monoclonal antibodies in type 1 poliovirus can be located outside or inside the antibody binding site. *J. Virol.* 57:81–90
10. Bordet, J. 1925. Le problème de l'autolyse microbienne transmissible et du bactériophage. *Ann. Inst. Pasteur Paris* 39:711–63
11. Borrel, A. 1903. Epithélioses infectieuses et épithéliomas. *Ann. Inst. Pasteur Paris* 17:81–118
12. Bouloy, M., Vialat, P., Girard, M., Pardigon, N. 1984. A transcript from the S segment of the bunya virus Germiston is uncapped and codes for the nucleoprotein and a non-structural protein. *J. Virol.* 49:717–23
13. Brenner, S., Jacob, F., Meselson, M. 1961. An unstable intermediate carrying information from genes to ribosomes for protein synthesis. *Nature* 190:576–81
14. Cash, E., Chamorro, M., Brahic, M. 1986. Quantitation with a new assay of Theiler's virus capsid proteins in the central nervous system of mice. *J. Virol.* 60:558–63
15. Chakrabarti, L., Guyader, M., Alizon, M., Daniel, M. D., Desrosiers, R. C., et al. 1987. Sequence of simian immunodeficiency virus from macaque and its relationship to other human and simian retroviruses. *Nature* 328:543–47
16. Charbit, A., Boulain, J. C., Ryter, A., Hofnung, M. 1986. Probing the topology of a bacterial membrane protein by genetic insertion of a foreign epitope; Expression at the cell surface. *EMBO J.* 5:3029–37
17. Charnay, P., Mandart, E., Hampe, A., Fitoussi, P., Tiollais, P., Galibert, F. 1979. Localization on the viral genome and nucleotide sequence of the gene coding for the major polypeptide of the hepatitis B surface antigen (HBs Ag). *Nucleic Acids Res.* 7:335–46
18. Clavel, F., Guetard, D., Brun-Vezinet, F., Chamaret, S., Rey, M. A., et al. 1986. Isolation of a new human retrovirus from West African patients with AIDS. *Science* 233:343–46
18a. Clavel, F., Guyader, M., Guetard, D., Salle, M., Montagnier, L., Alizon, M. 1986. Molecular cloning and polymorphism of the human immunodeficiency virus type 2. *Nature* 324:691–95
19. Crainic, R., Couillin, P., Blondel, B., Cabau, N., Bove, A., Horodniceanu, F. 1983. Natural variation of poliovirus neutralization epitopes. *Infect. Immun.* 41:1217–25
20. Danos, O., Katinka, M., Yaniv, M. 1982. Human papillomavirus type 1 complete DNA sequence and novel type of genome organization among papovaviruses. *EMBO J.* 1:233–36
21. Dejean, A., Bourgeleret, L., Grzeschik, K. H., Tiollais, P. 1986. Hepatitis B virus DNA integration in the sequence homologous to v-erb A and steroid receptor gene in a hepatocellular carcinoma. *Nature* 322:70–72
22. Delpeyroux, F., Chenciner, N., Lim,

A., Malpiece, Y., Blondel, B., et al. 1986. A poliovirus neutralization epitope expressed on hybrid hepatitis B surface antigen particles. *Science* 233:472–74

23. De Maeyer-Guignard, J., De Maeyer, E., Montagnier, L. 1972. Interferon messager RNA: translation in heterologous cells. *Proc. Natl. Acad. Sci. USA* 69:1203–7

24. Despres, P., Cahour, A., Dupuy, A., Deubel, V., Bouloy, M., et al. 1987. High genetic stability of the region coding for the structural proteins of yellow fever virus strain 17D. *J. Gen. Virol.* 68:2245–47

25. d'Herelle, F. 1921. *Le bactériophage.* Paris: Masson

26. Dubois, M. F., Pourcel, C., Rousset, S., Chany, C., Tiollais, P. 1980. Excretion of hepatitis B surface antigen particles from mouse cells transformed with cloned viral DNA. *Proc. Natl. Acad. Sci. USA* 77:4549–53

27. Duckworth, D. H. 1976. Who discovered bacteriophage? *Bacteriol. Rev.* 40:793–802

28. Duclaux, E. 1986. *Pasteur, histoire d'un esprit.* Sceaux, France: Charaire

29. Enders, J. F., Weller, T. H., Robbins, F. C. 1949. Cultivation of the Lansing strain of poliovirus in cultures of various human embryonic tissues. *Science* 109: 85–87

30. Gallo, R. C., Salahuddin, S. Z., Popovic, M., Shearer, G. M., Kaplan, M., et al. 1984. Frequent detection and isolation of cytopathic retroviruses (HTLV-III) from patients with AIDS and at risk of AIDS. *Science* 224:500–3

30a. Gallo, R., Sarin, P. S., Belmann, E. P., Robert-Guroff, M., Richardson, E., et al. 1983. Isolation of human T-cell leukemia virus in acquired immune deficiency syndrome (AIDS). *Science* 220:865–69

31. Girard, M. 1969. In vitro synthesis of poliovirus ribonucleic acid: role of the replicative intermediate. *J. Virol.* 3: 376–84

32. Guy, B., Kieny, M. P., Rivière, Y., Le Peuch, C., Dott, K., Girard, M., et al. 1987. The HIV F *(3'orf)* product is a phosphorylated GTP binding protein. *Nature* 330:266–69

33. Guyader, M., Emerman, M., Sonigo, P., Clavel, F., Montagnier, L., Alizon, M. 1987. Genome organization and transactivation of the human immunodeficiency virus type 2. *Nature* 326:662–69

34. Heerman, K. H., Goldmann, U., Schwartz, W., Seyfarth, T., Baumgarten, H., Gerlich, W. H. 1984. Large surface proteins of hepatitis B virus containing the pre-S sequence. *J. Virol.* 52:396–402

35. Herbomel, P., Bourachot, B., Yaniv, M. 1984. Two distinct enhancers with different cell specificities coexist in the regulatory region of polyomavirus. *Cell* 39:653–62

36. Horaud, F., Crainic, R., Van der Werf, S., Blondel, B., Wychowski, C., et al. 1987. Identification and characterization of a continuous neutralization epitope (C3) present on type 1 poliovirus. *Prog. Med. Virol.* 34:129–55

37. Hovanessian, A. G., Rivière, Y., Rubert, N., Svab, J., Chamaret, S., et al. 1981. Protein kinase (pp67-IFN) in plasma and tissues of mice with high levels of circulating interferon. *Ann. Inst. Pasteur Virol.* 132:175–88

38. Hovanessian, A. G., Wood, J. N. 1980. Anticellular and antiviral effects of $pppA(2'p5'A)_n$. *Virology* 101:81–90

39. Jacob, F. 1971. La belle époque. In *Of Microbes and Life,* ed. J. Monod, E. Borek, pp. 98–104. New York/London: Columbia Univ. Press

40. Jacob, F., Monod, J. 1961. Genetic regulatory mechanisms in the synthesis of proteins. *J. Mol. Biol.* 3:318–56

40a. Jacob, F., Wollman, E. L. 1961. *Sexuality and the Genetics of Bacteria.* New York/London: Academic

41. Kourilsky, P., Chaouat, G., Rabourdin-Combes, C., Claverie, J. M. 1987. Working principles in the immune system implied by the peptidic self model. *Proc. Natl. Acad. Sci. USA* 84:3400–4

42. Landsteiner, K., Levaditi, C. 1909. La transmission de la paralysie infantile aux singes. *C. R. Soc. Biol.* 67:594, 787–89

43. Lepine, P. 1954. Notions récentes concernant le virus poliomyélitique et leur application. *Rev. Prat.* 4:1919–26

44. Lepine, P. 1959. Etudes antigéniques sur le vaccin antipoliomyélitique français. *Ann. Inst. Pasteur* 97:780–93

45. Lepine, P. 1985. Evolution de la vaccination antirabique depuis Pasteur jusqu'à nos jours. In *La rage 100 ans après la découverte de Louis Pasteur,* ed. Fond. Marcel Mérieux, pp. 67–70. Lyon: Bosc Frères

45a. Letvin, N. L., Daniel, M. D., Sehgal, P. K., Desrosiers, R. C., Hunt, R. D., et al. 1985. Induction of AIDS-like disease in macaque monkeys with T-cell tropic retrovirus STLV-III. *Science* 230: 71–73

46. Levaditi, C. 1913. Virus de la poliomyélite et cultures de cellules *in vitro. C. R. Soc. Biol.* 75:202–5

47. Levaditi, C. 1913. Virus rabique et culture de cellules *in vitro*. *C. R. Soc. Biol.* 75:505

48. Levy, J. A., Hoffman, A. D., Kramer, S. M., Lanois, J. A., Shimabukuro, J. M., Oskiro, L. S. 1984. Isolation of lymphocytopathic retrovirus from San Francisco patients with AIDS. *Science* 225:840–42

49. Lwoff, A. 1953. Lysogeny. *Bacteriol. Rev.* 17:269–337

50. Lwoff, A. 1957. The concept of virus. *J. Gen. Microbiol.* 17:239–53

51. Lwoff, A. 1966. The prophage and I. In *Phage and the Origins of Molecular Biology*, ed. J. Cairns, G. S. Stent, J. D. Watson, pp. 88–89. Cold Spring Harbor, NY: Cold Spring Harbor Lab.

52. Lwoff, A., Horne, R., Tournier, P. 1962. A system of viruses. *Cold Spring Harbor Symp. Quant. Biol.* 27:51–55

53. Lwoff, A., Lwoff, M. 1961. Les évènements cycliques du cycle viral. I. Effets de la température. II. Les effets de l'eau lourde. III. Discussion. *Ann. Inst. Pasteur* 101:469–504

54. Lwoff, A., Tournier, P. 1966. The classification of viruses. *Ann. Rev. Microbiol.* 20:45–73

55. Maupas, P., Goudeau, A., Coursaget, P., Drucker, J., Bagros, P. 1976. Immunization against hepatitis B in man. *Lancet* 1:1367

56. Michel, M. L., Pontisso, P., Sobzak, E., Malpiece, Y., Streek, R. E., Tiollais, P. 1984. Synthesis in animal cells of hepatitis B surface antigen particles carrying a receptor for polymerized human serum albumin. *Proc. Natl. Acad. Sci. USA* 81:7708–12

57. Michelson, S., Tardy-Panit, M., Barzu, O. 1985. Catalytic properties of a human cytomegalovirus induced protein kinase. *Eur. J. Biochem.* 49:393–99

58. Montagnon, B., Fournier, B., Vincent-Falquet, J. C. 1985. Un nouveau vaccin antirabique à usage humain. In *Rabies in the Tropics*, ed. E. Kuwert, C. Mérieux, H. Koprowsky, K. Bögel, pp. 138–43. Berlin/Heidelberg/New York/Tokyo: Springer-Verlag

59. Neurath, A. R., Kent, S. B. H. 1988. The pre-S region of hepadnaviruses envelope proteins. *Adv. Virus Res.* In press

60. Nicolle, C., Laigret, J. 1935. La vaccination contre la fièvre jaune par le virus amaril vivant, desséché et enrobé. *C. R. Acad. Sci.* 201:312

61. Okamoto, T., Wong-Staal, F. 1985. Demonstration of virus-specific transcriptional activator(s) in cells infected with human T cell lymphotropic virus type III by *in vitro* cell-free system. *Cell* 47:29–35

62. Orth, G., Favre, M., Breitburd, F., Croissant, O., Jablonska, S., et al. 1980. Epidermodysplasia verruciformis: a model for the role of papillomaviruses in human cancer. *Cold Spring Harbor Conf. Cell Proliferation* 7:259–82

63. Pasteur, L. 1951. *Correspondance générale 4. 1885–1895.* Paris: Flammarion

63a. Porter, J. A. 1973. Some correspondence of Pasteur. *ASM News* 39:8–15

64. Roux, E. 1903. Sur les microbes dits "invisibles." *Bull. Inst. Pasteur Paris* 1:7–12; 49–56

65. Sarthou, J. L., Budkowska, A., Sharma, M. D., L'Huillier, M., Pillot, J. 1986. Characterization of an antigen-antibody system associated with epidemic non A non-B hepatitis in West Africa and experimental transmission of an infectious agent to primates. *Ann. Inst. Pasteur Virol.* 137E:225–32

66. Sellards, A., Laigret, J. 1932. Vaccination de l'homme contre la fièvre jaune. *C. R. Acad. Sci.* 194:1609–11

67. Sodroski, J. G., Patarca, R., Rosen, C. A., Haseltine, W. A. 1985. Location of the *trans* activating region on the genome of human T cell lymphotropic virus type III. *Science* 229:74–77

68. Sonigo, P., Alizon, M., Staskus, K., Klatzmann, D., Coles, S., et al. 1985. Nucleotide sequence of the visna lentivirus: relationship to the AIDS virus. *Cell* 92:369–82

68a. Stanley, W. M. 1935. Isolation of a crystalline protein possessing the properties of tobacco-mosaic virus. *Science* 81:644–54

69. Sureau, P. 1987. Rabies vaccine production in animal cell cultures. *Adv. Biochem. Eng. Biotechnol.* 34:11–123

70. Tiollais, P., Charnay, P., Vyas, G. N. 1981. Biology of hepatitis B virus. *Science* 213:406–11

71. Tiollais, P., Pourcel, C., Dejean, A. 1985. The hepatitis B virus. *Nature* 317:489–95

72. Tordo, N., Poch, O., Ermine, A., Keith, G., Rougeon, F. 1986. Walking along the rabies genomes: Is the large G-L intergenic region a remnant gene? *Proc. Natl. Acad. Sci. USA* 83:3914–18

73. Tsiang, H. 1985. An *in vitro* study of rabies pathogenesis. *Ann. Inst. Pasteur Virol.* 83:41–56

74. Van der Werf, S., Bregegere, F., Kopecka, H., Kitamura, N., Rothberg, P. G., et al. 1981. Molecular cloning of

the genome of poliovirus type 1. *Proc. Natl. Acad. Sci. USA* 78:5983–87

75. Wain-Hobson, S., Sonigo, P., Danos, O., Cole, S., Alizon, M. 1985. Nucleotide sequence of the AIDS virus, LAV. *Cell* 40:9–17

76. Waterson, A. P., Wilkinson, L. 1978. *An Introduction to the History of Virology.* Cambridge, UK: Cambridge Univ. Press

77. Wiktor, T. J. 1985. Historical aspect of rabies treatment. In *World's Debt to Pasteur, Wistar Symp. Ser. 3,* ed. H. Koprowski, S. Plotkin, pp. 141–51. New York: Liss

78. Wollman, E. L. 1966. Bacterial conjugation. See Ref. 51, pp. 216–25

79. Wollman, E. L., Jacob, F. 1955. Sur le mécanisme du transfert de matériel génétique au cours de la recombinaison chez *E. coli* K-12. *C. R. Acad. Sci.* 240:2449–51

80. Wollman, E., Wollman, E. 1938. Recherches sur le phénomène de Twort d'Herelle (bacteriophagie ou autolyse hérédo-contagieuse). *Ann. Inst. Pasteur Paris* 60:15–57

81. Wychowski, C., Van der Werf, S., Siffert, O., Crainic, R., Bruneau, P., Girard, M. 1983. A poliovirus type 1 neutralization epitope is located within amino acid residues 93 to 104 of viral capsid polypeptide VP1. *EMBO J.* 2:2019–24

SUBJECT INDEX

A

Absidia spp.
  nosocomial infection and, 527
Acetate
  formation of
    anaerobic protozoa and,
      470-71
Acetobacterium woodii
  anaerobic fermentative degra-
    dation by, 301
Acetosyringone
  Agrobacterium vir gene ex-
    pression and, 585
Acetyl coenzyme A
  oxidation of pyruvate to, 469-
    70
Acetylene
  nitrous oxide reduction and,
    249
Achromobacter cycloclastes
  nitric oxide reductase of, 244-
    46
  nitrite reductase of, 238
Acinetobacter calcoaceticus
  nosocomial infections and,
    457
  phosphotyrosine in, 106
Acinetobacter spp.
  pathogenicity of, 457-58
  phenol hydroxylase of, 272
  skin colonization by, 445
Acne
  Propionibacterium acnes and,
    457
Acquired immunodeficiency syn-
    drome (AIDS)
  animal models for, 610-11
  infection susceptibility and,
    452
  lentiviruses and, 609-10
  vaccine for, 759
  viruses, 751-54
Actinobacillus spp.
  coaggregation of, 634
*Actinobacillus actinomycetemco-
    mitans*
  periodontal disease and, 638
Actinomyces israelii
  coaggregation of, 631, 636
Actinomyces naeslundii
  adherence to teeth, 637
  coaggregation of, 631-34,
    636, 641, 644
  fimbriae of, 643
Actinomyces odontolyticus
  coaggregation of, 631-34

Actinomyces parabifidus
  coaggregation of, 631
Actinomyces spp.
  fimbriae of, 642-43
Actinomyces viscosus
  adherence to teeth, 637
  coaggregation of, 628, 631-
    34, 636, 641, 644
  fimbriae of, 642-43
Adamowicz, Philippe, 757
Adenylate kinase
  anaerobic protozoa and,
    478
Adhesins
  bacterial adherence to skin
    and, 450
  oral bacterial, 644
Aerobactin
  Shigella spp. and, 142-43
Aerobiosis
  anaerobic protozoa and, 474
Aeromonas formicans, 398
Aeromonas hydrophila, 396-98
  antimicrobial susceptibilities
    of, 413
  diarrheal disease and, 402-5
  endocarditis and, 406-7
  enterotoxins of, 410-12
  epidemiology of, 399-401
  extracellular enzymes of, 410
  eye infections and, 407
  immune response to, 409-10
  meningitis and, 405-6
  osteomyelitis and, 406
  prophylaxis and, 413
  septic arthritis and, 406
  skin infections and, 405
  urinary tract infections and,
    407
Aeromonas liquefaciens, 398
  hemorrhagic septicemia and,
    401-2
Aeromonas punctata, 398
  infections caused by, 408
Aeromonas salmonicida, 398
  furunculosis and, 401-2
Aeromonas sobria, 400
  diarrheal disease and, 403
  extracellular enzymes of, 410
  infections caused by, 408
Aeromonas spp., 395-413
  antimicrobial susceptibilities
    of, 413
  cultural characteristics of, 401
  enterotoxins of, 410-12
  extracellular enzymes of, 410
  immune response to, 409-10

infections caused by, 401-2
  human, 402-9
  prophylaxis and, 413
Age
  normal flora of skin and,
    447-48
Agrobacterium rhizogenes
  crown gall and, 575
  host range of, 422, 424
  Ri plasmid of
    vir region of, 585
Agrobacterium spp.
  chvA and *chvB* mutants of,
    580-81
  colonization and attachment
    of, 576-82
  host range of, 593-96
  infection and transformation
    of plants by, 575-97
  plant cell receptor for, 578-79
  plant wounding and, 582-84
  pscA mutants of, 581-82
  vir gene of
    expression of, 585-87
  virulence mutants of, 582
Agrobacterium tumefaciens
  crown gall and, 422-24, 575
  host range of, 422
  site-specific attachment of,
    578
  tumor induction by
    monocots and, 595-96
  T-DNA expression and,
    591-93
  T-DNA transfer and, 584-
    91
Alcaligenes eutrophus
  cadmium resistance in, 723
  haloaromatic degradation and,
    270
  phenol hydroxylase of, 272
Alcaligenes faecalis
  anaerobic growth of
    nitrous oxide-dependent,
      249
  nitric oxide reductase of, 245
  nitrite reductase of, 238
Alcaligenes spp.
  nitrite reductases of, 238-40
Alcaligenes strain A7
  methanol-degrading, 282
Alcaligenes xylosoxidans subsp.
    denitrificans
  benzoate metabolism by, 296
Alcohol dehydrogenase
  ethanol formation in anaerobic
    protozoa and, 472

CUMULATIVE INDEXES

CONTRIBUTING AUTHORS, VOLUMES 38–42

CHAPTER TITLES, VOLUMES 38–42

Annual Reviews Inc.

A NONPROFIT SCIENTIFIC PUBLISHER

ᗺ 4139 El Camino Way
P.O. Box 10139
Palo Alto, CA 94303-0897 • USA

Annual Reviews Inc. publications may be ordered directly from our office by mail or use our Toll Free Telephone line (for orders paid by credit card or purchase order, and customer service calls only); through booksellers and subscription agents, worldwide; and through participating professional societies. Prices subject to change without notice.   ARI Federal I.D. #94-1156476

- **Individuals:** Prepayment required on new accounts by check or money order (in U.S. dollars, check drawn on U.S. bank) or charge to credit card — American Express, VISA, MasterCard.
- **Institutional buyers:** Please include purchase order number.
- **Students:** $10.00 discount from retail price, per volume. Prepayment required. Proof of student status must be provided (photocopy of student I.D. or signature of department secretary is acceptable). Students must send orders direct to Annual Reviews. Orders received through bookstores and institutions requesting student rates will be returned. You may order at the Student Rate for a maximum of 3 years.
- **Professional Society Members:** Members of professional societies that have a contractual arrangement with Annual Reviews may order books through their society at a reduced rate. Check with your society for information.
- **Toll Free Telephone orders:** Call 1-800-523-8635 (except from California) for orders paid by credit card or purchase order and customer service calls only. California customers and all other business calls use 415-493-4400 (not toll free). Hours: 8:00 AM to 4:00 PM, Monday-Friday, Pacific Time.

Regular orders: Please list the volumes you wish to order by volume number.
Standing orders: New volume in the series will be sent to you automatically each year upon publication. Cancellation may be made at any time. Please indicate volume number to begin standing order.
Prepublication orders: Volumes not yet published will be shipped in month and year indicated.
California orders: Add applicable sales tax.
Postage paid (4th class bookrate/surface mail) by **Annual Reviews Inc.** Airmail postage or UPS, extra.

ANNUAL REVIEWS SERIES	Prices Postpaid per volume USA & Canada/elsewhere	Regular Order Please send:	Standing Order Begin with:
		Vol. number	Vol. number
Annual Review of ANTHROPOLOGY			
Vols. 1-14 (1972-1985)	$27.00/$30.00		
Vols. 15-16 (1986-1987)	$31.00/$34.00		
Vol. 17 (avail. Oct. 1988)	$35.00/$39.00	Vol(s). _____	Vol. _____
Annual Review of ASTRONOMY AND ASTROPHYSICS			
Vols. 1-2, 4-20 (1963-1964; 1966-1982)	$27.00/$30.00		
Vols. 21-25 (1983-1987)	$44.00/$47.00		
Vol. 26 (avail. Sept. 1988)	$47.00/$51.00	Vol(s). _____	Vol. _____
Annual Review of BIOCHEMISTRY			
Vols. 30-34, 36-54 (1961-1965; 1967-1985)	$29.00/$32.00		
Vols. 55-56 (1986-1987)	$33.00/$36.00		
Vol. 57 (avail. July 1988)	$35.00/$39.00	Vol(s). _____	Vol. _____
Annual Review of BIOPHYSICS AND BIOPHYSICAL CHEMISTRY			
Vols. 1-11 (1972-1982)	$27.00/$30.00		
Vols. 12-16 (1983-1987)	$47.00/$50.00		
Vol. 17 (avail. June 1988)	$49.00/$53.00	Vol(s). _____	Vol. _____
Annual Review of CELL BIOLOGY			
Vol. 1 (1985)	$27.00/$30.00		
Vols. 2-3 (1986-1987)	$31.00/$34.00		
Vol. 4 (avail. Nov. 1988)	$35.00/$39.00	Vol(s). _____	Vol. _____